Lecture Notes of the Institute for Computer Sciences, Social Informatics and Telecommunications Engineering 349

More information about this series at http://www.springer.com/series/8197

Honghao Gao · Xinheng Wang ·
Muddesar Iqbal · Yuyu Yin ·
Jianwei Yin · Ning Gu (Eds.)

Collaborative Computing: Networking, Applications and Worksharing

16th EAI International Conference, CollaborateCom 2020
Shanghai, China, October 16–18, 2020
Proceedings, Part I

Springer

Editors
Honghao Gao
Shanghai University
Shanghai, China

Xinheng Wang
Xi'an Jiaotong-Liverpool University
Suzhou, China

Muddesar Iqbal
London South Bank University
London, UK

Yuyu Yin
Hangzhou Dianzi University
Hangzhou, China

Jianwei Yin
Zhejiang University
Hangzhou, China

Ning Gu
Fudan University
Shanghai, China

ISSN 1867-8211 ISSN 1867-822X (electronic)
Lecture Notes of the Institute for Computer Sciences, Social Informatics
and Telecommunications Engineering
ISBN 978-3-030-67536-3 ISBN 978-3-030-67537-0 (eBook)
https://doi.org/10.1007/978-3-030-67537-0

This Springer imprint is published by the registered company Springer Nature Switzerland AG
The registered company address is: Gewerbestrasse 11, 6330 Cham, Switzerland

Preface

We are delighted to introduce the proceedings of the 16th European Alliance for Innovation (EAI) International Conference on Collaborative Computing: Networking, Applications and Worksharing (CollaborateCom 2020). This conference brought together researchers, developers and practitioners around the world who are interested in fully realizing the promise of electronic collaboration with special attention to the aspects of networking, technology and systems, user interfaces and interaction paradigms, and interoperation with application-specific components and tools.

The technical program of CollaborateCom 2020 selected 77 papers from 211 paper submissions, comprising 61 full papers, 13 short papers and 3 workshop papers in oral presentation sessions in the main conference tracks. The conference sessions were: Collaborative Applications for Network and E-Commerce; Optimization for Collaborative Systems; Cloud and Edge Computing; Artificial Intelligence; AI Application and Optimization; Edge Computing and CollaborateNet; Classification and Recommendation; Internet of Things; Collaborative Robotics and Autonomous Systems; Resource Management; Smart Transportation; Resource Management in Artificial Intelligence; Short paper Track and Workshop Track. Apart from high-quality technical paper presentations, the technical program also featured two keynote speeches and two technical workshops. The two keynote speeches were delivered by Dr Fumiyuki Adachi from Tohoku University and Dr. Deke Guo from National University of Defense Technology. The two workshops organized were Securing IoT Networks (SITN) and Collaborative Networking Technologies towards Future Networks (CollaborateNet). The SITN workshop aims to bring together expertise from academia and industry to build secure IoT infrastructures for smart society. The CollaborateNet workshop aims to facilitate all efforts to advance current networks towards content-centric future networks using collaborative networking technologies.

Coordination with the steering chair, Imrich Chlamtac, was essential for the success of the conference. We sincerely appreciate his constant support and guidance. It was also a great pleasure to work with such an excellent organizing committee team for their hard work in organizing and supporting the conference, in particular, the Technical Program Committee, led by our General Chairs and TPC Co-Chairs, Dr. Ning Gu, Dr. Jianwei Yin, Dr. Xinheng Wang, Dr. Honghao Gao, Dr. Yuyu Yin and Dr. Muddesar Iqbal, who completed the peer-review process of the technical papers and made a high-quality technical program. We are also grateful to the Conference Manager, Karolina Marcinova, for her support and to all the authors who submitted their papers to the CollaborateCom 2020 conference and workshops.

We strongly believe that the CollaborateCom conference provides a good forum for all researchers, developers and practitioners to discuss all scientific and technical aspects that are relevant to collaborative computing. We also expect that the future CollaborateCom conferences will be as successful and stimulating, as indicated by the contributions presented in this volume.

December 2020

Honghao Gao
Xinheng Wang

Conference Organization

Steering Committee

Chair

Imrich Chlamtac	Bruno Kessler Professor, University of Trento

Members

Song Guo	The University of Aizu, Japan
Bo Li	The Hong Kong University of Science and Technology
Xiaofei Liao	Huazhong University of Science and Technology
Xinheng Wang	Xi'an Jiaotong-Liverpool University
Honghao Gao	Shanghai University

Organizing Committee

International Advisory Committee

Velimir Srića	University of Zagreb, Croatia
Mauro Pezze	Università di Milano-Bicocca, Italy
Yew-Soon Ong	Nanyang Technological University, Singapore

General Chairs

Ning Gu	Fudan University
Jianwei Yin	Zhejiang University
Xinheng Wang	Xi'an Jiaotong-Liverpool University

TPC Chair and Co-chairs

Honghao Gao	Shanghai University
Yuyu Yin	Hangzhou Dianzi University
Muddesar Iqbal	London South Bank University

Local Chairs

Zhongqin Bi	Shanghai University of Electric Power
Yihai Chen	Shanghai University

Workshops Chairs

Yusheng Xu	Xidian University
Tasos Dagiuklas	London South Bank University
Shahid Mumtaz	Instituto de Telecomunicações

Publicity and Social Media Chairs

Li Kuang	Central South University
Anwer Al-Dulaimi	EXFO Inc
Andrei Tchernykh	CICESE Research Center
Ananda Kumar	Christ College of Engineering and Technology

Publications Chairs

Youhuizi Li	Hangzhou Dianzi University
Azah Kamilah Binti Draman	Universiti Teknikal Malaysia Melaka

Web Chair

Xiaoxian Yang	Shanghai Polytechnic University

Technical Program Committee

CollaborateNet Workshop

Amando P. Singun, Jr.	Higher College of Technology
BalaAnand Muthu	V.R.S. College of Engineering & Technology
Boubakr Nour	Beijing Institute of Technology
Chaker Abdelaziz Kerrache	Huazhong University of Science and Technology
Chen Wang	Huazhong University of Science and Technology
Chi-Hua Chen	Fuzhou University
Fadi Al-Turjman	Near East University
Muhammad Atif Ur Rehman	Hongik University
Rui Cruz	Universidade de Lisboa/INESC-ID
Suresh Limkar	AISSMS Institute of Information Technology

Collaborative Robotics and Autonomous Systems

Craig West	Bristol Robotics Lab
Inmo Jang	The University of Manchester
Keir Groves	The University of Manchester
Ognjen Marjanovic	The University of Manchester
Pengzhi Li	The University of Manchester
Wei Cheah	The University of Manchester

Internet of Things

Chang Yan	Chengdu University of Information Technology
Fuhu Deng	University of Electronic Science and Technology of China
Haixia Peng	University of Waterloo
Jianfei Sun	University of Electronic Science and Technology of China

Kai Zhou	Sichuan University
Mushu Li	University of Waterloo
Ning Zhang	Texas A&M University-Corpus Christi
Qiang Gao	University of Electronic Science and Technology of China
Qixu Wang	Sichuan University
Ruijin Wang	University of Electronic Science and Technology of China
Shengke Zeng	Xihua University
Wang Dachen	Chengdu University of Information Technology
Wei Jiang	Sichuan Changhong Electric Co., Ltd
Wen Wu	University of Waterloo
Wen Xhang	Texas A&M University-Corpus Christi
Wu Xuangou	Anhui University of Technology
Xiaojie Fang	Harbin Institute of Technology
Xuangou Wu	Anhui University of Technology
Yaohua Luo	Chengdu University of Technology
Zhen Qin	University of Electronic Science and Technology of China
Zhou Jie	Xihua University

Main Track

Bin Cao	Zhejiang University of Technology
Ding Xu	Hefei University of Technology
Fan Guisheng	East China University of Science and Technology
Haiyan Wang	Nanjing University of Posts & Telecommunications
Honghao Gao	Shanghai University
Jing Qiu	Guangzhou University
Jiwei Huang	China University of Petroleum
Jun Zeng	Chongqing University
Lizhen Cui	Shandong University
Rong Jiang	Yunnan University of Finance and Economics
Shizhan Chen	Tianjin University
Tong Liu	Shanghai Univerisity
Wei He	Shandong University
Wei Du	University of Science and Technology Beijing
Xiong Luo	University of Science and Technology Beijing
Yu Weng	Minzu University of China
Yucong Duan	Hainan University
Zijian Zhang	University of Auckland, New Zealand

SITN Workshop

A. S. M. Sanwar Hosen	Jeonbuk National University
Aniello Castiglione	Parthenope University of Naples
Aruna Jamdagni	Western Sydney University

Chunhua Sun	The University of Aizu, Japan
Deepak Puthal	Newcastle University
Jinguang Han	Queen's University Belfast
Julio Hernandez-Castro	University of Kent
Kashif Saleem	King Saud University
Md Zakirul Alam Bhuiyan	Fordham University
Mian Jan	University of Technology Sydney
Mohiuddin Ahmed	Canberra Institute of Technology
Nikolaos Pitropakis	Edinburgh Napier University
Qingchen Zhang	St. Francis Xavier University
Qingru Li	Hebei Normal University
Saurabh Singh	Dongguk University
Shancang Li	UWE Bristol
Syed Bilal Hussain Shah	Dalian University of Technology
Weizhi Meng	Technical University of Denmark
Yang Xu	Hunan University
Yongjun Zhao	Nanyang Technological University
Zhihong Tian	Guangzhou University

Contents – Part I

Cloud and Edge Computing

Artificial Intelligence

Contents – Part II

Smart Transportation

Resource Management in Artificial Intelligence

Collaborative Applications for Network and E-Commerce

Towards Accurate Search for E-Commerce in Steel Industry: A Knowledge-Graph-Based Approach

Maojian Chen[1,2,3], Hailun Shen[4], Ziyang Huang[4], Xiong Luo[1,2,3](✉), and Junluo Yin[1,2,3]

[1] School of Computer and Communication Engineering, University of Science and Technology Beijing, Beijing 100083, China
xluo@ustb.edu.cn
[2] Beijing Key Laboratory of Knowledge Engineering for Materials Science, Beijing 100083, China
[3] Shunde Graduate School, University of Science and Technology Beijing, Foshan, Guangdong 528399, China
[4] Ouyeel Co., Ltd., Shanghai 201999, China

Abstract. Mature artificial intelligence (AI) makes human life more and more convenient. However, in some application fields, it is impossible to achieve the satisfactory results only depending on the traditional AI algorithm. Specifically, in order to avoid the limitations of traditional searching strategies in e-commerce field related to steel, such as the inability to analyzing long technical sentences, we propose a collaborative decision making method in this field, through the combination of deep learning algorithms and expert systems. Firstly, we construct a knowledge graph (KG) on the basis of steel commodity data and expert database, and then train a model to accurately extract steel entities from long technical sentences, while using an advanced bidirectional encoder representation from transformers (BERT), a bidirectional long short-term memory (Bi-LSTM), and a conditional random field (CRF) approach. Finally, we develop an intelligent searching system for e-commence in steel industry, with the help of the designed KG and entity extraction model, while improving the searching performance and user experience in such system.

Keywords: Steel E-commerce · Knowledge graph (KG) · Entity extraction · Bidirectional encoder representation from transformers (BERT)

1 Introduction

Living in the era of big data, more and more data are generated on the Internet, and more choices are available for people, which makes it difficult for executives to make decisions [1]. In order to find the best strategy, the collaborative decision making (CDM) process is developed. It is a flexible process considering every aspect to obtain the best

H. Gao et al. (Eds.): CollaborateCom 2020, LNICST 349, pp. 3–18, 2021.
https://doi.org/10.1007/978-3-030-67537-0_1

benefit [2]. During the last decade, CDM is applied in many fields [3–8]. However, as far as the steel which is used as an essential material in our life, there is little research on CDM applied to the steel e-commerce field.

Among the e-commerce field in the steel industry, the search system is directly facing users to service. It emphasizes efficiency which means that the assigned tasks are completed with the shortest time while considering stability. Meanwhile, the accuracy of searching results and the efficiency of search system are largely influencing user experience, and user's satisfaction with the system reflects the pursuit of the steel e-commerce field.

Traditional steel search engines retrieve information on the Internet through keywords, and return relevant webpages containing strings to users. With the increasing complexity of business data, this retrieval method cannot accurately satisfy users' diverse demands, and it will greatly affect user' experience. For example, because some non-standard steel commodity information is filled in when uploading commodity information to the trading platform by traders, these existing search engines cannot retrieve and analyze complex information with multiple attributes well. Meanwhile, there are some informal vocabularies in user's inquiries, which may lead to an unsatisfactory ability to analyze steel daily inquiries. Then, they do not support the retrieval of the commodities' nickname, nor the retrieval of steel daily inquiry. In order to avoid the limitations of traditional steel search engines, new search engines on the basis of knowledge graph (KG) have attracted extensive attention from relevant researchers [9–12]. Hence, the introduction of KG will provide a new way for search engines of e-commerce trading platforms in steel industry.

Currently, there are some works on the metal and materials related to the steel industry. For example, it has been verified that constructing KG of steel enterprise operation and maintenance domain can obtain dispersed and heterogeneous information in a consistent way, while simplifying the process of obtaining information and improving the efficiency of obtaining information for engineers [13]. Then, by using DBpedia and Wikipedia, a method was developed to construct metallic materials KG [14]. A KG was constructed, where it included 115 materials properties and 69 relationships. This KG can represent arbitrarily complex property and relationships [15].

With the development of steel industry, it has accumulated amount of steel data forming expert databases. However, the above methods are using accurate algorithms or convenient tools to achieve the sort of steel data, rather than utilizing expert databases to improve the user experience and user satisfaction for the steel e-commerce.

Hence, in consideration of the above background, and to further effectively handle those difficulties, we combine steel commodity data and expert databases to construct a KG in steel industry, and standardize commodity information. Furthermore, through the use of bidirectional encoder representation from transformers (BERT) [16], bidirectional long short-term memory (Bi-LSTM) [17], and conditional random field (CRF) approach [18], we further optimize and improve the performance of existing search engine, so as to make the match between users' demands and commodities more accurate, make steel trading simpler and more effective, and improve user's experience.

The rest of this paper is organized as follows. Section 2 will introduce the related work, including KG, entity extraction and entity alignment model. In Sect. 3, we detail the method on the search system of steel e-commerce platform using KG. Then, we will introduce the experiments, including data sets, experimental results, and some discussions in Sect. 4. Finally, the conclusion and future work will be summarized in Sect. 5.

2 Related Work

In this section, we will introduce some key technologies in related to our method.

2.1 Knowledge Graph (KG)

In 2012, KG was proposed by Google, in an effort to optimize existing search engines [19]. Different from traditional search engines, the KG-based search engines can comprehend user's intentions from the semantic level and extract complicated information better, thus effectively improving the search performance.

Essentially, KG is a semantic network. Its nodes represent entities or concepts, and edges represent various semantic relationships between entities and concepts [20, 21]. With the full use of visual technology, KG can not only describe the knowledge resources and carriers, but also analyze and the relations between them [22]. In recent years, KG becomes one of the basic technologies in intelligent services, such as semantic search, intelligent question answering, and decision support [23].

The core technologies of KG involve knowledge extraction, knowledge representation, knowledge fusion, and some others. Knowledge extraction is used to extract entities, concepts, attributes, and relationships from various data sources. Knowledge representation means that the semantic information of entities can be expressed with dense low dimensional vectors, and then entities, relationships, and complicated semantic associations among entities can be efficiently calculated in a low dimensional space. Knowledge fusion includes semantic computing and data integration, which are to eliminate contradictions and ambiguities [24, 25].

Currently, there are many KG-based applications in different fields. In respect of search engines, whether Google, Bing, or Baidu, they all implement intelligent search on the basis of KG. For instance, if users want to retrieve information about "Yao Ming" in Bing, it will return a knowledge card of "Yao Ming" instead of some webpages containing the string "Yao Ming". The knowledge card shows Yao's date of birth, height and weight, the names of his spouse and children. This returned method can improve retrieval efficiencies greatly.

2.2 Unstructured Daily Queries Analysis

When a customer provides a daily query, it needs to be parsed to understand his requirements. Daily queries can be divided into three categories, i.e., a mixture of multiple keywords, Chinese and English mixed keywords, and sentences. If the query is just a number of keywords, it is relatively simple to parse. If it is a sentence, it needs to be deeply analyzed, which involves semantic analysis, entity extraction, and entity alignment. In this paper, we focus on the daily queries analysis with a sentence.

Entity extraction technique was proposed in 1996 [26], and it is automatically extracting various entities from text or sentence, like persons, time, locations, money [27]. Now, entity extraction is a basic technology of Chinese text analysis, and there are many machine learning methods applied to it [28, 29], such as hidden markov models (HMM), support vector machines (SVM).

The entity extracted from the daily queries may be a nickname, that is not consistent with the standard entity. Thus, these extracted entities cannot be used directly in other processes, and they need to be aligned.

There are some methods on entity alignment, most of which calculate the similarity between two entities to decide whether they are the same entity. IF-IDF algorithm counts the frequency of each character in an entity, and then transforms those frequencies into vectors, lastly calculates the similarity of the two vectors through cosine distance [30–32]. Word2Vec is to map each entity into a low-dimensional space to form a shorter vector, and then calculates the similarity of the two vectors through cosine distance [33, 34]. This method takes the semantic of the entity into account, and it is better than TF-IDF. In addition, there are string-based, corpus-based, knowledge-based, and other methods that can be used to calculate the similarity of two entities [35–39].

3 Methodology

In this section, we will introduce how to construct a KG in the steel industry with CDM process, and then develop a search system of e-commerce field. The whole framework is shown in Fig. 1.

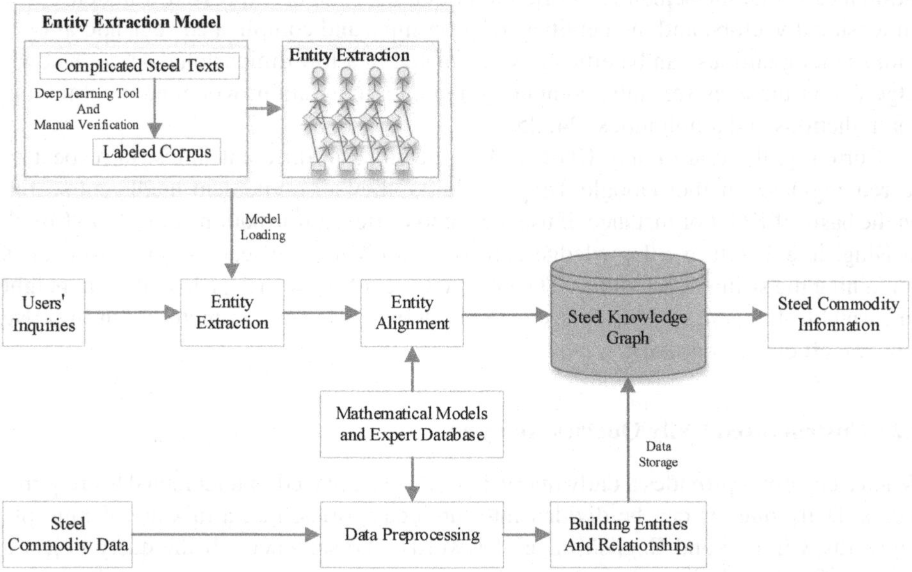

Fig. 1. The framework of establishing a KG-based search system for e-commerce platform.

Firstly, we construct a steel KG. In this processing, since the data of steel commodity are structured, we need to use mathematical models and expert database to extract important data. Then, according to the attributes of steel commodities, we establish entities and relations between entities, and those entities and relations will be stored in graph database. Secondly, in order to improve the computing performance, we use a third-party tool to label the unstructured steel data, and then manually verify them. Thirdly, to accurately comprehend users' intentions, some advanced deep learning algorithms are used to train these data to obtain an entity extraction model for the steel data. Fourthly, the trained entity extraction model is used to extract users' inquiries for getting entities and align entities. Finally, the aligned entities are matched with the entities in the graph database, and the corresponding steel commodity information is returned to users.

3.1 Preprocessing for the Steel Commodity Data

Firstly, we collect a lot of the steel commodity data from the Internet. Secondly, we clean and filter those data through mathematical models and expert database. Thirdly, we select some important data as entities, and construct relationships between those entities. Then, we store those data into graph database to construct a KG.

3.2 Preprocessing for the Steel Inquiry Data

Our purpose is to parse users' daily inquiries, thus we use the historical users' daily inquiry data as dataset. However, these data are unlabeled and unstructured. Therefore, we need to label those data first.

Among these steel data, there are many types of entities, such as numeral, technical term, name of a person, place name, and many others. Hence, we segment those data and tag them with their part-of-speech (POS). Then, the BMEO principle is used to further label the position of each character in the entity. Here, "B" means that the character is the first word in the entity, "M" represents that the character is the internal word in the entity, "E" shows that the character is the last word in the entity, and "O" expresses that the word is not an entity. Finally, we divide these data into three parts, including training set, validation set, and test set.

3.3 Entity Extraction Model

In this section, we will introduce how to use unstructured steel texts to train an entity extraction model.

Word Embedding with BERT. The data of the steel industry are text, while the input of training model must be numeral, therefore, we need to normalize those data. Here, we use BERT model to generate word embedding. Generally, the dimension of word embedding is less than the number of words.

Entity Extraction via Bi-LSTM+CRF Model. Bi-LSTM is a special recurrent neural network (RNN), which is mainly used to deal with the issue of vanishing gradient and exploding gradient during long sequence training.

The main idea of Bi-LSTM is to retain historical information and remember current information, by introducing a gate mechanism and controlling the degree of each unit. In so doing, it can retain important features and discard unimportant features. Totally, Bi-LSTM performs better in longer sequences than common RNN.

In this paper, the output of Bi-LSTM is the scores for each tag of the character. For example, if there are 7 tags, and the output of Bi-LSTM is shown as Fig. 2. For input w_0, the score of tag "B-Person" is 1.624, the score of tag "M-Person" is 0.819, the score of tag "E-Person" is 0.203, the score of tag "B-Organization" is 0.765, the score of tag "M-Organization" is 0.050, the score of tag "E-Organization" is 0.101, and the score of tag "O" is 0.035. It can select the tag with the highest score in each character as the result. But, in some cases, such as it described in Fig. 3, the result of two characters' tags in the entity "w_0w_1" tagged with "M-Organization" and "M-Person" is wrong clearly, since it is impossible for two middle characters of different categories to be adjacent.

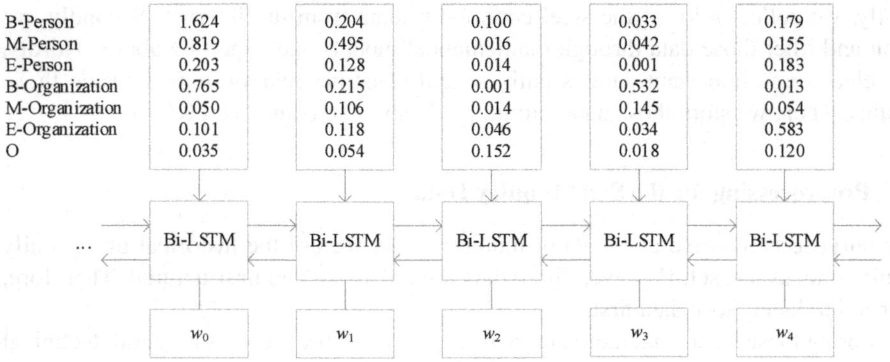

Fig. 2. An example of the output in Bi-LSTM.

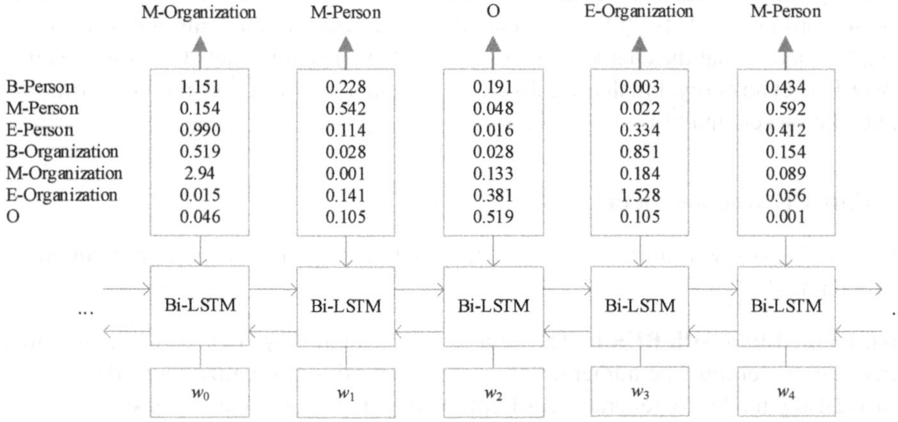

Fig. 3. An example of the result of Bi-LSTM.

CRF can automatically learn some restrictions during training to ensure that the predicted tags are valid. There are some restrictions showing as follows.

(1) The first character's tag in a sentence should start only with "B-" or "O", rather than others.
(2) For the tag result as "B-tag1 M-tag2 M-tag3 E-…", here tag1, tag2, tag3 should be the same tag. For example, "B-Person E-Person" is valid, and "B-Person M-Organization" is invalid.
(3) "O-tag" is illegal. Entities should start with "B-", not "M-" or "E-".

When these constraints are learned, the number of invalid tags in the final results is reduced dramatically. The structure of Bi-LSTM+CRF model is shown in Fig. 4.

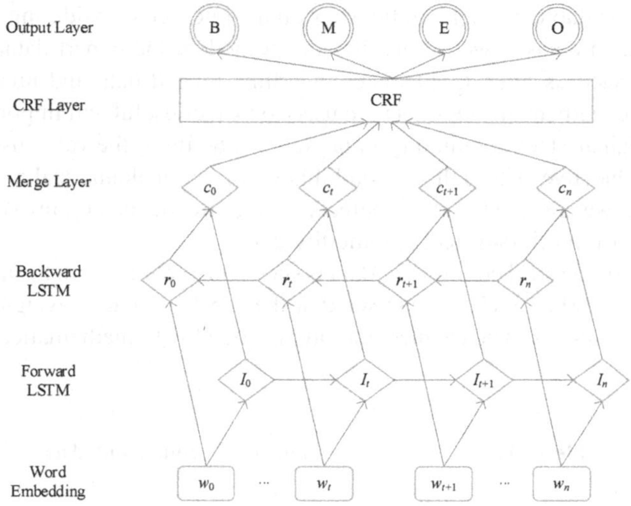

Fig. 4. The structure of Bi-LSTM+CRF model.

3.4 The KG-Based Search System

After obtaining the entity extraction model and KG, we accordingly develop a search system in relation to the steel e-commerce field. The whole process is as follows.

Firstly, when the customer inputs a daily inquiry, the system extracts entities from the inquiry according to the trained model. Because the extracted entities contain many professional terms in the steel industry, the expert database in the steel industry is used to align entities. Then, the entities obtained in the previous step are matched with the KG. Furthermore, we extract the IDs of the commodity, and they contain all the entities. Finally, the steel commodity information corresponding to the commodity IDs are returned to the customer.

In so doing, a KG-based search system for the steel e-commerce can be constructed.

4 Experiments

Different from the common industries, the data of the steel industry are more complex and professional. Therefore, it is necessary to combine expert knowledge to preprocess the data and construct KG. Generally, there are many graph databases, such as ArangoDB, Neo4j, and JanusGraph. In our experiments, we select Neo4j 4.1.1 as our graph database. Meanwhile, our experiments are conducted in the Python 3.6.8 environment running on Ubuntu 18.04.1.

4.1 Datasets and Data Preprocessing

Here, two datasets are used in our experiments, i.e., the steel commodity data, and the daily inquiry data in steel industry.

There are 20,366 steel commodity data, including more than 30 entity categories. Those data are available through public ways on the Internet. Considering that there are many dirty data in dataset, we use mathematical models and expert database to clean up those data, such as filtering attributes, handling invalid data and missing data. In this process, the mathematical models are used to extract useful and important attribute value from original steel commodity data, such as defining the rules used to extract the weight of objective. Through the combination of the mathematical models and the expert database, we are able to extract defects, surface treatment, coating type, and other attribute values from original steel commodity data.

The sample of the steel commodity data is shown in Table 1. From the attribute value of "zero spangle Z80 cr free", we can see that the "Z80" is coating weight, rather than coating typc, so it needs to be cleaned and preprocessed with mathematical models and expert database.

Table 1. The samples of the original steel commodity data.

Grade	Specification	Place	Resource ID	Coating type
SGC570	0.65*1240*C	Tang Steel	1398950769	Zero spangle Z80 cr free
HC420LAD+Z	1.65*1405*C	Ben Steel	1432699197	
DX53D+Z	0.8*1650*C	Shou Gang	1433283775	Light oiling

It is easy to obtain the steel commodity data, however, it is relatively difficult to collect unstructured steel inquiry data, since the inquiry data exist in the communication tools between customers and suppliers, such as email, social software. Through our cooperation partner, we got 2,600 daily inquiry data, which are from customers' historical purchase information. Therefore, all data we used in this paper are real and reasonable. After cleaning low available and repeated data, there remain 1,217 useful inquiry data.

The sample of the unstructured steel inquiry data is shown in Table 2. In this table, the expressions of customer's demands are informal. For example, in first inquiry data, "DC54D+Z" means that the grade of steel plate is "DC54D", and coating type is "Zinc". Similarity, for the "SCGA270D-45" in the third data, its grade is "DX52D", and coating type is "Zinc-Fe". After translating Table 2 from Chinese into English, the updated version is shown in Table 3.

Table 2. The samples of the unstructured steel inquiry data.

Index	非结构化的钢铁询单数据
1	求购:谁家有 DC54D+Z 2.5*66*1250的,我要一张
2	2.0的电镀锌SECCN5的样板,求购一块
3	求购SCGA270D-45,要2吨
4	上海地区求购 0.5*1000环保钝化无花

Table 3. The updated version of Table 2 after translating it from Chinese into English.

Index	Unstructured steel inquiry data
1	Who has DC54D+Z and 2.5*66*1250 steel plate, I want to buy one
2	Purchase SECCN5 steel plate with 2.0 electric galvanizing
3	Purchase 2 tons SCGA270D-45 steel plate
4	Purchase 0.5*1000 environmental passivation and zero zinc coil in Shanghai

4.2 Experiments and Result Analysis

In addition to data preprocessing mentioned above, there are other three parts in our experiments.

Constructing KG Based on Commodity Data in Steel Industry. After preprocessing the data, we select five important attributes of the steel commodity data as entities used in Neo4j, with the help of experts in steel industry. They are "Grade", "Coating Type", "Coating Weight", "Surface Structure", and "Surface Treatment". Moreover, we define the commodity ID, which is the central entity to connect those five types of entities. Then, the KG of steel industry is accordingly constructed. The storage architecture of steel commodity data in KG is shown as Fig. 5, and nodes with different colors represent different categories entities. For example, the red entity means "Grade", the green entity represents "Coating Type", and the orange entity is "Surface Structure". After translating Fig. 5 from Chinese into English, the updated version is shown in Fig. 6. There are 1,057 entities and 1,778 relationships stored in the Neo4j database, and a part of KG of steel commodity data is shown in Fig. 7. From this figure, we can easily find some information, e.g., for the Commodity ID "SP0009", whose surface structure is "Zero Spangle", and its coating weight is "40/40". After translating Fig. 7 from Chinese into English, the updated version is shown in Fig. 8.

Word Embedding. We use the BERT model to achieve word embedding of steel inquiry data. Here, in consideration of the characteristics of steel commodity data, we convert each word to a 128-dimensional vector. The sample of word embedding in steel industry is shown in Table 4.

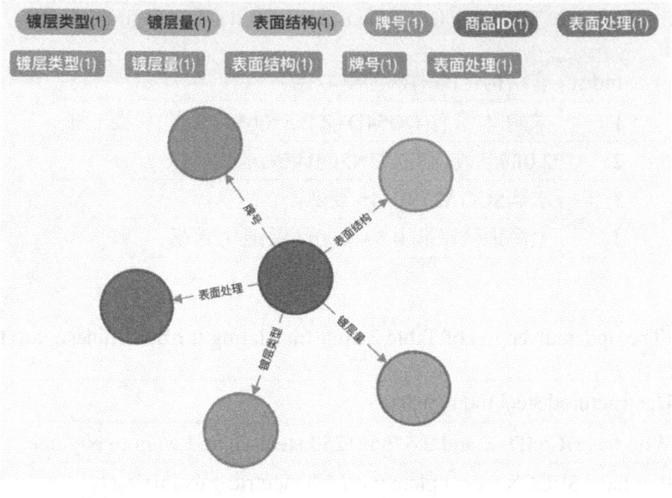

Fig. 5. The architecture of steel commodity data in KG.

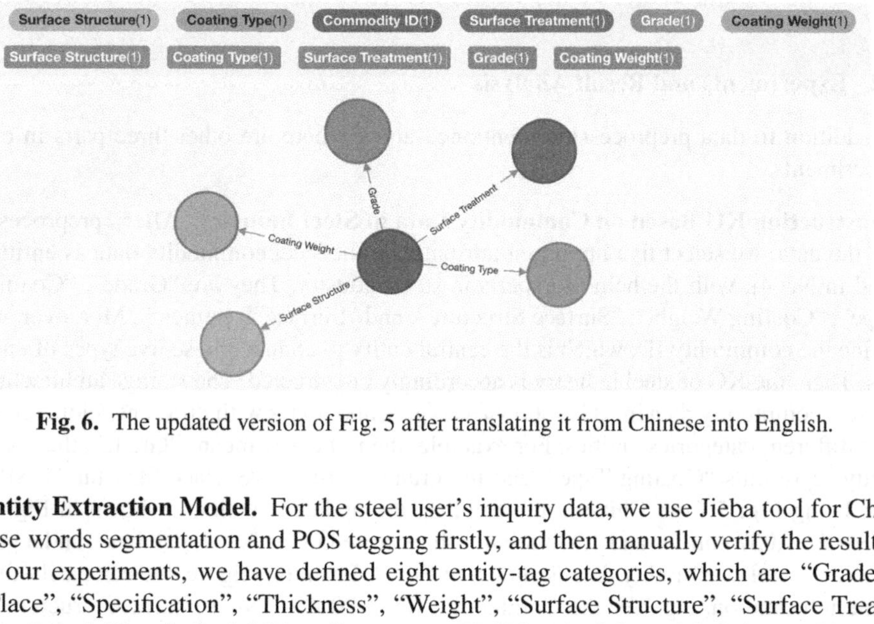

Fig. 6. The updated version of Fig. 5 after translating it from Chinese into English.

Entity Extraction Model. For the steel user's inquiry data, we use Jieba tool for Chinese words segmentation and POS tagging firstly, and then manually verify the results. In our experiments, we have defined eight entity-tag categories, which are "Grade", "Place", "Specification", "Thickness", "Weight", "Surface Structure", "Surface Treatment", and "Species". Additionally, we use BMEO principle to label the position of each character in the entity. Through this method, we can get 25 different POS tags. The example of above result is shown in Fig. 9. Firstly, the inquiry data "Purchase hot-dip zinc coating coil, cr free, zero spangles" has segmented and tagged simply with the Jieba tool and manual intervention. Here, each character in the entity whose tag belonging to the above eight tags will be further tagged with the BMEO principle.

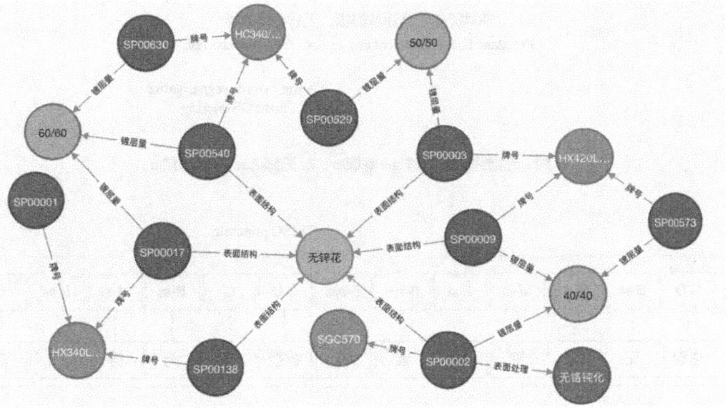

Fig. 7. A part of KG in displaying steel commodity data.

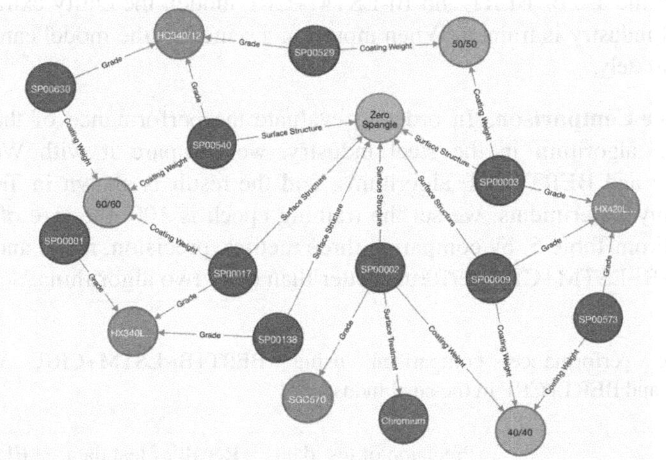

Fig. 8. The updated version of Fig. 7 after translating it from Chinese into English.

Table 4. The samples of word embedding obtained by BERT in the steel industry.

Chinese character	Word embedding
热	−0.28838387 0.5150681 …−0.4132588
镀	−0.44589975 0.41127107 …0.04480578
锌	−0.17859079 0.52611357 …0.04148377
C	−0.39107212 0.18615262 …−0.41394806
料	−0.25390878 0.8519357 …−0.42595372

Note: "热镀锌C料" appeared in this table means "GI Coil Grade DX51D".

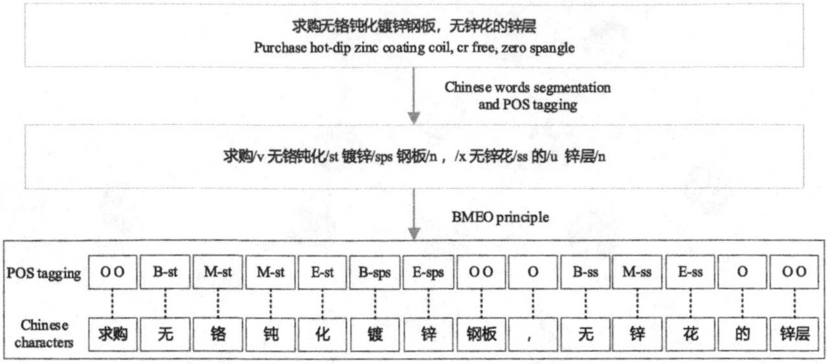

Fig. 9. The result of Chinese words segmentation, POS tagging and BMEO principle.

Through the use of BERT and Bi-LSTM+CRF model, the entity extraction model for the steel industry is trained. When inputting a sentence, the model can extract steel entities accurately.

Performance Comparison. In order to evaluate the performance of the BERT+Bi-LSTM+CRF algorithm in the steel industry, we compare it with Word2Vec+Bi-LSTM+CRF and BERT+CRF algorithms, and the result is shown in Table 5. Here, through many experiments, we set the training epoch is 100, the size of the training batch is 8. From Table 5, by comparing three metrics, precision, recall and F1, we find that BERT+Bi-LSTM+CRF performs better than other two algorithms.

Table 5. The performance comparison using BERT+Bi-LSTM+CRF, Word2Vec+Bi-LSTM+CRF and BERT+CRF in the steel industry.

Method	Precision of test data	Recall of test data	F1 of test data
Word2Vec+Bi-LSTM+CRF	87.31%	64.31%	0.74
BERT+CRF	87.50%	90.25%	0.89
BERT+Bi-LSTM+CRF	**89.71%**	**91.83%**	**0.91**

The KG-Based Search System for Steel E-Commerce. Based on the KG and entity extraction model, we develop a search system used in the steel e-commerce field. When a customer inputs a sentence as Fig. 10, there are 7 entities that can be extracted. After aligning entity with mathematical models and expert database, there are only 4 important entities related to the steel industry. By matching the 4 entities with the KG and comprehensively considering the expert system, we can find 43 related steel commodity IDs. Finally, we return the corresponding information of steel commodity data to the customer through these IDs. After translating Fig. 10 from Chinese into English, the updated version is shown in Fig. 11.

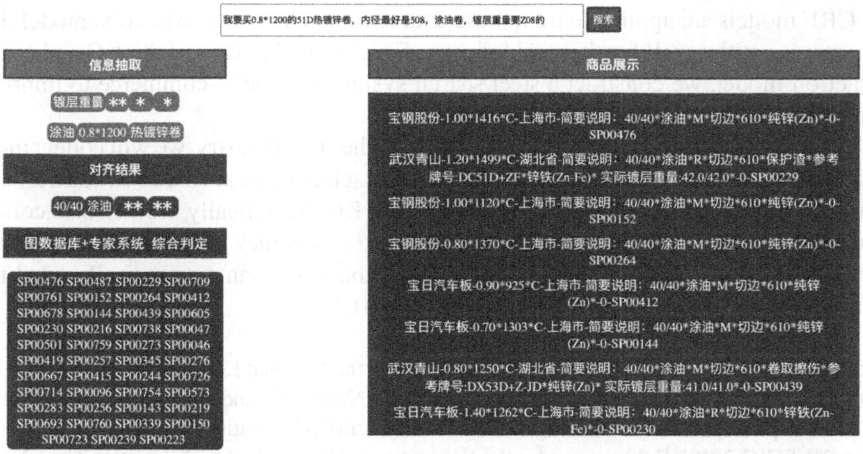

Fig. 10. The result of KG-based search system for steel industry.

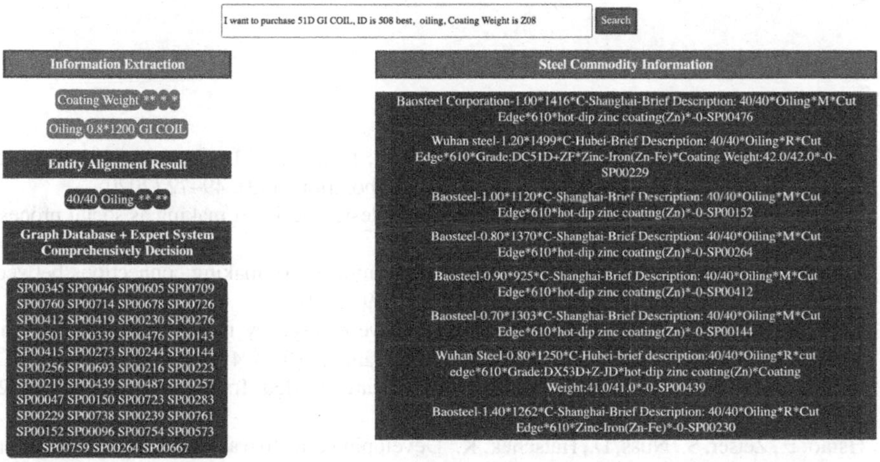

Fig. 11. The updated version of Fig. 10 after translating it from Chinese into English.

5 Conclusion

In this paper, through the incorporation of the CDM process into the steel e-commerce field, we construct a KG for the steel industry, train an entity extraction model which can accurately extract entities about steel from sentences, and develop a search system for the steel e-commerce, to serve the purpose of improving the system performance and optimizing user experience. Firstly, we collect steel commodity data from the Internet, and clean those data with some mathematical models and expert database in the steel industry to filter some important attributes. Secondly, these attributes as entities are stored in Neo4j graph database. Thirdly, Jieba tool, Chinese words segmentation, POS tagging, and BMEO principle are used for user inquiry data. Fourthly, BERT, Bi-LSTM, and CRF models are applied to train these data to obtain an entity extraction model that can extract entities well for the steel industry. Finally, on the basis of the KG and entity extraction model, we construct a steel search system for steel e-commerce to improve searching performance and user experience.

In the future, for expanding the scale of KG in the steel industry, we will collect more and more steel commodity data from the Internet, select more attributes as entities, and optimize the storage architecture of the data in the KG. Specifically, we will also collect more daily inquiry data about steel to improve the accuracy of the entity extraction model. In addition, we will further train a new model that can automatically label the unstructured steel data through BERT+Bi-LSTM+CRF.

Acknowledgment. This work was supported in part by the National Key R&D Program of China under Grant 2016YFC0600510, in part by the Beijing Natural Science Foundation under Grant 19L2029, in part by the Beijing Intelligent Logistics System Collaborative Innovation Center under Grant BILSCIC-2019KF-08, in part by the Scientific and Technological Innovation Foundation of Shunde Graduate School, USTB, under Grant BK19BF006, and in part by the Fundamental Research Funds for the University of Science and Technology Beijing under Grant FRF-BD-19-012A.

References

1. Benali, M., Ghomari, A.R., Zemmouchi-Ghomari, L., Lazar, M.: Crowdsourcing-enabled crisis collaborative decision making. Int. J. e-Collaboration **16**(3), 49–72 (2020)
2. Campbell, C., Roth, W., Jornet, A.: Collaborative design decision-making as social process. Eur. J. Eng. Educ. **44**(3), 294–311 (2019)
3. Chun, S., Shulman, S., Sandoval, R., et al.: Government 2.0: making connections between citizens, data and government. Inf. Polity **15**(1), 1–9 (2010)
4. Kapucu, N., Arslan, T., Demiroz, F.: Collaborative emergency management and national emergency management network. Disaster Prev. Manage. **19**(4), 452–468 (2010)
5. Lyndon, M., Angela, C.: Collaboration in health care. J. Med. Imaging Radiat. Sci. **48**(2), 207–216 (2017)
6. Hsiao, F., Zeiser, S., Nuss, D., Hatschek, K.: Developing effective academic accommodations in higher education: a collaborative decision-making process. Int. J. Music Educ. **36**(2), 244–258 (2018)
7. MacDonald, A., Clarke, A., Huang, L.: Multi-stakeholder partnerships for sustainability: designing decision-making processes for partnership capacity. J. Bus. Ethics **160**(2), 409–426 (2019)
8. Peng, P., Li, Y., Zhou, L.: Research on interactive collaborative decision-making method of equipment support task planning. In: 5th International Conference on Computer and Communication Systems, pp. 533–537. IEEE, Shanghai (2020)
9. Ebisu, T., Ichise, R.: Generalized translation-based embedding of knowledge graph. IEEE Trans. Knowl. Data Eng. **32**(5), 941–951 (2020)
10. Liu, Q., Li, Y., Duan, H., Liu, Y., Qin, Z.: Knowledge graph construction techniques. J. Comput. Res. Dev. **53**(3), 582–600 (2016)
11. Van Luijt, B., Verhagen, M.: Bringing semantic knowledge graph technology to your data. IEEE Softw. **37**(2), 89–94 (2020)
12. Zhou, J., Sun, X., Yu, X., Bian, X.: Knowledge graph and data application: intelligent recommendation. Telecommun. Sci. **35**(8), 165–172 (2019)

13. Jin, G., Lü, F., Xiang, Z.: Enterprise information integration based on knowledge graph and semantic web technology. J. SE Univ. (Nat. Sci. Edn.) **44**(2), 250–255 (2014)
14. Zhang, X., Liu, X., Li, X., Pan, D.: MMKG: An approach to generate metallic materials knowledge graph based on DBpedia and Wikipedia. Comput. Phys. Commun. **211**(2), 98–112 (2017)
15. Mrdjenovich, D., Horton, M., Montoya, J.H., Legaspi, C.M., Dwaraknath, S., Tshitoyan, V., Jain, A., Persson, K.A.: Propnet: a knowledge graph for materials science. Matter **2**(2), 464–480 (2020)
16. Devlin, J., Chang, M., Lee, K., Toutanova, K.: BERT: pre-training of deep bidirectional transformers for language understanding. In: Conference of the North American Chapter of the Association for Computational Linguistics: Human Language Technologies, pp. 4171–4186. ACL, Minneapolis, MN, USA (2018)
17. Hochreiter, S., Schmidhuber, J.: Long short-term memory. Neural Comput. **9**(8), 1735–1780 (1997)
18. Lafferty, J., Mccallum, A., Pereira, F.: Conditional random fields: Probabilistic models for segmenting and labeling sequence data. In: 18th International Conference on Machine Learning, pp. 282–289. Morgan Kaufmann, San Francisco (2002)
19. Steiner, T., Verborgh, R., Troncy, R., Gabarro, J., Van De Walle, R.: Adding realtime coverage to the Google knowledge graph. In: ISWC Posters and Demonstrations Track, pp. 65–68. CEUR-WS, Boston (2012)
20. He, P.: Counter Cyber Attacks by Semantic Networks: Emerging Trends in ICT Security. Morgan Kaufmann, Boston (2014)
21. Rahman, A.: Knowledge representation: a semantic network approach. Handbook of Research on Computational Intelligence Applications in Bioinformatics (2016)
22. Zhang, Y., Liu, X., Bai, X., Yin, J.: Collaborative research on intelligence perception and characterization of search engines. J. Beijing Inf. Sci. Technol. Univ. **34**(6), 19–24 (2019)
23. Huang, H., Yu, J., Liao, X., Xi, Y.: Review on knowledge graphs. Comput. Syst. Appl. **28**(6), 1–12 (2019)
24. Paulheim, H., Cimiano, P.: Knowledge graph refinement: a survey of approaches and evaluation methods. Seman. Web **8**(3), 489–508 (2017)
25. Zhang, Y., Dai, H., Kozareva, Z., Smola, A.J., Song, L.: Variational reasoning for question answering with knowledge graph. In: 32nd AAAI Conference on Artificial Intelligence, pp. 6069–6076. AAAI Press, New Orleans (2018)
26. Grishman, R., Sundheim, B.: Message understanding conference-6: a brief history. In: 16th International Conference on Computational Linguistics, pp. 466–471, Copenhagen (1996)
27. Yadav, V., Bethard, S.: A survey on recent advances in named entity recognition from deep learning models. In: 27th International Conference on Computational Linguistics, pp. 2145–2158. ACL, New Mexico (2018)
28. Goyal, A., Gupta, V., Kumar, M.: Recent named entity recognition and classification techniques: a systematic review. Comput. Sci. Rev. **29**, 21–43 (2018)
29. Li, J., Sun, A., Han, J., Li, C.: A survey on deep learning for named entity recognition. arXiv: 1812.09449 (2018)
30. Tata, S., Patel, J.: Estimating the selectivity of tf-idf based cosine similarity predicates. ACM SIGMOD Rec. **36**(4), 7–12 (2007)
31. Albitar, S., Fournier, S., Espinasse, B.: An effective TF/IDF-Based text-to-text semantic similarity measure for text classification. In: Benatallah, B., Bestavros, A., Manolopoulos, Y., Vakali, A., Zhang, Y. (eds.) WISE 2014. LNCS, vol. 8786, pp. 105–114. Springer, Cham (2014). https://doi.org/10.1007/978-3-319-11749-2_8
32. De Boom, C., Van Canneyt, S., Bohez, S., Demeester, T., Dhoedt, B.: Learning semantic similarity for very short texts. In: IEEE International Conference on Data Mining Workshop, Atlantic City, NJ, USA, pp. 1229–1234. IEEE, New York (2015)

33. Mikolov, T., Sutskever, I., Chen, K., Corrado, G., Dean, J.: Distributed representations of words and phrases and their compositionality. In: 26th International Conference on Neural Information Processing Systems, Nevada, pp. 3111–3119 (2013)
34. Wu, C., Wang, B.: Extracting topics based on Word2Vec and improved Jaccard similarity coefficient. In: IEEE Second International Conference on Data Science in Cyberspace, Shenzhen, China, pp. 389–397. IEEE, New York (2017)
35. Kenter, T., Rijke, M.: Short text similarity with word embeddings. In: 24th ACM International on Conference on Information and Knowledge Management, pp. 1411–1420. ACM (2015)
36. Huang, G., Guo, C., Kusner, M., Sun, Y., Weinberger, K.Q., Sha, F.: Supervised word mover's distance. In: 30th Conference on Neural Information Processing Systems. Neural Information Processing Systems Foundation, Barcelona, Spain, pp. 4862–4870 (2016)
37. Blanco, E., Moldovan, D.: A semantic logic-based approach to determine textual similarity. IEEE/ACM Trans. Audio Speech Lang. Processing **23**(4), 683–693 (2015)
38. Smarandache, F., Colhon, M., Vlăduţescu, Ş., Negrea, X.: Word-level neutrosophic sentiment similarity. Appl. Soft Comput. **80**, 167–176 (2019)
39. Lee, Y., Ke, H., Yen, T., Huang, H., Chen, H.: Combining and learning word embedding with WordNet for semantic relatedness and similarity measurement. J. Assoc. Inf. Sci. Technol. **71**(6), 657–670 (2020)

WSN Coverage Optimization Based on Two-Stage PSO

Wei Qi[1], Huiqun Yu[1,2(✉)], Guisheng Fan[1], Liang Chen[1], and Xinxiu Wen[1]

[1] Department of Computer Science and Engineering, East China University of Science and Technology, Shanghai, China
{yhq,gsfan}@ecust.edu.cn

[2] Shanghai Key Laboratory of Computer Software Evaluating and Testing, Shanghai, China

Abstract. Wireless Sensor Networks (WSN) coverage perception is an important basis for communication between the cyber world and the physical world in Cyber-Physical Systems (CPS). To address the coverage redundancy, hole caused by initial random deployment and the energy constraint in redeployment, this paper proposes a multi-objective two-stage particle swarm optimization algorithm (MTPSO) based on coverage rate and moving distance deviation to improve coverage efficiency. This algorithm establishes a multi-objective optimization model for above problems, and determines the candidate deployment scheme by reducing its local convergence probability through improved inertia weight, and then introduces virtual force mechanism to adjust the relative position between nodes. This paper mainly analyzes the influence of different initial deployment category and mobile nodes proportion on multi-objective optimization performance, and gives the corresponding algorithm implement. Simulation experiments show that compared with MVFA, SPSO and OPSO algorithms, MTPSO algorithm has a better redeployment coverage performance, which fully demonstrates its effectiveness.

Keywords: Hybrid WSN · Multi-objective optimization · Two-stage mechanism · Area coverage · Energy consumption balance

1 Introduction

With the development of Internet and artificial intelligence technology, CPS, a multi-dimensional complex system that integrates computing, network and physical environments, is being widely used in smart cities, smart manufacturing, national defense security and other fields [20]. WSN is located in the data perception layer of the smart city's technical architecture. It is responsible for the necessary data collection and measurement functions and participates in the important work of realizing the interaction between the CPS system and the physical environment [1].

H. Gao et al. (Eds.): CollaborateCom 2020, LNICST 349, pp. 19–35, 2021.
https://doi.org/10.1007/978-3-030-67537-0_2

WSN is composed of a series of sensing nodes and data storage receivers. Coverage awareness service is the main indicator for evaluating WSN service quality. Due to the limitations of some practical application environments, WSN mostly adopts random spraying for initial deployment. Although this method is more convenient, it often causes problems with coverage holes and redundancy. WSN sensing nodes are mainly powered by their own batteries [2]. Therefore, how to reduce and balance the energy consumption of redeployment is an important consideration for improving the working life of WSN.

From the perspective of perception range and energy, WSN redeployment is a multi-objective problem, and all feasible solutions need to be evaluated to determine the optimal solution. Particle swarm algorithm has the advantages of fewer parameters and fast convergence, but it's easy to fall into a locally optimal state. This paper proposes a multi-objective optimization model of WSN redeployment coverage based on the particle swarm algorithm, which aims to improve the area coverage and balance the mobile energy consumption among nodes, finally extend the overall operation time of the network. The main contributions of this article are as follows.

- A novel method is proposed to reduce the redeployment coverage problem in the hybrid WSN environment to a multi-objective optimization problem for area coverage rate and moving energy consumption deviation. Simultaneously, based on two randomly distributed scenarios, the above problem is analyzed.
- A two-stage method is proposed to optimize the multi-objective fitness function. On the premise of high-quality coverage, let as many mobile nodes as possible spread the deployment optimization work equally. At the same time, we regard the node movement as the result of the common action of various forces in a physical filed.

The rest of this paper is structured as follows: Sect. 2 introduces related work. Section 3 describes the WSN coverage analysis model and detailed redeployment optimization problems. Section 4 shows the design and implementation of our proposed algorithm. Section 5 describes the experimental environment settings and comparative experimental analysis. Section 6 summarizes this paper and looks forward to future work.

2 Related Work

Coverage perception is a research hotspot in the field of WSN. WSN can be divided into three states according to the mobility of nodes: static, mobile and hybrid. For different node types, there are different ways of coverage optimization.

The movement constraints of static nodes make it more restrictive in redeployment. Static WSN mostly adopts the node state rotation mechanism to expand the coverage and increase the working life of the network. Semprebom [14] proposed a (m, k)-Gur strategy, which provides a unified coverage monitoring application. The node autonomously performs a self-regulating choice, sends a message to the base station or sleeps to the next period. Habib Mostafaei [12] proposed a PCLA sleep scheduling algorithm that relies on an automatic learning machine to minimize the number of active sensors covering the monitoring area in order to improve the scalability and life cycle of the network. Ying Tian [17] made the nodes which have more remaining energy become work state to achieve the alternation of the multiple coverage sets. J. Sahoo [13] proposed a greedy algorithm based heuristic to schedule the sensors states in order to increase the network lifetime. Riham Elhabyan [5] combined the coverage-aware sleep scheduling with the clustering and proposed an discrete energy consumption model which depends on the status of sensors rather than distance.

Compared with static nodes, there are more research methods for mobile node optimization redeployment. Tongxin Shu [16] proposed a coverage optimization algorithm combining Voronoi Diagram and geometric center to maximize area coverage and minimize energy consumption under the constraints of communication range. Yaobing [10] proposed to adjust virtual force parameters based on the energy of neighbor nodes, and measure the uniformity as an experimental indicator. In addition, biologically heuristic intelligent optimization algorithms have also received extensive attention in this field. Mihoubi [11] proposed an algorithm that uses hybridization to improve the speed of bats and combines the Doppler effect to promote positioning performance. Bin Yu [8] proposed a periodic hybridization method to enhance the global search capability of the PSO algorithm and obtained higher convergence speed and area coverage rate. Amulya Anurag [3] proposed a negative speed PSO algorithm to improve the local search ability and obtain higher coverage efficiency. Tingli Xiang [19] proposed an optimization based on Cuckoo Search (CS) which divided the algorithm into two stages. The former is responsible for improving coverage rate, while the latter aims to reduce the average moving distance of mobile nodes and reduce energy consumption. S.T. Hasson [7] utilized the angle between two adjacent nodes to assign an optimal deployment scheme.

Now researchers are not only pursuing high coverage rate, but also tend to consider redeployment problem from the perspective of energy consumption. This paper describes the redeployment coverage problem as a multi-objective optimization, and proposes an improved virtual force particle swarm algorithm based on the characteristics of WSN mobile nodes. This paper uses weight coefficients to balance the goals, aiming to increase coverage rate and reduce mobile energy consumption deviation, so as to improve the coverage perception ability and network running time of WSN.

3 Analysis Model and Problem Statement

3.1 Network Model

For the convenience of research, this paper divides the two-dimensional monitoring area into $L_x \times L_y$ grids, and analyzes the N mobile nodes and M static nodes scattered in the binomial disc model shown in Fig. 1. Except for mobility, other

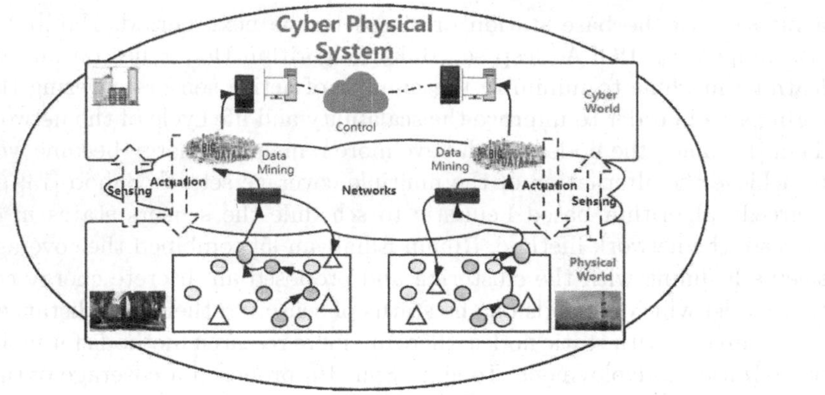

Fig. 1. The system architecture

properties of sensing nodes are the same. The perceived probability of nodes S_i to grid point q is shown in Eq. (1).

$$C(S_i, q) = \begin{cases} 1 \; d(S_i, q) \leq r \\ 0 \quad d(S_i, q) > r \end{cases} \tag{1}$$

Where r is the perception radius. $d(S_i, q)$ is the Euclidean distance S_i to the grid nodes q. Therefore, the node perception coverage rate Cov of the monitoring area is shown in Eq. (2).

$$Cov = \sum_{q=1}^{L_x \times L_y} \sum_{i=1}^{N+M} \frac{C(S_i, q)}{L_x \times L_y} \tag{2}$$

The moving distance deviation of all nodes is expressed as:

$$Std\_S = \sqrt{\frac{(\sum_{i=1}^{N} (dis_i - avg\_dis)^2)}{N \times (N-1)}} \tag{3}$$

dis_i is the moving distance of node S_i, avg_dis is the average moving distance of all nodes in the monitoring area. This paper focuses on the coverage optimization of hybrid WSN redeployment, which aims to improve area coverage rate and reduce the energy consumption deviation of nodes. Reference [10] knows that the energy consumption of a node movement is directly related to its distance. Therefore, the problem of moving energy consumption deviation can be converted into a moving distance deviation. Now, the multi-objective optimization model is shown in Eq. (4).

$$\begin{cases} max\_Cov(x) \\ min\_Std\_dis(x) \\ x \in L \end{cases} \tag{4}$$

In Eq. (4), the former two terms are respectively the maximum optimization of area coverage rate and the minimum optimization of movement deviation. The last one represents the boundary constraints of node movement. We refer to [6] normalize the targets value and utilize α to balance their weight. Based on the calculation of coverage rate and movement deviation in the redeployment optimization process, the fitness function of our algorithm is shown in Eq. (5).

$$fitness(x) = \alpha \times f(Cov(x)) + (1-\alpha) \times f(Std\_dis(x)) \tag{5}$$

We consider the WSN redeployment coverage that maximizes the fitness function $fitness(x)$ under boundary constraints.

3.2 Problem Statement

Based on the above analysis model, we mean to find a WSN redeployment scheme x that maximizes the fitness under boundary constraints. The problem is described formally by Eq. (6).

$$\begin{cases} find \quad x = (x_1, x_2 \cdots, x_N) \\ which\, max(\alpha \times f((Cov(x)) + (1-\alpha) \times f(Std\_dis(x)) \\ s.t. \quad x_i \in L \end{cases} \tag{6}$$

This is a D-dimensional decision vector. The model parameters and corresponding descriptions are summarized as shown in Table 1.

Table 1. Summary of model parameters

parameter	description
L	Monitoring area boundary
$L_x \times Ly$	Area grid number
r	Perception radius
R	Communication radius
$d(i, j)$	Euclidean distance between i and j
N, M	Number of mobile and static nodes
x	Node location for redeployment
Sum_dis	Total travel distance
Std_dis	Node moving distance deviation
Cov_i	Area coverage of deployment solution i
$fitness$	Multi-objective function fitness

4 Multi-objective Two-Stage PSO Algorithm

4.1 Algorithm Parameters in WSN Deployment

Although there are many other swarm intelligence algorithms. They have some inapplicability based on the current scene. For example, cuckoo search algorithm(CS) [19] uses Levy flight mechanism for birds, which is a random walk consisting of short-distance flight with small step length and occasional long-distance flight with large step length. So it is easy to appear in the area around the global optimal shock phenomenon, lower algorithm efficiency, etc. Besides, genetic algorithm (GA) is also a common swarm intelligence algorithm. However, it need coding and has crossover, mutation operation. So GA has many parameters and is complicated to implement [4].

Based on the observation of animal by individuals in the group to make the swarm activity behaviors, the particle swarm algorithm (PSO) uses the sharing of information movement of the entire group produce an evolutionary process from disorder to order in the problem solving space to obtain the global optimal solution [15]. On the one hand, PSO algorithm may converge to the optimal solution more rapidly. On the other hand, it does not need coding, does not have many parameters, only through the current search to share information, and uses internal speed to update, the principle is simpler, and the implementation is easier. Last but not least, PSO algorithm is mainly applied to continuous problem. Therefore, PSO algorithm is suitable for WSN redeployment coverage problem.

$$V_i{}^k = w \times V_i{}^{k-1} + c_1 \times r_1 \times (pbest_i - X_i{}^{k-1}) + c_2 \times r_2 \times (gbest_i - x_i{}^{k-1}) \quad (7)$$

$$X_i{}^k = X_i{}^{k-1} + V_i{}^k \quad (8)$$

This paper mainly analyzes the coverage problem of mobile node redeployment in the two-dimensional monitoring area. Assuming that the population size Num is the number of candidate schemes, the search space dimension is related to the coordinate dimension of mobile nodes in the monitoring area. The speed and position update equation of the particle swarm algorithm are shown in Eq. (7) and Eq. (8) which represent the nodes' movement change and the redeployment scheme currently. In Eq. (7), c_1 and c_2 are the weight coefficients of the particle tracking the historical local and global optimization redeployment solution respectively. In the original particle swarm algorithm, the inertia weight decreases linearly with the number of the iterations increases, as shown in Eq. (9). $Maxiter$ is the maximum number of iterations, and k is the current number of iterations.

$$w^k = w_{max} - (w_{max} - w_{min}) \times \frac{k}{Maxiter} \quad (9)$$

4.2 Algorithm Design

The original particle swarm algorithm is prone to fall into the dilemma of local optimal. Therefore, it must be improved to obtain the global optimal solution. This paper refers to the idea of [19] and divides the important into two parts. Firstly, set the inertia weight to a non-linear decreasing form to better balance the global and local optimization capabilities of searching particles. Secondly, abstract the monitoring area as a physical field, and introduce a virtual force mechanism to promote the uniformity of node distribution. The algorithm model is shown in Fig. 2.

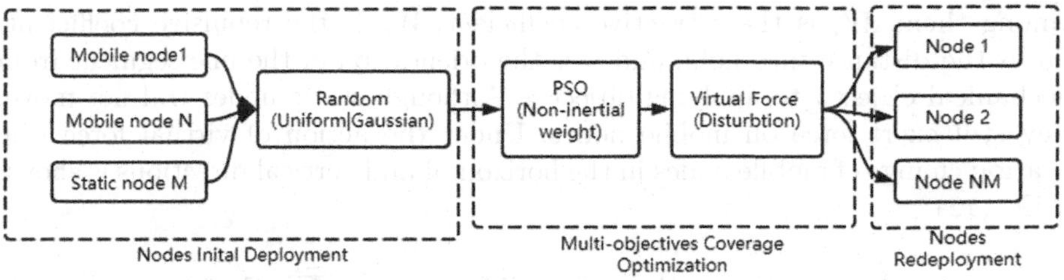

Fig. 2. The architecture of our algorithm

Non-inertial Weight

$$w = w_{min} + (w_{max} - w_{min}) \times exp(-20 \times (\frac{k}{Maxiter})^5) \tag{10}$$

Because the slope of the linearly decreasing inertia weight is fixed, the change of speed always remains at the same level. If the initial iteration does not produce good results, then as the iteration increases and the speed decays, it is likely to fall into a local optimum in the end. As shown the $w1$ in Fig. 3, the nonlinear inertia weight proposed in this paper has a larger parameter value in the early stage, which improves the global search ability. Keeping a smaller value in the later stage is beneficial to local search and accelerate the convergence of the algorithm.

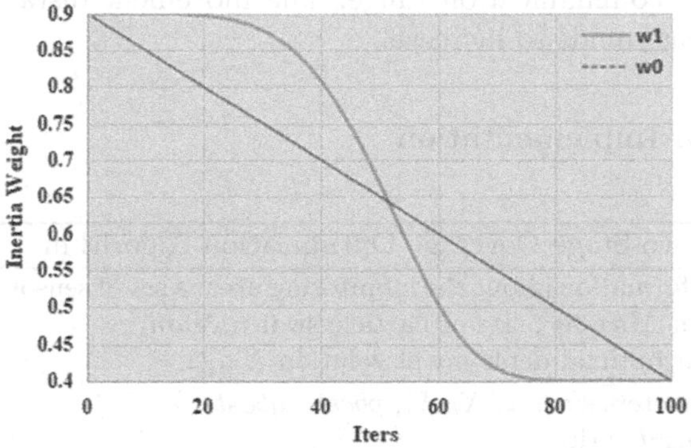

Fig. 3. The change of inertia weight with the number of iterations

Virtual Force Mechanism. This paper abstracts the entire monitoring area as a physical field, and the change of node location is regarded as the result of force objects under the influence of various forces. It refers to [18] to calculate the repulsive force of a mobile node by its neighbor nodes and the attractive force of surrounding uncovered grid points. The specific force analysis is shown in Eq. (11).

$$\boldsymbol{F}_{ij} = \begin{cases} W_a \times d_{ij}, \alpha_{ij} & D_{th} < d_{ij} < R \\ 0, & d_{ij} = D_{th} || d_{ij} > R \\ W_r \times d_{ij}, \alpha_{ij} + \pi & otherwise \end{cases} \tag{11}$$

Among them, W_a is the attractive coefficient, W_r is the repulsive coefficient, D_{th} is the distance threshold, and α is the orientation of the line segment from mechanical object i to applying object j. Although static nodes will not move, they still exert force on mobile nodes. Under the action of virtual force, the location change of mobile nodes in the horizontal and vertical directions is shown in Eq. (12).

$$\begin{cases} x_i(k+1) = x_i(k) + \frac{F_x}{F_{xy}} \times MaxStep \times e^{\frac{-1}{F_{xy}}}, F_x \neq 0 \\ y_i(k+1) = y_i(k) + \frac{F_y}{F_{xy}} \times MaxStep \times e^{\frac{-1}{F_{xy}}}, F_y \neq 0 \end{cases} \tag{12}$$

Among them, $x_i(k)$ and $y_i(k)$ are the original position coordinates of node S_i before the change, F_x and F_y are the horizontal and vertical components of the resultant force on node S_i. When the calculation result of the node position change exceeds the scope of the monitoring area, the latest position is set as the boundary value closest to the result, as shown in Eq. (13).

$$\begin{cases} x_i(k+1) = L_x,\ x_i(k+1) \notin L_x \\ y_i(k+1) = L_y,\ y_i(k+1) \notin L_y \end{cases} \tag{13}$$

Compared with other interference strategies [8], the virtual force method calculates the moving distance and direction of the target node by analyzing the relative distance between the target nodes and other nodes and uncovered grid points within the communication range. The movement distance is small and the overall coverage tends to increase.

4.3 Algorithm Implementation

Algorithm 1. Two-Stage Coverage Optimization Algorithm

Input: Related information about the monitoring area, a set of sensor nodes, maximum number of iteration $Maxiter$, size of particle swarm Num.
Output: The near optimal deployment solution X_{best}.

```
Initialize parameters such as X_i, V_i, pbest_i, gbest_i
for j=1 to Maxiter do
    for i=1 to Num do
        Update V_i, X_i
        Compute Cov_i, Std_dis, F_i corresponding to X_i
        F_i with pbest_i and making the higher be pbest_i.
    for i=1 to Num do
        Compute the force on mobile sensor nodes
        Update the location of mobile nodes
        Compare the current F_i forX_i and update the X_best
    Selecting the best near-optimal one of maximum fitness from Num deployment
  solutions as X_best
return X_best
```

Based on the above-mentioned inertial weight improvement and virtual force mechanism, we design a Multi-Objective Two-Stage PSO algorithm, which can be abbreviated as MTPSO, to reduce WSN deployment coverage problem. The specific implementation is shown in the pseudo code of Algorithm 1. MTPSO firstly initializes the position of the particles (that is the randomly distributed node deployment plan), speed and other necessary parameter settings (see 1). Then it utilizes a two-stage iterative optimization within the $Maxiter$ iterations constraint, of which the first stage is updating the speed and position, providing candidate target deployment (see 4–6); the second stage is adjusting the relative position among nodes according to the virtual force mechanism to optimize the candidate target position (see 8–10). Finally, it updates the global optimal deployment plan of the current particle swarm, according to the value of $fitness$ (see 11).

5 Experimental Evaluation

5.1 Environment Parameter

The experimental environment of this article uses MATLAB 2014R. The size of the monitoring area is $L_x \times L_y = 20 \times 20$, where the grid size is $px \times py = 1 \times 1$, the node perception radius r=3, and the communication radius $R = 2 \times r$. In practical applications, the initial deployment of WSN nodes affected by the physical environment mostly presents the data characteristics of Gaussian distribution [9]. This paper analyzes two initial deployments: Gaussian randomness and uniform randomness. Experiment parameters are shown in Table 2.

Table 2. Summary of model parameters

parameter	value
Num	100
$Maxiter$	100
w_{min}	0.4
w_{max}	0.9
c_1, c_2	2,2
r_1, r_2	[0,1]
D_{th}	$1.4 \times r$
W_a, W_r	1,1
$MaxSensor, MaxStep$	1,0.5
Pr	[0,0.2,0.4,0.6,0.8,1]

5.2 Determination of Target Weight

To determine α in Eq. (5), which ranges in 0.1–0.9. We conduct comparative experiments for different values of α when the proportion of mobile nodes is 1. The experimental results are obtained by 50 independent replicates, as shown in Table 3. It is obtained through experiments that when the coverage rate is used for single-objective optimization, the worst *Std_dis* of the MVFA algorithm is 9.0647. Therefore, 10 is regarded as the empirical maximum value of the movement deviation. It can be seen from Table 3 that with the coverage rate target coefficient increases, the value of *fitness* gradually increases.

Table 3. Summary of model parameters

α	0.1	0.2	0.3	0.4	0.5	0.6	0.7	0.8	0.9
Fitness	0.0778	0.1705	0.2610	0.3548	0.4482	0.5448	0.6474	0.7434	0.8469
Cov	0.8479	0.9195	0.9372	0.9432	0.9488	0.9496	0.9543	0.9509	0.9517
Std_dis	0.1551	0.3342	0.5750	0.7502	1.0488	1.2466	1.3750	1.7349	1.9248

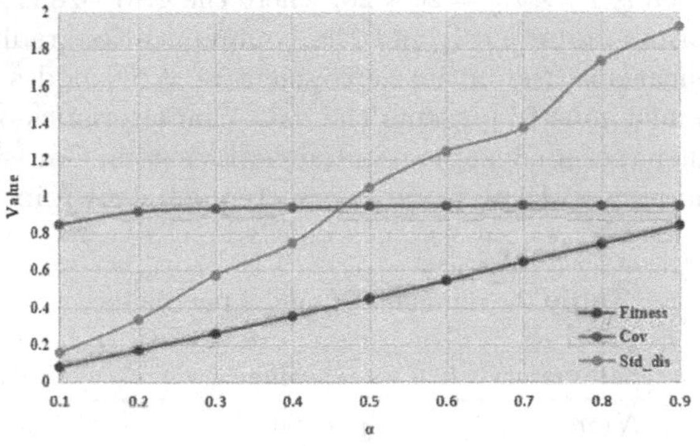

Fig. 4. Changes in fitness and optimization goals under different weights

As the weight coefficient α increases, the optimization target value of the coverage rate in the fitness function becomes more and more important, while the importance of the moving distance deviation decreases. Both targets tend to be optimized with larger weight coefficients. However, two targets cannot be maximized at the same time. The variance of the coverage rate is 0.0060, while the variance of the movement distance deviation is 0.5140, indicating that as the weight increases, the coverage rate fluctuates less than the deviation as shown in Fig. 4. In order to obtain the near-optimal solution, we set a largest weight for the coverage rate target, which is 0.9.

5.3 Algorithm Comparison

We compare MTPSO with typical coverage optimization algorithms, including OPSO, EPSO in [8] and MVFA in [10]. OPSO is the original particle swarm algorithm, and its inertia weight decreases linearly with the increase of iterations. EPSO uses periodic mutation to improve the global search ability. MVFA is based on a traditional virtual force algorithm, which adjusts the attractive force coefficients.

Comparison of Goals and Fitness Values. The performance of the four algorithms is developed from multiple perspectives: *fitness*, *Std_dis* and *Cov*. In addition, the energy consumption in the redeployment process is directly related to the movement of the node. Therefore, we also compare the total moving distance and the deviation of the moving distance among nodes. We present experimental configurations of five different scale networks with a ratio of mobile nodes ranging from 0.2 to 1 with an interval of 0.2 under two different initial deployments.

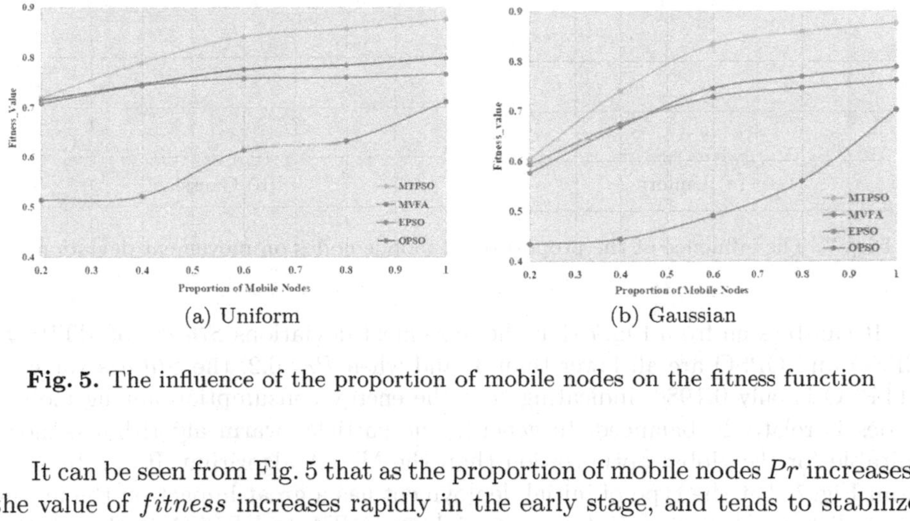

(a) Uniform (b) Gaussian

Fig. 5. The influence of the proportion of mobile nodes on the fitness function

It can be seen from Fig. 5 that as the proportion of mobile nodes Pr increases, the value of $fitness$ increases rapidly in the early stage, and tends to stabilize in the later stage. This shows that the mobility of the node has a very positive effect on the optimization, but the final effect will gradually decrease with the increase of Pr. Since the target weight α selected in this experiment is 0.9, the change trend of $fitness$ and Cov is roughly same as shown in Fig. 5 and Fig. 6. At the same time, MTPSO has the best coverage rate under both random initial deployments as shown in Fig. 6.

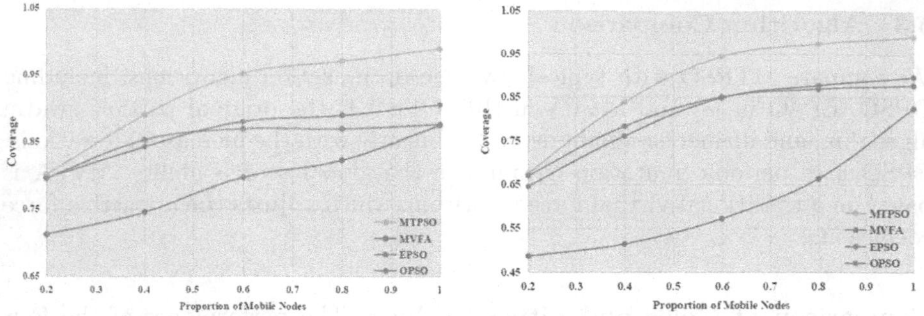

(a) Uniform (b) Gaussian

Fig. 6. The influence of the proportion of mobile nodes on coverage

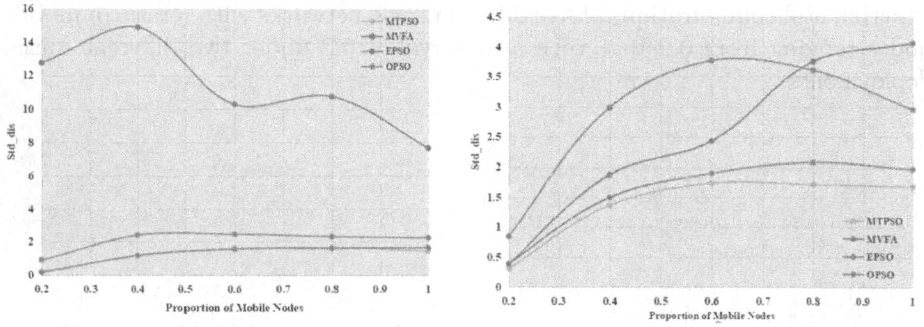

(a) Uniform (b) Gaussian

Fig. 7. The influence of the proportion of mobile nodes on movement deviation

It can be seen from Fig. 7 that the movement deviations *Std_dis* of MTPSO, EPSO and OPSO are all lower than 3, and when Pr=0.2, the *Std_dis* value of MTPSO is only 0.1986, indicating that the energy consumption among mobile nodes is relatively balanced. In general, the particle swarm algorithm is more suitable for the global optimization than the MVFA algorithm. It can be seen from Fig. 7 that the type of initial deployment has a great impact on the movement distance deviation *Std_dis*, especial for MVFA and EPSO. Although the *Sum_dis* of MVFA indicators are both maximum in both types, but the one in Gaussian is much smaller than that in the uniform type. However, the *Std_dis* of EPSO performs better in uniform initial deployment. This is because of the two optimization measures are based on different objects. MVFA uses local virtual force to promote the local optimization, while EPSO directly uses random arrangement of the global mutation.

As shown in Fig. 8, the total moving distance *Sum_dis* of MVFA is largest. Other algorithms are always kept at 350 or less. When Pr=1, the *Sum_dis* of EPSO is 305.9710. Under the premise that the Sum_{dis} difference of particle swarm-based algorithms is small. So, the smaller the value of *Std_dis*, the more

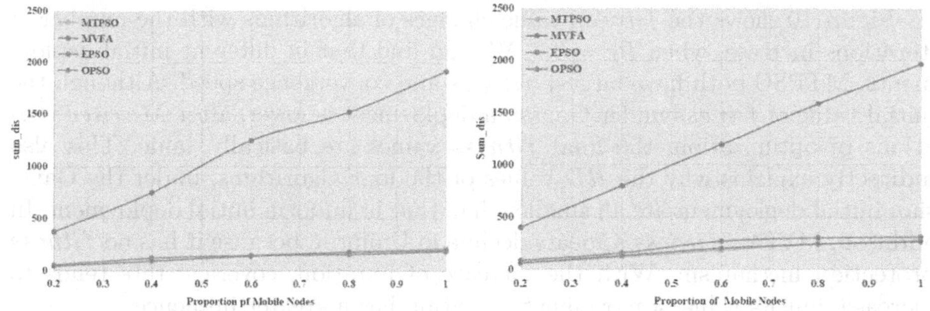

(a) Uniform (b) Gaussian

Fig. 8. The influence of the proportion of mobile nodes on the total moving distance

balance of the moving energy consumption, and the coverage holes are less likely to occur.

Performance Optimization Comparison. In this paper, the index RD is used to compare the coverage efficiency of the redeployment scheme, which is the ratio of the coverage rate to the average moving distance, shown in Eq. (14).

$$RD = \frac{Cov}{Sum\_dis/(NM \times Pr)} \tag{14}$$

Under the premise that other conditions remain unchanged, the coverage efficiency RD will increase as the proportion of mobile nodes Pr increases, but the RD change trend of different algorithms is different. As shown in Fig. 9, MTPSO has the best coverage efficiency when $Pr < 0.8$. However, as Pr increases in the later stage, the algorithm advantage will gradually decrease.

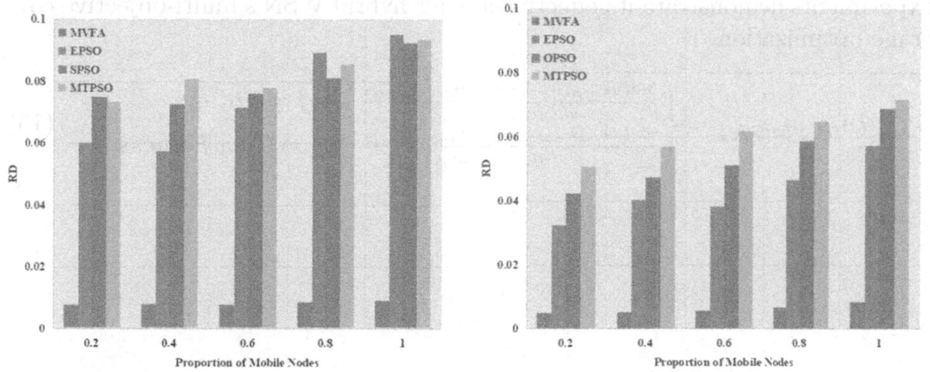

(a) Uniform (b) Gaussian

Fig. 9. The influence of the proportion of mobile nodes on RD

Figure 10 shows the $fitness$ value changes of algorithms with the number of iterations increase, when $Pr = 0.8$. We can find that in different initial deployments, MTPSO both have largest $fitness$ and convergence speed. Although the initial value of $fitness$ under Gaussian deployment is lower, after $Maxiter$ iterations of optimization, the final $fitness$ values are basically same. This also indirectly explains why the RD values of the four algorithms, under the Gaussian initial deployment are all smaller than that in uniform initial deployment. In addition, MVFA's $fitness$ appears decline in Uniform, because it has no $fitness$ protection mechanism. With the increase of iteration, coverage rate tends to increase, but its value of movement deviation has a greater influence.

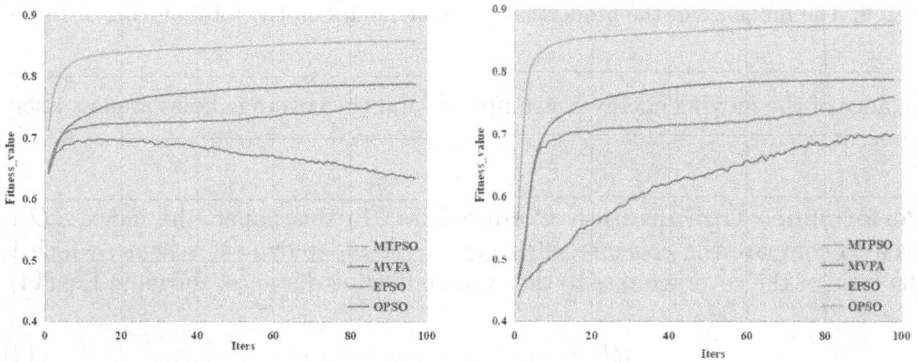

(a) Uniform (b) Gaussian

Fig. 10. The changes in fitness with the number of iterations

We calculate the performance improvement rate Imp of MTPSO compared with OPSO, EPSO and MVFA, as shown in Eq. (15). It is demonstrated that MTPSO has a significant improvement on algorithms, as shown in Table 4. The experiments demonstrate its effectiveness for hybrid WSN's multi-objective coverage optimization.

$$Imp_{(other,obj)} = \begin{cases} \frac{value_{other,obj} - value_{(ours,obj)}}{value_{(other,obj)}} & obj = Std\_dis \\ \frac{value_{ours,obj} - value_{(other,obj)}}{value_{(other,obj)}} & obj = Cov||Fitness \end{cases} \tag{15}$$

Table 4. The performance improvement rate of this algorithm compared with other algorithms

Initial Deployment	Uniform					Gaussian				
Pr	0.2	0.4	0.6	0.8	1	0.2	0.4	0.6	0.8	1
$Imp_{(MVFA,Cov)}$	12.51%	19.42%	19.64%	18.08%	13.14%	38.75%	62.93%	65.26%	46.61%	19.97%
$Imp_{(MVFA,Std\_dis)}$	98.44%	92.11%	84.76%	84.52%	80.82%	19.48%	27.44%	28.73%	54.31%	58.69%
$Imp_{(MVFA,Fitness)}$	40.07%	51.30%	37.15%	35.57%	23.27%	39.26%	66.75%	69.93%	53.38%	24.51%
$Imp_{(ESPO,Cov)}$	0.70%	4.34%	9.59%	11.68%	12.99%	0.85%	7.02%	11.02%	11.91%	12.75%
$Imp_{(ESPO,Std\_dis)}$	77.99%	50.96%	36.10%	27.66%	34.26%	63.60%	54.49%	53.93%	52.47%	43.38%
$Imp_{(ESPO,Fitness)}$	1.71%	6.12%	11.07%	12.88%	14.37%	1.76%	9.75%	14.38%	15.03%	14.93%
$Imp_{(OSPO,Cov)}$	0.69%	5.64%	8.26%	9.15%	9.28%	4.59%	10.17%	11.26%	10.90%	10.37%
$Imp_{(OSPO,Std\_dis)}$	-36.43%	1.61%	1.47%	-2.57%	11.89%	22.26%	8.90%	8.69%	17.62%	14.83%
$Imp_{(OSPO,Fitness)}$	0.62%	5.76%	8.46%	9.29%	9.72%	4.77%	10.59%	11.77%	11.67%	11.00%

6 Summary

In this paper, the redeployment problem based on hybrid WSN is described as a multi-objective optimization problem, and a two-stage coverage optimization algorithm is proposed to reduce it, which improves the area coverage rate and reduce the moving energy consumption deviation of nodes. Meanwhile, in order to reduce the problem that PSO is easy to fall into the local optimum and promote the convergence speed, MTPSO utilizes the nonlinear inertia weight as shown in Eq. (10) to balance the global and local search capabilities and obtain better candidate target positions. Then it introduces the virtual force mechanism to adjust the relative position among nodes. It can be found from Table 4 that compared with OPSO, EPSO and MVFA algorithms, it performs better in most cases, which proves its effectiveness for solving multi-objective redeployment coverage problem in hybrid WSN. In the future, we plan to take other optimization targets into consideration and analyze the hybrid WSN redeployment coverage problem with obstacles and hot spots, in order to optimize redeployment performance based on the above scenarios.

Acknowledgements. This work was partially supported by the National Natural Science Foundation of China under Grant nos. 61702334 and 61772200, Shanghai Municipal Natural Science Foundation under Grant nos. 17ZR1406900 and 17ZR1429700, the Planning Project of Shanghai Institute of Higher Education under Grant no. GJEL18135.

References

1. Abid, A., Kachouri, A., Mahfoudhi, A.: Data analysis and outlier detection in smart city. In: 2017 International Conference on Smart, Monitored and Controlled Cities (SM2C), pp. 1–4 (2017)
2. Amutha, J., Sharma, S., Nagar, J.: WSN strategies based on sensors, deployment, sensing models, coverage and energy efficiency: review, approaches and open issues. Wireless Pers. Commun. **111**(2), 1089–1115 (2019). https://doi.org/10.1007/s11277-019-06903-z

3. Anurag, A., Priyadarshi, R., Goel, A., Gupta, B.: 2-D coverage optimization in WSN using a novel variant of particle swarm optimisation. In: 2020 7th International Conference on Signal Processing and Integrated Networks (SPIN), pp. 663–668 (2020)
4. Chaturvedi, Y., Kumar, S., Bansal, P., Yadav, S.: Comparison among APSO, PSO GA for performance investigation of SEIG with balanced loading. In: 2019 9th International Conference on Cloud Computing, Data Science Engineering (Confluence), pp. 459–463 (Jan 2019). https://doi.org/10.1109/CONFLUENCE.2019.8776887
5. Elhabyan, R., Shi, W., St-Hilaire, M.: A full area coverage guaranteed, energy efficient network configuration strategy for 3D wireless sensor networks. In: 2018 IEEE Canadian Conference on Electrical Computer Engineering (CCECE), pp. 1–6 (2018)
6. Fan, G., Chen, L., Yu, H., Qi, W.: Multi-objective optimization of container-based microservice scheduling in edge computing (in press). Computer Science and Information Systems
7. Hasson, S.T., Finjan, A.A.R.: A suggested angles-based sensors deployment algorithm to develop the coverages in WSN. In: 2018 2nd International Conference on Inventive Systems and Control (ICISC), pp. 547–552 (2018)
8. Kong, H., Yu, B.: An improved method of WSN coverage based on enhanced pso algorithm. In: 2019 IEEE 8th Joint International Information Technology and Artificial Intelligence Conference (ITAIC), pp. 1294–1297 (2019)
9. Kumar, V., et al.: Optimal cluster count and coverage analysis in a Gaussian distributed WSNs using TESM. In: Satapathy, S.C., Bhateja, V., Nguyen, B.L., Nguyen, N.G., Le, D.-N. (eds.) Frontiers in Intelligent Computing: Theory and Applications. AISC, vol. 1014, pp. 335–345. Springer, Singapore (2020). https://doi.org/10.1007/978-981-13-9920-6_35
10. Li, Y., Zhang, B., Chai, S.: An energy balanced-virtual force algorithm for mobile-WSNs. In: 2015 IEEE International Conference on Mechatronics and Automation (ICMA), pp. 1779–1784 (2015)
11. Mihoubi, M., Rahmoun, A., Lorenz, P., Lasla, N.: An effective bat algorithm for node localization in distributed wireless sensor network. Secur. Priv. **1**, e7 (2018)
12. Mostafaei, H., Montieri, A., Persico, V., Pescap, A.: An efficient partial coverage algorithm for wireless sensor networks. In: 2016 IEEE Symposium on Computers and Communication (ISCC), pp. 501–506 (2016)
13. Sahoo, J., Sahoo, B.: Solving target coverage problem in wireless sensor networks using greedy approach. In: 2020 International Conference on Computer Science, Engineering and Applications (ICCSEA), pp. 1–4 (2020)
14. Semprebom, T., Montez, C., Arajo, G., Portugal, P.: A sleep-scheduling scheme for enhancing QoS and network coverage in IEEE 802.15.4 WSN. In: 2015 IEEE World Conference on Factory Communication Systems (WFCS), pp. 1–4 (2015)
15. Şenel, F.A., Gökçe, F., Yüksel, A.S., Yiğit, T.: A novel hybrid PSO–GWO algorithm for optimization problems. Eng. Comput. **35**(4), 1359–1373 (2018). https://doi.org/10.1007/s00366-018-0668-5
16. Shu, T., Dsouza, K.B., Bhargava, V., de Silva, C.: Using geometric centroid of Voronoi diagram for coverage and lifetime optimization in mobile wireless sensor networks. In: 2019 IEEE Canadian Conference of Electrical and Computer Engineering (CCECE), pp. 1–5 (2019)
17. Tian, Y., Wang, X., Jiang, Y., You, G.: A distributed probabilistic coverage sets configuration method for high density WSN. In: 2017 Chinese Automation Congress (CAC), pp. 2312–2316 (2017)

18. Wei, D., Huang, S., Bu, X.W.: A sensor deployment approach using improved virtual force algorithm based on area intensity for multisensor networks. Math. Probl. Eng. **2019**, 1–9 (2019). https://doi.org/10.1155/2019/8015309
19. Xiang, T., Wang, H., Shi, Y.: Hybrid WSN node deployment optimization strategy based on CS algorithm. In: 2019 IEEE 3rd Information Technology, Networking, Electronic and Automation Control Conference (ITNEC), pp. 621–625 (2019)
20. Yetis, H., Karakose, M.: A cyber-physical-social system based method for smart citizens in smart cities. In: 2020 24th International Conference on Information Technology (IT), pp. 1–4 (2020)

A Covert Ultrasonic Phone-to-Phone Communication Scheme

Liming Shi[1], Limin Yu[1], Kaizhu Huang[1], Xu Zhu[2], Zhi Wang[3], Xiaofei Li[4], Wenwu Wang[5], and Xinheng Wang[1(✉)]

[1] Xi'an Jiaotong-Liverpool University, Suzhou, China
liming.shi19@student.xjtlu.edu.cn,
{limin.yu,kaizhu.huang,xinheng.wang}@xjtlu.edu.cn
[2] University of Liverpool, Liverpool, UK
xuzhu@liverpool.ac.uk
[3] Zhejiang University, Hangzhou, China
zjuwangzhi@zju.edu.cn
[4] Westlake University, Hangzhou, China
lixiaofei@westlake.edu.cn
[5] University of Surrey, Guilford, UK
w.wang@surrey.ac.uk

Abstract. Smartphone ownership has increased rapidly over the past decade, and the smartphone has become a popular technological product in modern life. The universal wireless communication scheme on smartphones leverages electromagnetic wave transmission, where the spectrum resource becomes scarce in some scenarios. As a supplement to some face-to-face transmission scenarios, we design an aerial ultrasonic communication scheme. The scheme uses chirp-like signal and BPSK modulation, convolutional code encoding with ID-classified interleaving, and pilot method to estimate room impulse response. Through experiments, the error rate of the ultrasonic communication system designed for mobile phones can be within 0.001% in 1 m range. The limitations of this scheme and further research work are discussed as well.

Keywords: Smartphone · Aerial ultrasonic communication

1 Introduction

With smartphones being the most popular personal devices and the platforms for supporting smart services, the demands for reliable wireless communications rise rapidly. Communications based on electromagnetic (EM) waves are the main stream of wireless communications. Various protocols have been developed over the years to cover short-range and long-range communications, such as Wi-Fi, Bluetooth, and the latest 5G mobile communications. However, the spectra of electromagnetic waves are becoming scarce and congested. Under this circumstance, exploiting new transmission media and increasing bandwidth efficiency are two mainstream solutions.

H. Gao et al. (Eds.): CollaborateCom 2020, LNICST 349, pp. 36–48, 2021.
https://doi.org/10.1007/978-3-030-67537-0_3

Underwater navigation and communications have utilized acoustic wave since the last century [1,2]. Meanwhile, acoustic wave, especially ultrasonic component, has an abundant untapped bandwidth capacity in the aerial environment, but was not yet fully exploited. This work intends to utilise this resource and design an applicable scheme in face-to-face transmission scenarios. For example, a customer can get his ticket from the cashier without being in contact in a low-EM-signal-quality environment by using this scheme. This is particularly useful under the emergency situations, such as the outbreak of Covid-19 pandemic. Typically, the transmission rate of this kind of small-data-sized information is low, but security requirements bring challenges to the communication scheme.

Smartphone ownership has increased rapidly in the past decade, and smartphone is expected to become the technological product with the highest ownership rate in modern life. Currently, the most widely used phone-to-phone wireless communication systems are based on radio frequency (RF) and infrared (IR) transmission [3–5]. One significant merit of EM wave is its low decay rate in the air medium so that EM wave can propagate over a long distance. In addition, EM wave is transverse wave that has a propagation velocity of approximately 2.9×10^8 m/s, so EM signal under GHz frequency has a long enough wavelength to pass by small-sized solid barriers [6]. In terms of the acoustic wave, the attenuation rate of its propagation in the air is proportional to the wave frequency [7], which causes trouble to the wideband acoustic communications. Therefore, the usable bandwidth of EM wave communication is commonly more extensive than that of acoustic communications, and the data rate of the electromagnetic wave is higher too.

A critical challenge of the ultrasonic communication arises from wave reflection, mechanical energy loss, and Doppler shift: the reflection is mostly caused by multi-path propagation results in significant phase shift; the mechanical energy loss increases distance loss of sound propagation; the Doppler shift causes frequency deviation. Another significant challenge stems from mobile phones: the maximum sampling rate of mainstream smartphone is 48 kHz, and the frequency response capabilities vary widely. These conditions limit the coherent bandwidth, data rates, and also the transmission distance.

In the existing research on aerial acoustic communication systems, most work focuses on ultrasonic spectrum for a better signal-to-noise ratio (SNR) [8–12]. A Dual-In-Dual-Out acoustic communication system using frequency shifting keying (FSK) was proposed in [8]. A multi-channel aerial communication scheme using distance-selected binary phase shifting keying (BPSK) and binary amplitude shifting keying (BASK) was proposed in [9]. A mesh network structure based on Multiple Input, Multiple Output (MIMO) acoustic channel communication was proposed in [10]. The above work [8–10] were implemented on the PC or the specific-manufactured speaker&microphone platform, where these platforms all support a higher sampling rate and resolution than the smartphone platform. For research related to the smartphone platform, a low-rate chirp-based BPSK scheme to realize an up-to-25 m communication range was proposed in [12]. Researchers in [9] pointed out that the aerial ultrasonic phase characteristics change significantly with distance. Inspired by research work in [9,12], we have designed a covert aerial ultrasonic communication scheme that leverages chirp-like signal BPSK with ID-classified interleaving.

The main contributions of this work are summarized as follows: (1) We have proposed a short-distance covert aerial ultrasonic communication scheme for phone-to-phone communications; (2) A convolutional code plus ID-classified interleaving is used; (3) we leverage the phase noise characteristics of aerial ultrasonic wave to limit the effective communication range in the indoor environment.

The remainder of the paper is arranged as following: The system design is described in Sect. 2, with emphasis on selection of waveform and modulation scheme, estimation of room impulse response and its compensation, channel coding, and receiver design. Extensive experiments and performance evaluation of various aspects of this communication scheme are conducted and investigated in Sect. 3. Section 4 concludes the paper with discussions of limitations of this scheme and further research work.

2 System Design

Figure 1 shows a schemetic diagram of the proposed ultrasonic communication system. The transmitter block has two main parts, the encoder and the modulator respectively. The encoder functions to improve the channel gain via applying forward-error-checking code and interleaver; the core challenge in encoder block comes from the trade-off between code rate and channel coding gain. The modulator block takes challenges in the choice of modulation keying, which requires to balance the bit rate and bandwidth usage, communication distance and reliability. On the receiver block, the Room Impulse Response (RIR) estimation block functions to predict the channel state information and then leverage the predicted channel state model to restore the transmitted signal. The blocks in the receiver are paired to that in transmitter respectively: decoder matches the encoder, and demodulator matches the modulator; these blocks in the receiver perform reverse steps of the transmitter blocks to recover the information.

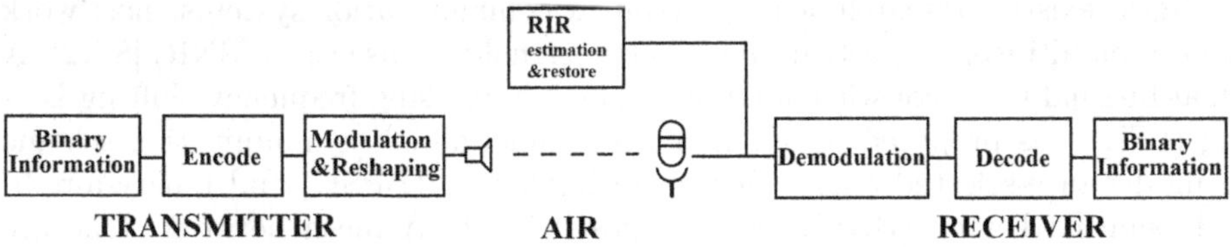

Fig. 1. Overview of the acoustic communication system.

2.1 Waveform and Modulation

The sound collection capabilities of current smartphones are limited to sampling rate of 44.1 kHz on iOS phones or 48 kHz on Android phones. That is equivalent to maximum signal frequency of 24 kHz, which means near ultrasonic. In order to improve the reliability of ultrasonic communications on smartphones, the chirp based BPSK is applied. Compared with frequency shifting keying used in [10,12], BPSK is less sensitive to the diversity of non-flat frequency response and is more suitable for signal modulation between various types of smartphones. The chirp signal is relatively insensitive to Doppler shift and power-efficiency, and the system uses a couple of orthogonal chirp-signals, which are noted as up and down chirps, to spread the frequency band. The expressions of up and down chirps are given in Eqns. (1) and (2), where s_{up} is the up chirp signal, s_{down} is the down chirp signal, f_{low} is the lowest frequency bound of the chirp signal, f_{high} is the highest frequency bound of the chirp signal, t is the time, τ_{chirp} is the duration of the chirp signal.

$$s_{up}(t) = \cos[2\pi(f_{low}t + \frac{f_{high} - f_{low}}{\tau_{chirp}}t^2)] \qquad 0 < t < \tau_{chirp} \tag{1}$$

$$s_{down}(t) = \cos[2\pi(f_{high}t - \frac{f_{high} - f_{low}}{\tau_{chirp}}t^2)] \qquad 0 < t < \tau_{chirp} \tag{2}$$

Figure 2 demonstrates an example of frame structure to send binary message '110100', which consists of a preamble, guard interval, data payload and trailing bits in sequence from the front to the back. The preamble is a down chirp modulated signal and is partly reshaped by a Blackman window (Fig. 3.(b)), which is used to synchronize the start position of the signal and channel state assessment. The guard interval serves as a transition between the load and the preamble to reduce the influence of inter-symbol interference. The chirp modulation uses up and down chirp to represent the binary '1' and '0' respectively, and the modulated signal of payload data is fully reshaped by the Blackman window (Fig. 3.(c)). A 2-bit trail bit is modulated by up and down chirps sequentially and part reshaped, which indicates the ending of the signal sequence.

In the process of demodulation, we use the matched-filter method to compare the received symbol's correlation coefficient. A Hilbert-transform method is used to detect the envelope of the correlation function. Demonstrated in Eq. (3), the matched-filter calculates the likelihood of the received symbol corresponding to its correlation to up and down chirps, and the value of likelihood is used in the following maximum likelihood (ML) decoder. In Eq. (3), the symbol *likelihood* describes the variable that represents the probability of the received bit to become the Boolean value '1', and the symbols $Correlation_{upchirp}$ and $Correlation_{downchirp}$ describe the normalized correlation corresponding to up and down chirps respectively.

$$likelihood = \frac{Correlation_{upchirp}}{Correlation_{downchirp} + Correlation_{upchirp}} \tag{3}$$

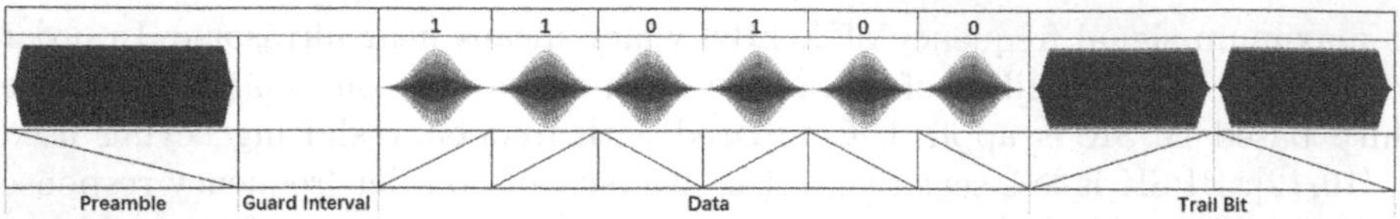

Fig. 2. Signal frame structure

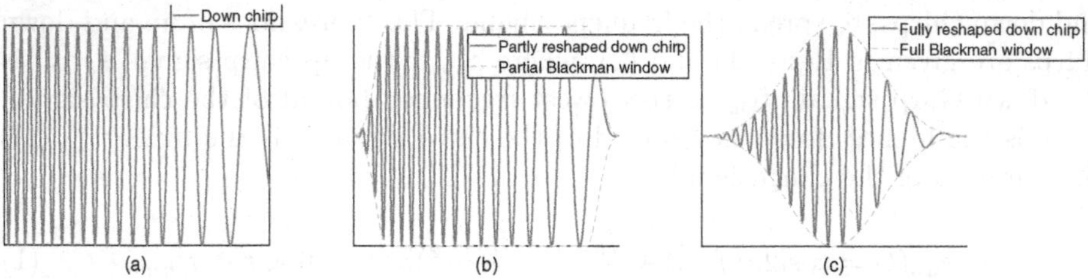

Fig. 3. (a). Down chirp signal; (b). Partly reshaped down chirp signal and partial Blackman window; (c). Fully reshaped down chirp signal and full Blackman window.

2.2 RIR Estimation and Compensation

RIR describes the acoustic channel state between the speaker and receiver. In an indoor scenario, RIR is typically determined by the environmental factors, which vary temporally and spatially. In the situation of face-to-face communications, the relative motion speed between the speaker and receiver is considered to be low enough compared with airborne sound speed, which means the Doppler shift can be ignored. Therefore, the RIR is primarily affected by multipath and shadow fading. In this work, we use a least-square-error method to restore an approximate RIR model from the received preamble.

The simplified channel model is demonstrated in Fig. 4. The received preamble $y[n]$ can be considered as the superposition of the convolution of transmitted preamble $x[n]$ with channel frequency response (the room impulse response) $h[n]$ and additive interference $i[n]$, and the equations of this process can be written as Eqns. (4), (5). The matrix expression of Eq. (5) can be rewritten as Eq. (6), and Eq. (7) simplifies Eq. (6) by substituting the x-elements matrix by matrix A. The error function E that defines the difference between the received preamble and the transmitted preamble is given in Eq. (8), and the square error is obtained in Eqns. (9) and (10), where σ is the variance function of environmental interference and the superscript ^{T} denotes the transposition. To minimize the square error, the partial derivative of the square error function with respect to variable H is shown in Eq. (11), and the optimal solution can be obtained when Eq. (11) equals to zero. Finally, we can obtain an approximation of the RIR model only depending on the transmitted and received preamble according to Eq. (12), where the superscript $^{-1}$ denotes the inverse matrix. Figure 5 shows the steps of RIR compensation: the algorithm leverages the Cepstrum analysis, using the reciprocal of the fast Fourier transform (FFT), to generate the inverse-RIR filter. The algorithm convolutes the inverse-RIR filter with the received data signal to restore and compensate the data signal.

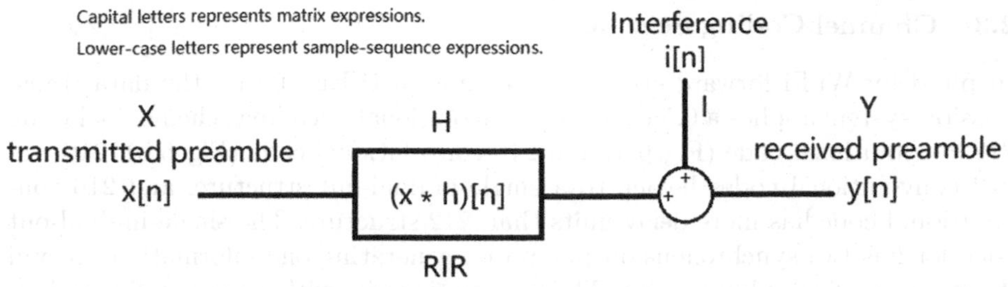

Fig. 4. Simplified channel model.

$$y[n] = x[n] * h[n] + i[n] \tag{4}$$

$$y[n] = \sum_{k=0}^{n} x[n-k]h[k] + i[n] \tag{5}$$

$$\begin{bmatrix} y[0] \\ y[1] \\ \vdots \\ y[n] \end{bmatrix} = \begin{bmatrix} x[0] & x[-1] & \cdots & x[-n] \\ x[1] & x[0] & \cdots & x[1-n] \\ \vdots & \vdots & \ddots & \vdots \\ x[n] & x[n-1] & \cdots & x[0] \end{bmatrix} \begin{bmatrix} h[0] \\ h[1] \\ \vdots \\ h[n] \end{bmatrix} + \begin{bmatrix} i[0] \\ i[1] \\ \vdots \\ i[n] \end{bmatrix} \tag{6}$$

$$Y = AH + I \tag{7}$$

$$E = Y - AH - I \tag{8}$$

$$|e|^2 = (Y - AH)(Y - AH)^T + \sigma \tag{9}$$

$$|e|^2 = Y^T Y - H^T A^T Y - Y^T AH + H^T A^T AH + \sigma \tag{10}$$

$$\frac{\partial}{\partial H}|e|^2 = -2A^T Y + 2A^T AH \tag{11}$$

$$\hat{H} = (A^T A)^{-1} A^T Y \tag{12}$$

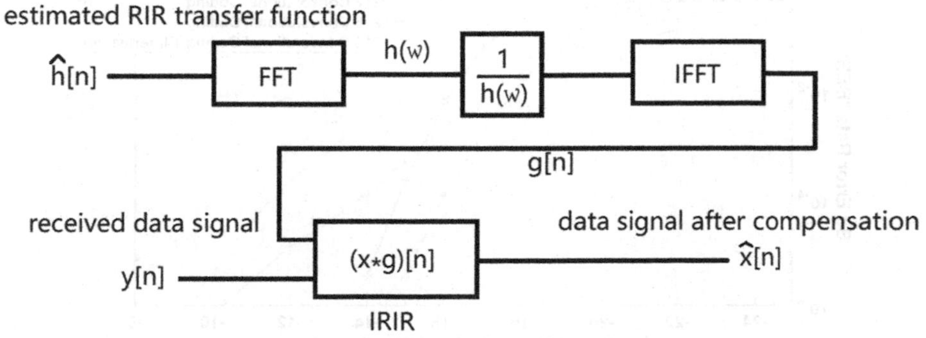

Fig. 5. RIR compensation method.

2.3 Channel Coding Scheme

Inspired by Wi-Fi forward error check coding in IEEE 802.11, the data transmission system applies a 0.5-code-rate convolutional encoding scheme, including 212 convolutional code (Fig. 6.a) and 216 convolutional code (Fig. 6.b). 216 and 212 convolutional codes belong to a single-in-dual-out structure, and 216 convolutional code has more delay units than 212 structure. The single-in-dual-out encoder has two synchronous output ports, generating one information bit and its corresponding redundant bit. The information bits with corresponding redundant bits are reordered into a one-dimensional bitstream by interleaving. We use the Matlab function '*randintrlv*' as the interleaver, the function performs data permutation according to the rand seed. In this coding scheme, the receiver is assigned a unique rand seed as its ID, and the receiver can only use it to recover the information. The decoder uses maximum likelihood (ML) algorithm based on Viterbi [13] algorithm. Figure 7 shows the theoretical bits error rate performance bound and coding gain of 212 and 216 structure, and the channel coding scheme that leverages 216 convolutional code plus interleaver can implement a zero-error communication under the same condition that the error rate of non-coding communication is around 10^{-2}. Table 1 displays the simulation parameters in Fig. 7.

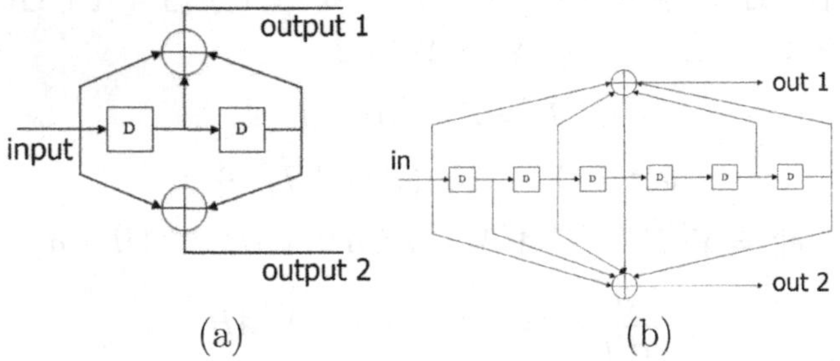

Fig. 6. (a). 212 convolutional code structure, (b). 216 convolutional code structure.

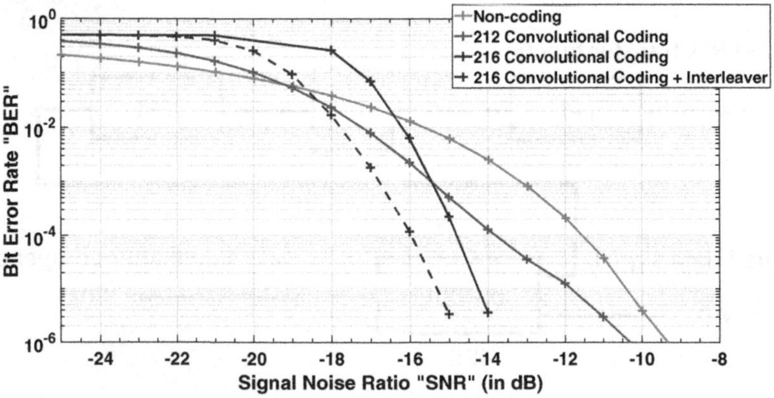

Fig. 7. Comparison of different coding schemes in terms of BER performance

Table 1. Simulation setup

Bit rate	100 bits/s
Frequency band	20∼22 kHz
Number of random bits	10^7
Sampling rate	48000
Channel	Gaussian noise channel

2.4 Receiver

The diagram of the process of demodulation and decoding is given in Fig. 8. The received signal firstly passes through the frame detector to extract the preamble and trail bit frames. The preamble frame is used to estimate the approximate RIR model by using the least square error method, and the RIR model is applied to generate the restore filter to compensate for the data signal. Following the detection of a preamble, the data sequence is divided into segments corresponding to symbols. Then, the symbol segments pass through the matched-filter to calculate the corresponding likelihood, and the receiver uses its unique deinterleaving seed to deinterleave the matched-filtering result. After deinterleaving, symbols are decoded by the ML decoder according to likelihood, and the receiver uses the checksum bits to verify the information.

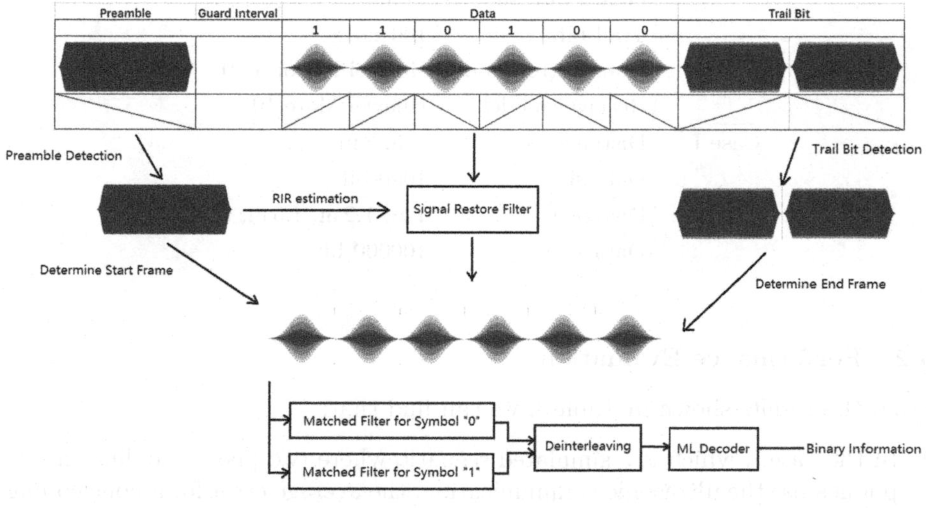

Fig. 8. Demodulation and decoding framework.

3 Experiments

3.1 Experimental Settings for Data Transmission

In our experiments, we tested and measured the error rate via simulating a scene (Case I in Table 2): two people holding smartphones exchange messages face-to-face, where the message can be a payment information link, QR code, and hyper-link. The data size of these messages is typically about 1000 bits, so we used a 1000-bits random binary data as the payload of the transitted data. We used Matlab R2019a to encode and modulate these data, and convert the processed data into '.wav' audio format. In the measurement, we used two smartphones to play and record the transmitted signal, and we used Matlab R2019a to demodulate and decode the received signal. In addition, we also tested fixed-point to fixed-point communication as a reference (Case II in Table 2). The detailed experiment parameters are displayed in Table 2.

Table 2. Experiment parameters

	Parameter	Value
Common	Room size	$5 \times 3 \times 2.7\,\text{m}^3$
	Bit rate	100 bits/s
	Frequency band	20∼22 kHz
	Preamble	85.3 ms
	Guard interval	21 ms
	Trail bits	100 ms
	Transmitter model	Huawei Honor V20
	Receiver model	Huawei Mate 10
Case I	Distance	1 m, 3 m
	Data size	1000-bit
Case II	Distance	1 m, 1.2 m, 1.5 m, 3.5 m
	Data size	100000-bit

3.2 Performance Evaluation

From the results shown in Table 3, we can find that:

1. In the case I, which is a simulated scenario where two people holding smartphones use the ultrasonic communication, the average error for a none coding communication is 0.05% at a distance around 1 m, and zero error rate for two coding cases. This result can be explained: when two people are holding mobile phones, the speakers and microphones on the two smartphones are in unpredictable motion, which leads to the incoherence of the channel stage. Overall, both schemes using channel coding are capable to achieve the short-distance face-to-face communication.

Table 3. Experimental result

Case	Coding sheme	Distance (m)	Average error rate
I	None	0.5–1.5	0.05%
	212 + Interleaver	0.5–1.5	0
	216 + Interleaver	0.5–1.5	0
	None	2–4	21.5%
	212 + Interleaver	2–4	48.8%
	216 + Interleaver	2–4	50.1%
II	None	1	0.014%
	212 + Interleaver	1	0
	216 + Interleaver	1	0
	None	1.2	0
	212 + Interleaver	1.2	0
	216 + Interleaver	1.2	0
	None	1.5	0.023%
	212 + Interleaver	1.5	0
	216 + Interleaver	1.5	0
	None	3.5	0.978%
	212 + Interleaver	3.5	0
	216 + Interleaver	3.5	0.001%

2. In the case I, all schemes fail to implement a zero-error communication when the distance is around 3 m. Compared with the same distance condition in the case II, which tests the fixed-point to fixed-point ultrasonic communication, the average error rates of the two channel coding schemes are approximately zero, where the ultrasonic communication quality is much better than that in the case I. This result can be explained: the current RIR estimation and compensation scheme is only applicable to coherent channels, while the channel state in the case I is frequently changing due to the motion between speaker and microphone.

3.3 Distance Loss

To evaluate the distance loss of airborne channel, we play 70 s single-symbol modulated signal in a $4.8 \times 8.8 \times 2.7\,m^3$-sized sitting room. Two testing smartphones are placed on stable supports, which are vertically 0.9m above the ground. We compare the average power spectral density of the received signals tested at 1 m, 5 m, and 10 m distance, and we found the linear distance loss for smartphones is approximately 1dB/m. Based on this estimate, the power density of the signal transmitted by tested smartphones will touch the noise level at 23 m, where the signal is undetectable over the distance (Fig. 9).

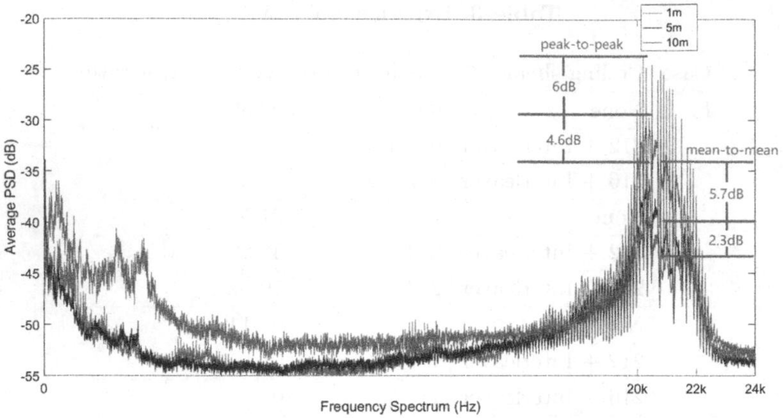

Fig. 9. Average PSD of modulated signal

3.4 Phase Noise

In the indoor room-sized environment, multi-path propagation causes significant phase shift that causes confusion in the demodulation of BPSK. Phase degradation characteristics are significant factors that determine the PSK modulation quality. To evaluate the characteristics in indoor environments, we conducted a chirp signal transmission test. Table 4 shows the experiment parameters, and we used a matched-filter method to evaluate the experimental result.

The convolution of matched-filter and signal outputs the correlation function between the signal and matched-filter-corresponding symbol. Theoretically, the matched-filter synchronizes the phase spectrum of the signal and can reshape the signal into impulse waveform. Under this circumstance, if the phase degradation is significant, the matched-filtering result will not exhibit an impulse waveform.

Figure 10 demonstrates the matched-filtering result of randomly picked down chirps symbol from results tested in 1 m and 5 m. In 1m-distance condition, Fig. 10.(a) exhibits a clear impulse in the correlation function of down chirp. However, in the 5 m-distance condition, Fig. 10.(c) and Fig. 10.(d) do not exhibit a clear impulse envelope, and two figures both exhibit a low correlation level. This phenomenon shows that the phase noise increases significantly with distance, and the Phase Shifting Keying method is difficult to recover information at a long distance.

Table 4. Experiment parameters

Parameter	Value
Room size	$5 \times 3 \times 2.7\,m^3$
Down chirp duration	10 ms
Frequency band	20∼22 kHz
Data size	1000
Distance	1 m, 5 m
Transmitter model	Huawei Honor V20
Receiver model	Huawei Mate 10

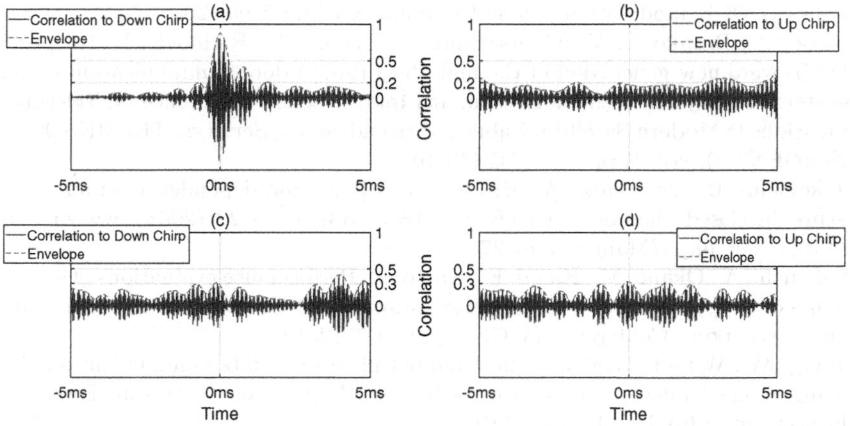

Fig. 10. The correlation functions of the down-chirp signal received at 1 m and 5 m. (a). Correlation to down-chirp symbol in 1 m-distance condition; (b). Correlation to up-chirp symbol in 1 m-distance condition; (c). Correlation to down-chirp symbol in 5 m-distance condition; (d). Correlation to up-chirp in 5 m-distance condition.

4 Conclusion and Future Work

In conclusion, we have developed a chirp-based BPSK modulation communication system on the smartphone platform to realize short-distance indoor communication. The implementation of convolutional coding enables the 1 m indoor data transmission error rate to be within 0.001%. The experiments on a face to face scenario proved that the work is of practical significance, and the technology can be applied in short-range communication scenarios such as the exchange of payment information. However, the current data transmission efficiency is low, and the system cannot respond to the RIR change in the case of sudden non-line of sight (NLOS). In further work, we plan to design a real-time RIR estimation and compensation scheme to improve data transmission stability over longer distances and take Doppler effect into consideration.

References

1. Lichte, H.: The influence of horizontal temperature layers in sea water and the range of underwater sound signals, Tracor Sciences & Systems (1919)
2. Estrada, R.F., Starr, E.A.: 50 years of acoustic signal processing for detection: coping with the digital revolution. IEEE Ann. Hist. Comput. **27**(2), 65–78 (2005)
3. Yu, Y., Zheng, L., Zhu, J., Cao, Y., Hu, B.: Technology of short-distance wireless communication and its application based on equipment support. In: AIP Conference Proceedings, vol. 1955, p. 040135. American Institute of Physics Inc., Apr 2018. https://doi.org/10.1063/1.5033799
4. Kraemer, R., Katz, M.: Design rules for future short-range communication systems, pp. 9–26 (2009)
5. Du, H., Xu, G.: Infrared indoor wireless MIMO communication system using 1.2 GHz OOK modulation. China Commun. **16**(5), 62–69 (2019)
6. Nesic, A., Brankovic, V., Oberschmidt, G., Dolle, T., Radnovic, I., Krupezevic, D.: Toward new generation of the high data rate in-door communication systems-system and key RF technologies. In: 4th International Conference on Telecommunications in Modern Satellite, Cable and Broadcasting Services. TELSIKS'99 (Cat. No.99EX365), vol. 1, pp. 232–235 (1999)
7. Jakevičius, L., Demčenko, A.: Ultrasound attenuation dependence on air temperature in closed chambers. ULTRAGARSAS **63**(1), 18–22 (2008). www.ndt.net/search/docs.php3?MainSource=27
8. Sakaushi, A., Okano, M., Kanai, K., Katto, J.: Performance evaluations of software-defined acoustic MIMO-OFDM transmission. In: IEEE Wireless Communications and Networking Conference (WCNC), pp. 1–6 (2018)
9. Jiang, W., Wright, W.M.D.: Multichannel ultrasonic data communications in air using range-dependent modulation schemes. IEEE Trans. Ultrason. Ferroelectri. Freq. Control **63**(1), 147–155 (2016)
10. Hanspach, M., Goetz, M.: On covert acoustical mesh networks in air. Communications **8**(11), 758-767 (2013)
11. Holm, S., Hovind, O.B., Rostad, S., Holm, R.: Indoors data communications using airborne ultrasound. In: Proceedings. (ICASSP'05). IEEE International Conference on Acoustics, Speech, and Signal Processing, 2005, vol. 3, pp. iii/957–iii/960 (2005)
12. Lee, H., Kim, T.H., Choi, J.W., Choi, S.: Chirp signal-based aerial acoustic communication for smart devices. In: IEEE Conference on Computer Communications (INFOCOM), pp. 2407–2415 (2015)
13. Viterbi, A.: Error bounds for convolutional codes and an asymptotically optimum decoding algorithm. IEEE Trans. Inf. Theory **13**(2), 260–269 (1967)

An Efficient and Truthful Online Incentive Mechanism for a Social Crowdsensing Network

Lu Fang[1], Tong Liu[1,2,3,4](✉), Honghao Gao[1], Chenhong Cao[1,3], Weimin Li[1], and Weiqin Tong[1,3,4]

[1] School of Computer Engineering and Science, Shanghai University, Shanghai, China
{shanhefl,tong_liu,gaohonghao,caoch,wmli,wqtong}@shu.edu.cn
[2] Shanghai Key Laboratory of Data Science, Shanghai, China
[3] Shanghai Institute for Advanced Communication and Data Science, Shanghai University, Shanghai, China
[4] Shanghai Engineering Research Center of Intelligent Computing System, Shanghai, China

Abstract. Crowdsening plays an important role in spatiotemporal data collection by leveraging ubiquitous smart devices equipped with sensors. Considering rational and strategic device users, designing a truthful incentive mechanism is a crucial issue. Moreover, another key challenge is that there may not exist adequate participating users in reality. To encourage more users to participate, the social relationship among them can be leveraged, as users may be significantly influenced by their social friends. In this paper, we assume recruited users to diffuse uncompleted sensing tasks to their friends, and propose an efficient and truthful online incentive mechanism for a such social crowdsensing network. Specially, we model the time-varying social influence of a user by extending two metrics of node centrality used in social networks. In order to maximize the accumulated social welfare achieved by the network, we design a user selection algorithm and a payment determination algorithm respectively, in which payments given to participants not only depend on data qualities but also related with social influences. We theoretically prove that our mechanism achieves properties of computational efficiency, individual rationality, and truthfulness. Extensive simulations are conducted, and the results show the superiority of our mechanism.

Keywords: Crowdsensing · Incentive mechanism · Truthfulness · Social influence

1 Introduction

In recent years, mobile smart devices like smartphones have been widespread in urban life, which are embedded with various sensors such as camera, microphone,

H. Gao et al. (Eds.): CollaborateCom 2020, LNICST 349, pp. 49–63, 2021.
https://doi.org/10.1007/978-3-030-67537-0_4

and accelerometer. Spatiotemporal sensing data can be collected over time in a large urban area by a mass of distributed mobile smart devices, which is referred as mobile crowdsensing. Some practical applications have been implemented based on mobile crowdsensing, such as passenger flow monitoring and environmental noise mapping. Providing sufficient incentives to participate device users is a crucial issue in mobile crowdsensing, in order to compensate their costs incurred by sensing, especially when rational and strategic users are considered. A number of incentive mechanisms [3,4,8] have been proposed, in which different facets of a crowdsensing network are taken into consideration, such as the quality or spatiotemporal coverage of sensing data and the dynamic arrivals of users and requests.

Most of these incentive mechanisms assume there are adequate participants in a crowdsensing network. However, it is not always true in reality. According to the statistics data published by TalkingData [1], the ratios of active users in May, 2019 in two popular traffic-monitoring applications, Tencent map and Google map, are only 2.99% and 0.48% respectively.

To solve the problem of insufficient participating users in a crowdsensing network, some researchers have tried to employ the social relationship among users to attract more participants. Incentive mechanisms with considering the social influences of users have been proposed in recent works [5,6,9,10,12–15]. Specially, in [10] and [9], the in-degree and out-degree of a user in a social network are calculated to measure the social influence of the user. Bayesian game and Stackelberg game are respectively employed to model the crowdsensing system. Social-aware incentive mechanisms are proposed in [15] and [6], where rewards given to participants depend on their social friends, to stimulate them attracting more users. Xu *et al.* [14] propose a two-tiered architecture, in which agents are selected as middlemen, who are employed to recruit their social neighbors as participants. However, most of these works ignore the social influence of a user is time-varying, which is related with the locations and reliability of his/her social friends.

In this work, we consider the problem of designing an incentive mechanism for strategic and unreliable users in a social crowdsensing network, aiming to maximize the social welfare of the network accumulated over time and guarantee the truthfulness of users. To be specific, we consider the quality of sensing data collected by different users are discrepant, which is decided by the reliability of each user. Moreover, we also consider some users in the network are inactive in sensing. Thus, participants are recruited not only for collecting sensing data but also for stimulating their inactive friends to become active. By taking advantage of the social network among users, we assume that uncompleted sensing tasks can be informed to the friends by a recruited user, and more users will be influenced by their social friends to participate in sensing.

Although some incentive mechanisms taking social influences of users into account have been designed in previous works, there still exist several challenges to overcome. *Firstly*, the social influence of a user is hard to be measured in a real-time crowdsensing network, which is not only determined by his/her centrality in the social network, but also related with the reliability of his/her social friends and the locations they may appear. *Secondly*, users can dynamically become active or inactive over time, which is unknown in advance to the platform. However, the platform should select participants and pay them incentives in real time, to support spatiotemporal data collection with high quality-of-service. *Thirdly*, there exists an inherent tradeoff between selecting users with high reliability and selecting users with high social influences. High-quality sensing data can be collected by users with high reliability on condition that there are sufficient users participating in sensing. *Last but not least*, we consider users are rational and strategic, who will not perform tasks with negative rewards and may submit false information for maximizing their own utilities.

To meet the challenges, we propose an efficient and truthful online incentive mechanism for a social crowdsensing network, which comprises of a user selection algorithm and a payment determination algorithm. Firstly, we measure the time-varying social influence of an active user by extending two metrics of node centrality used in social networks. Then, we model the interactions between users and the platform as a repeated reverse auction. In each auction, the platform collects the bids of active users and then greedily selects participants with high reliability, large social influences, and low costs one by one. Inactive users may be stimulated by selected users and then turn into active, if they are social friends of selected users. After receiving sensing data from selected users, the platform will provide incentives to them according to their data qualities and social influences. Finally, we theoretically prove that our proposed incentive mechanism achieves the properties of computational efficiency, individual rationality, and truthfulness.

The rest of the paper is organized as follows. We review the state-of-art related works in Sect. 2. Section 3 presents the network model and the formulated problem. In Sects. 4 and 5, we describe our proposed incentive mechanism in details and theoretically analyze the properties achieved by our mechanism, respectively. We present the simulation results in Sect. 6, and conclude this paper in Sect. 7.

2 Related Work

A lot of efforts [3,4,8] have been paid to design incentive mechanisms for mobile crowdsensing, encouraging rational smart device users to participate in sensing. To attract more users, social networks are also extensively introduced in mobile crowdsensing. In [2], rewarding social influences of participants in crowdsourcing systems are firstly put forward.

In [10] and [9], authors use the in-degree and out-degree of a user in a social network to measure the network effect of the user, which can quantify the influence of his/her behavior on social friends. Specially, a Bayesian game-based social-aware incentive mechanism is proposed in [10] for a crowdsensing system, in which the platform has incomplete information of users. And a two-stage Stackelberg game-based incentive mechanism are proposed for a crowdsensing system in [9], with analyzing the participation level and social influence of users. Yang *et al.* [15] propose a social-aware incentive mechanism in which the incentive given to a user depends on his/her social friends. Thus, users are motivated to drive social friends to provide high-quality services, with the objective to achieve higher utilities. In [5], a VCG-based incentive mechanism for multi-resource sharing in crowdsourcing is proposed, engaging users with low cost and high reputation as participants, in which the reputation of each user is updated according to his/her social influence. In all the above related works, an offline crowdsensing system is considered.

Differently, some other works consider an online crowdsensing system with tasks or users arriving in real time. Jiang *et al.* [6] consider time-sensitive tasks arriving to the platform in real time. In order to stimulate users to spread sensing tasks as soon as possible, extra rewards are paid to users for attracting others and the rewards are decreasing with the participating time of the attracted users. In [13], Xu *et al.* initially select participants according to their numbers of common social friends between participating users and their social friends via a greedy hill-climbing algorithm. Then, reverse auction is employed for selecting more users and providing incentives. [14] proposes a two-tiered crowdsourcing architecture, in which agents are recruited as middlemen between the platform and users. Agents are selected according to their time coverage and social influences which are measured by the similarity of interests between the agents and their social neighbors. Sun *et al.* [12] propose an incentive mechanism with considering the heterogeneity of users in social networks. To ensure users to participate in sensing continuously, participants are selected based on their social states and current workloads.

In these works, different metrics in social networks are employed to measure the social influences of users to provide social-aware incentives to participants. However, most of works ignore that the social influence of a user is time-varying, not only determined by the social network, but also related to the reliability and locations of his/her social friends.

3 Network Model and Problem Formulation

In this section, we first introduce the model of a social crowdsensing network considered in this work, and then mathematically formulate the problem of incentive mechanism design and user selection in such a network.

3.1 Network Model

In this work, we firstly consider a typical crowdsensing network, which comprises of a central platform located at the cloud and a plenty of mobile smartphone users distributed in an urban area. In such a crowdsensing network, a mass of fine-grained spatio-temporal sensing data could be collected for applications like real-time environment monitoring. For convenience, we divide time into equal-interval time slots (e.g., 20 min), and the set of time slots is denoted by

$\mathcal{T} = \{t_1, t_2, \cdots, t_s, \cdots\}$. We consider several points of interest (POIs) in an urban area need recruit smartphone users nearby (e.g., within 500 m) to collect sensing measurements continuously. We denote the set of POIs as $\mathcal{L} = \{l_1, l_2, \cdots, l_K\}$, where K is the number of POIs in the urban area. Moreover, the task of collecting measurements around POI l_k in time slot t_τ is denoted by φ_k^τ. Each measurement collected for task $\varphi_k^\cdot$ is associated with a preliminary value v_k obtained by the platform, which is related to the importance of POI l_k.

We consider there exists a universal set of registered smartphone users, denoted by $\mathcal{U} = \{u_1, u_2, \cdots, u_N\}$, where N represents the number of users. Due to different hardware, the measurements collected by different users may have diverse qualities. We model the qualities of measurements collected by user u_i via considering his/her reliability $q_i \in (0, 1]$, which could be derived from the historical measurements. The larger reliability of user u_i, the higher quality of his/her measurements, as well as the value achieved by the platform. Thus, we calculate the value of the measurement collected by user u_i for task φ_k^* as $q_i \cdot v_k$.

At the beginning of each time slot, *active users* will participate in performing tasks forwardly via a reverse auction with the platform. Specially, if user u_i becomes available in time slot t_τ, he/she will submit a bid $\beta_i^\tau = (\alpha_i^\tau, \Phi_i^\tau, b_i^\tau)$ to the platform. Here, α_i^τ represents the length of time window in which u_i will keep available, and Φ_i^τ is the set of tasks that u_i will perform during time window $[t_\tau, t_\tau + \alpha_i^\tau]$. As energy and bandwidth resources are consumed for doing tasks Φ_i^τ, a certain amount of costs is incurred on user u_i, denoted by c_i^τ. Considering the rationality and selfishness of users, b_i^τ is the bidding price claimed by u_i to the platform for completing tasks Φ_i^τ, which may be unequal to c_i^τ. In addition, the set of active users in time slot t_τ is represented by $\mathcal{U}^\tau \subseteq \mathcal{U}$.

To indicate whether a user is selected or not in time slot t_τ, we define an index vector $\mathbf{I}^\tau = [I_1^\tau, I_2^\tau, \cdots, I_N^\tau]$, in which $I_i^\tau \in [0, 1], \forall u_i \in \mathcal{U}^\tau$. Obviously, $I_i^\tau = 0$ if $u_i \notin \mathcal{U}^\tau$. Specially, $I_i^\tau = 1$ means user u_i is selected by the platform, otherwise u_i is not selected. If user u_i is recruited in t_τ, a certain amount of payment should be provided to u_i by the platform to compensate the cost of the user, which is denoted by p_i^τ. Given the decisions of user selection in period $[t_1, t_{\tau-1}]$, the total value of the measurements collected by these selected users for task φ_k^τ can be calculated as $V_k^\tau = \sum_{t=1}^{\tau-1} \sum_{I_i^t=1 \wedge \varphi_k^\tau \in \Phi_i^t} q_i v_k$. According to the law of diminishing marginal utility [7], we define the utility achieved by collected measurements increases with the total value of the measurements, while the marginal utility decreases. Thus, we define the utility achieved for task φ_k^τ as $\log(1 + V_k^\tau)$. Then, the utility of collected measurements in time slot t_τ is

$$U(\mathbf{I}^\tau) = \sum_{k=1}^{K} \sum_{t=\tau}^{\tau+\max_i\{\alpha_i^\tau\}} \left(\log(1 + V_k^t + \sum_{\varphi_k^t \in \Phi_i^\tau} I_i^\tau q_i v_k) - \log(1 + V_k^t) \right).$$

To promote more users (i.e., *inactive users*) to participate in sensing, the users selected to perform tasks are also responsible for diffusing sensing tasks to their inactive social friends through a social network. In this work, we consider the social network containing all the registered users is represented by a directed graph $G(\mathcal{U}, \mathcal{E})$, in which each node is a user. A directed edge from node u_i to node u_j, denoted by $e_{i,j} \in \mathcal{E}$, means user u_j is a friend or follower of user u_i. In addition, each edge $e_{i,j}$ is associated with a weight $w_{i,j}$, indicating the social influence made by u_i to u_j.

Intuitively, the larger social influence made by a user to his/her social neighbours, the higher effect achieved by the user to diffuse sensing tasks. Here, we employ two metrics, *weighted degree centrality* and *closeness centrality*, to evaluate the social influence made by a user to others. These two metrics are widely used to measure the node centrality in a social network. Generally, given a social network $G(\mathcal{V}, \mathcal{E})$, the weighted degree centrality and the closeness centrality of an arbitrary node $v_i \in \mathcal{V}$ are defined as $\sum_{v_j \in \mathcal{N}(v_i)} w_{i,j}$ and $\frac{|\mathcal{V}|-1}{\sum_{v_j \in \mathcal{V}} d_{i,j}}$, respectively. Here, $\mathcal{N}(v_i)$ is the set of social neighbours of v_i, and $d_{i,j}$ is the shortest distance between v_i and v_j. Obviously, the larger values of these two metrics, the higher centrality a node has.

Inspired by the two metrics of node centrality in social networks, we extend them to evaluate the social influence of each user in our social crowdsensing network. As some measurements have been collected in previous time slots, different POIs have various levels of urgency to collect more measurements at present. Therefore, we also consider various contributions made by different inactive users if they turn into participants, which are highly relevant with the their locations. We denote the probability of an inactive user $u_i \in \mathcal{U} \setminus \mathcal{U}^\tau$ appearing around POI l_k by $o_{i,k} \in [0, 1]$, which can be derived from his/her historical trajectories. Considering both the urgency of task φ_k^τ and the probability distribution of the location of inactive user u_i, we define the *potential contribution* of u_i as $O_i^\tau = \sum_{l_k \in \mathcal{L}} (o_{i,k} \cdot e^{-V_k^\tau})$. Intuitively, the less value of task φ_k^τ and the higher probability of u_i appearing around l_k, the more contributions achieved by u_i.

Then, we define two extended metrics to measure the social influence of a recruited user, named as *extended weighted degree centrality* and *extended closeness centrality*, respectively. Specially, given $\mathbf{I}^\tau$, the extended weighted degree centrality achieved by all recruited users is calculated as

$$\omega(\mathbf{I}^\tau) = \sum\nolimits_{u_j \in \bigcup_{I_i^\tau = 1} \mathcal{N}_i^\tau} \max\{I_i^\tau O_j^\tau w_{i,j} | \forall u_i \in \mathcal{U}^\tau\},$$

where $\mathcal{N}_i^\tau$ represents the set of inactive social neighbours of user u_i. Since an inactive user may be influenced by more than one recruited users, we only take the maximal influence into account, to avoid double counting. Note that we take the potential contribution of an inactive user as the auxiliary weight to evaluate the influence of recruited users. In addition, the extended closeness centrality achieved by all recruited users is calculated as

$$\upsilon(\mathbf{I}^\tau) = \sum_{u_i \in \mathcal{U}^\tau} \frac{I_i^\tau \cdot |\mathcal{U} \setminus \mathcal{U}^\tau|}{\sum_{u_j \in \mathcal{U} \setminus \mathcal{U}^\tau} (O_j^\tau)^{-1} d_{i,j}}.$$

Note that the extended closeness centrality of a recruited user is time-varying with the locations of his/her inactive social friends, introduced by taking their potential contributions into consideration. In summary, the utility obtained by the social influences of recruited users also follows the diminishing marginal law, which is defined as

$$W(\mathbf{I}^{\tau}) = \log\left(1 + \omega(\mathbf{I}^{\tau}) + \upsilon(\mathbf{I}^{\tau})\right).$$

3.2 Problem Formulation

In this work, we study the problem of online user selection and payment determination in a social crowdsensing network, aiming to maximize the social welfare of the network accumulated over time and maintain that rational and selfish users are truthful in participating in sensing.

Definition 1 (User Selection Problem). *In each time slot t_τ, a set of active users are selected by the platform to collect sensing data, indicated by $\mathbf{I}^{\tau}$, with the objective that the accumulated social welfare achieved by the crowdsensing network is maximized, i.e.,*

$$\max_{\{\mathbf{I}^{\tau}\}} \sum\nolimits_{\tau=1}^{\infty} [U(\mathbf{I}^{\tau}) + W(\mathbf{I}^{\tau}) - C(\mathbf{I}^{\tau})], \tag{1}$$

where $C(\mathbf{I}^{\tau}) = \sum_{u_i \in \mathcal{U}^{\tau}} I_i^{\tau} c_i^{\tau}$.

Definition 2 (Payment Determination Problem). *An incentive mechanism should be designed to determine the payment p_i^{τ} given to user u_i by the platform in time slot t_τ, if u_i is selected to collect sensing data in t_τ. Moreover, the incentive mechanism should achieve three properties, i.e., computational efficiency, individual rationality, and truthfulness.*

4 Incentive Mechanism Design

In this section, we propose a truthful online incentive mechanism for a social crowdsensing network. In particular, both user selection algorithm and payment determination algorithm are designed, to decide which users are recruited and how many payments are provided to them. In the following, we first present an overview of our proposed incentive mechanism, and then describe the two algorithms in detail, respectively.

4.1 Overview

At the beginning of each time slot, all active users will *firstly* submit bids to the platform. *Then*, after receiving all bids, the platform conducts user selection, to choose participants with high reliability, large social influences, and low costs. *Next*, all selected users are notified to collect sensing data and diffuse a message containing urgent uncompleted tasks to their inactive social friends. *Finally*, the platform will give proper payments to the selected users.

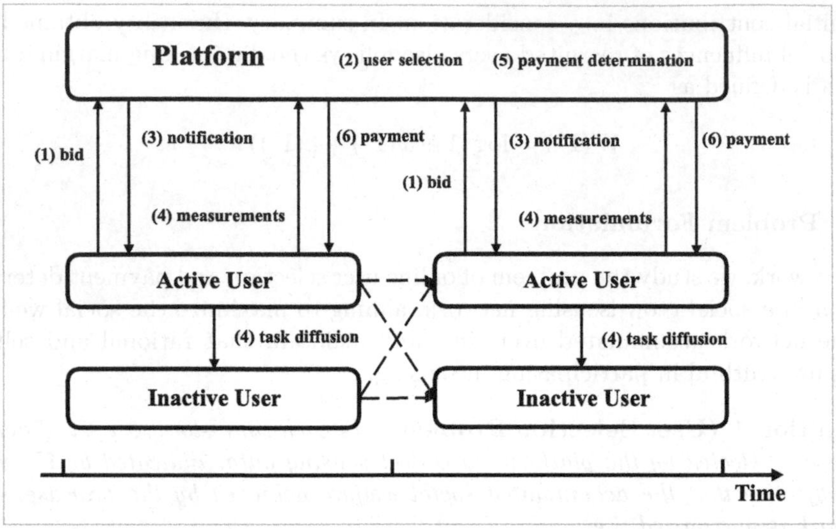

Fig. 1. The interactions between the platform and users in each time slot in the social crowdsensing network.

Note that users can dynamically switch between the active and inactive states. Specially, an inactive user, influenced by an active friend, may turn into active and further diffuse the message containing uncompleted tasks to his/her social friends. Thus, the message can spread to all users quickly through their social relationship (Fig. 1).

4.2 User Selection Algorithm

We first introduce the details of the online user selection algorithm, with assuming all users are truthful (i.e., the bidding price of each user equals to the cost). In each time slot, participating users are selected by the platform one by one. The main idea of this algorithm is to greedily choose the user who achieves the most *per-cost marginal utility*.

Definition 3 (User marginal utility). *Given the set of users already selected by the platform in the current time slot (indicated by* $\mathbf{I}^\tau$*), the marginal utility achieved by user* u_i *if he/she is selected by the platform in the next is defined as*

$$\Delta_i(\mathbf{I}^\tau) = U_i(\mathbf{I}^\tau) + W_i(\mathbf{I}^\tau), \tag{2}$$

where $U_i(\mathbf{I}^\tau)$ *and* $W_i(\mathbf{I}^\tau)$ *are calculated as*

$$U_i(\mathbf{I}^\tau) = U(\mathbf{I}^\tau \wedge I_i^\tau = 1) - U(\mathbf{I}^\tau \wedge I_i^\tau = 0), \tag{3}$$

$$W_i(\mathbf{I}^\tau) = W(\mathbf{I}^\tau \wedge I_i^\tau = 1) - W(\mathbf{I}^\tau \wedge I_i^\tau = 0). \tag{4}$$

Note that $U_i(\mathbf{I}^\tau)$ and $W_i(\mathbf{I}^\tau)$ represent the marginal utility achieved by the collected measurements and the social influences of user u_i, respectively.

At the beginning of each time slot, the platform first collects the bids of all active users, and initializes each element of indicator vector $\mathbf{I}^\tau$ as 0. Then, users are iteratively selected by the platform. In each iteration, the platform aims to find the user with the largest per-cost marginal utility, denoted by u_{i^*}. Specially, the platform calculates the marginal utility $\Delta_i(\mathbf{I}^\tau)$ of each unselected active user u_i. And hence the user $u_{i^*} = \arg\max_{u_i \in \mathcal{U}^\tau}\{\Delta_i(\mathbf{I}^\tau)/b_i^\tau | I_i^\tau = 0\}$ is selected as a participant (i.e., $I_{i^*}^\tau = 1$), unless the utility of his/her measurements is lower than his/her bidding price (i.e., $U_{i^*}(\mathbf{I}^\tau) < b_{i^*}^\tau$). The user selection iteration ends when the marginal utility of any unselected user is less than his/her bidding price, i.e., $\max_{u_i \in \mathcal{U}^\tau}\{\Delta_i(\mathbf{I}^\tau)/b_i^\tau | I_i^\tau = 0\} < 1$. Finally, the user selection algorithm outputs the indicator vector $\mathbf{I}^\tau$.

4.3 Payment Determination Algorithm

In order to ensure the strategic users honestly report their bids, we design a payment determination algorithm based on the rule of critical payment [11]. According to [11], to determine the payment given to a selected user u_{i^*}, we need to obtain the critical bid of u_{i^*} firstly. The critical bid is defined as the bid that makes the bid of user u_{i^*} fail (i.e., u_{i^*} can not be selected if the per-cost marginal utility of bid $\beta_{i^*}^\tau$ is less than the per-cost marginal utility of its critical bid), and the highest payment is achieved by u_{i^*} at the same time.

In what follows, we show how to find the critical bid of user u_{i^*}. We first initialize a virtual set of unselected users by excluding u_{i^*}, i.e., $\mathcal{U}' \triangleq \mathcal{U}^\tau \setminus \{u_{i^*}\}$, and a virtual user selection result $\mathbf{I}' \triangleq \mathbf{I}^\tau$. Moreover, the payment given to u_{i^*} is initialized as 0. Then, we virtually perform the user selection process as described in the previous subsection. In each iteration, user $u_{j^*} = \arg\max_{u_j \in \mathcal{U}'}\{\Delta_j(\mathbf{I}')/b_j^\tau | I_j' = 0\}$ is virtually selected (i.e., $I_j' = 1$), as long as $U_{j^*}(\mathbf{I}^\tau) \geq b_{j^*}^\tau$. We compare $\Delta_{i^*}(\mathbf{I}' \wedge I_{i^*}' = 1)/b_{i^*}^\tau$ and $\Delta_{j^*}(\mathbf{I}' \wedge I_{j^*}' = 1)/b_{j^*}^\tau$. When there exists $\Delta_{i^*}(\mathbf{I}' \wedge I_{i^*}' = 1)/b_{i^*}^\tau \geq \Delta_{j^*}(\mathbf{I}' \wedge I_{j^*}' = 1)/b_{j^*}^\tau$, user u_{i^*} would be selected instead of user u_{j^*}, if u_{i^*} were in the set of unselected users. Then, the payment given to u_{i^*} can be updated according to the per-utility cost of u_{j^*} as

$$p_{i^*|j^*}^\tau \triangleq \min\left(b_{j^*}^\tau \cdot \frac{\Delta_{i^*}(\mathbf{I}' \wedge I_{i^*}' = 1)}{\Delta_{j^*}(\mathbf{I}' \wedge I_{j^*}' = 1)}, U_{i^*}(\mathbf{I}' \wedge I_{i^*}' = 1)\right).$$

After all iterations, the critical bid of user u_{i^*} is obtained, which makes u_{i^*} achieve the highest payment. Thus, the payment given to user u_{i^*} can be calculated as

$$p_{i^*}^\tau = \max_{\left\{j^* | \frac{\Delta_{i^*}(\mathbf{I}' \wedge I_{i^*}' = 1)}{b_{i^*}^\tau} \geq \frac{\Delta_{j^*}(\mathbf{I}' \wedge I_{j^*}' = 1)}{b_{j^*}^\tau}\right\}} (p_{i^*|j^*}^\tau).$$

Note that if $U_{i^*}(\mathbf{I}' \wedge I_{i^*}' = 1)/b_{i^*}^\tau \geq 1$ after the iteration process of virtual user selection ends, which indicates there is no critical bid of user u_{i^*}, we set $p_{i^*}^\tau = \max\left(p_{i^*}^\tau, U_{i^*}(\mathbf{I}' \wedge I_{i^*}' = 1)\right)$.

5 Mechanism Analysis

In the following, we theoretically analyze that our mechanism can achieve the three desired properties, i.e., computational efficiency, individual rationality, and truthfulness.

Theorem 1. *Our incentive mechanism is computationally efficient. Specially, the computation complexity of the user selection algorithm and the payment determination algorithm is at most* $O(N^2)$.

Proof. First we prove that the user selection algorithm has polynomial-time computation complexity. The complexity for calculating the per-cost marginal utility $\frac{\Delta_i(\mathbf{I}^\tau)}{b_i^\tau}$ of all users in user selection algorithm is $O(|\mathcal{U}^\tau|)$. The operation of finding the user with largest per-cost marginal utility in current time slot is $O(|\mathcal{U}^\tau|)$. There are at most N bids in time slot t_τ because each user can only submit at most one bid in his/her active time window. So we have $|\mathcal{U}^\tau| \leq N$ and the operation analysed above will repeat at most N times. Then, it is easy to compute the total computation complexity of user selection algorithm which is at most $O(N^2)$.

Then we prove that the payment determination has polynomial-time computation complexity. The complexity for calculating the per-cost marginal utility $\frac{\Delta_i(\mathbf{I}^\tau)}{b_i^\tau}$ of all users in payment determination algorithm is $O(|\mathcal{U}^\tau|)$. The operation of choosing the user with highest per-cost marginal utility is $O(|\mathcal{U}^\tau|)$. There are at most N bids in time slot t_τ because each user can only submit at most one bid in his/her active time window. So we have $|\mathcal{U}^\tau| \leq N$ and the operation analysed above will repeat at most N times. So, it is easy to compute the total computation complexity of payment determination algorithm which is at most $O(N^2)$.

Theorem 2. Our mechanism guarantees that each user in the social crowdsensing network is individually rational.

Proof. Users are individual rational means each participating user will have a non-negative utility: $p_i^\tau - c_i^\tau \geq 0$. So we should prove that for each bid β_i^τ of participating users, the corresponding payment p_i^τ satisfies $p_i^\tau - c_i^\tau \geq 0$.

Since the payment of a selected user u_{i^*} for his/her bid $\beta_{i^*}^\tau$ is determined by two values which are $U_{i^*}(\mathbf{I}')$ and $b_{j^*}^\tau \frac{\Delta_{i^*}(\mathbf{I}')}{\Delta_{j^*}(\mathbf{I}')}$ according to payment determination algorithm. So we should prove that these two values are both not less than the bidding price $b_{i^*}^\tau$ of $\beta_{i^*}^\tau$.

First of all, it is obviously that we have $\frac{U_{i^*}(\mathbf{I}')}{b_{i^*}^\tau} \geq 1$ if u_{i^*} with bid $\beta_{i^*}^\tau$ is the m-th user selected in time slot t_τ. According to the judgement condition in payment determination algorithm, if there exists a m-th or m'-th $(m' > m)$ selected user u_{j^*} and $\frac{\Delta_{i^*}(\mathbf{I}')}{b_{i^*}^\tau} - \frac{\Delta_{j^*}(\mathbf{I}')}{b_{j^*}^\tau} \geq 0$ when u_{i^*} does not participate in time slot t_τ, we will update the payment of u_{i^*} with bid $\beta_{i^*}^\tau$.

Next, we move the per-cost marginal utility of user u_{j^*} to the right side of the inequation above and multiply $b_{i^*}^{\mathcal{T}}$ on both sides of the inequation. Now we have $\Delta_{i^*}(\mathbf{I}') \geq \frac{\Delta_{j^*}(\mathbf{I}')}{b_{j^*}^{\mathcal{T}}} b_{i^*}^{\mathcal{T}}$, then we have $b_{j^*}^{\mathcal{T}} \frac{\Delta_{i^*}(\mathbf{I}')}{\Delta_{j^*}(\mathbf{I}')} \geq b_{i^*}^{\mathcal{T}}$ since $\frac{\Delta_{j^*}(\mathbf{I}')}{b_{j^*}^{\mathcal{T}}}$ is a positive number.

As the prove process shown above, we have $p_{i^*}^{\mathcal{T}} = \min(b_{j^*}^{\mathcal{T}} \frac{\Delta_{i^*}(\mathbf{I}')}{\Delta_{j^*}(\mathbf{I}')}, U_{i^*}(\mathbf{I}')) \geq b_{i^*}^{\mathcal{T}}$, i.e., in the payment determination algorithm, we have $p_{i^*}^{\mathcal{T}} - b_{i^*}^{\mathcal{T}} \geq 0$. Due to the rationality of users, the real cost $c_{i^*}^{\mathcal{T}}$ of each user must lower than or equal to his/her bidding price $b_{i^*}^{\mathcal{T}}$. Thus, we successfully prove that users in our crowdsensing system are individual rational.

According to [11], our mechanism can achieve truthful if and only if it satisfies the following two conditions: (1) the user selection algorithm is monotonic, and (2) each selected user is paid the critical value. We first give the specific definitions of the two conditions in our crowdsensing network as follows.

- *The user selection algorithm is monotonic*, if user u_i with bid $\beta_i^{\mathcal{T}} = (\alpha_i^{\mathcal{T}}, \Phi_i^{\mathcal{T}}, b_i^{\mathcal{T}})$ selected as a participant will be still selected, if the user reports bid $\widetilde{\beta_i^{\mathcal{T}}} = \left(\widetilde{\alpha_i^{\mathcal{T}}}, \widetilde{\Phi_i^{\mathcal{T}}}, \widetilde{b_i^{\mathcal{T}}}\right)$ with $\widetilde{\alpha_i^{\mathcal{T}}} \geq \alpha_i^{\mathcal{T}}, \widetilde{b_i^{\mathcal{T}}} \leq b_i^{\mathcal{T}}, \Phi_i^{\mathcal{T}} \subseteq \widetilde{\Phi_i^{\mathcal{T}}}$.
- *The critical value of a user* indicates that if user u_i submits a bidding price lower than the critical value, u_i will be selected; otherwise, u_i will not be selected.

Theorem 3. Our mechanism guarantees that each user in the social crowdsensing network is truthful, including cost-truthful, time-truthful, and task-truthful..

Proof. We first prove that the user selection rule is monotonic.

Obviously, the $\frac{\Delta_i(\mathbf{I}^{\mathcal{T}})}{b_i^{\mathcal{T}}}$ of user u_i with $\beta_i^{\mathcal{T}}$ at each time slot will increase with the decrease of his/her bidding price $b_i^{\mathcal{T}}$. If user u_i with bid $\beta_i^{\mathcal{T}}$ is selected in time slot $t_{\mathcal{T}}$, u_i can also be selected if he/she submits a lower bidding price.

It is easily found that the $\frac{\Delta_i(\mathbf{I}^{\mathcal{T}})}{b_i^{\mathcal{T}}}$ of user u_i will not decrease if user u_i reports a wider active time window (the marginal utility of the measurements and social influence of users will not change when users report wider active time windows according to Eqs. (3) and (4)). So if user u_i is selected by submitting the bid $\beta_i^{\mathcal{T}}$ in time slot $t_{\mathcal{T}}$, u_i can also be selected by submitting a bid with wider active time window.

The $\frac{\Delta_i(\mathbf{I}^{\mathcal{T}})}{b_i^{\mathcal{T}}}$ of user u_i will increase if it bids a larger task set $\widetilde{\Phi_i^{\mathcal{T}}} \subseteq \Phi_i^{\mathcal{T}}$ ($\Phi_i^{\mathcal{T}}$ is the real task set that $u_i^{\mathcal{T}}$ can do during his/her active time window this time), so user u_i will submit his/her real task set for maximizing his/her own utility. Therefore, we can easily prove that if user u_i is selected with the bid $\widetilde{\beta_i^{\mathcal{T}}} = (\alpha_i^{\mathcal{T}}, \widetilde{\Phi_i^{\mathcal{T}}}, b_i^{\mathcal{T}})$ in time slot $t_{\mathcal{T}}$, u_i can also be selected if he/she submits a bid with a larger task set $\widehat{\Phi_i^{\mathcal{T}}} \subseteq \Phi_i^{\mathcal{T}}$ including tasks within his/her ability. Thus, we have proved that the user selection rule of our mechanism is monotonic.

Secondly, we verify that the payment $p_i^{\mathcal{T}}$ computed by payment determination algorithm is the critical value to user u_i with bid $\beta_i^{\mathcal{T}}$. If selected user u_i submits a bid $\widetilde{\beta_i^\tau} = (\alpha_i^\tau, \phi_i^\tau, \widetilde{b_i^\tau})$ whose $\widetilde{b_i^\tau} < p_i^\tau$, u_i will obviously be selected in time slot t_τ when $U_i(\mathbf{I}^\tau) \geq \widetilde{b_i^\tau}$, because there exists at least one user u_j who satisfies the inequation $\frac{\Delta_i(\mathbf{I}^\tau)}{\widetilde{b_i^\tau}} \geq \frac{\Delta_j(\mathbf{I}^\tau)}{b_j^\tau}$ and $U_j(\mathbf{I}^\tau) \geq b_j^\tau$ according to payment determination algorithm. Or user u_i must be selected if $U_i(\mathbf{I}^\tau) \geq \widetilde{b_i^\tau}$ when there is no other user u_j has $U_j(\mathbf{I}^\tau) \geq b_j^\tau$ in time slot t_τ. Thus, user u_i with bid $\widetilde{\beta_i^\tau}$ would be selected instead of user u_j with bid β_j^τ. On the contrary, if $\widetilde{b_i^\tau} > p_i^\tau$, user u_i with bid $\widetilde{\beta_i^\tau}$ can not be selected in time slot t_τ since there is no user u_j with $U_j(\mathbf{I}^\tau) \geq b_j^\tau$ satisfies $\frac{\Delta_i(\mathbf{I}^\tau)}{\widetilde{b_i^\tau}} \geq \frac{\Delta_j(\mathbf{I}^\tau)}{b_j^\tau}$ or the bidding price $\widetilde{b_i^\tau}$ is larger than $U_i(\mathbf{I}^\tau)$. Therefore, we prove that p_i^τ is the critical value to user u_i with bid β_i^τ.

6 Performance Evaluation

In this section, we conduct extensive simulations to show the performance achieved by our proposed incentive mechanism, in terms of the social welfare of the system.

6.1 Simulation Setup

We compare our proposed mechanism with the following four baselines.

- *Only-Bid selection Algorithm (OBA)*: Users are selected only based on the marginal utilities of their measurements. Specially, the user with the largest marginal utility of his/her measurements will be selected in each iteration, unless the marginal utility is lower than the bidding price of the user. The selected users will diffuse tasks to their social friends.
- *Only-Social selection Algorithm (OSA)*: Users are selected only based on the marginal utilities of their social influences. Specially, the user with the largest marginal utility of his/her social influence will be selected in each iteration, unless the marginal utility is lower than the bidding price of the user. The selected users will diffuse tasks to their social friends.
- *Random Selection Algorithm (RSA)*: A user is randomly selected in each iteration, as long as his/her marginal utility is not lower than his/her bidding price. The selected users will diffuse tasks to their social friends.
- *No-Social selection Algorithm (NSA)*: Users are selected only based on the marginal utilities of their measurements. Moreover, the selected users in NSA will not diffuse tasks to their social friends.

The default setup is illustrated as follows. The default number of POIs K in the mobile crowdsensing system is 20. The preliminary value of each measurement collected in location l_k, v_k, is uniformly distributed over $[2, 7]$. We set the number of registered users N is 60. The reliability of each user q_i is uniformly sampled in $(0, 1]$. In each time slot, we consider that 20% users randomly become active users, who will submit a bid to the platform. In each bid β_i^τ, the length of active time window α_i^τ and the bidding price b_i^τ of user u_i are uniformly distributed in $[1, 5]$, respectively. In each time slot during the active time window, a task $\varphi_k^t \in \Phi_i^\tau$ is randomly generated in location l_k, which will be performed by u_i if u_i is selected as participant. To simulate the social network, we randomly generate a directed edge between an arbitrary pair of users, to represent they are social friends or not. The weight of the directed edge $w_{i,j}$ (i.e., the social influence from u_i to u_j) follows a uniform distribution $\mathcal{U}(0, 1)$. An inactive social friend of a selected user may turn into active in the next time slot, according to the weight of the edge between them in the social network. The probability of each user appearing in each POI is uniformly sampled in $[0, 1]$. We run 50 time slots in each setting up of our simulations, with each time slot equals to 10 min. All the simulation results are the average of 20 runs under the same setting up.

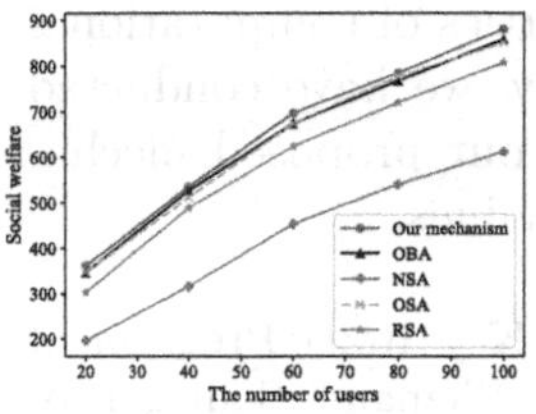

Fig. 2. Social welfare versus number of users

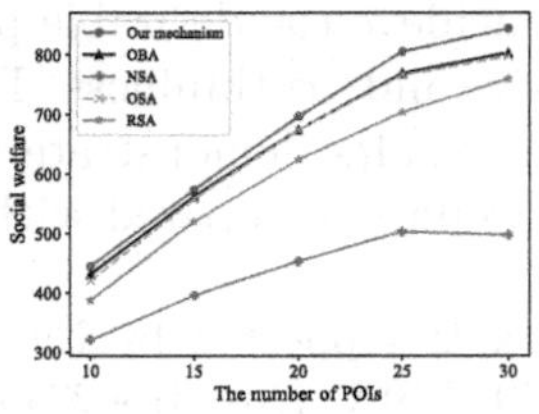

Fig. 3. Social welfare versus number of POIs

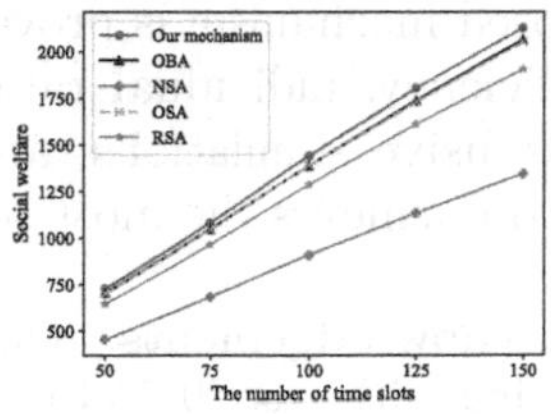

Fig. 4. Social welfare versus number of time slots

6.2 Evaluation Results

We compare the performance achieved by our proposed mechanism and the four baselines, in terms of the social welfare they obtain, i.e., $\sum_{\tau=1}^{|T|}[U(\mathbf{I}^\tau) + W(\mathbf{I}^\tau) - C(\mathbf{I}^\tau)]$. Specially, we vary the number of users, the number of POIs, and the number of time slots, to evaluate the performance of all algorithms under different setups. The evaluation results are shown in Figs. 2, 3, and 4, respectively.

We can easily observe that NSA achieves the worst performance, which is much lower than other algorithms. It is because NSA is the only algorithm without employing recruited users to diffuse sensing tasks to their social friends. This observation validates that leveraging the social relationship among users to attract more participants can make significant improvement to the performance of an incentivized crowdsensing system. Moreover, RSA performs worse than other algorithms except NSA under different setups, which can be seen as the lower bound of all feasible user selection solutions with considering the rationality of users. Furthermore, we can find that our proposed algorithm outperforms OBA and OSA, no matter how the parameters vary. It validates that taking both utilities achieved by measurements and social influences into the consideration of user selection is meaningful. Specially, when there are 30 POIs need to collect sensing data, our proposed mechanism can obtain 4.8% and 5.5% more social welfare than OBA and OSA, respectively.

7 Conclusion

In this paper, we consider the online incentive mechanism design for crowdsensing, in which the social relationship among users is leveraged to attract more participants. We define two novel metrics to evaluate the social influences of users, based on the definitions of node centrality in social networks. In addition, we propose an efficient and truthful online incentive mechanism, consisting of a user selection algorithm and a payment determination algorithm. Users are greedily selected based on their reliability, costs, and social influences. Our proposed mechanism is proved to achieve the desirable properties of computational efficiency, individual rationality, and truthfulness. Finally, we have conducted extensive simulations, and the results demonstrate that our proposed mechanism achieves the most social welfare, compared with baselines.

Acknowledgements. This work is supported by National Key R&D Program of China with No. 2017YFE0117500, NSFC (No. 61802245), the Shanghai Sailing Program (No. 18YF1408200), and STSCM (No. 19511121000). This work is also supported by the Open Project Program of Shanghai Key Laboratory of Data Science (No. 2020090600002).

References

1. TalkingData. http://mi.talkingdata.com/app-rank.html?type=105060
2. Chen, Y., Li, B., Zhang, Q.: Incentivizing crowdsourcing systems with network effects. In: IEEE INFOCOM 2016-The 35th Annual IEEE International Conference on Computer Communications, pp. 1–9. IEEE (2016)
3. Feng, Z., Zhu, Y., Zhang, Q., Ni, L.M., Vasilakos, A.V.: Trac: truthful auction for location-aware collaborative sensing in mobile crowdsourcing. In: IEEE INFOCOM 2014-IEEE Conference on Computer Communications, pp. 1231–1239. IEEE (2014)
4. Feng, Z., et al.: Towards truthful mechanisms for mobile crowdsourcing with dynamic smartphones. In: 2014 IEEE 34th International Conference on Distributed Computing Systems, pp. 11–20. IEEE (2014)
5. Gan, X., Li, Y., Wang, W., Fu, L., Wang, X.: Social crowdsourcing to friends: an incentive mechanism for multi-resource sharing. IEEE J. Sel. Areas Commun. **35**(3), 795–808 (2017)
6. Jiang, L., Niu, X., Xu, J., Wang, Y., Wu, Y., Xu, L.: Time-sensitive and sybil-proof incentive mechanisms for mobile crowdsensing via social network. IEEE Access **6**, 48156–48168 (2018)
7. Kauder, E.: History of Marginal Utility Theory, vol. 2238. Princeton University Press, Princeton (2015)
8. Lin, J., Li, M., Yang, D., Xue, G.: Sybil-proof online incentive mechanisms for crowdsensing. In: IEEE INFOCOM 2018-IEEE Conference on Computer Communications, pp. 2438–2446. IEEE (2018)

9. Nie, J., Luo, J., Xiong, Z., Niyato, D., Wang, P.: A stackelberg game approach toward socially-aware incentive mechanisms for mobile crowdsensing. IEEE Trans. Wirel. Commun. **18**(1), 724–738 (2018)
10. Nie, J., Luo, J., Xiong, Z., Niyato, D., Wang, P., Guizani, M.: An incentive mechanism design for socially aware crowdsensing services with incomplete information. IEEE Commun. Mag. **57**(4), 74–80 (2019)
11. Nisan, N., Roughgarden, T., Tardos, E., Vazirani, V.V.: Algorithmic Game Theory. Cambridge University Press, Cambridge (2007)
12. Sun, J.: An incentive scheme based on heterogeneous belief values for crowd sensing in mobile social networks. In: 2013 IEEE Global Communications Conference (GLOBECOM), pp. 1717–1722. IEEE (2013)
13. Xu, J., Bao, W., Gu, H., Xu, L., Jiang, G.: Improving both quantity and quality: Incentive mechanism for social mobile crowdsensing architecture. IEEE Access **6**, 44992–45003 (2018)
14. Xu, J., Guan, C., Wu, H., Yang, D., Xu, L., Li, T.: Online incentive mechanism for mobile crowdsourcing based on two-tiered social crowdsourcing architecture. In: 2018 15th Annual IEEE International Conference on Sensing, Communication, and Networking (SECON), pp. 1–9. IEEE (2018)
15. Yang, G., He, S., Shi, Z., Chen, J.: Promoting cooperation by the social incentive mechanism in mobile crowdsensing. IEEE Commun. Mag. **55**(3), 86–92 (2017)

CPNSA: Cascade Prediction with Network Structure Attention

Chaochao Liu[1], Wenjun Wang[1], Pengfei Jiao[2], Yueheng Sun[1](✉), Xiaoming Li[3], and Xue Chen[4]

[1] College of Intelligence and Computing, Tianjin University, Tianjin 300350, China
{chaochaoliu,wjwang,yhs,lxm696}@tju.edu.cn
[2] Center of Biosafety Research and Strategy, Tianjin University, Tianjin 300072, China
pjiao@tju.edu.cn
[3] School of International Business, Zhejiang Yuexiu University of Foreign Languages, Zhejiang, China
lxm696@tju.edu.cn
[4] Law School, Tianjin University, Tianjin 300072, China
xuechen@tju.edu.cn

Abstract. Online social medias provide convenient platforms for information spread, which makes the social network structure plays important role on online information spread. Although online social network structure can be obtained easily, few researches use network structure information in the cascade of the resharing prediction task. In this paper, we propose a cascade prediction method (named by CPNSA) involves the network structure information into cascade prediction of resharing task. The method is based on the recurrent neural network, and we introduce a network structure attention to incorporates the network structure information into cascade representation. In order to fuse network structure information with cascading time series data, we use network embedding method to get the representations of nodes from the network structure firstly. Then we use the attention mechanism to capture the structural dependency for cascade prediction of resharing. Experiments are conducted on both synthetic and real-world datasets, and the results show that our approach can effectively improve the performance of the cascade prediction of resharing.

Keywords: Cascade prediction · Deep learning · Network structure influences · Recurrent neural network · Cascade behavior

1 Introduction

Online social networks provide a new type of means for information spreading, which makes information spreading is affected by the online social networks. The exposed users are more likely to spread information than the not exposed one,

H. Gao et al. (Eds.): CollaborateCom 2020, LNICST 349, pp. 64–79, 2021.
https://doi.org/10.1007/978-3-030-67537-0_5

and weak ties play a more important role than strong ties [2]. The user more likely takes retweet action when he received information from multiple social neighbors, and the information spread farther and faster across clustered-lattice networks than across random networks [5].

The research of modeling information spreading data can be used in many fields, such as online social topic detection [4], online post's influence evaluation [37], public opinion monitoring [22], and marketing [23]. Information cascade modeling and prediction is one of these important researches. In this paper, we focus on the research of modeling information cascade, and improve the precision of cascade prediction.

There are already so many methods have been proposed from different perspectives to solve the cascade modeling issues. The wide used models are independent cascade model [6,8], linear threshold model [13], and their variants [3,9]. These models assume that the underlying diffusion pattern is known a priori, while the real-world data may not be like this. The optimization methods of the models are so complex that these models handle large-scale data difficultly. Point process based models are proposed to learn the dependency between the users who spread information [35,41], those models suppose the information spread cascade sequence data following the point process. However, the effectiveness of these models heavily depends on the carefully designed expressions of the point process, and the optimization methods are complex and difficult to compute in parallel. To solve these issues, researchers proposed neural network based models, which do not require an explicit priori assume and can process large-scale data by using GPU Computing. According to the researches [19,24,29,34,40], the neural network based models achieve better performances than the non-neural-network based sequential approaches. Thus, we propose our model based on the neural network to modeling and predicting information spreading cascade sequence data in this paper.

Recently, the neural network based model have been proposed to process the information spreading cascade modeling and predicting task. Wang *et al.* [29] proposed an attention-based recurrent neural network model to learn the cross-dependence influences between the users in information cascades' sequence data. But they did not actually use social network structure information. Wang *et al.* [30] proposed a sequential neural network based model with social network structure attention, to incorporate the social network structural information in information spreading cascade predicting. But they only consider users' neighborhood information. Due to the sparsity of the network, many legitimate links are missing [28]. As a result, their model is not sufficient to involve the network structure information. And because their model make social network adjacency matrix as input, the model can hardly process information spreading cascade data across large-scale social network.

To solve these issues, in this paper, we use the social network embedding method SDNE [28] to get the representations of nodes firstly. SDNE is proposed by Tang *et al.*, which can well maintain the local and global network structure by using the first-order and second-order proximity [27]. In this way, we can solve the missing links of social network issue, and reduce the calculations of training by reducing the representation dimension of social network nodes. Drawing on previous researches, we use the recurrent neural network to model the information spreading cascade sequence data. Thus, we can get the hidden representations of information spreading cascade sequences. To merge the social network structure information with information spreading cascade sequential information, we introduce a social network structure attention neural network layer before the output neural network layer. The proposed model is recognized as CPNSA, which is based on the recurrent neural network for cascade prediction with social network structure attention.

In the proposed model, we use the obtained users' social network embeddings by the SDNE as query vectors. Thus, the model not only considers the historical sequential state of activated users but also various types of social network structure information is transmitted during the cascade behavior modeling. In this way, the model can predict the next activated user precisely. We apply the proposed method on 6 synthetic datasets and the Digg real-world dataset with three baseline methods to compare the prediction performance of next activated users.

It is worthwhile to highlight several contributions of the CPNSA model here:

- We proposed a new model to deal with the modeling and predicting cascades of information resharing tasks, the model is robust by using the neural network.
- We consider social network structure information via attention mechanism in cascades of information resharing prediction task, which makes the method be able to capture nodes' dependency in network structure.
- Recently proposed network embedding method SDNE [28] is used, which makes the method be able to involve the first-order and second-order proximity between nodes. Thus, our model can deal with the missing links in social network.
- Our model utilizes the result of social network embedding, which reduces space complexity and more precise than the model using edges directly.
- CPNSA performs better than the other compared algorithms.

2 Related Work

In this part, we present recent researches related to our work, which mainly contains the researches of network structure's affection on the information diffusion and neural network based models for information spreading cascade modeling. Then we show our point of view on the existing information spreading cascade modeling researches. All these observations indeed motivate the work of this paper.

Social network structure plays important role on the information spreading, recent researchers attempt to involve social network structure information in the information spreading modeling tasks. Huang *et al.* [12] proposed a method utilizes community information for influence maximization task, the experiments' results validate social network structure's influence on the information spreading. Li *et al.* [17] investigate the close correspondence between social tie in information spreading process. Zhang *et al.* researched on the social influence and found that the social users' retweet actions are influenced by their close friends in their social ego networks [39]. Weng *et al.* [31] used social network structure information to predict memes on the social medias, their proposed model showed good performance on the prediction task of future popularity of a meme given its early spreading patterns by using the social network community structure. Su *et al.* [26] worked on the social contagious research and showed that there is an optimal social network community structure can maximizes spreading dynamics. Nematzadeh *et al.* [21] investigated the social network community structure effects on information spreading, and they found that global information spreading speed can be enhanced by the strong communities in online social network. Wu *et al.* [32] worked on the social network multi-community structure effects on information spreading research, and they found that the social network multi-community structure can facilitate the online social information spreading process. Qiu *et al.* [24] designed an end-to-end framework for feature representation learning to predict social influence. However, there are few types of research consider social network structure's influence in online social network information spreading cascade prediction models.

Types of research have been proposed to modeling the social network information spreading, such as Matchbox model [25,38], multiple additive regression trees (MART) model [33], maximum entropy classifier model [1], autoregressive-moving-average (ARMA) model [20], factor graph model [36], conditional random fields model [23] and so on. Because of the specific assumptions of the models, these models can hardly generalized to other datasets. And these models process large-scale data difficultly because of the complex optimization methods. Recently, researchers attempt to use deep neural networks to address above issues inspired by the recent success of deep neural networks in many other applications. These works mostly use the recurrent neural network (RNN) to learn the hidden dependences between the retweet users and time patterns of spreading action. Finally, the models can get the representations of information spreading cascade. Since these models do not require knowing the underlying information spreading model, they can process real-world data robust. In this paper, we also use the deep neural networks to build our model for dealing with the cascade prediction task.

There are already few models based on the neural networks. Xiao *et al.* [34] used two recurrent neural networks to build their model, one recurrent neural network is used to interprets the conditional intensity function of a point process, and the other recurrent neural network is used to learn the time patterns.

Zhang *et al.* [40] used neural network embedding method to transport the tweet contents, the social network user interests, the similarity information between the tweet content and social user interests, social user information and author of tweet information into neural network representations. They proposed an attention mechanism to encode the interests of the social users. Their model finally to predict whether a tweet will be retweeted by a user. Wang *et al.* [29] showed that each cascade generally corresponds to a diffusion tree, causing cross-dependence in cascade, so they proposed an attention-based recurrent neural network to capture the cross-dependence in cascade. Liu *et al.* [19] followed independence cascade model, they defined parameters for every social user with a latent influence vector and a susceptibility vector. The proposed model can be used to learn information spreading cascade dynamics. However, these models not actually use social network structure information. Wang *et al.* [30] proposed a diffusion model based on recurrent neural network, and involved the social network structure information by proposed a social network structure attention. But their model take social network adjacent matrix as input, which make the model can hardly process large-scale social network and difficult to handle the missing links in social networks. Liu *et al.* [18] proposed a cascade prediction model with community structure enhanced, but their model only focused on the community structure information.

Thus, in this paper, we use social network embedding tool to get a low dimensional representation for users, which can solve the missing links issue. We choose SDNE [28] to get social network embedding, since SDNE can learn local and global social network structure information. To merge the social network structure information with the information spreading cascade sequence data, we propose an social network structure attention layer to restrict the information spreading cascade representation.

3 Proposed Method

3.1 Problem Definition

In this paper, we focus on the task of further take retweet action users prediction and retweet time of the user prediction. The input data of our proposed model are social network structure data and information spreading cascade sequence data. The social network and information spreading cascade are defined as the following:

Definition 1. **Social Network.** A social network contains nodes and edges, and the edges represent the relationships between the nodes. We denote the set of nodes as V, the set of edges as E and the social network as $G = (V, E)$. Thus the number of nodes is $|V|$ and the number of edges is $|E|$, where $|\cdot|$ represent the size of the set.

Definition 2. **Information Spreading Cascade.** A information spreading cascade is a set of sequence information spreading data, it contains retweet users and retweet time. We denote the cascade as $S = \{(t_i, v_i)|v_i \in V, t_i \in [0, +\infty), t_i \leq t_{i+1}, i = 1, 2, ..., N\}$, it start from a original post user and the post time, and the retweet users and the corresponding retweet time are ascendingly ordered by time. We denote N as the number of users take part in the information spreading of the post, and denote the retweet data at time t_i as (t_i, v_i), where v_i represents the $i-$th retweet user.

Given the above definitions, we formulate our problem in the following part. The input data of our proposed model contains a social network $G = (V, E)$, and a collection of F information spreading cascades denote as $Q = \{S_f\}_{f=1}^{F}$. For one of the observed cascade, we denote the sequence data up to the k−th retweet behavior as $S_{\leq k}$. Our proposed model take the data input and learn the information spreading patterns. The trained model can be used to calculate the probability of the next take retweet action user v_{k+1} and the action time $t_{v_{k+1}}$. Thus, the probability of the next take retweet action user can be represented as $p(v_{k+1}|S_{\leq k})$.

3.2 Model Framework

Our proposed model is based on recurrent neural network, which can process the large-scale data easily. And we proposed an new social network structure attention to merge the social network structure information with the information spreading cascade sequence data. The proposed model is named by CPNSA, to solve the information spreading cascade prediction problem. The rationale of our model is that we use the recurrent neural network to learn the historical cascade sequential state of retweet users and retweet time. At the same time, we use the social network structure attention to learn the social network structure's effects on the information spreading actions. Based on these ideas, we propose a new deep learning-based cascade behavior modeling framework containing social network structure information. The system framework of the method is shown in Fig. 1. CPNSA mainly uses recurrent neural network (RNN) to model sequence dependence. In order to incorporate the impact of network structure, we propose a social network structure attention model to involve both local and global network structure.

Sequence Modeling. Our model employs two recurrent neural networks (RNNs) to model the user sequence and time sequence respectively. In each RNN, a hidden state is used to memorize the summarized history. In each step k of a cascade, the node v_k is represented as a low-dimensional vector $\mathbf{v}_k \in R^{d_v}$ through a mapping matrix $\mathbf{W}_{emv}$. The node embedding vector is represented as $\mathbf{v}_k = \mathbf{W}_{emv}^T v_k$ with the dimension d_v. Then, the hidden state representation of nodes' activity at step k can be $\mathbf{h}_k^{(0)} = RNN(\mathbf{v}_k, \mathbf{h}_{k-1}^{(0)})$ by using RNN. In addition, for the timing input t_k, we using inter-event duration $t_k - t_{k-1}$ as the temporal features $\mathbf{t}_k$. The hidden state representation of time sequence at step k can be $\mathbf{h}_k^{(1)} = RNN(\mathbf{t}_k, \mathbf{h}_{k-1}^{(1)})$ by using RNN. $\mathbf{h}_0^{(0)}$ and $\mathbf{h}_0^{(1)}$ are initialized as all zero vector.

Social Network Structure Attention. Considering that the influences of nodes in a social network are different, it is important to identify those key users and help extract representations of cascades. Wang *et al.* [30] proposed a network structure attention based on RNN. However, the query vector in [30] was

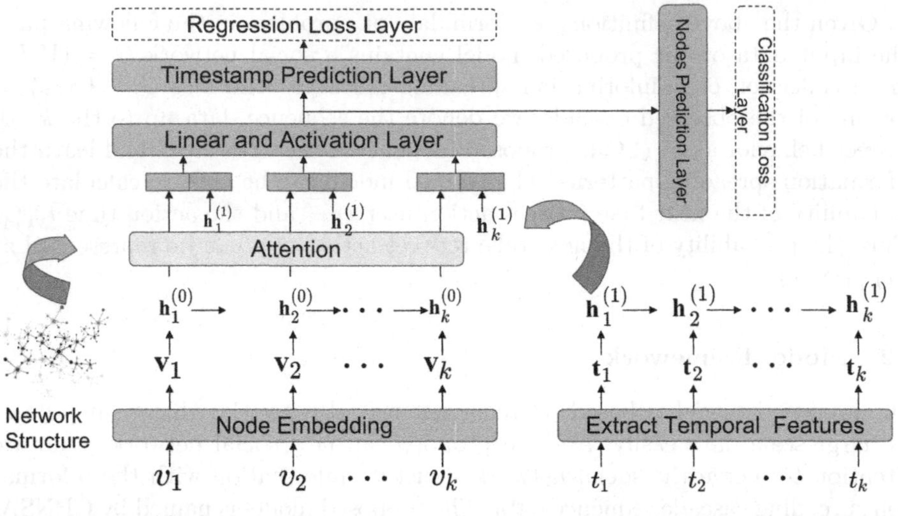

Fig. 1. Our recurrent neural network based model can be trained end-to-end. Users are converted to low-dimensional vectors by the user embedding layer, then feed into RNN network, and we can get hidden state vectors $\mathbf{h}_1^{(0)}, \mathbf{h}_2^{(0)}, ..., \mathbf{h}_h^{(0)}$. Time series are extracted as inter-event duration, and feed into recurrent neural network, finally we can get hidden state vectors $\mathbf{h}_1^{(1)}, \mathbf{h}_2^{(1)}, ..., \mathbf{h}_h^{(1)}$. We can get users' embedding vectors via SDNE [28]. Then the hidden state vectors $\mathbf{h}_1^{(0)}, \mathbf{h}_2^{(0)}, ..., \mathbf{h}_h^{(0)}$ can be transformed by using the social network structure attention layer, and concatenated with the hidden state vectors of time features feed into linear and activation layers, we can get the cascade's representation. Then two prediction layers are used to output the predicted activated node, and the associated timestamp. Cross-entropy and square loss are respectively used for event type and timestamp prediction.

computed complexity and can only utilize oversimplified neighbor information of network structure. Also that, many legitimate links are missing due to the sparsity of networks, which causing the Wang's model hardly model the data precisely. As a result, our model CPNSA is developed to consider both local and global social network structure information.

Thus, we use the social network embedding method SDNE [28] to represent the frist-order and second-order proximity of users into low-dimensional vectors. In this way, the global and local social network structure information can be involved in the social network embeddings. Given the network G of the training set and testing set, composed of N nodes and E edges, SDNE can learn highly non-linear network structure by using multiple nonlinear mapping functions to map the input data to a highly nonlinear latent space. SDNE also uses both the second-order and first-order proximity to capture the global and local network structure. Due to the sparsity of networks, SDNE imposes more penalty to the reconstruction error of the non-zero elements than that of zero elements. Following SDNE, we minimize the loss function to obtain embeddings of nodes $\mathbf{H}^{(e)}$:

$$\mathcal{L} = \mathcal{L}_{2nd} + \alpha\mathcal{L}_{1st} + \beta\mathcal{L}_{reg} \tag{1}$$

following the SDNE's definitions, we denote the $\mathcal{L}_{2nd}$ as the second-order proximity loss, and the $\mathcal{L}_{1st}$ as the first-order proximity loss. We denote the $\mathcal{L}_{reg}$ as an $\mathcal{L}2$-norm regularizer term in this paper. The α and the β represent the balance parameters.

Given the social network nodes' embeddings, we compute the effection of node at step i on the node at step k by using attention with the embedding of $\mathbf{h}_k^{(0)}$ as query vectors, like the Eq. (2):

$$\gamma_{k,i} = \frac{exp(\mathbf{h}_k^{(0)} A(\mathbf{h}_i^{(e)})^T)}{\sum_{j=1}^{k} exp(\mathbf{h}_k^{(0)} A(\mathbf{h}_j^{(e)})^T))} \tag{2}$$

where $\mathbf{h}_i^{(e)}$ is the social network structure embedding of node at step i, A is a parameter matrix, $\gamma_{k,i}$ is effection of node at step i on node at step k. We denote $\mathbf{h}_k^{(2)}$ as the final hidden representation of node sequence at step k, just like Eq. (3):

$$\mathbf{h}_k^{(2)} = \sum_{i=1}^{k} \gamma_{k,i}\mathbf{h}_i^{(0)} \tag{3}$$

Next Activated Node and Time Generation. The hidden representation of cascade at step k is given by merge the hidden representations $\mathbf{h}_k^{(1)}$ and $\mathbf{h}_k^{(2)}$:

$$\mathbf{h}_k^{cas} = \delta(\mathbf{W}_h^T(\mathbf{h}_k^{(1)} \oplus \mathbf{h}_k^{(2)}) + \mathbf{b}_h) \tag{4}$$

where $\mathbf{W}_h^T$ is the weight, $\mathbf{b}_h$ is the bias, $\oplus$ is the connection operation and $\delta()$ represents the activation function, and in this paper we use the PReLU activation function.

Next Activated Node Generation. The model adds a linear layer and an activation layer to project the hidden representation into the same space with the node embedding showed in Eq. (5):

$$\mathbf{h}_k^{node} = \delta(\mathbf{W}_{node}^T\mathbf{h}_k^{cas} + \mathbf{b}_{node}) \tag{5}$$

where $\mathbf{W}_{node}^T$ is the weight, and $\mathbf{b}_{node}$ is the bias. Finally, we calculate cosine similarities between the hidden vector $\mathbf{h}_k^{node}$ with the embedding vectors of all the nodes, and use a softmax layer to generate the probability distribution of next infected node as follow:

$$\mathbf{p}_k^{node} = softmax(\mathbf{h}_k^{node}\mathbf{W}_{emv}^T) \tag{6}$$

where $\mathbf{p}_k^{node} \in R^N$, N is the number of nodes.

Next Activated Time Generation. Based on the hidden representation of cascade at step k, we can generate the time inter-event duration between step $k+1$ and step k by adding a linear layer following Eq. (7)

$$t_{k+1} - t_k = \mathbf{W}_t^T\mathbf{h}_k^{cas} + \mathbf{b}_t \tag{7}$$

where $\mathbf{W}_t^T$ is the weight, and $\mathbf{b}_t$ is the bias.

3.3 Optimization

We introduce our learning process of the model as bellow. Given a collection of cascades $Q = \{S_f\}_{f=1}^{F}$, we treat the cascades are independent on each other. Thus, we can learn the model by maximizing the joint log-likelihood of observing Q in Eq. (8):

$$\mathbf{Loss}(Q) = \sum_{f=1}^{F} \sum_{i=1}^{N_f-1} logp((t_{k+1}, v_{k+1})|c_{v_k}, S_{\leq k}) \tag{8}$$

which is the sum of the logarithmic likelihood for all the individual cascades. We exploit backpropagation through time (BPTT) for training our model. In each training iteration, we vectorize activated nodes' information as inputs, including nodes' embedding, nodes' embedding calculated from network structure information and inter-event duration temporal features. At last, we apply stochastic gradient descent (SGD) with mini-batch and the parameters are updated by Adam [14]. To speed up the convergence, we use orthogonal initialization method in training process [10]. We also employ clips gradient norm for the parameters to prevent overfitting.

4 Experimental Setup

In this section, we introduce the data sets, the comparison methods, and the evaluation metrics used in the experiments to quantitatively evaluate the proposed framework.

4.1 Data Sets

Our experiments are conducted on two types of data sets—synthetic data and real-word data.

Synthetic Data. The data generation consists of two parts: network generation and cascade generation. We use two network generation tools to generate networks. The first network generation tool is following from previous work [29,30], we apply Kronecker graph model [16] to generate random network (RD) with the parameter matrix [0.5 0.5; 0.5 0.5]. We construct a network with default 1024 users and avenge 20°. The second network generation tool is the LFR benchmark proposed by Lancichinetti *et al.* [15], which is the popular used synthetic networks containing community structure generator. We set the average degree of nodes to be 20, the maximum degree of nodes to be 50, power-law exponent for the degree distribution to be 2, power-law exponent for the community size distribution to be 1. Then we generate two networks contain 500 nodes (LFR500) and 1000 (LFR1000) nodes separately. In the cascade generation part, for each activated node, we set the activation time of an activated user following a certain time distribution. Similar with Wang's setup, we choose two-time distributions for sampling: 1) mixed exponential (Exp) distributions, controlled by rate parameters in [0.01, 10]; 2) mixed Rayleigh (Ray) distribution, controlled by scale parameters in [0.01, 10] [29]. The cascade generation progress uses breadth-first to search for next activated node, and the progress will stop until the overall time exceeds the threshold ω or no node is activated. We set $\omega = 100$. Finally, six synthetic data sets are generated by different combinations of network scale and propagation time distributions, denoted by (RD, Exp), (RD, Ray), (LFR500, Exp), (LFR500, Ray), (LFR1000, Exp), and (LFR1000, Ray). We generate 20 cascades per node in each dataset and randomly pick up 80% of cascades for training and the rest for validation and test.

Real World Data. The Digg dataset proposed by Nathan *et al.* is used in this paper. The dataset contains diffusions of stories as voted by the users, along with friendship network of the users [11]. We drop the cascades with size larger than 1,000, as the large cascade rarely occurs in practice and may dominate the training process [29]. We randomly pick up 80% of cascades for training and the rest for validation and test.

4.2 Comparison Methods

For comparison with the proposed model, we evaluate the following methods on the data sets.

RNNPP [34]: The method of RNNPP takes a recurrent neural network (RNN) perspective to point process and models its background and history effect. The model can be used to predict event timestamp, main-type event and sub-type event. In this paper, we consider nodes as main-types and do not use the sub-type event prediction layer.

Recurrent Marked Temporal Point Processes (RMTPP) [7]: The method of RMTPP views the intensity function of a temporal point process as a nonlinear function of the history, and uses a recurrent neural network to automatically learn a representation of influences from the event history. The RMTPP can be applied in activated nodes timestamp and activated nodes prediction for information cascade.

Sequential Neural Network with Structure Attention (SNNSA) [30]: The SNNSA is a recently proposed method to model information diffusion, which can capture the structural dependency among users by using attention mechanism. However, the method only considers local network structure.

4.3 Evaluation Metrics

Our task is predicting next activated node, and next activated timestamp, given previously cascade information. Since the number of potential nodes is huge, we can regard the prediction task as a ranking problem with users' transition probabilities as their scores [29]. Each model outputs the infection probability distribution over all users and the actual infected user is expected to get the highest probability [30]. Thus, we evaluate the proposed method and comparison method by accuracy on top k (Acc@k) and mean reciprocal rank (MRR) which are the widely used metrics. For timestamp prediction, we use the root-mean-square error (RMSE) which measures the difference between the predicted time point and the actual one.

5 Experimental Results

5.1 Synthetic Data Results

Table 1, Table 2, and Table 3 show the prediction comparisons of next activated node on baselines and our proposed model respectively for ACC@5, ACC@10, and MRR metrics. As we can see, our proposed method CPNSA performs consistently and significantly better than other baselines on Acc@5, Acc@10, and MRR in all datasets. The results indicate that our proposed method can better predict next activated node. It is interesting to see that SNNSA performs better on (Rd, Exp) and (RD, Ray) datasets. This phenomenon demonstrates that although SNNSA involves network structure information via attention mechanism, the method does not fully consider network structure information, such as community structure.

Table 1. Predictive performance ACC@5 for predictions of next activation node on baselines and our proposed model named by CPNSA.

Method	500, Exp	500, Ray	1000, Exp	1000, Ray	Rd, Exp	Rd, Ray
RNNPP	0.0793	0.1796	0.0756	0.1388	0.0800	0.1337
RMTPP	0.3486	0.3308	0.3443	0.3094	0.6807	0.7144
SNNSA	0.3466	0.2851	0.2835	0.2357	0.7135	0.6566
CPNSA	**0.8268**	**0.7912**	**0.8930**	**0.8903**	**0.7428**	**0.8569**

Table 2. Predictive performance ACC@10 for predictions of next activation node on baselines and our proposed model named by CPNSA.

Method	500, Exp	500, Ray	1000, Exp	1000, Ray	Rd, Exp	Rd, Ray
RNNPP	0.1253	0.2142	0.0937	0.1604	0.0395	0.1389
RMTPP	0.4600	0.4508	0.4533	0.4117	0.7208	0.7925
SNNSA	0.4790	0.4217	0.4090	0.3515	**0.8116**	0.7792
CPNSA	**0.8367**	**0.8144**	**0.8950**	**0.8938**	0.7825	**0.8773**

Table 4 shows the prediction comparisons of next activation timestamp on baselines and our proposed model. The evaluation metric is RMSE. We can see that all the methods perform similarity. The available implementation does not allow SNNSA to compute the next activation time.

Table 3. Predictive performance MRR for predictions of next activation node on baselines and our proposed model named by CPNSA.

Method	500, Exp	500, Ray	1000, Exp	1000, Ray	Rd, Exp	Rd, Ray
RNNPP	0.0768	0.1566	0.0757	0.1215	0.0254	0.1312
RMTPP	0.2303	0.2208	0.2263	0.2063	0.4982	0.4736
SNNSA	0.2339	0.1960	0.1910	0.1579	0.4764	0.4309
CPNSA	**0.8222**	**0.7921**	**0.8958**	**0.8896**	**0.7055**	**0.8359**

Table 4. Predictive performance RMSE for predictions of next activation time on baselines and our proposed model named by CPNSA.

Method	500, Exp	500, Ray	1000, Exp	1000, Ray	Rd, Exp	Rd, Ray
RNNPP	7.0813	0.2359	3.1739	0.1958	12.1832	0.6288
RMTPP	**6.0980**	**0.2344**	**2.7998**	**0.1944**	11.6363	**0.5842**
SNNSA	-	-	-	-	-	-
CPNSA	6.5452	0.2829	3.0073	0.2317	**10.9942**	0.6866

5.2 Real Data Results

Table 5 shows the prediction comparisons of next activated node on Digg dataset. We perform all the algorithms on NVIDIA Tesla 32G GPU server. SNNSA algorithm runs out of memory. As we can see, CPNSA performs consistently best. SNNSA uses the whole adjacency matrix of the network, while our model uses the result of network embedding. As we know, the dimension of network embedding is far less than the adjacency matrix of the network. Thus, the space complexity of SNNSA is much larger than ours.

Table 5. Predictive performance of next activated node on Digg dataset.

Method	ACC@5	ACC@10	MRR
RNNPP	0.0088	0.0107	0.0076
RMTPP	0.0179	0.0298	0.0161
SNNSA	-	-	-
CPNSA	**0.7712**	**0.7644**	**0.7437**

Table 6 shows the prediction comparisons of next activated time on Digg dataset. We use RMSE as the evaluation metric. All the methods perform similarity. The results of SNNSA is not shown because of the not available implementation.

Table 6. Predictive performance of next activated time on Digg dataset.

Method	RNNPP	RMTPP	SNNSA	CPNSA
	30863.83	30862.12	-	30862.53

6 Conclusion

In this paper, we work on the cascade prediction task involving network structure information in the recurrent neural network framework (RNN) via attention mechanism. Different from traditional modeling methods, RNN is a convenient and effective tool for cascade predicting, avoiding strong prior knowledge on diffusion model and being flexible to capture complex dependence in cascades. Besides, recent researches find that network structure, such as community structure, always effects cascade behaviors. Thus we first embedding local and global network structure into nodes' representation vectors and using an attention mechanism in RNN to involve network structure information for capture the network structure effects in cascade.

We evaluate the effectiveness of our proposed model on both synthetic and real datasets. Experimental results demonstrate that our proposed model outperforms state-of-the-art modeling methods at the next activated node prediction task. Additionally, CPNSA performs better than SNNSA on both synthetic and real datasets, implying that our method not just involves neighborhood structure information, our method can also capture community structure effects.

Acknowledgement. This work was supported in part by the National Key Research and Development Plan of China under grant 2018YFC0831005 and the Supreme People's Court 2019 Judicial Research Major Project (ZGFYZDKT201916-01).

References

1. Artzi, Y., Pantel, P., Gamon, M.: Predicting responses to microblog posts. In: Proceedings of the 2012 Conference of the North American Chapter of the Association for Computational Linguistics: Human Language Technologies, pp. 602–606. Association for Computational Linguistics (2012)
2. Bakshy, E., Rosenn, I., Marlow, C., Adamic, L.: The role of social networks in information diffusion. In: Proceedings of the 21st International Conference on World Wide Web, pp. 519–528 (2012)
3. Barbieri, N., Bonchi, F., Manco, G.: Topic-aware social influence propagation models. Knowl. Inf. Syst. **37**(3), 555–584 (2013). https://doi.org/10.1007/s10115-013-0646-6
4. Cataldi, M., Di Caro, L., Schifanella, C.: Emerging topic detection on twitter based on temporal and social terms evaluation. In: Proceedings of the Tenth International Workshop on Multimedia Data Mining, p. 4. ACM (2010)
5. Centola, D.: The spread of behavior in an online social network experiment. Science **329**(5996), 1194–1197 (2010)

6. Cheng, S., Shen, H., Huang, J., Zhang, G., Cheng, X.: Staticgreedy: solving the scalability-accuracy dilemma in influence maximization. In: Proceedings of the 22nd ACM International Conference on Information and Knowledge Management, pp. 509–518. ACM (2013)
7. Du, N., Dai, H., Trivedi, R., Upadhyay, U., Gomez-Rodriguez, M., Song, L.: Recurrent marked temporal point processes: Embedding event history to vector. In: Proceedings of the 22nd ACM SIGKDD International Conference on Knowledge Discovery and Data Mining, pp. 1555–1564. ACM (2016)
8. Gomez-Rodriguez, M., Leskovec, J., Schölkopf, B.: Modeling information propagation with survival theory. In: International Conference on Machine Learning, pp. 666–674 (2013)
9. Goyal, A., Bonchi, F., Lakshmanan, L.V.: Learning influence probabilities in social networks. In: Proceedings of the third ACM International Conference on Web Search and Data Mining, pp. 241–250. ACM (2010)
10. Henaff, M., Szlam, A., LeCun, Y.: Recurrent orthogonal networks and long-memory tasks. arXiv preprint arXiv:1602.06662 (2016)
11. Hogg, T., Lerman, K.: Social dynamics of digg. EPJ Data Sci. **1**(1), 5 (2012)
12. Huang, H., Shen, H., Meng, Z., Chang, H., He, H.: Community-based influence maximization for viral marketing. Appl. Intell. **49**(6), 2137–2150 (2019). https://doi.org/10.1007/s10489-018-1387-8
13. Kempe, D., Kleinberg, J., Tardos, É.: Maximizing the spread of influence through a social network. In: Proceedings of the Ninth ACM SIGKDD International Conference on Knowledge Discovery and Data Mining, pp. 137–146. ACM (2003)
14. Kingma, D.P., Ba, J.: Adam: A method for stochastic optimization. arXiv preprint arXiv:1412.6980 (2014)
15. Lancichinetti, A., Fortunato, S., Radicchi, F.: Benchmark graphs for testing community detection algorithms. Phys. Rev. E **78**(4), 046110 (2008)
16. Leskovec, J., Chakrabarti, D., Kleinberg, J., Faloutsos, C., Ghahramani, Z.: Kronecker graphs: an approach to modeling networks. J. Mach. Learn. Res. **11**, 985–1042 (2010)
17. Li, K., Lv, G., Wang, Z., Liu, Q., Chen, E., Qiao, L.: Understanding the mechanism of social tie in the propagation process of social network with communication channel. Front. Comput. Sci. **13**(6), 1296–1308 (2019). https://doi.org/10.1007/s11704-018-7453-x
18. Liu, C., Wang, W., Sun, Y.: Community structure enhanced cascade prediction. Neurocomputing **359**, 276–284 (2019)
19. Liu, S., Zheng, H., Shen, H., Cheng, X., Liao, X.: Learning concise representations of users' influences through online behaviors. In: Proceedings of the Twenty-Sixth International Joint Conference on Artificial Intelligence, IJCAI-17, pp. 2351–2357 (2017)
20. Luo, Z., Wang, Y., Wu, X.: Predicting retweeting behavior based on autoregressive moving average model. In: Wang, X.S., Cruz, I., Delis, A., Huang, G. (eds.) WISE 2012. LNCS, vol. 7651, pp. 777–782. Springer, Heidelberg (2012). https://doi.org/10.1007/978-3-642-35063-4_65
21. Nematzadeh, A., Ferrara, E., Flammini, A., Ahn, Y.Y.: Optimal network modularity for information diffusion. Phys. Rev. Lett. **113**(8), 088701 (2014)
22. Nip, J.Y., Fu, K.W.: Networked framing between source posts and their reposts: an analysis of public opinion on china's microblogs. Inf. Commun. Soc. **19**(8), 1127–1149 (2016)

23. Peng, H.K., Zhu, J., Piao, D., Yan, R., Zhang, Y.: Retweet modeling using conditional random fields. In: 2011 11th IEEE International Conference on Data Mining Workshops, pp. 336–343. IEEE (2011)
24. Qiu, J., Tang, J., Ma, H., Dong, Y., Wang, K., Tang, J.: DeepInf: modeling influence locality in large social networks. In: Proceedings of the 24th ACM SIGKDD International Conference on Knowledge Discovery and Data Mining (KDD 2018) (2018)
25. Stern, D.H., Herbrich, R., Graepel, T.: Matchbox: large scale online bayesian recommendations. In: Proceedings of the 18th International Conference on World Wide Web, pp. 111–120. ACM (2009)
26. Su, Z., Wang, W., Li, L., Stanley, H.E., Braunstein, L.A.: Optimal community structure for social contagions. New J. Phys. **20**(5), 053053 (2018)
27. Tang, J., Qu, M., Wang, M., Zhang, M., Yan, J., Mei, Q.: Line: large-scale information network embedding. In: Proceedings of the 24th International Conference on World Wide Web, pp. 1067–1077. International World Wide Web Conferences Steering Committee (2015)
28. Wang, D., Cui, P., Zhu, W.: Structural deep network embedding. In: Proceedings of the 22nd ACM SIGKDD International Conference on Knowledge Discovery and Data Mining, pp. 1225–1234. ACM (2016)
29. Wang, Y., Shen, H., Liu, S., Gao, J., Cheng, X.: Cascade dynamics modeling with attention-based recurrent neural network. In: Proceedings of the 26th International Joint Conference on Artificial Intelligence,pp. 2985–2991. AAAI Press (2017)
30. Wang, Z., Chen, C., Li, W.: A sequential neural information diffusion model with structure attention. In: Proceedings of the 27th ACM International Conference on Information and Knowledge Management, pp. 1795–1798. ACM (2018)
31. Weng, L., Menczer, F., Ahn, Y.Y.: Predicting successful memes using network and community structure. In: ICWSM (2014)
32. Wu, J., Du, R., Zheng, Y., Liu, D.: Optimal multi-community network modularity for information diffusion. Int. J. Modern Phys. C **27**(08), 1650092 (2016)
33. Wu, Q., Burges, C.J., Svore, K.M., Gao, J.: Ranking, boosting, and model adaptation. Technical report, Microsoft Research (2008)
34. Xiao, S., Yan, J., Yang, X., Zha, H., Chu, S.M.: Modeling the intensity function of point process via recurrent neural networks. In: AAAI, vol. 17, pp. 1597–1603 (2017)
35. Xu, H., Wu, W., Nemati, S., Zha, H.: Patient flow prediction via discriminative learning of mutually-correcting processes. In: 2017 IEEE 33rd International Conference on Data Engineering (ICDE), pp. 37–38. IEEE (2017)
36. Yang, Z., Guo, J., Cai, K., Tang, J., Li, J., Zhang, L., Su, Z.: Understanding retweeting behaviors in social networks. In: Proceedings of the 19th ACM International Conference on Information and Knowledge Management, pp. 1633–1636. ACM (2010)
37. Ye, S., Wu, F.: Measuring message propagation and social influence on twitter.com. Int. J. Commun. Netw. Distrib. Syst. **11**(1), 59–76 (2013)
38. Zaman, T.R., Herbrich, R., Van Gael, J., Stern, D.: Predicting information spreading in twitter. In: Workshop on computational Social Science and the Wisdom of Crowds, nips, vol. 104, pp. 17599–601. Citeseer (2010)

39. Zhang, J., Liu, B., Tang, J., Chen, T., Li, J.: Social influence locality for modeling retweeting behaviors. In: IJCAI, vol. 13, pp. 2761–2767 (2013)
40. Zhang, Q., Gong, Y., Wu, J., Huang, H., Huang, X.: Retweet prediction with attention-based deep neural network. In: Proceedings of the 25th ACM International on Conference on Information and Knowledge Management, pp. 75–84. ACM (2016)
41. Zhou, K., Zha, H., Song, L.: Learning social infectivity in sparse low-rank networks using multi-dimensional Hawkes processes. In: Artificial Intelligence and Statistics, pp. 641–649 (2013)

Optimization for Collaborate System

Speech2Stroke: Generate Chinese Character Strokes Directly from Speech

Yinhui Zhang(✉), Wei Xi, Zhao Yang, Sitao Men, Rui Jiang, Yuxin Yang, and Jizhong Zhao

School of Computer Science and Technology, Xi'an Jiaotong University, Xi'an, China
manli0826@gmail.com, weixi.cs@gmail.com, zhaoyang9425@gmail.com, sitao.men@gmail.com, ruijiang.jerry@gmail.com, yangdx6@gmail.com, zjz@xjtu.edu.cn

Abstract. Chinese character is composed of spatial arrangement of strokes. A portion of these strokes combines to form phonetic component, which provides a clue to the pronunciation of the entire character, the others combine to form semantic component, which indicates semantic level information for speech context. How closely the connection between the internal strokes of Chinese characters and speech? In this paper, we propose Speech2Stroke, a end-to-end model that exploits the phonetic and morphologic level information of pictographic words. Specifically, we generate strokes directly from the speech by Speech2Stroke. The performance of Speech2Stroke is evaluated by the specific stroke error rate(SER). The SER of the optimal model can achieve 20.61%. Through the experiments and analysis, we show that our model has the ability to capture the alignment between audio and the internal structures of pictographic characters.

Keywords: Deep learning · Stroke of Chinese character · Pictographic word

1 Introduction

When Chinese people encounter a previously unseen character without any pinyin or phoneme, but one can still pronounce the character according to its graphical shape, and guess the meaning. There is a closely connection between speech and internal graphical components, part of which is a direct result of the mechanisms of Chinese formation system.

The traditional six writings were first described by Xu Shen in his dictionary Shuowen Jiezi in the ancient Han dynasty [13], which consists of six kinds of principles of Chinese formation. One of six writings, pictophonetic compound, is

This work was supported by NSFC Grant No. 61832008, 61772413, 61802299, and 61672424.

H. Gao et al. (Eds.): CollaborateCom 2020, LNICST 349, pp. 83–94, 2021.
https://doi.org/10.1007/978-3-030-67537-0_6

by far the most numerous characters. Pictophonetic characters are composed of at least two components, characters with the same phonetic components share similar pronunciation. As shown in Table 1, all these simplified Chinese characters have the same right-hand side part, and their pronunciation is extremely similar except the tone.

Chinese characters have been regarded as a logographic language, the logograms of Chinese characters convey fruitful information of their meanings. Recent researches have proved that internal structural features are beneficial in the field of natural language processing (NLP): radicals, also as semantic components in most cases, are useful in learning Chinese word embedding and language understanding task[19,20]. Cao et al.[4] decompose the character into units of smaller granularity, and propose stroke n-grams for Chinese word embedding. Su and Lee et al. [16] use a convolutional auto-encoder to extract character feature from bitmap to represent the character glyphs, which shows that glyph embedding enhance word analogy and word similarity tasks. Meng et al. [12] utilize the Tianzige-CNN structures to extract the semantic glyph-vectors with historical Chinese scripts, which is proved to improve a wide range of Chinese NLP tasks due to the rich pictographic information in Chinese character. Unfortunately, the aforementioned methods only focus on the signific component of character, however, they neglect the phonetic component.

In this work, given the rich phonetic components in Chinese characters, we ask a question: can machine learn the relationship between speech and internal structures of Chinese characters? In our proposed model, we use a similar idea with [4], which decomposes the character into a sequence of strokes. With such stroke order information, we design a model that can exploit the sequential correction with audio input. Specificially, we use a neural network model that takes a short speech signal as input and predicts a stroke vector representing the internal structure. To train our model, we use the self-built stroke order data set, comprised of 32 different types of Chinese strokes, which has more diversity compare to [4]. Our model simply exploits the natural advantages of Chinese speech and internal structures of character, without requiring any other phoneme, e.g., pinyin.

The contributions of this paper are as follows: 1) To the best of our knowledge, our work is the first to explore a model for generating the internal structural strokes directly from the mandarin speech. 2) a new task learning for multi-task learning or transfer learning in mandarin speech recognition and speech synthesis.

The rest of this paper is organized as follows. In Sect. 2, we begin by introducing the architecture of Speech2Stroke model in detail. We describe the experimental setup in Sect. 3. We evaluate our models on the test set, and compare the different architectures and decoding methods in Sect. 4. The paper is concluded in Sect. 5.

Table 1. Pictophonetic characters

Character	Radical	Phonetic	Pronunciation
油 (grease)	氵 (liquid)	由	yóu
柚 (pomelo)	木 (plants)	由	yóu
铀 (uranium)	钅 (metal)	由	yóu
釉 (glaze)	釆 (distinguish)	由	yòu

2 Speech2Stroke Model

We provide an overview of our Speech2Stroke model. As shown in Fig. 1, our proposed model consists of two main pipelines: 1) character decompose, which takes Chinese characters as input, and produces a sequence of stroke orders that correspond to the associated character; 2) a speech encoder, which takes a spectrogram of power normalized audio clips as input, and predicts a stroke sequence. During the training stage, we train our speech encoder with Connectionist Temporal Classification (CTC) loss function in an end-to-end way. During predicting, we decode the stroke order from the output probability of speech encoder. In the following sub-sections, we describe each of these components in detail.

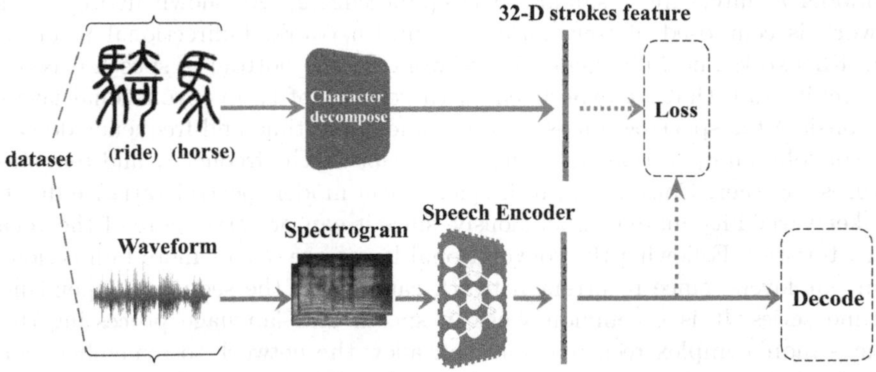

Fig. 1. The overall architecture of Speech2Stroke model

2.1 Character Decompose

A Chinese word is composed of a sequence of characters, and a Chinese character is composed of several graphical components. The characters with the same component share a similar pronunciation. The shape of the graphical components can be modeled by the stroke orders, which have finer granularity than radical. In our work, the strokes are categorized into 32 different types. As shown in Fig. 2, we use various strokes to depict the curved structure of character, therefore the structure information of character can be extracted completely from the stroke sequences, and we mark each stroke an integer ID from 1 to 32 respectively. The procedure of decomposing a word into IDs is shown in Fig. 3. First, we divide the simplified word into characters; then we retrieve the stroke sequence from every character, and add a blank space into the sequence of the strokes after finish a character. [1] provides an official guideline about the order for strokes involved in each simplified Chinese character; finally, we use integer IDs to represent the stroke sequence.

2.2 Speech Encoder

Let us define the sequence of acoustic feature vectors as $X = [X_1, X_2, X_3, ..., X_T]$, and $Y =$ [丶, 一, ㇍,......] denotes a stroke sequence. The goal of speech encoder is to convert an input sequence X into a transcription Y, the most likely stroke sequence Y^* is given by:

$$Y^* = \arg\max_{Y} p(Y|X) \tag{1}$$

Our speech encoder is a neural network fed with spectrogram of power. At each output time-step t, the speech encoder predicts the probability, $p(y_t|x)$, where y_t is the stroke of Chinese or the blank symbol at time-step t. The overall model architecture is similar to DeepSpeech2 [2]. As shown in Fig. 4, the network is composed of convolutional neural network, bidirectional recurrent neural network, and fully connected network. In the bottom of speech decoder, the architecture that we experiment with consists of two convolutional layers. We consider the spectrogram as a picture and apply time-and-frequency domain 2-D convolution operations [15], which can capture the frequency and temporal features together. Convolution in frequency can model spectral variance due to speaker variability more compendiously, since it can preserve more of the vocal characteristics. Following the convolutional layers are one or more bidirectional recurrent layers, since recurrent network can exploit the sequence dependence of time series. It is a common sense in speech and language processing that using a more complex recurrent unit can allow the network to remember more time dependence, therefore the Long Short-Term Memory (LSTM) units [9] and the Gated Recurrent Units (GRU) [6] are applied respectively in bidirectional recurrent layers, though it is computationally expensive to train them. After the bidirectional recurrent layers, we stack one fully connected layer which can adjust the output dimension conveniently. The last layer is a softmax layer that computes a probability distribution over the strokes. The parameters of the speech encoder are updated through the backpropagation through time algorithm [18] . The cost function of training is CTC loss function. In the following section, we will detail the CTC loss function and how to decode the best transcription from the output probability distribution.

Stroke Name	Dot	Horizontal	Horizontal-hook	Horizontal-left-falling	Horizontal-left-falling-bend-hook	Horizontal-slant-hook	Horizontal-turning	Horizontal-turning-vertical-hook
Shape, ID	㇔,1	㇐,2	㇖,3	㇇,4	㇌,5	㇠,6	㇕,7	㇆,8
Stroke Name	Horizontal-turning-rise	Horizontal-turning-bend	Horizontal-turning-bend-hook	Horizontal-turning-turning	Horizontal-turning-turning-left-falling	Horizontal-turning-turning-turning	Horizontal-turning-turning-turning-hook	Right-falling
Shape, ID	㇊,9	㇍,10	㇈,11	㇅,12	㇋,13	㇎,14	㇡,15	㇏,16
Stroke Name	Left-falling	Left-falling-dot	Left-falling-turning	Vertical	Vertical-hook	Vertical-rise	Vertical-bend	Vertical-bend-horizontal-hook
Shape, ID	㇒,17	㇛,18	㇜,19	㇑,20	㇚,21	㇙,22	㇄,23	㇟,24
Stroke Name	Vertical-turning	Vertical-turning-left-falling	Vertical-turning-turning	Vertical-turning-turning-hook	Rise	Bend-hook	Recline-hook	Slant-hook
Shape, ID	㇗,25	ㄣ,26	㇞,27	㇉,28	㇀,29	㇁,30	㇃,31	㇂,32

Fig. 2. Shapes of Chinese strokes

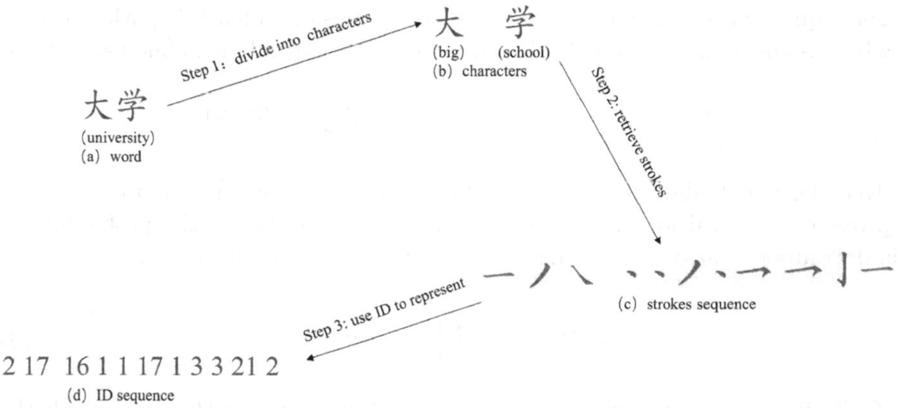

Fig. 3. Procedures of the generation of strokes from a word

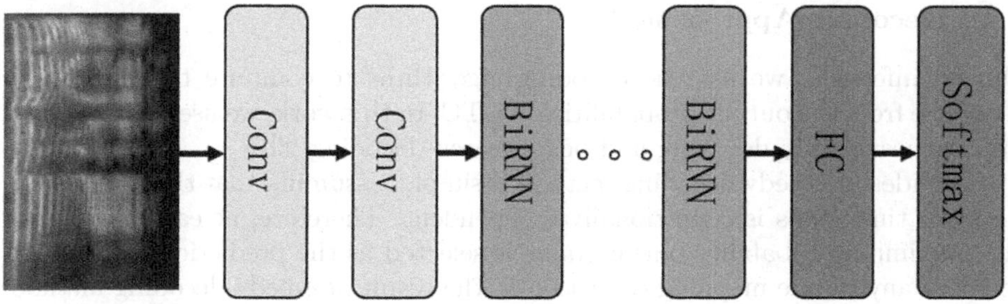

Fig. 4. Architecture of speech encoder

2.3 Connectionist Temporal Classification

Connectionist Temporal Classification (CTC) [7] is a popular method for end-to-end sequence learning, which usually applies in an RNN-based model to optimize predictions of the transcription sequence, it enables the end-to-end model training without pre-defined alignment information of training dataset. In the theory of CTC, let us denote x as the spectrogram frame sequence and y as the label sequence. CTC proposed a dynamic constrained procedure that adds a blank symbol ϕ as an additional label, the blank symbol can model the situation of pause and adjacent repeated labels in label sequence. The output of softmax of speech encoder has the same time-steps as the input sequence x, since the length of y should be shorter than the length of x due to the blank symbol. A CTC path π is defined as the concatenation of softmax observed labels at all time-steps. For the propose of mapping the path π to the target sequence y, CTC defines a many-to-one mapping operation β. β firstly collapses repeating tokens, then removes all blank symbols from the sequence processed in the previous step. For example, a sequence of 9 labels such as "aϕ ppϕ pϕ le", the sequence turns to "apple" after processing the operation β. The conditional probability of a given target sequence is defined as the sum of probabilities of all feasible paths through β, which is the definition of CTC loss function. The formula is defined as follows:

$$L_{CTC} = -\ln P(\mathbf{y}|\mathbf{x}) = -\ln \sum_{\pi \in B^{-1}(\mathbf{y})} P(\pi|\mathbf{x}) \tag{2}$$

In order to calculate the $P(\pi|x)$, CTC makes an assumption that every time-step output is conditionally independent of previous outputs, the probability of π is decomposed into the product of posteriors from each frame as:

$$P(\pi|\mathbf{x}) = \prod_{t=1}^{T} P(\pi_t|\mathbf{x}) \tag{3}$$

CTC uses a dynamic programming algorithm to efficiently sum up all the feasible paths. Therefore, the CTC technique provides practicable guidelines for training the end-to-end model by directly optimizing the CTC loss function.

2.4 Decoding Approaches

During inference, we use the decoding algorithms to compute the final label sequence from the output probabilities of CTC. In this work, we use two decoding approaches: greedy decoding and beam search decoding [8].

The idea of greedy decoding method is simple, assuming that the probability between time steps is conditionally independent. Therefore, at each time step, the maximum probability output value is selected as the prediction result. Due to the many-to-one mapping operation β, the result of greedy decoding method is an approximate solution, but it is very fast, which reduces the exponential time complexity to the length-dependent linear time complexity.

In this work, we also use beam search decoding algorithm, which is one of the popular decoding techniques for CTC decoding. Actually the beam search decoding is an improvement on greedy decoding. Beam search is a heuristic search algorithm that explores a breadth-first search tree by expanding the most promising node in limited sets. When the beam width is equal to 1, beam search decoding degrades to greedy decoding. The greater the beam width, the fewer states of search tree are pruned. Therefore, we can improve the performance of beam search decoding by increasing the beam width, but the side effects are increasing decoding time and memory consumption. In practice, it usually needs to find a trade-off between the beam width and time consumption. Due to the beam search sacrifices completeness by pruning, the solution of beam search is also not optimal.

3 Experimental Setup

3.1 Dataset

Our experiments are conducted on self-build Chinese strokes dataset and AISHELL-1 [3] open source mandarin speech corpus. Based on our observation, this is no publicly available simplified Chinese strokes dataset that has 32 different types of strokes, so we create the dataset by crawling webpage and manual correction.

The simplified Chinese strokes dataset contains 21964 Chinese characters and the stroke order of the corresponding Chinese characters. In addition, we find some interesting things that the length of the stroke order of character conforms to the normal distribution, and the lengths of most Chinese characters are distributed between 20–30.

We use the AISHELL-1 corpus dataset which comprises approximately 178 hours of recording from 400 speakers of different accent areas in China. All utterances of AISHELL-1 are encoded at the 16 kHz sampling rate with 16-bit resolution. All experiments use spectrogram of power as input features, computed with a 20ms window and shifted every 10ms. To reduce frequency leakage, we use a Hamming window function to reduce frequency leakage during the speech signal pre-processing.

3.2 Evaluation Metric

Since there is no standard metric for evaluating the stroke-based model, in order to compute the similarity between label strokes order sequence and prediction sequence, we propose the Stroke Error Rate (SER) which is inspired by the Word Error Rate (WER) of measuring the performance of Automatic Speech Recognition (ASR). A standard definition of the SER is as follows:

$$SER = \frac{I_s + S_s + D_s}{N_s} \tag{4}$$

where N_s is the number of strokes in the reference stroke order text, S_s is the number of strokes substituted, D_s is the number of strokes deleted and I_s is the number of strokes inserted required to transform the output text into the target. Fortunately, the numerator of SER is computed with the minimum number of operations required to convert the output prediction text into reference, which is known as Levenshtein distance [11]. The larger the number, the more different between the two stroke text.

3.3 Implementation Details

We reconstruct the speech dataset by replacing the label text with stroke sequences using the method of Sect. 2.1, the reconstructed dataset is used for all of our experiments in this work.

The architecture of convolutional layers of speech encoder is the same for all training and testing experiments, the convolutional block consists of a convolutional layer, batch normalization(BN) [10] layer and Hardtanh activation function layer, the hyperparameters of the first convolutional layer are as follows: the spectrogram is treated as one channel picture by convolutional operations, the number of out channels is 32, size of filters is 41×11 with a stride of 2×2, and the size of padding is 20×5. Similarly, the hyperparameters of the second layer are as follows: the numbers of input channels and output channels are 32, size of filters is 21×11 with a stride of 2×1, and the size of padding is 10×5. In the Bi-RNN blocks, we test different types of recurrent networks, including common RNN, GRU, and LSTM. In addition, we explore the impact of different layers on model performance. Each layer of Bi-RNN blocks has 800 cells. Batch normalization is also applied on each RNN layer. The last one fully connected layer has 34 hidden units, which represents the number of labels in the training set.

All the models are implemented with the PyTorch library [14], we use stochastic gradient descent (SGD) with a momentum of 0.9 [17] to train our model with CTC loss function described in Sect. 2.3, and set the batch size to 32 due to the GPU memory limitations. The initial learning rate is at 3e–4, The learning rate during training adopts 1/t annealing strategy after finishing every epoch, where t is set at 1.1. To accelerate the online learning, we train our model with SortaGrad [2] learning strategy. we also apply early stopping [5] to halt the training of model when the performance is degraded. During inference, we evaluate the test set using the checkpoint that produced the best SER on validation set.

4 Results

4.1 Evaluation of Various Architectures

Table 2 gives the SER obtained with our model using different types of recurrent networks in the Bi-RNN block. All results are obtained by beam search decoding method with a fixed beam size of 32. The convolutional layers and fully connected layers are fixed in these experiments. This is because CNN is used to extract the features of spectrogram, and the last layer, a fully connected network, is used for dimensional adaptation. The core of model is the Bi-RNN blocks, which model the temporal connection. As shown in Table 2, experiment E2 indicates that the GRU-based model has better performance than other RNN types, and we notice that there is an extremely big gap between E0 and E1, E2. it is revealed that the RNN-based model cannot handle the long time dependence, then the problem of vanishing gradient occurs, which makes it impossible to update the weights and ultimately leads to poor generalization.

Table 2. SER of Speech2Stroke with various RNN types

Exp ID	Architecture	SER %
E0	8-layer, 5 Bi-RNN	52.915
E1	8-layer, 5 Bi-LSTM	26.183
E2	8-layer, 5 Bi-GRU	20.609

Moreover, Table 3 gives the SER obtained with our model using different layers of recurrent networks in the Bi-RNN block. As Table 2 indicates GRU-based model performs best, we fix the RNN type as GRU and change the number of layers. The results of Table 3 show that experiment D1 has the lowest SER, when increasing the number of layers from 5 to 7, the performance of D2 is degraded, which seems to be a symptom of overfitting.

Table 3. SER of Speech2Stroke with various RNN layers

Exp ID	Architecture	SER %
D0	6-layer, 3 Bi-RNN	27.456
D1	8-layer, 5 Bi-LSTM	20.609
D2	10-layer, 7 Bi-GRU	22.216

4.2 Evaluation of Various Decoding Methods

Table 4 shows the effects of the decoding beam width and the decoding method on the SER for the test set. The architecture of experiment D1 is applied to this evaluation. As we see, experiment C0 with SER of 20.857%, which natively chooses the most likely label at each time step, as expected, beam search decoding method gives better results than greedy decoding. On the other hand, greedy decoding is much faster than any beam search decoding, 6.4 times faster than, C1, the fastest beam search based experiment. We can observe extra benefits when further increasing the beam width. But the price is increasing time consumption. In practice, we need to find a balance between them in different scenarios.

Table 4. SER and decoding time consuming of Speech2Stroke with various decoding methods. BS-32 denotes beam search with width of 32

Exp ID	Decoding method	SER %	Time (ms/item)
C0	Greedy Decoding	20.857	17
C1	BS-32	20.609	109
C2	BS-128	20.573	458
C3	BS-512	20.547	4677
C4	BS-1024	20.549	8308

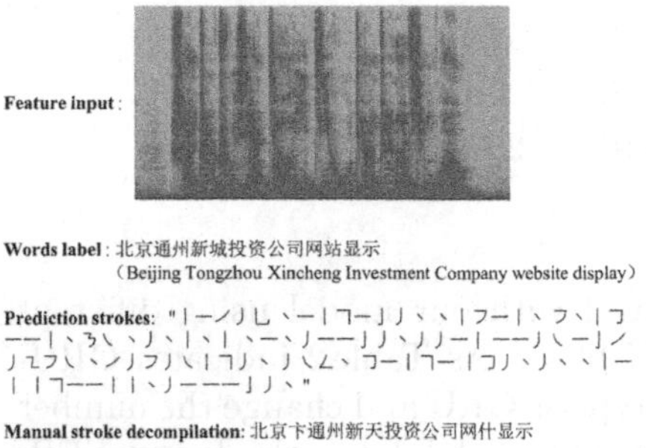
Feature input:
Words label: 北京通州新城投资公司网站显示
(Beijing Tongzhou Xincheng Investment Company website display)
Prediction strokes:
Manual stroke decompilation: 北京卞通州新天投资公司网什显示

Fig. 5. Qualitative results of case 1

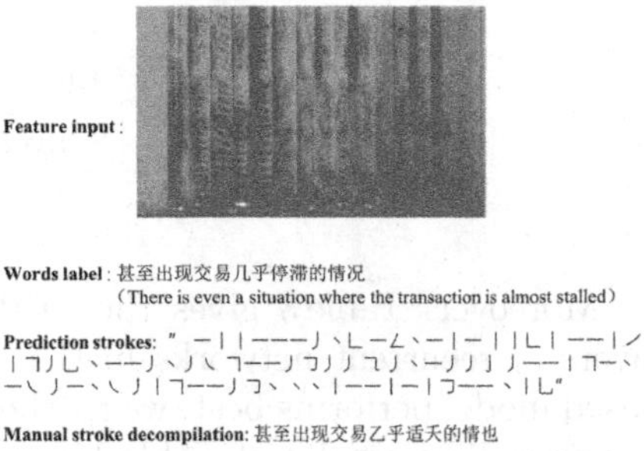
Feature input:
Words label: 甚至出现交易几乎停滞的情况
(There is even a situation where the transaction is almost stalled)
Prediction strokes:
Manual stroke decompilation: 甚至出现交易乙乎适天的情也

Fig. 6. Qualitative results of case 2

4.3 Qualitative Analysis

To better comprehend the quality of the connection between speech and strokes, we conduct qualitative analysis by illustrating some case studies of decoding results. As shown in Figs. 5 and 6, we randomly select two examples in the test set and input the features into Speech2Stroke model, then the model outputs the sequence of strokes. Due to it is complex to figure out the difference between strokes, we manual decompile the strokes to characters. The results of Figs. 5 and 6 show that our model can capture the connection between strokes and speech. Another interesting observation is that words can be determined without an additional language model, which is an advantage of exploiting the internal structural information. As an example of Fig. 5, "投资" (investment) is a pictophonetic word, whose internal structures can indicate the pronunciation and context. Therefore, it can determine the "公司" (company), this is because investment is semantically related to company.

5 Conclusion

In this paper, we propose a novel study of generating strokes of Chinese characters from the audio of speech. The experiments have proved that the internal structures of Chinese characters are strongly connected to speech. We believe that generating strokes gives a more comprehensive view of speech-strokes correlation and can exploit the new tasks with internal structure in Chinese speech research. In the future, we plan to explore the pix-based internal structural information from speech.

References

1. Standardization of stroke order for modern Chinese homepage (2005). http://www.moe.gov.cn/s78/A19/yxs_left/moe_810/s230/201001/t20100115_75615.html, Accessed 2 May 2020
2. Amodei, D., et al.: Deep speech 2: end-to-end speech recognition in English and Mandarin. In: International Conference on Machine Learning, pp. 173–182 (2016)
3. Bu, H., Du, J., Na, X., Wu, B., Zheng, H.: Aishell-1: an open-source mandarin speech corpus and a speech recognition baseline. In: 2017 20th Conference of the Oriental Chapter of the International Coordinating Committee on Speech Databases and Speech I/O Systems and Assessment (O-COCOSDA), pp. 1–5. IEEE (2017)
4. Cao, S., Lu, W., Zhou, J., Li, X.: cw2vec: learning Chinese word embeddings with stroke n-gram information. In: Thirty-Second AAAI Conference on Artificial Intelligence (2018)
5. Caruana, R., Lawrence, S., Giles, C.L.: Overfitting in neural nets: backpropagation, conjugate gradient, and early stopping. In: Advances in Neural Information Processing Systems, pp. 402–408 (2001)
6. Cho, K., et al.: Learning phrase representations using RNN encoder-decoder for statistical machine translation. arXiv preprint arXiv:1406.1078 (2014)
7. Graves, A., Fernández, S., Gomez, F., Schmidhuber, J.: Connectionist temporal classification: labelling unsegmented sequence data with recurrent neural networks. In: Proceedings of the 23rd International Conference on Machine Learning, pp. 369–376 (2006)
8. Hannun, A.Y., Maas, A.L., Jurafsky, D., Ng, A.Y.: First-pass large vocabulary continuous speech recognition using bi-directional recurrent DNNs. arXiv preprint arXiv:1408.2873 (2014)
9. Hochreiter, S., Schmidhuber, J.: Long short-term memory. Neural Comput. **9**(8), 1735–1780 (1997)
10. Ioffe, S., Szegedy, C.: Batch normalization: Accelerating deep network training by reducing internal covariate shift. arXiv preprint arXiv:1502.03167 (2015)
11. Levenshtein, V.I.: Binary codes capable of correcting deletions, insertions, and reversals. Soviet Phys. Doklady **10**, 707–710 (1966)
12. Meng, Y., Wet al.: Glyce: glyph-vectors for Chinese character representations. In: Advances in Neural Information Processing Systems, pp. 2742–2753 (2019)
13. Chinese. BLS. Macmillan Education UK, London (1999). https://doi.org/10.1007/978-1-349-27306-5_9
14. Paszke, A., et al.: Pytorch: an imperative style, high-performance deep learning library. In: Advances in Neural Information Processing Systems, pp. 8024–8035 (2019)
15. Sainath, T.N., Mohamed, A.R., Kingsbury, B., Ramabhadran, B.: Deep convolutional neural networks for LVCSR. In: 2013 IEEE International Conference on Acoustics, Speech and Signal Processing, pp. 8614–8618. IEEE (2013)

16. Su, T.R., Lee, H.Y.: Learning Chinese word representations from glyphs of characters. In: Proceedings of the 2017 Conference on Empirical Methods in Natural Language Processing, pp. 264–273 (2017)
17. Sutskever, I., Martens, J., Dahl, G., Hinton, G.: On the importance of initialization and momentum in deep learning. In: International Conference on Machine Learning, pp. 1139–1147 (2013)
18. Werbos, P.J.: Backpropagation through time: what it does and how to do it. Proc. IEEE **78**(10), 1550–1560 (1990)
19. Yin, R., Wang, Q., Li, P., Li, R., Wang, B.: Multi-granularity Chinese word embedding. In: Proceedings of the 2016 Conference on Empirical Methods in Natural Language Processing, pp. 981–986 (2016)
20. Yu, J., Jian, X., Xin, H., Song, Y.: Joint embeddings of Chinese words, characters, and fine-grained subcharacter components. In: Proceedings of the 2017 Conference on Empirical Methods in Natural Language Processing, pp. 286–291 (2017)

TAB: CSI Lossless Compression for MU-MIMO Network

Qigui Xu, Wei Xi, Lubing Han, and Kun Zhao(✉)

School of Computer Science and Technology, Xi'an Jiaotong University, Xi'an 710049, China
qiguixu@gmail.com, weixi.cs@gmail.com, 254215143@qq.com, pandazhao1982@gmail.com

Abstract. Multi-user MIMO (MU-MIMO) is an important technology to improve data transmission efficiency for future network, such as 5G and WiFi 6, due to its ability of enabling multi-users' concurrent transmissions. To achieve concurrent diversity gains, MU-MIMO network resource allocation relies on the feedback of Channel State Information (CSI) from multiple clients. CSI feedbacks from large user population, however, heavily degrade the throughput of a MU-MIMO network. Pursuing smart CSI feedback, we present a CSI timeliness-aware balanced mechanism, named TAB. It is a novel MU-MIMO protocol to eliminate unnecessary feedback overhead and improve CSI utilization within channel coherence time. TAB is fully compatible with the WiFi 5/6 standard and most state-of-the-art CSI feedback strategies, and is easy to be deployed on existing WiFi systems. Our software-radio based implementation and testbed experimentation demonstrate that TAB substantially improves the throughput of both downlink and uplink MU-MIMO network by 1.5× at least.

Keywords: MU-MIMO network · Channel State Information · Channel overhead

1 Introduction

Wireless network has been facing the growing demand of higher speed and more efficiency for the mass real-time data transmission among interconnected and intercommunicated multi-users. To deal with more resource consumption and better spectrum coverage, Multi-user Multiple-Input Multiple-Output (MU-MIMO) technology is being employed in many standard wireless protocols and infrastructures for future network, such as WiFi 6 [1] and 5G [9]. By allowing concurrent transmissions between a multi-antenna access point (AP) and multi-users, MU-MIMO holds the enormous potential to substantially improve

This work was supported by National Key R & D Program of China 2019YFB2102201, NSFC Grant No. 61772413, 61802299, and 61672424.

H. Gao et al. (Eds.): CollaborateCom 2020, LNICST 349, pp. 95–114, 2021.
https://doi.org/10.1007/978-3-030-67537-0_7

spectrum efficiency and network throughput. In MU-MIMO, users can report the estimation of their corresponding Channel State Information (CSI) to AP, and the AP selects the concurrent users with strong channel orthogonality accordingly to maximize the total channel capacity. CSI is a comprehensive metric of the energy attenuation and phase delay of wireless signals in multi-path channel with frequency selective fading [6]. For the AP in MU-MIMO, its data pre-coding before sent and data decoding after received rely on the accurate CSI estimation.

However, CSI feedback introduces high superfluous channel overhead, which constrains the network performance gain from MU-MIMO technology [14]. First, CSI report frame contains so many channel matrix elements and defined parameters that multi-user CSI feedback can occupy much system time and spectrum resource originally available for data transmission. Second, channel state may often fluctuate over time due to the channel mobility, dynamic interference and noise, thus most standard protocols require CSI feedback before each data transmission, regardless of the validity period for real-time CSI. Even worse, the CSI feedback overhead will increase to an intolerable degree with the growing user population in MU-MIMO network, overwhelming the channel resource utilization of normal data transmission. Besides, to measure real-time CSI, calculate the beamforming matrix and select the best transmission strategies, MU-MIMO network has to sacrifice lots of superfluous computation overhead and energy consumption. Thus it is crucial to reduce the CSI feedback overhead in MU-MIMO network.

To address the performance degradation of MU-MIMO network, existing solutions try to reduce unnecessary CSI feedbacks. On the one hand, some CSI compression methods are proposed to reduce the volume of complicated CSI matrix [3,10,12,19]. Without enough specific elements, however, inaccurate CSI matrix can not profile multi-path channel feature efficiently. As a result, the data transmission and parsing relying on accurate CSI may suffer from more efficiency loss. On the other hand, many novel channel sounding mechanisms are designed to reduce CSI feedback times. Some approaches support CSI share among location-related users [7,21], some [2,11,20] reuse CSI according to the similarity of CSI samples before and after, and some utilize simple CSI or zero CSI feedback strategies [15,16,18,22,23] to access channel. Lacking timely discriminative CSI feedback, however, fuzzy channel estimation can neither resist the effect of multi-path channel fading, nor avoid the accumulative error drift problem over time. Not to mention whether the real-world channel quality can afford for high efficient data transmission in MU-MIMO network. To sum up, if not considering the variance and fluctuation of real-time CSI effectiveness, blindly reducing feedback is often counterproductive and even aggravate the network performance deterioration.

In this paper we propose a CSI timeliness-aware balanced mechanism (TAB) to compress CSI while keeping its quality at the same time. First, based on the significant difference between adjacent training symbol samples, user can reply a lightweight ACK to AP, indicating that the real-time CSI has no significant change. Second, based on that CSI can be utilized to sense the fluctuation of wireless channel, AP accumulates CSI packet samples to estimate the variant range of channel state so as to predict the current CSI valid period related to each user, *i.e.*, channel coherence time (CCT). Finally, according to the CCT magnitude, the AP groups the multi-users with diverse priority, and schedules them via Round Robin with the corresponding constraint of maximum service time slice. In this way, the unnecessary CSI feedback overhead and the consequent resource consumption could be reduced significantly meanwhile the channel efficiency and network performance get improved efficiently.

The contributions of this paper can be summarized as follows.

1. We use both theoretical and experimental results to show that CSI feedback overhead is the cause of the performance of MU-MIMO limitation. This is the first work to target on comprehensive CSI sample analysis to realize CSI lossless compression via timeliness-aware balanced mechanism.

2. We develops an adaptive CCT-based multi-user scheduling scheme at AP and a CSI report occasion estimation method at users to eliminates unnecessary CSI feedback by replying a lightweight ACK frame to reuse historical CSI, so that the system throughput can be significantly improved.

3. We conduct extensive real-world experiments under practical settings, with prototype implementation using software-radio devices. Results show that TAB significantly improves the throughput of both uplink (UL) and downlink (DL) MU-MIMO network and decrease superfluous resource consumption.

2 Related Work

MU-MIMO technology has been put around for some time to resolve the insatiable demand for wireless capacity. Many standard protocols also have been exploring the application of MU-MIMO technology, such as LTE [4], IEEE 802.11ac (WiFi 5) [5], as well as the ongoing 5G [9] and 802.11ax (WiFi 6) [1]. They all employ Beamforming technology to improve channel Signal Noise Ratio (SNR) by concentrating the transmitted energy on the target receiver. Meanwhile, CSI [6,17] is introduced in MU-MIMO network to resist the effects of multi-path and frequency selective fading. CSI is the key component of MU-MIMO communication, but frequent CSI feedback also produces much superfluous channel overhead, which constrains and even cancels out the network throughput gain using MU-MIMO technology. To maintain this type of channel overhead in MU-MIMO system, Current CSI feedback reduction methods mainly fall into two categories: channel feedback volume compression or CSI feedback times reduction.

On the one hand, different compression methods are available to reduce the volume of CSI matrix. CSI-SF [3] uses the CSI value of a single data stream to predict the CSI value of multiple data streams, thereby reducing the CSI's oversampling. AFC [19] adaptively selects the compression level according to the Channel feature, quantizes or compresses CSI from 3 dimensions: time, frequency and numerical values. CSIFit [12] achieves high compression ratio by finding the sine signal of CSI and piecewise fitting it, based on Orthogonal Frequency Division Multiple (OFDM) technology. EliMO [10] adopt a two-way channel estimation to allow AP to accurately estimate DL CSI without explicit CSI feedback. Such compression methods sacrifice the CSI quality to save channel overhead, but inaccurate CSI may degrade the network performance in turn when faced with multi-path effect.

On the other hand, various CSI feedback mechanisms are designed to reduce CSI feedback times. Some schemes advise to share CSI in multi-user clusters. NEMOx [21] organizes a network into practical-size clusters with opportunistically synchronized APs in DL to balance their cooperation gain and spatial reuse. GCC [7] allows the location-related users to share a CSI matrix to limit the feedback quantity. Nevertheless, the premise of launching these schemes is that proper transmission mode and user groups are selected. Some strategies advocate reuse CSI to transmit more data packets according to CSI feature. Gabriel [2] reuses CSI by checking its validity period, which is obtained by F-distribution test of two CSI samples. RoFi [11] senses the rotation of device using Power Delay Profile (PDP) similarity and achieves rotation-aware CSI feedback while maintaining high throughput. QUICK [20] tries to reuse the CSI within its coherence time and schedule multi-users fairly. Unfortunately, these strategies can not prevent the real-time CSI from accumulative error drifting over time. There are also some approaches try to utilize simple CSI or zero CSI to access channel directly. OPUS [18] reduces CSI overhead via AP's iterative probing and users' competition report. Signpost [23] achieves scalable MU-MIMO signaling with zero CSI feedback. NURA [16] utilizes a lightweight UL user access mechanism with partial CSI feedback. TOUSE [15] employs Dynamic Time Warping (DTW) algorithm to evaluate the data rate and realizes the adaptive user selection with zero CSI feedback. Guidepost [22] combines the key ideas of NEMOx, Signpost and NURA, builds on a novel principle of indirection channel orthogonality evaluation to decouple and simplify the complicated computational and contention interaction among users. Although these methods can eliminate CSI feedback overhead significantly, they can not obtain a good Signal Interface Noise Ratio (SINR) without enough CSI in MU-MIMO transmission because of multi-path channel fading problem as well. In a word, these approaches have not consider the necessity for lossless CSI feedback or the role of historical CSI samples. Hence, this paper combines the current CSI sample analysis with the historical CSI sample, so as to reduce the unnecessary CSI feedback overhead when consider the effectiveness of the real-time CSI.

3 Channel Sounding Overhead

Channel State Information (CSI) is the key component for AP to allocate or parse data in MU-MIMO network. Considering the multi-path effect and frequency selective fading of wireless signals, CSI is a comprehensive measure of the energy attenuation and phase delay of signals from different paths, which reflects the diversities between different channels. In order to achieve accurate and efficient data transmission, it is essential for AP to obtain real-time CSI feedback via channel sounding. Nevertheless, CSI feedback will occupy a large number of network resources and bring much superfluous overhead to the MU-MIMO system. We must be aware that CSI reporting is time-consuming and frequency-spectrum-consuming while the concurrent capacity of AP can be maintained at the same time is limited. According to the channel sounding process, the influence of its channel overhead on system data throughput mainly depends on the size of CSI report frame and the feedback frequency of real-time CSI.

On the one hand, the CSI report frame in wireless protocol is very complicated, including many CSI matrix elements and specified preambles. For the scenario with a 4-antenna AP, the size of CSI matrix with 52 subcarriers estimated by each user with one antenna is $1 \times 4 \times 52 \times (32/8 \times 2) = 1664$ bytes, which requires 32 OFDM symbols including the Real and imaginary parts and it takes 4 µs to transmit each symbol. Then, the data of CSI vector is embedded in the data field of PHY Protocol Data Unit (PPDU) as shown in Fig. 1(a). In addition, the VHT-LTF (Long Training Field) in preamble is mainly applied to channel estimation and correction. Its length depends on the number of transmitted streams, and here it equals 4 symbols. As described above, reporting a CSI vector by a user may take up to 180 µs by sending the aggregation frame. If there are 20 users in the MU-MIMO network, it would take 3.6 ms to report all CSIs. Not to mention that during channel sounding, different users has various CSI valid periods which would gives rise to plenty of waiting and turnaround time for CSI request and report. According to the 802.11ac standard [5], however, the length of data field should not exceed 5.5 ms. That is, with the growing of user population, the CSI feedback overhead may approach or even overwhelm the channel resources available for data transmission. Even in 802.11ax standard [1], multi-user CSIs are fed back via UL MU-MIMO, the system time overhead is merely replaced with the channel spectrum resource consumption, rather than partially eliminated. Fortunately, it is found that a user may simply reply an ACK to indicate no change of the current CSI, which only costs 4 µs as shown in Fig. 1(b).

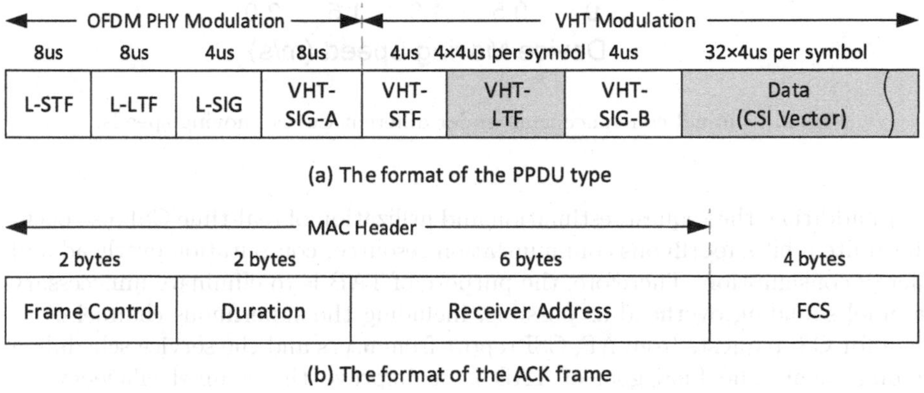

Fig. 1. CSI report frame analysis

On the other hand, real-time CSI on wireless channels has a certain period of validity, i.e., Channel Coherence Time (CCT). During this time period, the channel state remains almost constant, and all packets can be decoded using the same CSI. To analyze the CCT corresponding to different SNRs and user device moving speeds, the AP is allowed to continuously send packets to the user. The CCT is defined as the CSI update interval that leads to a 3 dB SNR loss for the decoding results of packets [13]. As Fig. 2, these stationary users have a relatively longer CCT exceeding 250 ms. Even at 1 m/s, their CCTs are much larger than the maximum packet transmission time 5.5 ms specified by 802.11ac. Only with extremely high moving speeds do the users need a much faster frequency to update CSI, which seems to be impractical because the human speeds on foot generally vary from 0.5 m/s to 2.0 m/s. It is evident that if the device in MU-MIMO network keeps stationary, the surrounding environments and system parameters of which remain the same during a long time, it is unnecessary to require the user to report CSI to AP before each transmission, like the 802.11ac protocol. Frequent CSI updates tend to many redundant operations, of which the resulting channel overhead undoubtedly restricts the throughput capacity of the MU-MIMO network.

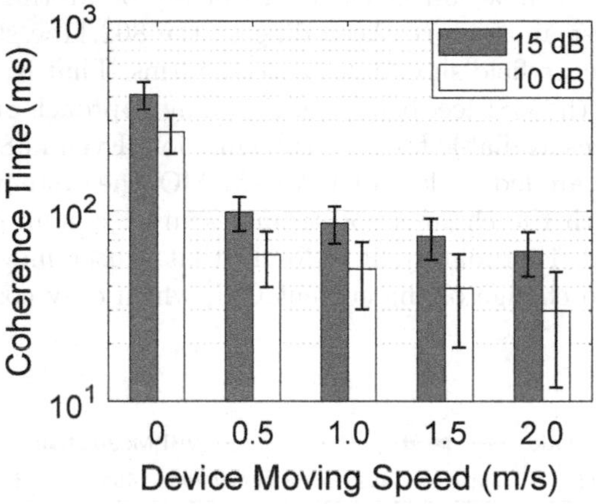

Fig. 2. Channel coherence time under different device moving speeds.

In addition, the request, estimation and utilization of real-time CSI also occupies quite a bit superfluous communication resource, computation overhead and energy consumption. Therefore, the purpose of TAB is to eliminate unnecessary channel sounding overhead as possible, including the superfluous channel overhead for CSI requests from AP, CSI report from users and the service scheduling among them. The final goal of TAB is to improve the channel efficiency for MU-MIMO data transmission.

4 TAB Protocol Design

In this section, we will introduce the details of TAB, which aims to resolve the performance degradation of MU-MIMO network. To reduce unnecessary channel sounding overhead, TAB combines CSI reporting occasion estimation method, channel coherence time (CCT) prediction method and multi-user balanced scheduling method. First, based on the significant difference between adjacent LTF symbols sample in time, unnecessary CSI report from users can be reduced. Second, according to the fluctuation state of CSI packets for a while, the CCT can be predicted to indicate the valid period of current CSI. Finally, with the priority grouping and time slice cycling, the multi-user balanced scheduling component determines which group of users should be scheduled and whether the current CSI needs updating.

4.1 Work Flow of TAB

TAB is compatible with 802.11ac/ax standard and QUICK protocol [20], except that TAB leverages a novel update operation on CSI Matrix List (CML). As Fig. 3, TAB includes both AP and client protocol adjustment, while the sampling form and window size are different, and the operation purpose is different.

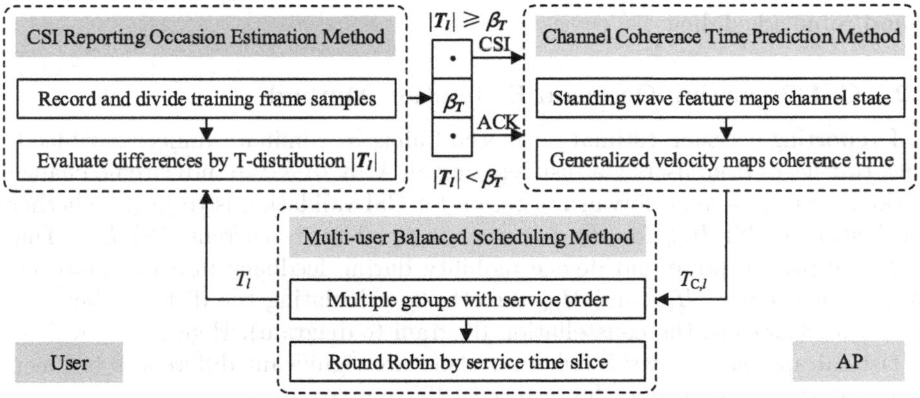

Fig. 3. The work flow of TAB

User conducts the short window sampling on the received training frame symbols, in order to determine whether to feedback the complete CSI estimation results. This process mainly consists of the following four steps:

(i) The user receives the Null Data Packet (NDP) frame including Long training Field (LTF) from the AP, and then analyzes and records the LTF as an evaluation sample for subsequent channel conditions.

(ii) The current user constructs a T-distribution hypothesis test to estimate the significant difference $|T_l|$ between the current LTF sample and the historical

LTF sample. If there is no significant difference between the samples before and after (*i.e.*, $|T_l| < \beta_T$), then go to step (**iii**); Otherwise, then go to step (**iv**).

(**iii**) The current user only reports a short ACK frame to inform the AP that the previous CSI can be reused, and accumulates the current LTF sample behind the historical LTF sample record as the new historical sample record.

(**iv**) The current user uses the current LTF sample to estimate the real-time CSI and immediately reports it to the AP, and then overwrites the cleared historical LTF sample with the current LTF sample.

AP conducts the long window sampling of effective CSI packets to estimate the corresponding channel coherence time of each user and schedule them fairly. This process mainly consists of the following three steps:

(**i**) AP records the effective CSI packet samples through QUICK protocol and starts analysis after reaching a certain length of time window, which is set to 30 ms here to ensure that the existing packets can effectively reflect the quasi-standing wave characteristic of CSI amplitude during device movement.

(**ii**) For cumulative CSI packet samples, pre-treatment of CSI, power distribution analysis, characteristic frequency estimation, device generalized speed estimation and CCT prediction via interpolation mapping are performed successively in order to achieve the CCT $T_{C,l}$ related to each user l.

(**iii**) According to the magnitude of predicted CCT, AP groups each user, sets the corresponding maximum service time slice T_{max} as system constraint and the suitable scheduling order as well as fair service time T_l during multi-user round robin scheduling.

4.2 CSI Reporting Occasion Estimation Method

CSI reporting occasion estimation method aims to eliminate unnecessary feedback overheads from users via just replying an ACK to AP to indicate no change of their CSIs. A straightforward criterion for CSI validation is to judge whether the historical CSI $H_{l,h}$ for the l-th user is equal to its current CSI $H_{l,c}$. Due to the impact of noise and device mobility during feedback interval, however, we cannot compare $H_{l,h}$ and $H_{l,c}$ directly by calculating the distance between their coordinates in the constellation diagram (c-diagram). Hence, we leverage statistical approach to verify whether there are significant differences between $H_{l,h}$ and $H_{l,c}$ by analyzing the received LTF samples in user.

We can reasonably assume that if the CSI changes, the mean of $H_{l,c}$ would be different from the mean of $H_{l,h}$. Whereupon, we can construct the following hypothesis:

$$\mathrm{H}_0 : \overline{H}_{l,c} = \overline{H}_{l,h}, \mathrm{H}_1 : \overline{H}_{l,c} \neq \overline{H}_{l,h}$$

where $\overline{H}_{l,c}$ and $\overline{H}_{l,h}$ is the expectation of $H_{l,c}$ and $H_{l,h}$ respectively. If H_0 holds, the historical CSI $H_{l,h}$ is still qualified in this turn. Considering different lengths of $H_{l,c}$ and $H_{l,h}$, TAB performs this hypothesis testing of H_0 via the *T-distribution test* and we can construct the tests statistic T as follows:

$$T_l = \frac{(\overline{H}_{l,c} - \overline{H}_{l,h})}{S_w\sqrt{\frac{1}{N_{l,c}} + \frac{1}{N_{l,h}}}} \sim t\left(N_{l,c} + N_{l,h} - 2\right)$$

where

$$S_w = \frac{(N_{l,c} - 1)\, S_{l,c}^2 + (N_{l,h} - 1)\, S_{l,h}^2}{(N_{l,c} + N_{l,h} - 2)}$$

and $S_{l,c}^2$ and $N_{l,c}$ denotes the variance and sample size of $H_{l,c}$, respectively and similarly for $H_{l,h}$. Hence, with the significance level α, the rejection region is

$$|T_l| = \left| \frac{(\overline{H}_{l,c} - \overline{H}_{l,h})}{S_w \sqrt{\frac{1}{N_{l,c}} + \frac{1}{N_{l,h}}}} \right| \geq t_{\alpha/2}\,(N_{l,c} + N_{l,h} - 2) \tag{1}$$

In TAB, we set $\alpha = 0.01$ in which $t_{\alpha/2} \geq 2.576$ when $N_{l,c} + N_{l,h} - 2 \leq \infty$. Thus the decision threshold is set as $\beta_T = 2.576$. As long as $|T_l| < \beta_T$, we can believe that H_0 holds 99 % confidence.

In this way, partial CSI report strategy can be adopted to reduced the length of CSI report frame. If most CSI vectors of all subcarriers have significant change, full CSI should be reported as usual. If partial CSI vectors of these subcarriers have significant change, the corresponding CSI vectors of continuous adjacent subcarriers should be reported. In contrast, the other CSI vectors making no evident difference can be replaced with some flag bits indicating sub-ACK. Once most CSI vectors of all subcarriers have no significant change, the user can reply an ACK frame to instead to indicate that its historical CSI can be reused for the AP. In addition, after a user report current CSI, the corresponding recorded LTF samples must be refreshed. Otherwise, the user should continue to record to generate longer historical samples, for avoiding CSI error drift over time.

4.3 Channel Coherence Time Prediction Method

Channel coherence time (CCT) is the time validity period T_C within which the data packets can be decoded by the same CSI without increased bit error ratio (BER). This feature can help us to constrain the operation of TAB system. For one, the CCT can guide AP reduce unnecessary channel sounding overhead including CSI request and feedback. For the channels within CCT, it can consider transmitting longer aggregate packets by reusing CSI. For another, CCT can provide effective basis for the AP to schedule users. The concurrent capacity of AP can be maintained at the same time is limited. Once faced with a large number of users, orderly user scheduling can reduce channel overhead caused by frequent channel sounding. Therefore, reasonable prediction of CCT will greatly maintain the effectiveness of real-time CSI and the fairness of user scheduling, and reduce unnecessary channel sounding overhead.

With the CSI-quality protection of QUICK, here CCT is mainly determined by both device motion state and environmental changes related to channel mobility. It has been proved that [8,20], when a wireless device is moving indoor with a low speed v, though the Doppler shift is negligible, some periodical ripples with frequency f_o in continuous CSI amplitude sample still appears. This quasi-standing wave phenomenon of wireless signals can satisfy

$$f_o = \frac{v}{\lambda/2} \tag{2}$$

where λ is the wavelength of original signal and $\lambda = 12.5\ cm$ in 2.4GHz WiFi. That is, for a mobile channel, its moving speed can be inferred by analyzing the characteristic frequency of the quasi standing wave taken on in CSI sample. Based on the above relationships, the CCT prediction method in TAB is a two-level mapping process, specific as follows.

1. *Pre-treatment of CSI sample.* The phase of CSI $\mathbf{H} = [H_1, H_2, \ldots, H_N]$ for N selected subcarriers in multi-path environment is uniformly distributed between $[0, 2\pi]$, which provides no discriminative information. Hence, we only use its amplitude $\mathbf{A} = |\mathbf{H}|$ to estimate the comprehensive channel state during sampling. Since the instantaneous reception rate of packets is unstable, $\mathbf{A}$ is resampled to a stable reception frequency f_w, denoted as $\mathbf{A_{re}}$.

2. *Power distribution Analysis via CSI.* We utilize the short-time Fourier transformation (STFT) with 50% overlapping Hanning window to obtain the power spectral density (PSD) of $\mathbf{A_{re}}$ in different subcarriers, in which the channel noise and motion state are both reflected significantly. Take device moving speed $v = 1.20\,\text{m/s}$ for example as Fig. 4, specific moving speed will make the received samples present a certain quasi-standing wave characteristic frequency. Without loss of generality, the accumulative length of $\mathbf{A_{re}}$ is set to 30 ms.

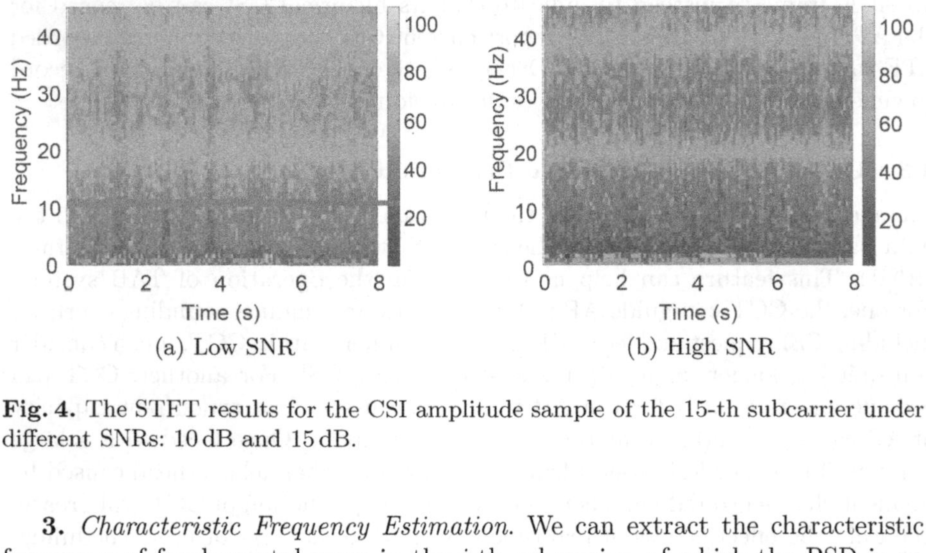

Fig. 4. The STFT results for the CSI amplitude sample of the 15-th subcarrier under different SNRs: 10 dB and 15 dB.

3. *Characteristic Frequency Estimation.* We can extract the characteristic frequency of fundamental wave in the i-th subcarrier, of which the PSD is no less than 1/3 of the maximum PSD, as follows.

$$f_o^i = \frac{\sum_{f_{\min} < f_j < f_{\max}} f_j \times \omega_j}{\sum \omega_j} \tag{3}$$

where the weight ω_j represents the corresponding PSD value and the frequency range should be selected between $f_{\min} = 8$ Hz and $f_{\max} = 32$ Hz according to the Eq. (2). Without loss of generality, the final characteristic frequency f_o of the standing wave is estimated by the median of N selected subcarriers' characteristic frequency as follows

$$f_o = median\left(f_o^1, f_o^2, ..., f_o^N\right) \tag{4}$$

where $N = 30$ in our work.

4. *Device Generalized Speed Estimation.* To comprehensively measure the channel state fluctuation degree, we use the custom device generalized speed as the evaluation metric. Hence, the device generalized speed can be expressed as

$$\bar{v} = \frac{\lambda \times f_o}{2} \tag{5}$$

For $v = 1.2\,\mathrm{m/s}$, the estimated results are $1.19\,\mathrm{m/s}$ (15 dB) and $1.24\,\mathrm{m/s}$ (10 dB) respectively, very close to the truth.

5. *CCT Prediction via Interpolation Mapping.* By employing Piecewise Cubic Hermite Interpolation (PCHIP), plenty of statistical results in 10 dB can be used to fit the relationship between $\bar{v}$ and T_C, as Fig. 5. A simple lookup table is constructed at the AP, so that once the generalized speed is estimated, the corresponding CCT can be obtained immediately. TAB takes 0.05 as the step size, and constructs a lookup table containing 41 data pairs, which replaces a large amount of computing overhead with a little fixed memory.

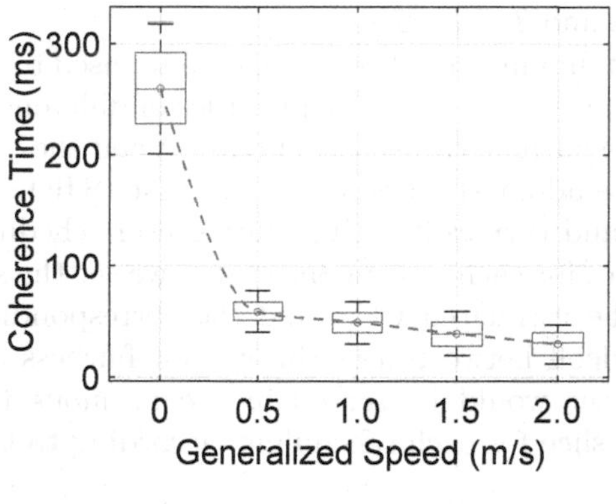

Fig. 5. PCHIP fitting between channel coherence time and Device Generalized Speed.

Note that the CCT prediction method is a process of estimating the overall CCT by partial cumulative CSI samples within the CCT. Under common human walking and channel noise, CSI samples would generally not suffer Cliff-like variations. The corresponding CCT is far greater than the specified maximum data packet length, as well as the sample time used to estimate the overall CCT. The historical CSI packets are recorded merely when satisfying QUICK protocol with affordable channel quality and often plays a auxiliary role. Only when there are extreme changes does the AP need to start over with CSI accumulating.

In addition, CCT only helps the AP select and schedule users, and limit CSI calibration timeout. Hence, a coarse-gained CCT prediction is enough efficient for TAB to work stably. Occasional inaccurate CCT values will only increase some redundant operations of feedback, record and refresh, but not obvious impacts on the whole MU-MIMO system.

4.4 Multi-user Balanced Scheduling Method

There is no doubt that the previous proposed method has reduced CSI feedback overhead before data transmission as much as possible. Nevertheless, the unfair service for multi-users will still lead to serious throughput imbalance problems between different users. Since AP prefers to schedule these users whose CSIs is still usable in CML, the user with a longer CSI lifecycle based on CCT would obtain a longer service time. Especially, the CCT of stationary user is much longer than that of mobile user. As a result, other users with short CCT will have to wait for service in a long time and even forever. Therefore, it is very necessary to group the users based on different priorities and schedule them with the maximum service time slice.

On the one hand, for the CCTs of vary length from different users, we adopt multi-group multi-priority service strategy. Based on the empirical results, we divides the users with different CCTs into 3 categories:

(i) $T_C \leq 48$ ms and $T_{max} = 24$ ms.

(ii) $48 \text{ ms} < T_C \leq 144$ ms and $T_{max} = 72$ ms.

(iii) $T_C > 144$ ms and $T_{max} = 216$ ms.

where T_{max} is the maximum time slice in each group, used to limit the maximum service time for each group of users. The priorities of different groups are divided by the urgency of the current transmission task. When the priority level of task urgency is similar, we adopt short run-time job first (SJF) algorithm to reduce the average waiting and turnaround time during user scheduling.

On the other hand, for each user in one group, we set the specific user service time according to the estimated CCT and the corresponding maximum time slice. To make a tradeoff between user throughput fairness and network global throughput, every user would be served in a continuous time of as close as possible to this time slice for each scheduling, according to the formulation

$$T_l = \left\lfloor \frac{T_{\max}}{T_{C,l}} + \frac{1}{2} \right\rfloor \times T_{C,l} \tag{6}$$

where T_l is the maximum continuous service time available to each user l, $T_{C,l}$ is the CCT related to the l-th user and $\lfloor x \rfloor$ is the floor function of x.

So far, the continuous service time of each user can be as close as possible to the maximum time slice in its group, which avoids that some individual users have been occupying channel resources. All the service time of users is integer times of their own CCTs, which can fully improve the utilization rate of real-time CSI and reduce unnecessary channel overhead. For a network connected with 20 users, each user can be served for twice or thrice at least in every second. Such service conditions will not affect the experience of users to use the Internet.

5 Implementation and Evaluation

In this section, we present and analyze the experimental results using a implemented TAB prototype.

5.1 Experimental Setup

We have implemented TAB on USRP-N210 and USRP-X310 radio platforms with corresponding UHD software packages, as shown in Fig. 6(a). An AP with multiple antennas is built with one USRP-X310 plus multiple SBX daughterboards. Each concurrent user is a USRP-N210 equipped with a SBX daughterboard, providing 40 MHz bandwidth. To allow multiple users to transmit concurrently, we connect the USRP-N210 devices to a laptop and control their transmissions by an instruction script. A similar instruction script is also installed in the two SBX daughterboards of USRP-X310. For precise time synchronization, an external clock model is adopted as a common clock source to connect the AP and users.

The entire system is compatible with the 802.11ac/ax protocol, QUICK protocol and standard OFDM-MIMO specifications, including the modulations (16-QAM, 64-QAM, 256-QAM) and code rates. Hence, TAB can be easily implemented on Commercial Off-The-Shelf Network Interface Cards (COTS NICs) without hardware modification, such as Intel 5300 NIC. For the user, it's supported by the NICs that the device could calculate the CSI upon receiving NDP from AP, and place the CSI into data field of a frame and feed the frame back to the AP. For the AP, constellation diagram is supported for NICs to decode received symbols. Given essential information, we could implement TAB on the COTS NICs through programming its network card driver. In addition, we simulate the different SNR conditions by tuning the transmission power within the range of [5 dBm, 20 dBm].

TAB implements OFDM modulation, packet detection, channel estimation and symbol demodulation. We use LabVIEW to achieve OFDM and the channel estimation in the prototype system based on USRP. Besides, we have obtained some micro-benchmark results and the overall performance using Intel 5300NIC with 4-axis motion testbed in Fig. 6(b). Moreover, to verify our methods, we examined TAB in three different scenarios with general multi-path as shown in Fig. 7: (1) corridor, (2) office and (3) workshop.

5.2 CSI Reusability Within CCT

In this section, we evaluated the effectiveness of reusing CSI within CCT. To illustrate this point, we first verify the performance of TAB in reducing CSI feedback overhead. Figure 8 show the average number of CSI feedback times for continuous 100 rounds of UL and DL MU-MIMO transmission in TAB, respectively. Figure 8(a) shows that the average number of CSI feedback times in UL increases with the growth of the number of users and AP's antennas. Obviously, feedback overhead can be significantly reduced by leveraging CSI reuse, as they

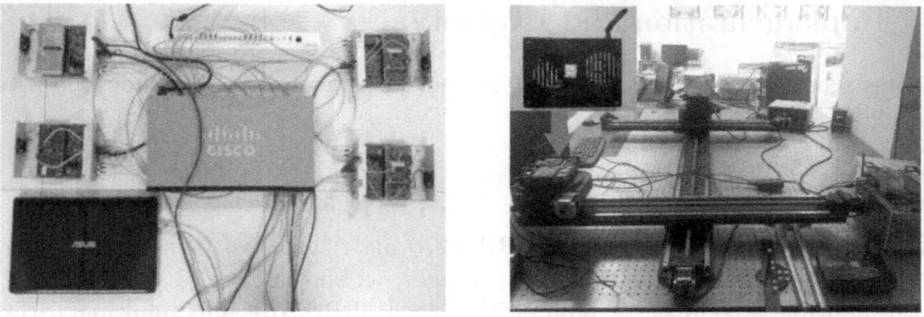

(a) USRP with LabVIEW (b) 5300NIC on 4-axis testbed

Fig. 6. TAB Prototype system and experimental devices

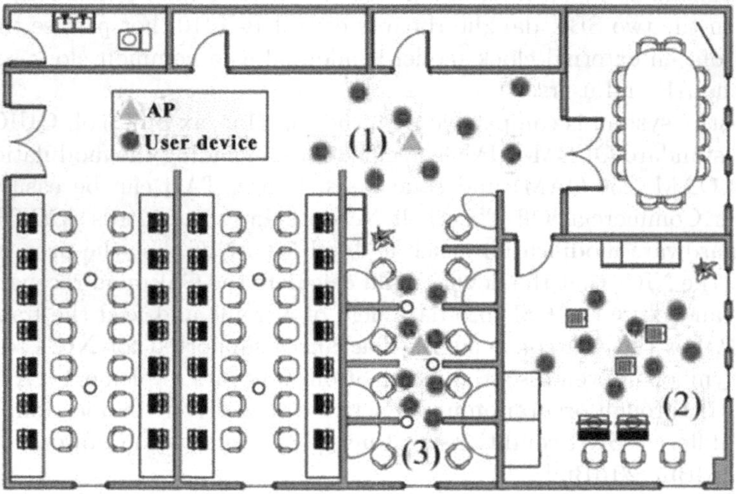

Fig. 7. Distribution of devices in different experimental scenarios

only need less than 4 times of feedback for 100 rounds. When the user population reached 5 times of the number of AP's antennas, the average CSI feedback times would be close to 70% of the number of AP's antennas, which means the performance of TAB begins to degrade to the traditional MU-MIMO network. Figure 8(b) shows that the average number of CSI feedback times in DL is even smaller. It slightly increases while the number of AP's antenna increases, finally fluctuating around 1. That means TAB has a great performance in reducing CSI feedback for both UL and DL MU-MIMO transmission.

Longer CCT does not mean higher reusability of CSI, which is also influenced by distributed user selection and scheduling. To measure CSI utilization, we introduce "Packet Length Per CSI" (PLPCSI) as the indicator to illustrate the average packets length decoded by different users' CSI in MU-MIMO net-

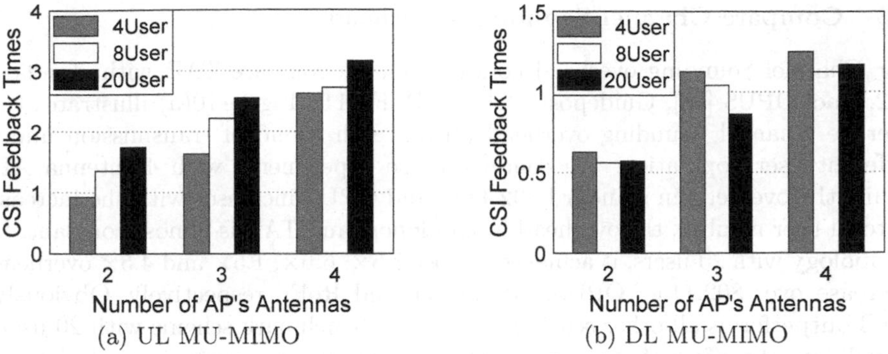

Fig. 8. The count of CSI feedback for continuous 100 rounds of transmission in TAB.

work. Figure 9 presents the PLPCSI values for each user after 100 consecutive concurrent transmissions in a 8-user MU-MIMO network: (a) 2-AP scenario; (2) 4-AP scenario. Since the 802.11ac has specified the maximum data packet length, the PLPCSI of traditional network is not more than 5.5 ms. In TAB, all the PLPCSI values of different users have been improved to varying degrees. Moreover, when faced with the same scale of users, the PLPCSI values will increase significantly with the growing of AP's antenna, *i.e.*, CSI utilization is higher.

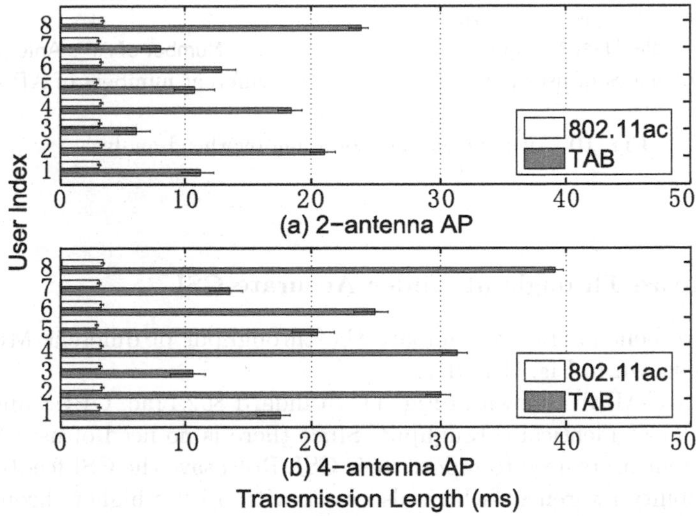

Fig. 9. PLPCSI values for each user after 100 consecutive concurrent transmissions.

5.3 Compare Channel Sounding Overhead

For Channel Sounding overhead comparison, we compare TAB with standard 802.11ac, OPUS [18], Guidepost [22] and RoFi [11]. Figure 10(a) illustrates the average Channel Sounding overhead during each round of transmission under different user population. We conduct these experiments with 4-antenna AP. While the overhead in standard 802.11ac and OPUS increases with the increase of total user number, the overhead in Guidepost and TAB is almost constant. In a topology with 20 users, it achieves almost 7.5×, 6.6×, 1.5× and 4.5× overhead decrease over 802.11ac, OPUS, Guidepost and RoFi, respectively. Obviously, TAB outperforms all other schemes. We run a benchmark scheme with 20 users to validate the effect of the number of AP's antenna on overhead, and plot the result in Fig. 10(b). It is clearly that the overhead increases when the number of AP's antenna grows in all these four MU-MIMO systems. However, the growth rate in TAB is still in a tolerable range. Hence, TAB can be scalable easily.

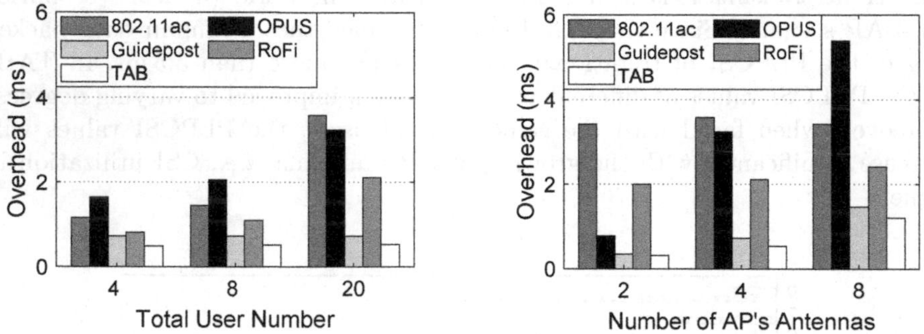

(a) Different scales of user population (b) Different numbers of AP antenna

Fig. 10. Average channel sounding overhead analysis

5.4 Compare Throughput Under Accurate CSI

In this micro-benchmark, we compare the throughput of different MU-MIMO systems, as shown in Fig. 11 and 12.

For DL MU-MIMO shown in Fig. 11, Standard 802.11ac, OPUS and Guidepost have almost identical throughput. Since there is no need of user selection, only concurrent users need to report their CSI. RoFi save the CSI feedback overhead by mobility-awareness, which also work with a little higher throughput in mobile scenarios. TAB achieves the highest throughput among all schemes. It is because that each concurrent users within CCT can continuously receive data from AP without CSI feedback. Moreover, the performance of TAB in static scenario is better than it in mobile scenario due to longer CCT.

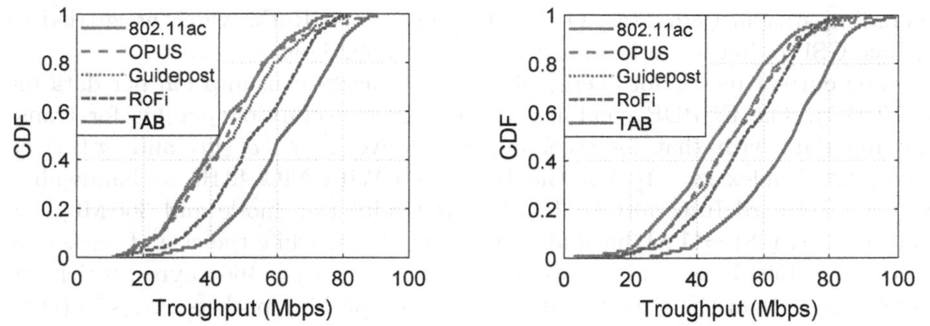

(a) Mobile scenarios (b)Static scenarios

Fig. 11. The throughput comparison among different systems in DL MU-MIMO transmission under different device motion states.

For UL MU-MIMO shown in Fig. 12, RoFi, OPUS, Guidepost and TAB have too much higher throughput than 802.11ac. In order to select the orthogonal users, AP needs all the users to report their CSI in 802.11ac. The four novel MU-MIMO protocol have reduced much channel sounding overhead and user access time to different extent. Especially, TAB utilize multi-user balanced scheduling method to provide higher CSI utilization and stable data transmission availability, so that the throughput of TAB is the 1.5 5 times of 802.11ac.

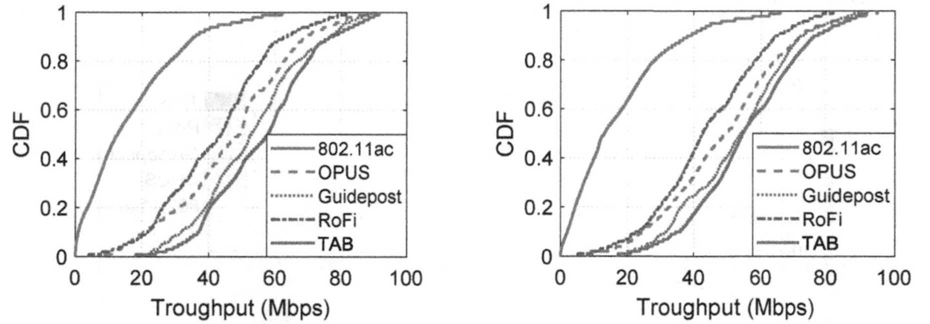

(a) Mobile scenarios (b) Static scenarios

Fig. 12. The throughput comparison among different systems in UL MU-MIMO transmission under different device motion states.

5.5 Compare Energy Consumption

In this micro-benchmark, we approximately evaluate the energy consumption about CSI feedbacks of TAB compared with four existing MU-MIMO systems, including standard 802.11ac, OPUS, Guidepost and RoFi, which have tried to reduce CSI feedback overhead except standard 802.11ac.

The current metric for energy efficiency is energy consumption per data bit (ECPDB). The ECPDB combines the energy consumption per bit for transmitting data with that for receiving, respectively, *i.e.*, $et(m)$ and $er(m)$ as using MCS index m [11]. For the Intel 5300 WiFi NIC 40 Hz as bandwidth, 800 ns as Guard Interval(GI), 16QAM as modulation mode and 360 Mbps as date rate, $er(28)$ =11 nJ/bit and $et(0)$ =90 nJ/bit, while the size of packets is $size(p_i) = 1500$ bytes and the size of CSI is $size(csi_i) = 1664$ bytes, which are much longer than the size of control frame and probing packets, *i.e.* $size(ctr_i)$ and $size(pro_i)$. Hence, the energy consumption of CSI feedback would almost cost $90 \times 1664 \times 8/(90 \times 1664 \times 8 + 11 \times 1500 \times 8) = 90.1\%$ of total energy consumption for transmission in standard 802.11ac. Fortunately, TAB has significantly reduced the number of CSI feedbacks $\sum_{i=1}^{N} size(csi_i)$ as much as possible.

Figure 13 shows that the results of energy consumption in different methods under both mobile and static scenarios. It is clearly that OPUS, Guidepost, RoFi and TAB can reduce much higher energy consumption than the standard 802.11ac. Besides, OPUS and Guidepost have reduced CSI feedback and even do not need CSI feedback, but they have to add some other information and steps to represent the CSI report. Moreover, zero CSI cannot profile the channel state accurately which degrades the efficiency of data transmission. Thereby RoFi and TAB have consumed quite less energy than them based on the direct reduction of CSI feedback times. In addition, the fluctuation of the energy consumption in TAB is lower than all other schemes, benefiting from the comprehensive and balanced guarantee for the channel efficiency.

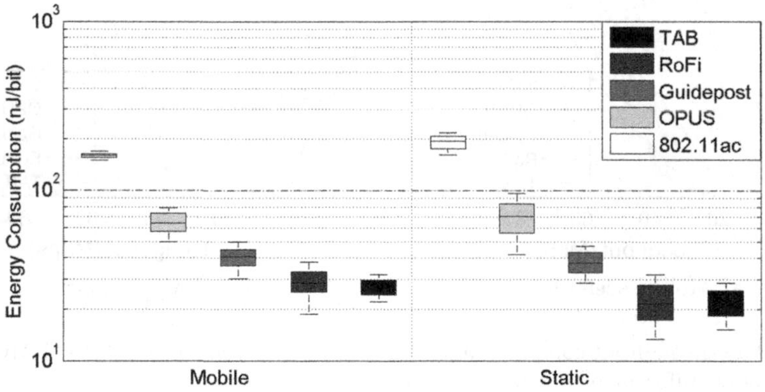

Fig. 13. Energy consumption analysis of different MU-MIMO systems.

6 Conclusion

Our work demonstrates that the unnecessary CSI feedback from concurrent users will lower the throughput in MU-MIMO network, due to the high superfluous overhead. Hence, we propose TAB to reduce unnecessary channel overheads for MU-MIMO WLANs. TAB designs a CSI timeliness-awareness balanced mechanism to determine whether a user's scheduling and CSI feedback is necessary, which significantly improve the CSI utilization and channel throughput. We implement the prototype TAB over software-radio devices. Extensive experiment results show that TAB can substantially improve the throughput for MU-MIMO networks and reduce superfluous resource consumption.

References

1. Afaqui, M.S., Garcia-Villegas, E., Lopez-Aguilera, E.: IEEE 802.11ax: challenges and requirements for future high efficiency WiFi. IEEE Wirel. Commun. **24**(3), 130–137 (2017)
2. Cai, Y., et al.: CSI feedback reduction by checking its validity period: poster. In: Proceedings of the 22nd Annual International Conference on Mobile Computing and Networking (MobiCom 2016), pp. 469–470. ACM (2016)
3. Crepaldi, R., Lee, J., Etkin, R., Lee, S.J.: CSI-SF: estimating wireless channel state using CSI sampling & fusion. In: Proceedings of the 32nd IEEE Conference on Computer Communications (INFOCOM), pp. 154–162 (2012)
4. Duplicy, J., et al.: MU-MIMO in LTE systems. EURASIP J. Wirel. Commun. Netw. **2011**(1), 1–13 (2011). https://doi.org/10.1155/2011/496763
5. Gast, M.: 802.11ac: A Survival Guide. O'Reilly Media, Sebastopol (2013)
6. Halperin, D., Hu, W., Sheth, A., Wetherall, D.: Tool release: gathering 802.11n traces with channel state information. ACM SIGCOMM Comput. Commun. Rev. **41**(1), 53–53 (2011)
7. Fang, J., Wang, L., Qin, Z., Liu, J., Lu, B.: GCC: group-based CSI feedback compression for MU-MIMO networks. Mobile Netw. Appl. **23**(3), 407–418 (2018). https://doi.org/10.1007/s11036-018-1015-1
8. Jiang, Z.P., et al.: Communicating is crowdsourcing: Wi-Fi indoor localization with CSI-based speed estimation. J. Comput. Sci. Technol. **29**(4), 589–604 (2014). https://doi.org/10.1007/s11390-014-1452-7
9. Li, S., Li, D.X., Zhao, S.: 5G Internet of Things: a survey. J. Ind. Inf. Integr. **10**, 1–9 (2018)
10. Ma, Y., Gang, Z., Shan, L.: ELiMO: eliminating channel feedback from MIMO. In: 2017 IEEE International Conference on Smart Computing (SMARTCOMP), pp. 1–8 (2017)
11. Ma, Y., Zhou, G., Lin, S., Chen, H.: RoFI: rotation-aware WiFi channel feedback. IEEE Internet Things J. **4**(5), 1684–1695 (2017)
12. Mukherjee, A., Zhang, Z.: Channel State Information Compression for MIMO systems based on curve fitting, pp. 1–9 (2016)
13. Shen, W., Lin, K.C., Chen, M., Tan, K.: SIEVE: scalable user grouping for large MU-MIMO systems, pp. 1975–1983 (2015)
14. Sur, S., Pefkianakis, I., Zhang, X., Kim, K.H.: Practical MU-MIMO user selection on 802.11 ac commodity networks. In: Proceedings of the 22nd ACM Annual International Conference on Mobile Computing & Networking (MobiCom 2016), pp. 122–134. ACM (2016)

15. Tang, Z., Qin, Z., Zhu, M., Fang, J., Ma, H.: TOUSE: a fair user selection mechanism based on dynamic time warping for MU-MIMO networks. KSII Trans. Internet Inf. Syst. **11**(9), 4398–4417 (2017)
16. Teng, W., Zhang, X.: Random access signaling for network MIMO uplink. In: IEEE Conference on Computer Communications (INFOCOM), pp. 1–9 (2016)
17. Xi, W., Ma, R., Cai, Y., Zhao, K.: Prevent CSI spoofing in uplink MU-MIMO transmission. In: Proceedings of the 1st Workshop on Context Sensing & Activity Recognition (CSAR 2015), pp. 13–18 (2015)
18. Xie, X., Zhang, X.: Scalable user selection for MU-MIMO networks. In: Proceedings of the 33rd IEEE Conference on Computer Communications (INFOCOM), pp. 808–816 (2014)
19. Xie, X., Zhang, X., Sundaresan, K.: Adaptive feedback compression for MIMO networks. In: Proceedings of the 19th Annual International Conference on Mobile Computing & Networking (MobiCom 2013), pp. 477–488 (2013)
20. Zhang, S., Xi, W., Xu, Q., Zhao, K., Cai, Y.: Accurate CSI estimation to eliminate unnecessary transmission for MU-MIMO networks. In: Proceedings of the 2019 International Conference on Embedded Wireless Systems and Networks (EWSN 2019), pp. 166–177. ACM, Junction Publishing, USA (2019)
21. Zhang, X., Sundaresan, K., Khojastepour, M.A.A., Rangarajan, S., Shin, K.G.: NEMOx: scalable network MIMO for wireless networks. In: Proceedings of the 19th Annual International Conference on Mobile Computing & Networking (MobiCom 2013), pp. 453–464. ACM (2013)
22. Zhou, A., Teng, W., Zhang, X., Ma, H.: Guidepost: scalable MU-MIMO user selection via indirect channel orthogonality evaluation. IEEE Trans. Mob. Comput. **18**(7), 1556–1570 (2018)
23. Zhou, A., Wei, T., Zhang, X., Liu, M., Li, Z.: Signpost: scalable MU-MIMO signaling with zero CSI feedback. In: Proceedings of the 16th ACM International Symposium on Mobile Ad Hoc Networking and Computing (MobiHoc 2015), pp. 327–336 (2015)

Towards Mobility-Aware Dynamic Service Migration in Mobile Edge Computing

Fangzheng Liu, Bofeng Lv, Jiwei Huang(✉), and Sikandar Ali

Beijing Key Laboratory Petroleum Data Mining,
China University of Petroleum-Beijing, Beijing 102249, China
2019310704@student.cup.edu.cn, lvbofeng@foxmail.com,
{huangjw,sikandar}@cup.edu.cn

Abstract. Mobile edge computing is beneficial to reduce service response time by pushing cloud functionalities to the network edge. However, it is necessary to consider whether to conduct service migration to ensure the quality of service as users migrate to new locations. It is challenging to make migration decisions optimally due to the mobility of the users. To address this issue, we propose a mobility-aware dynamic service migration scheme for mobile edge computing. In order to predict a mobile user's movement behavior in terms of boundary crossing probability, we use a new approach for modeling user mobility and formulate the service migration problem as a Markov Decision Process (MDP). This policy can effectively weigh the relationship between delay and migration costs. Our methods capture general cost models and provide a mathematical framework to design optimal service migration policies. Experimental evaluations based on real-world mobility traces of Beijing taxis show superior performance of the proposed solution.

Keywords: Mobile edge computing · Service migration · Markov Decision Process (MDP) · User mobility

1 Introduction

With the prevalence of mobile terminals and the Internet of Things (IoT), Mobile Edge Computing (MEC) has emerged as a novel architecture where cloud computing services are extended to the edge of networks leveraging mobile base stations [1–3]. It integrates the techniques of cloud computing and mobile computing and pushes part of the applications, data and services away from centralized cloud data centers to the logical extremes of a network where edge servers are deployed. Since the local edge servers are located closer to the users and IoT devices than the centralized cloud data centers, the quality of service (QoS, e.g. response time and throughput) and privacy can be improved, and the overhead can be reduced as well [4]. Therefore, MEC has become increasingly popular for supporting a variety of innovative applications and services in mobile environments.

H. Gao et al. (Eds.): CollaborateCom 2020, LNICST 349, pp. 115–131, 2021.
https://doi.org/10.1007/978-3-030-67537-0_8

In most of MEC scenarios, the locations of users and devices are time-varied and dispersed in a wide area. Devices and users on the edge site can only access to the services within the signal coverage of edge base stations (or MEC servers). When they move out, they can choose to continue to let the service process at the original edge node, and ensure the continuity of service through the data transmission between the edge nodes; however, too long network distance may increase the delay of data transmission between the user and the edge server that hosts the service, and affect the service quality perceived by users [5]. To address the issue brought by the user mobility, dynamic service migration techniques have been put forward for improving user experience under MEC [6].

The basic idea of dynamic service migration in edge computing is to migrate the services from one edge server to another edge server according to the movements of the users being invoking the services. Figure 1 illustrates a typical scenario. In service migration, we have to solve the following two problems. The first one is whether or not to migrate a service at a certain time point, and if yes the second one is where to migrate the service. Migrating a service may cause service interruption and bring in network overhead, whereas not migrating a service may increase the data transmission delay between the user and the edge server that hosts the service when the user moves away from its original location. It is quite challenging to make an optimal migration decision due to the uncertainty of user mobility as well as the complex trade-off between the costs related to migration and distant data transmission.

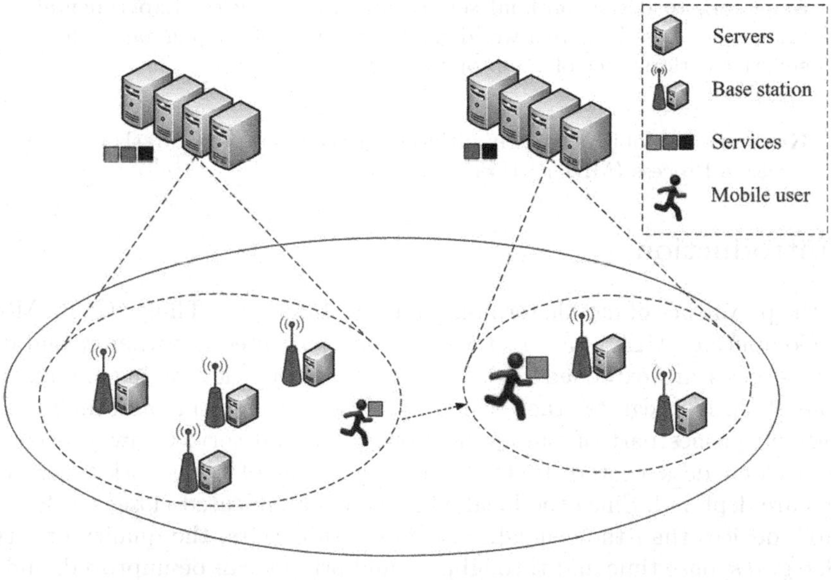

Fig. 1. Service migration in mobile edge computing

For a smooth service migration in MEC, user mobility as one of the most important factors that should be taken into account. There have been existing works on mobility-driven service migration dedicating to model or predict user mobility patterns. The performance of MEC in the presence of user mobility is first studied in [7] using a Markovian mobility model, but decisions on whether and where to migrate the service are not considered. A preliminary work on mobility-driven service migration based on Markov Decision Processes (MDPs) is given in [8,9], which mainly considers one-dimensional (1-D) mobility patterns and takes the uniform random walk migration model as modeling hypothesis. But in fact individual users do not necessarily follow a uniform random walk. To the best of our knowledge, real user mobility has not been considered in the literature, which is a much more realistic scenario compared to the uniform random walk model and we consider in this article.

To address this problem, this paper presents a mobility-aware dynamic service migration scheme for mobile edge computing. Based on the analysis of the trajectory data from users, we propose a geometry-based user mobility model for predicting the probability of a user to move out from the coverage of the current edge server to another one. Considering the trade-off between QoS and cost, the service migration is formulated by a Markov decision problem, and a mobility-aware dynamic migration (MODEM) algorithm is designed. Finally, with trajectory data set from real-life applications, extensive simulation experiments are conducted to validate the effectiveness of our MODEM algorithm.

The remainder of the paper is organized as follows. Section 2 reviews the representative research efforts relevant to our work. Section 3 presents user mobility model to predict the next moving area of the user. Section 4 introduces general cost models and provides a mathematical framework to design optimal service migration policies. Section 5 reports detailed experimental results. Finally, Sect. 6 concludes this work and discusses future research directions.

2 Related Work

Mobile edge computing is an extension of cloud computing, with the benefits of reduced delay. Due to the edge nodes coverage is small, the user's mobility will have a great impact on service quality. Choosing a reasonable service migration strategy based on the predicted results of user mobility is crucial to ensuring service quality. There has been extensive work devoted to user mobility prediction and service migration.

Research directions related to service migration mainly focuses on the load balancing of distributed data center. Ouyang *et al.* [10] has proposed an Lyapunov optimization technique to incorporate the long-term budget into a series of real-time optimization problems, which achieve a desirable balance between time-averaged user-perceived latency and migration cost. Similarly, Chowdhury *et al.* [11] has proposed to use the load information of each node in a period of time in the data center to predict the load and change trend at the next time point. On this basis, the allocation of services is determined to avoid unnecessary overhead caused by frequent service migrations [12]. However, these methods are mainly based on network load, user requests and other information to make decisions, without further consideration of user mobility [13]. As the user moves, the distance between the user and the MEC server where the service is located also changes relatively, and the original connection scheme may no longer be optimal.

In order to make accurate location prediction, the work [14] and [15] extracted features of multiple dimensions from users' historical information, such as network status features and social frequency features at different times, and effectively integrated them into a unified framework by using the factor graph (FG) model. All the above work on predicting mobility in service migration has a common assumption that we have perfect information about users over a period of time. However, in the actual environment, it is difficult to accurately predict the above users and network information. At the same time, for each decision moment, due to the lack of understanding of network environment parameters, users will consume additional communication costs in collecting system information. There are many research areas related to user mobility, such as context switching in cellular networks [16] and wireless ATM networks [17]. Nevertheless, these studies cannot be directly applied to service migration scenarios due to different decision spaces.

Many studies, e.g., [18–21], migrate the service to the vicinity of the current location of the user by means of the virtual machine dynamic migration technology, so as to ensure a low delay when the user use the service. However, this will entail significant service migration costs (such as additional network bandwidth usage and power resource consumption). To solve this problem, the change of network connection state is modeled by introducing user mobility have been investigated in [22–24]. Nevertheless, most of these schemes adopt the random walk model, rarely exploring the user's trajectory data and predicting the user's movement. In addition, these work pay less attention to the influence of QoS (such as network delay and migration cost) on edge server selection in service migration, so it is difficult to choose the optimal service migration strategy. In contrast, the method proposed in this paper can predict the location of the user at the next time, and considering the limitations of long-term prediction, it can make more intelligent decisions at the current time to avoid the cost of frequent service migration.

Different from the existing work, we proposed a mobile-aware dynamic service migration scheme, aiming at making more intelligent service migration decisions through location prediction, so as to reduce the service delay perceived by users and improve their service quality. We solved this problem by establishing a geometry-based user mobility prediction model and describing the service migration problem as a markov decision process. Finally, we developed a motion-aware dynamic migration algorithm.

3 User Mobility Model

In order to obtain the optimal solution of mobility-aware service migration, the foundation is to capture the dynamics of user mobility and try to precisely

Table 1. List of notations.

Notation	Description
$(x(t), y(t))$	The user's position coordinates
v_t	The speed of user at timeslot t
v_{max}	The maximum moving speed of user
D	The distance between the cellular network and TA boundary
$d(t)$	The distance between user and service at timeslot t
$P(cell_i/X_t)$	The probability that the user moves to cell i given the state X_t
$(v_{x(t)}, v_{y(t)})$	The velocity vector in $x(t)$ and $y(t)$ direction
Δv	The change of velocity between time $t+1$ and t
$u(t)$	The position of user at timeslot t
$w(t)$	The position of service at timeslot t
$cost_{com}(t)$	Communication cost at timeslot t
$cost_{mig}(t)$	Migration cost at timeslot t
$A(s)$	action space
$Size$	The size of the service to be migrated
$s(t)$	Initial state at timeslot t
$s^{'}(t)$	Intermediate state at timeslot t
S	The state space, including the location of the mobile user and the location of the base station where the service is located
π	Decision policy
$C_a(s_0)$	The sum of costs when taking action a in state s_0
$V^*(s_0)$	Discount sum cost when starting at state s_0
$P[a(s_0), s_1]$	Transition probability from state s_0 to the next initial state s_1
$\varphi_c, \varphi_l, \beta, \delta_c, \delta_l, \mu$	Parameters related to the service migration model
Dis	Maximum distance between user and service to maintain connection

predict the movement of users among the base stations in the near future. In this section, we propose user mobility model for prediction. To facilitate presenting the model in a formal way, we summarize all the notations used in the following discussions of this paper in Table 1.

In mobile edge computing, a user moves under the coverage of the base stations. An edge server is deployed to process the requests submitted by users from one or multiple base stations. Without loss of generality, we assume that each base station is equipped with an edge server. For the cases when multiple base stations share an edge server, we focus on the coverage area of the edge server and regard the base stations as one cell.

To solve the migration problem effectively and efficiently, we consider a time-slotted model for formulating the user mobility, and optimal policies are obtained at the beginning of each time slot according to the *state* of the user. The length of the time-gap interval is denoted by t, and the state of a user at time t is expressed as Eq. (1) where $(x(t), y(t))$ are the coordinates of the user and $v(t)$ is the velocity.

$$X_t = \{x(t), y(t), v(t)\} \tag{1}$$

Service migration is more likely to occur when users move near or across the cell boundaries. However, how to define the boundaries depends on the velocity of the user mobility. Therefore, we dynamically define a *Target Area (TA)* according to the upper bound of the velocity which is denoted by v_{max}, and the width of a TA is given by Eq. (2).

$$D = v_{max} \cdot t \tag{2}$$

At each decision epoch, only the users in the TA are possible to move to another cell, which may trigger a service migration. All the users in the central area, otherwise, will remain in the cell during the time interval. Consequently, we need to calculate the possibility of a user to move to another cell only when the user is located in the TA, which can significantly reduce the computational overhead of calculating the optimal solution during the service processes.

Figure 2 shows the coverage areas of TAs in a cellular network. Conventionally, we use a hexagon to represent the coverage area of a cell, and multiple cells constitute a cellular network. Dashed lines illustrate the boundaries of the TA, and the coverage area of a TAs is the hexagon ring between the TA boundary and cell boundary.

Specifically, we illuminate our mobility model when a user is moving in the TA of a cell as Fig. 3. The center coordinates of the cell are denoted by (a_0, b_0), and the distance between the user and the cell center is expressed by $d(t) = \sqrt{(x(t) - a_0)^2 + (y(t) - b_0)^2}$. We define a stochastic variable θ to represent the moving direction of the user during the time slot at time t, and let θ_i denote the direction to the i-th vertex of the cell for $i = 0, 1, \ldots, 5$.

At time t, the probability of the user moving to cell i given the current state X_t has a general form expressed as Eq. (3), where $f(\cdot)$ is the probability density function of the moving direction.

$$P\left(cell_i | X_t\right) = \int_{\theta_i}^{\theta_{(i+1)\%6}} f\left(\theta | X_t\right) d\theta \tag{3}$$

Additionally, we define the stochastic variate of the velocity as $v_t = (v_{x(t)}, v_{y(t)})^T$ in x and y direction respectively, and thus θ can be simply calculated by the following equation.

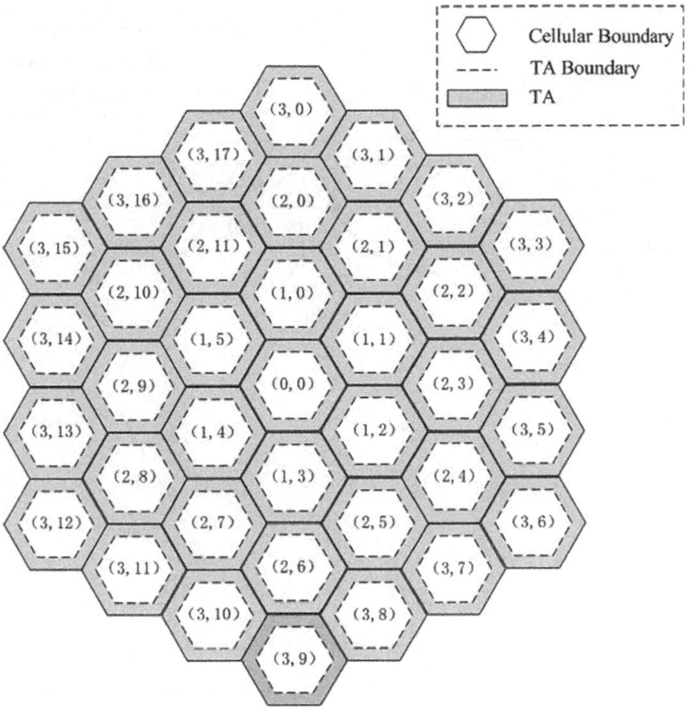

Fig. 2. Target areas in a cellular network.

$$\theta = g\left(v_t\right) = \arctan \frac{v_{y(t)}}{v_{x(t)}} \tag{4}$$

In most of the general cases, the probability density function of the velocity and its moving direction can be approximated by a Gaussian distribution [17]. Formally, we have

$$f(\theta|X_t) \sim N(\mu_\theta, \sigma_\theta^2) \tag{5}$$

where μ_θ is the mean and σ_θ is the standard deviation.

Afterward, the variate θ can be approximately expressed by Eq. (6).

$$\theta \approx g\left(v_t\right) + G\left(\Delta v\right) \tag{6}$$

where

$$G = \left.\frac{\partial g}{\partial v}\right|_{v=v_t} = \left[\frac{-v_{y(t)}}{v_{x(t)}^2 + v_{y(t)}^2}, \frac{v_{x(t)}}{v_{x(t)}^2 + v_{y(t)}^2}\right] \tag{7}$$

and

$$\Delta v = v_{t+1} - v_t \tag{8}$$

Δv is the change of velocity between time $t+1$ and t. Since v_t can be assumed to confrom to Gaussian distribution, and thus Δv also has a Gaussian

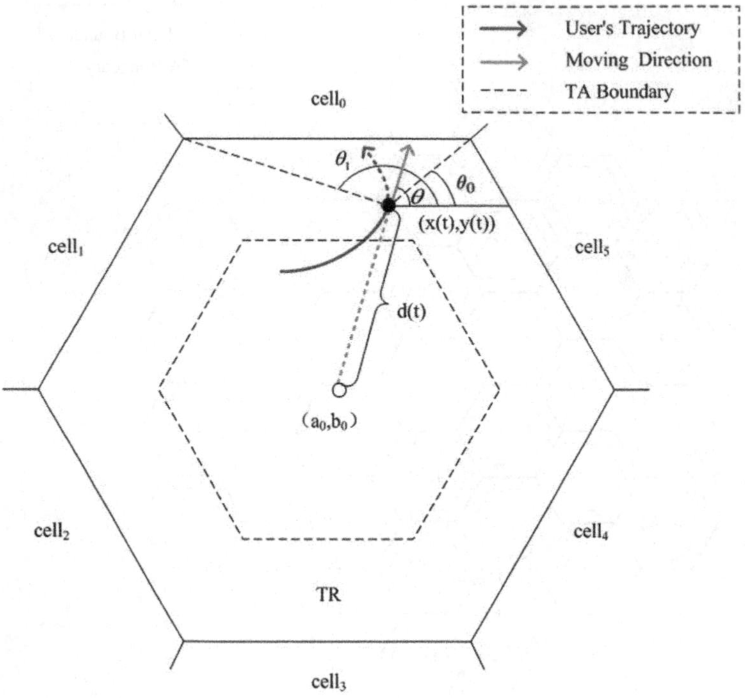

Fig. 3. User mobility model based on geometry.

distribution with the mean of $\mu_{\Delta v}$ and the variance of $\sigma^2_{\Delta v}$. With Eq. (6), we have

$$\mu_\theta = g(v_t) + \mu_{\Delta v} \tag{9}$$

$$\sigma_\theta = \sigma_{\Delta v} \tag{10}$$

In summary, one can conclude that the probability of a user moving to the i-th cell at the time slot $t+1$ can be calculated using the following expression.

$$P\left(cell_i|X_t\right) = \Phi\left(\frac{\theta_{i+1} - \mu_\theta}{\sigma_\theta}\right) - \Phi\left(\frac{\theta_i - \mu_\theta}{\sigma_\theta}\right) \tag{11}$$

In addition, in the cases when the speed of the mobile user is very slow, i.e., $\mu_{v_t} \approx 0$, $f(\theta|X_t)$ becomes a simple uniform distribution over $[0, 2\pi)$. Thus, the probability $P\left(cell_i|X_t\right)$ can be calculated by Eq. (12).

$$P\left(cell_i|X_t\right) = \frac{\theta_{i+1} - \theta_i}{2\pi} \tag{12}$$

4 Model and Algorithm of Service Migration

The user mobility model presented in the previous section can formulate the dynamic movements of users among different coverage areas of edge servers. With such model, in this section, we analyze the cost brought by migrating a service, and the formulate a dynamic optimization problem of service migration between two edge servers. An algorithm for solving the optimization problem is presented.

4.1 Cost Model

When predicting that a user will move to the coverage of another edge server with a high probability, a reasonable decision should be made whether to migrate the service being used by the user to the other edge server. The objective is to balance the trade-off between the QoS degradation and migration cost. For both of them, the distance should be fully taken into account for analysis, while other factors such as the size of the service should also be considered.

Let $u(t)$ and $w(t)$ denote the locations of the user and the service at time t, respectively. We assume that $w^{'}(t)$ is the location of the service that we try to migrate, according to the prediction result obtained form the mobility model. First, we analyze the communication cost which is closely related to the distance between the user and the service. The communication cost model has been studied by several existing literature, and hence we apply an exponential model [4], [18]. Specifically, we calculate the communication cost between the user at location $u(t)$ and the service located at $w^{'}(t)$ using the following expression.

$$cost_{com}(t) = \begin{cases} \varphi_c + \varphi_l \beta^{\|u(t)-w^{'}(t)\|} & u(t) \neq w^{'}(t) \\ 0 & u(t) = w^{'}(t) \end{cases} \tag{13}$$

where φ_c, φ_l and β are commonly non-negative parameters defined by the service provider, and β should be greater than 1.

The decision on triggering a service migration may bring in additional migration cost, and this type of cost depends on the distance between the locations of before and after the service migration, denoted by $w(t)$ and $w^{'}(t)$ respectively. Moreover, the size of the service *Size* also has a certain impact on the cost of service transfer, since transferring a big service may consume considerable networking resources. Similar to the communication cost model, we define the migration cost using Eq. (14), where $\delta_c \geq 0$, $\delta_l \geq 0$, and $\mu > 1$.

$$cost_{mig}(t) = \begin{cases} \delta_c + \delta_l \mu^{Size \cdot \|w(t)-w^{'}(t)\|} & w(t) \neq w^{'}(t) \\ 0 & w(t) = w^{'}(t) \end{cases} \tag{14}$$

We define that the system state at the beginning of the time slot t is denoted by $s(t) = (u(t), w(t))$. Considering both the communication cost and migration cost, the total cost of our model is given by Eq. (15).

$$C(s(t)) = cost_{com}(t) + cost_{mig}(t) \tag{15}$$

4.2 Service Migration

Dynamic service migration is to make optimal decision at each time slot according to the system state. The user mobility model is applied to predict the location of a moving user in the near future, while the cost model is used to calculate the migration cost for supporting the optimal solution.

We let π denote a sequence of decision, which is a mapping between a state $s(t)$ and an action $a \in A(s)$. Let $a_\pi(s(t))$ denote the action taken by strategy π when the system is in state $s(t)$. This control action will trigger a state transition of the system from the current state $s(t)$ to an intermediate state $s^{'}(t) = ((u(t), w^{'}(t)) = a_\pi(s(t))$. Let $C_{a_\pi}(s(t))$ represent the sum of migration and communication costs incurred by taking the control action $a_\pi(s(t))$ in the time slot t, and we have $C_{a_\pi}(s(t)) = cost_{com}(t) + cost_{mig}(t)$. Starting from an initial state $s(0) = s_0$, the long term expected cost given a policy π is expressed as Eq. (16)

$$V_\pi(s_0) = \lim_{t\to\infty} E\left\{\sum_{\tau=0}^{t} \gamma^\tau C_{a_\pi}(s(\tau)) \,|s(0) = s_0\right\} \tag{16}$$

where $0 < \gamma < 1$ is the discount factor used to distinguish short-term costs from long-term costs. The long-term cost is multiplied by the discount factor, which means that in this model, the current short-term cost is more important than the uncertain long-term cost.

In this paper, the ultimate objective of service migration is to minimize the total cost given an initial state s_0, i.e.,

$$V^*(s_0) = \min_\pi V_\pi(s_0) \tag{17}$$

Equation (17) can be precisely formulated by a Markov Decision Process (MDP) with an infinite horizon discounted cost. The Bellman's equation of the MDP is shown by Eq. (18).

$$V^*(s_0) = \min_a \left\{C_a(s_0) + \gamma \sum_{s_1\in S} P[a(s_0), s_1] \cdot V^*(s_1)\right\} \tag{18}$$

where $P[a(s_0), s_1]$ denotes the probability of the system to transfer from state $s^{'}(0) = s_0^{'} = a(s_0)$ to $s(1) = s_1$. The transition probability here is calculated by the mobility model presented in Sect. 3. With the relationship of state transitions, we have $s(t+1) = (u(t+1), w^{'}(t)) = (u(t+1), w(t+1))$, where $w(t+1) = w^{'}(t)$. In the following statement, the time symbol t will be omitted if not specified.

The solutions of the optimality equations include the minimum expected discounted total cost $V^*(s)$ and the optimal policy π. The optimal policy π indicates the migration target of the service given the state s. The procedures of our mobility-aware dynamic migration (MODEM) algorithm in mobile edge computing are summarized as follows:

- The one-step cost at the k-th step is given by $C_a(s)$ when the system is in the state s and a control action a is selected. In this work, it is determined by the communication cost $cost_{com}(t)$ and the service migration cost $cost_{mig}(t)$.
- The state transfer mechanism is probabilistic which is controlled by the transition probabilities $P[a(s), s^{'}]$ of all states s, and the control action a is selected from all the feasible actions of state s. In this work, the transition probabilities are calculated by the user mobility model.
- The minimum cost function $V^*(s)$ includes the cost of one step and the minimum expected cost of all possible state transitions at the $(k+1)$-th step.

With all the calculations presented above, one can apply some well-known existing algorithms to solve the MDP problem. Considering the computational complexity of the algorithms, we select policy iteration algorithm, which is often able to find the optimal solution in the minimal number of iterations. The procedures of the policy iteration algorithm is shown in Algorithm 1. The output is the service migration strategy π in a certain state (that is, in the case of service migration, which server should be migrated from the current server to the surrounding server), which can minimize the total migration cost $V^*(s)$.

Algorithm 1. Policy-iteration algorithm based on mobility-aware

Input: $C_a(s)$, $P[a(s), s^{'}]$;
Output: $\pi*$, $V^*(s)$;

```
π ∈ A(s) arbitrarily for all s ∈ S;
for k = 0, 1, ... do
    Compute the probability of transition P[a(s), s'] by solving Eq. (11);
    Compute cost C_a(s) by solving Eq. (15);
    Solve V*(s) = C_a(s) + γ Σ_{s'∈S} P[a(s), s'] · V*(s'),
    find V*(s) for π;
    Update the policy according to V*(s),
    π_{k+1} ← argmin_a { C_a(s) + γ Σ_{s'∈S} P[a(s), s'] · V*(s') };
    if π_{k+1} = π_k then
        π* ← π_k;
        return π*, V*(s) ;
    end if
end for
```

It is worth noting that in practice, the maximum allowable distance between the mobile user and the service to maintain communication is usually bounded, that is, although users can move in unbounded space, the state space controlled by the service is limited. Therefore, we assume that the maximum connection distance between the user and the service is Dis, then our policy only needs to focus on the state of $(u(t) - w(t)) \in [0, Dis]$. If the user moves to another service location where the distance exceeds the threshold Dis, the service migration is automatically triggered. Instead, we need to consider the mobility of users and the cost of service migration to make a reasonable migration decision.

5 Simulation Results

To validate the our approach, we conduct simulation experiments in MATLAB. A data set containing the GPS trajectories of 10,357 taxis in the city of Beijing is applied [25]. It has been released by Microsoft Research Asia, and covers the dates from Feb. 2 to Feb. 8 in the year of 2008. In our experiments, each active taxi is regarded as a mobile user in the MEC system, and its location varying with time is obtained from the longitude, latitude and time data from the data set.

The base stations are placed randomly, covering all the active area of the taxis. We simply assume that each taxi connects to its nearest base station measured by Euclidean distance. The base station connected to the taxi collects the taxi's service request, location and other parameter information. The length of a time slot is 60 s, and the base station calculates parameters such as the moving speed of the taxi based on the information obtained.

The relationship between the GPS track of the taxi and the cellular network is shown in Fig. 4. Subfigure (a) is the display of the GPS track of the taxi in the map, and subfigure (b) is the display of the GPS track of the taxi in the cellular network. Among them, the center of the hexagon is the base station, and the size of each hexagon is the coverage area of the base station. According to the survey, it is found that the 5G base station in densely populated areas is kept at about 200 m, and the suburban area is kept at about 500 to 1,000 m, so the radius of the hexagon in the experiment is set at about 400 m.

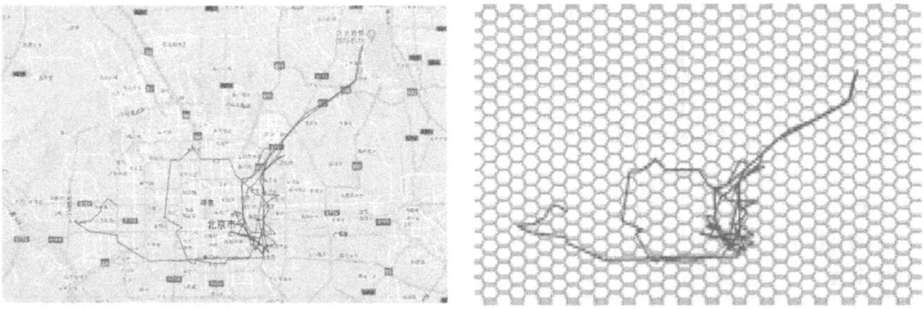

(a) (b)

Fig. 4. The relationship between the GPS trajectories of a taxi and cellular network

In order to validate the effectiveness of our MODEM approach, we select another three schemes for comparison as follows.

- **Non-Migration (NM)**: The service will not be migrated no matter how the user moves, unless the distance between them exceeds the maximum threshold Dis. Formally, in the area of $d(t) < Dis$, $a = 0$ always holds.
- **Always Migration (AM)**: We always migrate the service as soon as the user moves to another cell. Thus, the action variable a is always set to 1.
- **Random Walk Model (RWM)**: We still use the MDP algorithm to find the optimal solution, but replace our mobility model with a random walk model which has been widely applied in service migration [23]. We assume that the probability of the user leaving a cell is p, and thus the probability of staying inside the cell is $1 - 6p$.

5.1 Migration Cost Parameter Analysis

We analyze the algorithm by changing the migration cost parameters and the results are shown in Figs. 5 and 6.

As shown in Fig. 5, the total discounted cost of the NM approximates the cost of our MODEM algorithm when φ_l is small. The result can be explained by Eq. (13). the communication cost is relatively small when φ_l is small, and the policy is more in favor of NM. On the contrary, when φ_l is large, the total discounted cost of our MODEM algorithm approximates to the AM scheme. That is because, the communication cost is relatively large when φ_l is large, and the policy is more in favor of AM.

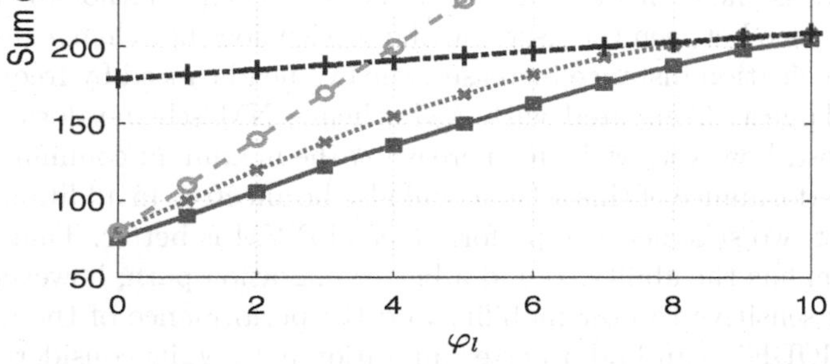

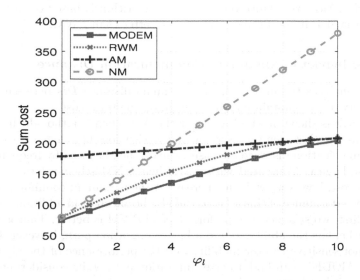

Fig. 5. Parameter analysis: the communication function parameter φ_l.

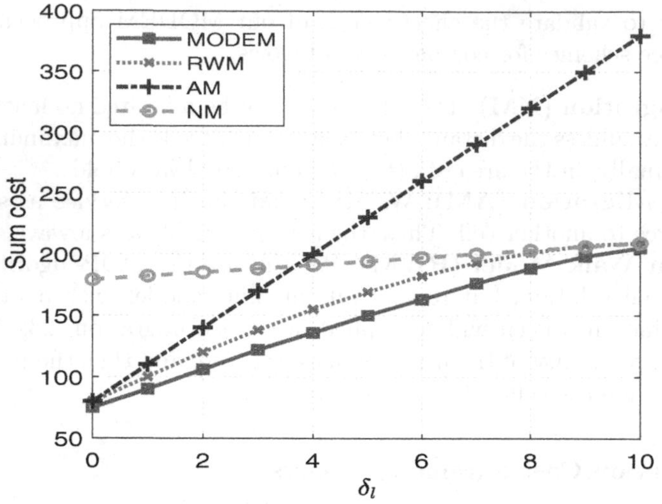

Fig. 6. Parameter analysis: the migration cost parameter δ_l.

In contrast, as shown in Fig. 6, the total discounted cost of the MODEM approximates the AM when δ_l is small. The result can be explained by Eq. (14), the migration cost is relatively small when δ_l is small, and the policy is more in favor of AM. On the contrary, when δ_l is large, the total discounted cost of our MODEM algorithm approximates to the NM scheme. That is because, the migration cost is relatively large when δ_l is large, and the policy is more in favor of NM. Moreover, since our MODEM algorithm is based on the mobility prediction model, its performance is always better than RWM.

5.2 The Impact of Maximum Communication Distance

This part analyzes the impact of the maximum distance *Dis* between the user and the service to maintain communication on the total cost.

The result is shown in the Fig. 7, for the AM scheme, although the communication distance between the user and the service does not change, as the maximum communication distance increases, the overhead caused by frequent migration is the largest. Compared with AM scheme, NM scheme does not require migration cost, however, with the increase of the maximum communication distance, the performance of this scheme will also be affected. In addition, compared with the first two schemes, the performance of RWM is better. That is because, the algorithm has the ability to find a better migration path, however, the algorithm is not sensitive to user mobility and the performance of the algorithm is mediocre. MODEM can find a better migration path while considering the user mobility, so it has better performance.

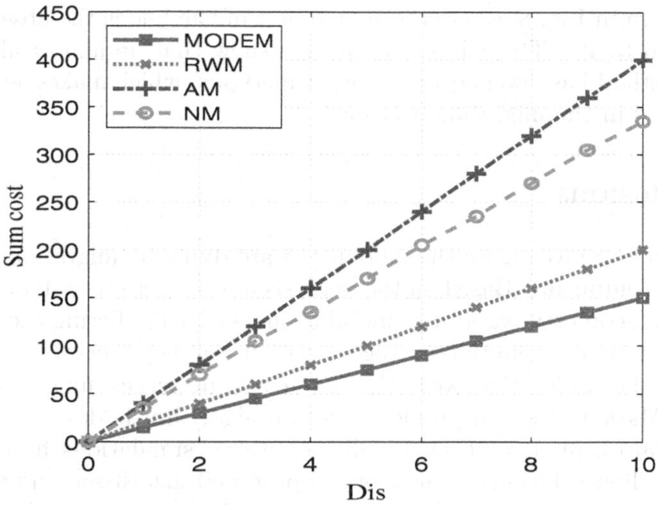

Fig. 7. The impact of the maximum distance *Dis* between the user and the service to maintain communication on the total cost.

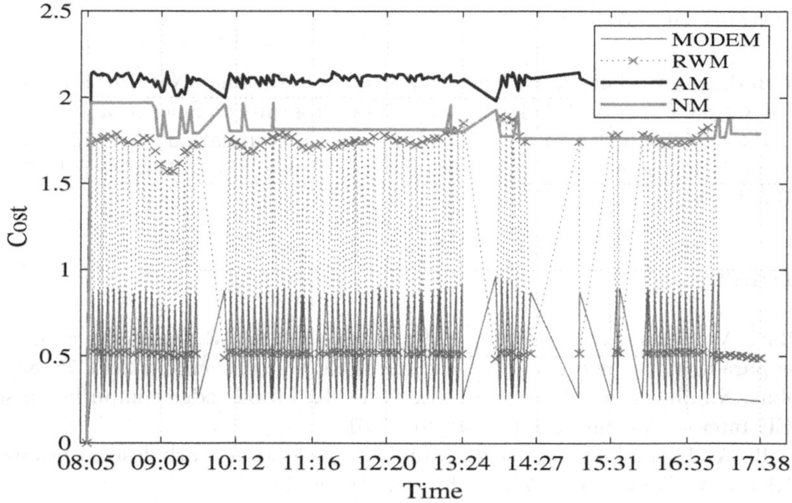

Fig. 8. Cost compared to alternative algorithms in trace-driven simulation.

5.3 Comparison of Simulation Results

The costs of the four migration strategies are compared by tracking the driver simulation and the results are shown in Fig. 8.

The track data of a certain taxi driver on a certain day was randomly selected (taxi data from 8:05 to 17:38 were selected in the experiment because taxi tracks were relatively active in the daytime) to compare the service migration cost. The result is shown in Fig. 8, where the sparse part of the line is the absence of data (e.g. 14:27 to 15:40). The comparison results show that, in almost all cases, the proposed method has lower cost than other methods, which makes our algorithm better verified in the migration decision.

6 Conclusions

In this paper, we investigate the mobility-aware dynamic migration problem in mobile edge computing. Based on the analysis of the trajectory data from users, we propose a geometry-based user mobility model for predicting the probability of a user to move out from the coverage of the current edge server to another one. Considering the trade-off between QoS and cost, the service migration is formulated by a Markov decision problem, and an algorithm is designed for finding the long-term optimal solution. Finally, extensive simulations have been conducted to evaluate the effectiveness of the proposed algorithm, and the impacts of the model parameters are further analyzed. The experimental results show that our approach can dynamically find the optimal service migration decisions by reducing the cost while improving the QoS. This work is expected to provide a theoretical model and a practically solution of optimal service migration in MEC systems for mobile users.

Acknowledgment. This work was supported by National Natural Science Foundation of China (No. 61972414), Beijing Nova Program of Science and Technology (No. Z201100006820 082), Beijing Natural Science Foundation (No. 4202066), and Fundamental Research Funds for Central Universities (Nos. 2462018YJRC040 and 2462020YJRC001).

References

1. Villari, M., Fazio, M., Dustdar, S., Rana, O., Ranjan, R.: Osmotic computing: a new paradigm for edge/cloud integration. IEEE Cloud Comput. **3**(6), 76–83 (2016)
2. Abbas, N., Zhang, Y., Taherkordi, A., Skeie, T.: Mobile edge computing: a survey. IEEE Internet Things J. **5**(1), 450–465 (2018)
3. Ceselli, A., Premoli, M., Secci, S.: Mobile edge cloud network design optimization. IEEE/ACM Trans. Netw. **25**(3), 1818–1831 (2017)
4. Satyanarayanan, M.: The emergence of edge computing. Computer **50**(1), 30–39 (2017)
5. Peng, Q., et al.: Mobility-aware and migration-enabled online edge user allocation in mobile edge computing. In: 2019 IEEE International Conference on Web Services (ICWS), pp. 91–98. IEEE (2019)
6. Wang, S., Xu, J., Zhang, N., Liu, Y.: A survey on service migration in mobile edge computing. IEEE Access **6**, 23511–23528 (2018)
7. Taleb, T., Ksentini, A.: An analytical model for follow me cloud. In: Proceedings of IEEE GLOBECOM 2013, December 2013
8. Ksentini, A., Taleb, T., Chen, M.: A Markov decision process-based service migration procedure for follow me cloud. In: 2014 IEEE International Conference on Communications (ICC), pp. 1350–1354. IEEE (2014)

9. Wang, S., Urgaonkar, R., He, T., Zafer, M., Chan, K., Leung, K.K.: Mobility-induced service migration in mobile micro-clouds. In: Proceedings of IEEE MILCOM 2014, October 2014
10. Ouyang, T., Zhou, Z., Chen, X., et al.: Follow me at the edge: mobility-aware dynamic service placement for mobile edge computing. IEEE J. Sel. Areas Commun. **36**(10), 2333–2345 (2018)
11. Chowdhury, M., Rahman, M.R., Boutaba, R.: ViNEYard: virtual network embedding algorithms with coordinated node and link mapping. IEEE/ACM Trans. Netw. **20**(1), 206–219 (2011)
12. Minarolli, D., Mazrekaj, A., Freisleben, B.: Tackling uncertainty in long-term predictions for host overload and underload detection in cloud computing. J. Cloud Comput. **6**(1), 1–18 (2017). https://doi.org/10.1186/s13677-017-0074-3
13. Wang, S., Zhang, X., Zhang, Y., Wang, L., Yang, J., Wang, W.: A survey on mobile edge networks: convergence of computing, caching and communications. IEEE Access **5**, 6757–6779 (2017)
14. Kschischang, F.R., Frey, B.J., Loeliger, H.-A.: Factor graphs and the sum-product algorithm. IEEE Trans. Inf. Theory **47**(2), 498–519 (2001)
15. Wu, Q., Chen, X., Zhou, Z., Chen, L.: Mobile social data learning for user-centric location prediction with application in mobile edge service migration. IEEE Internet Things J. **6**(5), 7737–7747 (2019)
16. Xenakis, D., Passas, N., Merakos, L., Verikoukis, C.: Mobility management for femtocells in LTE-advanced: key aspects and survey of handover decision algorithms. IEEE Commun. Surv. Tutorials **16**(1), 64–91 (2013)
17. Liu, T., Bahl, P., Chlamtac, I.: Mobility modeling, location tracking, and trajectory prediction in wireless ATM networks. IEEE J. Sel. Areas Commun. **16**(6), 922–936 (1998)
18. Ceselli, A., Premoli, M., Secci, S.: Mobile edge cloud net- work design optimization. IEEE/ACM Trans. Netw. 25(3), 1818–1831 (2017)
19. Nelson, M., Lim, B.-H., Hutchins, G., et al.: Fast transparent migration for virtual machines. In: USENIX Annual Technical Conference, General Track, pp. 391–394 (2005)
20. Zhou, A., Wang, S., Ma, X., Yau, S.S.: Towards service composition aware virtual machine migration approach in the cloud. IEEE Trans. Serv. Comput. **13**, 735–744 (2019)
21. Sung, J.-W., Han, S.-J., Kim, J.-W.: Virtual machine provisioning for computation offloading service in edge cloud. In: 2019 IEEE 12th International Conference on Cloud Computing (CLOUD), pp. 490–492. IEEE (2019)
22. Plachy, J., Becvar, Z., Strinati, E.C.: Dynamic resource allocation exploiting mobility prediction in mobile edge computing. In: IEEE 27th Annual International Symposium on Personal, Indoor, and Mobile Radio Communications (PIMRC), pp. 1–6. IEEE (2016)
23. Wang, S., Urgaonkar, R., Zafer, M., He, T., Chan, K., Leung, K.K.: Dynamic service migration in mobile edge computing based on Markov decision process. IEEE/ACM Trans. Netw. **27**(3), 1272–1288 (2019)
24. Machen, A., Wang, S., Leung, K.K., Ko, B.J., Salonidis, T.: Live service migration in mobile edge clouds. IEEE Wirel. Commun. **25**(1), 140–147 (2017)
25. Yuan, J., Zheng, Y., Xie, X., Sun, G.: Driving with knowledge from the physical world. In: Proceedings of the 17th ACM SIGKDD International Conference on Knowledge Discovery and Data Mining, pp. 316–324 (2011)

Research on Debugging Interaction of IoT Devices Based on Visible Light Communication

Jiefan Qiu[1(✉)], Chenglin Li[1], Yuanchu Yin[1], and Mingsheng Cao[2]

[1] Zhejiang University of Technology, Hangzhou 310023, China
qiujiefan@zjut.edu.cn

[2] University of Electronic Science and Technology of China, Chengdu 610054, China

Abstract. Wireless sensors normally deployed in inaccessible areas. Once the wireless communication of sensor node fails, the lost node cannot be repaired by debugging interaction which depends on this communication. Visible light communication (VLC) is a supplement of the traditional radio-wave wireless communication by and needs a dedicated device or module. Thus, VLC is hard to be applied in low-cost sensor nodes. We implement a hybrid duplexing debugging interaction system (HDDIS) based on VLC in general smartphone and sensor node. The smartphone is taken as a VLC gateway to send debugging codes to the sensor node. In order to improve the VLC transmission rate of debugging codes, we propose a novel debugging code compression method for data source and channel coding. With regard to the data source, we analyze the binary instructions and reuse opcodes and leverage a bit-mask technique to compress operands. The average compression rate of binary instructions reaches 84.11%. For channel coding, we optimize the dual-header pulse interval modulation (DH-PIM) and propose the overlap DH-PIM (ODH-PIM) by introducing a LED half-on state. The LED half-on state can improve the representation ability of each symbol. The experiment results illustrate that our modulation reduces transmission time by 10.71% compared with DH-PIM.

Keywords: Sensor nodes · Debugging · Visible light communication · Instruction · Modulation

1 Introduction

The wireless sensor network is composed of decentralized and self-organized sensor nodes. Once the network deployment, it is difficult to maintain the nodes, especially the nodes are deployed in the inaccessible area, such as in a museum show in cases [1]. The reparation and debug of sensor nodes depend on the built wireless network infrastructure.

However, some failures of the sensor network are related to unreliable wireless communication. If the sensor nodes on the critical network path are temporarily or permanently lost, and they cannot be repaired and work, the performance of the network system is possibly degenerated, and even invalid the entire system of the network. Meanwhile, it is difficult to employ an alternatives interaction method to complete a debugging node

H. Gao et al. (Eds.): CollaborateCom 2020, LNICST 349, pp. 132–147, 2021.
https://doi.org/10.1007/978-3-030-67537-0_9

due to the limited resources. In practice, developers have to abandon these lost nodes and replace them by backup nodes to which the tasks running on lost nodes are migrated [2–4]. The problem is that developers rely on experience and intuition to determine whether the tasks should be migrated, and that is usually inaccurate. In addition, developers cannot figure out the reason for the node lost and prevent node loss. Even if task migration is successful, the current status of a task is not stored. That finally affects the performance of entire networks. To this end, we hope to find an alternative debugging interaction method without modifying the local hardware of sensor nodes.

In recent years, with the development of visible light technology, the general sensor node, and smartphone equip various visible light components such as LED, camera, and ambient light sensor. These components provide a potential capacity for visible light communication (VLC). More and more traditional sensor devices have begun to use VLC as a complement of radio wave communication (RWC) for traditional wireless communication. VLC adopts unidirectional propagation different from the traditional RWC with broadcast way. This kind of propagation is not easy to be intercepted or eavesdropped and improves transmission security. In addition, VLC does not occupy the bandwidth resources of current sensor networks and affects the wireless communication between sensor nodes [5]. Therefore, we try to apply VLC in debugging interaction.

However, current VLC requires dedicated visible light components or modules. For example, Fan L et al. applied VLC to the access control system. Therefore, it is necessary to modify the existing system by adding a special VLC circuit. The circuit composes of the photodiode, amplifier, comparator, and MCU [6]. In order to improve the transmission rate and solve the problem of multi-channel transmission, Wang Y et al. designed a duplexing indoor communication system based on visible light RGB-LED. The duplexing communication system employs RGB light emitting diodes, low pass filters, electronic amplifiers, and photodiodes [7]. VLC components are sensitive to environmental optical noise. Adiono T et al. proposed a solution to reduce the influence of optical noise. This solution requires an analog filter which is taken as the front-end receiver [8]. Such extra hardware modification increase cost and not suitable for deployed sensor nodes, which is a bottleneck of the VLC application in IoT.

In this paper, we implemented a hybrid duplexing debugging interaction system (HDDIS) based on VLC applied in general sensor nodes. We tap the potential capacity of VLC components in smartphone and sensor nodes and regard the smartphone as a VLC-based gateway for debugging node in field. The debugging interaction uplink mainly transmits debugging information from the malfunction sensor node. It leverages the smartphone's optical camera as the signal receiver, and the node's LED as the signal transmitter. The debugging interaction downlink mainly transmits updating codes to fix nodes. It leverages the smartphone's flashlight as the signal transmitter and the node's ambient light sensor as a signal receiver. In this system, since the data transmitted by downlink are relative to binary code for an update, and its size is large compared with debugging information data. In addition, the general ambient light sensor equipped in a node has a serious light sensitivity latency, which prolongs data transmission and lower transmission rate.

To this end, we have proposed a set of compression schemes in source and channel coding. In source coding, we respectively compress opcode and operand of one instruction. In practice, a large number of redundancies exist in opcodes. The same part of the opcodes of the instruction can be reused. In addition, a part of the operand, such as a base address, is also represented by bit-mask to further shorten the length of each instruction. In channel coding, we designed the overlap dual-header pulse interval modulation (ODH-PIM). DH-PIM is an isochronous pulse time modulation in which data are encoded as discrete time slots between adjacent pulses. The width of the low power is used to represent a symbol. We add a LED half-on state as an overlapping mark in DH-PIM. This mark realizes an overlap representation to compress two data into one symbol and further reduces the number of light pulses.

Besides, since the optical camera completes shooting in units of image frames, the interval at which the node LED sends data needs to be consistent with the optical camera's shooting time to transmit debugging information successfully. For this reason, we further propose a feedback frame synchronization scheme for rolling shutter. In this synchronization solution, the smartphone communicates with the node to coordinate signal transmission and reception synchronization. Real-time detection of the collected images to ensure synchronous communication.

Finally, we conducted a serial of experiments to verify our method from compression rate, bit error rate (BER), transmission time, and energy consumption. The experiment results illustrate our compression method reduce 80% of redundant codes. At the same time, the ODH-PIM owns the lower bit error rate than PWM, and the less transmission time than DH-PIM at the cost of an extra 17.36% energy overhead.

The rest of this paper is structured as follows: Sect. 2 shows relative work; Sect. 3 gives the overview of the HDDIS; Sect. 4 describes the design of instruction compression based on bitmask and ODH-PIM. Section 5 describes the uplink synchronization scheme. Section 6 describes experimental and evaluates our approach and Sect. 7 closes with a conclusion.

2 Related Work

Chen P Y et al. [9] introduced a mask-based compression method and used it in real-time embedded systems. Kumar R N et al. [10] proposed a method of combining a lookup table (LUT) and a mask, which increases the matching range of the lookup table. Early compression work [11, 12] used masks for code compression in embedded systems, which significantly improves the code compression rate. However, the above articles need to transfer some dictionaries or lookup tables, which increases the amount of data transferred. We expect to judge the type of data only when compressing and perform data compression without transferring the dictionary.

At present, intensity modulation/direct detection (IM/DD) is mainly used in VLC to communicate. DPIM modulation transmits data in units of groups, and Ghassemlooy et al. [13] applied DPIM modulation for the first time in the field of optical communications. Subsequently, Aldibbiat et al. [14] proposed double-head pulse interval modulation (DH-PIM) based on DPIM. Kaili Yao et al. [15] proposed a novel scheme using pulse position to improve the BER performance instead of amplitude of Carrier-less Amplitude

Phase. Agha Yasir et al. [16] specifically combined the PPM and pulse shape modulation, this scheme provided has improved in system performance along with computational complexity by increasing the bandwidth and the number of pulses. However, there are few modulations designed to compress transmission data.

CMOS sensors also have the function of receiving and converting optical signals into electrical signals, so many researchers have begun to study the role of optical cameras in VLC. T. S. Fuzile and R. Heymann et al. [17] implemented a three-color LED VLC system on the mobile terminal, but the data transmission rate was low due to the global shutter and the simple modulation and demodulation method.

In response to the above research work, we have designed a HDDIS based on visible light using the existing components on the existing nodes and the general sensor module of the Android device.

3 System Overview

To repairing the wireless communication lost node, we design HDDIS based on VLC. As shown in Fig. 1, in HDDIS, the smartphone is taken as VLC gateway, and send binary codes to a sensor node. The node downloads codes and preforms the repair process. Finally, debugging results is reported bake to smartphone. The two links use different sensor components and communication methods to achieve node debugging. The downlink uses traditional VLC technology to update and repair the sensor node. It uses a smartphone flashlight as the signal transmitter and the ambient light sensor on the sensor node as the signal receiver. The sensor node then feeds back the debugging

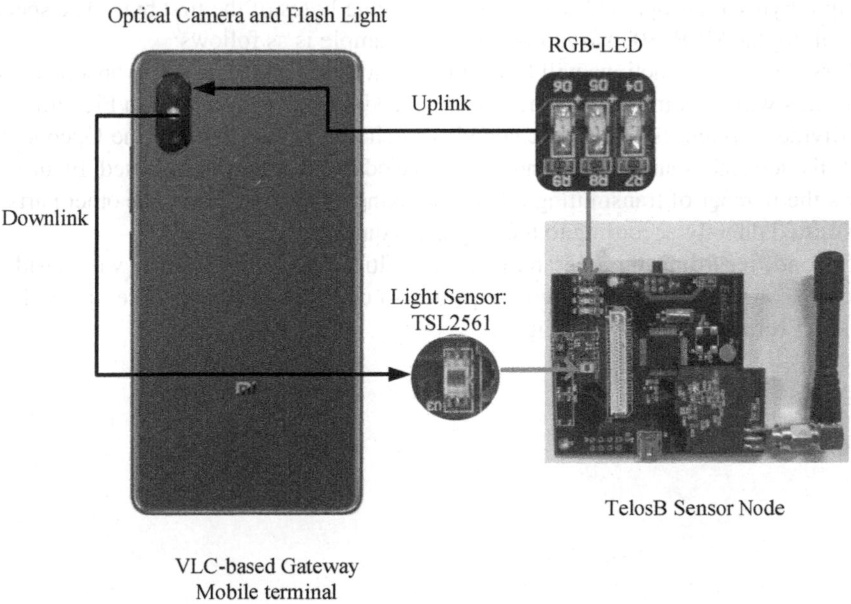

Fig. 1. Hardware interaction diagram

information to the smartphone from the uplink. In the uplink, the LED on the node acts as a signal transmitter, and the optical camera of the smartphone acts as a signal receiver using OCC to collect LED light information.

4 Downlink: Instruction Compression Based on Bitmask and Pulse Compression Method

The downlink transmits a large amount of binary data to repair the sensor node. However, the ambient light sensor on the node has low accuracy, resulting in a long transmission time of the data unit and a low transmission rate of the overall codes. To this end, we respectively optimize the instruction representation for source coding and the symbol representation for channel coding.

4.1 Compress the Binary Instruction

When transmitting the new code, we only transmit the modified function, which can effectively reduce the size of the transmitted data. These functions will be converted by the compiler into instructions for the next analysis, and finally into compressed binary data. In this paper, instructions are divided into three types: single operand instructions, double operand instructions, and jump instructions. Each type of instruction has the same part in the opcode. And single operand instructions, double operand instructions may occupy different numbers of bytes due to different addressing modes. Therefore, we have analyzed the instruction structure and remove some fixed opcode binary data, and further compressed the operand of multiple byte instructions. When instructions occupy multiple bytes, their operand will occupy following bytes of the first byte. The specific plan taking the MSP430 instruction set as an example is as follows:

First, binary instructions will be divide into a lot of sequences. The opcode data in sequences will be compressed. The opcode encoding format is shown in Fig. 2(a). Sort flag divides sequences into four categories as shown in Fig. 2(b). In the Opcode segment, fix sequences are omitted and only changed sequences is transmitted. Figure 2(c) shows the number of transmitting bits in each kind of opcode. Then, the other parts are transmitted directly according to the original sequences.

Second, regarding the operand data of multiple bytes instruction, we considered a method of replacing the all-zero sequence to compress the data. The operand data encoding format as it shows in Fig. 3.

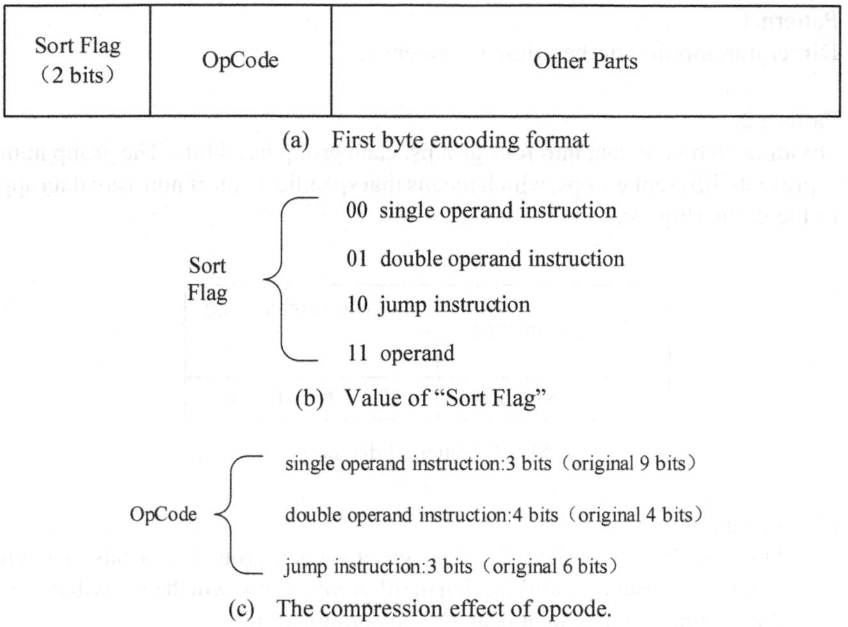

Fig. 2. Opcode compression coding diagram

Sort Flag	Pattern	Group Number	Non-zero Data
(11) 2 bits	2 bits	2 bits	4 bits/8 bits

Fig. 3. Operand compression coding diagram

The sort flag corresponds to that in Fig. 2(a), indicates that this is an operand sequence. The pattern means that different schemes are used for compression. The group number indicates the group number where non-zero data appear. And the non-zero data shows specific non-zero data content. Next, we introduce the schemes adopted by each pattern shown in Table 1. 16 bits of data are divided into four groups, each group has 4 bits.

Table 1. Four different pattern

Pattern	Value	Data size and number of groups
Pattern 1	00	-, -
Pattern 2	01	4 bit, 1 group
Pattern 3	10	8 bits, 2 groups
Pattern 4	11	8 bits, 2 groups

(1) Pattern 1:
Direct transmission of the original sequence;

(2) Pattern 2:
Dividing 16 bits of data into four groups, each group has 4 bits. The group number represents different groups, which means that specific 4 bits of non-zero data appear in the group (Fig. 4);

16 bits sequence	0000 0000 0000 0000
Group Number	00 01 10 11

Fig. 4. Pattern 2 diagram

(3) Pattern 3:
Dividing 16 bits of data into four groups, each group has 4 bits. The group number represents a combination of different groups, which means that specific 8 bits of non-zero data appears in the combination;

16 bits sequence	0000 0000 0000 0000	0000 0000 0000 0000	0000 0000 0000 0000	0000 0000 0000 0000
Group Number	00	01	10	11

Fig. 5. Pattern 3 diagram

(4) Pattern 4:
Dividing 16 bits of data into two groups, each group has 8 bits. The group number" represents different groups, which means that specific 4 bits of non-zero data appears in the group (Fig. 6).

16 bits sequence	0000 0000 0000 0000	0000 0000 0000 0000
Group Number	00	11

Fig. 6. Pattern 4 diagram

In some cases, the sequence transmitted includes 16 bits of data whose bits are all-zero. At this time, since two group number are consumed in pattern 4, group number can be set to 10 to indicate that this is a group of all-zero sequences. By analyzing the original binary sequence of the code, it can be found that the frequency of the sequence is not low.

Take the instructions in the MSP430 instruction set as an example as shown in Fig. 7. The "Push" instruction is a single-byte single operand instruction, so there is no need to consider additional operand compression. The "Call" instruction is also a single-operand instruction, but it occupies two bytes due to different addressing modes, so the operand of the second byte can be compressed.

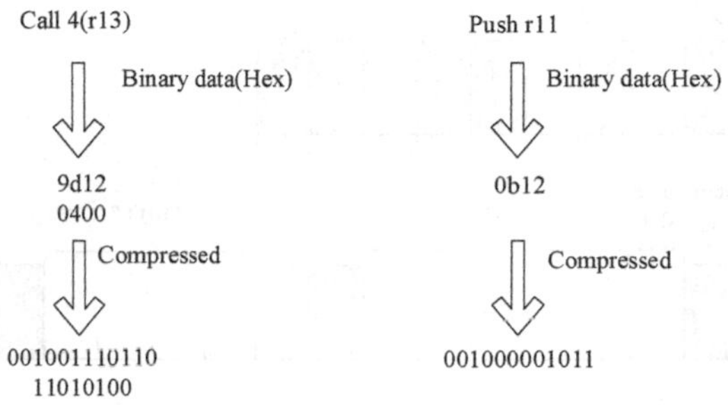

Fig. 7. Example of instruction compression

4.2 Introduce LED Half-on State for Compression Modulation

After the first step of compressing the binary instruction, a better modulation method can be designed to further compress the code. However, the commercial off-the-shelf device usually has low accuracy, making it difficult to achieve high-speed data transmission. Therefore, there is necessary to improve existing modulation methods to solve this problem.

Using the ambient light sensor, the Measuring process of light intensity needs a duration of light-on state. If we intend to shorten the duration, the sensor captures a new LED state different from LED On/Off state. We mark the new state as LED half-on state. Moreover, we designed the ODH-PIM based on the existing DH-PIM modulation method by adding LED half-on state. DH-PIM is an isochronous pulse time modulation in which data are encoded as discrete time slots between adjacent pulses. A symbol that encodes M bits of data is represented by k slots of low power and followed by one pulse of constant power, where $0 \leq k \leq L/2-1$ and $L = 2^M$. The pulse of constant power will last t_r or $2 * t_r$ for two numbers which are a radix-minus-one complement. For example, the number "0100" has 4 slots of low power and 1 slot pulse of constant power. The number "1011" is a radix-minus-one complement of "0100" and has 4 slots of low power and 2 slots pulse of constant power. A symbol of the DH-PIM usually contains 4 bits of data. However, the ODH-PIM can compress 8 bits of data into one symbol because two sets of data can be overlapped.

Because different sets of data have different time slot sizes when two sets of data pulses are overlapped together for modulation, and special pulses need to be used to separate them. At this time, the LED half-on state plays an important role. As shown in Fig. 8, given M = 4, after representing the contents of two sets of data, a time slot pulse is introduced as an order symbol, indicating the order of the two sets of data. And a set of inverted marks is used to determine which set of data needs to be complemented. In ODH-PIM modulation, the order flag indicates that the order of the two sets of data needs to be changed, and if there is no flag, the data set with a shorter period is first. Besides, there are four cases of the reverse flag.

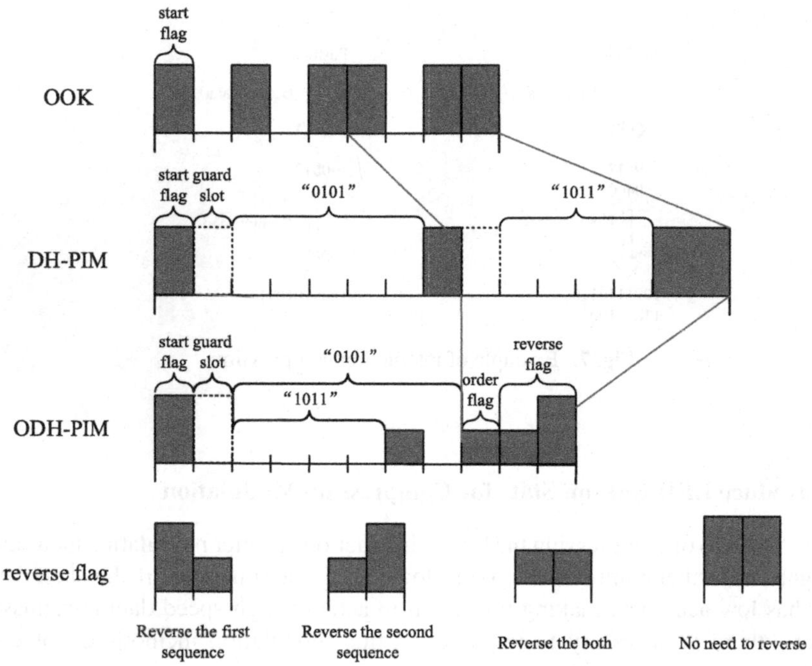

Fig. 8. Three modulation comparison. Bit series "01011011"

5 Uplink: Synchronization Scheme for Rolling Shutter Camera

In optical camera communication, synchronization has always been one of the most important issues. First, the sampling of the optical camera is performed randomly. It may occur during any symbol of the signal, so any symbol may be lost at any time. Secondly, the frame sampling interval is variable, which depends on the characteristics of the image sensor and optical channel. When the camera collects node LED stroboscopic information, the data symbols will be lost due to the lack of uniform turn-on time at the transmitter and receiver and the optical camera's unstable frame rate.

To this end, we propose a frame synchronization scheme for the rolling shutter camera. In this synchronization solution, the smartphone communicates with the node to coordinate signal transmission and reception synchronization. In the communication process, due to the smartphone's unstable frame rate, the data stripes' position in the collected image frame will shift. Due to the hardware device's limitation, the time accuracy that can be controlled is low. It is impossible to guarantee that a complete data frame is collected in the effective area. To solve this problem, we take the green light as the indicator for detecting the synchronization status, and the yellow and red lights continuously send the same data frame within the duration of the image frame to ensure that one image frame can successfully receive one data frame. Due to the instability of the optical cameras' frame rate, two adjacent data frames are simultaneously collected in one image frame. Therefore, after sending a data frame, let the green light send a low pulse to indicate a data frame's end. If the dark stripes in the green light are detected in the image frame, the image frame contains two data frames. Furthermore, synchronization has expired. At this time, the feedback information needs to be sent by the smartphone's flashlight, and the node end receives the feedback information and resynchronizes. The program consists of two parts:

(1) Initial synchronization at the beginning of transmission;
(2) Resynchronization after transmission interruption (Fig. 9).

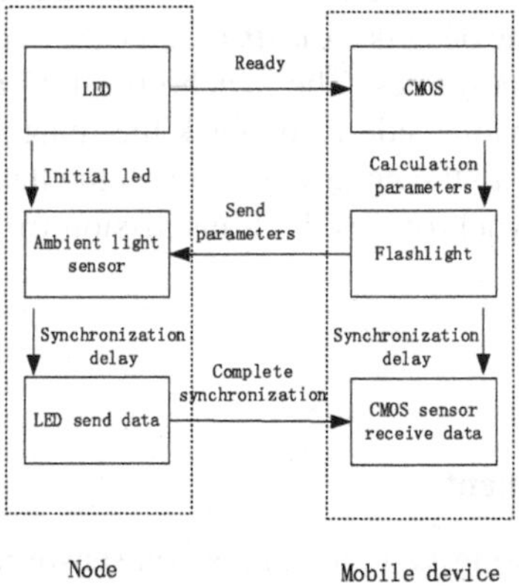

Fig. 9. Initial synchronization diagram

In the initial stage, we have a lot of preparations to do. When the node is in the ready state, the three LEDs are always on, the smartphone takes pictures and collects and determines the value of the synchronization parameter according to the result of the image processing. The synchronization parameter is shown in Fig. 10. The smartphone calculates the effective area width according to the position of the LED light in the image, which is marked as X. Then the smartphone will send the parameter code to the node end, and the node end can calculate the LED emission frequency suitable for the effective area width according to the parameter.

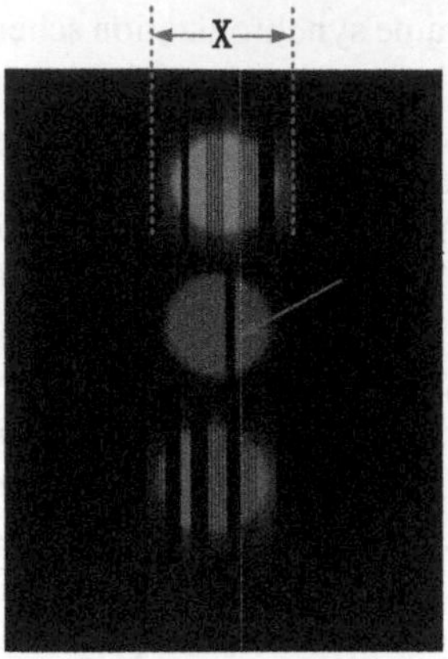

Fig. 10. Transmission parameter diagram

After the initial synchronization, the node LED repeatedly sends the corresponding data frame within each image frame's duration. However, due to the unstable frame rate, the data stripes will shift in the image frame. When the offset exceeds one image frame's duration, the same image frame may contain two adjacent data frames. The green light is used as the detection of the synchronization state. Whenever a data frame is sent, the green light will send a low pulse to separate two adjacent data frames. When the green light in the image frame detects dark stripes, the data collected in an effective area containing two adjacent data frames. The synchronization has failed and needs to be resynchronized. The smartphone side feeds back the image frame's sequence number with synchronization failure to the node side through the flashlight. After the node side receives and decodes, it resynchronizes the transmission from the data frame indicated by the sequence number.

6 Evaluation

6.1 Experiment Environment

Based on the Android operating system and general sensor nodes, we realize a HDDIS using a commercial off-the-shelf Android smartphone and popular sensor nodes TelosB. In this system, smartphone need to be equipped with a flashlight, optical camera, and its installed Android support for Camera2 API. The TelosB owns RGB-LED and ambient light sensor.

6.2 Experiment Content

We conduct a series of experiments about updating code blocks from five different functions shown in Table 2. The Timer and LS Ctrl function relative to controlling sensor node. Bubble sort, Dijkstra, and Horspool is sort of algorithms.

As shown in Fig. 11, we compared the compressed code size with the original code, and the compression rate of the five code blocks is given in Table 2. It is obviously that the size of the code block is reduced as least 8%, especially in LS Ctrl Fun with 20% reduction. Due to more reused bits from the opcode of single operand instructions than multi-byte instructions, it can achieve better effect that applying our compression scheme in the code block including more single operand instructions, such as Timer Fun and Bubble Sort.

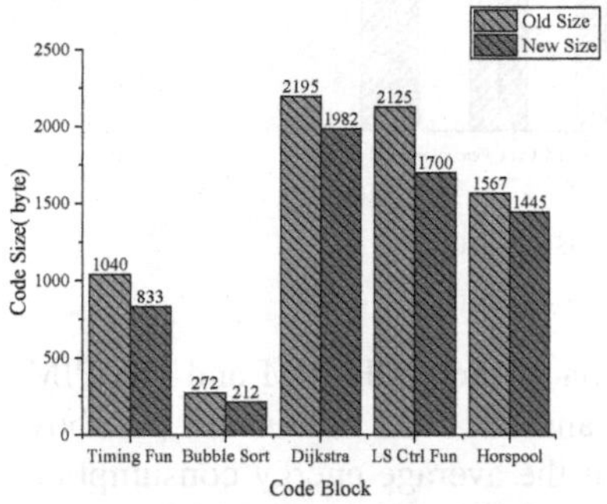

Fig. 11. Compressed Code Size

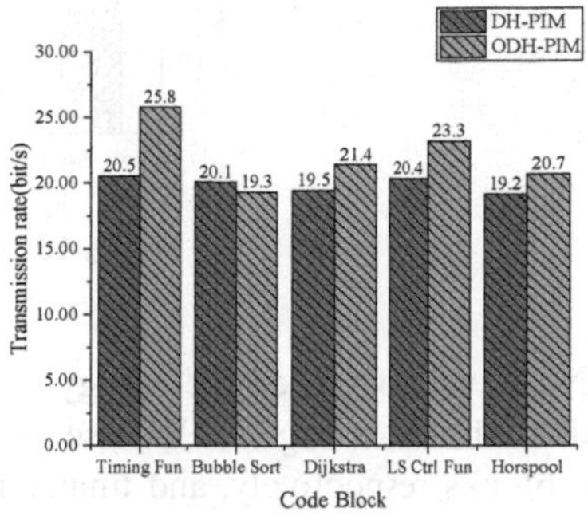

Fig. 12. Transmission rate

The transmission rate of ODH-PIM is given in Fig. 12. We compared ODH-PIM with DH-PIM. In order to neatly compares, these two methods are used to transmit origin code without compression. In the most cases of transmitting code blocks, our method is better than DH-PIM. Especially in Timing Fun case, ODH-PIM is 25.8% faster than DH-PIM because ODH-PIM can effectively reduce the length of a symbol. In Bubble Sort case, we found a large number of consecutive zero existing in binary codes which ODH-PIM hard to reduce, and the transmission rate of ODH-PIM remains about 3.9% below DH-PIM.

Table 2. Compression Rate

Code Block	Brief	Size(bit)	Compression Rate	Complete time(s)
Timer Fun	Control timer	9360	80.10%	364.2
Bubble Sort	Data sorting	2128	77.94%	118.7
Dijkstra	Find the shortest path	17560	90.30%	821.5
LS Ctrl Fun	Control light sensor	17000	80%	732.3
Horspool	Character match	12536	92.21%	606.4

We also apply the binary code compression and ODH-PIM at the same time, and record the transmission completion time shown in Table 2. Table 2 list the transmission completion time of each code block. The instruction compression based on bitmask reduces the size of the code block and ODH-PIM shorten the length of each symbol. Both methods guarantee that the commercial off-the-shelf devices complete the debugging interaction by VLC.

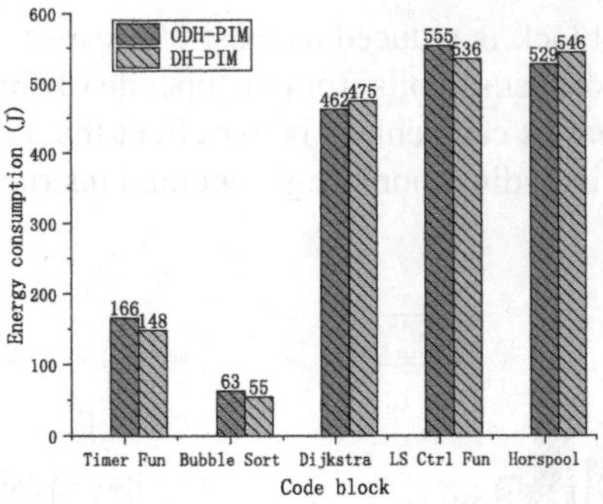

Fig. 13. Energy consumption

Next, we compared the energy consumption of the ODH-PIM and DH-PIM. The result is shown in Fig. 13. We use ODH-PIM and DH-PIM to transmit the above five code blocks respectively, and finally figure out the average energy consumption. The DH-PIM consumes 352 J of energy. However, ODH-PIM consumes 355 J energy. Since realizing the LED half-on state in ODH-PIM requires flashlight remaining a duration of high-speed flash, it will pay extra 7.36% energy consumption than DH-PIM. However, the energy consumption of flashlight is from smartphone and the extra part is slight comparing with the transmission rate improvement by LED half-on state.

The bit error rate (BER) is an important fact to evaluate performance of ODH-PIM. Figures 14 and 15 demonstrate the BER of ODP-PIM with different angles and distances. In the downlink, different distances between the flashlight to the ambient light sensor result in different light intensities captured by the sensor. From Fig. 14, it is obviously that BER is very high if the transmission distance is over 20 cm. The reason is that the sensor is difficult to detect the LED half-on state once exceeding 20 cm and BER boost due to lack of the state.

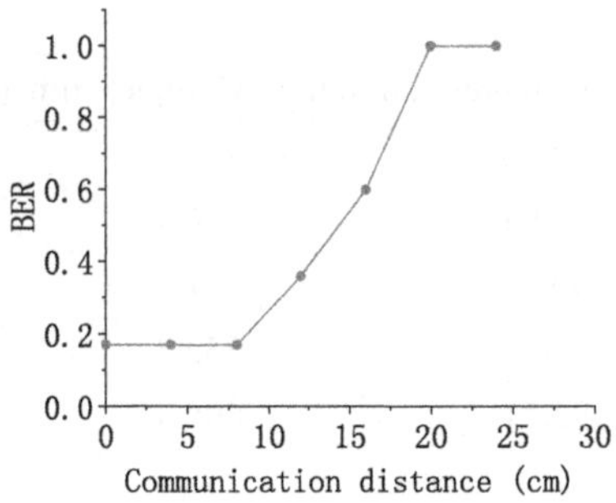

Fig. 14. BER under different distances

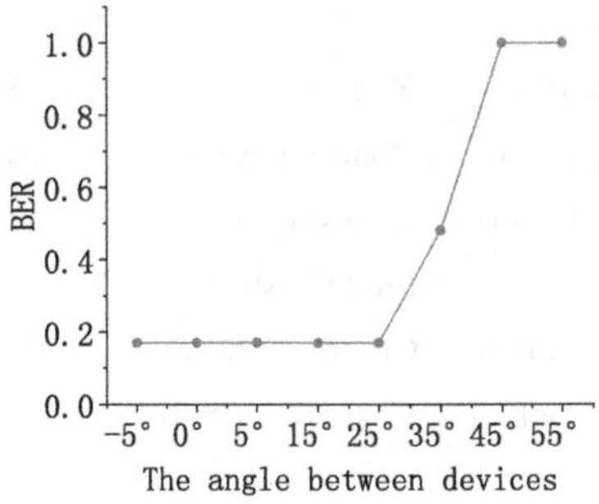

Fig. 15. BER under different angles

Because VLC owns the directionality character, the transmission angle is also factor to impact performance of transmission. As shown in Fig. 15, when the angle between both devices exceeds 45°, VLC is unavailable using smartphone and TelosB sensor. The angle also affects the light intensity collected by the sensor. When the angle is less than 25°, the sensor is able to capture enough light intensity from the flashlight, so the VLC can work with a low BER. When the angle is between 25° and 45°, the flashlight provide enough light intensity for communication, and that leads to boost BER.

The debugging information stroboscopic sequence captured by the optical camera is saved as an image for decoding in the uplink. However, its own hardware parameters, such as exposure time, will also affect imaging. The experimental results are shown in Fig. 16. When the exposure time is the minimum exposure time of 10 microseconds because each line's exposure time is short, the transition zone around the effective area is narrow. The high-frequency frame header is easy to identify decode. When the exposure time is increased to 100 microseconds, it can be clearly observed that the light information is also collected outside the LED's actual light-emitting area, and the transition band is wider, which affects the identification and detection of the high-frequency frame head. On the other hand, larger exposure time can increase the width of the effective area, and at the same time increase the data capacity of a single frame of data. So the appropriate exposure time can be set according to actual hardware conditions to achieve the best communication effect.

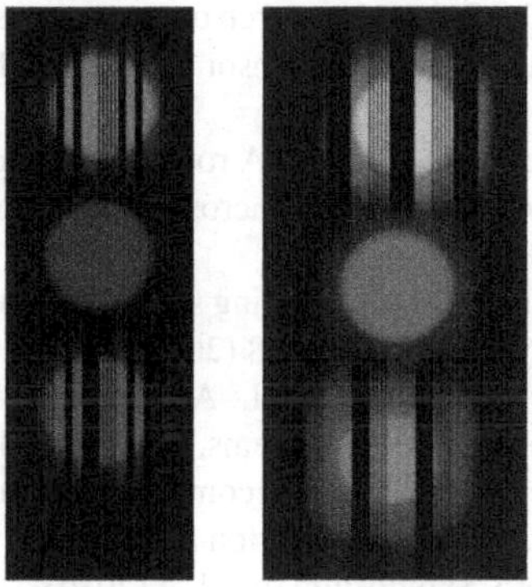

Fig. 16. Comparison of imaging under different exposure time conditions

7 Conclusion

In this paper, we apply VLC-based debugging interaction applying in the sensor nodes which malfunction caused by radio communication failure, and further design a hybrid duplexing debugging interaction system based on commercial off-the-shelf device. This system adopts different visible light sensing means in the physical layer of uplink and downlink. In view of the unstable frame rate of general CMOS optical cameras, we also designed a frame synchronization scheme to ensure fast and reliable transmission of debugging data. Due to transmitting debugging binary code with downlink, two methods from source and channel coding are proposed to improve downlink transmitting performance. For source coding, the character of transmitted binary code makes bitmask-based instruction compression available to reduce the size of transmitted data.

For channel coding, we optimize DH-PIM by introducing LED half-on state, and put forward Overlap DH-PIM to increase the transmission rate. Experimental results demonstrate the validity of the above two methods. In future work, by adding error correction in VLC, the transmission rate can be improved. In addition, the drone equipping flashlight and camera owns the potential capacity of VLC. By sending the drone to place nearby the lost sensor node, it extends VLC-based debugging interaction to most of field-deployed application.

Acknowledgment. This research work is supported by Zhejiang Provincial Natural Science Foundation of China under Grant (LY20F020026).

References

1. Li, D., Liu, W., Cui, L.: EasiDesign: an improved ant colony algorithm for sensor deployment in real sensor network system. In: Proceedings of IEEE GLOBECOM, pp. 1–5 (2010)
2. Du, W.Z., Chen, H.M., Li, D., Cui, L.: Research on a gateway switching mechanism based on gateway's energy-harvesting for wireless sensor networks. Chin. High Technol. Lett. **26**(6), 631–642 (2016)
3. Shah, P.A., Awan, K.M., Rehman, Z., et al.: A route optimized distributed IP-based mobility management protocol for seamless handoff across wireless mesh networks. Mob. Networks Appl. **23**(4), 752–774 (2018)
4. Liu, J., Chung, S.H.: An efficient load balancing scheme for multi-gateways in wireless mesh networks. J. Inf. Process. Syst. **9**(3), 365–378 (2013)
5. Shao, S., Khreishah, A., Rahaim, M.B., et al.: An indoor hybrid WiFi-VLC internet access system. In: Mobile Ad Hoc and Sensor Systems, pp. 569–574 (2014)
6. Fan, L., Liu, Q., Jiang, C., et al.: Visible light communication using the flash light LED of the smart phone as a light source and its application in the access control system. In: 2016 IEEE MTT-S International Wireless Symposium, pp. 1–4 (2016)
7. Das, S., Chakraborty, A., Chakraborty, D., Moshat, S.: PC to PC data transmission using visible light communication. In: Computer Communication and Informatics, pp. 1–5 (2017)
8. Adiono, T., Pradana, A., Putra, R.V.W., et al.: Analog filters design in VLC analog front-end receiver for reducing indoor ambient light noise. In: IEEE Asia Pacific Conference on Circuits and Systems, pp. 581–584 (2017)
9. Chen, P.Y., Wu, C.C., Jiang, Y.J.: Bitmask-based code compression methods for balancing power consumption and code size for hard real-time embedded systems. Microprocess. Microsyst. **36**(3), 267–279 (2012)
10. Kumar, R.N., Chandran, V., Valarmathi, R.S., et al.: Bitstream compression for high speed embedded systems using separated split look up tables (LUTs). J. Comput. Theor. Nanosci. **15**(5), 1719–1727 (2018)
11. Seong, S.W., Mishra, P.: Bitmask-based code compression for embedded systems. IEEE Trans. Comput. Aided Design Integrated Circuits Syst. **27**(4), 673–685 (2008)
12. Haider, S., Nazhandai, L.: A hybrid code compression technique using bitmaskand prefix encoding with enhanced dictionary selection. In: Proceedings of Conference on Compilers, Architecture, and Synthesis for Embedded Systems, pp. 58–62. ACM, New York (2007)
13. Ghassemlooy, Z., Hayes, A.R., Seed, N.L., et al.: Digital pulse interval modulation for optical communications. IEEE Commun. Mag. **36**(12), 95–99 (1998)

14. Aldibbiat, N.M., Ghassemlooy, Z., McLaughlin, R.: Error performance of dual header pulse interval modulation (DH-PIM) in optical wireless communications. IEEE Proc Optoelectron **148**(2), 91–96 (2001)
15. Yao, K., Wu, N., Wang, X., et al.: A novel power efficient modulation scheme for VLC systems. In: 2016 IEEE/CIC International Conference on Communications in China (ICCC). IEEE (2016)
16. Ali, A.Y., Zhang, Z., Zong, B.: Pulse position and shape modulation for visible light communication system. In: International Conference on Electromagnetics in Advanced Applications. IEEE (2014)
17. Fuzile, T.S., Heymann, R.: Investigating a visible light communication channel with an LED transmitter and a smartphone receiver. In: IEEE Africon 2017 Proceeding, pp. 354–358 (2017)

Attacking the Dialogue System at Smart Home

Erqiang Deng[1,2](✉), Zhen Qin[1,2,3], Meng Li[1,2,3], Yi Ding[1,2,3,4], and Zhiguang Qin[1,2,3]

[1] School of Information and Software Engineering, University of Electronic Science and Technology of China, Chengdu, China
dylandeq@outlook.com

[2] Network and Data Security Key Laboratory of Sichuan Province, University of Electronic Science and Technology of China, Chengdu, China

[3] Institute of Electronic and Information Engineering of UESTC, Chengdu, China

[4] Institute of Electronic and Information Engineering of UESTC in Guangdong, Guangdong 523808, China

Abstract. Intelligent dialogue systems are widely applied in smart home systems, and the security of such systems deserves concern [1, 2]. In this paper, we design a threatening scenario of dialogue systems at a smart home. A trojan robot is disguised as one part of the whole system but generates dialogue adversarial examples to attack the normal robots according to the information of users. To achieve the goal in such a scenario, the responding speed, the correctness of the grammar, and the consistency of semantic is necessary. Based on these requirements, we propose a novel method named Attention weight Probability Estimation Attack (APE) to allocate the keys words in dialogue and substitute these words with synonyms in real-time. We perform our experiments on popular classification datasets in the DNN model, and the result shows that APE effectively attacks the system with low responding time and a high success rate.

Keywords: Smart home · Security · Dialog system · Adversarial example

1 Introduction

In the scenario of smart home, voice interaction has been the optimal choice, because of the natures of short responding time, fluency of interaction and the convenience of operation [3]. These voice interaction systems are composed of speech signal processing system (SSP) and natural language processing system (NLP). In the front end, the SSP system samples the original information and convert the frequency signals to text information [4], and in the back end, the NLP system recognizes the text information and generates corresponding feedback to users, with comprehending the semantic information of input. In both SSP and NLP parts, deep neural network (DNN) is mainstream technology.

However, previous works have shown the vulnerability of DNN. Szegedy et al. [5] firstly added small perturbations to the input images, which led to the misjudgment of the DNN model with high classification confidence. Jia and Liang [6] deceived the reading

H. Gao et al. (Eds.): CollaborateCom 2020, LNICST 349, pp. 148–158, 2021.
https://doi.org/10.1007/978-3-030-67537-0_10

comprehension model by changing a few words in a paragraph. With the deployment of the dialogue system at smart homes, the latent risk of security is inevitable and deserves concern. Based on this consideration, we design this paper to research the security issue and try to expose the vulnerability of such systems from the perspective of an attacker.

To achieve this goal, we design a Trojan scenario [7] at a smart home. As shown in Fig. 1, robot No.1 is a Trojan invader. It receives the input of the user and generates an adversarial example to attack the normal robot No.2, leading it misclassify the information, although only a few words are changed and the semantic information seems consistent with the original input. The existence of the Trojan invader is possible if the number of robots at a smart home is large.

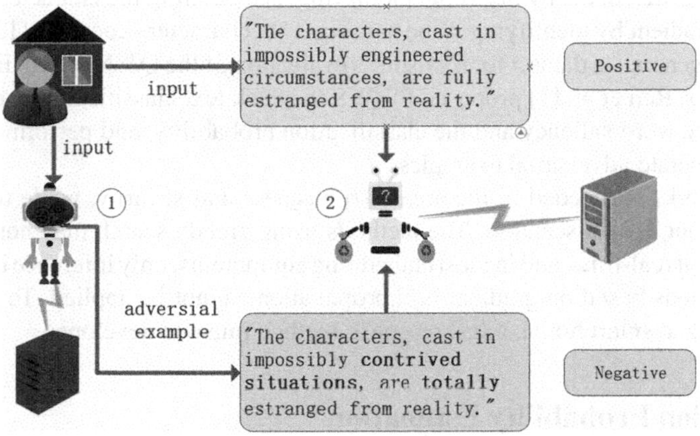

Fig. 1. The Trojan scenario at smart home. Robot No. 1 is a invader, it receives the information of user and tries to attack the normal robot No. 2 by crafting adversarial examples.

In this paper, only the NLP part of the dialogue system is involved, because the complexity and variability of language are contained in text, which makes the NLP system more vulnerable to be attacked. We performed our experiments on IMDB sentiment classification tasks and generated adversarial examples to attack the DNN models. In such a Trojan scenario, the responding speed is critical to perform an attack, and the grammar correctness and the semantic consistency are also important to disguise itself from being perceived by humans. To satisfy these requirements, we propose a novel method based on the attention mechanism, which indicates the distribution of importance of the words in a paragraph. We estimate the probability of a successful attack by attention distribution and select the words with high attention weights to be substituted. Our main contributions are:

Proposing a novel method to estimate the probability of successful attack in text.

Proposing an algorithm to search possible adversarial words in short time.

Designing a possible scenario of adversarial attack in physical world and craft adversarial examples in real-time.

2 Related Work

Szegedy et al. [5] proposed the notion of adversarial attack in the area of computer vision (CV). They added perturbation into images which were negligible to human but misguided the DNN. Szegedy developed this branch of security in DNN, and many works were engaged in this area. The attack methods of FGSM [8], JSMA [9], and C&W [10] were proposed and the defensive methods of adversarial training and distillation followed in CV.

In text area, Papernot et al. [11] calculated the forward derivative, i.e. Jacobian to generate adversarial text sequences on RNN. TextFool [11] used the concept of FGSM and USES backpropagation to calculate the cost gradient of each training sample. The classifier error was made by inserting, modifying, and deleting the character containing the largest gradient by identifying it and naming it hot character. Xue et al. [13] proposed a method to generate a dataset for robustness evaluation of the Q&A system in the black-box scenarios. Ren et al. [1] proposed PWWS to attack text classification models. They calculated the word saliency and the classification probability, and performed a greedy search to generate adversarial examples.

These works succeeded in the area of robustness and security, while they are not suitable for our Trojan scenario. The methods using greedy search may not satisfy the requirement of real-time, and in most embedding equipments, only inference is supported and the methods based on gradient backpropagation cannot be applied. To perform an in time attack at smart home, a fast response method must be developed.

3 Attention Probability Estimation

The attention mechanism [14] and its upgraded technologies [15, 16] have boosted performance on a range of NLP tasks. The attention weights explicitly reflect representations of input components, and these weights can be used to identify the importance of the input from the perspective of statistical distribution. Based on this feature of the attention mechanism, we approximate the attention weights distribution as the saliency map of input when we deploy a real-time attack, to reduce the expense of gradient calculation of backpropagation.

3.1 Adversarial Attack in Text

Suppose a text sequence $x = \{\omega_1, \omega_2 \ldots \omega_n\}$, where $\omega_i \in D$, and D is the space of dictionary. A DNN model receives the input x and outputs the prediction of classification of x, $F(x) = y$.

An adversarial example is defined as:

$$F(x) = y \text{ and } F(x') \neq y \tag{1}$$

where $x \approx x'$. It indicates that the adversarial example x' is similar to x and the difference between x and x' is negligible for humans, but it makes model F misjudge the information.

3.2 Attention Based Classification

Attention Mechanism
The attention mechanism is designed to focus on the key parts of a whole sequence, to strengthen the ability to abstract the most important information. Mathematically, the attention mechanism is a weights-refreshed affine transformation. It can be expressed as:

$$f(x) = \tilde{A}x + b = \sum_{\mathrm{i}} \alpha_{\mathrm{i}} x_i + b \tag{2}$$

Different from the parameters of full connection layer in DNN, the α weights are not directly updated by backpropagation, but are determined by query-key calculation:

$$\alpha_{\mathrm{i}}(Q, K) = \mathrm{softmax}(q_i^T k_i) \tag{3}$$

The query-key pair with a higher product of multiplication will be allocated a higher weight in affine transformation. The attention mechanism shows its effectiveness in the scenario of long sequences to solve the gradient vanishing problem (Fig. 2).

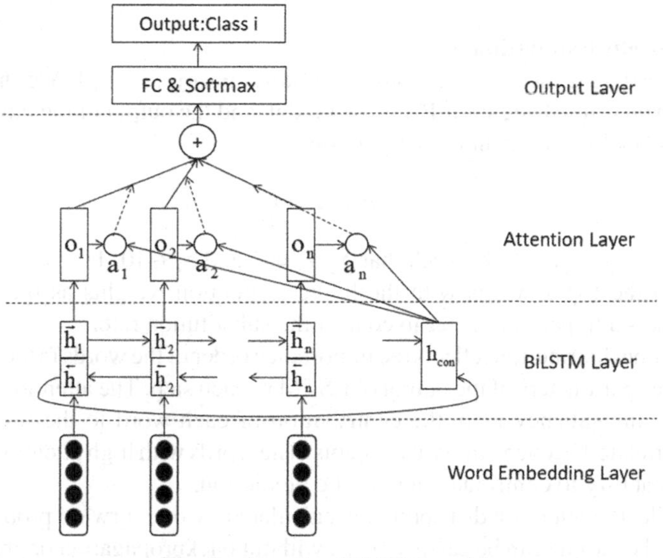

Fig. 2. Attention BiLSTM model.

Self-attention in LSTM Model
A self-attention BiLSTM model is applied to extract features. $\overrightarrow{\mathrm{h}}_{\mathrm{i}}$ and $\overleftarrow{\mathrm{h}}_{\mathrm{i}}$ is the hidden vectors of step i, o_i is the output of step i. h_{con} is the concatenated vector of $\overrightarrow{\mathrm{h}}_{\mathrm{n}}$ and $\overleftarrow{\mathrm{h}}_1$,

where n means the numbers of the input sequence. a_i is set as the weight of vector o_i, and is calculated by the function:

$$\alpha_i = \frac{\exp(o_i^T h_{con})}{\sum_{j=1}^{n} \exp(o_j^T h_{con})} \tag{4}$$

The hidden layer has the information of whole input, the output contributes more can be allocated a higher attention weight, because of the larger product of the vector multiplication. The final output of LSTM layer is:

$$output = \sum_{i=1}^{n} \alpha_i o_i \tag{5}$$

3.3 Word Substitution

In this paper, we perform the attack based on the synonym substitution and estimating the replacement order by the weight of attention. Our method is named Attention Weight Probability Estimation Attack (APE).

Attention Weight Estimation
We define a text sequence with n words $x = \{\omega_1, \omega_2, \omega_3, \ldots, \omega_n\}$. We input the text sequence to a pretrained attention-BiLSTM model, and the output of the model contains the predicted label y and attention distribution:

$$y, a = F(x) \tag{6}$$

where $a = \{\alpha_1, \alpha_2, \ldots, \alpha_n\}$ such that $\sum \alpha_i = 1$ and $\alpha_i \in [0, 1]$.

We select the top k words with the highest attention weights as the substitution objects, and k is a hyper-parameter to control the substitution rate.

The attention distribution reflects the importance order of the words in the sequence. It is the refreshed parameters of the output of LSTM in each step. The attention distribution doesn't equal the saliency map, the contribution of each word to the result, but it is positively correlated between the two concepts. The words with high attention coefficient has high probability to contribute more to the prediction.

Meanwhile, the attention distribution is calculated by one forward propagation, and the adversarial example can be easily crafted without backpropagation or greedy search. The requirement of real-time is the reason why we use attention distribution to determine the substitution order.

Word Substitution Strategy. Assuming that ω_i is the substitution word in x, we use the WordNet tool to build a set of synonyms of ω_i, and we define the set as:

$$S_i = \{\omega'_{i1}, \omega'_{i2}, \ldots, \omega'_{im}\} \tag{7}$$

Where m is the number of synonyms. A substitution candidate could be a single word or a short phrase and S_{ω_i} could be empty if no synonym is found. We use two strategies to perform the substitution, the first one is Optimal Greedy Substitution and the second one is probability estimation based on attention distribution and beam search.

Optimal Greedy Substitution.
When k substitution words in a text x are determined by attention distribution, we replace the words in the order of attention weights. For one word ω_i, we try to find the synonym which makes the text change most when substitution is performed. It can be expressed as:

$$\omega_i* = \underset{\omega_i' \in S_i, y' \neq y}{\arg\max} \left(P(y'|x')\right) \tag{8}$$

Where:

$$x' = \{\omega_1, \omega_2, \ldots \omega_i' \ldots \omega_n\} \tag{9}$$

When ω_i is replaced by ω_i*, the same strategy is repeated on the next candidate word. After all the k words are substituted, an adversarial example is crafted:

$$x* = \{\omega_1, \omega_2, ..\omega_i * ..\omega_j * ..\omega_k * ..\omega_n\} \tag{10}$$

This new text sequence has a relatively high probability of a successful attack, but it's not time efficient enough because of the greedy search.

Algorithm 1

Require: Generating adversarial text x' in real time.
1. Input the original text x to the Trojan model F(self-attention BiLSTM)
2. Get the predicted label of original input: F(x)=y
3. Get the attention weight distribution of input of each word.
4. Select k words with high attention weight
5. **for** i=1 to k:
6. Choose the word with i-th highest attention weight, which is ω_i.
7. Get the substitution set $S_i = \{\forall \omega'_{ij}\}$, where ω'_{ij} are m synonyms of ω_i.
8. Combine the synonym words in as a string ss_i and input it into F.
9. Get the attention weight distribution of ss_i.
10. Select q words with high attention weight $\omega*_i$
11. Generate $x'(i)$ by replacing ω_i in $x'(i-1)$ with $\omega*_i$
12. **if** $F(x'(i)) \neq y$:
13. $x' = x'(i)$
14. **end for**
15. **end for**

Attention Based Substitution Search
This method is also based on attention distribution. When a synonym set of a word is built, we connect these synonym words into a new string and put it into the attention BiLSTM model. Then the output of substitution combination is:

$$y_s, A_s = F(S_{\omega_i}) \tag{11}$$

where $S_{\omega_i} = \{\omega'_{i1}, \omega'_{i2}, \ldots, \omega'_{im}\}$, y_s is the prediction of the model, and A_s is the attention distribution of these words. The combination of these words is meaningless but the attention distribution probably reflects the priority of their contribution to the predicted result y_s. If $y_s = y = F(x)$, the words with low attention weight are selected first to perform an attack, and if $y_s \neq y$, the words with high attention weight are prior choices.

Then we use a heuristic search to find a locally optimal solution. Every circular is a synonym of the word ω_i, and the darker color means a better attack expectation. In each first step, we select the top q synonym words for substitution. Except for the first and last steps, there are q^2 choices to replace the relative words in original input x. We select q best choice to calculate the next step. At last, one best sequence is determined and the adversarial examples are crafted. An example with q = 2 is shown in Fig. 3, and the procedure is shown in Algorithm 1.

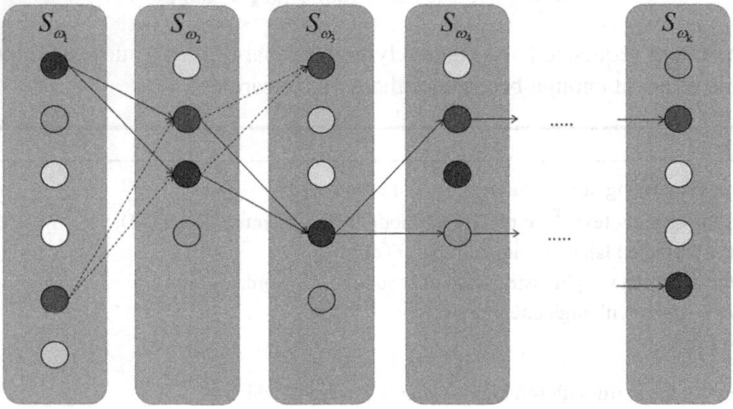

Fig. 3. Synonym Words Attention based heuristic search.

4 Experiment

4.1 DataSets and Deep Neural Models

IMDB. IMDB sentiment is a binary sentiment(positive and negative) classification dataset. It collects highly polar movie reviews from the IMDB, the world's most popular and authoritative website for movies. In this dataset, a set of 25,000 comments is provided for training, and 25,000 for testing.

Attn Bi-LSTM. The word embedding dimension is set as 200, and 128 units in BiLSTM layer. An attention layer is added to refresh the weights of output. A full connection layer with softmax is used to get the result. The framework of Attn-BiLSTM is shown in Fig. 1, and the original accuracy of Attn-BiLSTM is 90.6%.

4.2 Attention Weight V.S. Saliency Map

The saliency map is the contribution of the words to the predicted class. Previous work WS and PWWS estimated the order of importance by saliency map. Attention weight is an inner parameter of the BiLSTM, and the feature with high attention weight will contribute more to the prediction result. Figure 4 and Fig. 5 show that the saliency map and attention weight have a positive correlation.

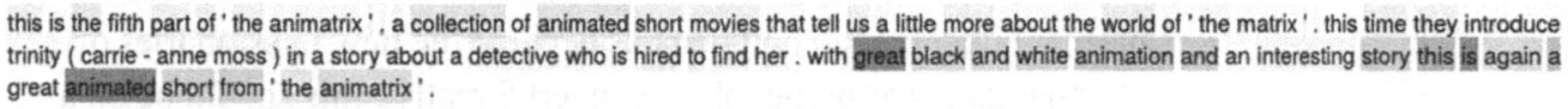
this is the fifth part of ' the animatrix ' , a collection of animated short movies that tell us a little more about the world of ' the matrix ' . this time they introduce
trinity (carrie - anne moss) in a story about a detective who is hired to find her . with great black and white animation and an interesting story this is again a
great animated short from ' the animatrix ' .

Fig. 4. The effect of Saliency Map.

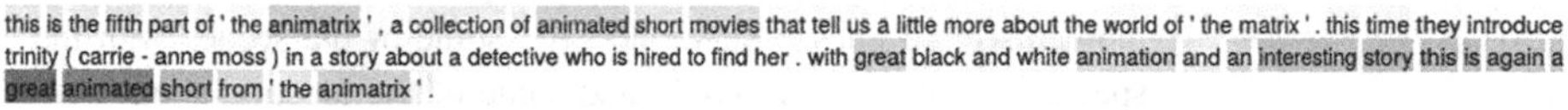
this is the fifth part of ' the animatrix ' , a collection of animated short movies that tell us a little more about the world of ' the matrix ' . this time they introduce
trinity (carrie - anne moss) in a story about a detective who is hired to find her . with great black and white animation and an interesting story this is again a
great animated short from ' the animatrix ' .

Fig. 5. The effect of Attention weight distribution.

4.3 Attacking Result

We experimented by controlling the number of substitution words. In some cases, even a text is long, a modification of one word changed the result. Examples are shown in Table 1. In 'Text information' column of this table, the back words are the original material of the dataset. The green ones are the selected words by our method and the red ones in brackets are the substitution counterparts. When the adversarial examples are input to the same model, the polarities of these texts are changed, as shown in the first two columns.

Table 1. APE attack results.

Original Result	Adversarial Result	Text information
Negative	Positive	i was very excited when paranormal state first came on a&e. i thought that it may bring some more interesting ghostly evidence . the production value looked good and i really love the logo . then , after about few episodes in , i started to feel that this show may not be looking for evidence but had a strong religious agenda.it seems like every case they investigate has some big powerful evil demon that can't even make a teacup move on camera , yet everyone is terrified . then comes some power of christ ritual that saves every-one.also , there is very little focus on other members of the team . the entire show focuses on ryan and he feels like one of those people that hands you pamphlets about his church on the street.has para-normal phenomenon and demons become the new missionaries of christianity , scaring people to convert ? really , this should be on a christian network . i was very disappointed(defeated).
Negative	Positive	Primary(chief) plot!primary direction! poor interpretation
Positive	Negative	i saw this movie(pic) last night and thought it was decent . it has it 's moments(bit) i guess you would say . some of the scenes with the special ops forces were cool , and some of the location shots were very authentic . i wo n't be putting this movie in my dvd collection but it is fair enough to recommend for renting . i guess nothing set the movie at another level compared to others of the same genre . the action is good , the acting is decent , the women are extremely seductive and exotic in my opinion , and the story is pret-ty(middling) interesting . 7 out of ten "

We compare four substitution method. The first one is the Random method which replaces K words in a text randomly. PWWS [1] method calculate the saliency map and replace the word in the order of saliency value. APE + GS is our method attention probability estimation, and the synonyms are tested by a greedy search. APE + ABS means the Attention-based Substitution Search mentioned in Sect. 3.3. K is the max number of substitution words in a text. The results are shown in Table 2. It can be concluded that the attack effect of APE + GE is significant especially when the K is large.

Table 2. Substitution attack comparison.

Prediction accuracy	K = 1	K = 5	K = 10	K = 15	K = 20
Self-attn BiLSTM	90.6%				
Random	88.1%	84.2%	82.4%	78.8%	75.5%
PWWS	**83.4%**	72.7%	57.9%	45.0%	42.0%
APE + GS	83.8%	**63.1%**	**38.8%**	**37.1%**	**15.9%**
APE + ABS	85.3%	72.8%	65.1%	55.1%	52.3%

And the curves of more tested data are shown in Fig. 6.

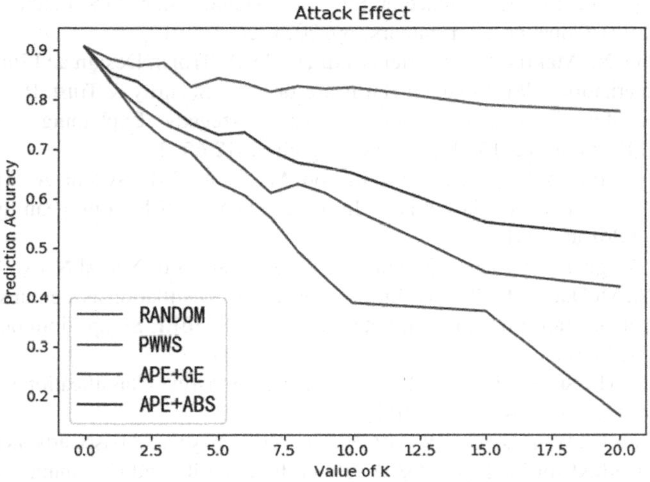

Fig. 6. The comparison of attack effect of four methods.

5 Conclusion

In this paper, we study the security issue of smart home. We design a possible trojan scenario of attacking the dialogue devices or robots at smart home, and we propose a novel method based on attention distribution to craft textual adversarial examples and an algorithm to improve the responding speed. The result shows that it works and exceeds the previous substitution method. However, the transferibility of the method should be considered and the substitution distribution could be predicted in a more precise way, these issues should be concerned in our future work.

Acknowledgement. We thank the anonymous reviewers for their insightful comments on the preliminary version of this paper. This work is supported by the Natural Science Foundation of Guangdong Province (Grant No. 2018A030313354). Any findings, opinions, or conclusions in this paper are those of the authors and do not reflect the views of the funding agency.

References

1. Ren, S., Yihe, D., Kun, H., Che, W.: Generating Natural Language Adversarial Examples through Probability Weighted Word Saliency. ACL (2019)
2. Copos, B., Levitt, K., Bishop, M., Rowe, J.: Is Anybody Home?. Inferring Activity From Smart Home Network Traffic. Security & Privacy Workshops, IEEE (2016)
3. Kwabena, O.-A., Qin, Z., Zhuang, T., Qin, Z.: MSCryptoNet: Multi-Scheme privacy-preserving deep learning in cloud computing. IEEE Access **7**, 29344–29354 (2019)
4. Awni, H., Case, C., Casper, J., Catanzaro, B., Diamos, G.: Deep speech: scaling up end-to-end speech recognition. Computer Science (2014)
5. Szegedy, C., Zaremba, W., Sutskever, I., Bruna, J.: Intriguing properties of neural networks. In: Proceedings of the 2nd International Conference on Learning Representations (ICLR 2014) (2014)
6. Jia, R., Liang. P.: Adversarial examples for evaluating reading comprehension systems. In Proceedings of the 2017 Conference on Empirical Methods in Natural Language Processing (EMNLP 2017). Copenhagen, Denmark, pp. 2021–2031 (2017)
7. Jin, Y., Kupp, N., Makris, Y.: Experiences in Hardware Trojan Design and Implementation. In: IEEE International Workshop on Hardware-oriented Security & Trust. IEEE (2009)
8. Ian, J.: Goodfellow, Jonathon Shlens, Christian Szegedy. Explaining and Harnessing Adversarial Examples (2015). https://arxiv.org/abs/1412.6572
9. Papernot, N., McDaniel, P., Jha, S., Fredrikson, M., Celik, Z.B., Swami, A.: The limitations of deep learning in adversarial settings. In: Proceedings of IEEE European Symposium on Security and Privacy (2016)
10. Carlini, N., Wagner, D.: Towards Evaluating the Robustness of Neural Networks (2016)
11. Papernot, N., McDaniel, P., Wu, X., Jha, S., Swami, A.: Distillation as a defense to adversarial perturbations against deep neural networks. In: 2016 IEEE Symposium on Security and Privacy (SP) (2016)
12. Liang, B., Li, H., Su, M., Bian, P., Li, X., Shi, W.: Deep Text Classification Can be Fooled. arXiv preprint arXiv:1704.08006 (2017)
13. Xue, M., Yuan, C., Wang, J., Liu, W.: DPAEG: a dependency parse-based adversarial examples generation method for intelligent Q&A robots. In: Security and Communication Networks 2020 (2020): 5890820:1-5890820:15
14. Bahdanau, D., Cho, K., Bengio, Y.: Neural machine translation by jointly learning to align and translate. arXiv preprint arXiv:1409.0473 (2014)
15. Devlin, J., Chang, M.W., Lee, K., et al.: Bert: Pre-training of deep bidirectional transformers for language understanding. arXiv preprint arXiv:1810.04805 (2018)
16. Vaswani, A., et al.: Attention is all you need. In: Advances in Neural Information Processing Systems, pp. 5998–6008 (2017)

Boosting the Performance of Object Detection CNNs with Context-Based Anomaly Detection

Jan Blaha(✉), George Broughton, and Tomáš Krajník

Artificial Intelligence Center, FEE CTU, Prague, Czech Republic
{jan.blaha,george.broughton,tomas.krajnik}@fel.cvut.cz

Abstract. In this paper, we employ anomaly detection methods to enhance the ability of object detectors by using the context of their detections. This has numerous potential applications from boosting the performance of standard object detectors, to the preliminary validation of annotation quality, and even for robotic exploration and object search. We build our method on autoencoder networks for detecting anomalies, where we do not try to filter incoming data based on anomality score as is usual, but instead, we focus on the individual features of the data representing an actual scene. We show that one can teach autoencoders about the contextual relationship of objects in images, i.e. the likelihood of co-detecting classes in the same scene. This can then be used to identify detections that do and do not fit with the rest of the current observations in the scene. We show that the use of this information yields better results than using traditional thresholding when deciding if weaker detections are actually classed as observed or not. The experiments performed not only show that our method significantly improves the performance of CNN object detectors, but that it can be used as an efficient tool to discover incorrectly-annotated images.

Keywords: Anomaly detection · Object detection · Context-aware neural networks · Explainable neural networks · Autoencoders

1 Introduction

The development of deep-learning (DL)—as a subset of the machine learning field (ML)—in the last few years has brought great advances in the field of artificial intelligence (AI), and with it the spread of such deep neural networks into life-critical domains. That places a great deal of pressure onto the reliability, efficiency, and also on the careful preparation of datasets for a large number of individual applications. Moreover, these requirements are even stronger when the implementation is deployed in systems which interact with the physical world, since they can cause injuries to people around them as well as cause damage to property and the robot itself.

H. Gao et al. (Eds.): CollaborateCom 2020, LNICST 349, pp. 159–176, 2021.
https://doi.org/10.1007/978-3-030-67537-0_11

We believe that addressing the problem of contextual understanding can be useful when considering the above-stated concerns when dealing with perception. A system capable of taking into account context can often be much more effective in its tasks. This has some biological foundations, with strong evidence implying that humans similarly depend on this kind of contextual reasoning when performing tasks such as object recognition[3]. Greater reliability comes with greater accuracy, but it is also connected to context modelling by research [9,11], and also with the explainability of decisions that the system makes.

The complexity of deep neural networks has meant that the question of explainability is still open to research, but explainable context modelling would necessarily help the system to present its decisions in a human understandable way. In this paper, we present the idea that an important part of context modelling is the ability to spot anomalies, which is a well studied problem, and deep learning has been applied to it in the past [32].

Apart from not being able to reason about the decisions of the employed system, the training of large deep architectures is tedious and time-consuming. This is mainly because of the preparation required for the diversity, relevance and size of the training datasets, and most importantly, the quality of their annotation. We show that the use of anomaly detection addresses both of the aforementioned problems: it can spot potential detection errors of the networks, as well as identify incorrectly annotated images. If the system is able of spotting anomalies in the incoming data, it can flag wrongly annotated data in automatic dataset annotations, which then puts less pressure on human annotators or people that manually check the results. Even datasets completely annotated by humans are prone to errors, so double-checking of suspicious results can be useful as well. Better datasets inevitably lead to better models, and possibly to positive impacts on the length of the learning process, which can lead to money-saving due to the high energy consumption of datacenters used for the training of such huge models.

The ability to detect potential mistakes in one's output is a necessary precursor to automatically correcting one's mistakes, which, in turn, leads to higher accuracy and efficiency in the task performed. In another scenario, we might assume that our detection systems output is reliable, and the knowledge of anomalous outputs can be used for different tasks like robotic spatial or spatio-temporal exploration as indicated in [24,25]. Here, a high anomaly score indicates the possibility of object presence at a given location even if it is not currently visible, or can indicate that the detected object is at an unusual location. The work presented at [25] shows that the ability to detect these situations can be used to trigger robot's exploratory behaviours, which results in efficient spatio-temporal exploration [25].

This is in contrast to other currently existing DL-based models, which incorporate context modelling into their object detecting architectures, like [15]. Our approach is based on two standalone methods, one of which detects objects and the other is handling the context modelling through anomaly detection. This brings not only advantages in terms of being able to reason about the decisions of such a combined model, but also minimal overhead to the training and deployment of the context-aware detection system.

Our method for anomaly detection builds upon existing research of using autoencoders for this task. We present a proof-of-concept for how these could be used as an extension of existing capabilities of state-of-the-art neural network models for object detection. In our experiments, we look at how the autoencoder can be attached to the output of the YOLOv3 object detection neural network [21]. Unlike the anomaly detection methods [1,22,33], who focus solely on the ability to detect anomalous images, our method is designed to indicate which actual feature causes the anomaly. We also provide a link to a repository containing all the code used for reproduction of the described methods and experiments [13].

To test the abilities of our method, we perform two experiments for finding positive anomalies, that is objects that are present but anomalous in the context, and negative anomalies, i.e. objects that are expected to be present but are not. These experiments are performed over annotations present in the Microsoft COCO dataset [18], and they resulted in flagging up several potentially wrong annotations, such as those shown in Fig. 1. Finally, we combine our method for anomaly detection with the YOLOv3 network and test its accuracy against the standalone YOLOv3, with our system determining whether low probability detections should be boosted up in confidence given the context of the surrounding detections. This experiment is based on comparing the mentioned methods by their accuracy on images in Microsoft COCO dataset, and we take the included annotations as ground truth.

2 Related Work

During the last few years, there has been much research conducted in the field of deep learning, particularly for the task of computer vision, making it impossible to cover the entire field in a short paper. In this section, we focus on two main topics relevant to this work—object detection networks and autoencoders with their use for anomaly detection.

2.1 Object Detection

Computer vision has advanced rapidly in the last decades due to the introduction of deep-learning into the field. Starting with AlexNet in 2012 [17], the progress quickly lead to the introduction of the exponentially bigger VGG architecture [27], GoogLeNet [28], and ResNet [14] networks. Once whole image classification was achieved, research efforts focused at the task of multiple object detection and localisation, such as You Only Look Once (YOLO) [21].

Of course, the task of training these ever-growing networks with more and more training parameters placed a huge requirement for an ever greater number of training examples. With this, a number of public image datasets were made available, providing annotations for many millions of images [8].

Fig. 1. Our system can be used for the cleaning or correcting of DL/ML image datasets, by flagging unlikely objects. This Figure presents a selection of images from the Microsoft COCO dataset [18] with objects that our system identified as being anomalous given the context of the other objects detected. In the top-left the presence of a surfboard on the wall given that it is surrounded by kitchen objects was flagged to be anomalous. On the right, we have an image from the output of the YOLO neural network, where a horse was detected (without the understanding of it being a painting) when all the rest of the objects are typically found indoors, and according to the context, this was strange. Finally, in the bottom left, we have the annotations provided by the COCO dataset. Our system suggested both that a bicycle annotated outside the window was strange given the indoors nature of the rest of the image, and that there should probably be some chairs in this scene, where instead it was annotated using a sofa.

These datasets have powered the training of these huge networks, but it is not without its problems. Annotating datasets using object detection networks, trained from other datasets, leads to the amplification of errors. Therefore, the image datasets must be hand-annotated, and with the current architectures datasets must be tailored for a specific domain, as for example the prior on the distribution of objects in different environments can not be easily changed for the already-trained network. A tragic example of how a wrong choice of prior dataset distribution can cause a system failure is the Uber autonomous car fatality, where the dataset used to train the pedestrian detector was biased towards humans at crosswalks [5].

Even with software aimed at streamlining the task of annotating datasets [10], the task is still intensely tedious, repetitive, and time-consuming. This inevitably leads to errors creeping into the dataset, even when the annotations are also checked by other humans. This can be in the form of inconsistent annotating across different people, such as edge cases where different people label the object as different classes, to whether they choose to annotate very small objects in the background of an image or not. These errors are then manifested in the trained networks that learn from bad data. They may then be then be biased to make the same kind of mistake the annotators made, whether by rejecting a good detection, or misclassifying.

Therefore tools that are better able to understand the images, for example, that can look for bad annotations in a dataset are of immense importance. The authors of [31] have shown that understanding beyond the individual parts of an image, i.e. awareness of the global context, is highly beneficial. However, the most popular state-of-the-art detectors, such as YOLO, do not make use of this. Instead, they work by looking for regions of interest, and then classifying them separately. Thus, there is no interplay between these classification results within the same image. For human vision, context is a strong cue – people can detect objects that are expected to be seen in a particular context even if they are largely occluded, poorly illuminated or anomalously-shaped, see Fig. 1. However, a context-unaware method might reject a detection simply because poor illumination or occlusion causes the detection probability to fall below a fixed threshold. As shown in [31] this has a potentially negative effect on the overall accuracy.

In last few years there has been quite some research into how to incorporate context inherently into the architecture of these object detectors [2,6,15, 20,26,34]. All these new architectures present a unique approach to context modelling ranging from applying spatial Recurrent Neural Networks (RNNs) over the whole image [2] to modelling objects and relationships between them as graphs [20]. The context is then learned from the training dataset together with the object detection and further augments the learning process. We present a simpler approach, which is easier to train. It presents minimal overhead to the standard object detection pipelines, and its separation from the detection engine results in the ability to indicate the cause of the anomaly, i.e. the result is explainable. In this paper, we will address how contextual information can be added to traditional detectors through anomaly detection, and assess to what extent it is or is not beneficial.

2.2 Autoencoders and Their Use for Anomaly Detection

One of the first efforts to use a network with a layer smaller than its input (bottleneck) to learn compact representations of input data goes to 1991 [16]. This approach is based on creating a network capable of learning a mapping for nonlinear component analysis, which reduces the dimensionality of presented data while preserving their most significant features. The result is a general, versatile method that learns a generalised representation of the inputs.

It has been shown that two-layer bottleneck architectures are equivalent to standard Principal Component Analysis (PCA) when trained with a squared error loss criteria [4]. Nowadays applications of this architecture range from denoising by extraction of important features from images [29] or recorded speech [12] to anomaly detection through dimensionality reduction of general data [23].

Autoencoders (AE), as they became known, are nowadays an interesting but niche form of neural networks that can learn by themselves how to perform lossy compression on the inputs, no matter the type of data they are given. During the training phase, they would learn this compression-by-generalisation by training through comparison of the network inputs against its outputs generated by the hidden 'bottleneck' layer. When the network had been sufficiently trained, it could be split in half at the bottleneck, with the first half becoming the encoder, responsible for data compression, and the second half becoming the decoder, used for recreating the original data from the encoded form. While being interesting for learning how to encode any kind of data, 'vanilla' autoencoders did not reach widespread use for a number of reasons; the compression ratio is poor, they require time to learn for every data kind used, and the relationship between datasets and the optimal configuration of the network is not obvious.

However, one of their popular applications is anomaly detection. For example, [22] showed how the network architecture can be used for anomaly detection in telemetry data. Furthermore, others [33,35] have extended the technique to deep-autoencoders. Their research has shown how deep autoencoders can further boost the performance of autoencoder anomaly detection by capturing the characteristics of the underlying processes that generate the data. Another related architecture is the variational autoencoder (VAE) [1], which looks at the use of reconstruction probabilities rather than simply minimising loss.

The way these networks work is that essentially they learn a generalised distribution of the data. Once trained, the inputs to the network can be compared to its outputs to locate the parts of the input data that the network failed to recreate. The inability to reproduce a particular input implies that its presence in the training data is rare, i.e. that the input is anomalous.

In this paper, we will look at combining the problems of the context in deep object detectors with the use of autoencoders to spot anomalies. We hypothesise that through anomaly detection, the autoencoders can quantify the consistency of scene context, which can then be used to detect wrong classifications as well as to refine the output of the object detectors.

3 Method

This section concerns the use of autoencoders for anomaly detection itself. After the problem definition, we describe the training process and the design choices influencing the parameters used in the autoencoders.

The core idea we are elaborating upon is that the autoencoders are trained to reproduce their inputs with a constrained intermediate representation, which forces them to model the underlying processes that generate the input data. We argue that in the case of object detection, the underlying process that generates the visible classes relates to the scene context. We hypothesise that training dataset classes exhibit strong correlation in their occurence within similar scenes represented by the individual images. Therefore, one would expect that the images would often contain similar combinations of objects typical for common scenes, and that the autoencoders would be able to learn these frequent combinations. Subsequently, the autoencoders should be able to learn if an object is missing or redundant in an observed combination. To evaluate this hypothesis, we utilise the Microsoft COCO dataset, which covers images of common objects in their natural context [18].

3.1 Conventions

The following convention will be used for the rest of the method description:

1. Input vector—Given the architecture of simple linear autoencoders, their input is a list of numbers which we call the input vector. We use each position of this fixed-length vector to encode the presence of an object class. The class-position map can be arbitrary, we use the ordering induced by COCO dataset class ids.
2. Output vector—The output of the autoencoder is of the same dimension and ordering as the input.
3. Annotation—The COCO dataset we used for both training and evaluation of our method comes in the form of images and segmentation masks of annotated objects visible in those images. For our purposes, we only use the class labels attached to individual boxes, which we refer to as the annotation. We use the term annotations to indicate the set of annotations present in one image, as well as the set of all annotations present in a dataset grouped by the image they belong to.

3.2 Design of Autoencoder

We created a model with one hidden layer fully connected to both the input and output layers. The size of the hidden layer was a tested parameter with regards to anomaly detection performance, and the size of the input-output layers was equal to the number of dataset classes. To investigate the method properly we tried several different sizes of the encoding layer, which effectively determines the relationship between how simplified and therefore how generalised the inputs become in the inner, encoded representation.

The autoencoders were trained to reproduce a set of input vectors, where for every set of annotations A from the set of all sets of annotations the input vector was created as

$$i[class] = \begin{cases} 1 & class \in A \\ 0 & \text{otherwise.} \end{cases} \tag{1}$$

As the autoencoder size is linked to the number of classes in the output vector of the object detector, it is extremely small in comparison to the object detector. For example, when paired with YOLO using the COCO image dataset, the autoencoder is comprised of an input layer of 80 neurons, then a hidden layer of in the region of 10 neurons, then an output layer of 80 again. Therefore with less than 200 neurons, the run-time performance is extremely fast on low-end modern hardware, in the sub-millisecond range to perform inference on a GPU.

3.3 Anomaly Detection

For the task of detecting anomalies using such a trained network, multiple approaches can be employed. The universal autoencoder-based anomaly detecting methods usually use a certain measure of anomality of a data sample. This measure is then computed for the data at hand, and based on its value the sample is either flagged as an anomaly or not. Such a measure can be the mean squared sum of errors between the input and output vector—the reconstruction error—or for example the reconstruction probability, employed by Variational AutoEncoders (VAEs) [1]. The reason why we can't adopt the same approach is due to a different formalisation of the task. These methods usually consider anomalies to be generated by a completely different process than the normal data, so their use case is then to filter anomalous inputs out of the incoming data stream.

In contrast, we know that all of our data was created by the same process and it is only the context that makes part of the input vector anomalous or not. We therefore emphasise the ability to detect anomalous components of the input vector, which not only provide us with the information in the form of an anomality score, but also allows for an explanation of the results. In certain scenarios, it can enable us to correct mistakes in the data, if that is the expected cause of anomalies.

The way we trained our autoencoders to reproduce vectors representing object class presence in some plausible scene with limited resources, leads to a behaviour where presence or absence of a given feature in the input vector affects with varying strength all of the features of the output vector. In other words when the context—the present features—strongly suggests the presence of another, currently absent object, the output of the network will be higher for that missing object. Furthermore, this also applies the other way round, when the context disagrees with the presence of some object, its value will be lowered in the output vector.

Although this gives us the general idea about how to read the output of the AE, it is yet to be determined how large must be the difference between the value for a given feature in the input and the output vector, so that it is reasonable to flag the feature value as anomalous. We do not investigate this problem much as we think it is a large problem worthy of its own investigation. Instead we limit ourselves to consider only a static thresholding value, which we set accordingly to the task at hand, as can be seen in the section about experiments.

We consider something as an anomaly, if the output of the autoencoder for this class differed from the input by more than 0.5.

4 Experiments

We designed a set of experiments to test the performance of our method in several scenarios, corresponding to two of the presented potential applications—the verification of potentialy false annotations and boosting of standard object detection networks.

First of all, we test the ability of the network to detect anomalies—both positive and negative. Then we present an experiment to prove the applicability of our method to the problem of adding contextual awareness to deep-learning object detectors to boost their performance.

One of the problems with quantifying anomalous objects in a scene is that standard clusters of objects often seen together may only semi-regularly include certain classes. For example, in a kitchen scene, the presence of cutlery is not always expected as it may be tidied away and out of view. Therefore, the system should not identify the positive or negative identifications of these as anomalous. When coming up with a method for evaluation, we had to take this into account.

4.1 Training of the Autoencoders

Before the actual experiments we must discus how the anomaly detecting autoencoders were trained.

As a training set for the autoencoder, we processed the annotations from the COCO dataset in a way where we created an input vector for every image annotation, set A, according to Eq. 1. We excluded the object class *person* from all annotations, because as we found out during preliminary experiments, it is vastly over-represented in the dataset and thus is not reasonably represented for anomaly analysis. For the breakdown of the dataset, we used the COCO validation set to determine how good our methods were performing, and we broke the training set for autoencoder into two sets for training and validation, in a ratio of 10:1. We trained all models for 120 epochs with the binary cross entropy loss function [7] and using the Adadelta optimizer [30].

With a careful study of the training graphs, we determined what shape of autoencoder would work best for us. Counterintuitively, the goal of training isn't to minimise loss; that would defeat the point of generalisation and would always favour a larger intermediate layer. Instead, some loss is what is helping us generalise. In validation some loss will always be present as there will be less common groups of objects in the evaluation dataset than in the training dataset. In fact if one considers the total number of possible subsets, it is a certainty. The important point here though is that the less likely a cluster is, the more likely it is less common because it contains some anomaly.

After taking this into consideration, as can be seen in Fig. 2, we decided that the best point that simplifies the model enough to generalise, but also still has relatively low loss, lies somewhere around the range of 15–25 neurons. Below this number, the loss increases massively. Above this number, each additional neuron has little affect on the end model. Therefore we ended up using a model for our experiments with 20 neurons, in the centre of the sweet-spot.

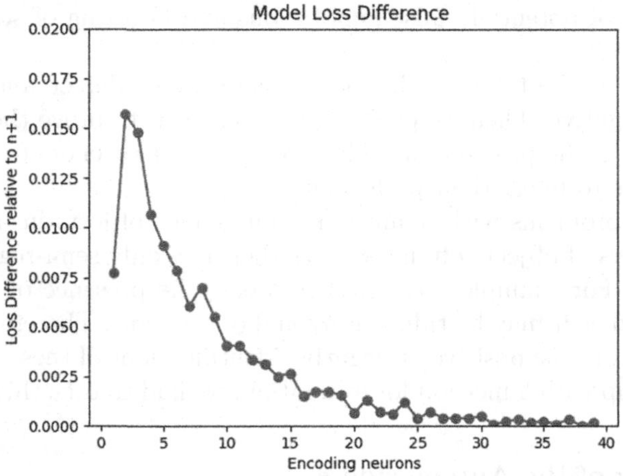

Fig. 2. Here we can see how much loss each model gained relative when it's final neuron was removed. On the X-axis is the number of neurons in the model, and on the Y-axis how much additional loss the model gained compared to a model one neuron larger. We can see that removing neurons from very large intermediate layers has little effect on the loss. This is relatively steady down to about the range of 15–25 neurons, where the removal of each neuron begins to introduce more and more loss. Below 15 neurons, removing each extra neuron adds significant extra loss to the model, before dropping for one neuron as the loss is already very high.

4.2 Anomaly Detection

Our first two experiments focus on assessing the capabilities of the autoencoder networks to detect positive and negative anomalies. The motivation for this experiment is the application to verification of automatically or otherwise annotated datasets. If one has an idea about the spatial distribution of objects that are being annotated in some images, then our method would be able to detect these annotations that violate such a distribution. The actual performance is presented as a result of the two following experiments, which consist of artificially adding and removing object classes to a set of manually annotated images and seeing how well our system can detect these. We trained several autoencoders that differed in the number of neurons in the encoding layer, effectively setting the compression ratio.

For these experiments, we used the images and annotations from the Microsoft COCO dataset [19], which are already split into training and validation sets. As mentioned before, this data was preprocessed to remove annotations of 'people', and we also imposed some limits regarding the minimum number of annotations required in an image, which we discuss later, as looking for anomalies in images with single classes makes little sense.

Positive Anomalies. In this experiment, we wanted to investigate how well autoencoders can detect anomalous classes artificially added to images. We took all the images with at least three different annotations present, and then added an extra class that was not already present. We then looked at how well our system was able to flag this added class as being anomalous.

Out of all 80 possible classes, we would select randomly 10, which we would then add one by one to all the images, where this class was not already present in the annotations. Adding an object consists of setting to 1 the respective value in the autoencoders input vector. Then, if the output of the autoencoder for this class differed from the input by more than 0.5, we would classify it as a correct anomaly detection. Because of the random nature of the experiment, we re-ran it for 20 times to get more statistically significant results.

Negative Anomalies. Regarding the negative anomalies, we designed a similar experiment, where instead of adding random object classes we instead removed classes randomly from those that appeared in a given image.

We wanted to show the autoencoders detecting the objects we removed from the scene, so as a part of the preprocessing, we filtered out the images the autoencoders considered anomalous in the dataset itself. The images with less than five annotations were filtered as well, so that after removal of an object class, there would be at least four classes left for the autoencoder to understand the context.

For every remaining image, randomly five of its annotations were selected and one at a time was removed from the set, so their respective position set to 0 in the input vector. If the output of the autoencoder differed at the corresponding position of this class from the input by more than 0.25, we would classify it as a correct negative anomaly detection. The reason for a threshold lower for negative anomalies is that a presence of single object is highly fluctuating in real-world environments, so we settle with a lower confidence—the group of objects, after we removed one class, might actually be perfectly normal.

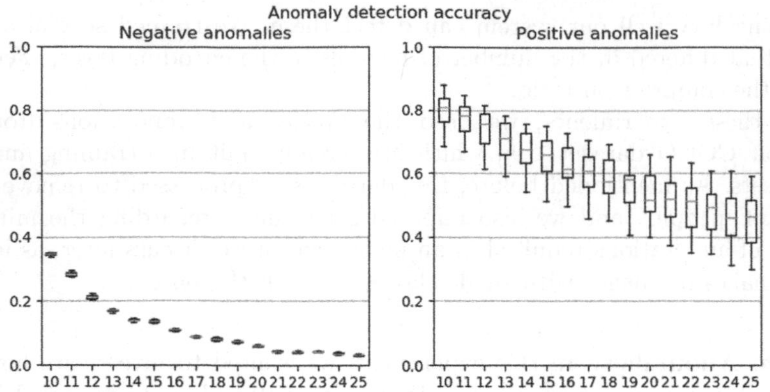

Fig. 3. Results of the first two experiments to test the ability of the autoencoder to spot anomalies, artificially introduced to a scene. On the left we removed object classes from the scene, on the right we added them. Every box represents a model with a different number of neurons in the encoding layer.

Results. All the results are depicted in Fig. 3.

Detecting negative anomalies is clearly a much harder problem than detecting the positive ones, as discussed already in the section about the experiment. These results are predictably worse, because removing objects from the scene might not result in an anomalous scene, but only another valid setting, by the nature of the negative anomalies.

The best accuracy in the proposed experiment was achieved by the autoencoder with the smallest number of neurons and was around 35%. That the more complex models are able to better capture the fluctuating nature of single objects and removing one from a valid set might have only created another valid set, probably led to decrease in accuracy with the number of encoding neurons. However, we expect the negative anomaly detection to perform better in real applications with more complex models, because they should exhibit a lower false-positive rate.

For positive anomalies, we show that our method is very good at detecting these, with an accuracy rate mostly between 75 and 85% for the best model. Also, as we can see from the graphs, the lower the number of neurons in the encoding layer, the better the results.

This last result is due to the fact that more complex models with more neurons are then likely to pick up anomalies during their training from the training dataset. That in turn is caused by anomality not being an absolute property, but rather a context-related measure. And last, since we added random classes, some of them had to be generally less common objects as well, which small encoders might not have reflected.

4.3 Boosting Object Classification by Context-Awareness

In this experiment, we wanted to show how well our method is able to dynamically asses the output of a neural network regarding it's confidence in individual detections. We used the pretrained YOLOv3 network for object detections, which we then run on the images in the COCO dataset, against which it was trained.

Using this network one normally has to set a thresholding parameter, which determines at what level of confidence one will accept the output of the network as a valid object detection. Our intention was to use the negative anomaly detection to strengthen the confidence in objects that were detected below this threshold, but given the context of the scene, they are likely to be observed.

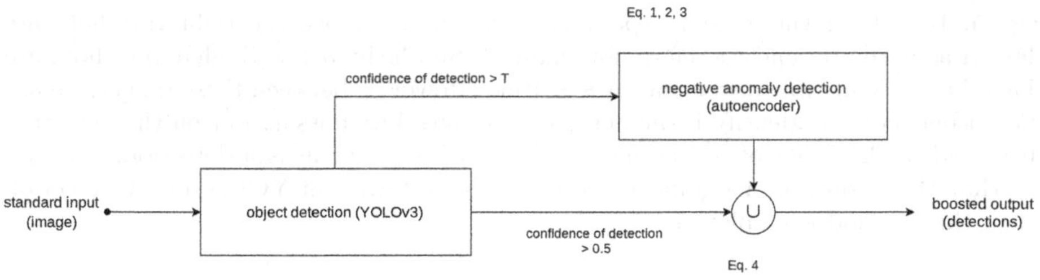

Fig. 4. Diagram describing the application of anomaly detection to boosting of the object detection. First, the input image is analysed by an object detector. The negative anomalies, that are detected are then combined with the standard output of the detector, for boosting the accuracy of detections.

The whole experiment was set up to examine the difference in accuracy between a standard object detecting network and one enhanced with an autoencoder. We needed to tailor the experiment to account for two basic properties of the situation. We know that with a high threshold the network performs reasonably well [21], but the threshold also determines how many objects will be detected, so it can be effectively used for setting the accuracy versus extracted information ratio. Second, we are limited by the ground truth available, basically the annotated image dataset.

The general idea of the experiment is to count how many times the network was wrong on the validation dataset against which it was trained. Denoting the decision threshold T, we compare how well the YOLOv3 model performs with respect to T, to a case where we set the threshold to 0.5 as a reasonable default and let the autoencoder detect other objects in the image, that should be included as well, but their confidence lies in the interval $<T,\ 0.5>$. The process is depicted in Fig. 4.

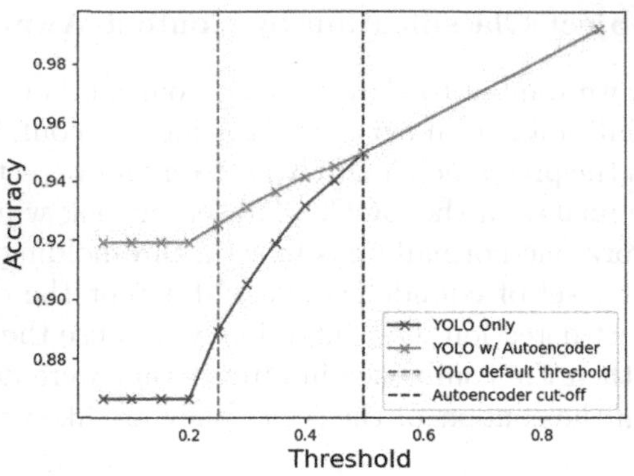

Fig. 5. Results of the second experiment. Detections above the right threshold are classed as absolute, and therefore not changed. Similarly, below the left threshold are classed as so weak not to warrant investigation. However, between these ranges, we use the autoencoder to identify if the detection, despite just missing out on the detection threshold, makes a lot of contextual sense. If so, we count it as a detection. One can see that this results in a significant boost to the accuracy of YOLO object detection when this method is employed

Results. Specifically, the procedure is as follows:

1. We analyse the image in the dataset with YOLOv3 network, which results in a set of detections *Det* and we create a set

$$D = \{d \mid conf(d) > T, d \in Det\}, \quad (2)$$

   where $conf(d)$ is the level of confidence the network has in particular detection.
2. We calculate the accuracy, comparing the set D with the set of annotations we have from COCO dataset. This is the accuracy of standard YOLO itself. We only compare whether the class was present in annotations or not - we don't compare the number of detections or bounding box precision.
3. Next, we create an input vector i for the autoencoder:

$$i[c] = max\{conf(d) \mid class(d) = c, d \in D\}, \quad (3)$$

   where c is the particular object class, $class(d)$ is the class of given detection d and in the case that no object of a given class was detected, we set the input to 0.
4. The output vector o of the autoencoder is then transformed into a set

$$O = \{c \mid o[c] > 0.1, c \in C\} \cup \{class(d) \mid conf(d) > 0.5, d \in Det\}, \quad (4)$$

   where C is the set of all classes and 0.1 was selected as a reasonable threshold for autoencoder output from our experience with experiments for anomaly detections. We then compare the set O with the set of annotations of given image the same way we did with the simple YOLO output.

Results of this experiment are depicted in Fig. 5. We show that on the interval <0.2, 0.5> our method clearly outperforms the standard YOLOv3 network. At the level of 0.2 by almost 5%. The reason both curves converge to the same accuracy value beyond a threshold of 0.5 is that we always include everything with detection confidence above this value as a given, a certain detection used for gathering the context.

5 Conclusion

We investigated the use of autoencoder-based anomaly detections for the task of image-based object detection. The core idea of our approach is to use autoencoders to capture the contextual information present in images, which affects the joint occurrence of the detected object classes. Simply put, we use the autoencoders to answer two questions: 'What does not belong here?' and 'What is missing here?'. We demonstrate that autoencoders can quantify the aforementioned questions and use this quantification to detect potentially incorrect detections and annotations in the training data. We also show that knowledge of the context can boost the precision of state-of-the-art object detection frameworks.

Our work shows that decoupling the context-aware autoencoder from the context-unaware object detector results in a system that can be re-trained for a particular object distribution in short time while being computationally inexpensive. In the experiments performed, we demonstrate that using a ~200-neuron-sized autoencoder can not only significantly improve performance of the YOLOv3 detector, but also detect errors artificially injected into the training set. Thus, the method presented addresses two prominent issues troubling practical deployments of machine learning methods – deployment in scenarios with different distributions of the detected classes and dataset annotation quality.

In the future we would like to extend our work in several directions. We would like to investigate different, more complicated linear or deep autoencoders capable of capturing not only co-occurrence, but also the spatial and temporal relations between the detected classes. We will compare the decoupled approach to the newest context-aware deep neural networks, that incorporate different levels of contextual information to achieve better performance. Finally, we believe that the ability to spot potential errors in the detector's output is an important step towards introspective, explainable learning methods and we will pursue their deployment in systems capable of long-term autonomous operation.

Acknoledgement. The authors acknowledge the support of the Czech Science Foundation project "Towards long-term autonomy through introduction of the temporal domain into spatial representations used in robotics "20-27034J". The calculations were performed using computational resources provided by the OP VVV MEYS funded project CZ.02.1.01/0.0/0.0/16_019/0000765 "Research Center for Informatics".

References

1. An, J., Cho, S.: Variational autoencoder based anomaly detection using reconstruction probability. Spec. Lect. IE **2**(1), 1–18 (2015)
2. Bell, S., Lawrence Zitnick, C., Bala, K., Girshick, R.: Inside-outside net: detecting objects in context with skip pooling and recurrent neural networks. In: Proceedings of the IEEE Conference on Computer Vision and Pattern Recognition, pp. 2874–2883 (2016)
3. Biederman, I.: Recognition-by-components: a theory of human image understanding. Psychol. Rev. **94**(2), 115 (1987)
4. Bishop, C.M.: Pattern Recognition and Machine Learning. Information Science and Statistics. Springer, New York (2006)
5. Board, T.N.T.S.: Collision between vehicle controlled by developmental automated driving system and pedestrian, Tempe, Arizona, p. 78, 18 March 2018
6. Chen, X., Gupta, A.: Spatial memory for context reasoning in object detection. In: Proceedings of the IEEE International Conference on Computer Vision, pp. 4086–4096 (2017)
7. Creswell, A., Arulkumaran, K., Bharath, A.A.: On denoising autoencoders trained to minimise binary cross-entropy. arXiv:1708.08487 [cs, stat] October 2017
8. Deng, J., Dong, W., Socher, R., Li, L.J., Li, K., Fei-Fei, L.: ImageNet: a large-scale hierarchical image database. In: 2009 IEEE Conference on Computer Vision and Pattern Recognition, pp. 248–255. IEEE (2009)
9. Divvala, S.K., Hoiem, D., Hays, J.H., Efros, A.A., Hebert, M.: An empirical study of context in object detection. In: 2009 IEEE Conference on Computer Vision and Pattern Recognition, pp. 1271–1278. IEEE (2009)
10. Dutta, A., Zisserman, A.: The VGG image annotator (via). arXiv preprint arXiv:1904.10699 (2019)
11. Galleguillos, C., Belongie, S.: Context based object categorization: a critical survey. Comput. Vis. Image Understand. **114**(6), 712–722 (2010)
12. Gehring, J., Miao, Y., Metze, F., Waibel, A.: Extracting deep bottleneck features using stacked auto-encoders. In: IEEE International Conference on Acoustics, Speech and Signal Processing, pp. 3377–3381, May 2013. DOIurl10.1109/ICASSP.2013.6638284. ISSN: 2379-190X
13. George Broughton, J.B.: Auto image anomaly (2020). https://github.com/broughtong/Auto-Image-Anomaly
14. He, K., Zhang, X., Ren, S., Sun, J.: Deep residual learning for image recognition. In: Proceedings of the IEEE Conference on Computer Vision and Pattern Recognition, pp. 770–778 (2016)
15. Hu, H., Gu, J., Zhang, Z., Dai, J., Wei, Y.: Relation networks for object detection. In: Proceedings of the IEEE Conference on Computer Vision and Pattern Recognition, pp. 3588–3597 (2018)
16. Kramer, M.A.: Nonlinear principal component analysis using autoassociative neural networks. AIChE J. **37**(2), 233–243 (1991). https://doi.org/10.1002/aic.690370209. https://aiche.onlinelibrary.wiley.com/doi/abs/10.1002/aic.690370209
17. Krizhevsky, A., Sutskever, I., Hinton, G.E.: ImageNet classification with deep convolutional neural networks. In: Advances in Neural Information Processing Systems, pp. 1097–1105 (2012)
18. Lin, T.Y., et al.: Microsoft COCO: common objects in context. arXiv:1405.0312 [cs] February 2015

19. Lin, T.-Y.: Microsoft COCO: common objects in context. In: Fleet, D., Pajdla, T., Schiele, B., Tuytelaars, T. (eds.) ECCV 2014. LNCS, vol. 8693, pp. 740–755. Springer, Cham (2014). https://doi.org/10.1007/978-3-319-10602-1_48
20. Liu, Y., Wang, R., Shan, S., Chen, X.: Structure inference net: object detection using scene-level context and instance-level relationships. In: Proceedings of the IEEE Conference on Computer Vision and Pattern Recognition, pp. 6985–6994 (2018)
21. Redmon, J., Divvala, S., Girshick, R., Farhadi, A.: You only look once: unified, real-time object detection. In: Proceedings of the IEEE Conference on Computer Vision and Pattern Recognition, pp. 779–788 (2016)
22. Sakurada, M., Yairi, T.: Anomaly detection using autoencoders with nonlinear dimensionality reduction. In: Proceedings of the MLSDA 2014 2nd Workshop on Machine Learning for Sensory Data Analysis, p. 4. ACM (2014)
23. Sakurada, M., Yairi, T.: Anomaly detection using autoencoders with nonlinear dimensionality reduction. In: Proceedings of the MLSDA 2014 2nd Workshop on Machine Learning for Sensory Data Analysis - MLSDA 2014, Gold Coast, Australia QLD, Australia, pp. 4–11. ACM Press (2014). https://doi.org/10.1145/2689746.2689747. http://dl.acm.org/citation.cfm?doid=2689746.2689747
24. Santos, J.M., Krajník, T., Fentanes, J.P., Duckett, T.: Lifelong information-driven exploration to complete and refine 4-d spatio-temporal maps. IEEE Robot. Autom. Lett. **1**(2), 684–691 (2016). https://doi.org/10.1109/LRA.2016.2516594
25. Santos, J.M., Krajník, T., Duckett, T.: Spatio-temporal exploration strategies for long-term autonomy of mobile robots. Robot. Auton. Syst. **88**(C), 116–126 (2017)
26. Shrivastava, A., Gupta, A.: Contextual priming and feedback for faster R-CNN. In: Leibe, B., Matas, J., Sebe, N., Welling, M. (eds.) ECCV 2016. LNCS, vol. 9905, pp. 330–348. Springer, Cham (2016). https://doi.org/10.1007/978-3-319-46448-0_20
27. Simonyan, K., Zisserman, A.: Very deep convolutional networks for large-scale image recognition. arXiv preprint arXiv:1409.1556 (2014)
28. Szegedy, C., et al.: Going deeper with convolutions. In: Proceedings of the IEEE Conference on Computer Vision and Pattern Recognition, pp. 1–9 (2015)
29. Vincent, P., Larochelle, H., Bengio, Y., Manzagol, P.A.: Extracting and composing robust features with denoising autoencoders. In: Proceedings of the 25th International Conference on Machine Learning - ICML 2008, Helsinki, Finland, pp. 1096–1103. ACM Press (2008). https://doi.org/10.1145/1390156.1390294. http://portal.acm.org/citation.cfm?doid=1390156.1390294
30. Zeiler, M.D.: ADADELTA: an adaptive learning rate method. arXiv:1212.5701 [cs] December 2012
31. Zhao, R., Ouyang, W., Li, H., Wang, X.: Saliency detection by multi-context deep learning. In: Proceedings of the IEEE Conference on Computer Vision and Pattern Recognition, pp. 1265–1274 (2015)
32. Zhou, C., Paffenroth, R.C.: Anomaly detection with robust deep autoencoders. In: Proceedings of the 23rd ACM SIGKDD International Conference on Knowledge Discovery and Data Mining - KDD 2017, Halifax, NS, Canada, pp. 665–674. ACM Press (2017). https://doi.org/10.1145/3097983.3098052. http://dl.acm.org/citation.cfm?doid=3097983.3098052
33. Zhou, C., Paffenroth, R.C.: Anomaly detection with robust deep autoencoders. In: Proceedings of the 23rd ACM SIGKDD International Conference on Knowledge Discovery and Data Mining, pp. 665–674. ACM (2017)

34. Zhu, Y., Urtasun, R., Salakhutdinov, R., Fidler, S.: segDeepM: exploiting segmentation and context in deep neural networks for object detection. In: Proceedings of the IEEE Conference on Computer Vision and Pattern Recognition, pp. 4703–4711 (2015)
35. Zong, B., et al.: Deep autoencoding gaussian mixture model for unsupervised anomaly detection (2018)

Cloud and Edge Computing

The Design and Implementation of Secure Distributed Image Classification Model Training System for Heterogenous Edge Computing

Cong Cheng[1], Huan Dai[2(✉)], Lingzhi Li[1(✉)], Jin Wang[1], and Fei Gu[1]

[1] School of Computer Science and Technology, Soochow University, Suzhou 215006, China
lilingzhi@suda.edu.cn
[2] School of Electronics and Information Engineering, Suzhou University of Science and Technology, Suzhou 215009, China
daihuanjob@163.com

Abstract. Deep learning provides many new and efficient solutions for edge computing. We study training image classification models on edge devices in this paper. Although there have been many researches on deep learning in edge computing. Most of them did not consider the impact of the limited service capabilities of edge devices, the problem of straggler and insecurity of training data on the system. We design a new distributed computing system to train image classification models on edge devices. To be more specific, we vectorize the convolutional neural network (CNN) to transform it to a lot of matrix multiplications. These matrix multiplications can be arbitrarily cut into many smaller matrix multiplications suitable for computing on edge devices. Besides, our system utilizes codes to ensure the stability and security of distributed matrix multiplications on edge devices. In the performance evaluation, we test the performance of matrix multiplications and a CNN model training in our system with uncoded and coded strategies. The evaluation results show that the system with code strategies perform better than with uncoded strategies on the edge devices having the problem of straggler. In summary, we design a secure distributed image classification model training system for heterogenous edge computing.

Keywords: Edge devices · Distributed computing · Deep learning

This work was supported in part by the National Natural Science Foundation of China (62072321, 61672370, 61702354), "Six Talent Peak Project" of Jiangsu Province (XYDXX-084), CERNET Innovation Project (NGII20190314), China Postdoctoral Science Foundation (2020M671597), Jiangsu Postdoctoral Research Foundation (2020Z100) Scientific Research Project of Suzhou University of Science and Technology (XKZ2017004), Postgraduate Research and Practice Innovation Program of Jiangsu Province (SJCX19_0801), Tang Scholar of Soochow University and the Priority Academic Program Development of Jiangsu Higher Education Institutions (PAPD).

H. Gao et al. (Eds.): CollaborateCom 2020, LNICST 349, pp. 179–198, 2021.
https://doi.org/10.1007/978-3-030-67537-0_12

1 Introduction

In recent years, edge computing is developing steadily and getting popular. The computing systems and applications based on edge computing are emerging in endlessly. They have provided lots of solutions for many practical problems. For instance, edge computing has outstanding performance in smart health [1], transportation system [6] and object detection [19]. Meanwhile, edge computing is superior to traditional cloud computing in four aspects: latency, data security, scalability and reliability [5]. Computing at the edge instead of computing on the cloud, which has lower latency and facilitates the protection of data security. Hence, we decide to train the image classification model on edge devices.

As well as edge computing, deep learning is booming. Nowadays, efficient image processing technologies are based on deep learning. And it has wide applications in practice. For example, deep learning helps analyze medical images [14,17], construct image [2,3,9,10,16] and classify image [8]. We train the image classification model using convolutional neural networks (CNNs) on edge devices. Because CNNs are important and popular networks of deep learning. They are mainly composed of the convolutional layer, pooling layer and fully connected layer. And convolutional layer and fully connected layer are primary computing layers in CNNs [4].

In general, the edge devices are responsible for collecting data and the cloud utilizes it to train the deep learning model as we can see in Fig. 1. However, this scheme has several drawbacks. First, all the collected data needs to be transmitted to the cloud through the network, which requires a large amount of bandwidth resources. Second, once the cloud is attacked, the whole system will not work well because the architecture of it is cloud-centric. Third, the security of data is easily threatened in the process of uploading to the cloud. Finally, the computing resources of cloud are expensive. In contrast, training CNN models on edge devices has the following advantages: low latency, cheaper computing resources. To sum up, it is necessary to train the deep learning models on edge devices.

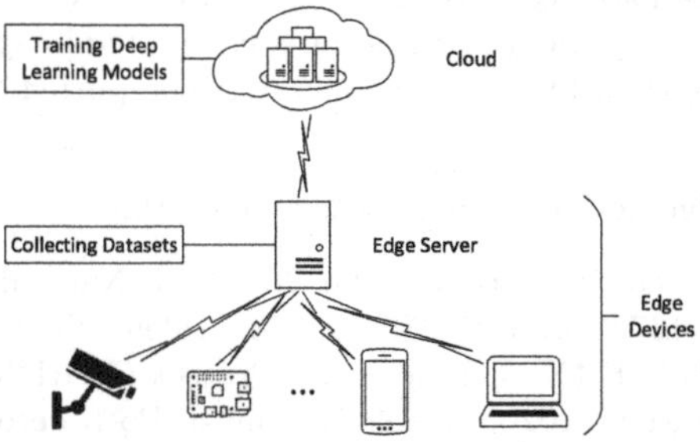

Fig. 1. The edge devices are responsible for collecting datasets and the cloud is responsible for training deep learning models.

However, training the deep learning model on the edge devices has several challenges [5]. First, communication latency may affect the time it takes for the master node to transmit data to the worker nodes. Second, network bandwidth determines the amount of data transmitted from the mater node to the worker nodes per unit time. Third, the compute capability of the worker nodes may affects the computing power of the entire computing system. The computing power of the edge devices is weaker than the cloud. And training a CNN model in a single edge device is scarcely possible. Therefore, we have to use distributed computing method to combine multiple edge devices. Data parallelism and model parallelism are two existed methods of distributed deep learning. Data parallelism is to copy the untrained model to many computing equipment so that they have the same untrained model, and then divide the dataset to them for parallel training. So data parallelism can accelerate the training speed of a CNN model because the dataset is consumed in parallel. However, data parallelism requires the subordinate computing equipment to be able to independently train the entire model. Obviously, ordinary edge devices cannot meet this requirement. Model parallelism is the process of cutting huge models into different parts and deploying them to many computing equipment so that a huge model which cannot be trained on a single computer can be trained in the distributed environment. But the problem of straggler easier occurs to edge devices, which makes the model training using model parallelism method failure. Hence, the two distributed deep learning methods above are not suitable for edge devices. So, we have to design a new distributed computing method to train the CNN models on edge devices.

We transform the computing load in convolutional and fully connected layers to many subtasks, which is equivalent to divide the big task that training a CNN model to lots of subtasks. The convolution transforming to matrix multiplication mentioned in [4,18] can help vectorize CNN. The computing load of training a CNN model transforms to lots of matrix multiplications with it. Then these matrix multiplications can be cut arbitrarily to fit the operation on an edge device. Hence, we can train the CNN models on edge devices.

Nevertheless, distributed computing on the edge devices has a few drawbacks need to be an addressed. The problem of straggler and data security problem are the main issues of distributed computing on edge devices. The edge worker devices distributed computing tasks may not timely return all computed results due to the problem of straggler. And the security of computing data could be threatened when there are edge worker devices being attacked. So we utilize the redundancy coding strategy to handle these problems. After the computing tasks which are matrix multiplications have been coded, we get redundancy of the tasks. Then the redundant computing tasks are assigned to edge worker nodes to deal with the negative impact of the straggler problem. Meanwhile, the details of them are protected because the computing tasks are coded.

Overall, the contributions of this paper are as follows:

- We find a new CNN distributed training method, which is more suitable for running on edge devices with limited service capabilities and unstable states than data parallel and model parallel training methods.
- In this paper, we design and implement a distributed computing system running on edge servers and edge work nodes. It effectively combines and utilizes various computing resources of the edge, which further enhances the computing power of the edge.
- In addition, we apply coded strategies in our distributed computing system. Then we conduct comparative experiments on edge devices simulated by distributed processes with different latencies. The experimental results show that the distributed computing system using coded strategies can not only protect the computing data, but also deal with the negative impact caused by the problem of stragglers.

2 Related Work

2.1 Machine Learning for Edge Computing

There have been many researches on the cooperation between edge computing and machine learning. [21] proposed a scheme for distributed training of machine learning models based on gradient descent on the multiple edge nodes. The training data is collected and stored by the edge node and used for the training of the machine learning model on the edge node instead of being sent to the central location for training the model, which saves network bandwidth. However, [21] does not consider the impact of the security of training data and the limited service capabilities of edge nodes on the system. [19] designed a distributed and efficient target detection system in which end devices, edge servers and cloud cooperate with each other and have a clear division of labor. The end devices are responsible for data collection, compression and transmission to edge server, where the edge servers train the local model, and the cloud server is responsible for aggregating the global model and updating the model on the edge servers. This division of labor makes the entire system more efficient. But, the security of training data is also not taken into consideration in [19]. [13] proposed a deployment strategy of deploying the lower layers of CNN on the edge server and the higher layers on the cloud. Compared with deploying all layers of CNN in the cloud, deploying the lower layers of CNN on edge servers close to the data source can save network bandwidth and ease the computing pressure of the cloud. However, it only uses the computing resources of the edge server, and does not make full use of the computing resources of the edge worker nodes under the edge server. In addition, the security of training data has not been taken seriously.

2.2 Coded Distributed Matrix Computing

There are a lot of matrix computing in computer applications. Coded distributed matrix convolution is studied in [7], which uses the MDS coding strategy to complete the matrix convolution within the deadline on the worker nodes with the problem of stragglers. However, the limited number of worker nodes is not considered in [7], so it may achieve its theoretical effects in practical applications. Coded distributed matrix multiplication is also studied in [15], which compares the performance of the uncoded, task replication, and coded strategies in a distributed computing environment with the problem of straggler. The experimental results show that coded strategies are better than other strategies under the same experimental environment, and the rateless coded strategy is superior than the MDS code strategy in terms of experimental performance and time complexity. On this basis, we apply the rateless coding strategy and MDS coding strategy to our CNN distributed training system instead of simple matrix multiplication.

2.3 Coded Distributed Machine Learning

The distributed machine learning is already a mature technology. In order to deal with the problem of straggler, code strategies are introduced in distributed machine learning. The linear regression algorithm based on coding distributed computing is studied in [12], which can therefore run well on the distributed computing system with the problem of straggler. However, the linear regression algorithm is just a simple algorithm in machine learning. For deep learning algorithms with non-linear operations like CNN, training methods in [12] cannot work. In this paper, we design a new distributed training method that can be used for CNN model training to provide the problem a solution.

3 Design of the Secure Distributed Image Classification Models Training System Based on Coding

In order to train CNN models on edge devices, we design a secure distributed training system for an image classification model that can run on edge devices. Firstly, we introduce the system framework and show the components of the system and their general functions from a macro perspective. Secondly, we introduce the vectorization of CNN, which converts the complex computation of CNN into matrix multiplication so that we can perform the coded distributed matrix multiplication on this basis. Thirdly, we introduce several distributed computing strategies which are applied to distributed computing systems and help deal with the problem of straggler. Finally, we introduce the design of an edge distributed computing system which is an important part of our system.

3.1 System Framework

From the Fig. 2, we can see the detailed framework of the secure distributed image classification models training system. It consists of an edge server and a number of edge worker nodes. In the left part of the figure is the network topology of our system that every edge worker node independently communicates with the edge server. Specifically, the edge worker nodes obtain the task from the edge server and then return the task result to the edge server in this network topology. In the right part of the figure, we can see what the edge server is responsible for and will be introduced in detail next.

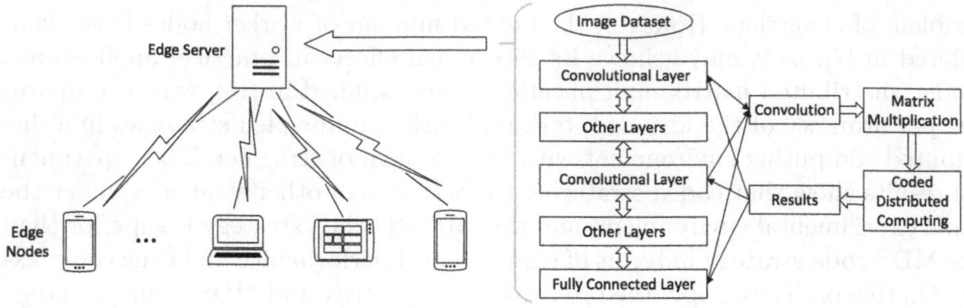

Fig. 2. The framework of the secure distributed image classification models training system.

Edge Server. The edge server maintains a CNN model on it. So it is responsible for updating the parameters of convolutional and fully connected layers. In addition to this, the edge server also do jobs of making coded computing tasks based on matrix multiplication and integrating the returned task results. The method of convolution transforming to matrix multiplication which is introduced in 3.2 helps edge server extract computing tasks from the CNNs. Then, these computing tasks are distributed to the edge worker nodes. Finally, the results computed by edge worker nodes are collected by the edge server and then a round of coded distributed computing ends.

Edge Worker Nodes. There are many kinds of edge worker nodes, such as raspberry pies and smart phones. Nevertheless, they all have the ability of communicating with the edge server and have a few computing power. Edge worker nodes constitute the basis of edge computing.

3.2 Vectorization of Convolutional Neural Network.

Although CNNs have different layers, the computational load is mainly concentrated in the convolutional layer and the fully connected layer. And the total computational load of convolutional layers and fully connected layers is nearly equal to the computational load of CNN. For example, the VGGNet which is a classical CNN. In the Table 1 of [20], we get the configurations of VGGNet. As can be seen from the configurations, the parameters of the CNN are concentrated in the convolutional layer and the fully connected layer. So the entire computational load of VGGNet is composed of the computational load of the convolutional layer and fully connected layer obviously. Meanwhile, the distributed training of the CNN can be transformed into the distributed execution of the computational load of the convolutional layer and the fully connected layer.

Our system is based on the vectorized CNN and vectorization has been proved to accelerate the speed of training CNN models [4]. As mentioned above, the convolutional layer and the fully connected layer are the main computational load layers in CNNs and the parameters and bias that need to be updated only exist in them. Therefore, the vectorization of the convolutional neural network is the vectorization of the convolutional layer and the fully connected layer.

As shown in Fig. 3, there are forward and backward propagation of CNN on the edge server. In the forward propagation, the input image is processed by the convolutional layer and the fully connected layer to obtain the predicted value. The deviation between the predicted value of the image and the label value becomes the product of the forward propagation i.e. the gradient. In the backward propagation, the gradient is passed back to update the parameters in CNN.

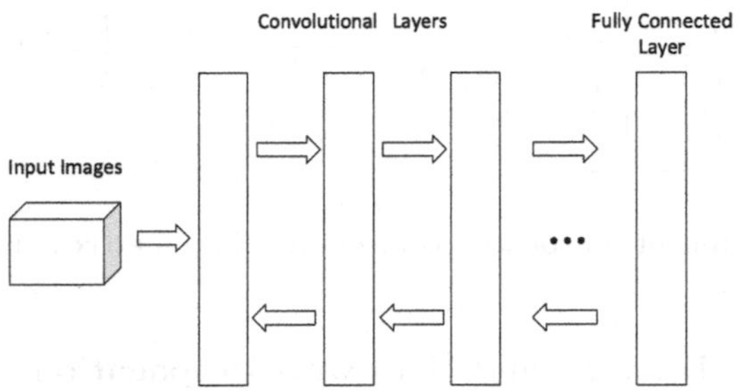

Fig. 3. Forward and backward propagation of CNN on the edge server.

The forward and backward propagation in a convolutional layer can be converted to the forward and backward propagation in a fully connected layer by the method of convolution transforming to matrix multiplication. Next we will introduce the vectorization of forward propagation and the backward propagation in CNNs, i.e. the vectorization of the forward propagation and the backward propagation in the fully connected layer.

Convolution Transforms into Matrix Multiplication Scheme. It is hard to implement a distributed CNNs training system if we do not transform the convolution. Then, a method in [4] can transform convolution to matrix multiplication, which has been proved to speed up image convolution operation in practical application. And it brings us another advantage that we can do distributed computation easier because the image convolution is transformed to matrix multiplication.

An example is shown in Fig. 4, which uses a unique way of matrix unfolding to implement that image convolution turns to matrix multiplication. As we can see, four 2×2 matrices are derived from the 3×3 image matrix after the convolution kernel window scans the image matrix with step size 1. Then, they respectively unfold to a single row and convolution kernel unfold to a single col. Finally, four rows make up a big matrix which multiplies with convolution kernel col. And then the result of the matrix multiplication is converted to the result of the convolution operation.

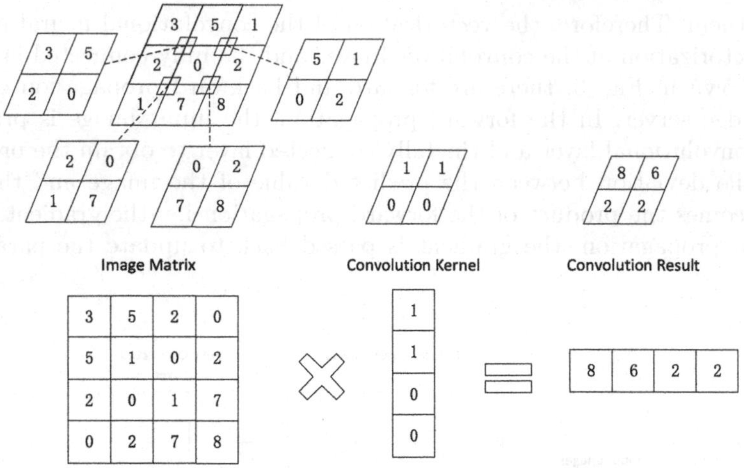

Fig. 4. Example of convolution transforming into matrix multiplication.

Vectorization of Forward and Backward Propagation. In the [4], there is an initial introduction to vectorization of forward propagation and backward propagation, the following is a more detailed vectorization of forward and backward propagation that we have improved on this basis. The forward propagation of the CNNs is to calculate the predicted value of the training samples. The vectorization process of the forward propagation is as follows:

$$Z^{[l]} = A^{[l-1]}.W^{[l]} + b^{[l]}, \tag{1}$$

$$A^{[l]} = g^{[l]}(Z^{[l]}). \tag{2}$$

In Eq. 1, $W^{[l]}$ is the parameter matrix of the lth layer of CNN and $b^{[l]}$ is the bias of the lth layer of CNN. $A^{[l-1]}$ is the output of $l-1$th layer and also the input of lth layer. In Eq. 2, $g^{[l]}$ is the activation function of lth layer.

The backward propagation of the CNNs is to transmit the difference between the predicted value and the real value of the training sample back to convolutional layers and fully connected layers of the CNNs and update the parameters in them. The vectorization process of backward propagation is as follows:

$$\mathrm{d}Z^{[l]} = \mathrm{d}A^{[l]}.[g^{[l]}(Z^{[l]})]', \ \mathrm{d}A^{[l-1]} = \mathrm{d}Z^{[l]}.W^{[l]T}, \tag{3}$$

$$\mathrm{d}W^{[l]} = A^{[l-1]T}.\mathrm{d}Z^{[l]}, \tag{4}$$

$$\mathrm{d}b^{[l]} = sum(\mathrm{d}Z^{[l]}), \tag{5}$$

$$W^{[l]} := W^{[l]} - \alpha \mathrm{d}W^{[l]}, \ b^{[l]} := b^{[l]} - \alpha \mathrm{d}b^{[l]}. \tag{6}$$

In the equations above, d[ ] is the gradient of [ ] and [ ]′ is the derived function of [ ] and $[\,]^T$ is the transposition of [ ] during backward propagation. In Eq. 5, $\mathrm{d}b^{[l]}$ is the sum of $\mathrm{d}Z^{[l]}$ in rows. In Eq. 6, α is the learning rate of the CNNs.

3.3 Distributed Matrix Multiplication Strategies

The Uncoded Strategy. As a comparison benchmark, we keep the simplest strategy that the uncoded strategy. When do the distributed matrix multiplication $A \times B$ with uncoded strategy, we cut the matrix A evenly along the rows into n matrix blocks $\{a_1, a_2, \ldots, a_n\}$. Then, we distribute n tasks $\{a_1 \times B, a_2 \times B, \ldots, a_n \times B\}$ to p workers. In this strategy, the matrix multiplication $A \times B$ has been completed when the master node received all n computation results.

The *r*-Replication Strategy. In the r-replication strategy, we do the multiplication of matrix A with matrix B. The matrix A is repeated r times, $A = A_1 = A_2 = \cdots = A_r$. So, the computing task $A \times B$ is also repeated r times. Then we get r computing tasks $\{A_1 \times B, A_2 \times B, \cdots, A_r \times B\}$. In distributed computing, the $A \times B$ task is divided into n subtasks and the total number of subtasks is rn. When the rn subtasks are distributed to worker nodes and at least n results returned or at most $rn - n + 1$ results returned, the result of $A \times B$ can be got.

The *(m,n)* MDS Code Strategy. In the MDS code strategy, we use the Vandermonde matrix to encode the matrixes which come from CNNs. Any n row vectors of the $m \times n$ Vandermonde matrix can form an $n \times n$ invertible matrix. The process of matrix encoding and returned result decoding is as follow:

$$E_{m\times t} = M_{m\times n} \times A_{n\times t}, \tag{7}$$

$$R_{m\times s} = E_{m\times t} \times B_{t\times s}, \tag{8}$$

$$O_{n\times s} = M_{n\times n}^{-1} \times R_{n\times s}. \tag{9}$$

The matrix $M_{m\times n}$ in Eq. 7 is a Vandermmonde matrix, which can encode the matrix $A_{n\times t}$ redundancy to get redundant encoded matrix $E_{m\times t}$. Then the redundant encoded matrix $E_{m\times t}$ multiplied by the matrix $B_{t\times s}$ in the distributed environment. In the equation 9, the matrix $R_{n\times s}$ from the matrix $R_{m\times s}$ can be decoded with the inverse of its encoding matrix to get the result of $A_{n\times s} \times B_{s\times t}$ finally.

The LT-Code Strategy. Luby Transform (LT) code is one of rateless codes that the number of encoded matrix blocks is not fixed and it increasing with the need of decoding end for decoding out the whole result. Comparing with the MDS code, the LT code has simpler encoding way based on $\oplus$. The operation $\oplus$ is exclusive or. In our system, we use addition and subtraction operations instead of the exclusive or operation and they have the same performance in practice.

In the encoding phase, a degree d is selected to represent the number of original matrix blocks participating in the encoding. The d original matrix blocks are selected from n original matrix blocks randomly and the degree $d \in \{1, 2, \ldots, n\}$. The choice of degree d conforms to the probability distribution $\rho(d)$ which is the robust soliton degree distribution and can be found in [15]. The set of the subscripts of d original matrix blocks is S_d and $S_d \subseteq \{1, 2, \ldots, n\}$. The process

Algorithm 1: Matrix Encoding Algorithm Using LT Code

Input:
The original matrix blocks M;
The number of original matrix blocks n_m;
The robust soliton degree distribution $\rho(d)$;
The redundancy ratio r;
Output:
The encoded matrix blocks E;
The number of encoded matrix blocks n_e;

$n_e = r \times n_m$;
for $i = 0$ *to* n_e **do**
  $d \leftarrow$ get a degree randomly according to $\rho(d)$;
  for $j = 1$ *to* d **do**
    $m \leftarrow$ pick an original matrix block randomly from M;
    $e = e + m$;
  $E[i] = e$;
return E, n_e.

of encoding is shown in Algorithm 1. In the first step of Algorithm 1, we get a degree d according to the robust soliton degree distribution $\rho(d)$. For the original n matrix blocks, the set of values of degree d is $\{1, 2, \ldots, n\}$. Then, we pick d matrix blocks from the original matrix blocks and get the sum of them. The sum of the d matrix blocks is an encoded matrix block. By repeating the above operation, we can theoretically obtain an endless stream of encoded matrix blocks. However, we set a redundant parameter r to determine the number of encoded matrix blocks produced in practice. So, we get rn encoded matrix blocks finally.

In the decoding phase, the decoding starts with blocks of degree 1 and the neighbor blocks of them as we can see in Fig. 5. With the decoding carries on, there are more and more blocks of degree 1. And the blocks of degree 1 are the decoded blocks actually. When all original blocks are decoded, the decoding phase is finished.

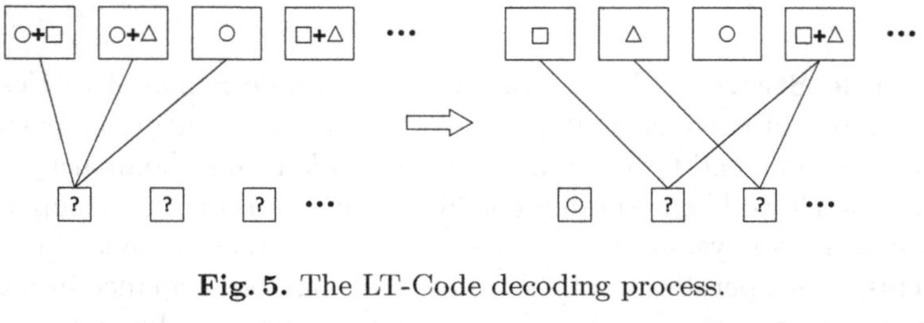

Fig. 5. The LT-Code decoding process.

3.4 The Scheme of Coded Strategies Protecting Data Security

In addition to maintaining the forward and backward propagation of CNNs, the edge server is also responsible for distributing compute tasks to edge worker nodes and collecting computed results from them. After vectorizing the CNNs, we get lots of large matrix multiplications. In a forward propagation of convolutional and fully connected layers, the image data are transform to matrix A and the weight parameters are transform to matrix B. So, protecting the security of the matrix A is equivalent to protecting the security of the image data. As we can see from Fig. 6, matrix A multiplying with matrix B is done by distributed computing according to our design. Firstly, divide the matrix A into many small matrix blocks evenly and then encode them using coding matrix. The size of the small matrix block is determined according to the size of the largest matrix multiplication that the worst-performing edge worker node can undertake. Secondly, make computing tasks. Thirdly, assign the computing tasks to edge worker nodes. Finally, collect the results returned from edge worker nodes and combine them into the result of $A \times B$.

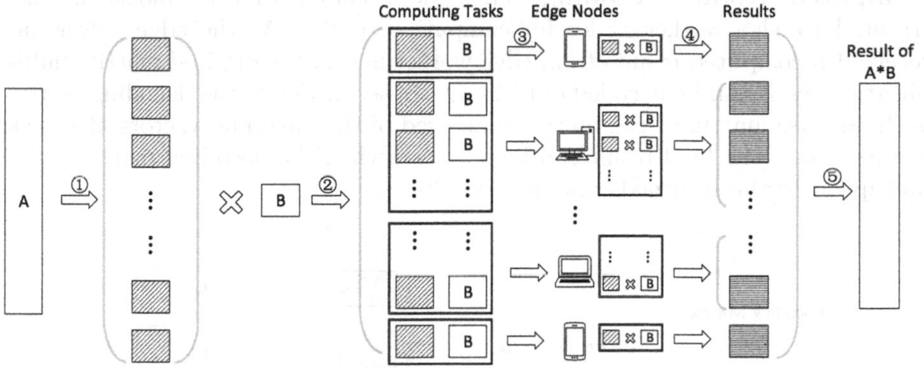

Fig. 6. Make, distribute computing tasks and collect results.

Matrix Redundancy Encoding. The redundant encoding has been proven to perform well in distributed computing [12]. To cope with the problem of stragglers, redundant encoding of the original matrix is inevitable. Because it encodes the original matrix to ensure data security and makes the original matrix redundant at the same time. We use the MDS code strategy as an example. The detailed process is shown in Fig. 7. The Vandermonde matrix can encode the original matrix redundancy in the MDS code strategy, but the encoding vector in the Vandermonde matrix will increase as the size of the encoded matrix increase, which also means that the computational load getting larger. In order to reduce the computational load and low the complexity of encoding, we uniformly partition the original matrix to n smaller blocks at first. Next, the large matrix is encoded with redundancy based on these matrix blocks. The $m \times n(m > n)$ coding matrix is used to encode n small matrix blocks. Finally, we get m encoded matrix blocks.

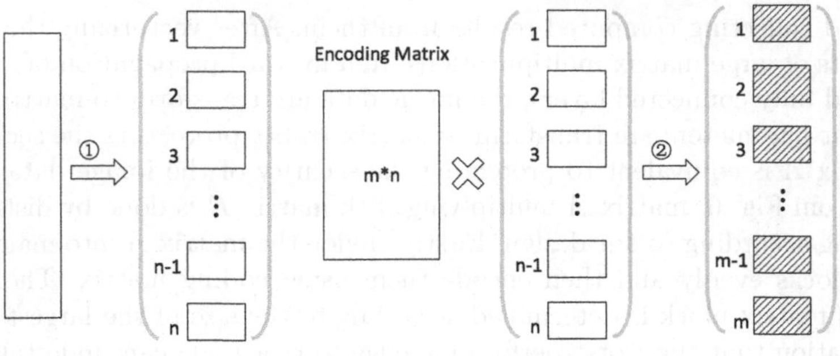

Fig. 7. Divide the big matrix into n blocks, then use the $m \times n (m > n)$ Vandermonde matrix to encode the matrix blocks with redundancy.

Computed Results Decoding. Then the m encoded matrix blocks are distributed to edge worker nodes for computing results. As the edge server has received n computed results from edge worker nodes, the original matrix multiplication result can be decoded out. As we can see in Fig. 8, the decoding matrix is the inverse matrix of the matrix composed of the encoding vectors of n task results. Then, the final result can be obtained when the decoding matrix is left multiplied by the returned computed results.

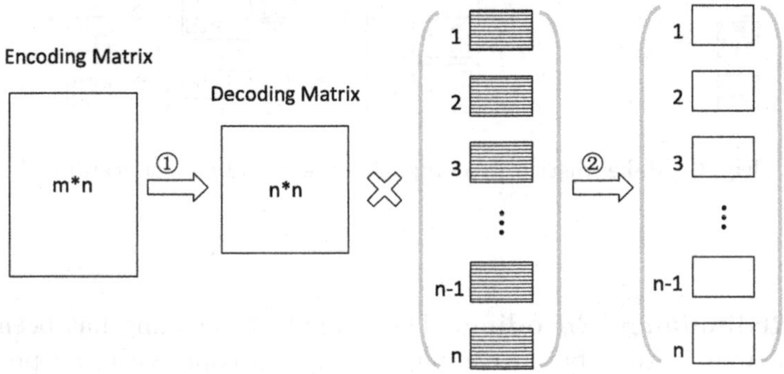

Fig. 8. The process of decoding n result matrix blocks.

Data Security Protecting. In this paper, we protect the security of image data during the transmission from the edge server to the edge worker nodes and when the edge worker nodes perform computing tasks. As we can see in the Fig. 6, the matrix A that needs protection is transformed from the image data. If matrix A is directly transmitted in plaintext from the edge server to the edge worker nodes, the content of matrix A is vulnerable to be leaked during transmission and calculation on the edge worker nodes. Therefore, we use the coding matrix to encode the matrix blocks of matrix A. After that, the content of the encoded matrix blocks has been protected during transmission and calculation. Even if the attacker collects all the encoded matrix blocks by intercepting the data packets of the transmission or attacking the edge worker nodes, he cannot get the content of matrix A. Because the coding matrix is stored in the edge server and it does not participate in the calculation on the transmission and edge worker nodes. In the end, we protect the security of matrix A, which is to protect the security of image data.

3.5 Design of Edge Distributed Computing System

In our system, we do the matrix multiplications with distributed computing. Hence, the scheme of distributing and recycling computing tasks plays an important role in the secure distributed image classification model training system. Because the edge worker nodes have different performance at work, and it is difficult for an edge server to actively contact all edge worker nodes. How to assign the computing tasks to edge worker nodes determines the performance of the entire system. As we can see in Fig. 9, we create two pools on the edge server: the task pool and the result pool. Due to the two pools, we divide a round of computation into three specific steps. In the first step, every edge worker node randomly gets a task from the task pool. In the second step, after completing the computation of the task, the edge worker node puts the result into the result pool. Finally, the edge server takes all results out from the result pool and integrates them into a final result.

We consider the different performance of the edge worker nodes in the edge distributed computing system. Although these edge worker nodes get computing tasks of the same size meanwhile, they return computed results at different times. And it may lead to the problem of straggler. Previously, we introduced redundant coding of computing tasks to address the problem of straggler in distributed computing. Specifically, the final result can be successfully decoded if first n computed results in $m(m > n)$ computing tasks are returned to the result pool. At the same time, we find that if the return of the first n computed results is accelerated, the time to complete this round of calculation will be shorter. The edge distributed computing system we designed can do this job. It has three advantages:

Advantage 1: The computing tasks are properly assigned. In the edge server, we create two pools: the task pool and the result pool. Hence, the edge worker nodes are not assigned tasks passively, but actively takes them from the task pool. Under this mechanism, the more powerful the edge worker node is, the more computing tasks it can accomplish. From a macro point of view, the edge worker nodes with powerful ability undertake a lot of work, while the edge worker nodes with weak ability undertake little work. This reasonable and healthy way of the division of labor makes our system run efficiently and thus speeds up the return of the first n computed results.

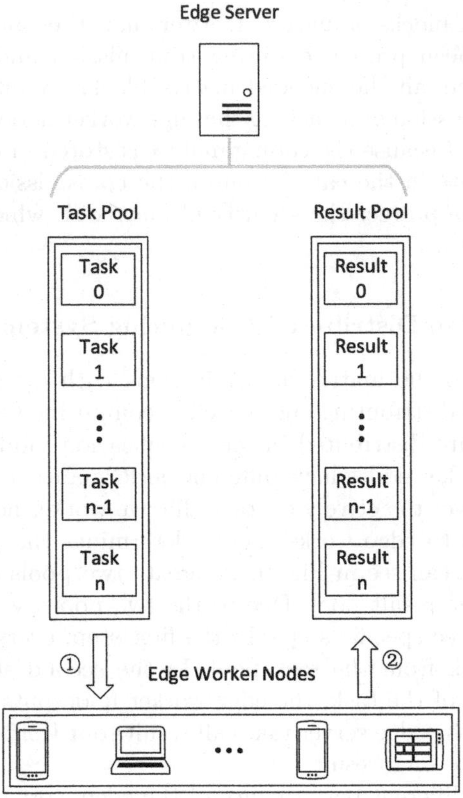

Fig. 9. The edge distributed computing system.

Advantage 2: Easy implementation of the system. The system does not need to set tasks of different sizes for edge worker nodes with different performance. And there is no need to determine which edge worker nodes are the straggler nodes. In the traditional distributed computing system, the edge worker nodes with powerful ability are assigned large computing tasks for reducing the time it takes to complete the all computing tasks. In our system, the edge worker nodes actively fetch tasks from the task pool. So, we just need to partition the original task evenly at the beginning of a distributed computing that the tasks in the task pool are the same size.

Advantage 3: The contribution of straggler edge worker nodes is not wasted. The straggler edge worker nodes take more time to return the computed results than others. However, it does not mean that their work must be abandoned. Especially in our system, the original computing task is divided into many sub-tasks. During a round of distributed computing, the powerful nodes do lots of tasks and the straggler nodes do few tasks rather than nothing. This design of the edge distributed computing system successfully utilizes the computational resources existing in the straggler nodes and increases the computing resources of the whole system.

4 Performance Evaluation

In this section, we compare the performance of distributed matrix multiplication strategies in the edge distributed computing system which is running on edge devices.

4.1 Experimental Settings

In uncoded distributed matrix multiplication strategies, we have uncoded strategy and 2-replication strategy. Relatively, we have the MDS strategy and the LT-code strategy in coded strategies. Before the experiments, we have set a delay model to simulate the latency on the edge worker nodes during the distributed computing. In this paper, we use the distributed processes to simulate the edge worker nodes. In order to simulate the problem of stragglers of edge worker nodes, we add latency T to the processes using a *Sleep*() function. The latency T is a set $\{t_1, t_2, \ldots, t_n\}$ relative to n distributed processes. Then, we make the latency T sequentially complies with the uniform, normal and exponential distribution to get three different kinds of latency sets. For testing the performance of the edge distributed computing system, we perform the coded distributed matrix multiplication experiments on it in the first place. Finally, we conduct a distributed training experiment of an image classification model on it to evaluate the performance of our secure distributed image classification model training system.

4.2 Coded Distributed Matrix Multiplication Experiments

In the first place, we need to evaluate the performance of the secure edge distributed computing system above. Uncoded schemes and coded schemes in the system are compared. Moreover, the performance of different coded schemes are also compared. Then, we make the computing tasks for the evaluation of the system and design serval experimental schemes of matrix multiplication. Before each round of distributed computations, we split the matrix A into ten blocks $\{a_1, a_2, \ldots, a_{10}\}$ along the horizontal direction. So, we get 10 original tasks $\{a_1 \times B, a_2 \times B, \ldots, a_{10} \times B\}$.

The 10 original tasks were processed by uncoded, 2-replication, MDS and LT-code strategies. Then, we put them in the task pool. In the scheme A, we test the computing time required for computing tasks of different sizes using uncoded, 2-Replication, MDS code, and LT-code strategies on the ten distributed processes. And, the latency set on the distributed processes obeys uniform, normal or exponential distribution. In the scheme B, we test the performance of the system completing a same computing task when the latency sets that obey uniform, normal or exponential distribution.

Scheme A: We consider different sizes of matrix multiplication computing on the 10 distributed processes which have latencies that obey uniform, normal or exponential distributions. The 6 kinds of matrix multiplications are from 1000×1000 matrix A multiplying with 1000×10 matrix B to 6000×1000 matrix A multiplying with 1000×10 matrix B. So we have six sizes of computing tasks carrying on the 10 distributed processes with 3 kinds of distributed latencies.

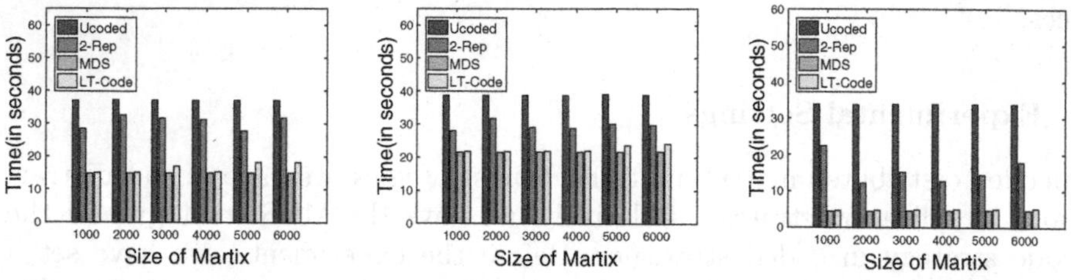

(a) Uniform distribution. (b) Normal distribution. (c) Exponential distribution.

Fig. 10. The matrix multiplications of different sizes carrying on the 10 distributed processes with the latencies that obey the uniform, exponential and normal distribution.

As we can see in Fig. 10(a), Fig. 10(b) and Fig. 10(c), the system with the coded distributed matrix multiplication strategies takes significantly less time to complete the six computing tasks of different sizes than the system with uncoded and 2-Replication strategies whether the latency on the 10 distributed processes obeys uniform, normal or exponential distribution. Among coded strategies, the performance of MDS strategy and LT-code strategy are close. We can also see that the 2-Replication strategy outperforms the uncoded strategy in our experiments. The experiments of the scheme A show that in the distributed computing environment with the problem of straggler, the 2-Replication strategy with backup computing tasks can alleviate the problem of straggler, but the actual performance is inferior to coded strategies.

Scheme B: We consider the multiplication of a 6000×1000 matrix A with 1000×10 matrix B computing on the 10 distributed processes with different latencies. The ranges of values of the latencies that obey uniform distribution are from range between 1 and 10 seconds to range between 1 and 90 seconds. The normal distribution latencies have same expectation 20 and different variances from 12 to 20. And the exponential distribution latencies have different expectations from 12 to 20.

Figure 11(a) shows that the edge distributed computing system adopting different distributed computing strategies takes different times to complete a same computing task when the latencies of 5 ranges on 10 distributed processes obey uniform distribution. From the Fig. 11(a), we can see that coded schemes

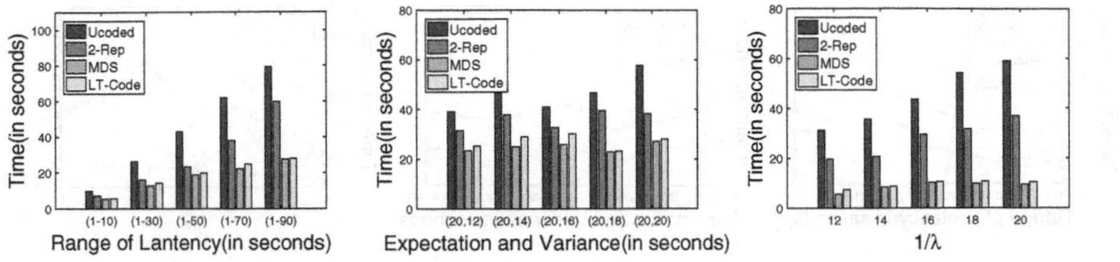

(a) Uniform distribution. (b) Normal distribution. (c) Exponential distribution.

Fig. 11. The multiplication of 6000×1000 matrix A with 1000×10 matrix B carrying on the 10 distributed processes with the latencies that obey the uniform, exponential and normal distribution and have different ranges, different variances and different expectations.

are obviously superior to the uncoded schemes in time spent no matter which range of latencies is set on the distributed processes. And with the range of latencies increases, the performance gap between uncoded and coded schemes is widening, which shows the stability of coded schemes from another point of view. In the Fig. 11(b), the latencies we set on the distributed processes follow normal distribution. And these latency sets have same expectation and different variances. In this condition, we still find that the coded schemes are more excellent than the uncoded ones. In the Fig. 11(c), the latencies on the distributed processes obey exponential distribution. The latencies have different expectations from 12 to 20. As the expectation of the latencies increases, the time taken by the uncoded schemes to complete a same computing task increases linearly. The difference is that the coded schemes are less affected by rising expectations of the latencies.

4.3 Secure Distributed Image Classification Model Training Experiments

We train the vectorized CNN model in the edge distributed computing system and use uncoded schemes and coded schemes respectively. As alluded to above, the coded schemes can deal with the problem of stragglers of edge worker nodes and protect the security of the dataset. For evaluating the performance of our system ,we consider using the MNIST dataset [11] to train a VGGNet [20] model. Besides, we set three kinds of latency sets on the ten distributed processes. Three kinds of latencies respectively obey uniform, normal and exponential distribution. When model training runs out of a batch of images which are 50 images, we record the time spent by our system.

In the Fig. 12(a), the latency sets on the 10 distributed processes obey normal distribution. Whether they are uncoded schemes or coded schemes, the time spent by the system increases as the latency range increases as we can see from the Fig. 12(a). The coded schemes still can bring less time cost to our system. In the coded schemes, MDS strategy and LT-code strategy have comparable per-

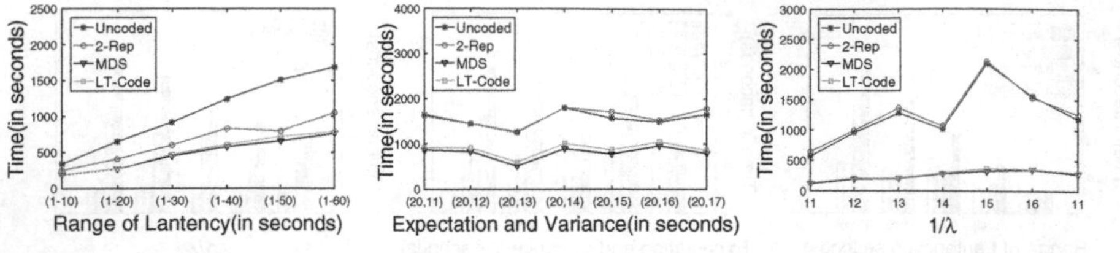

(a) Uniform distribution. (b) Normal distribution. (c) Exponential distribution.

Fig. 12. Training VGGNet using the MNIST dataset on the 10 distributed processes with the latencies that obey the uniform, exponential and normal distribution and have different ranges, different variances and different expectations.

formance. Then, the coded schemes also have better performance than uncoded schemes in Fig. 12(b) where the latency sets obey normal distribution. In the Fig. 12(c), the latency sets on 10 distributed processes obey exponential distribution. With the increasing of expectation, coded schemes perform more stable than uncoded schemes. The MDS strategy and LT-code strategy bring our system less time cost than uncoded strategy and 2-replication strategy.

The coded schemes not only spends less time in distributed model training than the uncoded schemes, but also improves the stability of the system. As the range of latency increases, the time spent in the coding schemes increase more slowly than the time spent in the uncoded schemes in Fig. 12(a). Similarly, we can also see from Fig. 12(c) that as the expectation of the latency increases, the time spent curve of the coded schemes fluctuates less and is smoother than that of the uncoded schemes. In summary, the coded schemes can improve the stability of the system when the image classification model is trained on heterogeneous edge worker nodes with different latencies.

5 Conclusion

In this paper, we designed a secure distributed image classification models training system for heterogenous edge computing. The system can run on edge devices and can deal with the problem of straggler of edge worker nodes. Besides, it has the ability to protect the privacy of computing data because our system adopts coded strategies. It brings two advantages. First, we use many idle edge devices to perform task computing to reduce the cost of the system. Second, we have implemented the secure training of CNN models on edge devices. The datasets collected at the edge no longer need to be uploaded to the cloud, which avoids the security problem during transmission and saves the transmission bandwidth. After the experiment, we demonstrated that the system in this paper can deal with the problem of straggler and can run smoothly on edge devices.

References

1. Abdellatif, A.A., Mohamed, A., Chiasserini, C.F., Tlili, M., Erbad, A.: Edge computing for smart health: context-aware approaches, opportunities, and challenges. IEEE Netw. **33**(3), 196–203 (2019)
2. Bahrami, K., Shi, F., Rekik, I., Shen, D.: Convolutional neural network for reconstruction of 7T-like images from 3T MRI using appearance and anatomical features. In: Carneiro, G., et al. (eds.) LABELS/DLMIA -2016. LNCS, vol. 10008, pp. 39–47. Springer, Cham (2016). https://doi.org/10.1007/978-3-319-46976-8_5
3. Benou, A., Veksler, R., Friedman, A., Riklin Raviv, T.: De-noising of contrast-enhanced MRI sequences by an ensemble of expert deep neural networks. In: Carneiro, G., et al. (eds.) LABELS/DLMIA -2016. LNCS, vol. 10008, pp. 95–110. Springer, Cham (2016). https://doi.org/10.1007/978-3-319-46976-8_11
4. Chellapilla, K., Puri, S., Simard, P.: High performance convolutional neural networks for document processing (2006)
5. Chen, J., Ran, X.: Deep learning with edge computing: a review. Proc. IEEE **107**(8), 1655–1674 (2019)
6. Chen, Y., Zhang, Y., Maharjan, S., Alam, M., Wu, T.: Deep learning for secure mobile edge computing in cyber-physical transportation systems. IEEE Netw. **33**(4), 36–41 (2019)
7. Dutta, S., Cadambe, V., Grover, P.: Coded convolution for parallel and distributed computing within a deadline. In: IEEE International Symposium on Information Theory (ISIT), pp. 2403–2407. IEEE (2017)
8. Ghafoorian, M., et al.: Deep multi-scale location-aware 3D convolutional neural networks for automated detection of lacunes of presumed vascular origin. NeuroImage Clin. **14**, 391–399 (2017)
9. Golkov, V., et al.: q-space deep learning: twelve-fold shorter and model-free diffusion MRI scans. IEEE Trans. Med. Imaging **35**(5), 1344–1351 (2016)
10. Hoffmann, N., Koch, E., Steiner, G., Petersohn, U., Kirsch, M.: Learning thermal process representations for intraoperative analysis of cortical perfusion during ischemic strokes. In: Carneiro, G., et al. (eds.) LABELS/DLMIA -2016. LNCS, vol. 10008, pp. 152–160. Springer, Cham (2016). https://doi.org/10.1007/978-3-319-46976-8_16
11. LeCun, Y., Bottou, L., Bengio, Y., Haffner, P., et al.: Gradient-based learning applied to document recognition. Proc. IEEE **86**(11), 2278–2324 (1998)
12. Lee, K., Lam, M., Pedarsani, R., Papailiopoulos, D., Ramchandran, K.: Speeding up distributed machine learning using codes. IEEE Trans. Inf. Theory **64**(3), 1514–1529 (2017)
13. Li, H., Ota, K., Dong, M.: Learning IoT in edge: deep learning for the internet of things with edge computing. IEEE Network **32**(1), 96–101 (2018)
14. Litjens, G., et al.: A survey on deep learning in medical image analysis. Med. Image Anal. **42**, 60–88 (2017)
15. Mallick, A., Chaudhari, M., Joshi, G.: Rateless codes for near-perfect load balancing in distributed matrix-vector multiplication. arXiv preprint arXiv:1804.10331 (2018)
16. Nie, D., Cao, X., Gao, Y., Wang, L., Shen, D.: Estimating CT image from MRI data using 3D fully convolutional networks. In: Carneiro, G., et al. (eds.) LABELS/DLMIA -2016. LNCS, vol. 10008, pp. 170–178. Springer, Cham (2016). https://doi.org/10.1007/978-3-319-46976-8_18

17. Pouyanfar, S., et al.: A survey on deep learning: algorithms, techniques, and applications. ACM Comput. Surv. (CSUR) **51**(5), 92 (2019)
18. Ren, J.S., Xu, L.: On vectorization of deep convolutional neural networks for vision tasks. In: Twenty-Ninth AAAI Conference on Artificial Intelligence (2015)
19. Ren, J., Guo, Y., Zhang, D., Liu, Q., Zhang, Y.: Distributed and efficient object detection in edge computing: challenges and solutions. IEEE Network **32**(6), 137–143 (2018)
20. Simonyan, K., Zisserman, A.: Very deep convolutional networks for large-scale image recognition. Computer Science (2014)
21. Wang, S., et al.: When edge meets learning: adaptive control for resource-constrained distributed machine learning. In: IEEE INFOCOM 2018-IEEE Conference on Computer Communications, pp. 63–71. IEEE (2018)

HIM: A Systematic Model to Evaluate the Health of Platform-Based Service Ecosystems

Yiran Feng, Zhiyong Feng, Xiao Xue, and Shizhan Chen(✉)

Department of Intelligence and Computing, Tianjin University, Tianjin 300350, China
{zyfeng,fengyiran,jzxuexiao,shizhang}@tju.edu.cn

Abstract. With the vigorous development of the platform-based service ecosystem represented by e-commerce, service recommendation is used as a personalized matching method. There exist some service recommendation strategies that mainly focus on high popularity services and ignore non-popular ones. This will not only lead to oligopoly, but also be detrimental to the health of the platform-based service ecosystem. In addition, the health evaluation indicators for this kind of ecosystems are mostly qualitative and single. In view of the above phenomenon and based on the system view of balance and health, a health index model (HIM) is proposed to measure the health from two aspects quantitatively: stability and sustainability. Specifically, the model includes the system activity and organizational structure reflecting stability, as well as the productivity and vitality reflecting sustainability, which helps to illustrate the health status of the platform-based service ecosystem from the perspective of multi-dimensional integration. Additionally, this paper analyzes the factors affecting the health of this ecosystem based on HIM. In this work, a platform-based service ecosystem simulation model is constructed by using the computational experiment method to verify the effectiveness of HIM. The simulation results show that the HIM can reasonably measure the health of such ecosystems, which has guiding significance for the overall management and sound development of e-commerce platforms.

Keywords: Health index model · Platform-based service ecosystem · Recommendation strategies · Computational experiment

1 Introduction

With the rapid development of e-commerce [11], a platform-based service ecosystem with service providers, services, consumers and service operators as the core is gradually formed. In the real world, Alibaba's Taobao is a very typical platform-based service ecosystem. In the ecosystem, service providers refer to those who are in possession of resources, and provide services to consumers. Services are products that can be provided by service providers. Consumers can

H. Gao et al. (Eds.): CollaborateCom 2020, LNICST 349, pp. 199–216, 2021.
https://doi.org/10.1007/978-3-030-67537-0_13

use the platform to choose services that meet their needs. The platform operators can connect multi-party services and formulate marketing strategies, thus to increase the efficiency of value creation.

As the development of the platform-based service ecosystem, the recommendation strategy as a marketing method plays an important role in it [16,22]. It promotes the healthy development of the platform-based service ecosystem while meeting the individual needs of users [14]. Nowadays, some recommendation strategies focus on services with high popularity, ignoring services with low popularity and cooperation between services [2,17], which is not conducive to the healthy development of platform service ecosystem. Studying its health is not only helpful in deepening the understanding of the ecosystem but also beneficial to its sound development. Therefore, measuring the health of the platform-based service ecosystem has become an important research topic.

In recent years, correlative research efforts have been posed to study the health of the ecosystem [21]. Gini coefficient [4] and Shannon-Weiner index [18] are recognized indicators that can reflect the state of the ecosystem, but they can only reflect one aspect of the system. The traditional Quality of Service [3,12] and system performance evaluation [10] also have the above problems, which are weak and single for measuring the health of the ecosystem. Therefore, how to measure the health of the platform-based service ecosystem from multiple dimensions is an urgent research issue to be solved.

To tackle the aforementioned issues, first of all, this paper proposes a health index model to describe the health of the platform-based service ecosystem from the two aspects of stability and sustainability [13,15]. System activity and organizational structure reflect system stability; Productivity and vitality reflect sustainability. Then, the computational experiment [20,24] is used to design models of consumers, service providers, and the platform operator to ensure the construction of the platform-based service ecosystem. The evolution of the platform-based service ecosystem under the three recommendation strategies of random, collaborative filtering and bundling are simulated. Finally, the proposed health index model is used to measure the health of the ecosystem dynamically.

The main contributions of this paper are as follows:

- A health index model is proposed, which can evaluate the health of the platform-based service ecosystem in multiple dimensions. This contributes to our understanding and the healthy development of the platform-based service ecosystem.
- By comparing the two indicators of Gini coefficient and service survival rate, the rationality of the health index model is verified through the computational experiment.
- The factors that affect the healthy development of the platform-based service ecosystem are studied and analyzed. This can provide new ideas for the analysis of the platform-based service ecosystem.

The structure of this paper is organized as follows. Section 2 shows the health index model of the platform-based service ecosystem. Sect. 3 presents the details

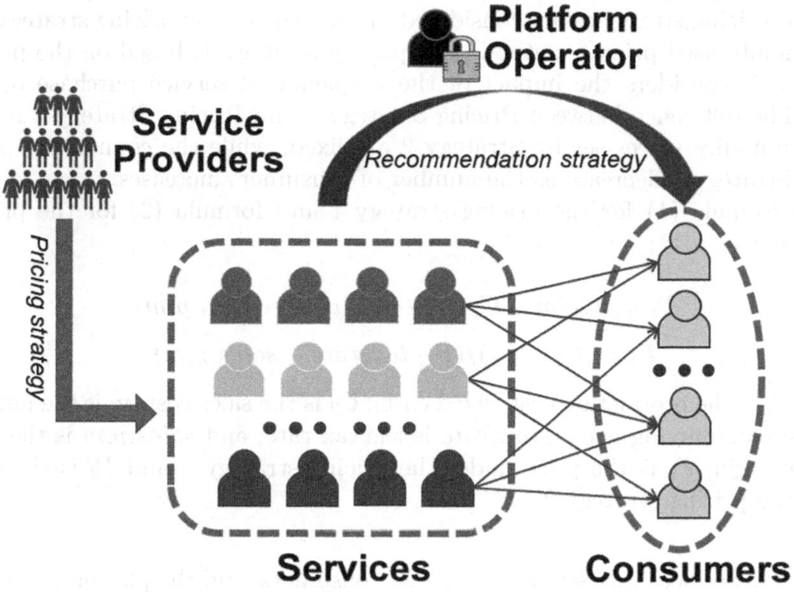

Fig. 1. The platform-based service ecosystem

and results of the experiments. Section 4 reviews the related work. Section 5 summarizes the whole paper.

2 The Health Index Model of Platform-Based Service Ecosystem

2.1 The Platform-Based Service Ecosystem

The platform-based service ecosystem is a complex and interdependent system mainly composed of the service providers, services, consumers and platform operator. Service providers can use the resources of the platform to improve their own value, continue to develop new products for consumers; Services are products that can be provided by service providers; Consumers can use the platform to choose services that meet their needs; The platform operators can connect multi-party services and formulate marketing strategies. In the real world, Alibaba's Taobao is consistent with the platform-based service ecosystem, so it is of considerable significance to study such ecosystems.

Based on the characteristics of the platform-based service ecosystem, this paper constructs the ecosystem from four main aspects: service providers, services, consumers and platform operators, as illustrated in Fig. 1.

Service Providers. A service provider is an enterprise or individual that provides services to consumers through the platform. The pricing behavior of service providers is mainly considered.

Two pricing strategies are considered in this paper. The pricing strategy 2 is a commonly used pricing method. The pricing strategy 1, based on the pricing strategy 2, considers the impact of the frequency of service purchase on the price. The difference between Pricing Strategy 1 and Pricing Strategy 2 is that the commodity prices set by Strategy 2 are fixed, while the commodity prices set by Strategy 1 decrease as the number of consumers increases.

See formula (1) for the pricing strategy 1 and formula (2) for the pricing strategy 2.

$$P_1 = (C_1/m + C_2)/(1 - tax\_rate - sales\_p\_m) \tag{1}$$

$$P_2 = (C_1 + C_2)/(1 - tax\_rate - sales\_p\_m) \tag{2}$$

where C_1 is the production cost (fixed cost); C_2 is the sales cost; m is the number of consumers buying goods; tax_rate is the tax rate, and $sales\_p\_m$ is the sales profit margin; P_1 is the price under the pricing strategy 1 and P_2 is the price under the pricing strategy 2.

Services. Services link service providers, consumers, and the platform operator to form a platform-based service ecosystem. The main characteristic properties of the service are shown in Table 1.

Consumers. In the experiment system, agent model is used to depict the behavior features of various consumers, which is the individual with independent decision and cooperative competition ability. The formal description of consumer agent is shown as formula (3).

$$Consumer =< S, Et, Yt > \tag{3}$$

where S represents consumer attributes; Et represents perceived events; Yt represents decision-making mechanism.

Consumer attributes S: The attributes of consumers includes four aspects.

(1) Willingness to pay(V_i): The willingness of consumers to pay refers to the valuation of the service goods or the price they are willing to pay, where the willingness to pay obeys a uniform distribution of [0, a].
(2) Consumer demands(Cd): Different consumers have different demands for service goods. The demand here corresponds to the type of service goods.
(3) Consumer satisfaction(Cs): Consumer satisfaction is the degree of satisfaction of consumers after enjoying the service, and the initial consumer satisfaction is set to a fixed value.
(4) Purchase history(Ph): Records of service goods purchased by consumers.

Table 1. The main characteristic properties of the service

Property name	Brief description	Remarks
P	The selling price	P is the selling price of the service. Different pricing strategies lead to different prices of services
C_1	Production cost	C_1 refers to various production costs incurred in providing services
C_2	Selling cost	C_2 refers to the production cost of the service sold or the labor cost of the labor service provided and the business cost of other sales
T	Type of service	T is the category of the service that the service provider can provide to the customer
Qos	Quality of service	Qos refers to the sum of the characteristics and characteristics of services that can meet the requirements and potential needs, and refers to the extent to which services can meet the needs of the served
R	Service reputation	R refers to the reputation of the service, which will not change during the experiment
Rt	Response time	Rt is the time that the customer has to wait for the service
L	Service life	The initial life value of the service is fixed. The rules for updating the life value are as follows. If the consumer currently chooses to use this service, the service life remains unchanged; If no consumer chooses the service, the service life decreases; If the service life value becomes 0, the service is dead
Re	Service revenue	Re refers to the benefits of services
Ce	The cross-elasticity coefficient of demand	Ce reflects the complementarity between services

Decision mechanism Yt: The consumer's decision mechanism mainly reflects the behavior of consumers. There are two main behaviors: Selecting recommended services and satisfaction evaluating of the purchased services.

(1) Select recommended services: Consumers decide whether to purchase a service based on their willingness to pay and the average satisfaction of the service.
(2) Satisfaction evaluation of services: Consumers rate services after using them. Satisfaction evaluation is related to service attributes (quality of service, response time and service reputation). Evaluate the service according to formula (4).

$$Satisfactory = \alpha \times Qos + \beta \times Rt + \gamma \times R \quad (\alpha + \beta + \gamma = 1) \tag{4}$$

where Qos is the quality of the service; Rt is the response time; R is the service reputation.

Perceived Events Et: The external events perceived by consumer can affect their decision-making behavior. The perceived events of consumers are as follows.

Perceive the Service Price and Average Satisfaction: Consumers can perceive the price and average satisfaction of the current candidate service.

Table 2. Three ways of the bundling strategy

Ways	Applicable conditions
Pure bundling	The absolute value of the demand cross-elasticity of bundling parties is small (the degree of complementarity is low) and the service quality is good
Mixed bundling	The absolute value of the demand cross-elasticity of bundling parties is at the center (the degree of complementarity is in the middle) and the service quality is average
Separation strategy	The absolute value of the demand cross-elasticity of bundling parties is high (the degree of complementarity is high)

The Platform Operator. The platform operator mainly regulates and controls the entire platform through marketing strategies, so as to make the whole ecosystem healthy and stable. The followings are common recommended strategies.

The Bundling Strategy: The bundling strategy is a marketing strategy commonly used by e-commerce platforms. The relevant content of this strategy is as follows.

(1) Price of bundled services: The price of the bundled service is shown in the following formula.

$$P < P1 + P2 \tag{5}$$

where $P1$ and $P2$ are the prices of two bundled services sold separately; P is the price of the bundle.

(2) Three ways of the bundling strategy are shown in Table 2.

(3) Profit distribution for bundled services: Game theory is used for profit distribution of bundled services. Formula (6) uses the Shapley's law in cooperative games for profit distribution.

$$\begin{aligned} \phi_i &= \sum_{i \in S \in I} W(|S|)[V(S) - V(S \setminus \{i\})] \\ W(|S|) &= \frac{(n-|S|)!(|S|-1)!}{n!} \end{aligned} \tag{6}$$

where I={1,2,...,n} is the set of all the people participating in the game; S is a subset of I; $|S|$ is the number of elements in the set S; $V(S)$ represents the profit of set S; $V(S\backslash\{i\})$ represents the profit of other individuals in the set S except i; ϕ_i represents the profit that individual i should earn.

The Collaborative Filtering Strategy: The collaborative filtering recommendation strategy is the currently popular recommendation algorithm. It is mainly divided into the following two steps:

(1) Building a service similarity matrix: Use the Pearson correlation coefficient to calculate the similarity between services based on the history of all users, and build a similarity matrix. This is shown in formula (7),

$$sim_1(i_1, i_2) = \frac{\sum_{u \in u_{i_1} \cap u_{i_2}} (r_{u_{i_1}} - \overline{r_{i_1}}) \times (r_{u_2} - \overline{r_{i_2}})}{\sqrt{\sum_{u \in u_{i_1} \cap u_{i_2}} (r_{u_{i_1}} - \overline{r_{i_1}})^2} \times \sqrt{\sum_{u \in u_{i_1} \cap u_{i_2}} (r_{u_2} - \overline{r_{i_2}})^2} \quad (7)$$

where $r_{u_{i_1}}$ and r_{i_1} indicate the rating of the items (i_1 and i_2) by user u; $\overline{r_{i_1}}$ and $\overline{r_{i_2}}$ represent the average of the scores of items i_1 and i_2; $sim_1(i_1, i_2)$ indicates the similarity of items i_1 and i_2.

(2) Make a recommendation: According to the similarity matrix of the service and the user's historical purchase record, the user's service recommendation is made.

The Random Recommendation Strategy: The random recommendation strategy is to randomly select services for consumers to recommend.

The dynamic change of the platform-based service ecosystem depends on the match between the service and the consumer driven by the recommendation strategy. In practice, the service is recommended to consumers. If the service cannot be selected by consumers for a long time, its life value will decrease and gradually die; Consumers select services from the list of candidate services based on service prices and service evaluations, and then evaluate the service. Under different circumstances, the different matching of consumers and services will affect the healthy development of the entire ecosystem.

2.2 The Health Index Model

Nowadays, with the rapid development of e-commerce, many platform-based service ecosystems with extensive influence have been formed. Assessing the health of the platform-based service ecosystem is a great help for its research methods and establishment approaches. A healthy platform-based service ecosystem is of high stability, which means it can resist changes in the external environment, and of high sustainability, which means it can continuously tap its inherent potential to achieve sustainable development. Based on the definition of platform-based service ecosystem, this paper draws on the health research of ecosystems, and

Fig. 2. The health index model

provides a platform-based service ecosystem health index model as shown in Fig. 2. The service ecosystem health index is expressed as formula (8).

$$HI = S_1 \times \alpha + S_2 \times \beta \tag{8}$$

where HI is the system health index; S_1 is the stability of the system; S_2 is the sustainability. The shares of S_1 and S_2 in the entire index formula are α and β, which are divided based on the importance of stability and sustainable development to the health of the service ecosystem. For a platform-based service ecosystem, stability determines the existence of the ecosystem, and sustainable development determines the development prospects of the system. This paper considers the significance of existence is more important than development. Therefore, $\alpha>\beta$ is set in the experimental part.

Stability. Platform-based service ecosystems' stability is referred to as the ability to maintain itself after being interfered, which means minor fluctuations and random disturbances would not change the system. Drawing on the relatively mature research on health in the ecosystem, this paper divides the factors that affect the stability of ecosystem into the following two factors: system activity and organization structure. The stability of the ecosystem can be expressed by the formula (9).

$$S_1 = O \times A \tag{9}$$

where S_1 is referred as the stability of the service ecosystem; O is the organization structure; A is the system activity.

(1) Organization Structure. It means the complexity of ecosystem's structure. This paper draws on the Shannon-Wiener diversity index in the biosphere to reflect the organizational structure. This method can reflect the types of services in the platform-based service ecosystem and the uniformity of the distribution

of various services, as shown in formula (10).

$$O = -\sum_{i=1}^{m}(Pi)(lnPi) \tag{10}$$

where m is the number of types of services; $P_i = n_i/N$; N means the total number of services in ecosystem, and n_i means the number of the i-th type of service.

(2) System Activity. In an ecosystem, activity refers to all the energy that can be measured according to the nutrient cycle and productivity. Drawing it into a platform-based service ecosystem, activity means production, that is, service coverage or resource utilization. This satisfies the formula (11).

$$A = \sum_{i=1}^{m}\frac{n_l}{N} \tag{11}$$

where m is the number of types of services; N means the total number of services in ecosystem, and n_l means the number of services used by consumers in the i-th service goods.

Sustainability. Service ecosystem's sustainability refers to the trend of constantly development with the change of time and environment. Considering the sustainability of service ecosystem, this paper divides the factors affecting sustainable development into the following two aspects: productivity and vitality. The following formula can express the sustainability of the ecosystem.

$$S_2 = P \times R \tag{12}$$

where S_2 is the sustainability of the system; P is the productivity; R is the vitality.

(1) Productivity. Productivity refers to the ability to transform the technology and other innovative raw materials to low cost and new productions. The simplest and most effective measure of this ability is the average return on investment. This paper uses the service average investment return rate to evaluate the service ecosystem's productivity. It satisfies the following formula.

$$P = \frac{1}{N} \times \sum_{i=1}^{N}\frac{Pr_i}{C_i} \tag{13}$$

where N is the total number of services; Pr_i is the ith service products' price; C_i is the ith service's cost; Pr_i/C_i is the investment rate.

(2) Vitality. For the platform-based service ecosystem, vitality refers to the ability to provide lasting benefits for services that depend on it. In the face of external shocks, the mutual relationship between the members of the ecosystem can play a certain role in buffering. Therefore, the most straightforward measure of vitality is the survival rate of ecosystem members.

$$R = \frac{n}{N} \tag{14}$$

where N is the total number of services; n is the number of the survived services.

The above is the platform-based service ecosystem health index model proposed in this paper.

3 Experiments and Evaluation

This paper studies a platform-based service ecosystem similar to Taobao. In reality, the relevant source data sets cannot be obtained. Therefore, the method of computational experiments is used to conduct simulation experiments.

The computational experiment has the advantage of accurate control, simple operation and repeatability, and has been widely used in analyzing various complex systems, such as transportation system, ecological environment, socioeconomic system, political ecosystems, etc. Hence, computational experiment can be used to design various experiment scenarios to compare the performance of different strategies. This paper aims to simulate and compare bundling strategy, collaborative filtering strategy and random recommendation strategy under the same experimental environment.

3.1 Experimental Parameter Settings

The computational experiment system is constructed as an artificial society laboratory, see Sect. 2.1 for details. In order to make the experiment easy for readers to understand, the prototype of this model is based on Taobao. The experimental parameters are shown in Table 3.

3.2 Experiment Results and Analysis

Based on the proposed evaluation model, the evolution of the platform-based service ecosystem under the three recommendation strategies is simulated. The service diversity index, resource utilization rate, average investment return rate and service survival rate in each state of the system are corresponded to the calculation of the health index formula, and are normalized to get the health value of the system. This paper uses a five-level evaluation scale to quantify the various evaluation results.

From Table 4, it is known that 0.4 is the alert value of the ecosystem health index. When the health index is above 0.4, ecosystem's health level is at least

Table 3. The experimental parameters

System variables	Experimental setup
System variable settings	50*50
Consumer number	It can be set up according to experiment acquirements
Consumers' Willingness to pay	V_i represents consumers' wiliness to pay. V_i follows uniform distribution in[0,a]. In this paper, it is set in [0, 100]
Consumers' shopping history	This paper generates consumer score data according to Weibull distribution
Service number	The service is divided into long tail service and popular service. According to the pareto principle, set the number of the two to 2: 8. The number of services varies from 20 to 2000 according to experimental needs
Type of service	The number of service types is positively correlated with the number of services. In the experimental part, we use Arabic numerals to mark the service types, and the same number means the same service type
Service cost	The service cost contains production cost and sale cost. The two individual cost are floats lying in [1, 6]
Service price	Service prices are set according to two pricing strategies
Quality of service	Service quality refers to the sum of the characteristics and characteristics of services that can meet specified and potential requirements, in other words, the extension that the service can meets. It follows a random distribution in [10, 20]. {(10∼12) means bad, (12∼15) means normal, (15∼18) means good, (18∼20) means perfect.}
Service reputation	It means the prestige of the service. It follows a random distribution in [10, 20]. {(10∼12) means bad, (12∼15) means normal, (15∼18) means good, (18∼20) means perfect.}
Service initial life value	Initial values: random from the range of [40,50]
Response time	It means the time to wait before enjoying the service. Usually, hot service's responding time is longer. It follows a random distribution in [10, 20]. {(10∼12) means bad, (12∼15) means normal, (15∼18) means good, (18∼20) means perfect.}
The cross-elasticity coefficient of demand	The absolute value of the cross-elasticity coefficient between services follows a random distribution in [0, 15].{(0∼5) means the value is small, (5∼10) means the value is centered, (10∼15) means the value is large.}
Service satisfaction score	Initial values: random from the range of [10,20]
Consumer satisfaction	Consumers will score the service in the end. A satisfactory score is related to service property. The calculation is shown in formula (4)

Table 4. Five-level evaluation scale

Health level	Very low	Low	Normal	High	Very high
Threshold	0.2	0.4	0.6	0.8	1

normal and even better; when the health index is below 0.4, ecosystem's health level is low or worse.

The experimental results in this paper include two aspects. First, the rationality of the health index model is verified by comparison with the Gini coefficient and service survival rate. Secondly, this paper analyzes the factors that affect the long-term healthy development of the platform-based service ecosystem.

The Rationality of the Health Index Model

The Relationship Between Gini Index and Health Index. The Gini coefficient is a commonly used indicator to measure the income gap and reflect the state of the system. It is introduced into this paper to measure the service income gap. Comparison with the accepted Gini coefficient can verify the rationality of the proposed health index model. The Gini coefficient is shown in formula (15).

$$G = 1 - \frac{1}{n} \times (2 \sum_{i=1}^{n-1} W_i + 1) \tag{15}$$

where n means that the service is equally divided into n groups; W_i reflects the percentage of total service income accumulated to group i as a percentage of total population income.

International practice regards below 0.3 as a relatively average income; 0.3–0.4 is regarded as a relatively reasonable income; 0.4 is a warning value, and when the Gini coefficient reaches 0.5 or more, it means income disparity.

In the simulated platform service ecosystem, the relationship between Gini index and health index under the three strategies can be found. As shown in the Fig.3, when service income Gini index is well, service ecosystem health index is high. Overall, with the continuous increase of the Gini coefficient in the service ecosystem, the health index has first increased and then decreased. When Gini coefficient is less than 0.4, which means service income gap is small, the health index of the ecosystem can be maintained at a relatively healthy value; When Gini coefficient is between 0.4 and 0.5, the health index will decline. However, it is still greater than the alert value; when Gini coefficient exceeds 0.5, the health index will be lower than the alert value, and with the continuous increase of the Gini coefficient, the health index will continue to decline.

From the above analysis, it can be found that the trend of the health index and the Gini coefficient is approximately the same, so it can explain the rationality of the health index model.

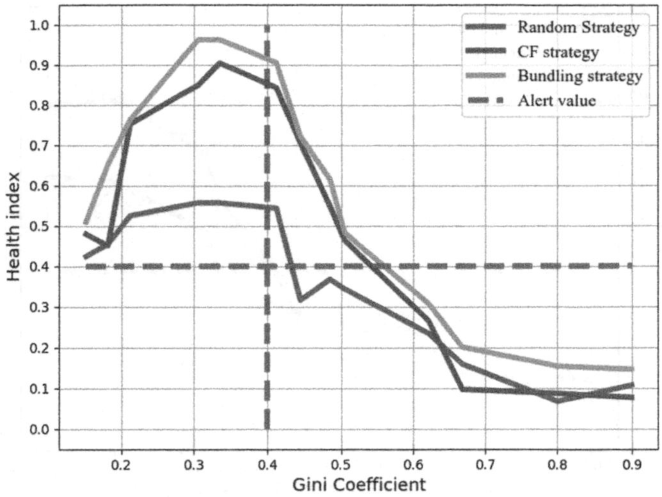

Fig. 3. Relationship between Gini coefficient and the health index

The Relationship Between the Survival Rate and the Health Index. The health index has a certain relationship with the survival rate of services in the ecosystem. The calculation of service survival rate is shown in formula (14). With the dynamic evolution of the platform-based service ecosystem, the service survival rate is constantly changing. As can be seen from Fig. 4, with the increase of service survival rate, the health index of the ecosystem also increases and then stabilizes. When the service survival rate is low, the health index is low; When the service survival rate is high, the health index is also high.

The relationship between health index and service survival rate, Gini index can reflect health index's reasonability to some extent. What's more, the health index proposed in this paper studies the service ecosystem from multiple dimensions, taking into account the ecosystem's service diversity, service survival rate, service coverage rate, and service investment return rate. Compared with a single indicator for measuring ecosystems, our proposed health index model covers multiple aspects of indicators that can more comprehensively and reasonably reflect the health of an ecosystem.

Factors Affecting the Health of the Platform-Based Service Ecosystem. The purpose of this paper is to verify the health index model by evaluating recommendation strategies. The reasons for choosing these three strategies are as follows. The random recommendation strategy is selected as a benchmark for evaluation; The collaborative filtering recommendation strategy is selected as the currently popular recommendation algorithm; The bundling strategy can increase the attention of non-popular products, but also it has brought about problems such as decreased consumer satisfaction. This strategy is currently a marketing strategy commonly used by e-commerce platforms, so it is chosen.

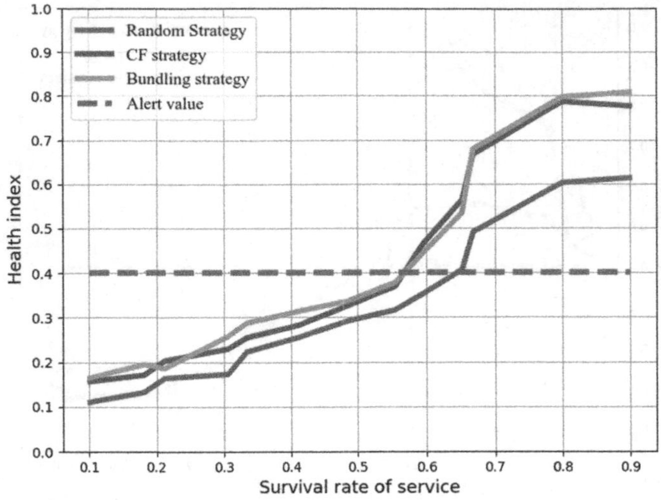

Fig. 4. Relationship between service survival rate and the health index

The Ratio of Consumers to Services and Different Recommendation Strategies. In the experiment, the platform-based service ecosystem health index for the three strategies is shown in Fig. 5, where the x-axis corresponds to the ratio of consumers to services, and the y-axis corresponds to the platform-based service ecosystem health index. On the whole, as the ratio continues to increase, the health indexes of the bundling strategy, collaborative filtering strategy, and random strategy increase sharply; When the ratio of consumers to services reaches a fixed value, the health index of the bundling strategy and collaborative filtering strategy grows slowly and steadily, while the health index of the randomly recommended strategy tends to stabilize; As the ratio continues to increase, the health index of the bundling strategy and collaborative filtering strategy gradually stabilizes.

It can be found from Fig. 5 that when the ratio of consumers to services is less than 3, the health index under the collaborative filtering strategy is higher than the other two strategies, so collaborative filtering is the current optimal strategy; When the ratio of consumers to services is greater than 3 and less than 14, the health index under the bundling strategy is higher than the other two strategies and the bundling strategy is the current optimal strategy; When the ratio of consumers to services is greater than 14, although the health index under the bundling strategy is slightly higher than the collaborative filtering strategy, the health index of the two are not much different, so both strategies are optimal strategies.

In summary, the quantitative relationship between consumers and services has an impact on the health of the platform-based service ecosystem. Moreover, different recommendation strategies should be used at different ratios to make the platform-based service ecosystem healthier.

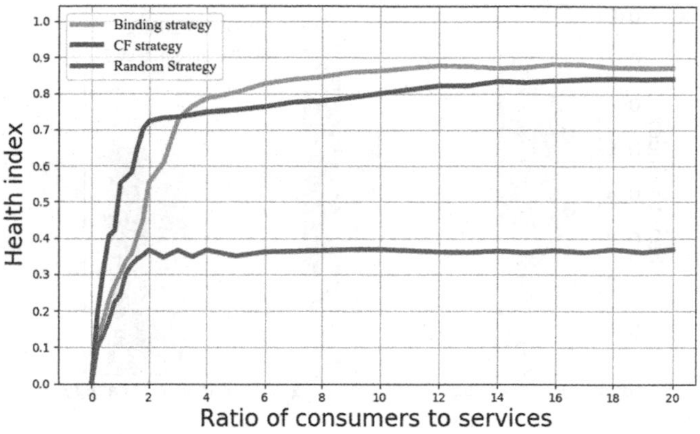

Fig. 5. Relationship between Ratio of consumers to services and the health index

Different Pricing Strategies and Recommendation Strategies. This experiment compares the platform-based service ecosystem health index under two pricing strategies and three recommended strategies. It is shown from the Fig. 6 that when the number of services is much larger than the number of consumers, the collaborative filtering strategy performs best, and the use of pricing strategy 2 is better. If the number of consumers is much larger than the number of services, the bundling strategies will perform best, and the use of pricing strategy 1 is better.

From the above analysis, different recommendation strategies will affect the health of the platform-based service ecosystem. And different pricing strategies should be used under different recommendation strategies to ensure a healthier platform-based service ecosystem.

4 Related Work

4.1 Study on the Health of Ecosystems

The new defined service system resembles the concept of business ecosystem largely. A service system is a value-coproduction configuration of people, technology, and other internal and external service systems [19]. The research on the health of service ecosystems can draw on the research on ecosystems in the biosphere. Constanza believes that a healthy ecosystem is stable and sustainable, has vitality, can maintain its organization and maintain its ability to operate itself, and is resilient to external pressures [7]; Rapport et al. Believed that ecosystem health refers to that the ecosystem has no pain reflection, is stable and sustainable development, that is, the ecosystem has vitality over time and can maintain its organization and autonomy, and is easy to recover under external stress [13]. Based on the former research, this paper proposes a health index model based on the service ecosystem.

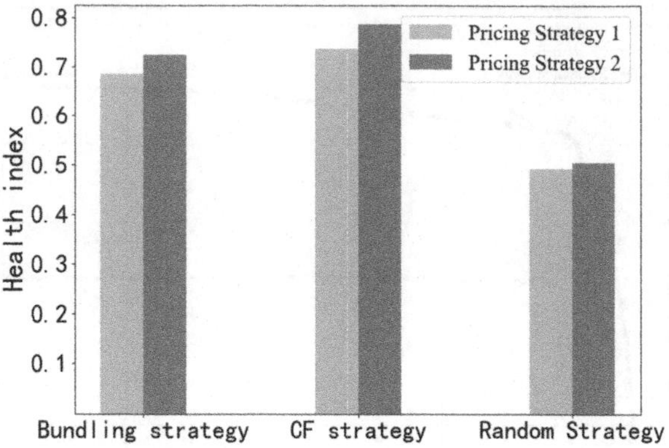

(a) The ratio of consumers to services is 0.8

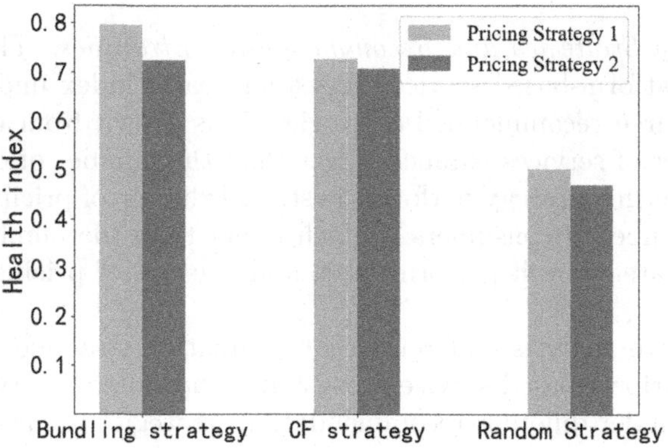

(b) The ratio of consumers to services is 5

Fig. 6. Impact of two pricing strategies and recommendation strategies on health index

4.2 Recommendation Strategy

The recommendation strategy is considered as one of the effective methods to alleviate the imbalance between information production and acquisition caused by the information overload problem [9]. In the field of traditional service recommendation, many researchers have done a lot of related work [5]. Collaborative filtering algorithm [6], which is widely used at present, has greatly promoted the development of recommendation systems. This recommendation algorithm can be divided into two forms: user-based recommendation and item-based recommendation [1]. The basic principle of user-based recommendations is as follows. Firstly, user preference is found by analyzing their historical data; secondly, it finds object service's neighbor according to their preference; finally, according to neighbor user's historical preference data, it recommends for object users [8].

The basic principles of item-based recommendations are similar to user-based recommendations. Firstly, the similarities between services rather consumer's similarity is computed; Secondly, the nearest number of the unmarked service is found; thirdly, according to the nearest neighbor's mark, the service's mark is predicted; finally, the service which has the highest predicted mark could be recommended to the users [23].

5 Conclusion

In order to study the health of the platform-based service ecosystem, first, this paper proposes a health index model to measure its health quantitatively. Secondly, a simulation model of the platform-based service ecosystem is constructed by means of computational experiments. Driven by the three recommendation strategies, the platform-based service ecosystem continues to evolve. Finally, this paper uses the proposed health index model to measure the health of the ecosystem dynamically. The results show that the proposed health index model is reasonable and can provide decision support for the management of the ecosystem. In addition, we have also studied and analyzed the factors affecting the healthy development of the platform-based service ecosystem. For future work, we look forward to adding more quantitative indicators to the health index model to measure the health of the platform-based service ecosystem more accurately.

Acknowledgment. This work is supported by the National Key R&D Program of China grant No.2017YFB1401201, the National Natural Science Key Foundation of China grant No.61832014 and No.61972276the Shenzhen Science and Technology Foundation grant JCYJ20170816093943197, and the Natural Science Foundation of Tianjin City grant No.19JCQNJC00200.

References

1. Ambulgekar, H.P., Pathak, M.K., Kokare, M.B.: A survey on collaborative filtering: tasks, approaches and applications. In: Chakraborty, M., Chakrabarti, S., Balas, V.E., Mandal, J.K. (eds.) Proceedings of International Ethical Hacking Conference 2018. AISC, vol. 811, pp. 289–300. Springer, Singapore (2019). https://doi.org/10.1007/978-981-13-1544-2_24
2. Bai, B., Fan, Y., Tan, W., Zhang, J.: DLTSR: a deep learning framework for recommendation of long-tail web services. IEEE Trans. Serv. Comput. **13**, 73–85 (2017)
3. Cardoso, J., Sheth, A.P., Miller, J.A., Arnold, J., Kochut, K.J.: Quality of service for workflows and web service processes. J. Web Semant. **1**(3), 281–308 (2004)
4. Chen, L., Huang, Q.: An exact calculation method for Gini coefficient and its application in China. J. Discrete Math. Sci. Crypt. **21**(6), 1235–1240 (2018)
5. Chen, X., Zheng, Z., Liu, X., Huang, Z., Sun, H.: Personalized QoS-aware web service recommendation and visualization. IEEE Trans. Serv. Comput. **6**(1), 35–47 (2011)

6. Chen, X., Zheng, Z., Lyu, M.R.: QoS-aware web service recommendation via collaborative filtering. In: Bouguettaya, A., Sheng, Q., Daniel, F. (eds.) Web Services Foundations, pp. 563–588. Springer, Heidelberg (2014). https://doi.org/10.1007/978-1-4614-7518-7_22
7. Costanza, R., et al.: The value of ecosystem services: putting the issues in perspective. Ecol. Econ. **25**(1), 67–72 (1998)
8. Fu, M., Qu, H., Moges, D., Lu, L.: Attention based collaborative filtering. Neurocomputing **311**, 88–98 (2018)
9. Jannach, D., Zanker, M., Felfernig, A., Friedrich, G.: Recommender Systems: An Introduction. Cambridge University Press, Cambridge (2010)
10. Kiinzli, S., Poletti, F., Benini, L., Thiele, L.: Combining simulation and formal methods for system-level performance analysis. In: Proceedings of the Design Automation and Test in Europe Conference, vol. 1, pp. 236–241 (2006)
11. Li, S., Fan, Y.: Research on the service-oriented business ecosystem. In: 2011 3rd International Conference on Advanced Computer Control, pp. 502–505. IEEE (2011)
12. Napitupulu, D., et al.: Analysis of student satisfaction toward quality of service facility. Phys. Educ. **954**(1), 012019 (2018). http://arxiv.org/abs/Physics
13. Rapport, D.J., Costanza, R., McMichael, A.: Assessing ecosystem health. Trends Ecol. Evol. **13**(10), 397–402 (1998)
14. Resnick, P., Varian, H.R.: Recommender systems. Commun. ACM **40**(3), 56–58 (1997)
15. Schaeffer, D.J., Herricks, E.E., Kerster, H.W.: Ecosystem health: I. measuring ecosystem health. Environ. Manag. **12**(4), 445–455 (1988)
16. Schafer, J.B., Konstan, J., Riedl, J.: Recommender systems in e-commerce (2014)
17. Singh, M., Matsui, Y.: Effect of long tail and trust on customer motivation behind online shopping use: comparative study between physical product and service product. In: PACIS, p. 221 (2017)
18. Spellerberg, I.F., Fedor, P.: A tribute to claude shannon (1916–2001) and a plea for more rigorous use of species richness, species diversity and the 'shannon-wiener' index. Glob. Ecol. Biogeogr. **12**(3), 177–179 (2003)
19. Spohrer, J., Maglio, P.P., Bailey, J., Gruhl, D.: Steps toward a science of service systems. Computer **40**(1), 71–77 (2007)
20. Wen, D., Yuan, Y., Li, X.R.: Artificial societies, computational experiments, and parallel systems: an investigation on a computational theory for complex socioeconomic systems. IEEE Trans. Serv. Comput. **6**(2), 177–185 (2012)
21. Xiao, J., Chen, S., He, Q., Feng, Z., Xue, X.: An android application risk evaluation framework based on minimum permission set identification. J. Syst. Softw. **163**, 110533 (2020)
22. Xie, F., Chen, Z., H., Feng, X., Hou, Q.: TST: Threshold based similarity transitivity method in collaborative filtering with cloud computing. Tsinghua Science and Technology (2013)
23. Xue, F., He, X., Wang, X., Xu, J., Liu, K., Hong, R.: Deep item-based collaborative filtering for top-n recommendation. ACM Trans. Inf. Syst. (TOIS) **37**(3), 1–25 (2019)
24. Xue, X., Kou, Y.M., Wang, S.F., Liu, Z.Z.: Computational experiment research on the equalization-oriented service strategy in collaborative manufacturing. IEEE Trans. Serv. Comput. **11**(2), 369–383 (2016)

SETE: A Trans-Boundary Evolution Model of Service Ecosystem Based on Diversity Measurement

Tong Gao, Zhiyong Feng, Shizhan Chen, and Xue Xiao(✉)

Tianjin University, No. 135 Yaguan Road, Haihe Education Park, Tianjin, China
{gt19960226,zyfeng,shizhan,jzxuexiao}@tju.edu.cn

Abstract. Trans-boundary and integration are important characteristics of the development of modern service industry. With the development of Internet technology, trans-boundary cooperation between domains constantly emerges, which drives the development of service ecosystem. Currently, there is a lack of an appropriate model for analyzing the impact of trans-boundary services on the entire service ecosystem. In this paper, we propose a service ecosystem trans-boundary evolution model (SETE). It analyzes the interactions between user needs and services, and focuses on the mechanism of trans-boundary services to promote the evolution of the service ecosystem. At the same time, we develop a diversity measurement algorithm for service ecosystems based on the theory of biodiversity in ecology. Based on these, a computational experimental system is established. It simulates the trans-boundary evolution mechanism of the service ecosystem and shows the characteristics of each stage of the service ecosystem evolution. At last, we verify the effectiveness of the SETE model through actual cases (the Alibaba Group). The results show that the SETE model can provide new ideas for the study of the trans-boundary evolution of the service ecosystem, and provide decision support for the development direction of the modern service industry.

Keywords: Diversity measurement · Service ecosystem · Trans-boundary service · Service evolution · Computational experiment

1 Introduction

The new social form of "Internet+" has promoted the trans-boundary and integration development between industries. Services are increasingly breaking traditional domain boundaries. Trans-boundary service provides users with innovative services through trans-domain cooperation, and creates value that a single-domain service cannot create [8]. Trans-boundary services and single-domain services work together to form a service ecosystem. As time goes by, dynamic cooperation between services has promoted the development of the service ecosystem

H. Gao et al. (Eds.): CollaborateCom 2020, LNICST 349, pp. 217–236, 2021.
https://doi.org/10.1007/978-3-030-67537-0_14

[20]. Therefore, studying how to characterize the trans-boundary evolution mechanism of the service ecosystem can help us better understand and manage it.

Before trans-boundary services were created, services came from different domains and service providers, and they did not have uniform development standards. Semantic conflicts make cross-domain collaboration between services very difficult. It not only limits the ability to create value of services, but also hinders the development of the service ecosystem. The trans-boundary service solves the problems. It establishes communication channels between domains, allows services between different domains to invoke each other, and provides users with a new user experience that cannot be provided by the single domain services. At the same time, the creation of trans-boundary services has a huge impact on the cooperation and competition relationships of services in the service ecosystem. The trans-boundary and integration of various domains has caused great changes in the diversity of the service ecosystem and promoted the evolution of the service ecosystem.

Research in this field is facing challenges. Firstly, the rapid changes in user needs and the rapid development of modern service industries make data volumes very large and difficult to obtain. Secondly, the complex and changeable service ecosystem makes it difficult to capture the regular pattern of the generation and evolution of trans-boundary services. In order to better understand the development of the service ecosystem, we attempt to build a model (SETE) based on the diversity measurement, and study the trans-boundary evolution mechanism of the service ecosystem. The main contributions of this article are as follows:

- Propose a model for trans-boundary evolution of the service ecosystem based on diversity measurement;
- The measurement algorithm designed to assess the diversity of service ecosystems;
- Prove the effectiveness of SETE through computational experiment and case study.

The rest of this article is organized as follows. Section 2 introduces the specific details of the SETE model. Section 3 introduces the mechanism of trans-boundary evolution. Section 4 describes the diversity measurement algorithm. Section 5 shows a computational experiment system and a case study to validate the model. Section 6 introduces the related work. Section 7 summarizes the research of this article.

2 Structure of SETE Model

As shown in Fig. 1, the service ecosystem can be viewed as a collection of services, the relationships between services, and other entities [7]. It is a system where multiple members create and share value together. It can be analyzed from four levels: service network level, value network level, domain level, and service ecosystem level. The service network level includes various web services, and the services form a network by invoking each other. The value network level contains value flow between services. The domain level contains resources such as specific value network their service providers, semantic rule bases, etc. The ecosystem level includes user needs and the environment. In order to better explain the trans-boundary evolution mechanism of the service ecosystem, we first define some basic concepts in the service ecosystem.

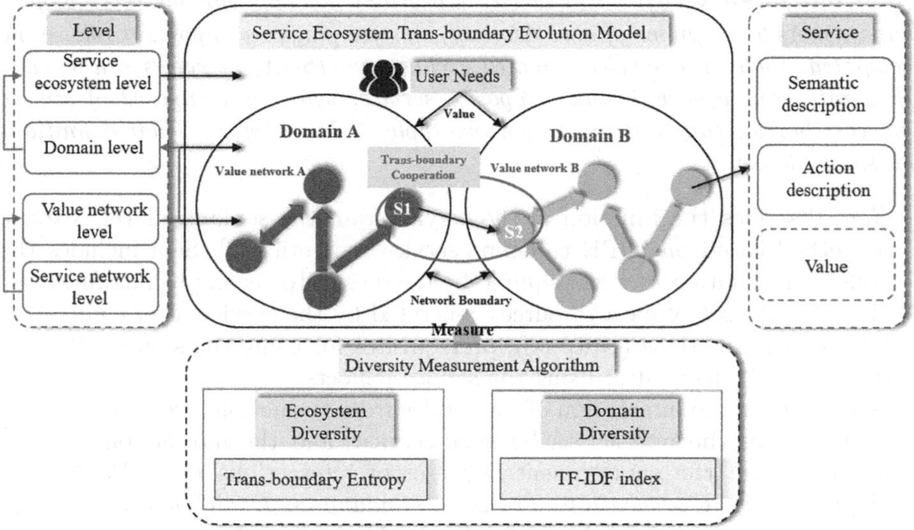

Fig. 1. Structure of the dynamic trans-boundary evolution model of the service ecosystem.

2.1 Service Network Level

Services provide users with solutions through cooperation, thus forming a network. The service network is the foundation of the domain structure and evolution of the service ecosystem and an important condition for trans-boundary evolution.

Theorem 1 (Service Network). *Facing diversified user demands, services need to invoke other services to complete the functions. The cooperative relationship between services influences each other and forms a service network. Services provide richer and more flexible solutions in the service network. The service network can be defined as a collection of cooperative relations in the domain*

$$SN = (V, E) \tag{1}$$

where V represents the set of points in SN and E is the set of edges in SN. And

$$V = \{S_1, S_2, \ldots, S_n\} \tag{2}$$

$$E = \{(S_i, S_j)\,|1 \leq i \leq n, 1 \leq j \leq n\} \tag{3}$$

where S represents a node in the service network, which is service. n represents the number of services in the service network.

Theorem 2 (Service). *The service ecosystem includes various types of participants. As the supplier of user demands, service plays an important role in the ecosystem. Under the service-oriented architecture (SOA), services can be classified into atomic service and composite service, which is composed of atomic services. Service can be defined by a two-tuple* $< Sem, Act >$*. Their definitions are shown below.*

- *Sem* describes the function of the service from the semantic level. We use a method based on IOPE to define service semantics [6]. Sem includes the collection of data resources required by the service to accomplish its functions (Inputs), the set of data resources generated by the service after completing its business functions (Outputs), prerequisites for using the service (Preconditions) and effects after using the service (Effects).
- *Act* describes the interaction of the service from the action level. The interaction includes the interaction between services and the interaction between a service and the environment. The set of interactions is $<$ *DealOrder*, *Reproduction*, *Cooperation*, $Trans - boundary >$. *DealOrder* means that the service creates value by responding to invocations from users and other services. *Reproduction* refers to the expansion of atomic services. When the value of the service reaches a certain threshold, it will decide whether to develop atomic services to expand the business scope. *Cooperation* means that services will implement collaboration through mutual invocation, service integration, and other unforeseen ways. Trans-boundary refers to the innovation or improvement of services through the cross-domain integration of resources and capabilities.

2.2 Value Network Level

The value network is the foundation of the value flow in the service ecosystem. It's a necessary condition to promote the evolution and development of the service ecosystem. Service is the carrier of value. The flow of value is realized by mutual invocation between services.

Theorem 3 (Value Network). *Values are transferred between services through invocation, which forms a value network. There are a large number of cooperation relationships and value exchanges in the domain. The upstream service can call the downstream service, and the downstream service can feedback information for the upstream service. The node set and edge set of the value network VN are consistent with the corresponding service network SN.*

Theorem 4 (Service Value). *When users interact with services, both parties involve the interaction of information flow and cash flow. Users choose services that meet their needs, and service suppliers profit by providing services. Service value is the value generated when the service interacts with users. Service value describes the value that services create when interacting. The value of service s at time t can be defined*

$$V_s(t) = QoS_s(t) + price_s(t) * fre_s(t) - cost_s(t) \quad (4)$$

where $QoS_s(t)$ is the quality of the service, and it indicates whether the performance or availability of the service s can meet the demand of users. $QoS_s(t)$ will affect the number of users attracted by the service and the maximum number of times the service can be invoked per unit time. $QoS_s(t)$ will increase when the service is perfected, and decrease when the service is overloaded. $price_s(t)$ represents the price charged by the service each time it is invoked. $fre_s(t)$ represents the frequency the service is invoked, and the product of $price_s(t)$ and $fre_s(t)$ represents the revenue of service. $cost_s(t)$ is the cost of the service, including development costs, invocation costs, maintenance costs, etc. When the service needs to develop new function or improve current functions, development costs will be incurred. When the service needs to invoke other services, the service need the invocation costs. When a service is defective, service providers need to pay the cost of maintaining the service.

Theorem 5 (Network Boundary). *Different service networks are affected by issues such as semantic conflicts and interface mismatches, and cannot be invoked by each other. In other words, there is a network boundary between service networks. Suppose two value networks* VN_A *and* VN_B. V_A *and* V_B *denote their node set respectively.* E_A *and* E_B *denote their edge set respectively. Then there will be*

$$V_A \cap V_B = \emptyset \wedge E_A \cap E_B = \emptyset \quad (5)$$

Although domain boundaries make it impossible for services between domains to interact, domains can exchange information and value through service providers, shared user groups, etc.

2.3 Domain Level

The complex relationships between services and services, services and environment lead to the community structures in the ecosystem. Domain describes concepts and their relationships in a particular community structure and is used to analyze semantics and build rule bases.

Theorem 6 (Domain). *Different services belong to different community structures, which is the domain. Services in the same domain follow unified development standards, rules, and semantic libraries and meet specific user needs. In addition to the value network, the domain also includes service providers and the rule base. The formal formula to express the domain is*

$$Dom = (VN, RuleBase, Supplier) \tag{6}$$

RuleBase represents the semantic specification and rule base in the domain. *RuleBase* stores the mapping relationships between semantic concepts, which is used to detect and resolve semantic inconsistencies between web services. The mapping relationships can be manually added and modified. A mapping relationship can be formally represented by a five-tuple $(C1, C2, mapType, rule, cons)$, where $C1$ and $C2$ represent semantic concepts, mapType represents the type of semantic mapping relationship (such as equivalence, inclusion, irrelevance, etc.), *rule* represents the conversion rule between $C1$ and $C2$, and *cons* represents the constraint rule of this mapping.

Supplier represents the service suppliers that provide the services of the domain. Atomic web services have their own service suppliers, users obtain services through the service providers, and the profit pays the operation and service maintenance fee.

2.4 Ecosystem Level

Trans-boundary cooperation between domains meets users' growing and changing needs. Fusion and dynamic collaboration between services drive the service ecosystem to evolve.

Theorem 7 (Service Ecosystem). *Service ecosystem is a logical collection of network services [1]. In the service ecosystem, services obtain value by meeting the needs of users. In addition to various domains, the service ecosystem also includes users and the environment.*

$$SerEcosystem = (User, Environment, Dom_1, Dom_2, ..., Dom_m) \tag{7}$$

where $User$ represents the users who use the services in the service ecosystem, environment represents the basic environmental information of the entire ecosystem, and m represents the number of domains in the ecosystem.

Theorem 8 (Trans-boundary Service). *Trans-boundary service refers to the service that integrates services by service fusion or service matching in order to sole problems such as data inconsistencies and business inconsistencies. Trans-boundary service can interact with services in two domains by normal invocations. After the service integrated with other services, the service value becomes*

$$V_i' = V_i - \mathrm{cost_{trans}} + \delta \tag{8}$$

where V_i' is the service value after trans-boundary, V_i is the service value before trans-boundary, $cost_{\mathrm{trans}}$is the value consumption generated by trans-boundary, and δ is the added value generated by trans-boundary. When $\delta \gg cost_{\mathrm{trans}}$, trans-boundary service can continue to develop.

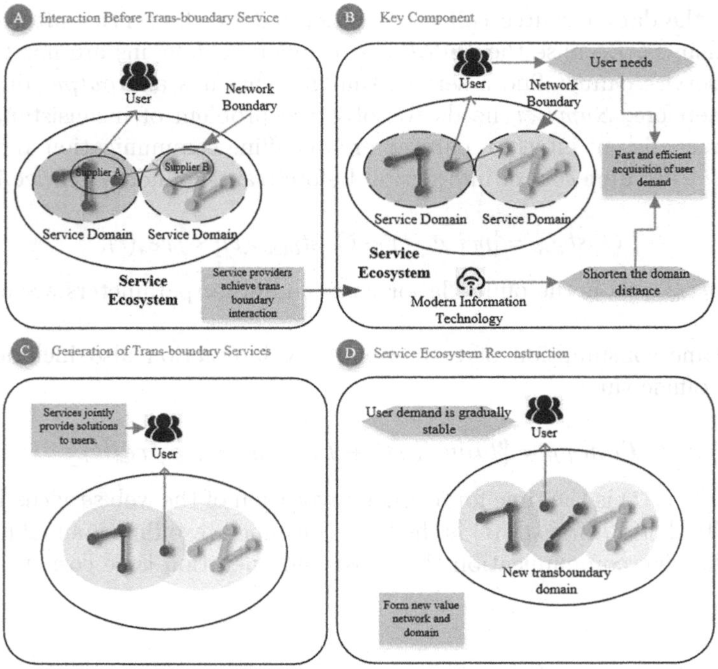

Fig. 2. Structure of trans-boundary evolution mechanism.

3 Mechanism of Trans-Boundary Evolution

Under the SETE model, we analyzed the differences between trans-boundary interactions before and after the trans-boundary service are generated and proposed the generation and mechanism of trans-boundary evolution. As shown in Fig. 2, part A shows the trans-boundary interaction before trans-boundary services are generated. Part B shows the key components. Part C and part D demonstrates the changes in the service ecosystem caused by trans-boundary services.

3.1 Interaction Before Trans-Boundary Service

Before the generation of trans-boundary service, services between domains cannot invoke each other. Interaction between domains needs to be completed by online web services and offline human behavior. The Trans-boundary interaction request before trans-boundary service generation can be defined as the following five-tuple:

$$TR_{pre} =< S_{id}, Dom, Supplier, des, output > \quad (9)$$

where S_{id} is the id of the service to request interaction, Dom is the domain of the service, des describes the function requested by the service, and $output$

describes the data resource collection generated by the service after completing its business. Because the *RuleBase* in different domains are not the same, the web service cannot find a service that matches *des* and *output* outside the *Dom*. Therefore, *Supplier* needs to solve the problem of inconsistent semantics and inconsistent interface parameters by offline communication and manual matching. The value cost of interaction before trans-boundary service is

$$Cost_{TR} = [price_s(t) + Cost_{labor}(t)] * fre_s(t); \tag{10}$$

where $Cost_{labor}(t)$ is the offline labor cost, and other parameters are similar to Eq. 4.

The time consumption of trans-boundary interaction also includes online time and offline time

$$Time_{TR} = [Time_{inv}(t) + Time_{labor}(t)] * fre_s(t) \tag{11}$$

where $Time_{inv}(t)$ is the time for a single invocation of the web service. This time is generally short. $Time_{labor}(t)$ is the time consumed by offline manual matching. Due to the high communication time cost and operation time cost, this time is long.

3.2 Key Components

The key components for the generation of trans-boundary services include the trans-boundary needs of users, modern information technologies, and the collection of services in the service ecosystem. User needs are rapidly changing and growing, which promotes cooperation between domains to propose innovative services. The modern information technologies have shortened the distance between users and services. Services can efficiently obtain user needs through methods such as big data analysis and cloud computing. The cross-domain interaction between services has provided a trans-boundary technical basis between domains. These help services to update and fuse to meet the evolving and diverse needs of users.

3.3 Generation of Trans-Boundary Service

Before the generation of trans-boundary service, the supply capacity of trans-boundary cooperation services is restricted by the rules of the two domains and the capabilities of offline entities. The increasing demand of users for high-quality and low-price services has promoted the integration of services and proposed innovative trans-boundary services. The service developers refactored the parameters between services in different domains and added the mapping rules of interface semantics to the *RuleBase* of the two domains. This solves the problem of inconsistencies in the service interface. And the services in different domains can invoke each other by S_{id} and *des*. Its supply capability is greatly improved.

After the generation of trans-boundary service, the cost of trans-boundary interaction is reduced to

$$Cost_{TR}' = price_s(t) * fre_s(t) \tag{12}$$

The time for trans-boundary interaction is also greatly shortened.

$$Time'_{TR} = Time_{inv}(t) * fre_s(t) \tag{13}$$

3.4 Service Ecosystem Reconstruction

Trans-boundary services have established a new domain system, and at the same time, it has made users break away from the original domains and enter the new. The new domain created by trans-boundary services attracts and stimulates user demand through its own service performance. After emerging domain occupies a certain share, the traditional domains are affected by new domains, resulting in changes in domain scale and value network structure. It leads to the complete disappearance of the traditional service ecosystem and the end of the trans-boundary evolution.

4 Diversity Measurement Algorithm

Because the boundaries of the service ecosystem are difficult to determine and the dynamic evolution mechanism is complex and changeable, how to measure the changes of the service ecosystem in the trans-boundary evolution is a difficult problem. We introduce the concept of biodiversity [12] in ecology into the service ecosystem, which is used to measure the stability of the service ecosystem and service capacity. Biodiversity believes that the stability of ecosystems is closely related to the number of species in the system and the size of interactions between species [5]. Studying the diversity of service ecosystems is important for maintaining their stability. The diversity measurement algorithm quantifies the characteristic indicators of the service ecosystem at each stage of trans-boundary evolution by measuring the trans-boundary entropy of the whole service ecosystem and the TF-IDF index of each domain.

4.1 Trans-Boundary Entropy

Entropy is a parameter representing the state of matter in thermodynamics, and it is a measure of the degree of chaos in the system. Shannon borrowed this idea from thermodynamics into information theory as a measure of the observation or measurement work involved in assessing the state of the system [13]. The higher the system's entropy value, the higher its overall complexity and diversity.

The evolving service ecosystem can be viewed as a network. IBM defines the state of the value network as the set of nodes participating in any given consumer interaction. Then, the information entropy is used to calculate the complexity of the value network [2]. We define the interaction set as the set of trans-boundary

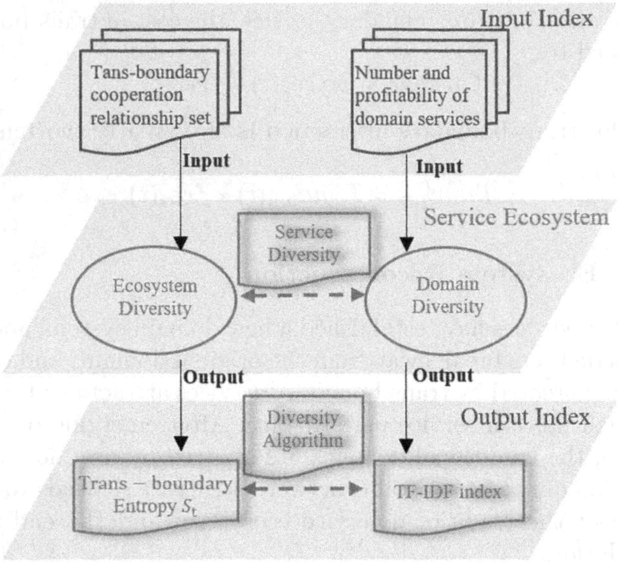

Fig. 3. Structure of diversity measurement algorithm.

types that may exist in the system. And calculate the trans-boundary entropy S_t of the service ecosystem (Fig. 3).

$$S_t = \sum_{i=1}^{D} \left[-\left(\frac{N_{it}}{N_t}\right) \log \left(\frac{N_{it}}{N_t}\right) \right] \tag{14}$$

where D is the trans-boundary set of the service ecosystem, N_{it} is the number of services that have i trans-boundary at t time, N_t is the number of all services at t time, and the base of the logarithm is 2.

4.2 TF-IDF Index

Domain diversity refers to the number of domains in the service ecosystem and the proportion of various domains in the system. Maintaining balance and stability in all areas of the service ecosystem is very important for the development of the service ecosystem. In this paper, the TF-IDF algorithm, a common weighting technology for information retrieval and data mining, is used to calculate the proportion of various domains in the service ecosystem. The higher the TF-IDF value in a domain, the larger its proportion in the service ecosystem. The more unstable the TF-IDF value of a system is, the less stable the development of the system is. Compared with counting the number of services in the service system, using a weighting technology can help us to observe the size of the domains in the ecosystem more intuitively and clearly. The TF-IDF index can be evaluated by the following equation

$$TFIDF_{ij} = TF_{ij} * IDF_i \tag{15}$$

where $TFIDF_{ij}$ is the TF-IDF index of the i domain at j time of the ecosystem, TF_{ij} is the proportion of the i domain at j time, and IDF_i is the weight value of the i domain.

TF_{ij} can be calculated by the following formula

$$TF_{ij} = \frac{v_{ij}}{v_j} \tag{16}$$

where v_{ij} refers to the sum of the total value of services in the i domain at j time, and v_j is the total value of all services at j time in the service ecosystem.

The weight value of domain i, IDF_i, can be calculated by this formula

$$IDF_i = \log_5 \left(\frac{\sum_{j=1}^{j=T} N_j}{\sum_{j=1}^{j=T} N_{ij}} \right) \tag{17}$$

where T is the time for the evolution of the service ecosystem, N_j is the total amount of all services at j time, $\sum_{j=1}^{j=T} N_j$ is the sum of the number of services in all periods, N_{ij} is the number of services in the i domain at j time, and $\sum_{j=1}^{j=T} N_{ij}$ is the sum of the number of services in i domain in all periods.

5 Construction of Computational Experiment and Result Analysis

By modeling the main elements of the service ecosystem, we can formulate a computational experiment platform to simulate the trans-boundary evolution mechanism of the service ecosystem [15]. In this section, we will analyze the impacts of trans-boundary services based on the experiment data and conduct a case study of Alibaba Group.

5.1 Construction of Computational Experiment

This experiment is based on the SETE model. In the service ecosystem, user demand and services are the main participants in interactive activities. Therefore, we built an agent-based model involving user demands and services on the Repast Symphony platform.

We set two experiments, the experimental group and the control group. Both of the two experiments set the same service ecosystem environments: each of the ecosystems comprises 3 domains, and the initial amount of services in each domain is 30, 20, 15 respectively. In order to facilitate the calculation and comparison of experimental data, all parameters are reduced in proportion to the actual data. Meanwhile, we randomly design invocation relationships in the domain. Apart from this, the experimental group added the trans-boundary evolution mechanism. When the service perceives a certain amount of trans-boundary demands in the system, the service will choose whether to try to trans-boundary integration based on its current income situation and value situation. By comparing the results of the two groups, we can observe the impact of the trans-boundary mechanism on the service ecosystem. Both ecosystems run 50 ticks.

Construction of Demand Agent. Demand is an important factor affecting the service evolution. And it can directly affect the survival of the service. We can make rules for demand agents (such as the number and type of demand agents) to model the trend of various market fluctuations. The structure of the demand agent is shown in Table 1.

Table 1. Attributes of the demand agent

Attributes	Description
DemandCategory	It represents the type of this demand. It describes the functions that the user needs the service to provide, including the input information and the desired service effect
DemandVolume(t)	It represents the current amount for this demand. The current volume of a kind of demand is affected by the growth trend of the demand and the supply ability of services
DemandPrice	It represents the market price of this type of demand. The value range for different types of services is $N(10, 4)$
IncreaseTrend(t)	It represents the increasing trend of demand. The growth trend of demand can be divided into rigid growth and non-rigid growth. The demand growth rate x for rigid growth per unit time follows a normal distribution: $x \sim N(\mu, \sigma^2)$, where μ determines the range of demand growth and σ^2 determines the intensity of demand growth. The demand growth rate y for non-rigid growth per unit time follows a uniform distribution: $y \sim U(a, a+b)$. In our experiment, the value range of μ and a is $[100, 200]$, and the value range of σ and b is randomly generated between $[10, 20]$

Construction of Service Agent. Service is the carrier of the value network in the service ecosystem and the basis for studying the mechanism of trans-boundary evolution. We construct the service agent according to the Definition 2.1. Service attributes include semantic attributes, value attributes, and interaction behaviors. The semantic and value attributes of the service agent are shown in Table 2, and the interaction behaviors are shown in Table 3.

5.2 Analysis of Computational Experiment Results

Analysis of Experimental Group Results

Table 2. Semantic and value attributes of service agent

Attributes	Description
SemDescription	It represents the semantic attributes of a service. It uses key words to describe the function-related parameters of the service, including inputs, outputs, preconditions, and effects. The service uses keywords to match corresponding needs and invoke other services
Cost	It represents the necessary cost of this service. It is distributed randomlylow(1), middle(2), high(3)
$QoS(t)$	It represents the current performance of this service and describes the quality of the service. The quality of the service range is $N(3, 1)$
$Value(t)$	It represents the value currently owned by the service agent, as shown in Formula 4
Vison	It represents the range of information available to the service agent. The larger the range, the more likely it is to capture user needs and partners
MovePace	It represents the moving distance of the service per unit time in the service ecosystem. The larger the distance, the more sensitive the service
MaxDemandNum	It represents the total number of calls to the service in a unit of time. This attribute may affect whether subsequent user choose this service

Table 3. Behaviors of service agent

Behaviors	Description
DealOrder	When the service agent finds a demand that meets its business scope, the service moves to the location of the demand, provides the service for the demand, and obtain value
Reproduction	When the service's value reaches a certain range, the service will generate sub-services to expand the scope of the service or improve its service capabilities. The attributes of the new sub-service will be affected by inheritance and mutation. After the parent service generates a child service at time t, its value will change as follows: $V_{parent}(t+1) = V_{parent}(t) - V_{child}(t)$ (18)
Cooperation	When the service itself cannot meet the user's needs, it will complete the current demand by invoking other services and pay the invoking fee
Transboundary	Services achieve innovation through trans-boundary integration. Trans-boundary service can create value that cannot be created by single-domain service. The value of trans-boundary service is shown in 8

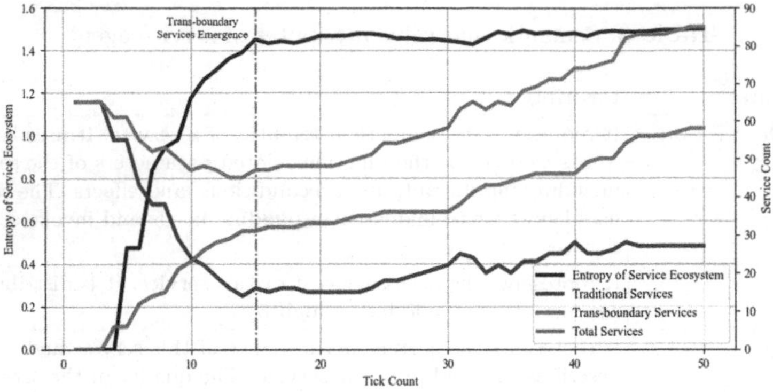

Fig. 4. Trans-boundary entropy and number of services in the experimental group.

Impact of Trans-Boundary on S_t Index. The impact of trans-boundary services on trans-boundary entropy S_t is shown in the black line in Fig. 4, and the axis is on the left. The other three polylines are the amount of single domain services and cross-domain services and the total number of services in the system over time, and the axis is on the left. In the trans-boundary service generation stage (Tick $\leq$ 15), services break through the domain boundaries to integrate and form trans-boundary services. This led to explosive growth in trans-boundary relationships in the service ecosystem. St of the ecosystem has improved significantly in a short time and reached a peak at tick = 15. After the generation of trans-boundary services ($15 < \text{tick} \leq 50$), the ecosystem adapts to the generation of trans-boundary services and continuously adjusts the relationship structure, but there is no longer a large number of new trans-boundary services produced. Therefore, the S_t index has not changed significantly. At the same time, the growth in the number of trans-boundary services has slowed, and the number of services in a single domain has also slowly recovered. This is because the supply of trans-boundary services in the service ecosystem has reached saturation, and the market structure is changing to accommodate these new services. In the end, the number of trans-boundary services and the number of single-domain services is balanced, and the service ecosystem structure is also stable.

Impact of Trans-Boundary on TF-IDF Index. It can be found in Fig. 5 that the time when the TF-IDF index of each domain and the trans-boundary entropy S_t of ecosystem reach stability is very different. In the initial stage of the service ecosystem (tick $\leq$ 3), there were only three primitive domains in the service ecosystem, each occupying a certain proportion. In the stage of trans-boundary service generation (tick = 15), in order to meet the growing trans-boundary demand of users, a large number of services tried to cross the boundary. The TF-IDF index of the new domain reached its peak in a short time, and the supply of trans-boundary services reached a saturated state. However, in the stage of service ecosystem reconstruction ($15 < \text{tick} \leq 45$), because the service supply is greater than the user demand, some services are difficult to make a

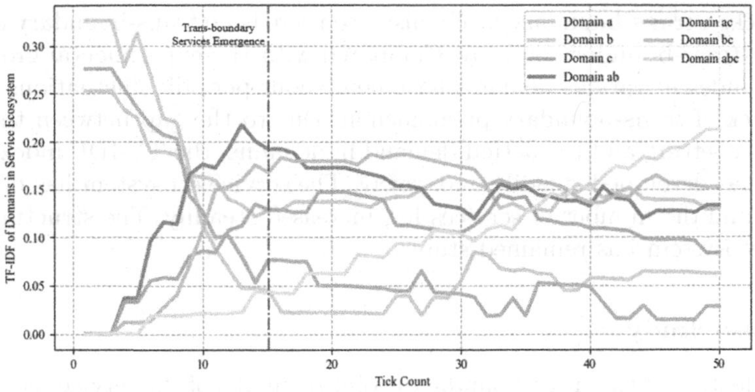

Fig. 5. TF-IDF index of domains in the experimental group.

long-term profit. Some services cannot continue to innovate, causing their development to stagnate or even die, which has caused the proportion of emerging trans-boundary domains in the market to decline. At last ($45 < \text{tick} \leq 50$), the ecosystem gradually generates a new stable value network, and the fluctuation of TF-IDF index in various domains has slowed. The supply of various services in the ecosystem and the needs of users have reached a balance, the structure of the ecosystem has also returned to stability, and the process of trans-boundary evolution has ended.

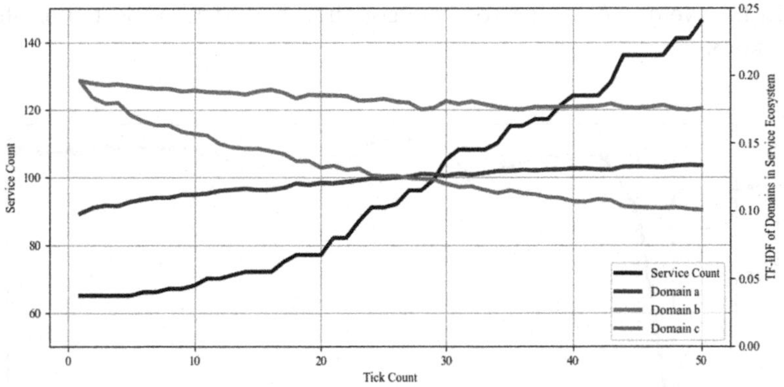

Fig. 6. Number of services and TF-IDF index of domains in the control group.

Analysis of Control Group Results. The black polyline in Fig. 6 is the number of services in the system, and the axis is on the left. The other lines are the TF-IDF index of the three domains, and the axis is on the right. The control group lacks a trans-boundary mechanism, so there is no trans-boundary relationship and trans-boundary entropy. Compared with the experimental group, The TF-IDF index in various domains have no obvious periodic fluctuations because of the lack of trans-boundary phenomenon. Due to the gap between the initial parameter setting and the actual demand in domains, the TF-IDF index in each domain has developed steadily and slowly. The service ecosystem has developed steadily and the number of services has increased steadily. The structure of the service ecosystem has remained stable.

5.3 Case Study

The key feature of the trans-boundary evolution model of the service ecosystem is that the S_t index and $TF-IDF$ index are asynchronously balanced. According to this characteristic, we can measure the degree of trans-boundary evolution of the service ecosystem. In order to verify the validity of the SETE model, we select a specific service ecosystem (Alibaba Group) as a case study. The data is based on the 2014–2019 quarterly performance reports released by Alibaba Group. The performance reports record the quarter revenue of each domain and the strategic investment of Alibaba Group over the four years. Alibaba Group began to enter the local customer service domain through a strategic joint venture in January 2015. And in May 2015, it was announced that the establishment of the Cainiao logistics service. Over the past four years, Alibaba Group has continuously adjusted its investment scale and business structure, and finally achieves positive income in these two domains, and has a certain share in the Chinese market. We calculated the S_t and $TF-IDF$ index of Alibaba Group in four years according to the algorithm in Part C of Sect. 3 as shown in Figs. 7 and 8.

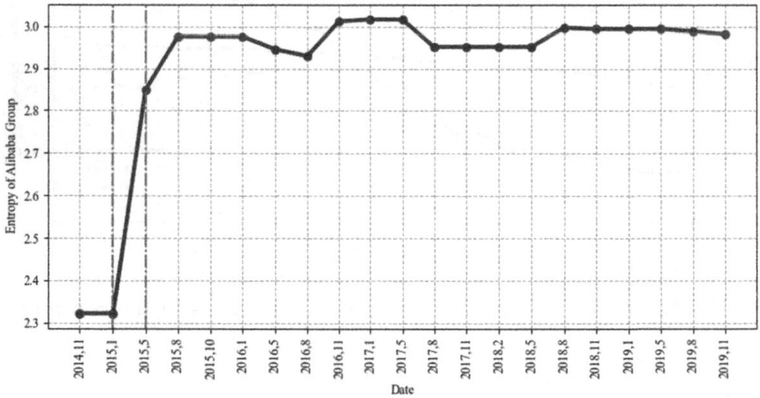

Fig. 7. Trans-boundary entropy of Alibaba Group.

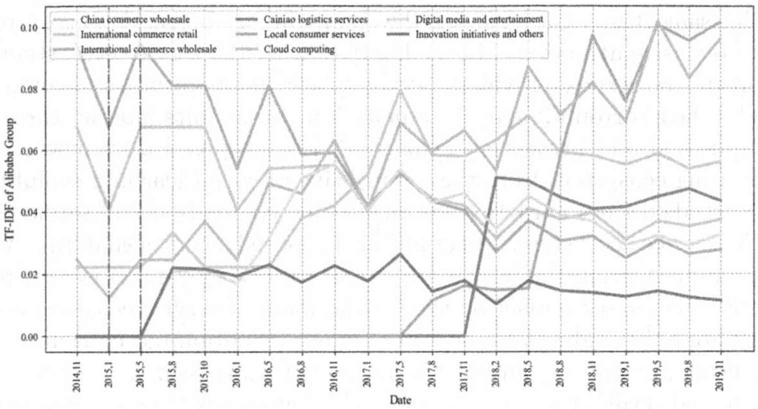

Fig. 8. TF-IDF index of Alibaba Group.

The results of Alibaba Group's diversity measurement are similar to the experimental group. Figure 7 is the change of the trans-boundary entropy of Alibaba Group in each quarter. It can be found that after the Alibaba Group announced its entry into the domain of local consumer services and Cainiao logistics services, its S_t index has increased significantly and remained at a high level since then. Figure 8 shows the TF-IDF index of the Alibaba Group in each quarter, except for the main retail sector in China. It shows that the TF-IDF values in all domains of the Alibaba Group have been affected after they began the trans-boundary industries. However, it was not until 2017 that the two trans-boundary domains began to occupy a certain weight and reached stability in 2019. Therefore, the SETE model is effective for studying the trans-boundary impact on the service ecosystem.

6 Related Work

In recent years, research on trans-boundary services has gradually emerged. Wu et al. proposed the theory of crossover service [17]. Crossover service emphasizes the phenomenon of crossover and integration between services in the modern service industry at the commercial level. Trans-boundary service emphasizes communication among domains by service integration at the web service level [4]. Service composition [9] strictly constrains the set of candidate services in order to accurately invoke services, which is a common method for service cooperation to solve user needs. Xue et al. [18] proposed the concept of "service bridge" to evaluate the trans-boundary impact of the Internet model. Some studies have used complex network to analyze the evolution of services and service relationships. Wang et al. [14] proposed an impact analysis model based on service dependency. Fokaefs et al. [3] analyzed the evolution of WSDL documents of Web services through the VTracker differentiation algorithm to discuss the impact of service changes and development on service system maintainability.

At the same time, the growing development of the service ecosystem has attracted inclusive attention. Liu et al. [10] defined the service ecosystem as services, consumers, service providers, service platform operators, and communities affected by their surrounding environment. The basic units interact through service composition and invocation. Xue et al. [19] proposed a framework for the manufacturing ecosystem from the perspective of social learning evolution. Wu et al. [16] studied the interaction between value co-creation and service innovation in the service ecosystem. Yin et al. [11] proposed a conceptual framework for defining and measuring the health of the software ecosystem. However, research on the trans-boundary impact on the development of service ecosystems is lack. Our work not only analyzes the trans-boundary evolution mechanism of the service ecosystem, but also quantifies the impact of trans-boundary services on the ecosystem, and verifies the effectiveness of the framework through examples. This not only helps to understand the emergence and development of trans-boundary services, but also helps to build a more realistic model of the evolution of the service ecosystem.

7 Conclusion

This paper introduces the trans-boundary evolution mechanism of the service ecosystem, develops a measurement algorithm to quantify the impact of trans-boundary services on the diversity of the service ecosystem, and conducts case studies of Alibaba Group. Using the computational experimental system, we explore the entire process of trans-boundary evolution of the service ecosystem.

In our research, we defined the overall architecture and cross-border evolution process of the service ecosystem. By measuring the diversity of service ecosystems, we find that the generation of trans-boundary services will cause the overall complexity of service ecosystems to increase immediately. However, the proportion of emerging domains in the ecosystem will not reach stability at the same time. It will continue to be turbulent during the evolution period of the ecosystem and will not be stable until the ecosystem structure is stable. The emergence of trans-boundary services can have an impact on the diversity of service ecosystems, and it will take some time for the ecosystem structure to stabilize again.

Although the service ecosystem is dynamic and changeable, by modeling its key elements, it can help us better analyze the activities and evolution mechanisms of the service ecosystem. In the future, we will try to build a more comprehensive theoretical framework to track and predict the generation and evolution of trans-boundary services, and provide constructive opinions for the development of the ecosystem.

Acknowledgement. This work is supported by the National Key R&D Program of China grant No. 2017YFB1401201, the National Natural Science Key Foundation of China grant No. 61832014 and No. 61972276, the Shenzhen Science and Technology Foundation grant JCYJ20170816093943197, and the Natural Science Foundation of Tianjin City grant No. 19JCQNJC00200.

References

1. Barros, A.P., Dumas, M.: The rise of web service ecosystems. IT Prof. **8**(5), 31–37 (2006)
2. Basole, R.C., Rouse, W.B.: Complexity of service value networks: conceptualization and empirical investigation. IBM Syst. J. **47**(1), 53–70 (2008)
3. Fokaefs, M., Mikhaiel, R., Tsantalis, N., Stroulia, E., Lau, A.: An empirical study on web service evolution. In: 2011 IEEE International Conference on Web Services, pp. 49–56. IEEE (2011)
4. Guo, S., Xu, C., Xue, X., Feng, Z., Chen, S.: Research on trans-boundary convergence of different service chains in health service ecosystem. J. Med. Imaging Health Inform. **10**, 1734–1745 (2020)
5. Hector, A., Bagchi, R.: Biodiversity and ecosystem multifunctionality. Nature **448**(7150), 188 (2007)
6. Hu, X., Feng, Z., Chen, S., Huang, K., Li, J., Zhou, M.: Accurate identification of ontology alignments at different granularity levels. IEEE Access **5**, 105–120 (2016)
7. Huang, K., Fan, Y., Tan, W.: An empirical study of programmable web: a network analysis on a service-mashup system. In: 2012 IEEE 19th International Conference on Web Services, pp. 552–559. IEEE (2012)
8. Kossinets, G., Watts, D.J.: Empirical analysis of an evolving social network. Science **311**(5757), 88–90 (2006)
9. Kukko, M.: Knowledge sharing barriers in organic growth: a case study from a software company. J. High Technol. Manag. Res. **24**(1), 18–29 (2013)
10. Liu, Y., Fan, Y., Huang, K.: Service ecosystem evolution and controlling: a research framework for the effects of dynamic services. In: 2013 International Conference on Service Sciences (ICSS), pp. 28–33. IEEE (2013)
11. Manikas, K., Hansen, K.M.: Reviewing the health of software ecosystems-a conceptual framework proposal. In: Proceedings of the 5th International Workshop on Software Ecosystems (IWSECO), pp. 33–44. Citeseer (2013)
12. Mori, A.S., Lertzman, K.P., Gustafsson, L.: Biodiversity and ecosystem services in forest ecosystems: a research agenda for applied forest ecology. J. Appl. Ecol. **54**(1), 12–27 (2017)
13. Shannon, C.E.: A mathematical theory of communication. Bell Syst. Tech. J. **27**(3), 379–423 (1948)
14. Wang, S., Capretz, M.A.: A dependency impact analysis model for web services evolution. In: 2009 IEEE International Conference on Web Services, pp. 359–365. IEEE (2009)
15. Wang, Y.B., School, B., University, H.: Modeling and simulation technology based on repast simphony. Comput. Syst. Appl. **24**(10), 17–22 (2015)
16. Wu, X., Ma, R., Shi, Y.: How do latecomer firms capture value from disruptive technologies? A secondary business-model innovation perspective. IEEE Trans. Eng. Manage. **57**(1), 51–62 (2010)
17. Wu, Z., Yin, J., Deng, S., Wu, J., Li, Y., Chen, L.: Modern service industry and crossover services: development and trends in China. IEEE Trans. Serv. Comput. **9**(5), 664–671 (2015)
18. Xue, X., Gao, G., Wang, S., Feng, Z.: Service bridge: transboundary impact evaluation method of internet. IEEE Trans. Comput. Soc. Syst. **5**(3), 758–772 (2018)

19. Xue, X., Wang, S., Zhang, L., Feng, Z., Guo, Y.: Social learning evolution (SLE): computational experiment-based modeling framework of social manufacturing. IEEE Trans. Industr. Inf. **15**(6), 3343–3355 (2018)
20. Yin, J., Zheng, B., Deng, S., Wen, Y., Xi, M., Luo, Z., Li, Y.: Crossover service: deep convergence for pattern, ecosystem, environment, quality and value. In: 2018 IEEE 38th International Conference on Distributed Computing Systems (ICDCS), pp. 1250–1257. IEEE (2018)

A DNN Inference Acceleration Algorithm in Heterogeneous Edge Computing: Joint Task Allocation and Model Partition

Lei Shi[1], Zhigang Xu[1(✉)], Yi Shi[2], Yuqi Fan[1], Xu Ding[3], and Yabo Sun[1]

[1] School of Computer Science and Information Engineering, Hefei University of Technology, Hefei 230009, China
xuzhig1995@163.com

[2] Intelligent Automation Inc., 15400 Calhoun Drive, Rockville, MD 20855, USA

[3] Institute of Industry and Equipment Technology, Hefei University of Technology, Hefei 230009, China

Abstract. Edge intelligence, as a new computing paradigm, aims to allocate Artificial Intelligence (AI)-based tasks partly on the edge to execute for reducing latency, consuming energy and improving privacy. As the most important technique of AI, Deep Neural Networks (DNN) has been widely used in various fields. And for those DNN based tasks, a new computing scheme named DNN model partition can further reduce the execution time. This computing scheme partitions the DNN task into two parts, one will be executed on the end devices and the other will be executed on edge servers. However, in a complex edge computing system, it is difficult to coordinate DNN model partition and task allocation. In this work, we study this problem in the heterogeneous edge computing system. We first establish the mathematical model of adaptive DNN model partition and task offloading. The mathematical model contains a large number of binary variables, and the solution space will be too large to be solved directly in a multi-task scenario. Then we use dynamic programming and greedy strategy to reduce the solution space under the premise of a good solution, and propose our offline algorithm named GSPI. Then considering the actual situation, we subsequently proposed the online algorithm. Through our experiments and simulations, we proved that our proposed GSPI algorithm can reduce the system time cost by at least 32% and the online algorithm can reduce the system time cost by at least 24%.

Keywords: Task allocation · Model partition · Edge computing · Edge intelligence

1 Introduction

As the most popular algorithm in deep learning, nowadays Deep Neural Networks (DNN) has achieved very good results in various fields, such as face recognition [1], autonomous driving [2], speech recognition [3], and so on. Especially

H. Gao et al. (Eds.): CollaborateCom 2020, LNICST 349, pp. 237–254, 2021.
https://doi.org/10.1007/978-3-030-67537-0_15

with the advent of the Internet of Everything [4], the number of the embedded devices (such as mobile phones, wearables, smart cameras), which tend to do some DNN calculation jobs, are increased dramatically. However, these embedded devices are resource-constrained, and they do not have enough computing ability to run these DNNs. The traditional approach is based on cloud computing solution [5], which means the input data generated from end devices is uploaded to the cloud center for executing, and then results are sent back to the end devices after the DNN calculation finished on the cloud. But with this cloud-centric approach, a large amount of data needs to be transmitted to the cloud through the wide area network, and this will cause high latency and network congestion.

Fortunately, in recent years, a new computing paradigm called edge computing [6] prompts us some new ways to deal with these issues. In edge computing environment, computation tasks will be allocated to edge servers which are close to end devices for guaranteeing real-time performance. For example, in [7], authors proposed an edge computing framework based on video processing to deal with the high latency and network congestion problems in traditional cloud computing. In addition, edge computing has the potential to play a huge role in smart homes [8] and smart cities [9]. Many scholars have done researches on the resource scheduling and task allocations on edge computing. For example, in [10], authors proposed an online algorithm named OnDoc, which can satisfy the task scheduling requested deadline as much as possible. In [11], authors investigated a green mobile-edge computing (MEC) system with energy harvesting devices and developed an effective computation offloading strategy.

The concept of the edge intelligence was proposed in [12], which can be considered as the combination of the artificial intelligence and the edge computing. Edge intelligence has the characteristics that edge computing is close to the user, which can effectively solve the high latency problem caused by running DNN in cloud computing mode. In edge intelligence, an important research hotspot is to further accelerate DNN inference. Different researchers have done different works on this hotspot. Some studies aim at making lighter DNN models so that they can be deployed on resource-constrained devices. For example, in [13], authors proposed the method of compressing deep neural network with pruning, trained quantization and Huffman coding, and making DNN models easier to be deployed on embedded systems with limited hardware resources. In [14–17], scholars design lightweight networks by improving the structure of the network model. But making lighter DNN models may reduce the accuracy, so other studies aim at dividing the DNN inference model into two parts: one part is executed on the end device, and the other is executed on the edge server. We call this method as model partition [18]. For example, in [19], authors proposed a DeepWear model, which applied the DNN model partition to DNN-based wearable device applications, and achieved a good acceleration effect. In [20], based on BranchNet [21], authors proposed the method which combining the model partition and the model early exit, for providing the low-latency edge intelligence. In [22], authors proposed a partition method based on graph min-cut method, and proposed the partition algorithms under the light workload and the heavy workload respectively. In [23], based on DNN model partition, authors further reduced the delay by encoding the feature of middle layer. In [24], authors proposed a cost-driven partition strategy for DNN-based applications over the cloud, edge and end devices, and the partition strategy effectively reduced the system cost within the corresponding deadline of each task. In [25], authors proposed the partition and offloading strategy which can make the optimal tradeoff between performance and privacy for battery-powered mobile devices. In Fig. 1, we show a simple DNN view (Fig. 1(a)) and its partition example (Fig. 1(b)).

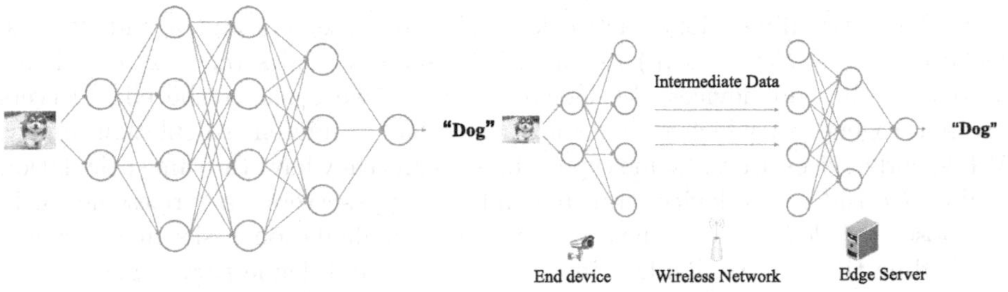

(a) A simple DNN (b) Partition of the DNN

Fig. 1. A simple DNN & partition of the DNN

Previous work has made some contributions in the field of model partition. However, most of them only considered the situation of the single end device to single edge server pattern. In reality, the optimal partition point under the case of the single end device to single edge server pattern may not be suitable for the multi-device multi-server system. For example, the optimal partition point for the same task is different under the different system loads. So in this paper, different from other researchers' previous works, in addition to get an optimal solution, we also consider some good suboptimal partition points in our system. More concretely, tasks will adopt several different available partition points. For further analysis, to minimize time cost for tasks in the heterogeneous edge system, the process of the model partition and task allocation are inseparable. So we jointly consider model partitioning and task allocation, and try to design the algorithm to minimize the average time cost of all tasks.

Specifically, the main contributions of the paper are as follows: 1) We establish the mathematical model of adaptive DNN model partition and task offloading under the heterogeneous edge computing; 2) We use dynamic programming to solve complex problems and propose the Greedy Strategy for Progressive Inference (GSPI) algorithm; 3) Considering the actual situation, it is necessary to make the selection of the partition point and the allocation of tasks in the online situation, and we propose the corresponding online algorithm. Through simulation experiments, we find that using DNN model partition in heterogeneous edge system can effectively reduce the time cost. And both offline and online algorithms can achieve good performance.

The rest of this paper is organized as follows: In Sect. 2, we introduce our system model and define our problem. In Sect. 3, we give the model partition algorithm and task allocation algorithm respectively. In Sect. 4, we give the simulation results and analyze them. In Sect. 5, we summarize this paper.

2 System Model and Problem Definition

2.1 DNN Model Partition and Network

Consider a two-dimensional network consists of u servers and n end devices. Denote $s_i(s_i \in S, i = 1...u)$ as one of the servers, and $e_i(e_i \in E, i = 1...n)$ as one of the end devices. As shown in Fig. 2, we consider the edge servers are heterogeneous, which means each server has a different calculation ability. While end devices are the same type, which means they have the same calculation ability. Consider end devices transmit data to edge servers via wireless network. Suppose end devices may have one or several calculation tasks in the whole scheduling time T. We divide T into h time slots, and define $\tau_i(\tau_i \in T, i = 1...h)$ as one of the time slot. In general, suppose the length of all time slot is the same, i.e., $\tau_1 = \tau_2 = ... = \tau_h$. Consider each task may start in a different time slot, and there are N tasks in the whole scheduling time T. Denote m_i^j as one of the task, where i means the task is from e_i, and j means the task starts from τ_j.

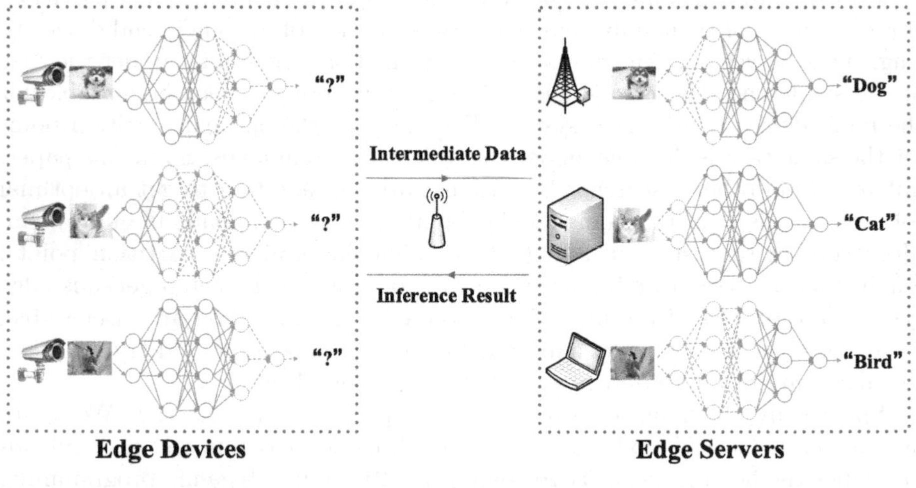

Fig. 2. Edge computing network with heterogeneous servers

Suppose all tasks use a same DNN model for calculating, and suppose this DNN model has v layers. Denote $r_i(r_i \in R, i = 1...v)$ as one of the layer, and denote d_0 as the initial input data and d_i as the output data for layer r_i.

For each server, only one task can be calculated at a time, which means when the server processes multiple tasks, the other tasks will not be executed until the previous task is completed.

By using model partition, the DNN calculation for each task can be divided into two parts. As shown in Fig. 1(b), the first part will be calculated on the end device, and the second part will be calculated on a server. Each task may use any partition point (as shown in Fig. 2). Suppose all devices can connect directly with all servers in this region. We want to decide each task's partition point and calculating server, so that all tasks can be calculated out as soon as possible.

2.2 Problem Formulation

Table 1. Notations

Variables	Meaning
B	The bandwidth of the system;
d_0	The input data of the DNN;
d_i	The output data of the layer r_i;
s_i	One of the edge server $i \in \{1, 2, 3, ..., u\}$;
e_i	One kind of end devices $i \in \{1, 2, 3, ..., n\}$;
τ_i	One of the time slots $i \in \{1, 2, 3, ..., h\}$;
r_i	One of the layer of the network $i \in \{1, 2, 3, ..., v\}$;
m_i^j	The task from the end device e_i at the j-th time slot;
N	The total number of the all tasks;
$t(m_i^j)$	The total time to complete the task m_i^j;
$t_d(m_i^j)$	The executing time of the end device part for task m_i^j;
$t_t(m_i^j)$	The transmission time of the intermediate data task m_i^j;
$t_w(m_i^j)$	The waiting time for executing the task m_i^j on the edge server;
$t_s(m_i^j)$	The executing time of the edge server part for task m_i^j;
t_D^α	The time cost for end device to execute the r_α layer of the DNN;
$t_{S_k}^\beta$	The time cost for edge server to execute the r_β layer of the DNN;
$J_{m_i^j}^k$	The executing time on the k-th edge server for task m_i^j;
$t_r(m_i^j)$	The time of returning result for task m_i^j;

For the task m_i^j, if the end device i does not generate a task at time slot j, then the task m_i^j does not exist. Here we use $E(m_i^j)$ to indicate whether there is a task m_i^j, and it can be expressed as (Table 1)

$$E(m_i^j) = \begin{cases} 1: \text{the end device } i \text{ generates a task at time slot } j; \\ 0: \text{otherwise.} \end{cases}$$

The total time to complete the task m_i^j can be expressed as the following

$$t(m_i^j) = t_d(m_i^j) + t_t(m_i^j) + t_w(m_i^j) + t_s(m_i^j) + t_r(m_i^j), \tag{1}$$

where $t_d(m_i^j)$ indicates the time cost for executing on the end device e_i, $t_t(m_i^j)$ indicates the time cost for data transmission to a server, $t_w(m_i^j)$ is the waiting time for execution, $t_s(m_i^j)$ is the execution time of task m_i^j on edge server, and $t_r(m_i^j)$ is the time for returning results. In the following we give the specific formulation for each item. We will first give the simple items for the first, the second, and the last item, then give the complex items for the forth and the third item.

For the first item $t_d(m_i^j)$, we use a binary variable $x_p(m_i^j)$ to indicate whether the task m_i^j is partitioned at the r_p layer of the DNN, i.e.

$$x_p(m_i^j) = \begin{cases} 1 : E(m_i^j) = 1 \text{ and } m_i^j \text{ is partitioned at the } r_p \text{ layer of the DNN;} \\ 0 : \text{otherwise.} \end{cases}$$

Note that $\sum_{p=1}^{v} x_p(m_i^j) = 1$. Then the first item $t_d(m_i^j)$ can be expressed as

$$t_d(m_i^j) = \sum_{p=1}^{v} (x_p(m_i^j) \cdot \sum_{\alpha=1}^{p} t_D^\alpha), \tag{2}$$

where t_D^α is the time cost for the end device to execute the r_α layer. Since all end devices are the same type, so different end devices have a same t_D^α.

For the second item $t_t(m_i^j)$, it can be calculated by

$$t_t(m_i^j) = \sum_{p=1}^{v} (x_p(m_i^j) \cdot \frac{d_p}{B}), \tag{3}$$

where d_p is the output data of layer r_p and B is the bandwidth.

For the last item $t_r(m_i^j)$, it is the time cost to return inference result and can be considered as a constant.

For the forth item $t_s(m_i^j)$, it will be decided by the server which executing task m_i^j eventually. Denote a binary scheduling variable $y_k(m_i^j)$ to indicate whether the task m_i^j is executed on the server s_k, i.e.,

$$y_k(m_i^j) = \begin{cases} 1 : E(m_i^j) = 1 \text{ and } m_i^j \text{ is executed on the server } s_k; \\ 0 : \text{otherwise.} \end{cases}$$

Apparent we have $\sum_{k=1}^{u} y_k(m_i^j) = 1$, and we have

$$t_s(m_i^j) = \sum_{k=1}^{u} (y_k(m_i^j) \cdot J_{m_i^j}^k), \tag{4}$$

where $J^k_{m_i^j}$ is the executing time on the k-th edge server for task m_i^j. $J^k_{m_i^j}$ can be expressed as

$$J^k_{m_i^j} = \sum_{p=1}^{v}(x_p(m_i^j) \cdot \sum_{\beta=p+1}^{v} t_{s_k}^{\beta}), \tag{5}$$

where $t_{s_k}^{\beta}$ is the time cost for edge server s_k to execute the r_β layer of the DNN.

For the third item $t_w(m_i^j)$, denote $AS(m_i^j)$ as the arriving time of the task m_i^j, we have

$$AS(m_i^j) = j + t_d(m_i^j) + t_t(m_i^j) = j + \sum_{p=1}^{v}(x_p(m_i^j) \cdot (\sum_{\alpha=1}^{p} t_D^{\alpha} + \frac{d_p}{B})). \tag{6}$$

Since when task m_i^j starts to run on the server, all tasks $m_{i'}^{j'}$ come to this server before $AS(m_i^j)$ should be accomplished. Here we use a binary variable $z_t^1(m_i^j)$ to judge the arrival time of a task, and $z^1_{AS(m_i^j)}(m_{i'}^{j'}) = 1$ means that the task $m_{i'}^{j'}$ arrives the edge server before the task m_i^j. For those tasks arriving at the edge server at the same time slot, we use another binary variable $z_t^2(m_i^j)$ to represent them. $z^2_{AS(m_i^j)}(m_{i'}^{j'}) = 1$ means that the task $m_{i'}^{j'}$ arrives the edge server at the same time slot as the task m_i^j. Note that in order to solve the execution order of the tasks satisfied $z_t^2(m_i^j)$, we adopt the first-generation-first-serving (FGFS) strategy for these tasks. These two binary variables are expressed as follows:

$$z_t^1(m_i^j) = \begin{cases} 1 : AS(m_i^j) < t; \\ 0 : \text{otherwise}. \end{cases}$$

$$z_t^2(m_i^j) = \begin{cases} 1 : AS(m_i^j) = t; \\ 0 : \text{otherwise}. \end{cases}$$

Then the third item $t_w(m_i^j)$ can be expressed as

$$\begin{aligned} t_w'(m_i^j) = \sum_{k=1}^{u}\Big(y_k(m_i^j) \cdot \big(\sum_{j'=1}^{h}\sum_{i'=1}^{n}(y_k(m_{i'}^{j'}) z^1_{AS(m_i^j)}(m_{i'}^{j'}) \cdot (J^k_{m_{i'}^{j'}}))\big) \\ + \big(\sum_{j'=1}^{j-1}\sum_{i'=1}^{n}(y_k(m_{i'}^{j'}) z^2_{AS(m_i^j)}(m_{i'}^{j'}) \cdot (J^k_{m_{i'}^{j'}}))\big)\Big) - (AS(m_i^j) - 1). \end{aligned} \tag{7}$$

In (7), it is composed of three parts. The first part is the total time cost of all tasks that arrive at the server before the task m_i^j, under the premise of being allocated to the same server as the task m_i^j. The second part is the total time cost of all tasks that arrive at the server at the same time as task m_i^j but were generated before the task m_i^j, under the premise of being allocated to the same server as the task m_i^j. The last part is the time that the system has been running until the task m_i^j arrived at the server. Apparently the value of $t_w(m_i^j)$ needs to be positive, i.e:

$$t_w(m_i^j) = \max\{0, t'_w(m_i^j)\}. \tag{8}$$

Then we can get the average time of all tasks in the system,

$$t_{Average} = \frac{1}{N}\sum_{i=1}^{n}\sum_{j=1}^{h} t(m_i^j). \tag{9}$$

Then our problems can be formulated as,

$$\begin{aligned}
&\min \quad t_{Average} \\
&\text{s.t.}(2)(3)(4)(5)(6)(7)(8)(9) \\
&\textstyle\sum_{p=1}^{v} x_p(m_i^j) = 1 \quad (\forall i \in E, j \in T\&E(m_i^j) = 1) \\
&\textstyle\sum_{k=1}^{u} y_k(m_i^j) = 1 \quad (\forall i \in E, j \in T\&E(m_i^j) = 1).
\end{aligned} \tag{10}$$

In (10), $E(m_i^j)$, d_i, d_p, and other symbols are all constants or determined values for specific network. $x_p(m_i^j)$ and $y_k(m_i^j)$ are binary variables. However, these binary variables almost appear in all items with different forms. These make the original problem model complex and hard to be solved directly. So for the problem to be solved in polynomial time, we need to give further analysis and find some way to reduce the complexity of the original problem.

3 Algorithms

In the last section, we give the original problem model and show that it is difficult to be solved directly. In this section, we will try to find some feasible solution for the whole network. Firstly, in order to get the best possible experimental result, we designed an offline algorithm. In the offline algorithm, all the $E(m_i^j)$ can be know in advance, that is, we know the generation time of all tasks. However, in a real network with multiple end devices and multiple edge servers, we may only know tasks that have been generated or are generating, but we do not know tasks that will be generated. In other words, for a time τ_k, we know all $E(m_i^j)$ where $(i = 1 \dots n)$ and $(j \leq k)$, but we know nothing about $E(m_i^j)$ where $(j > k)$. That means in reality we need to design an online algorithm with only the knowledge about the history and the current tasks. In the following, we first design the offline algorithm in Subsect. 3.1. Then in Subsect. 3.2, we design the online algorithm.

3.1 Offline Strategy

For the offline strategy, suppose we know all $E(m_i^j)$. When considering all tasks directly, the solution space is infinite, since there are $(u * v)^N$ combinations of N task strategies. To design an offline algorithm in polynomial time, we must reduce the solution space. Here we consider two aspects to reduce the solution space.

First, for a DNN model with v layers, each layer may be the position of the partition point. However, we find that most of layers have no potential to be partition points. Because the intermediate data for these layers are too large, and partition in these layers may cause high transmission cost. So we need to design an algorithm to exclude these layers. We call the algorithm as the Partition-Points-Selection (PPS) algorithm. Through PPS algorithm, we can reduce the solution space size in a certain extent.

Second, based on the dynamic programming, we try to reduce the solution space further. The main idea is first generated tasks will be first handled. We call the designed algorithm based on this idea as the Greedy Strategy for Progressive Inference (GSPI) algorithm.

Now we give the details.

The PPS Algorithm

For the layers that have no potential to be partition points, we can use our Partition-Point-Selection algorithm to exclude them in advance, that is, we will select a few good partition points. Then in order to determine which partition points will be selected, we need to define a time standard *Latency*. Here, if a partition point is a good partition point, then it needs to satisfy that the total time cost of the task $t(m_i^j)$ is less than *Latency* without considering the waiting time $t_w(m_i^j)$. In this way, we will get a suitable set of partition points SP_u for each server s_u.

But it is worth noting that there are multiple servers with different computing capabilities in our system, and for a same task, the total cost time at the same partition point is different when different servers are selected. So for the whole system, the suitable partition points should meet latency requirements on all servers. Therefore, we need to use the common suitable partition points on all servers as the final set of suitable partition points SP.

The detailed steps are in Algorithm 1. After get the all available partition points, each task can choose any partition point in SP. Through Partition-Points-Selection algorithm, we can greatly reduce the choice of unreasonable partition points, thereby reducing the solution space.

The GSPI Algorithm

Now we discuss the GSPI algorithm. After using the PPS algorithm, we have shrunk the solution space for $x_p(m_i^j)$. However, there are still lots of variables of $x_p(m_i^j)$, $y_k(m_i^j)$ in (10). Though it is impossible to determine all these variables once, it is possible to determine them step by step by greedy. In detail, suppose we need to allocate a task generated in the first time slot. As the first generated task, it can select any possible partition point and can select any edge server without calculating $t_w(m_i^j)$, since $t_w(m_i^j) = 0$ now. We can calculate all values and select the best one as this task's allocating strategies. We can continue doing this calculation in the following time slot based on the strategies of tasks in the previous time slots. The most difficult calculating item in (10) is $t_w(m_i^j)$. But by using the GSPI algorithm, when we try to allocate a task m_i^j, all $t_w(m_{i'}^{j'})$ where

---

Algorithm 1. Partition-Points-Selection algorithm

Input:
S: Optional Server number; B: Bandwidth; V: the total layer for the network; d_i: the output data of the layer $ri, i \in \{1, 2, 3, ..., v\}$; t_d^i; executing time for the layer ri on the end device; $t_{s_u}^i$: executing time for the layer ri on the end server s_u, $u \in \{1, 2, 3, ..., N\}$; *Latency*:latency requirement for the system;
Output: the optional partition points SP in the system
  for $u \in S$ **do**
    for $v \in V$ **do** $T \leftarrow \sum_{i=0}^{v} t_d^i + \frac{d_v}{B} + \sum_{k=v+1}^{V} t_{s_u}^k$
      if $Latency \geq T$ **then**
        Recode the partition point for server s_u in SP_u;
      end if
    end for
  end for
  $SP \leftarrow \bigcap_{u \in \{1,2,3,...,N\}} SP_u$
  return SP

$(j' < j)$ have already been determined. And the solution space is been reduced further.

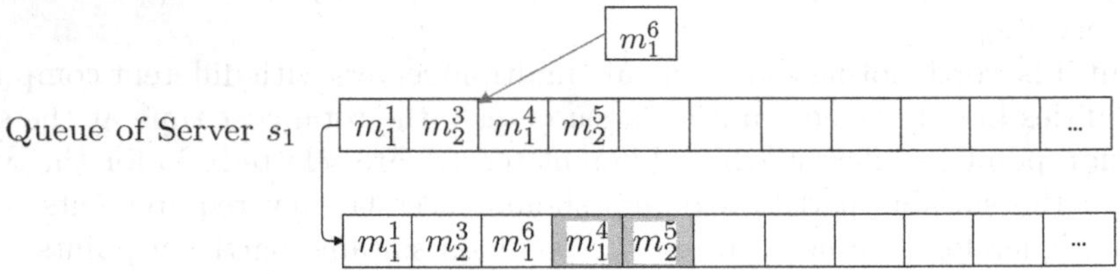

Fig. 3. Insert a task to the server's queue. A task with a gray background indicates that its strategy is improvable.

Notice that we may meet the situation that task m_i^j generated after task $m_{i'}^{j'}$, i.e., $(j > j')$, but it reach to the same edge server before $m_{i'}^{j'}$, i.e., $AS(m_i^j) < AS(m_{i'}^{j'})$. For example, in Fig. 3, the task m_1^6 is assigned to the server after the task m_1^4 and m_2^5, but will arrive before the task m_1^4, i.e. $AS(m_1^6) < AS(m_1^4) < AS(m_2^5)$. Obviously the strategies of the task m_1^4 and m_2^5 are improvable because m_1^6 was miss-considered when strategy center made their strategies. For this situation, the offline algorithm can remake the strategy for improving the performance. The task m_1^4 and m_2^5 will be removed from the queue of the server and their strategies will be remade. We will consider this situation in our GSPI algorithm.

Based on the above ideas, the GSPI algorithm can be summarized into the following three steps:

Algorithm 2. The GSPI Algorithm

Input: S:Optional servers; Q:Task Queue of edge servers; M:the tasks stack; J:cost Matrix; SP:Optional partition points.
Output: the partition and allocation strategy for the task in M.

1: **while** M!=Empty **do**
2: $M' \leftarrow M.pop()$;
3: **while** $M'! = NULL$ **do**
4: **for** $m_i^j \in M'$ **do**
5: get the strategy according to $\arg\min_{k \in S; p \in SP} t(m_i^j)$;
6: insert the tasks into Q;
7: update the statu of edge server Q.
8: **end for**
9: **end while**
10: **for** m_i^j in Q **do**
11: **if** the strategy of m_i^j is improvable **then**
12: add m_i^j into Q'.
13: **end if**
14: **end for**
15: **for** m_i^j in Q **do**
16: remake the strategy for m_i^j.
17: **end for**
18: update Q
19: **end while**

Step 1: Initialization. Put all tasks into the task stack M in units on their generated time. The first generated task is at the top of the stack, and the last generated task is at the bottom of the stack.
Step 2: Get the Tasks' Strategies. Take out all tasks on the top of the stack M. Calculate their optimal strategies one by one. Note that when the strategy of one task is generated, the strategy-making center will update the status of each edge server.
Step 3: Check the improvable Strategies. After the strategies of all tasks for a stack frame are completed, we need to check whether there is any task with improvable strategy, such as the task m_1^4 and the task m_2^5 in Fig. 3. For these tasks with improvable strategies, they should be removed from the servers' queues, and the strategies should be remade.

The step 2 and step 3 will be repeated until the stack M is empty. The GSPI algorithm is shown in Algorithm 2. Line 3–9 are the process of a single time slot task, and line 10–17 are the process of checking strategies and remaking the strategies.

3.2 Online Strategy

For the online strategy, we only know tasks that have been generated and are being generated. And when the tasks are generated, we should make their strategies in real time. Here we propose a strategy-making architecture as shown in

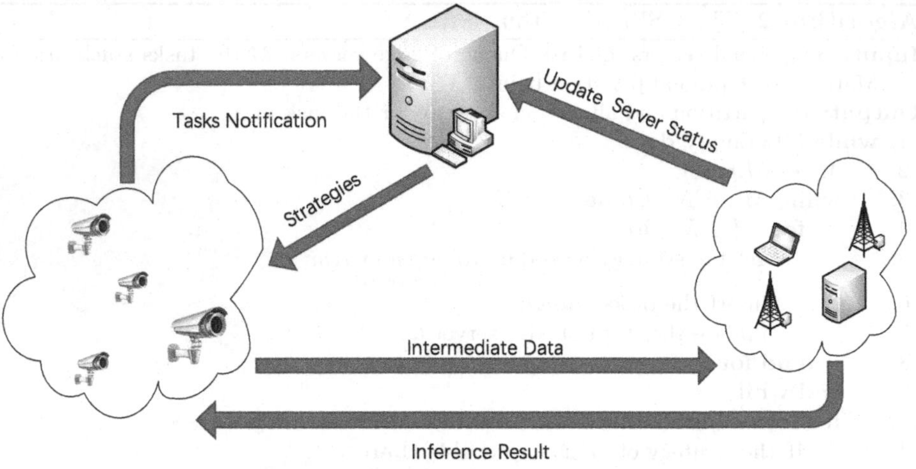

Fig. 4. Strategy model

Fig. 4. We use a much powerful edge server as the strategy-making center. As the end devices generate tasks, the strategy center will decide the tasks calculating strategies after receiving the notification. For the online strategy, the strategy center does not know the future tasks when making strategies. So formula (7) should be modified accordingly,

$$t'_w(m_i^j) = \sum_{k=1}^{u}\Big(y_k(m_i^j)\cdot\Big(\sum_{j'=1}^{j}\sum_{i'=1}^{n}(y_k(m_{i'}^{j'})\cdot(J^k_{m_{i'}^{j'}}))\Big)\Big)-(AS(m_i^j)-1). \quad (11)$$

Here the modified formula ignores the tasks generated after time slot j to meet the characteristics of real-time strategies-making.

Most steps of the online algorithm is the same as the offline algorithm. But unlike the offline algorithm, since we do not have the knowledge about the future tasks, and we need to calculate each task in-time, so we will not consider the situation that task m_i^j generated after task $m_{i'}^{j'}$ but reaches to the same edge server before $m_{i'}^{j'}$. That means, if this situation occurs, task $m_{i'}^{j'}$ will be added to the server stack directly and wait to be executed.

The heuristic algorithm of our online algorithm is shown in Algorithm 3. We use an auxiliary task queue Q' temporarily save the state of server queue. After the strategies for all tasks in a time slot are formulated, copy the status of Q' to Q. In this way, the strategy center can return the strategies of tasks at the same time.

3.3 Complexity Analysis

In this subsection, we will give the complexity analysis of three algorithms. For the Algorithm 1, there are u servers in the system, and the DNN model has v

Algorithm 3. Online Algorithm

Input: S:Optional servers; Q:Task Queue of edge servers; M_τ:Tasks generated at the current momentτ; J:cost Matrix; SP:Optional partition points.
Output: The strategies for the all tasks in M_τ and the updated Q.

1: $\kappa \leftarrow length(M_\tau)$
2: **if** $\kappa > 1$ **then**
3: $\quad Q' \leftarrow Q$
4: $\quad$ **while** M_τ!=NULL **do**
5: $\quad\quad m_i^j \leftarrow M_\tau.pop()$
6: $\quad\quad$ Get the best strategy according to $\underset{k \in S; p \in SP}{\arg\min}\ t(m_i^j)$;
7: $\quad\quad$ Update the Q'.
8: $\quad$ **end while**
9: $\quad Q \leftarrow Q'$
10: **else**
11: $\quad$ Get the best strategy according to $\underset{k \in S; p \in SP}{\arg\min}\ t(m_i^j)$;
12: **end if**
13: Update the Q.

layers. It takes $u * v$ times to calculate each possible combination, and $u * v$ can be treated as a constant. So the time complexity of Algorithm 1 is $O(1)$.

For the Algorithm 2, to get the strategy of a single task in line 5, we need to calculate the total time consumption of the task $t(m_i^j)$. From formula 1 and the corresponding expansion, we can get that except for $t_w(m_i^j)$, the other items are all constant items. To calculate $t_w(m_i^j)$, the formula 7 tell us that all tasks need to be considered. So for any task, the time complexity of computing $t_w(m_i^j)$ is $O(N)$, where N is the total number of the tasks. Then get the best strategy according to $\underset{k \in S; p \in SP}{\arg\min}\ t(m_i^j)$, the time complexity come to $O(a * N)$, where a is a constant related to the set of servers S and the set of available partition points SP. And there are N tasks, so the final time complexity is $O(a * N^2)$, which can be expressed as $O(N^2)$.

For the Algorithm 3, to get the strategy of a single task in line 6, it is the same as Algorithm 2, that is, the main complexity lies in $t_w(m_i^j)$. To calculate $t_w(m_i^j)$, the formula 11 tell us that all tasks generated before task m_i^j need to be taken into account, that is, to calculate the $t_w(m_i^j)$ of the N-th task, the strategies of the first N-1 tasks need to be considered. So for all tasks, their average time complexity of computing $t_w(m_i^j)$ is $O(N/2)$, where N is the total number of the tasks. Then get the best strategy according to $\underset{k \in S; p \in SP}{\arg\min}\ t(m_i^j)$, the time complexity come to $O(a * N/2)$, where a is a constant related to the set of servers S and the set of available partition points SP. Finally, there are N tasks in the entire process, so the above content has to be repeated N times, and the final time complexity is $O(a * N^2/2)$, which can be expressed as $O(N^2)$.

4 Simulation and Experiment

In this section, we introduced the related work of experiments and simulations. The DNN model used in our experiments is VGG16 [26]. The model is trained using the cifar-10 dataset, and the framework used is pytorch. All tasks in our system is the inference task based on VGG16. We use the computing power of a CPU to simulate the computing power of our end device, and use the GPUs on different computers as our servers. To get the constants in Sect. 2, such as t^{α}_{D}, $t^{\beta}_{S_k}$, d_i and so on. First we run a DNN inference on a CPU for 100 times and get the execution time of each layer of the DNN. Then we take the average result of these 100 experiments as our experimental data, i.e the value of t^{α}_{D}. The values of other constants are obtained by the same way. Notice that we set enough end devices and each end device have enough time to calculate the generated tasks, that is, there will be no waiting tasks on the device. We set 4 servers with computing capabilities in the system.

We first study these DNN-based tasks in Subsect. 4.1, and give experimental results of Partition-Points-Selection. Then based on these experimental results, we give further experimental setting. In Subsect. 4.2, we validate the proposed model and algorithm. We conducted experiments under different bandwidths and different number of tasks. Experiment result shows that compared to running the DNN in end-only mode, running the DNN in a multiple partition points mode can reduce system time cost by 32% on average. And compared to the server-only mode, the system time is reduced by 48% on average.

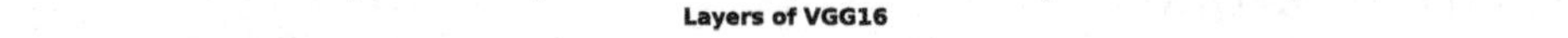

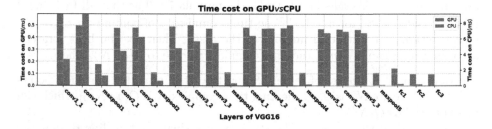

Fig. 5. Each layer's executing time of VGG16 on GPU and CPU

4.1 Results of Partition Points

First, we get the execution time of each layer of VGG16 on the end device and edge servers and get the amount of intermediate output data. Figure 5 shows the execution time of each layer of the model on CPU and one of the GPUs, and the time cost of running a task on the end device is more than ten times that on the edge server.

Then we get the total time cost of a DNN task at different partition points under the premise of ignoring the task waiting time. One of the result is shown as Fig. 6, and finally we get 6 partition points. Compared with the original 20 partition points, this can greatly reduce the amount of calculation. The red stars in Fig. 6 indicate the selected partition points.

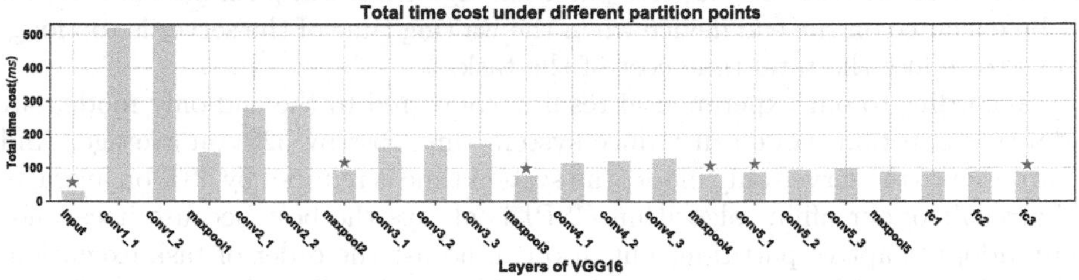

Fig. 6. The time cost of VGG16 at different partition points

Based on the above experiment result, we define 1 ms as one time slot. Then we use the time slot as our minimum time unit, and convert normal time representation into time slot representation. Table 2 shows the time cost for end device and edge servers at different partition points. The partition point $P1$ means the entire task will be completely uploaded to the server for calculation, and partition point $P6$ means that the entire task is calculated on the end device.

Table 2. Time cost of device and servers in different partition points

Partition schemes	End Device	Server1	Server2	Server3	Server4
P1	0	7	10	20	38
P2	25	5	7	13	27
P3	41	3	5	8	19
P4	62	2	3	4	9
P5	71	1	2	3	6
P6	88	0	0	0	0

4.2 Experiment Results

Firstly we set time slots $h = 100$ and the bandwidth $B = 4$ Mbps. Then we set different number of tasks in 100 time slots, and run these tasks in end-only mode, server-only mode and our adaptive partition mode. The result is shown in Fig. 7(a).

When the number of tasks is relatively small, we can see the results of our algorithm are very close to the server-only mode. The reason for this is that when there are few tasks, the waiting time of the queues on the servers are almost 0. This drives the system to run the entire task on the server. As the number of tasks gradually increases, the waiting time of the queue on the servers will increase greatly. As a result, the average task time cost in server-only mode will greatly increase. But our adaptive partition algorithm will push part of the task to be executed on the end device when the waiting time of the server is too long, so as to reduce the total time cost of the task.

According to our experimental results, compared to the end-only mode, our adaptive partition mode can reduce system time cost by 32% on average. And compared to the server-only mode, the system time is reduced by 48% on average. The result of our offline algorithm GSPI is always the best because it can not only adopt adaptive partition, but also can adjust the order of task execution. The online-algorithm we proposed also performs well. When the number of tasks is small, its performance is the same as the offline algorithm, and as the number of tasks increases, it cost at most 10% more system time than offline algorithms.

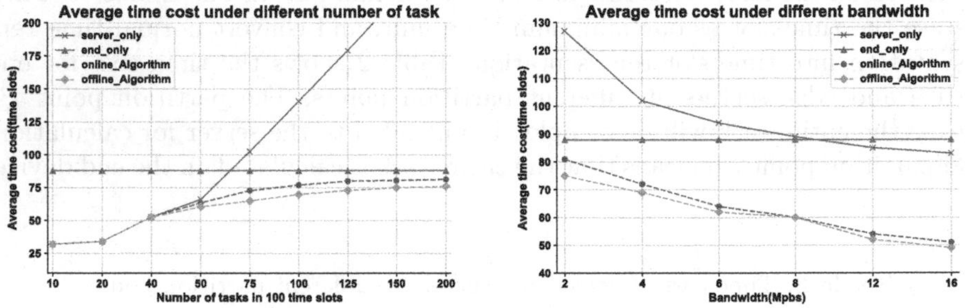

(a) Different number of tasks

(b) Different bandwidth

Fig. 7. (a) The average time cost for different number of tasks. The bandwidth $B = 4$ Mbps. (b) The average time cost under different bandwidth. Number of tasks is set as 75 in 100 time slots.

And Fig. 7(b) shows that we set 75 tasks in 100 time slots and bandwidth B from 2 Mbps to 16 Mbps. From the result, we can know that higher bandwidth will produce less time cost. But the bandwidth does not affect the end-only mode, because no intermediate data needs to be uploaded to the edge server. It is worth noting that the adaptive partition mode we proposed performs better than the server-only mode at any bandwidth.

5 Conclusion and Future Work

In this paper, we studied the DNN model partition and task allocation in heterogeneous edge system. We first establish the mathematical model. The model contains a large number of binary variables, and the solution space will be too large to be solved directly. Then we proposed the adaptive DNN model partition and task allocation algorithm named GSPI. Considering the real-time nature of the reality, we also proposed corresponding online algorithm. To the best of our knowledge, this is the first attempt to put the DNN model partition in a heterogeneous edge system, especially considering the waiting time of the server queue. Our simulation experiments proved that using model partition in heterogeneous edge systems can effectively reduce the time cost of DNN-based tasks.

This article is an attempt to introduce model partition into the heterogeneous edge system. In the future, we will consider that the end devices in the system have different computing capabilities, and the network environment from the device to the edge server is dynamically changing.

Acknowledgements. This article was supported by the National Key Research And Development Plan (Grant No. 2018YFB2000505), National Natural Science Foundation of China (Grant No. 61806067) and Key Research and Development Project in Anhui Province (Grant No. 201904a06020024).

References

1. Parkhi, O.M., Vedaldi, A., Zisserman, A.: Deep face recognition. In: BMVC (2015)
2. Chen, C., Seff, A., Kornhauser, A.L., Xiao, J.: DeepDriving: learning affordance for direct perception in autonomous driving. In: 2015 IEEE International Conference on Computer Vision (ICCV), pp. 2722–2730 (2015)
3. Chan, W., Jaitly, N., Le, Q.V., Vinyals, O.: Listen, attend and spell: a neural network for large vocabulary conversational speech recognition. In: 2016 IEEE International Conference on Acoustics, Speech and Signal Processing (ICASSP), pp. 4960–4964 (2016)
4. Snyder, T., Byrd, G.: The internet of everything. Computer **50**(6), 8–9 (2017)
5. Pandey, P., Singh, S., Singh, S.: Cloud computing. In: ICWET (2010)
6. Shi, W., Cao, J., Zhang, Q., Li, Y., Xu, L.: Edge computing: vision and challenges. IEEE Internet Things J. **3**, 637–646 (2016)
7. Long, C., Cao, Y., Jiang, T., Zhang, Q.: Edge computing framework for cooperative video processing in multimedia IoT systems. IEEE Trans. Multimedia **20**, 1126–1139 (2018)
8. Deschamps-Sonsino, A.: Smarter Homes. Apress, New York (2018)
9. Alba, E., Chicano, F., Luque, G. (eds.): Smart-CT 2016. LNCS, vol. 9704. Springer, Cham (2016). https://doi.org/10.1007/978-3-319-39595-1
10. Liu, L., Huang, H., Tan, H., Cao, W., Yang, P., Li, X.-Y.: Online DAG scheduling with on-demand function configuration in edge computing. In: Biagioni, E.S., Zheng, Y., Cheng, S. (eds.) WASA 2019. LNCS, vol. 11604, pp. 213–224. Springer, Cham (2019). https://doi.org/10.1007/978-3-030-23597-0_17
11. Mao, Y., Zhang, J., Letaief, K.B.: Dynamic computation offloading for mobile-edge computing with energy harvesting devices. IEEE J. Sel. Areas Commun. **34**, 3590–3605 (2016)
12. Zhou, Z., Chen, X., Li, E., Zeng, L., Luo, K., Zhang, J.: Edge intelligence: paving the last mile of artificial intelligence with edge computing. Proc. IEEE **107**, 1738–1762 (2019)
13. Han, S., Mao, H., Dally, W.J.: Deep compression: compressing deep neural network with pruning, trained quantization and Huffman coding. CoRR abs/1510.00149 (2015)

14. Howard, A.G., et al.: MobileNets: efficient convolutional neural networks for mobile vision applications. arXiv:abs/1704.04861 (2017)
15. Sandler, M., Howard, A.G., Zhu, M., Zhmoginov, A., Chen, L.C.: MobileNetV2: inverted residuals and linear bottlenecks. In: 2018 IEEE/CVF Conference on Computer Vision and Pattern Recognition, pp. 4510–4520 (2018)
16. Zhang, X., Zhou, X., Lin, M., Sun, J.: ShuffleNet: an extremely efficient convolutional neural network for mobile devices. In: 2018 IEEE/CVF Conference on Computer Vision and Pattern Recognition, pp. 6848–6856 (2018)
17. Ma, N., Zhang, X., Zheng, H.-T., Sun, J.: ShuffleNet V2: practical guidelines for efficient CNN architecture design. In: Ferrari, V., Hebert, M., Sminchisescu, C., Weiss, Y. (eds.) Computer Vision – ECCV 2018. LNCS, vol. 11218, pp. 122–138. Springer, Cham (2018). https://doi.org/10.1007/978-3-030-01264-9_8
18. Kang, Y., et al.: Neurosurgeon: collaborative intelligence between the cloud and mobile edge. In: ASPLOS 2017 (2017)
19. Xu, M., Qian, F., Zhu, M., Huang, F., Pushp, S., Liu, X.: DeepWear: adaptive local offloading for on-wearable deep learning. IEEE Trans. Mob. Comput. **19**, 314–330 (2020)
20. Li, E., Zeng, L., Zhou, Z., Chen, X.: Edge AI: on-demand accelerating deep neural network inference via edge computing. IEEE Trans. Wireless Commun. **19**, 447–457 (2020)
21. Teerapittayanon, S., McDanel, B., Kung, H.T.: BranchyNet: fast inference via early exiting from deep neural networks. In: 2016 23rd International Conference on Pattern Recognition (ICPR), pp. 2464–2469 (2016)
22. Hu, C., Bao, W.S., Wang, D., Liu, F.: Dynamic adaptive DNN surgery for inference acceleration on the edge. In: IEEE INFOCOM 2019 - IEEE Conference on Computer Communications, pp. 1423–1431 (2019)
23. Ko, J.H., Na, T., Amir, M.F., Mukhopadhyay, S.: Edge-host partitioning of deep neural networks with feature space encoding for resource-constrained Internet-of-Things platforms. In: 2018 15th IEEE International Conference on Advanced Video and Signal Based Surveillance (AVSS), pp. 1–6 (2018)
24. Lin, B., Huang, Y., Zhang, J., Hu, J., Chen, X., Li, J.: Cost-driven off-loading for DNN-based applications over cloud, edge, and end devices. IEEE Trans. Industr. Inf. **16**, 5456–5466 (2020)
25. Shi, C., Chen, L., Shen, C., Song, L., Xu, J.: Privacy-aware edge computing based on adaptive DNN partitioning. In: 2019 IEEE Global Communications Conference (GLOBECOM), pp. 1–6 (2019)
26. Qassim, H., Feinzimer, D., Verma, A.: Residual squeeze VGG16. arXiv:abs/1705.03004 (2017)

A Novel Probabilistic-Performance-Aware and Evolutionary Game-Theoretic Approach to Task Offloading in the Hybrid Cloud-Edge Environment

Ying Lei[1], Wanbo Zheng[2(✉)], Yong Ma[3], Yunni Xia[1(✉)], and Qing Xia[4]

[1] Software Theory and Technology Chongqing Key Lab, Chongqing University, Chongqing, China
xiayunni@hotmail.com

[2] School of Mathematics, Kunming University of Science and Technology, Yunnan 650500, China
zwanbo2001@163.com

[3] School of Computer and Information Engineering, Jiangxi Normal University, Nanchang, China

[4] Chongqing Key Laboratory of Smart Electronics Reliability Technology, Chongqing, China

Abstract. The mobile edge computing (MEC) paradigm provides a promising solution to solve the resource-insufficiency problem in mobile terminals by offloading computation-intensive and delay-sensitive tasks to nearby edge nodes. However, pure edge resources can be limited and insufficient for computational-intensive applications raised by multiple users, which calls for a hybrid architecture with a centralized cloud server and multiple edge nodes and smart resource management strategies in such hybrid environment. The problem is however challenging due to the distributed nature and intrinsic dynamicness of the environment. Existing researches in this direction mainly see that edge servers are with constant performance and consider the offloading decision-making as a static optimization problem. In this paper, instead, we consider that geographically distributed edge servers are with time-varying performance and introduce a dynamic offloading strategy based on a probabilistic evolutionary game-theoretic framework. To validate our proposed framework, we conduct experimental case studies based on a real-world dataset of cloud edge resource locations and show that our proposed approach outperforms traditional ones in terms of multiple metrics.

Keywords: Task offloading · Mobile edge computing · Evolutionary game theory · Probabilistic QoS

1 Introduction

Due to limited battery power, storage capacity and computational power, mobile devices face challenges in executing delay-sensitive and resource-hungry complex

H. Gao et al. (Eds.): CollaborateCom 2020, LNICST 349, pp. 255–270, 2021.
https://doi.org/10.1007/978-3-030-67537-0_16

applications such as augmented reality and online gaming. Mobile Edge Computing (MEC) is widely believed as a remedy to alleviate this problem [10,11]. In MEC, the mobile edge is enhanced with analysis and storage capabilities, possibly by a dense deployment of computational servers or by strengthening the already-deployed edge entities such as small cell base stations. Therefore, mobile devices are allowed to offload their computationally expensive tasks to the edge servers while meeting some specific quality of service (QoS) requirements [8]. This process, referred to as computation offloading, is feasible due to the fact that edge servers are usually deployed near mobile users, specifically in comparison to the remote cloud servers.

Its challenge lies in that offloading decision can't be made by using conventional centralized resource allocation schemes, due to the fact that such mechanism requires the availability of global knowledge of status of all involving nodes. As a result, distributed and autonomous approaches, where the individual mobile devices and mobile edge servers make autonomous offloading decisions, are in high need.

For the purpose described above, we develop a novel dynamic offloading approach by using a probabilistic-performance-aware game-theoretic model. Different from most existing methods in this direction, our proposed method considers that geographically distributed edge servers are with time-varying performance [19] instead of the constant one. To validate the proposed framework, we conduct experimental case studies based on a real-world dataset of cloud/edge resource locations. Experimental results clearly suggest that our proposed approach outperforms traditional ones in terms of multiple metrics.

2 Related Work

As an important theme of mobile edge computing (MEC) [10], mobile task offloading has attracted a lot of research interests recently. Existing studies can be classified into two categories in terms of scheduling fashion: online solutions and offline solutions [4,14,30]. For online solutions, the offloading decision for every mobile task will be made at once when it arrives. While the offline ones usually split the continuous-time into discrete time slots, requests which arrive in the same time slot will be scheduled together at the end of that time slot.

Most of the current studies are offline ones, for example, Chen *et al.* [8] formulate the task offloading problem at a certain time slot as a mixed integer non-linear program in software defined ultra dense network (SD-UDN), then transform this optimization problem into two sub-problems and develop a Lagrange-multiplier-based algorithm to solve them. Hosseinzadeh *et al.* [18] proposes an ANN-PSO algorithm which combines machine learning and metaheuristic algorithm, the behavior model in LTS is divided into two stages to evaluate the process of service selection and composition, in order to avoid premature convergence and guarantee the requested QoS factors. Alameddine *et al.* [4] define the dynamic task offloading and scheduling (DTOs) problem as a Mixed Integer Program (MIP), they also propose a novel thoughtful decomposition based on the technique of the Logic-Based Benders Decomposition to reduce the searching space resulted by massive requests. Alfakih *et al.* [5] proposed an offloading decision-based state-action-reward-state-action method (OD-SARSA) to sovle the question of developing an efficient resource management model for the selected MEC server in a multi-edge network by reducing system cost, including energy consumption and computing time delay. Zaw *et al.* [29] formulate the resource allocation problem in multi-edge network as a Generalized Nash Equilibrium Problem (GNEP) and prove the existence and uniqueness of the game, two distributed algorithms are proposed to update Lagrange Multipliers on client and MBS respectively. Fantacci *et al.*[15] establish a matching game with externalities between the application requested by the IIoT devices and the ECSs to raise the VRCs placement problem in a hybrid EC-Cloud network structure for an IIoT scenario. Xia *et al.* [26] regard the cloud-edge network as a graph, and propose an optimal approach named EDD-IP based on the Integer Programming technique to solve the problem of edge data distribution (EDD) in the edge computing environment. For a large-scale EDD problems, EDD-A algorithm they proposed could find the approximate solutions effectively.

There are also some studies focus on the online solution of this problem, for example, Liu *et al.* [22] and Du *et al.* [14] assume the task arrivals are i.i.d over time and their average arrival rates are pre-given. Chen *et al.* [7] and Xu *et al.* [28] consider task arrival follows Poisson process, and random arrival rate is used capture the time change of task arrival mode at different times. While Zhao *et al.* [30] assume the unloading demands of mobile users follow the binomial distribution of given probability. Based on these assumptions, a local edge double-layer queuing model is established, and the Lyapunov optimization technique is used to maintain the stability of the system. Xia *et al.* [24] considers the arbitrary arrival time of requests and propose a best-fit-decreasing-based method to yield offloading decision in real-time. However, their approach is a centralized one. Xia *et al.* [27] introduce data migration and data caching technology into task offloading in the edge environment, model the collaborative edge data caching problem (CEDC) as a constrained optimization problem, and propose an online algorithm based on Lyapunov optimization to achieve the goal of saving bandwidth cost, reducing network latency and minimizing access costs. Aslanpour *et al.* [6] comprehensively discusses the performance and metrics for cloud, fog and edge computing from multiple perspectives, so as to help researchers identify appropriate metrics for evaluating performance by analyzing specific scenarios.

3 Modeling and Formulation

3.1 System Model

In this paper, we consider a hybrid environment composed of a macro centralized cloud server is deployed, and K micro edge nodes. It is assume that the hybrid environment accomodates N users and each user is allowed to access both the centralized cloud server or one of the edge ones. The coverage of cloud servers is usually large enough to cover all users in a certain geographical area. On the contrary, edge nodes cover their near-by area, and are weaker than cloud one in terms of computing power, storage resources and other resource allocation. According [13], a large area is usually divided into J small service regions with such that the number of edge servers in each region is nearly equal. Users in each region form a population, and they have the stationary proportion to adopt the same strategy under the assumption of bounded-rationality.

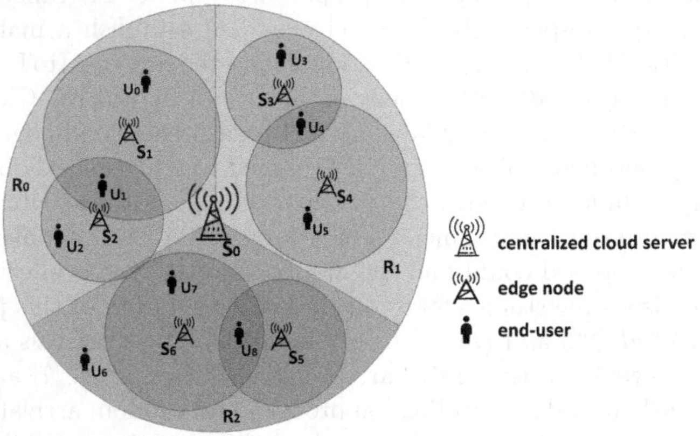

Fig. 1. Edge computing environment example

Figure 1 shows a good example of the hybrid environment described above. In this example, there exists a centralized cloud server (S_0), 6 edge ones ($S_1 \sim S_6$) and 9 users ($U_0 \sim U_8$). Coverage areas of different edge nodes may overlap and thus a user may have multiple candidates for task offloading [14]. For example, U_1 can be allocated to S_0, S_1 or S_2. In contrast, U_6 can only offload its task to S_0. We partition the area into three regions of the same size ($R_0 \sim R_2$), and two edge nodes are deployed in each region.

The notation used in this paper is shown in Table 1.

3.2 Computation Model

The tasks offloading time for U_i can be estimated as $T_i = \Delta_i^{ul} + \Delta_i^{dl} + \Delta_i^{bh} + \Delta_i^{exe}$ is composed of four parts [3], where Δ_i^{ul} indicates the uplink communication time, Δ_i^{dl} the downlink time, Δ_i^{bh} the backhaul link time and Δ_i^{exe} the backhaul link time. According to [9,20], Δ_i^{bh} can be infinitesimal, and the downlink time Δ_i^{dl} can usually be a constant ξ. Δ_i^{ul} and Δ_i^{exe} are calculated according to the network conditions and device performance. Therefore, the T_i can be expressed as:

$$\begin{aligned} T_i = t(i,k) &= \Delta_i^{ul} + \Delta_i^{dl} + \Delta_i^{bh} + \Delta_i^{exe} \\ &= \frac{b_i}{w_k \eta_i^k} + \xi + \frac{d_i}{f_k} \end{aligned} \tag{1}$$

Table 1. Key Notations

Notation	Description
S_k	The kth server
lat_k, lng_k	Latitude and longitude of S_k
R_k	Coverage radius of S_k
W_k	Maximum bandwidth of S_k
F_k	Maximum computing power of S_k
C_f^k	Unit price for computing of S_k
C_t^k	Unit price for communication of S_k
V_k	Rent of S_k
num_k	The number of users who select cloud service on S_k
w_k	Real-time bandwidth of the cloud service on S_k
f_k	Real-time computing power of the cloud service on S_k
U_i	The ith end-user
lat_i, lng_i	Latitude and longitude of U_i
b_i	Amount of data of U_i's task
d_i	Number of calculation instructions of U_i's task
t_i	The maximum tolerable delay of U_i
c_i	The maximum tolerable cost of U_i
a_i	The population to which U_i belongs
$dist(i,k)$	The distance between U_i and S_k
P_j	The jth population

where $\eta_i^k = \lambda/dist(i,k)$ is decided by the distance between U_i and S_k. As the distance increases, the bit error rate increases and the average transmission speed decreases [20]. w_k and f_k indicate the averaged computing power and averaged bandwidth of S_k, respectively. f_k can be calculated based on its corresponding historical distribution [19] according to:

$$f_k = \int_{-\infty}^{\infty} Pr(t)t\mathrm{d}t. \tag{2}$$

Acoording (3), monetary cost of U_i comprises three parts, namely the communication cost, the computation cost, and the fees for renting server. C_t^k and C_f^k indicate the price per unit transmission rate and the charge per unit computation resources of S_k, respectively. The rent V_k^{rent} is related to the current load of the server. If the server is used more efficiently and provides services for more users, the lower the rent shared by each user.

$$\begin{aligned} C_i = c(i,k) &= C_k^{commu} + C_k^{compu} + V_k^{rent} \\ &= C_t^k \frac{b_i}{w_k \eta_i^k} + C_f^k \frac{d_i}{f_k} + \frac{V_k}{num_k} \end{aligned} \tag{3}$$

We use US_i to represent U_i's user satisfaction, which can be expressed as the probability that both offloading time and monetary cost constraints are met [19]:

$$US_i = Pr(T_i \leq t_i) \cdot Pr(C_i \leq c_i) \tag{4}$$

US_{P_j} represents the satisfaction of the entire population, where num_j denotes the number of users in P_j:

$$US_{P_j} = \sum_{i=1}^{num_j} US_i \tag{5}$$

US is the satisfaction of the whole area and it is regarded as the ultimate optimization objective:

$$US = \sum_{j=1}^{J} US_{P_j} \tag{6}$$

4 Evolutionary Game Formulation

In this section, we first give the evolutionary game formulation for the problem. Due to the limitation of computation resource and bandwidth resource in each edge node, end-users in different regions have to compete for resources in clouds. Based on the replicator dynamics [25], we consider an evolutionary stable strategy (ESS) for the problem of hybrid-edge-cloud-based task offloading.

4.1 Game Formulation

A normal game includes three factors: the player set, the strategy set, and the payoff function of every player when choosing a strategy. In the context of an evolutionary game, the population is utilized to represent the group of players with the same properties [17]. We introduce the formulation of the evolutionary game as below:

Players: End-users are denoted as players in the game.

Population: Players are grouped into different populations by geographical locations. We denote the set of population as $\{\mathbb{N}_0, \mathbb{N}_1, \ldots, \mathbb{N}_J\}$. Players in each population are all located in the same geographical region.

Strategy: Each player's strategy refers to the selection of servers. In this game, there are $1 + K$ services for players to select. Accordingly, the server selection strategy can be denoted as $\mathbb{K} = \{0, 1, 2, \ldots, K\}$, which refers to the selection of the centralized cloud server and K edge nodes. $\mathbb{S}_j$ which notes the server selection set of P_j is the subset of $\mathbb{K}$.

Population Share: num_j^k denotes the number of end-users who selecting strategy S_k for population $\mathbb{N}_j$. Thus, $x_j^k = num_j^k / num_j$ is population share of S_k in population $\mathbb{N}_j$, where $x_j^k \in [0, 1]$.

Population State: The population shares of all servers constitute the population state denoted by a vector $\mathbf{x}_j = \left[x_j^0, x_j^1, x_j^2, \ldots, x_j^K\right]$. We have $\sum_{k=0}^{K} x_j^k = 1$. We use a matrix $\mathbb{X}$ to denote the population state space which contains all J population states.

$$\mathbb{X} = \begin{pmatrix} x_0^0 & x_0^1 & \ldots & x_0^K \\ x_1^0 & x_1^1 & \ldots & x_1^K \\ \vdots & \vdots & \ddots & \vdots \\ x_{J-1}^0 & x_{J-1}^1 & \ldots & x_{J-1}^K \end{pmatrix} \tag{7}$$

Payoff: The payoff of a player is decided by its net utility function. The net utility function is based on (4).

$$\pi_j^k(i) = Pr(T_i \leq t_i) \cdot Pr(C_i \leq c_i) \tag{8}$$

4.2 Evolutionary Stable Strategy (ESS)

In a traditional game theory, all players can achieve a stable state where no player can further obtain extra benefit by unilaterally changing its strategy. Such a state is called Nash equilibrium (NE). In this work, we call the game Γ as the game of service selection, where N is the player set, $\mathbb{S} = \{s_1, s_2, \ldots, s_N\}$ is the strategy set of all players, and $\Pi = \{\pi_1, \pi_2, \ldots, \pi_N\}$ is the set of payoff function. Let $\mathbb{S}_{-i} = \{s_1, \ldots, s_{i-1}, s_{i+1}, \ldots, s_N\}$ be a strategy profile of all players except player i, and $\pi(s_i, s_{-i})$ is set to be the payoff function of player i when this player selects the strategy s_i while others select s_{-i}. Then, the corresponding Nash equilibrium can be defined as below.

Definition 1. *A Nash equilibrium (NE) of a strategic game* $\Gamma =< \mathbb{N}, \mathbb{S}, \Pi >$ *is a profile* $s^* \in \mathbb{S}$ *of actions with the property that for every player* $i \in N$ *we have:*

$$\pi(s_i, s_{-i}^*) \leq \pi(s_i^*, s_{-i}^*), \forall s_i \in \mathbb{S} \tag{9}$$

The Nash equilibrium has a property of self-reinforcement that each player has no motivation to deviate from this equilibrium. The general solution to obtain the Nash equilibrium is on the assumption of complete rationality among all players. However, with a small perturbation, all players may change their strategies to reach another Nash equilibrium. In an evolutionary game theory [17], an equilibrium strategy is adopted among players with bounded rationality, which can resist small disturbances. This equilibrium strategy is called ESS and is defined as below.

Definition 2. *A strategy profile* $\mathbb{S}^* = \{s_1^*, s_2^*, \ldots, s_N^*\}$ *is an ESS iff* $\forall s_i \notin \mathbb{S}^*, s_{-i} \neq s_{-i}^*$:

$$\begin{aligned} &1. \quad \pi(s_i, s_{-i}^*) < \pi(s_i^*, s_{-i}^*) \\ &2. \quad \pi(s_i, s_{-i}^*) = \pi(s_i^*, s_{-i}^*), \pi(s_i, s_{-i}) < \pi(s_i^*.s_{-i}) \end{aligned} \tag{10}$$

Compared with Nash equilibrium, the condition (1) of Definition 2 ensures that ESS is a Nash equilibrium (NE). The condition (2) of Definition 2 ensures the stability of the game process. During the process of strategy evolution, players using mutation strategy will decrease until all players in the population asymptotically approach to the ESS. In our problem, end-users adapt their strategies among a finite set of strategies to get a better payoff. In each time, each end-user can have his/her own strategy set and the information of average payoff in the same population. Each end-user can repeatedly evolve his/her strategy over time for the cloud service selection. After sufficient repetitive stages, all end-users' strategy profile approaches to an ESS. The process of this strategy replication can be modeled by replicator dynamics, which is described in the next section.

4.3 Replicator Dynamics

In the dynamic evolutionary game [12], replicator dynamics provides a method to acquire the population information of others and converges towards an equilibrium selection. It is also significant to investigate the speed of convergence of strategy adaptation to reach evolutionary equilibrium (EE) that the population will not change its selection [23]. The basic idea is that in a population of players with bounded rationality, strategy with better results than average level will be gradually adopted by more players, and the proportion of players' strategies will change consequently. It is given as below:

$$\dot{x}_j^k(t) = \delta x_j^k(t)[\pi_j^k(t) - \bar{\pi}_j(t)] \tag{11}$$

where δ is used to control the convergence speed of strategy adaption for players in the same population. $\pi_j^k(t)$ is the current payoff of the individuals choosing strategy k in P_j, $\bar{\pi}_j(t)$ is the average payoff of $\mathbb{N}_j$. The growth rate $\dot{x}_j^k(t)$ is relevant to the difference between the payoff $\pi_j^k(t)$ for that selection strategy and the population's average payoff $\bar{\pi}_j^k(t)$ as well as the current size of population share $\pi_j^k(t)$. The average payoff $\bar{\pi}_j(t)$ of the population can be derived as:

$$\bar{\pi}_j(t) = \sum_{k=0}^{K} x_j^k(t)\pi_j^k(t) \tag{12}$$

Based on the replicator dynamics of strategy selection in P_j, the number of end-users that choose the strategy s has a positive growth trend in the population if their payoff is above the average payoff in the same population. Through setting

Algorithm 1. Evolutionary Game Algorithm

Input: N : number of end-users; K : number of servers; J : number of populations; ε : error factor.
Output: $\mathbb{X}_j^*$: The ESS of P_j

```
Initialize: b_i, d_i, t_i, c_i, W_k, F_k, V_k, R_k, C_f^k, C_t^k, λ, ξ, δ
t ← 0
X_j^* ← 0, ∀j ∈ J
Each player determines its available server set as the strategy set S_i based on its location information
Each player randomly selects server k from its strategy set S_i
Each servers acquires n_k and allocates w_k and f_k.
All servers calculate the resources they have allocated and send the information to the players.
Player calculates revenue π_j^k based on information returned by the servers
The server collects the revenue π_j^k returned by the players and calculates the π̄_j
while |π_j^k − π̄_j| ≥ ε do
    Each population computes ẋ_j^k,and update x_j^k = x_j^k + ẋ_j^k
    update X_j
    Player calculates revenue π_j^k based on information of X_j.
    The server collects the revenue π_j^k returned by the players and calculates the π̄_j
    Plays change their strategies with probability δ, when their payoff is less than the average payoff in the same population.
    Update t = t + τ
end while
```

$\sum_k \dot{x}_j^k(t) = 0$, we can get the fixed point of the replicator dynamics, in which the population state will not change and no player is willing to change its strategy since all players in the same population have the same payoff.

The algorithm based on replicator dynamics is described as:

5 Experimental Evaluation

To validate our proposed method, we carry out simulative experiments based on the Edge User Allocation (EUA) dataset [16] for positions of cloud servers and end-users, and a dataset of [32] for the performance of cloud servers. Figure 2 shows that the area contains 200 end-users, 1 centralized cloud server and 12 edge servers. The servers are marked red. End-users of the same population share the same color.

As for the performance data for the centralized cloud, we test a typical third-party commercial cloud service, *i.e.*, Tencent cloud [2]. Figures 3 and 4 show its measured thoughput and computing performance (in terms of MIPS, *i.e.*, Million Instructions Per Second) of the centralized cloud, and these data are partitioned into 24 consecutive windows [1].

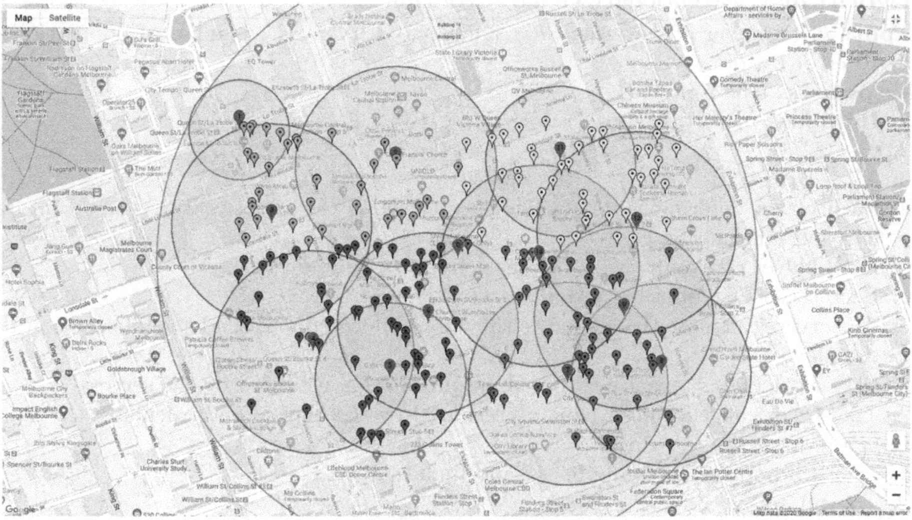

Fig. 2. Geographical distribution of experimental data

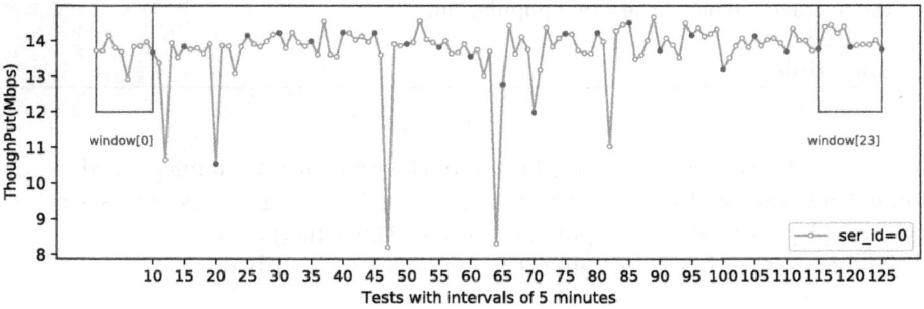

Fig. 3. Throughput for the centralized cloud

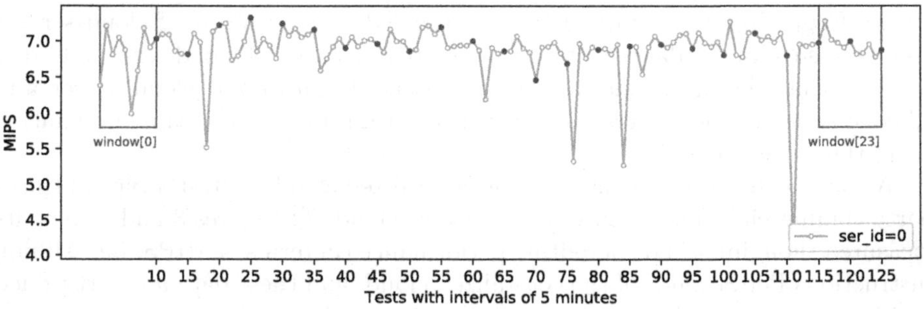

Fig. 4. MIPS for the centralized cloud

We compare our proposed method with three existing ones: Greedy [31], Genetic Algorithm (GA) [21], and Nash-based Game [9].

Greedy. Greedy algorithm refers to the algorithm that takes the best or optimal choice in each step when solving a problem, hoping to lead to the best or optimal algorithm. In this environment, each player chooses the server that can bring the highest US to offload based on the current resource allocation, regardless of whether other player's strategy changes will affect his next strategy selection.

Genetic Algorithm (GA). GA is a method to find the optimal solution by simulating the natural evolution process. We encode the players' strategy selection, get new populations through crossover and mutation, and then select through Roulette Wheel Selection.

Nash-Based Game. Game theory is another distributed algorithm based on Nash equilibrium, called Nash-based, which treats tasks equally without considering dynamically allocating computation resources to different types.

As can be seen from Figs. 5, 6, 7 and 8: 1) our method beats its peers at all 24 windows in terms of total user satisfaction, *i.e.*, US; 2) our method and Nash-based Game outperform others in terms monetary cost and our method shows more stable advantage, fewer iteration rounds, and offloading time than Nash-based Game through all windows.

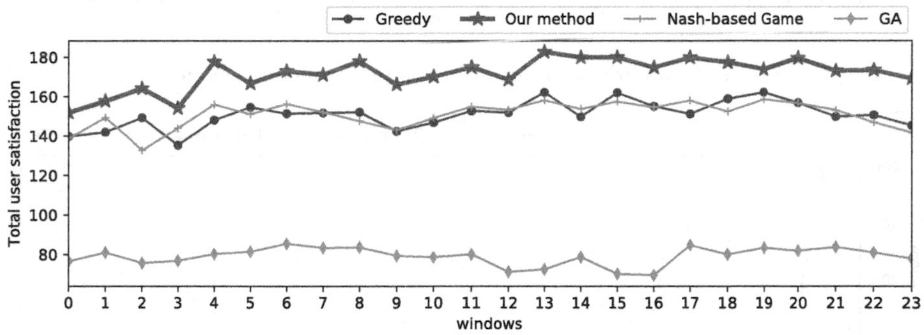

Fig. 5. Comparison of total user satisfaction at different windows

In this part, we compare the convergence of two game theory algorithms for the population after mutation. First of all, the population reached EE 1. After five iterations, a small number of users in the population will mutate randomly, showing that the size of user tasks changes, and the tolerable delay and cost will also change. After the mutation, the population readjust the offloading strategy to reach EE 2. The number of iterations to reach the new EE is taken as a measure of the population's convergence under a certain game theory strategy.

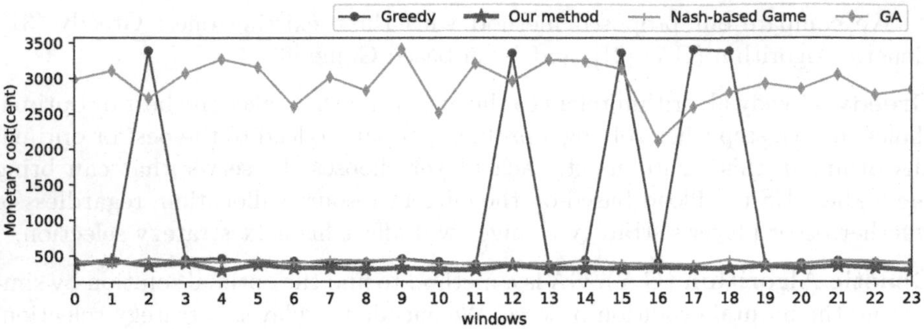

Fig. 6. Comparison of monetary cost at different windows

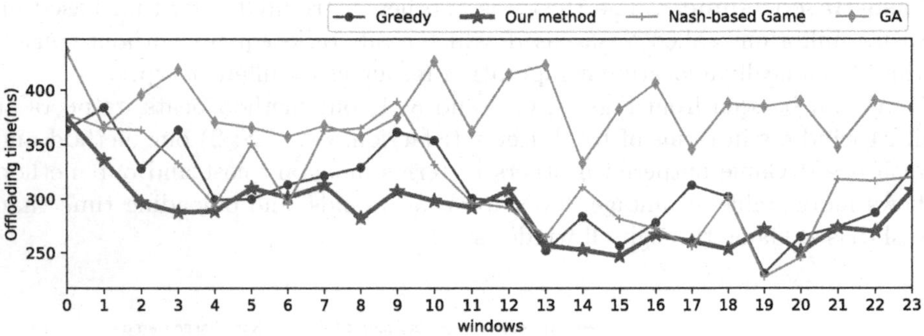

Fig. 7. Comparison of offloading time at different windows

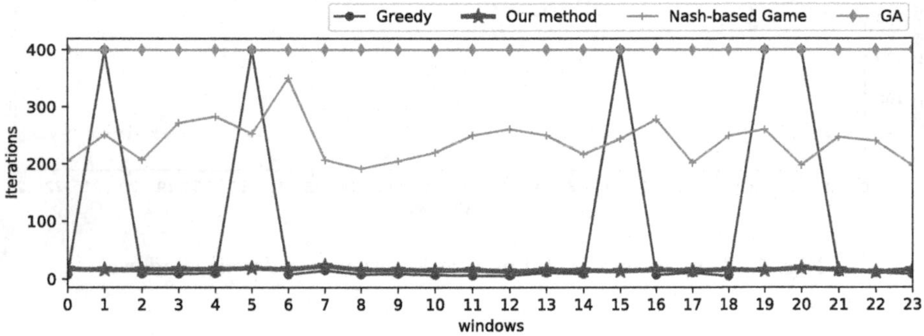

Fig. 8. Comparison of iterations at different windows

As shown in Fig. 10, players need to go through 10 iterations to reach new EE after mutation by using traditional game theory algorithm, while evolutionary game algorithm (our method) only needs one iteration to reach new EE in Fig. 9. Experimental results clearly suggest that our proposed approach outperforms traditional ones in terms of Anti-interference ability.

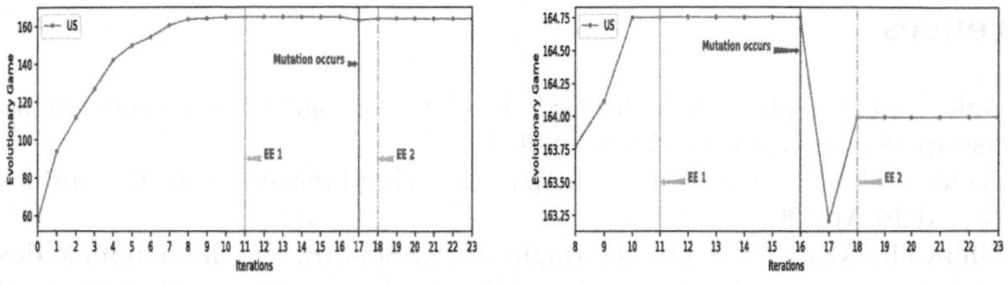

Fig. 9. Convergence of evolutionary game.

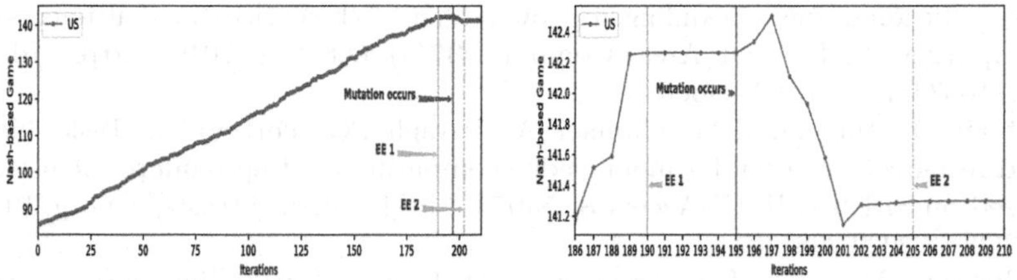

Fig. 10. Convergence of Nash-based game.

6 Conclusion

In this paper, we study the offloading problem for the cloud-edge-hybrid architecture and assume that geographically distributed edge servers are with fluctuating real-time performance. We develop a dynamic offloading strategy based on a probabilistic evolutionary game-theoretic model. For the model validation purpose, we test our method against other existing ones through simulations based on a well-known dataset of cloud/edge resource locations. It's clear to see that our proposed approach outperforms its peers in terms of multiple metrics.

In the future, we plan to carry out the following work: 1) We plan to consider the mobility of users and edge nodes, and study the strategy of multi-user task online offloading under irregular trajectory; 2) we intend to introduce a performance prediction model based on time series and neural network, and use trajectory prediction information and performance prediction information to drive the multi-user task offloading decision-making model; 3) We plan to consider the impact of task failure and transmission failure on task offloading under untrusted communication conditions, and design fault tolerant and fault-tolerant multi-user task offloading strategies and algorithms; 4) etc.

Acknowledgement. This work is supported in part by the Graduate Scientific Research and Innovation Foundation of Chongqing, China (Grant No. CYB20062 and CYS20066), and the Fundamental Research Funds for the Central Universities (China) under Project 2019CDXYJSJ0022.

References

1. Google maps platform. https://developers.google.cn/maps/documentation/javascript/tutorial. Accessed 6 May 2020
2. Tencent cloud platform. https://intl.cloud.tencent.com/zh/product/cvm. Accessed 16 Apr 2020
3. Al-Shuwaili, A.N., Simeone, O., Bagheri, A., Scutari, G.: Joint uplink/downlink optimization for backhaul-limited mobile cloud computing with user scheduling. CoRR abs/1607.06521 (2016). http://arxiv.org/abs/1607.06521
4. Alameddine, H.A., Sharafeddine, S., Sebbah, S., Ayoubi, S., Assi, C.: Dynamic task offloading and scheduling for low-latency IoT services in multi-access edge computing. IEEE J. Sel. Areas Commun. **37**(3), 668–682 (2019). https://doi.org/10.1109/JSAC.2019.2894306
5. Alfakih, T., Hassan, M.M., Gumaei, A., Savaglio, C., Fortino, G.: Task offloading and resource allocation for mobile edge computing by deep reinforcement learning based on SARSA. IEEE Access **8**, 54074–54084 (2020). https://doi.org/10.1109/ACCESS.2020.2981434
6. Aslanpour, M.S., Gill, S.S., Toosi, A.N.: Performance evaluation metrics for cloud, fog and edge computing: a review, taxonomy, benchmarks and standards for future research. Internet Things **12**, 100273 (2020). https://doi.org/10.1016/j.iot.2020.100273. http://www.sciencedirect.com/science/article/pii/S2542660520301062
7. Chen, L., Zhou, S., Xu, J.: Computation peer offloading for energy-constrained mobile edge computing in small-cell networks. IEEE/ACM Trans. Networking **26**(4), 1619–1632 (2018)
8. Chen, M., Hao, Y.: Task offloading for mobile edge computing in software defined ultra-dense network. IEEE J. Sel. Areas Commun. **36**(3), 587–597 (2018)
9. Chen, X., Jiao, L., Li, W., Fu, X.: Efficient multi-user computation offloading for mobile-edge cloud computing. IEEE/ACM Trans. Networking **24**(5), 2795–2808 (2016). https://doi.org/10.1109/TNET.2015.2487344
10. Chen, Z., Cheng, S.: Computation offloading algorithms in mobile edge computing system: a survey. In: Data Science - 5th International Conference of Pioneering Computer Scientists, Engineers and Educators, ICPCSEE 2019, Proceedings, Part I, Guilin, China, 20–23 September 2019, pp. 217–225 (2019). https://doi.org/10.1007/978-981-15-0118-0_17
11. Cong, P., Zhou, J., Li, L., Cao, K., Wei, T., Li, K.: A survey of hierarchical energy optimization for mobile edge computing: a perspective from end devices to the cloud. ACM Comput. Surv. **53**(2), 38:1–38:44 (2020). https://doi.org/10.1145/3378935
12. Cuong, D., Tran, N., Tran, D., Pham, C., Alam, M.G.R., Hong, C.S.: Toward service selection game in a heterogeneous market cloud computing. In: Proceedings of the 2015 IFIP/IEEE International Symposium on Integrated Network Management, IM 2015, pp. 44–52, June 2015. https://doi.org/10.1109/INM.2015.7140275
13. Dong, C., Wen, W.: Joint optimization for task offloading in edge computing: an evolutionary game approach. Sensors (Switzerland) **19**(3) (2019). https://doi.org/10.3390/s19030740
14. Du, W., Lei, T., He, Q., Liu, W., Lei, Q., Zhao, H., Wang, W.: Service capacity enhanced task offloading and resource allocation in multi-server edge computing environment. In: 2019 IEEE International Conference on Web Services, ICWS 2019, Milan, Italy,8–13 July 2019, pp. 83–90 (2019). https://doi.org/10.1109/ICWS.2019.00025

15. Fantacci, R., Picano, B.: A matching game with discard policy for virtual machines placement in hybrid cloud-edge architecture for industrial IoT systems. IEEE Trans. Ind. Informatics **16**(11), 7046–7055 (2020). https://doi.org/10.1109/TII.2020.2999880
16. He, Q., Cui, G., Zhang, X., Chen, F., Deng, S., Jin, H., Li, Y., Yang, Y.: A game-theoretical approach for user allocation in edge computing environment. IEEE Trans. Parallel Distrib. Syst. **31**(3), 515–529(2019). https://doi.org/10.1109/tpds.2019.2938944
17. Hofbauer, J., Sigmund, K.: Evolutionary game dynamics. Bull. Am. Math. Soc. **40**(4), 479–519 (2003)
18. Hosseinzadeh, M., Tho, Q.T., Ali, S., Rahmani, A.M., Souri, A., Norouzi, M., Huynh, B.: A hybrid service selection and composition model for cloud-edge computing in the internet of things. IEEE Access **8**, 85939–85949 (2020). https://doi.org/10.1109/ACCESS.2020.2992262
19. Hwang, S., Hsu, C., Lee, C.: Service selection for web services with probabilistic qos. IEEE Trans. Serv. Comput. **8**(3), 467–480 (2015). https://doi.org/10.1109/TSC.2014.2338851
20. Lan, Z., Xia, W., Cui, W., Yan, F., Shen, F., Zuo, X., Shen, L.: A hierarchical game for joint wireless and cloud resource allocation in mobile edge computing system. In: 10th International Conference on Wireless Communications and Signal Processing, WCSP 2018, Hangzhou, China, 18–20 October 2018. pp. 1–7 (2018). https://doi.org/10.1109/WCSP.2018.8555606
21. Li, W., Xia, Y., Zhou, M., Sun, X., Zhu, Q.: Fluctuation-aware and predictive workflow scheduling in cost-effective infrastructure-as-a-service clouds. IEEE Access **6**, 61488–61502 (2018). https://doi.org/10.1109/ACCESS.2018.2869827
22. Liu, C.F., Bennis, M., Debbah, M., Poor, H.V.: Dynamic task offloading and resource allocation for ultra-reliable low-latency edge computing. IEEE Trans. Commun. **67**(6), 4132–4150 (2019)
23. Niyato, D., Hossain, E.: Dynamics of network selection in heterogeneous wireless networks: an evolutionary game approach. IEEE Trans. Veh. Technol. **58**(4), 2008–2017 (2009)
24. Peng, Q., et al.: Mobility-aware and migration-enabled online edge user allocation in mobile edge computing. In: 2019 IEEE International Conference on Web Services (ICWS), pp. 91–98. IEEE (2019)
25. Taylor, P., Jonker, L.: Evolutionary stable strategies and game dynamics. Math. Biosci. **40**, 145–156 (1978). https://doi.org/10.1016/0025-5564(78)90077-9
26. Xia, X., Chen, F., He, Q., Grundy, J.C., Abdelrazek, M., Jin, H.: Cost-effective app data distribution in edge computing. IEEE Trans. Parallel Distrib. Syst. **32**(1), 31–44 (2021). https://doi.org/10.1109/TPDS.2020.3010521
27. Xia, X., Chen, F., He, Q., Grundy, J.C., Abdelrazek, M., Jin, H.: Online collaborative data caching in edge computing. IEEE Trans. Parallel Distrib. Syst. **32**(2), 281–294 (2021). https://doi.org/10.1109/TPDS.2020.3016344
28. Xu, J., Chen, L., Zhou, P.: Joint service caching and task offloading for mobile edge computing in dense networks. In: IEEE INFOCOM 2018-IEEE Conference on Computer Communications, pp. 207–215. IEEE (2018)
29. Zaw, C.W., Ei, N.N., Im, H.Y.R., Tun, Y.K., Hong, C.S.: Cost and latency tradeoff in mobile edge computing: A distributed game approach. In: IEEE International Conference on Big Data and Smart Computing, BigComp 2019, Kyoto, Japan, 27 February–2 March 2019, pp. 1–7 (2019). https://doi.org/10.1109/BIGCOMP.2019.8679304

30. Zhao, H., Deng, S., Zhang, C., Du, W., He, Q., Yin, J.: A mobility-aware cross-edge computation offloading framework for partitionable applications. In: 2019 IEEE International Conference on Web Services, ICWS 2019, Milan, Italy, 8–13 July 2019, pp. 193–200 (2019). https://doi.org/10.1109/ICWS.2019.00041
31. Zhao, P., Tian, H., Fan, B.: Partial critical path based greedy offloading in small cell cloud. In: IEEE 84th Vehicular Technology Conference, VTC Fall 2016, Montreal, QC, Canada, 18–21 September 2016, pp. 1–5 (2016). https://doi.org/10.1109/VTCFall.2016.7881145
32. Zheng, Z., Zhang, Y., Lyu, M.R.: Investigating QoS of real-world web services. IEEE Trans. Serv. Comput. **7**(1), 32–39 (2014). https://doi.org/10.1109/TSC.2012.34

Artificial Intelligence

Identification of Sequential Feature for Volcanic Ash Cloud Using FNN-LSTM Collaborative Computing

Lan Liu[1](✉), Cheng-fan Li[2,3], Xian-kun Sun[1], and Jiangang Shi[4]

[1] School of Electronic and Electrical Engineering, Shanghai University of Engineering Science, Shanghai 201620, China
liulan@sues.edu.cn

[2] School of Computer Engineering and Science, Shanghai University, Shanghai 200444, China

[3] Shanghai Institute of Advanced Communication and Data Science, Shanghai University, Shanghai 200444, China

[4] Shanghai Shangda Hairun Information System Co Ltd., Shanghai 200444, China

Abstract. Collaborative computing performs quickly and accurately the task via combining the multimedia, multi-methods, and multi-clients. Analyzing of traditional feedforward neural network (FNN), long short-term memory (LSTM) neural networks and remote sensing data, this paper proposes a new identification method of sequential feature based on FNN-LSTM collaborative calculation in the volcanic ash cloud monitoring. In this method, combining remote sensing data, the FNN network is used firstly to construct the identification model of volcanic ash cloud. Next, the LSTM network is used to identify the sequential feature of dynamic changes in volcanic ash cloud based on the text data of the volcanic ash report. And then the simulation and true volcanic ash cloud case is performed and analyzed. The experimental results show that: 1) the proposed method is high in training accuracy with 76.54% and testing accuracy with 77%, respectively, and has obvious advantages for small-scale data volumes; 2) the total accuracy and RMS of the simulation analysis reached 79.05% and 0.0149, respectively, it verified the feasibility and effectiveness in the prediction of spatiotemporal evolution; 3) the anti-noise property and the image segmentation effect is good, the obtained sequential feature of the volcanic ash cloud are closer to the actual diffusion. It can provide a reference for sequential feature extraction and dynamic monitoring of volcanic ash cloud in complex environments.

Keywords: Collaborative computing · Remote sensing · Volcanic ash cloud · Identification · Long Short-Term Memory (LSTM)

1 Introduction

The volcanic ash cloud formed by volcanic eruption has the characteristics of strong suddenness, high spatial distribution and long diffusion distance, which has attracted widespread attention [1, 2]. In fact, volcanic ash cloud is affected by meteorological

H. Gao et al. (Eds.): CollaborateCom 2020, LNICST 349, pp. 273–289, 2021.
https://doi.org/10.1007/978-3-030-67537-0_17

factors at the time of the eruption, such as the geological conditions, wind direction and wind speed, etc. The location, height and coverage range of volcanic ash cloud are often very uncertain [3–5]. As an important space-to-ground observation technology, satellite remote sensing can obtain dynamic change information of volcanic ash cloud in a timely, accurate and efficient manner [6–10]. There is an obvious sequential feature in the dynamic change of volcanic ash clouds, and the sequential feature extraction of volcanic ash is usual got by the volcanic ash diffusion models. However, on the one hand, most of these conventional simulation methods are based on numerical simulation and existing experience. On the other hand, the reference and calibration data are also mainly coming from several fixed ground observation points. The local characteristics are significant and are not conducive to the dynamic, real-time and large-scale monitoring of volcanic ash cloud [11].

Via analyzing and summarizing the existing works of volcanic ash cloud diffusion, there are three commonly used types of volcanic ash cloud diffusion models at present, to be specific, it contains: 1) Lagrange function-based method, e.g., smoke plume model and puff model, 2) Monte Carlo-based method, e.g., nuclear accident management model (NAME), Canadian emergency response model (CANERM), hybrid single particle Lagrangian integrated trajectory (HYSPLIT), volcanic ash forecast transmission and diffusion model (VAFTAD), 3) climate change prediction-based method, e.g., atmospheric circulation model. However, these models are not completely consistent in the actual volcanic ash cloud diffusion states because of the model own characteristics [12–16]. For example, the puff model is originally intended as a supplementary means when satellite remote sensing cannot monitor volcanic ash cloud, and the effective prediction time does not exceed 48 h; VAFTAD, CANERM and HYSPLIT models, as a tracking model, are suitable for diffusion prediction at short-term meteorological scales, and have large errors for long-term prediction; Atmospheric circulation model generally assume that volcanic ash aerosols are evenly distributed along the zonal direction.

Deep reinforcement learning is a learning framework that combines the advantages of reinforcement learning and deep neural networks [17]. Compared to the traditional learning methods, in a continuous state or huge space state environment, deep reinforcement learning can learn mapping functions by the neural network and automatically complete the online learning process by interacting with the actual environment. It can well adapt to dynamically changing environments [18]. The previous diffusion models based on numerical simulation, relied heavily on parameter settings, has a poor capability of self-learning and interaction and complicated calculation. In theory, deep reinforcement learning can automatically complete online learning by interacting with the actual environment and without human intervention training and can more realistically approach to the dynamic diffusion of volcanic ash cloud. The large-scale training and deep artificial neural networks (ANN) have been performed by computer automatically with the discovery of multilayer feedforward neural networks (FNN) algorithm since 2006 [19–21]. Due to the obvious sequential feature of volcanic ash cloud diffusion, the existing dense trajectory-based methods, and convolutional neural networks (CNN) cannot represent the motion information well. By introducing the a set of memory units, the long short term memory (LSTM) neural network is able to judge forgotten old information and learn new information by itself, and it effectively overcomes the space-time mapping relationship and gradient explosion of traditional models. Therefore, learning and mining dynamic evolution rule of dynamic diffusion using deep learning neural networks has become a hot and difficult point in volcanic ash cloud monitoring.

Via combining multi-technologies including computer networks, communications and multimedia, collaborative computing can complete a task together by researchers from different regions, different times and different cultural backgrounds. It has the advantages of high efficiency, accuracy, and simplicity. Based on the collaborative computing, neural network, remote sensing data and dynamic change of volcanic ash cloud, in this paper, a new identification method of sequential feature for volcanic ash cloud using FNN and LSTM (i.e., FNN-LSTM) collaborative computing method is proposed. The rest of the paper is constructed as follows: Sect. 2 describes the collaborative computing model of sequential feature identification for volcanic ash cloud. Section 3 presents related theoretical works including FNN, LSTM and FNN-LSTM collaborative computing. Section 4 elaborates the simulation experiment and analysis. Section 5 devote the true volcanic ash cloud case. Finally, the conclusions and discussions are summarized in Section 6.

2 Collaborative Computing Model of Sequential Feature Identification

Collaborative computing refers to the behavior of relatively independent computing subjects in space and time series to complete a certain computing task in accordance with pre-established interconnection modes, interaction methods and technologies and computing strategy. The mode of collaborative computing can be divided into business level, process level and service instruction level in terms of the granularity of calculation. The key to collaborative computing is the allocation mapping and decomposition mechanism, module service interface and data interaction specifications.

As a typical big data, remote sensing data has the characteristics of large amount of data, multi-types, and wide application. Meanwhile, the structure and type of auxiliary text data corresponding to the remote sensing data is diverse and complex. Identification and extraction of target information quickly and efficiently from the complicated data has become an important application of current collaborative computing. In the sequential feature identification of volcanic ash cloud based on remote sensing data, it can be performed by collaborative calculation mode with integrating multiple data sources, multiple methods and multiple computing resources [22].

2.1 Multi-temporal Data Collaborative Calculation

It is a specific manifestation of image-spectrum synergy. By comprehensive analysis (e.g., change detection and disaster monitoring) of multi-temporal remote sensing images and other auxiliary text data, the collaborative computing of volcanic ash cloud change at different stages (e.g., early, middle and late eruption) is formed based on the multi-temporal data.

2.2 Multi-knowledge Collaborative Computing

It can improve the identification accuracy and computational efficiency of sequential feature for volcanic ash cloud based on remote sensing data via the integration of multi-knowledge synergy. In the identification of sequential feature for volcanic ash cloud information, it is possible to improve the identification accuracy and reduce the interference of similar information by the multi-knowledge synergy models, such as prior knowledge, spatial configuration of ground features, causality and deduction, etc.

2.3 Multi-algorithms Collaborative Computing

For the same and given remote sensing and text data, different algorithms usual gets different information identification and extraction results. To some extent, it can significantly improve the reliability and identification accuracy of volcanic ash cloud sequential feature by integrating multi-algorithms.

3 Related Theoretical Basis

3.1 FNN

By arranging neurons in layers, each neuron of FNN is only connected to neurons in the previous layer [23–25] and only receive the output of the previous layer and then pass the result to the next layer. There is no feedback between neurons in each layer. So the FNN is also called multilayer perceptron sometimes. So far, FNN has become one of the most widely used neural networks.

(1) Single-Layer FNN
Single-layer FNN refers to a single-layer neural network that contains only one output layer and no hidden layers and is one of the simplest FNN structures. For any neuron in the output layer, the transformation from input to output can be expressed as:

$$\begin{cases} s_j = \sum_{i=1}^{n} w_{ji}x_i - \theta_j \\ y_j = f\left(s_j\right) = \begin{cases} 1, s_j \geq 0 \\ 0, s_j < 0 \end{cases} \end{cases} \tag{1}$$

where $x = [x_1, x_2, \ldots, x_n]^T$ is the feature vector inputted, $i, j \in [1, 2, \cdots, n]$,$w_{ji}$ is the weight from x_i to y_j.

(2) Multi-layer FNN
Compared with single-layer FNN, the structure of multi-layer FNN model is relatively complicated. It not only contains an input layer and an output layer, but also there may be one or more hidden layers in the middle. So sometimes the multi-layer FNN is also

called multilayer perceptron network. And then the transformation a given neuron from input to output canbe expressed as:

$$\begin{cases} S_i^{(q)} = \sum_{j=0}^{n_{q-1}} w_{ij}^{(q)} x_j^{(q-1)}, (x_0^{(q-1)} = \theta_i^{(q)}, w_{i0}^{(q-1)} = -1) \\ x_i^{(q)} = f\left(S_i^{(q)}\right) = \begin{cases} 1, S_i^{(q)} \geq 0 \\ -1, S_i^{(q)} < 0 \end{cases} \end{cases} \quad (2)$$

where $i = 1, 2, \ldots, n_q, j = 1, 2, \ldots, n_{q-1}, q = 1, 2, \ldots, n$. At this time, each layer of multi-layer FNN can be regarded as a separate single-layer FNN, and the complex classification of the multi-layer FNN can be carried out by nonlinearly classification for the input information of each layer.

3.2 LSTM

In the LSTM network, there are input gates, forget gates, and output gates in the cell processor. It can effectively avoid the long-term dependence issue in the traditional recurrent neural network (RNN) [26, 27]. The detailed information flow process of LSTM can be expressed as:

1) Determine the percentage of information is kept from the cell, which is done in the forgotten gate. The output of the network is a value in the range of (0, 1), which is used for subsequent calculation of the degree of retention of old memory.
2) Judge whether the information is updated to the next cell, which is mainly done in the input gate. It contains two steps, i.e., new information generation and usage of new information. The combination of the two can be performed by the production.
3) Combine the steps 1) and 2) to get the information of new cell status. The retention part of old information and the new learning information are combined to generate new cell status information now.
4) Via the new cell status information, the output value of classifier is calculated in the output gate, and the final output value is gotten.

3.3 FNN-LSTM Collaborative Computing

In the LSTM network, there are input gates, forget gates, and output gates in the cell processor. It can effectively avoid the long-term dependence issue in the traditional recurrent neural network (RNN) [26, 27]. The detailed information flow process of LSTM can be expressed as:

(1) Normalization of Training Set

In this study, data preprocessing mainly contains geometric correction and cloud removal. In the data processing, the parameters from geographic position data set were first extracted and used to perform the geometric correction, and then the calibration data set was obtained. Next, the respectively bands R(red), B(blue) and G(green) were selected from remote sensing dataset and further generated a sample dataset, and then a corrected total dataset was obtained by stacking. For the ETM and TM remote sensing data with thin cloud, the cloud removal was performed by cloud mask data.

For the given remote sensing image, the normalization is performed by the *premnmx* function for vary of channel data of satellite-borne sensor. And the formula of normalization can be expressed as:

$$[PN, \min p, \max p, TN, \min t, \max t] = premnmx\,(P, T) \quad (3)$$

where P is the vector matrix $R \times Q$ formed by the input vectors and input variables; T is the vector matrix $S \times Q$ formed by the target and output vectors; PN is the vector matrix $R \times Q$ of normalized input vectors; min$p$ is the minimum vector $R \times 1$ contains P; max$p$ is the maximum vector $R \times 1$ contains P; TN is the normalized target vector matrix $S \times Q$; min$t$ is vector $S \times 1$ of minimum values for each target value T; max$t$ is vector $S \times 1$ of minimum values for each target value T.

(2) FNN Model Construction

The simple FNN model can be constructed by the *newff* function, and the formula can be expressed as:

$$net = newff\,(PR, [S1, S2, \ldots, SN], \{TF1, TF2, \ldots, TFN\}, BTF, BLF, PF) \quad (4)$$

where PR is the matrix $R \times 2$, defines the minimum and maximum values of input vector R;S_i is the number of neurons in the i-layer; TF_i is the transfer function of i-layer, and the default function is $\tan sig$; BTF is the training function, and the default function is *trainlm*; BLF is the weight/threshold learning function, and the default function is *learngdm*; PF is the performance function, and the default function is *mse*.

(3) LSTM Model Construction

In the experiment, the LSTM and Dense functions are used in the detailed model construction.

The LSTM function can be expressed as:

```
keras.layers.recurrent.LSTM(units, activation='tanh',
recurrent_activation='hard_sigmoid', use_bias=True,
kernel_initializer='glorot_uniform',
recurrent_initializer='orthogonal',
bias_initializer='zeros', unit_forget_bias=True,
kernel_regularizer=None,
recurrent_regularizer=None, bias_regularizer=None,
activity_regularizer=None, kernel_constraint=None,
recurrent_constraint=None,
bias_constraint=None, dropout=0.0, recurrent_dropout=0.0)
```

where units are the output dimension; input_dim is the input dimension; return_sequences are the control return type, and generally defaults to False; input_length is the length of input sequence.

And the Dense function can be expressed as:

```
keras.layers.Dense(units,
activation=None,
use_bias=True,
kernel_initializer='glorot_uniform',
bias_initializer='zeros',
kernel_regularizer=None,
bias_regularizer=None,
activity_regularizer=None,
kernel_constraint=None,
bias_constraint=None)
```

where units are the number of the neurons; activation is the activation function; use_bias represents whether to add bias items; kernel_initializer is the initialization method of weight; bias_initializer is the initialization method of bias value; kernel_regularizer is the normalization function of weight; bias_regularizer is the normalization method of bias value; activity_regularizer is the normalization method of output; kernel_constraint is the limit function of weight change; bias_constraint is the limit function of bias value change.

4 Simulation Experiment and Analysis

To verify the effectiveness and feasibility of the proposed FNN-LSTM collaborative computing method in this paper, in the next, we compare and analyze the FNN-LSTM method in view of model performance, spatiotemporal evolution and anti-noise property.

4.1 Model Performance

In the test, the CNN, FNN and LSTM methods were introduced and used to compare and analysis of the proposed FNN-LSTM method in this work.

The loss rate comparison of the CNN, LSTM, FNN and the FNN-LSTM method is plotted in Fig. 1. When the fitting is reached, the proposed FNN-LSTM method requires the least number of training steps, followed by the FNN and CNN methods, and the last is LSTM method. Although At the beginning of training, the loss rate of FNN-LSTM method is larger than other three methods in the late stage, the difference is small and the loss rate in the early stage is lower, and the total performance of the proposed FNN-LSTM method is superior to the CNN, LSTM and FNN methods.

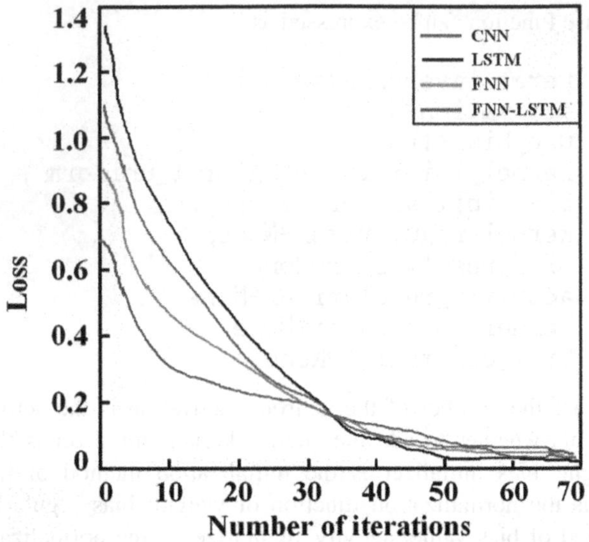

Fig. 1. Loss rate comparisons of training set.

It can be clearly seen from Fig. 1 that the above methods in the experiment can fit in fewer training iterations. Compared to other three methods, in total the proposed FNN-LSTM method in this paper has lower loss rate. And then the performance comparison of the above four methods is shown in Table 1.

Table 1. Performance comparison of different methods

Model	Training accuracy (%)	Testing accuracy (%)	Time overhead(s)
CNN	64.52	63	0.48
LSTM	70.08	69	0.60
FNN	73.21	73	0.72
FNN-LSTM	76.54	77	0.98

As shown in Table 1, the proposed FNN-LSTM collaborative computing method in this paper can effectively combine the advantages of FNN neural and LSTM's sequential feature, and its testing accuracy and training accuracy reached 76.54 and 77%, respectively. The testing accuracy of the above four methods is only slightly lower than that of the training accuracy in total. It shows that these methods have strong normalization ability. The proposed FNN-LSTM method in this paper has a large number of layers and a complex structure, resulting in a large time overhead and reached 0.98 s, which is much higher than 0.48 s, 0.60 s and 0.72 s of the CNN, LSTM and FNN methods, respectively. In short, compared to the traditional single neural network, the proposed FNN-LSTM collaborative computing method has obvious advantages in the training and testing accuracy and time overhead.

4.2 State Transition in Spatial Changes

In this section, taking the traditional CA model with three layers and nine nodes as an example, the simulation and analysis of FNN-LSTM method for the different data nodes via the setting the Moore as the structure of node. In our test, the assumption is that other cell state in the CA model is randomly changed with the data node information. The cells can be divided into three classes, such as normal cells, partially mutated cells, and completely mutated cells.

And then the state transition probability of the test cell was calculated by the FNN-LSTM method. And the state transition of data nodes at different times is plotted in Fig. 2.

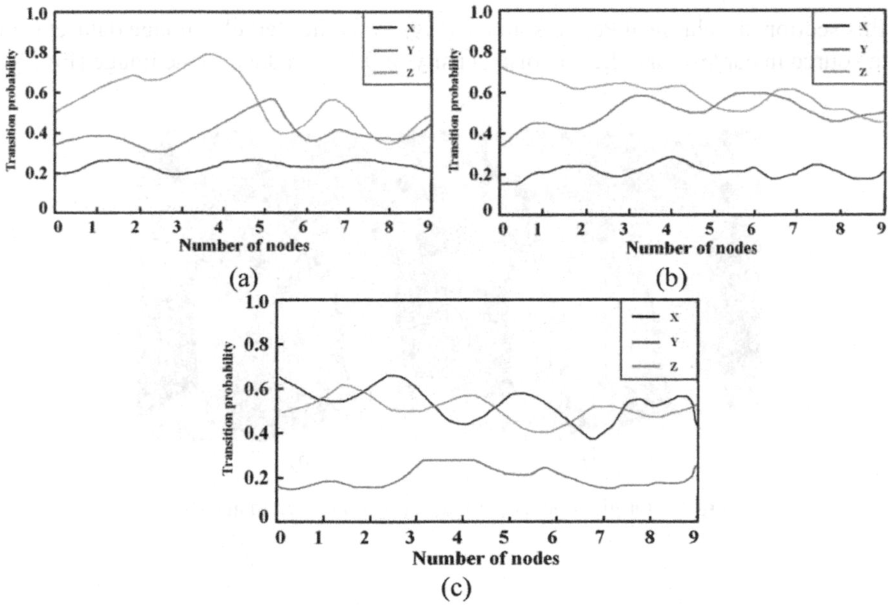

Fig. 2. State transition of data nodes, (a) t, (b) $t + 1$, (c) $t + 2$.

In view of the given learning times as 5000, the accuracy of the state transition was evaluated based on the precision and root mean square (RMS). And the results are shown in Table 2.

It can be seen from Fig. 2 and Table 2 that there are obvious randomness of state changes in data node, and the precision of the three types of data nodes reached 81.02%, 86.25% and 69.88%, and the RMS reached 0.0151, 0.0160 and 0.0136, respectively. Meanwhile, in the case of given learning times as 5000, the precision of the simulation results the proposed method reached 79.05% and the RMS reached 0.0149, respectively from the average value point of view. It indicates that the proposed method has a feasibility and effectiveness for the dynamic change questions, i.e., volcanic ash cloud monitoring and sequential feature extraction, etc.

Table 2. Accuracy evaluation

	Precision (%)	RMS
x	81.02	0.0151
y	86.25	0.0160
z	69.88	0.0136
Average value	79.05	0.0149

4.3 Anti-noise Property

In this section, the classical Peppers image (Fig. 3a) in the Berkeley image dataset is the data source in our test, and then a noise density of 0.1 was added to the image (Fig. 3b).

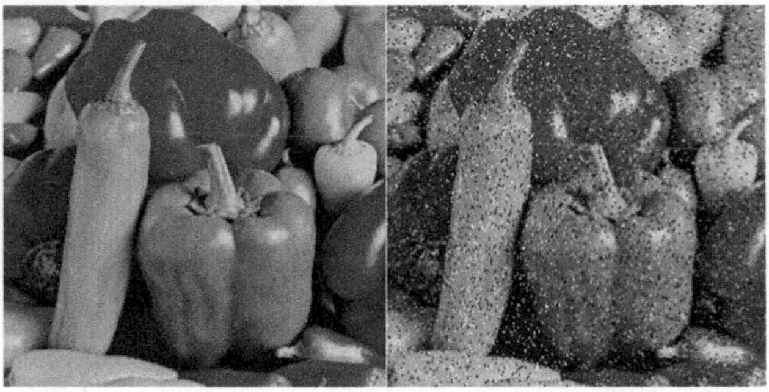

(a) (b)

Fig. 3. Original peppers images (a) and noised image (b).

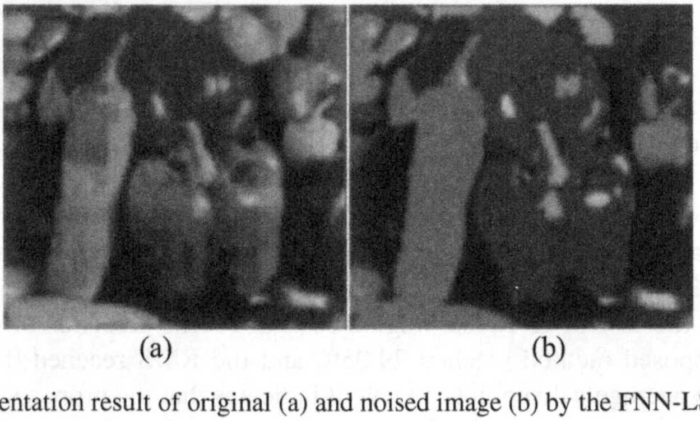

(a) (b)

Fig. 4. Segmentation result of original (a) and noised image (b) by the FNN-LSTM method.

Figure 4 clearly show the segmentation result of original image and noised image by the FNN-LSTM method proposed in this paper. The segmentation effect of original image and the noised image is very similar, and the local details are segmented successfully.

To evaluate the anti-noise property of the proposed FNN-LSTM method, the indexes including peak signal-to-noise ratio (PSNR) and CPU time-consuming were introduced to evaluate the image effect. And the result is shown that the FNN-LSTM method has the PSNR value with 23.05 and the time-consuming with 26.55 s when the segmentation threshold was setting as 0.5. To some extent, the proposed FNN-LSTM method has strong anti-noise property and achieved a good image segmentation result.

5 Sequential Feature Identification of Volcanic Ash Cloud Case

5.1 Data Source

(1) Remote Sensing Data

For the most widely used MODIS satellite-borne sensor in current volcanic ash cloud monitoring, after many tests, the following channel data of MODIS sensors are finally used in this experiment, and specific include:

1) Channels 31 and 32
   The spectral ranges of the channels 31 and 32 are in the range of 10.78–11.2 μm and 11.77–12.27 μm, respectively, the signal-to-noise ratio is 0.05 $NE\Delta t$, and the resolution is 1000 m. Within these two channels, the absorption characteristics of volcanic ash and water vapor and ice in the infrared band are significantly different [17, 28, 29]. The split window brightness temperature difference (SWTD) method currently widely used can distinguish ordinary meteorological cloud from volcanic ash cloud based on the difference between channels 31 and 32.
2) Channels 6 and 7
   The spectral ranges of the channels 6 and 7 are in the range of 1628–1652 nm and 2105–2155 nm, respectively, the signal-to-noise ratio is 275 $NE\Delta t$, and the resolution is 500 m. Within these two channels, the spectral reflection characteristic of volcanic ash detrital particles is significantly different with that of mineral components [30–33]. Then if the difference between channel 6 (center wavelength of 1.64 μm) and channel 7 (center wavelength of 2.13 μm) is negative, it is determined to be non-volcanic ash cloud samples, such as land, clouds and snow, etc.
3) Channel 28
   The spectral range of the channel 28 is in the range of 7.175–7.475 nm, the signal-to-noise ratio is 0.25 $NE\Delta t$, and the resolution is 1000 m. In view of the histogram of reflection spectrum characteristics of volcanic ash cloud on channels 28, 31 and 32 of the MODIS sensor [34], it can be seen that channel 28 has similar spectral characteristic changes as channels 31 and 32. Therefore, the channel 28 data is also selected in the training set data.
   In addition, be subjected to the different channel data of MODIS sensor used in the test has different resolutions (e.g., 250 m, 500 m and 1000 m), the resolution

of all channel data with not 1 km were resampled to 1 km resolution to facilitate calculation and reduce complexity.

(2) Text Data
In this experiment, via consulting a large number of literatures and many tests, at last the VAA Text data released by the Tokyo Volcanic Ash Consultation Center (Tokyo VAAC), one of the largest amounts of data storage centers, were selected as the model input data [35–37]. It overcomes the few samples in the training set when the input remote sensing data samples insufficient to some extent.

For the MODIS sensor, due to the time interval or revisit circle of the two sensors is too long, and the amount of data is too small, it is able to form an effective input time series of volcanic ash cloud diffusion. For VAA Text, VAAC monitors every volcanic eruption and gives forecast maps with 6, 12 and 18-h. VAA text can just make up for the too little time series compared with satellite-borne MODIS sensors. Therefore, the time series of samples in the training set of the FNN-LSTM collaborative computing method proposed in this paper can be composed of text data with 0 (current time), 6 and 12-h. Partial data of standard VAA text are shown in Fig. 5.

```
OBS VA CLD: SFC/FL130 N5644 E16122 - N5631 E16147 - N5619 E16137 -
N5636 E16114 MOV SE 20KT
FCST VA CLD +6 HR: 06/1720Z SFC/FL130 N5517 E16313 - N5439 E16322 -
N5424 E16308 - N5637 E16117 - N5643 E16126
FCST VA CLD +12 HR: 06/2320Z SFC/FL130 N5408 E16433 - N5226 E16453 -
N5219 E16441 - N5429 E16244 - N5636 E16117 - N5641 E16129
FCST VA CLD +18 HR: NO VA EXP
```

Fig. 5. Partial data of standard VAA text.

From Fig. 5, it can be seen that VAA Text contains not only the coverage of the current volcanic ash cloud (represented by latitude and longitude coordinates), but also the wind direction and wind speed (represented by SE 20KT) at the current moment.

5.2 Result and Comparison

Taking the MODIS image of the volcanic ash cloud of Etna on October 30, 2002 as the data source, the volcanic ash cloud information was identified by FNN-LSTM method from the MODIS satellite remote sensing image. And then the traditional SWTD method is used to evaluate and compare the identification accuracy of the volcanic ash cloud by FNN-LSTM method. MODIS false color image of Etna ash cloud on October 30, 2002 is shown in Fig. 6.

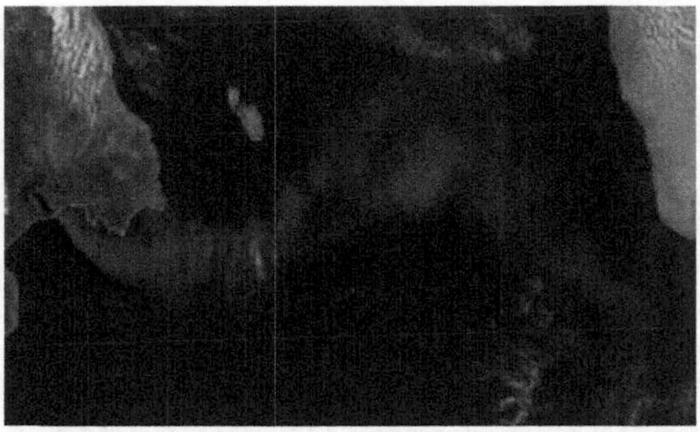

Fig. 6. MODIS false color image of Etna ash cloud on October 30, 2002.

Etna volcanic ash cloud on October 30, 2002 identified by the proposed FNN-LSTM method in this paper is shown in Fig. 7.

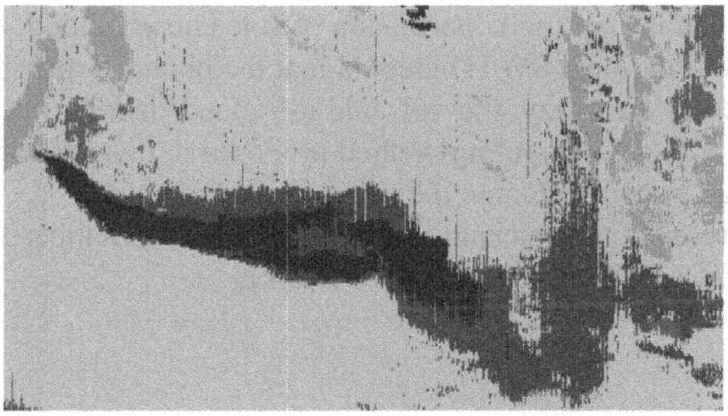

Fig. 7. Etna volcanic ash cloud on October 30, 2002 identified by the proposed FNN-LSTM method. The dark blue area represent the area with more volcanic ash particles, and the light blue area represent the area with less volcanic ash (i.e., volcanic ash mixed with water vapor); the yellow area represent the ocean; the brown-yellow area represent the land; the green area represent the ordinary meteorological cloud.

From Fig. 7, it can be seen that the main distribution area of the Etna volcanic ash cloud on October 30, 2002 contains the concentration area of volcanic ash particle and the volcanic ash mixed water vapor area was basically accurately identified and detected by the FNN-LSTM method, and achieved the good effect of volcanic ash cloud monitoring. However, there is also a lot of noise information in the identified volcanic ash cloud; for example, part of the seawater area at the edge of the volcanic ash cloud was misclassified into land area.

Etna volcanic ash cloud on October 30, 2002 identified by the traditional SWTD method in this paper is shown in Fig. 8.

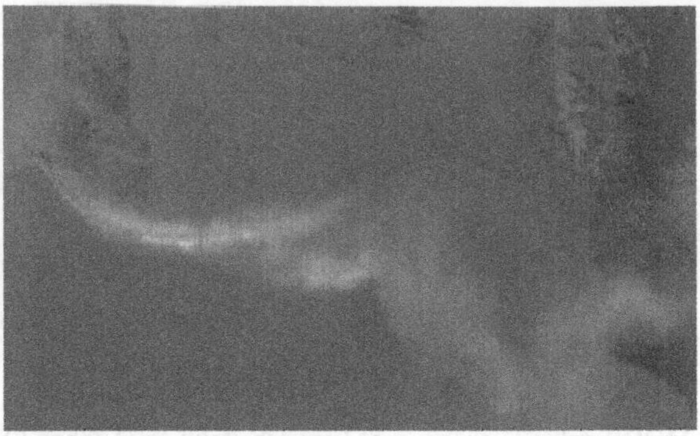

Fig. 8. Etna volcanic ash cloud on October 30, 2002 identified by SWTD method, the yellow area represents the volcanic ash cloud distribution.

From Figs. 7 and 8, it can be seen that the volcanic ash cloud information identified by the FNN-LSTM collaborative calculation method and SWTD method are basically consistent in the spatial shape and distribution of volcanic ash cloud. And what's more, compared with the traditional SWTD method that the obtained distribution boundaries of volcanic ash cloud are fuzzy, the volcanic ash cloud distribution identified by the FNN-LSTM collaborative computing method proposed in this paper has clear and neat boundaries, and can better overcome the above problems.

Sequential feature of Etna volcanic ash cloud diffusion on October 30, 2002 identified by the proposed FNN-LSTM method is shown in Fig. 9.

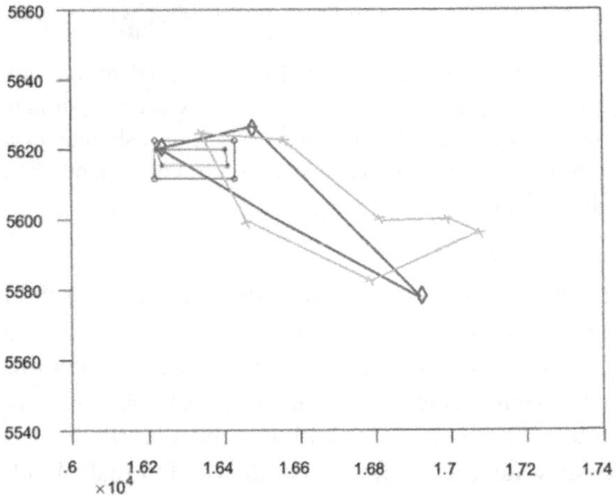

Fig. 9. Sequential feature of Etna volcanic ash cloud diffusion on October 30, 2002, the red, blue and carmine lines represent the coordinate of volcanic ash cloud at the first, second and third time point by the observation, the green lines represent the coordinate of volcanic ash cloud at the third time point simulated by the proposed FNN-LSTM method. (Color figure online)

From Fig. 9, it can be seen that the Etna ash cloud will spread eastward on October 30, 2002, and VAAC predicts that the volcanic ash cloud will spread southeast to east. Compared with the VAAC's report, the sequential feature of volcanic ash cloud identified by the FNN-LSTM collaborative calculation method proposed in this paper is closer to the actual diffusion.

6 Conclusions and Discussions

Satellite-borne remote sensing technology is capable of near-real-time monitoring of large-scale, fast-changing natural disasters, and has become an important support means for earth resources and environment monitoring and security work. Dynamic monitoring of volcanic ash cloud is a complex systematic project. Combining remote sensing data, to accurately identify and extract the sequential feature of volcanic ash cloud diffusion is the difficulty that determines the entire volcanic ash disaster monitoring and disaster prevention and reduction. Given the huge advantages of collaborative computing, many achievements including monitoring of volcanic ash cloud disaster have been made from multiple perspectives based on the latest information and communication technologies.

Via analyzing of the FNN, LSTM and sequential feature of volcanic ash cloud, the identification of sequential feature for volcanic ash cloud was carried out in this paper by proposed FNN-LSTM collaborative calculation method. The simulation experiment and true volcanic ash cloud case show that the FNN-LSTM collaborative computing method proposed in this paper has strong anti-noise property and high accuracy and achieved better dynamic monitoring effect of volcanic ash cloud.

Based on the current work, some possible related issues in the future may include:

1) Increase the usage of multi-temporal remote sensing data, and perform the more detailed sequential feature identification and analysis of volcanic ash cloud based on auxiliary data that have a greater impact on the dynamic monitoring, such as climate, temperature, humidity and rainfall, etc.
2) Optimize the proposed FNN-LSTM collaborative calculation method to improve the computational efficiency of sequential feature identification and extraction for volcanic ash cloud.
3) Carry out the identification of sequential feature for volcanic ash cloud based on different regions, multi-cases, and multi-source satellite-borne remote sensing data to improve the applicability of the proposed method.
4) Explore the boundary matching method between identified sequential feature by text data and image from sensors to improve the overall accuracy and dynamic monitoring effect of volcanic ash cloud diffusion.

Acknowledgements. This work was supported by the Science and Technology Development Foundation of Shanghai in China under Grant No. 19142201600 and Graduate Innovation and Entrepreneurship Program in Shanghai University in China under Grant No. 2019GY04.

References

1. Prata, F., Kristianesen, N., Thomas, H.E., et al.: Ash metrics for European and trans-atlantic air routes during the Eyjafjallajoukull eruption 14 April to 23 May 2010. J. Geophys. Res. Atmos. **123**(10), 5469–5483 (2018)
2. Krippner, J.B., Belousov, A.B., Belousova, M.G., Ramsey, M.S.: Parametric analysis of lava dome-collapse events and pyroclastic deposits at Shiveluch volcano, Kamchatka, using visible and infrared satellite data. J. Volcanol. Geoth. Res. **354**(1), 115–129 (2018)
3. Cheng, B., Liang, C.B., Liu, X.N., et al.: Research on a novel extraction method using deep learning based on GF-2 images for aquaculture areas. Int. J. Remote Sens. **41**(9), 3575–3591 (2020)
4. Mossop, S.C.: Volcanic dust collected at an altitude of 20 km. Nature **203**(4947), 824–827 (1964)
5. Farlow, N.H., Oberbeck, V.R., et al.: Size distributions and mineralogy of ash particles in the stratosphere from eruptions of Mount St. Helens Sci. **211**(4484), 832–834 (1981)
6. Hobbs, P.V., Radke, L.F., Eltgroth, M.W., et al.: Airborne studies of the emissions from the volcanic eruptions of Mount St. Helens Sci. **211**(4484), 816–818 (1981)
7. Krueger, A.J.: Sighting of El Chichon sulfur dioxide clouds with the Nimbus 7 total ozone mapping spectrometer. Science **220**(4604), 1377–1379 (1983)
8. Malingreau, J.P.: Kaswanda: monitoring volcanic eruptions in Indonesia using weather satellite data: the Colo eruption of July 28, 1983. J. Volcanol. Geoth. Res. **27**(1–2), 179–194 (1986)
9. Prata, A.J.: Observations of volcanic ash clouds in the 10–12 μm window using AVHRR/2 data. Int. J. Remote Sens. **10**(4–5), 751–761 (1989)
10. Krotkov, N., Torres, O., Seftor, C., et al.: Comparison of TOMS and AVHRR volcanic ash retrievals from the August 1992 eruption of Mt. Spur. Geophys. Res. Lett. **26**(4), 455–458 (1999)
11. Mccarthy, E.B., Bluth, G.J.S., Watson, I.M., et al.: Detection and analysis of the volcanic ash clouds associated with the 18 and 28 August 2000 eruptions of Miyakejima volcano. Jpn. Int. J. Remote Sens. **29**(22), 6597–6620 (2008)
12. Carey, S., Sigurdsson, H.: Influence of particles aggregation on deposition of distaltephra from the May 18, 1980, eruption of Mount St. Helens Volcano. J. Geophys. Res. Atm. **87**(B8), 7061–7072 (1982)
13. Stenchikov, G.L., et al.: Radiative forcing from the 1991 Mount Pinatubo volcanic eruption. J. Geophys. Res. Atmos. **103**(D12), 13837–13857 (1998)
14. Kirchner, I., Stenchikov, G.L., Graf, H.F., Roboc, A., Carlos Antuna, J.: Climate model simulation of winter warming and summer cooling following the 1991 Mount Pinatubo volcanic eruption. J. Geophys. Res. Atmos. **104**(D16), 19039–19055 (1999)
15. Durant, A.J., Villarosa, G., Rose, W.I., Delmelle, P., Prata, A.J., Viramonte, J.G.: Long-range volcanic ash transport and fallout during the 2008 eruption of Chaiten volcano. Chile. Phys. Chem. Earth **45–46**, 50–64 (2012)
16. Steensen, T., Stuefer, M., Webley, P., et al.: Qualitative comparison of Mount Redoubt 2009 volcanic ash clouds using the PUFF and WRF-Chem dispersion models and satellite remote sensing data. J. Volcanol. Geoth. Res. **259**(2), 235–247 (2013)
17. Ellrod, G.P.: Impact on volcanic ash detection caused by the loss of the 12.0 μm "split window" band on GIES imagers. J. Volcanol. Geoth. Res. **135**(1–2), 91–103 (2004)
18. Liu, Z.Y., Vaughan, M., Winker, D., et al.: The CALIPSO lidar cloud and aerosol discrimination: version 2 algorithm and initial assessment of performance. J. Atmos. Ocean. Tech. **26**(7), 1198–1212 (2009)

19. Hinton, G., Deng, L., Yu, D., et al.: Deep neural networks for acoustic modeling in speech recognition: the shared views of four research groups. IEEE Signal Proc. Mag. **29**(6), 82–97 (2012)
20. Krizhevsky, A., Sutskever, I., Hinton, G.: Image classification with deep convolutional neural network. Adv. Neural Inf. Process. Sys. **4**(4), 1097–1105 (2012)
21. Arel, I.: Deep reinforcement learning as foundation for artificial general intelligence. Artif. Gen. Intell. **33**(1), 89–102 (2012)
22. Shen, Z.F., Li, J.L., Yu, X.J.: Water information extraction of Baiyangdian wet land based on the collaborative computing method. J. Geo-inf. Sci. **18**(5), 690–698 (2016)
23. Floropoulos, N., Tefas, A.: Complete vector quantization of feedforward neural networks. Neurcomputing **367**, 55–63 (2019)
24. Shehu, G.S., Cetinkaya, N.: Flower pollination-feeodfoward neural network for load forecasting in smart distribution grid. Neural Comput. Appl. **31**(10), 6001–6012 (2019)
25. Guo, Y.M., Peng, H., Yong, Y.: Blind separation algorithm for non-cooperative PCMA signal based on feedfoward neural network. Acta Electron. Sin. **47**(2), 302–307 (2019)
26. Xu, X., He, H.G.: A gradient algorithm for neural-network-based reinforcement learning. Chin. J. Comput. **26**(2), 227–233 (2003)
27. Silver, D., Huang, A., Maddison, C.J., et al.: Mastering the game of Go with deep neural networks and tree search. Nature **529**(7587), 484–489 (2016)
28. Filizzola, C., Lacava, T., Marchese, F., et al.: Assessing RAT (robust AVHRR techniques) performances for volcanic ash cloud detection and monitoring in near real-time: the 2002 eruption of Mt. Etna (Italy). Remote Sens. Environ. **107**(3), 440–454 (2007)
29. Watson, I.M., Realmuto, V.J., Rose, W.I., et al.: Thermal infrared remote sensing of volcanic emissions using the moderate resolution imaging spectroradiometer. J. Volcanol. Geoth. Res. **135**(1–2), 75–89 (2004)
30. Gangale, G., Prata, A.J., Clarisse, L.: The infrared spectral signature of volcanic ash determined from high-spectral resolution satellite measurements. Remote Sens. Environ. **114**(2), 414–425 (2010)
31. Marzano, F.S.: Remote sensing of volcanic ash cloud during explosive eruptions using ground-based weather RADAR data processing. IEEE Signal Proc. Mag. **28**(2), 124–126 (2011)
32. Nakagawa, M., Ohba, T.: Minerals in volcanic ash 1: primary minerals and volcanic glass. Glob. Environ. Res. **6**(2), 41–51 (2003)
33. Cronin, S.J., Hedley, M.J., Neall, V.E., et al.: Agronomic impact of tephra fallout from 1995 and 1996 Ruapehu volcano eruptions. New Zealand Environ. Geol. **34**(1), 21–30 (1998)
34. Picchiani, M., Chini, M., Corradini, S., et al.: Volcanic ash detection and retrievals using MODIS data by means of neural networks. Atmos. Meas. Tech. **4**(12), 2619–2627 (2011)
35. Webley, P.W., Atkinson, D., Collins, R.L., et al.: Predicting and validating the tracking of a volcanic ash cloud during the 2006 eruption of Mt. Augustine volcano. B. Am. Meteorol. Soc. **89**(11), 1647–1658 (2008)
36. Winker, D.M., Tackett, J.L., Getzewich, B.J., et al.: The global 3D distribution of tropospheric aerosols as characterized by CALIOP. Atmos. Chem. Phys. **13**(6), 3345–3361 (2013)
37. Folch, A., Costa, A., Basart, S.: Validation of the FALL 3D ash dispersion model using observations of the 2010 Eyjafjallajökull volcanic ash clouds. Atmos. Environ. **48**(2), 165–183 (2012)

Hybrid CF on Modeling Feature Importance with Joint Denoising AutoEncoder and SVD++

Qing Yang[1], Heyong Li[1], Ya Zhou[2], Jingwei Zhang[2(✉)], and Stelios Fuentes[3]

[1] Guangxi Key Laboratory of Automatic Detection Technology and Instrument, Guilin University of Electronic Technology, Guilin 541004, China
[2] Guangxi Key Laboratory of Trusted Software, Guilin University of Electronic Technology, Guilin 541004, China
gtzjw@hotmail.com
[3] Leicester University, Leicester, UK

Abstract. AutoEncoder is an unsupervised learning approach that can maps inputs to useful intermediate features, which can be used to build recommendation. Intermediate features of different entities obtained by AutoEncoder may have different weight for predicting users behavior. However, existing research typically uses a uniform weight on intermediate features to make a fast learning algorithm, this general approach may lead to the limited performance of the model. In this paper, we proposes a novel approach by using SGD to dynamically learn the intermediate features importance, which can integrate the intermediate features into matrix factorization framework seamlessly. In the previous works, the entities intermediate features learned by AutoEncoder are modeled as a whole. On this basis, we proposes to use attention parameters in entity intermediate feature to dynamically learn the intermediate features importance and build fine-grained model. By learning unique attention unit for each entity intermediate feature, the entities intermediate features are integrated into the matrix factorization framework better. Extensive experiments conducted over two real-world datasets demonstrate our proposed approach outperforms the compared models.

Keywords: Denoising AutoEncoder · Collaborative Filtering · Attention unit

1 Introduction

With the explosive growth of online information, the recommendation system is gradually widely deployed in various terminal devices to meet the user's information screening needs [22]. Among the various recommendation strategies, Collaborative Filtering (CF) has been widely adopted due to its precision and efficiency [21]. However, the existing CF methods are insufficient to model nonlinear relation and side information of user and item entities [17,20]. With the widespread

H. Gao et al. (Eds.): CollaborateCom 2020, LNICST 349, pp. 290–304, 2021.
https://doi.org/10.1007/978-3-030-67537-0_18

application of deep learning, many studies apply neural networks to alleviate this problem [28].

Based on CF, [5] present a general framework named NCF, which use neural network to model latent features of users and items, on this basis, [1] adopt an attention mechanism to adapt the representation of a group, and recommendation items for group. [27] propose a dual channel hypergraph collaborative filtering (DHCF) method to model the representation of users and items so that these two types of data can be seamlessly interconnected while still keeping their specific properties unchanged. [11] proposed a new method named Field-weighted Factorization Machines (FwFMs) to model the different feature interactions between different fields in a much more memory-efficient way. [2] describe a novel Field-Leveraged embedding network to learn inter-field and intra-field feature interactions.

However, many existing works compute the feature interactions in a simple way and care less about the importance of features. Due to this, [10] propose AutoFIS to automatically identify important feature interactions for factorization models by training the target model to convergence. [25] propose an automated interaction discovering model for CTR prediction named AutoCTR. [7] propose a new model named FiBiNET, aim to model feature importance by using the the Squeeze-Excitation network (SENET) mechanism. [13] devise AutoInt to automatically learn the high-order feature interactions of input features, these models have made great progress.

Recent research have demonstrated that AutoEncoder have the capability of capturing the complex relationships within raw data, and compact representations in the hidden layers [16]. Based on this, many excellent recommendation models have been proposed [12,18,23]. These research can be divided into two categories [24]. The first category focuses on designing recommendation models based on AutoEncoder only, without using any components of traditional recommendation models. For example, [18] applies the Denoising AutoEncoder to model distributed representations of the users and items via formulating the user-item feedback data. [12] formulated Collaborative Filtering as a AutoEncoder. However, these methods do not employ any extra information of users and items, which leads to the second research category. It aims to use AutoEncoder to learn intermediate feature representations and embed them into classic CF models. For instance, [23] improve Collaborative Filtering (CF) by integrating intermediate features of item entities into matrix factorization framework. The intermediate features are obtained by AutoEncoder. [23] has achieved a good feedback. However, it unified the intermediate features weight to model, which may lead to limited improvement of model performance. In fact, not every dimension of the intermediate features of different item entities obtained by unsupervised approach has predictability [3,19]. Some feature elements contribute little to the prediction, and the useless feature elements even introduce noise, which can result in model to be hindered [6].

In this paper, we design a multi-dimensional attention parameter that seamlessly fuse the intermediate features of the item entity with the factorization framework, it can be learned by stochastic gradient descent. Compared with the previous research in this direction, the work of our paper is summarized as follows:

- We propose to use learn-able attention parameter to discriminate the importance of each dim of intermediate feature learned from AutoEncoder. More meaningful, the weight of intermediate feature is learned by stochastic gradient descent automatically.
- We conduct extensive experiments on two real-world datasets, and the results show that our model outperforms the compared methods significantly.

The rest of the paper is organized as follows. Section 2 provides the problem definition and revisits SVD-based latent factor models and AutoEncoder framework. Section 3 describes our method and learning algorithm in detail. Section 4.1 presents experimental results for the performance comparisons and components analysis. Section 5 summarizes our work and discusses our future directions.

2 Preliminaries

2.1 Problem Definition

Given a set of users $U = [1, ..., M]$, a set of items $I = [1, ..., N]$, we use r_{ui} to express the rating of the user u for the item i, and $r_{ui} \subset R \in \mathcal{R}^{M \times N}$, R is an incomplete matrix of ratings [14]. The known score is expressed as $I = \{(u,i)|r_{ui}\ is\ known\}$. We use $\hat{r_{ui}}$ represents the predicted value of r_{ui}, which captures the interaction score between the user u and the item i.

2.2 Denoising AutoEncoder

A traditional AutoEncoder takes a vector $x \in [0,1]^d$ as input, and transforms it into hidden representation $y \in [0,1]^{\acute{d}}$ through a deterministic mapping:

$$y = h_\theta(x) = s(W \cdot x + b) \tag{1}$$

Its parameter set is $\Theta = \{W, b\}$, where W is a $\acute{d} \times d$ weights matrix and b is a biases column vector. The hidden representation y is then mapped back to a reconstructed vector $z \in [0,1]^d$ through:

$$z = h_\theta^{'}(y) = s(W^{'} \cdot y + b^{'}) \tag{2}$$

With parameters set $\Theta^{'} = \{W^{'}, b^{'}\}$. The weight matrix $W^{'}$ of reverse mapping may be constrained by $W^{'} = W$ in an optional manner. The parameters of this model are determined by minimize the average reconstruction error:

$$\arg\min_{\Theta,\Theta'} \frac{1}{n} \sum_{i=1}^{n} L(x, z) \tag{3}$$

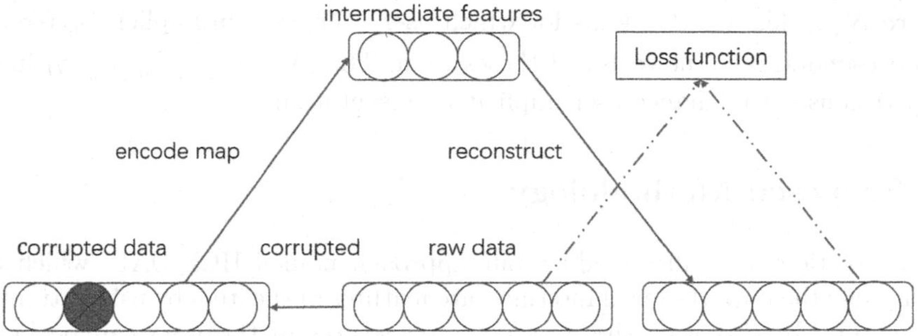

Fig. 1. The architecture of Denoising AutoEncoder

Where L is the traditional square error loss function $L(x - z) = \|x - z\|^2$.

As shown in see Fig. 1, the Denoising AutoEncoder [15] reconstruct a repaired input x from a corrupted version $\widetilde{x}$ by virtue of a stochastic mapping $\widetilde{x} \sim p(\widetilde{x}|x)$. There are two common corruption choices include additive Gaussian noise and the multiplicative mask-out/drop-out noise [18]. In this paper, we choose Gaussian noise to corrupt input data.

2.3 Latent Factor CF Models

Funk-SVD. Funk-SVD [4]is a classic CF model, which map features of users and items to a shared latent factor space of dimensionality k, each item i corresponds to a latent property vector $q_i \in \mathbb{R}^k$, and each user u corresponds to a hidden preference vector $p_u \in \mathbb{R}^k$. The prediction is done by taking an inner product $\hat{r_{ui}} = q_i^T \cdot p_u$, which denote the prediction score of user u to item i.

Biased SVD. It would be inadequate to explain the full rating value by latent factor interaction. Based on Funk-SVD, [8] propose a improved version named biased SVD, which introduces the concept of user and item deviation, the prediction equation is as follows:

$$\hat{r_{ui}} = \mu + b_u + b_i + q_i^T \cdot p_u \tag{4}$$

The parameter μ denotes the global average rating of each user, b_i indicates the bias term of item i, b_u is observed bias of user u, q_i and p_u inherit the definition of Funk-SVD.

SVD++. When explicit feedback is insufficient, implicit information such as browsing and purchase history information can be used to gain insight into user preferences and alleviate the cold start problem. SVD++ [8]is an improved model by minding the biased SVD. The model is more precise, the prediction formula is as follows:

$$\hat{r_{ui}} = \mu + b_u + b_i + q_i^T \cdot (p_u + |N(u)|^{-\frac{1}{2}} \cdot \sum_{j \in N(u)} y_j) \tag{5}$$

Where $N(u)$ denotes the items for which user u uttered an implicit preference, y_j represents the latent vector of theses item. The $|N(u)|^{-\frac{1}{2}} \cdot \sum_{j \in N(u)} y_j$ in the Eq. (5) is used to enhance user implicit representation.

3 Proposed Methodology

In this section, we will introduce our approach named HCF-DAE, which can distinguish the importance of intermediate features of the item by using attention parameters. In this way, the intermediate features of the item can be better integrated into SVD++.

3.1 Our Approach HCF-DAE

Based on autoencoder, [23] applies the item entity intermediate features to the SVD++ framework, constructs hybrid model named Auto-SVD++, and improves prediction performance. However, the Auto-SVD++ embed item entity intermediate features into SVD++ with a uniform weights, which can limit performance of model. For example, for user u, the intermediate features $CAE(i)$ of item entity i contributes to model performance improvement, we name it a useful feature, however, the intermediate features $CAE(j)$ of item entity j may be useless features, and even affect the predictive performance of the model. Based on the above analysis and [23], our paper proposes a new method named HCF-DAE, the architecture of our proposed model see Fig. 2, where p_u and q_i represent the latent features of users and items respectively. Attention unit in Fig. 2 is equivalent to our attention strategy, which can distinguish the importance of item entity intermediate features learn from DAE. To sum up, our model integrates each intermediate feature into SVD++ separately.

Our model can be split into two stages:pre-train stage and re-train stage. These two stage are independent with two loss function (see Eq. (3) and Eq. (7)). In the pre-train stage, we use Denoising AutoEncode to generate ItemFeature $dae(x)$ in hidden layer of DAE, the loss function is Eq. (3). While in the re-train stage, we will update all parameters. Among these parameters, we model the useful features in intermediate features via the attention parameters, the loss function of this step is Eq. (7). By doing this, we can model these intermediate features importance finer-grained and automatically. The prediction score of our method is calculated as:

$$\hat{r_{ui}} = \mu + b_u + b_i + (f_i \odot dae(x_i) + q_i^T) \cdot (p_u + |N(u)|^{-\frac{1}{2}} \cdot \sum_{j \in N(u)} y_j) \tag{6}$$

Where $f_i \in \mathbb{R}^{N \times k}$ represents the attention parameters of intermediate feature of the item entity i, $dae(x_i)$ denotes the intermediate features of item entity i learn from DAE. We use $\odot$ to denote the dot multiplication of matrix. The dot multiplication of f_i and $dae(x_i)$ in the Eq. (6) represents that we assign different weights to each element of the intermediate feature, in this way, we

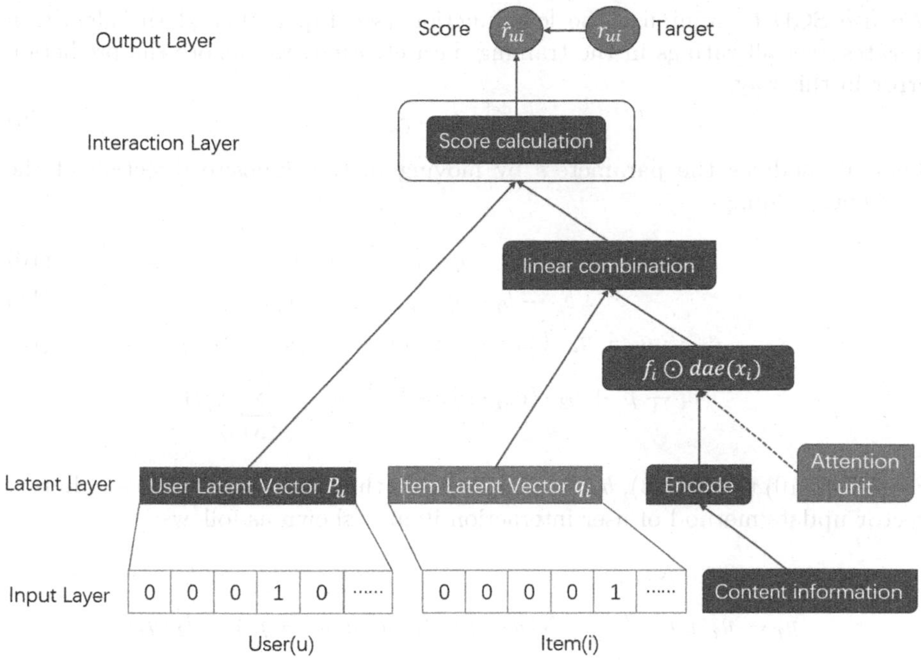

Fig. 2. The architecture of our proposed model HCF-DAE

name f_i as attention unit of intermediate features. We calculate f_i and $dae(x_i)$ in a dot-multiplication manner, so that while improving performance, we also take into account the complexity of our approach. We use stochastic gradient descent (SGD) to update f_i at the element level, and the derivation process is shown in the next section.

3.2 Optimization

Firstly, we use the DAE to obtain the intermediate features of the items the loss function is Eq. (3), and then, we optimize the model in the following way. We learns the parameters by minimizing the regular squared error loss function, which is defined as follows:

$$\underset{b^*,p^*,q^*,y^*}{\arg\min} \sum_{(u,i)\in I} (r_{ui} - \hat{r_{ui}})^2 + \Lambda \cdot f_{reg} \tag{7}$$

The first term of Eq. (7) inherits the error function construction of SVD++. Where Λ represents the weight of regularization parameters, we set it to a single value, f_{reg} is a regularization term of all updated accessories, we set it to avoid model overfitting. f_{reg} expands as follows:

$$f_{reg} = \|f_i\|^2 + b_i^2 + b_u^2 + \|q_i\|^2 + \|p_u\|^2 + \sum_{j\in N(u)} \|y_j\|^2 \tag{8}$$

We use SGD to minimize the loss function (see Eq. (7)), and the algorithm iterates over all ratings in the training dataset. First, we denote the prediction error in this way:

$$e_{ui} \stackrel{\text{def}}{=} r_{ui} - \hat{r_{ui}} \tag{9}$$

Then it modifies the parameters by moving in the opposite direction of the gradient, yielding:

$$b_u \leftarrow b_u + \eta_1 \cdot (e_{ui} - \beta_1 \cdot b_u) \tag{10}$$

$$b_i \leftarrow b_i + \eta_1 \cdot (e_{ui} - \beta_1 \cdot b_i) \tag{11}$$

$$q_u \leftarrow q_u + \eta_2 \cdot (e_{ui} \cdot (f_i \odot dae(x_i) + p_i) - \beta_2 \cdot q_u) \tag{12}$$

$$p_i \leftarrow p_i + \eta_2 \cdot (e_{ui} \cdot (q_u + |N(u)|^{-\frac{1}{2}} \cdot \sum_{j \in N(u)} y_j)) \tag{13}$$

From Eq. (10) to Eq. (13), b_u, b_i, q_u, p_i inherits the definition of SVD++. Latent vector update method of user interaction item is shown as follows:

$$\begin{aligned} &\forall j \in N(u): \\ &y_j \leftarrow y_j + \eta_2 \cdot (e_{ui} \cdot |N(u)|^{-\frac{1}{2}} \cdot (f_i \odot dae(x_i) + p_i) - \beta_2 \cdot y_j) \end{aligned} \tag{14}$$

Equation (14) is used to update y_j, where j belongs to user history click item. Similarly, we use the back propagation algorithm to derive the updating method of attention parameters f_i:

$$f_i \leftarrow f_i + \eta_3 \cdot (e_{ui} \cdot (p_i + f_i \odot dae(x_i)) \cdot (q_u + |N(u)|^{-\frac{1}{2}} \cdot \sum_{j \in N(u)} y_j)) \tag{15}$$

From Eq. (10) to Eq. (15), we use η_1, η_2 and η_3 to express the learning rates of parameters updating, β_1 and β_2 are the regularization parameters. We use Eq. (10) to (13) to update parameters b_u, b_i, q_u, p_i. Equation (15) is the process of updating the attention unit f_i of intermediate features. The above formula is updated along the opposite direction of gradient. We will introduce our experiments in the next part.

4 Experiments

When we optimize the parameters, both SGD and GD can be selected. When SGD is used to adjust parameters, one batch of data is used at a time. While GD is used for optimization, all training data are used in each iteration. Due to this, we employ SGD to optimize all parameters. Specific description, for items, we use DAE to option intermediate features. For each user u, we compute the average ratings μ. After that, we begin to update all parameters (see Algorithm 1).

4.1 Experiment Settings

Datasets Description. We evaluate the effectiveness of our proposed model on two public datasets. Movielens is the datasets of ratings to movie, which has been widely utilized to investigate the performance of many recommendation algorithms. This paper uses two stable benchmark Movielens datasets: Movielens-100k and Movielens-1M. Movielens-100k includes 100,000 ratings for 1,660 movies by 943 users, and Movielens-1M contains 1,000,209 ratings of 3,706 movies from 6,040 users. In addition to rating information, Movielens-100k also contains content information for items and users. In this paper, we use genres, year and items ID as content information of items [23]. In the next section, we would define the evaluation metrics of our model.

Algorithm 1. Training algorithm of our model HCF-DAE

```
procedure UPDATE PARAMETERS
    initial the parameters q_u, p_i, b_u, b_i, y_i, f_i.
    generate ItemFeature dae(x).
    for all user u do
        for all training samples of user u do
            compute sum ratings r_u of user u.
            compute num ratings n_r of user u.
        end for
        compute average ratings of user u:
        μ_u = r_u / n_r.
    end for

    repeat
        for all user u do
            for all training samples of user u do
                upadate parameters q_u, p_i, b_u, b_i:
                from Equation (10) to Equation (13).
                upadate parameters f_i:
                Equation (15).
            end for
            for all training samples of user u do
                upadate parameters y_j:
                Equation (14).
            end for
        end for
    until epoch>=epochs
end procedure
```

Evaluation Metrics. We use the Root Mean Square Error (RMSE) and the Mean Absolute Error (MAE) as the evaluation metrics of experiment. Which has been widely utilized to evaluate the performance of the CF recommendation. RMSE is define as:

$$RMSE = \sqrt{\frac{\sum_{(u,i)\in I}^{M,N}(r_{ui} - \hat{r_{ui}})^2}{|T|}} \tag{16}$$

Where $|T|$ is the number of ratings in the test dataset, and r_{ui} denotes the rating user u to item i. $\hat{r_{ui}}$ represents the corresponding prediction rating. Which represents the error between the predicted value and the true value. The MAE is expressed as:

$$MAE = \frac{\sum_{(u,i)\in I}^{M,N}|r_{ui} - \hat{r_{ui}}|}{|T|} \tag{17}$$

It denotes the average value of the absolute errors of the predicted and observed ratings. The definition of r_{ui} and $\hat{r_{ui}}$ are same as RMSE. Our goal is to reduce the value of RMSE and MAE.

Compared Methods. To verify the efficiency of our model, we compare the proposed model with the following models:

- Funk-SVD [4]. The process of singular value decomposition is simplified into two low-rank matrices, and the two matrices are applied to represent the latent factor of users and items respectively. Then use the two matrix inner products to predict the rating, and obtain the recommending item list.
- SVD++ [9]. This model extends Biased SVD, increases users and item offsets, global average value of rating and historical latent feedback information from users, which has achieved good results.
- AutoSVD++ [23]. This model use AutoEncoder to obtain item intermediate features, then, it integrates the item intermediate features into SVD++, thus modeling the rich content information of the item, and alleviating the cold start problem to a certain extent. Compared with SVD++, Auto-SVD++ has made better experimental results.
- AutoRec [12] utilizes the reconstruction characteristics of the AutoEncoder, proposes two collaborative filtering variants:user-based (U-AutoRec) and item-based (I-AutoRec) that respectively take the partially collected user vector or item vector as input. Our experiments only observe the performance of U-AutoRec.

Parameter Settings. We apply grid search to find the optimal parameters. We tuned the learning rates η_1, η_2, η_3 in the range of [0.001,0.002,...,0.009,0.01], and the value of β_1, β_2 was searched in [0.001,0.002,...,0.01]. For AutoSVD++, we inherit all parameters in [23]. After experiments, the parameters of our model (HCF-DAE) and compared method has be determined and showed in the Table 1.

Table 1. Parameter setting for comparison model and our method

Parameters	η_1	η_2	η_3	β_1	β_2
HCF-DAE	0.005	0.007	0.005	0.005	0.015
AutoSVD++	0.007	0.007	–	0.005	0.015
SVD++	0.005	0.008	–	0.005	0.015
Funk-SVD	–	0.01	–	–	0.01

4.2 Performance Comparison

In this subsection, we summarize the overall experiments of our model and compared models on two movielens test sets respectively. We conduct experiments to check the performance of our model. [26] suggests that sampling should be avoided for metric calculation, nevertheless, sampling is unavoidable in our experiments. In order to ensure the accuracy and credibility of the experimental results, we do five experiments in each group to find the average value of the evaluation metric. Figure 3 shows performance comparison between HCF-DAE and Auto-SVD++ on the Movielens-1M dataset. The experimental results show that AutoSVD++ iterative converges after 15 iterations, and our model iterates 20 times to converge. The final result of HCF-DAE is better than Auto-SVD++ and other two compared models. While the performance of HCF-DAE is not significantly improved compared to SVD++ in Movielens-100k dataset. We speculate that this is due to the small scale of Movielens-100k dataset. This also reflects that AutoEncoder is suitable for model large-scale and complex dataset, while it's difficult to take advantage of deep learning frameworks when AutoEncoder process small-scale datasets. We find that different datasets partition has an effect on the model performance. Figure 4-1 shows experiments on the Movielens-1M dataset. We divide the dataset into test and training dataset in different proportions, and on the premise that the HCF-DAE performs best, the model performs poorly with the training dataset becomes smaller. We analyze that the training dataset is small in size and the model does not have sufficient data for training, so the ideal result cannot be obtained on the test dataset.

4.3 Impact of Attention Unit

We use the Algorithm 1 to iterate the datasets for 30 times, then, we obtain the trajectory of the test error. In order to make the experimental results more reliable, we train five times on the datasets, and we take the average value of RMSE and MAE. Table 2 and Figs. 5, 6 shows the experimental results of our model and the compared methods on these two datasets. We found that the final performance of our approach on both two datasets are the best. Even compared with AutoSVD++, the performance of our approach has also been improved. In the macro point of view, we analyze that comparison with unified weight modeling intermediate features, we use attention unit to model intermediate features

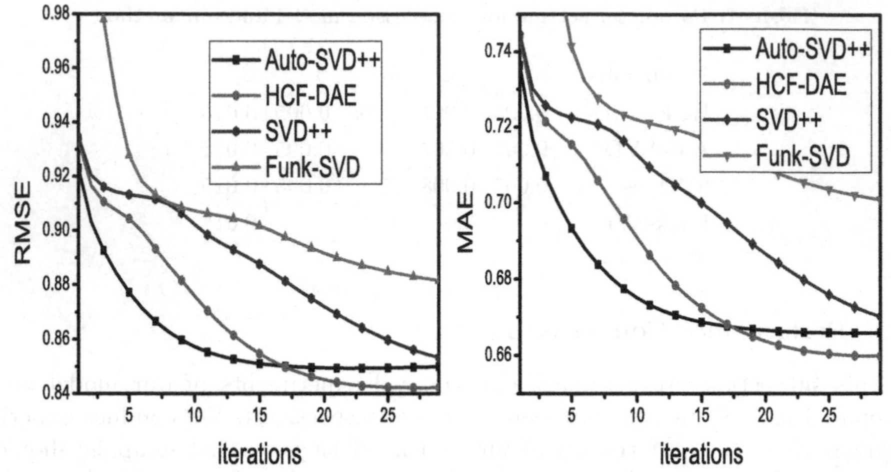

Fig. 3. 1). Test error (RMSE) comparison of each epoch of HCF-DAE and comparison models on Movielens-1M dataset. 2). Performance comparison of each epoch of HCF-DAE and comparison models with the evaluation metrics of average MAE.

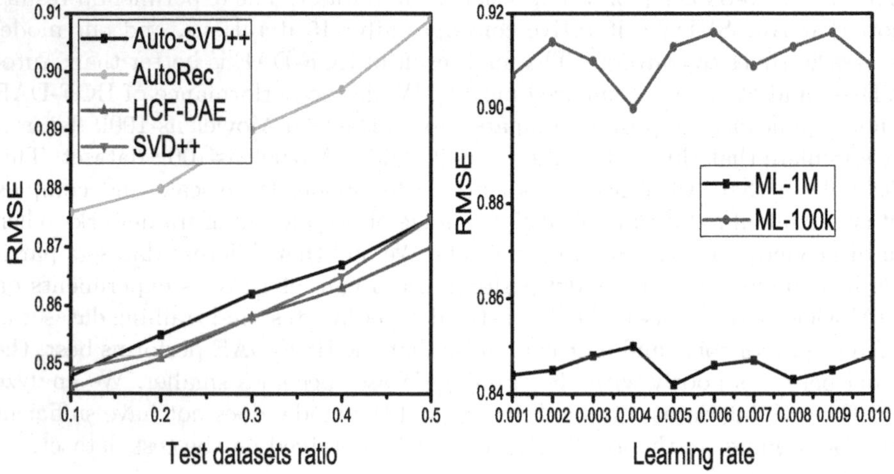

Fig. 4. 1). Average error (RMSE) of each epoch of HCF-DAE and comparison models on Movienens-1M by splitting test dataset with different proportions. 2). Effect of attention parameter learning rate on HCF-DAE performance.

importance is successful. From the micro-level, we analyze that the key advantage of HCF-DAE is ability in interpreting the attention weights of intermediate features. In conclusion, by introducing attention unit, our model can distinguish useful features from noise features, thus the intermediate features obtain from the AutoEncoder can seamlessly integrated with the SVD++ frame work.

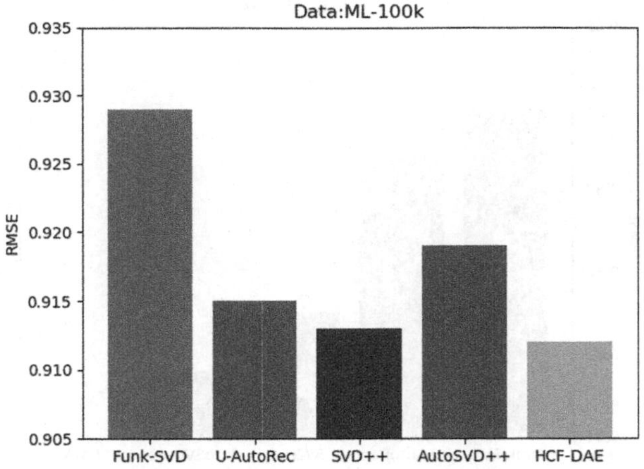

Fig. 5. Histogram of test error of HCF-DAE and comparison models on ML-100k dataset with the evaluation metrics of RMSE.

Table 2. Final performance comparison of our method and compared model for Movielens-1M and Movielens-100k.

	ML-1M		ML-100k	
Model name	RMSE	MAE	RMSE	MAE
Funk-SVD	0.880	0.700	0.929	0.733
U-AutoRec	0.874	0.687	0.915	0.720
SVD++	0.855	0.672	0.913	0.719
AutoSVD++	0.848	0.666	0.919	0.723
HCF-DAE	0.846	0.665	0.912	0.718

4.4 Impact of Different Learning Rates of Attention Unit

We observe the effect of different learning rates on model performance by changing the η_3, which is the attention parameters of the intermediate features. In the Fig. 4-2 we found that when the experiment is performed on ML-1M, the value of $\eta_3 = 0.005$ can let HCF-DAE achieve the best performance. While the experiment on ML-100k, the performance of HCF-DAE is best when $\eta_3 = 0.004$. This shows that the learning rate of attention parameters will fluctuate with the change of dataset. Meanwhile, experiments on ML-1M (see Fig. 4-2) indicated that no matter what value the learning rate of the attention parameters is set, the performance of HCF-DAE is better than SVD++, and in most cases, HCF-DAE performance is better than AutoSVD++. It shows that our model has made a better improvement, and the idea we put forward worked well in the model.

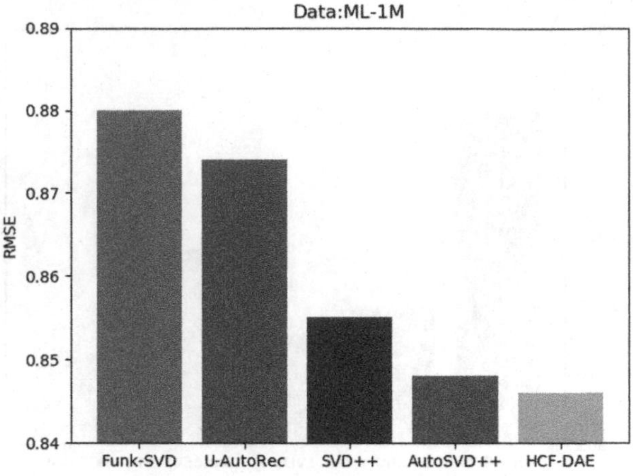

Fig. 6. Histogram of performance of HCF-DAE and comparison model on ML-1M dataset with the evaluation metrics of RMSE.

5 Conclusion

In this paper, we propose a new method to solve the problem of unified weight when AutoEncoder and SVD++ frame work are merged. Our approach uses a clever fusion strategy to learn the intermediate features importance. Instead of manually setting, the attention parameters of the intermediate features in our model are estimated independently, this can more effectively model item feature. We have conducted extensive comparison experiments on two public datasets. The results show that our approach obtains the better prediction results in the comparison model. In future, we plan to explore deep version for our work in the following two directions. First, we intend to further improve the fusion strategy of AutoEncoder and matrix factorization recommendation models. We try to use recurrent neural network to denotes attention strategy. Second, we will take measures to reduce the computational complexity of the algorithm to make model better applied to large-scale datasets.

Acknowledgements. This work was supported in part by the National Natural Science Foundation of China under Grant 61862013, 61662015, U1811264, U1711263, and in part by the Natural Science Foundation of Guangxi Province under Grant 2020GXNSFAA159117 and Grant 2018GXNSFAA281199.

References

1. Cao, D., He, X., Miao, L., An, Y., Yang, C., Hong, R.: Attentive group recommendation. In: The 41st International ACM SIGIR Conference on Research & Development in Information Retrieval, pp. 645–654 (2018)
2. Chen, W., Zhan, L., Ci, Y., Lin, C.: FLEN: leveraging field for scalable CTR prediction. arXiv preprint arXiv:1911.04690 (2019)

3. Cheng, Z., Ding, Y., He, X., Zhu, L., Song, X., Kankanhalli, M.S.: A^3NCF: an adaptive aspect attention model for rating prediction. In: IJCAI, pp. 3748–3754 (2018)
4. Funk, S.: Netflix update: try this at home. https://sifter.org/~simon/journal/20061211.html. Accessed 11 Dec 2006
5. He, X., Liao, L., Zhang, H., Nie, L., Hu, X., Chua, T.S.: Neural collaborative filtering. In: Proceedings of the 26th International Conference on World Wide Web, pp. 173–182 (2017)
6. He, X., Tang, J., Du, X., Hong, R., Ren, T., Chua, T.S.: Fast matrix factorization with nonuniform weights on missing data. IEEE Trans. Neural Netw. Learn. Syst. **31**, 2791–2804 (2019)
7. Huang, T., Zhang, Z., Zhang, J.: FiBiNET: combining feature importance and bilinear feature interaction for click-through rate prediction. In: Proceedings of the 13th ACM Conference on Recommender Systems, pp. 169–177 (2019)
8. Koren, Y.: Factorization meets the neighborhood: a multifaceted collaborative filtering model. In: Proceedings of the 14th ACM SIGKDD International Conference on Knowledge Discovery And Data Mining, pp. 426–434 (2008)
9. Koren, Y., Bell, R., Volinsky, C.: Matrix factorization techniques for recommender systems. Computer **42**(8), 30–37 (2009)
10. Liu, B., et al.: AutoFIS: automatic feature interaction selection in factorization models for click-through rate prediction. arXiv preprint arXiv:2003.11235 (2020)
11. Pan, J., et al.: Field-weighted factorization machines for click-through rate prediction in display advertising. In: Proceedings of the 2018 World Wide Web Conference, pp. 1349–1357 (2018)
12. Sedhain, S., Menon, A.K., Sanner, S., Xie, L.: AutoRec: autoencoders meet collaborative filtering. In: Proceedings of the 24th International Conference on World Wide Web, pp. 111–112 (2015)
13. Song, W., Shi, C., Xiao, Z., Duan, Z., Xu, Y., Zhang, M., Tang, J.: AutoInt: automatic feature interaction learning via self-attentive neural networks. In: Proceedings of the 28th ACM International Conference on Information and Knowledge Management, pp. 1161–1170 (2019)
14. Strub, F., Gaudel, R., Mary, J.: Hybrid recommender system based on autoencoders. In: Proceedings of the 1st Workshop on Deep Learning for Recommender Systems, pp. 11–16 (2016)
15. Vincent, P., Larochelle, H., Bengio, Y., Manzagol, P.A.: Extracting and composing robust features with denoising autoencoders. In: Machine Learning, Proceedings of the Twenty-Fifth International Conference (ICML 2008), Helsinki, Finland, 5–9 June 2008 (2008)
16. Vincent, P., Larochelle, H., Lajoie, I., Bengio, Y., Manzagol, P.A.: Stacked denoising autoencoders: Learning useful representations in a deep network with a local denoising criterion. J. Mach. Learn. Res. **11**, 3371–3408 (2010)
17. Wang, X., He, X., Cao, Y., Liu, M., Chua, T.S.: KGAT: knowledge graph attention network for recommendation. In: Proceedings of the 25th ACM SIGKDD International Conference on Knowledge Discovery & Data Mining, pp. 950–958 (2019)
18. Wu, Y., DuBois, C., Zheng, A.X., Ester, M.: Collaborative denoising auto-encoders for top-n recommender systems. In: Proceedings of the Ninth ACM International Conference on Web Search and Data Mining, pp. 153–162 (2016)
19. Xiao, J., Ye, H., He, X., Zhang, H., Wu, F., Chua, T.S.: Attentional factorization machines: learning the weight of feature interactions via attention networks. arXiv preprint arXiv:1708.04617 (2017)

20. Xin, X., He, X., Zhang, Y., Zhang, Y., Jose, J.: Relational collaborative filtering: modeling multiple item relations for recommendation. In: Proceedings of the 42nd International ACM SIGIR Conference on Research and Development in Information Retrieval, pp. 125–134 (2019)
21. Xue, F., He, X., Wang, X., Xu, J., Liu, K., Hong, R.: Deep item-based collaborative filtering for top-n recommendation. ACM Trans. Inf. Syst. (TOIS) **37**(3), 1–25 (2019)
22. Zhang, S., Yao, L., Sun, A., Tay, Y.: Deep learning based recommender system: a survey and new perspectives. ACM Comput. Surv. (CSUR) **52**(1), 1–38 (2019)
23. Zhang, S., Yao, L., Xu, X.: AutoSVD++ an efficient hybrid collaborative filtering model via contractive auto-encoders. In: Proceedings of the 40th International ACM SIGIR conference on Research and Development in Information Retrieval, pp. 957–960 (2017)
24. Zhang, S., Yao, L., Xu, X., Wang, S., Zhu, L.: Hybrid collaborative recommendation via semi-autoencoder. In: Liu, D., Xie, S., Li, Y., Zhao, D., El-Alfy, E.S. (eds.) Neural Information Processing, pp. 185–193. Springer, Cham (2017). https://doi.org/10.1007/978-3-319-70087-8_20
25. Song, Q., Cheng, D., Zhou, H., Yang, J., Tian, Y., Hu, X.: Towards automated neural interaction discovery for click-through rate prediction. In: Proceedings of the 26th ACM SIGKDD International Conference on Knowledge Discovery & Data Mining, pp. 945–955 (2020)
26. Krichene, W., Rendle, S.: On sampled metrics for item recommendation. In: Proceedings of the 26th ACM SIGKDD International Conference on Knowledge Discovery & Data Mining, pp. 1748–1757 (2020)
27. Ji, S., Feng, Y., Ji, R., Zhao, X., Tang, W., Gao, Y.: Dual channel hypergraph collaborative filtering. In: Proceedings of the 26th ACM SIGKDD International Conference on Knowledge Discovery & Data Mining, Ser. KDD 2020, pp. 2020–2029 (2020)
28. Hidasi, B., Karatzoglou, A.: Recurrent neural networks with top-k gains for session-based recommendations, pp. 843–852 (2017)

Sentiment Analysis of Film Reviews Based on Deep Learning Model Collaborated with Content Credibility Filtering

Xindong You[1], Xueqiang Lv[1], Shangqian Zhang[1], Dawei Sun[2(✉)], and Shang Gao[3]

[1] Beijing Key Laboratory of Internet Culture and Digital Dissemination Research, Beijing Information Science and Technology University, Beijing, China
youxindong@bistu.edu.cn

[2] School of Information Engineering, China University of Geosciences, Beijing, China
sundaweicn@cugb.edu.cncs

[3] School of Information Technology, Deakin University, Waurn Ponds Victoria, Geelong, Australia
shang.gao@deakin.edu.au

Abstract. Sentiment analysis of film reviews is the basis of obtaining the opinions of movie viewers. It has an important influence on movie public opinion control and stimulating potential viewers. Due to the natural openness and randomness of social media, there may exist a considerable amount of useless or false information in film review comments, making it challenging to analyze the credibility of the comments. This paper proposes a fine-grained sentiment analysis method based on the key-viewpoint sentences of Chinese film reviews, where a deep learning model is used to classify the fine-grained emotions in film reviews. Based on the analysis results, a method for calculating the credibility of review comments is proposed. Under the credibility criteria, corpus screened through credibility filtering algorithm, the overall sentiment classification can obtain 9% improvement on accuracy than the original corpus, which verifies the validity of the credibility algorithm. The higher quality corpus achieved by the credibility algorithm is benefit for improving the accuracy of the sentiment classification.

Keywords: Sentiment analysis · Film reviews mining · Natural language processing · Deep learning · Credibility algorithm

1 Introduction

With the rise of social media platforms and their widespread use, natural language processing for social network data has become a hot research topic. In terms of movie review, more and more people are publishing their review comments and opinions on film forums or relevant websites. These comments and opinions contain emotions, perceptions, attitudes, feelings, and behavioral tendencies towards films, cast members, the film industry, and its market. Internet word-of-mouth has a significant impact on

H. Gao et al. (Eds.): CollaborateCom 2020, LNICST 349, pp. 305–319, 2021.
https://doi.org/10.1007/978-3-030-67537-0_19

consumers' choice of watching movies. The higher the word-of-mouth evaluation, the higher is their box office revenue. Therefore, sentiment analysis of film evaluation helps understand the emotional tendency of the audience and timely acquire their opinions and attitudes [1]. It is very important for movie public opinion control, product marketing, and stimulation of potential viewers.

Text sentiment analysis, also known as opinion mining, is the process of analyzing, processing, summarizing and reasoning subjective texts with sentimental tendencies [2]. For example: "The rhythm is just to the point". This sentence expresses a positive view. In the sentence, "rhythm" is used as the characteristic word, and the corresponding emotional words are "just to the point", then "rhythm" and "just to the point" are pairs of characteristic viewpoint words. The short sentence of the viewpoint words is called a key viewpoint sentence. It can be seen from the sentiment words that the emotion of the sentence belongs to positive emotion. This paper uses a Bi-GRU model to perform a fine-grained sentiment analysis on key-viewpoint sentences. Based on the analysis result, a credibility algorithm for film review filtering is proposed. A Bi-LSTM is used to analyze the sentiment of the filtered reviews with higher credibility.

Due to its natural openness, virtuality, and randomness, especially the communication and word-of-mouth characteristics, social media have great commercial benefits for marketing and advertising. An 'internet water army' may be hired to publish a large number of false contents, including a considerable amount of useless or false information to influence the audience. It is important to analyze the content, identify these false comments, and evaluate the comment credibility. To accomplish the above objectives, we first analyze the characteristics of emotional tendency, sentence length and position of key-viewpoint sentences. Based on the analysis result, we then propose a credibility algorithm to filter film reviews. By classifying the overall sentiment on the credibility-filtered comments, 9% accuracy improvement is obtained than that of the unfiltered original corpus, which proves the validity of our credibility algorithm.

As a whole, the main contributions of this paper are listed as bellow.

(1) Corpus for extracting the fine-grained sentiment tendency is constructed, in which key viewpoint sentences and the sentiment tendency of the key viewpoint sentences are labelled.
(2) Comparison experiments are conducted on the corpus to analyze the fine-grained sentiment tendency with different deep learning model.
(3) Comments filtering algorithm is designed for obtaining higher quality corpus for overall sentiment analysis, which is based on the fine-grained sentiment analysis, length of the key viewpoint sentence and the position of the key viewpoint sentence.
(4) Extensive experiments are conducted on the corpus filtered by the different credibility with different mainstream deep learning model, which show that 9% improvement is achieved than on the un-filtered original corpus with Bi-LSTM employed.

2 Related Work

Sentiment Analysis can be divided into coarse-grained and fine-grained by granularity. Coarse-grained sentiment analysis only judges the overall emotional tendency at the text level. However, the overall sentimental tendency of a review text may be different or opposite to the emotional propensity of the individual evaluation opinions in the evaluation text [3]. Therefore, a finer granular analysis of sentiment is needed to identify the emotional tendencies of different opinions in the evaluation text. The aspect-level sentiment classification is a fine-grained sentiment classification task [4, 5]. The goal of this task is to infer the emotional polarity of corresponding evaluation words for the evaluation objects appearing in a given text [6–8]. Fine-grained sentiment analysis focuses on the extraction and analysis of attribute words and emotional words. Early scholars often used traditional machine learning and rule-based methods to extract emotion evaluation units in texts to perform fine-grained sentiment analysis. Liu Li et al. pruned the grammar tree by establishing a syntactic path library [9]. For texts containing multiple evaluation objects and evaluation words, they removed irrelevant evaluation objects regularly. The result of fine-grained sentiment analysis was obtained by extracting the correct evaluation unit. Tang et al. tried to improve the accuracy of microblogging fine-grained sentiment analysis by constructing a product feature ontology and extracting implicit product features by using feature word-emotional word pairs [10]. Traditional machine learning or rule methods only consider shallow semantic in-formation. Although this method is relatively simple and straightforward, it relies extremely much on information such as feature engineering and emotional dictionary. It requires knowledge support in related fields and the classification effect is not satisfactory.

Since it expresses textual semantic information with low-dimensional dense vectors, deep learning method contains deeper semantic information, which is convenient for mining and analyzing the relationship between words. Therefore, it has become a mainstream method for natural language processing related research. The TD-LSTM and TC-LSTM models proposed by Tang et al. enhanced the topic information based on the LSTM model, thereby improved the accuracy of sentiment classification for specific topics [11]. Ruder et al. implemented a hierarchical LSTM to model the correlation be-tween sentences in perspective sentiment analysis [12]. The hierarchical network structures worked better on multiple domains and language samples than models without levels. Ma et al. proposed an interactive attention network model to learn the target word and context representation separately, which completed the target word sentiment analysis task [13]. While paying attention to the overall text, they also introduced the characteristics of target words into the model. Xue et al. proposed a model based on the combination of CNN and gate mechanism [14]. The proposed structure is much simpler than the attention layer applied to the existing models, and the calculation of the model is easy to parallelize in training. Guan et al. used the attention mechanism to learn the weight distribution of each word on the sentiment tendency of the sentence directly from the word vector and used Bi-LSTM to learn the semantic information of text, and then improved the classification effect through parallel fusion [15]. Zhang et al. proposed a method of sentiment analysis of film reviews based on multi-feature fusion, which improved the sentiment analysis results on Chinese film reviews [16]. Deep learning has an excellent self-encoding ability and can well establish the mapping from low-level signals to high-level semantics. In this paper, a deep learning model Bi-GRU is selected as the basic model in the sentiment analysis experiments at the aspect-level, and a Bi-LSTM is selected as the basic model in the sentiment analysis experiments at the text level. Improvements obtained verify the effectiveness of the models.

It is important to evaluate the credibility of the comments. At present, the research on the quality of network information is more about the exploration of network opinion leaders. In essence, it is a method of judging high-quality users, mainly relying on external indicators such as citation, forwarding amount, and comment number to conduct credibility calculation [17]. This kind of method subjectively assumes that the content published by high-quality users is of high quality. While the number of high-quality users themselves is relatively few, the quality of content published by most users cannot be detected. Sheng et al. used Weibo users who had larger forwarding numbers and comments to build a domain-related vocabulary to calculate content similarity and evaluate the credibility of published contents [18]. This method essentially excavated the shallow mining of information, but did not fully consider the deep semantic information of the text, and it relied heavily on indirect information. Inspired by the afore-mentioned work, this paper considers three important characteristics of movie review comments: emotional tendency, sentence length and position of key viewpoint sentences. By considering the result of the fine-grained sentiment analysis text-level sentiment classification is conducted on these more credible comments filtered by the review credibility algorithm. The validity of the credibility algorithm is verified by substantial experiments.

3 Sentiment Analysis Based on Content Credibility Filtering

Figure 1 is a flow chart of the proposed fine-grained sentiment analysis based on the key viewpoint sentences. It consists of 4 main parts: corpus preprocessing and key viewpoint sentence labelling part, fine-grained sentiment classification model construction part, film review credibility algorithm part, and overall sentiment classification model part.

3.1 Corpus Preprocessing and Key Viewpoint Sentence Annotation

The experimental corpus adopted in this paper is the film reviews on 2017 China Film Lists on Douban website, covering comedy, thriller, horror, crime, action and other film genres. The film reviews are divided into 1 to 5 stars according to the star rating, corresponding to five rates of emotional tendency: highly recommended, recommended, just fine, poor, bad. There are 3755 movie reviews that have been manually labelled. The length of a review is the number of words included. The average sentence length is 47.465 words.

Corpus Samples are shown in Table 1, where the first column is the corpus number, followed by the emotional tendency rating score of the original corpus, the original corpus details, the manually labelled key viewpoint sentences and the emotional tendency of key viewpoint sentences, respectively.

According to the previous research work and our analysis, key viewpoint sentences can be divided into the following 3 categories.

Category 1: Sentence expression is an emotional tendency toward a single explicit evaluation object. That sentence contains a single explicit evaluation object and corresponding context-free emotion words. For example, the sentence "a" of the above

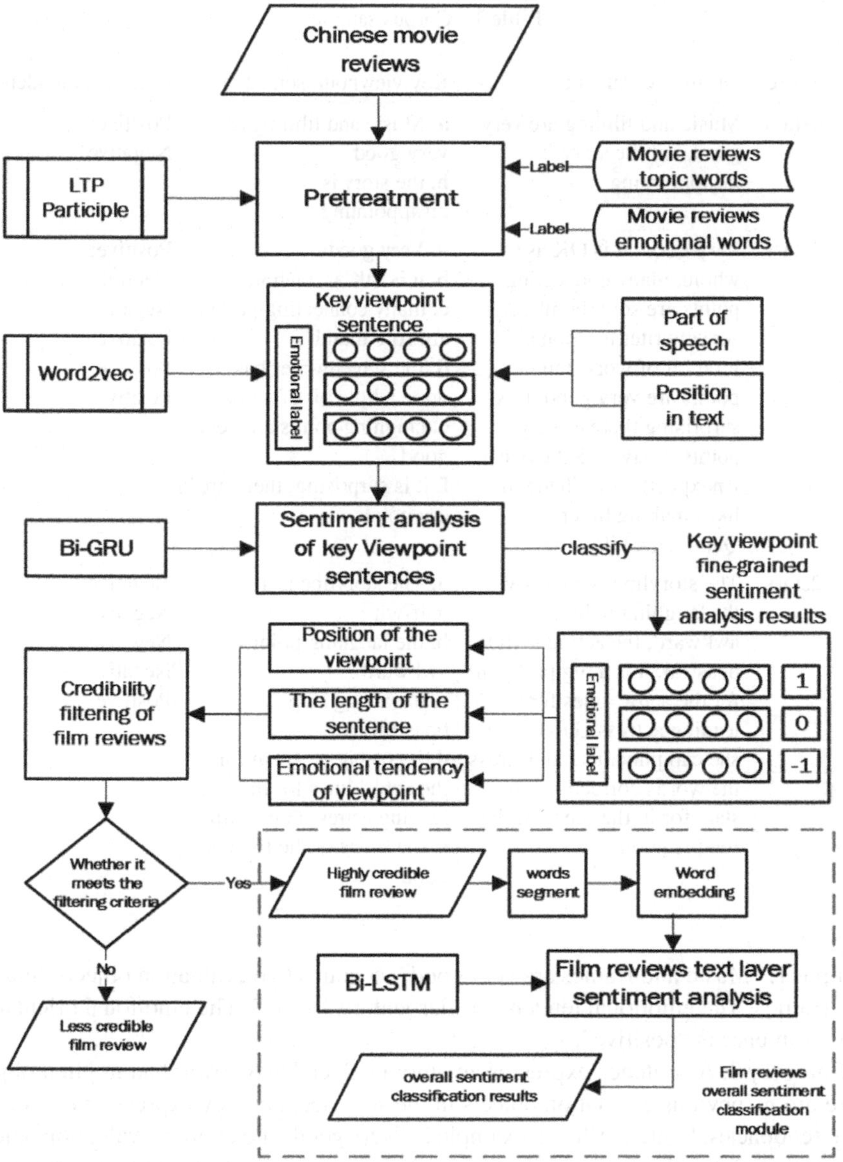

Fig. 1. Flow chart of sentiment analysis of film reviews based on content credibility filtering

example 3: "The storyline is a half-wit", this viewpoint sentence contains one single evaluation object "storyline" and the corresponding derogatory emotional tendency word "half-wit". The sentiment orientation of this sentence is "negative".

Category 2: Sentence expression is an emotional tendency toward multiple explicit evaluation objects. That sentence contains a plurality of explicit evaluation objects and corresponding context-free emotion words. For example, the sentence "a" of the above

Table 1. Corpus samples

No	Score	Original comment	Key viewpoint sentence	Emotional tendency
1	4stars	Music and filming are very good, but the story is disappointing	**a.** Music and filming are very good **b.** the story is disappointing	Positive; Negative
2	3stars	Very good, it is OK as a whole, many connecting points are sore thumb, the screenwriter has done a little bit of work, laughing points are very good, it is surprising there are cry points, I have cried twice. Unexpectly, Lu Zhengyu likes making bitter Qianlong Childhood!...	**a.** Very good **b.** it is OK as a whole **c.** many connecting points are sore thumb **d.** the screenwriter has done a little bit of work **e.** laughing points are very good **f.** it is surprising, there are cry points	Positive; Neutral; Negative; Positive; Positive; Positive
3	2stars	The storyline is a half-wit, the laughing point is awkward, the acting skill is boastful, it is a big problem for the actor' lines for the leading actress Guo Caijie, she can punctuate and stress the words correctly, two stars for it, the scene of the food is good.	**a.** The storyline is a half-wit **b.** the laughing point is awkward **c.** the acting skill is boastful **d.** it is a big problem for the actor' lines for the leading actress Guo Caijie **e.** the scene of the food is good	Negative; Negative; Negative; Negative; Positive

example 1: "Music and frames are very good", contains two evaluation objects "music" and "frame". The emotional tendency is derogatory, "good". The emotional orientation of this sentence is "positive".

Category 3: A sentence expresses an emotional tendency toward an implicit object. There are no obvious evaluation objects in the sentence, but they express obvious emotional tendencies. Sentence "b" in example 2, "very good", no explicit evaluation objects are mentioned, but there are emotional words - "very good". The sentiment orientation of this sentence is "negative".

According to the above analysis, it is found that there may be multiple key viewpoints in one comment, and each key viewpoint sentence may contain multiple opinions. After a statistical analysis of the 3,755 film reviews, the number of key viewpoint sentences included is 9019, and the number of views is 10,807.

3.2 Fine-Grained Sentiment Classification Model Based on Key Opinion Sentences

The overall architecture of the fine-grained sentiment classification model based on key viewpoint sentences is shown in Fig. 2. In this paper, a Word2vec word vector is used as the model input, a Bi-GRU model is used to learn the commentary dependency, and a Softmax layer is used as the output of the classification model to obtain the corresponding emotional tendency of the key opinion sentences. Over the corpus with key opinion sentences labelled and the emotion polarity of the key opinion sentences labelled. The key opinion sentences will be extracted and the categories of the key opinion sentences will be classified by the GRU neural network.

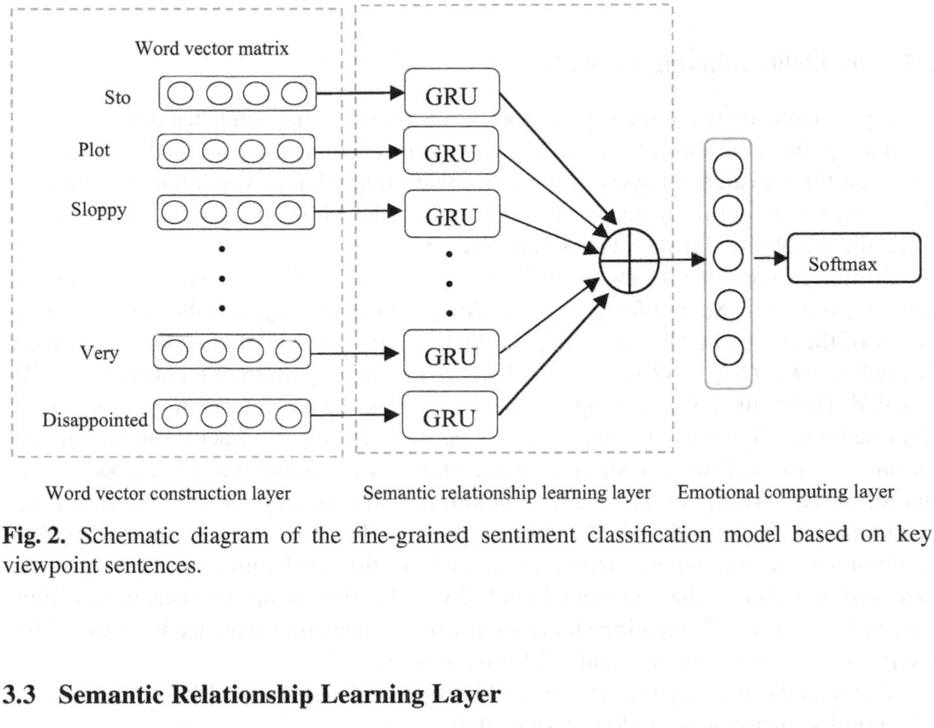

Fig. 2. Schematic diagram of the fine-grained sentiment classification model based on key viewpoint sentences.

3.3 Semantic Relationship Learning Layer

The sentiment analysis of the key opinion sentences needs to consider the semantic dependence between words. This paper uses a Bi-GRU model to obtain the semantic dependence information from film reviews. GRU (Gated Recurrent Unit) was proposed by Cho [19], which is a variant of the LSTM[20] model. It removes the forgetting gates, input gates, and output gates in the LSTM and only retains the reset gates and update gates. The advantage of LSTM model is kept but with a simplified structure.

The update gate of GRU controls how much information in the hidden state is passed to the current hidden state. The greater the value of the update gate, the more information is updated at the current time. This is very similar to the memory unit in a LSTM network, which helps GRU learn long-term dependencies. Since the training parameters are less than those of LSTM, its training speed is faster.

The Bi-GRU model is used to improve the accuracy of emotional classification results in key sentences. This structure provides complete forward and backward context information about the input sequence at the output layer.

3.4 Emotional Computing Layer

Define h_t to represent the feature of sentiment classification. Softmax classifier is used to classify key viewpoint sentences. The predicted sentiment category of a comment is y^, and y is the actual category. To train the model better, a cross-entropy cost function is used as the objective function to measure the difference between y distribution and y^ distribution. The back-propagation mechanism is used to train and update the parameters in the model.

3.5 Credibility Filtering Algorithm

For a given movie review, the key viewpoint sentence is a short sentence that represents the main point of the audience. It is a key part of the commentary emotional information. Through the statistical analysis of the sentiment orientation of key opinion sentences, the fine-grained sentiment analysis is conducted on the whole film reviews. At the same time, the credibility of user reviews is produced.

Splitting an original review by symbols,'，'、'。'、'；'、'！'、'？', into a set of n short sentence units Text Set = $\{Text_1, Text_2,\ldots,Text_m\}$; $Review_{orientation}$ is the emotional tendency of the comment. Opi Set = $\{Opi_1, Opi_2,\ldots,Opi_n\}$ is a set of viewpoint sentences; Ori Set = $\{Ori_1\ Ori_2,\ldots,Ori_n\}$ is a set of viewpoint polarity. Both of them are manually labelled; Define the collection OpinionSet to store all the key information of the viewpoint sentence: Opinion Set = $\{opinion_1, opinion_2,\ldots,opinion_n\}$. Each opinion sentence $opinion_i$ consists of unequal short sentence units $Text_j$. $opinion_i$ records the key information of each viewpoint sentence. It includes 3 attributes: $opinion_i = \{begin_i, end_i, orientation_i\}$. $begin_i$ indicates the start position of the short sentence unit of the i-th viewpoint sentence, and end_i indicates the end position of the ith viewpoint sentence. $orientation_i$ indicates the sentimental tendency of the ith opinion sentence, with a value range of $\{-1,0,1\}$. The sentimental tendency of the viewpoint sentence is set to +1 for the positive, 0 for the neutral, and −1 for the negative.

The specific steps of the review credibility calculation method based on the key viewpoint sentence are as follows (Algorithm 1).

Algorithm 1: Method for calculating review credibility based on key opinion sentences

---

Input: a user movie review, Review=$Text_1$, $Text_2$, ..., $Text_n$;
Output: The credibility of this film review, $Confidence_1$, $Confidence_2$, $Confidence_3$
Begin
1. for each piece $Opi_i \epsilon OpiSet$, determine
2. if Opi_i is an indivisible unit clause
3. if Opi_i is inseparable $f(i) = 0$;
4. else
$f(i)$ = number of unit sentences by segmenting − 1;
5. end if
6. end for
7. The position of the clause: pointer ← 1. Set the view collection OpinionSet to null
8. for i ← 1 to n do;
9. for k ← pointer to m do ;
10. if $Text_k$ can be found in Opi_i then
11. $begin_i$=k; end_i=k+ $f(i)$; $orientation_i$=Ori_i
12. $opinion_i$={ $begin_i$, end_i, $orientation_i$}
13. Add $opinion_i$ to OpinionSet
14. pointer ← k + $f(i)$;i = i+1;
15. end if
16 end for
17 end for
18. Confidence $_1$=$Review_{orientation}$ $* \sum_{i=1}^{n} orientation_i$;
19. $Confidence_2$ = $Review_{orientation} * \sum_{i=1}^{n} orientation_i *$ LengthWeigh(i);
20. $Confidence_3$=$Review_{orientation} * \sum_{i=1}^{n} orientation_i *$ Position$Weigh$(i) * LengthWeigh(i);
End

LengthWeigh(i) represents the length of the sentence of the ith opinion sentence $opinion_i$LengthWeigh(i)=$end_i - begin_i$. Position$Weigh$(i) is a normalized quadratic function that gives different weights to viewpoints at different positions:

$$\text{Position}Weigh(\text{i})=2*[(\frac{begin_i}{m})^2 - \frac{1}{2}]+1$$

3.6 Construction of the Overall Emotional Classification Model

The overall emotional classification experiment uses the film reviews with positive credibility as experimental data. It is designed to verify the validity of the credibility algorithm.

The overall sentiment classification model is similar to the key sentence sentiment classification model in Sect. 3.2, except that the Bi-GRU model is replaced by the Bi-LSTM model. Because the data for sentiment analysis at the text layer is a complete review, the length of the sentence is longer than the key sentence. The model of subsequent classification layer is Softmax.

4 Experiments and Results Analysis

4.1 Experimental Parameters and Evaluation

In the experiments, 0.9G movie review data crawled from Douban Website is used as input, and the word2vec's skip-gram model to train word embedding vectors. Each word vector is 200-dimensional. For unregistered words, a uniform distribution of U (−0.01,0.01) is used to randomly initialize the word vector. Other parameters use the default value of the model, and use all-zero initialization < PAD> to terminate a sentence. During the neural network training period, the word vector is allowed to be fine-tuned. After training, a word vector dictionary containing 149,195 words are obtained. The model is trained by 64 samples in each batch, setting Adam's learning rate to 0.001, the cost function to 0.001, Dropout to 0.5, and the number of Bi-GRU units is 256. The hidden layer of the Bi-GRU model is set to 200. If the accuracy does not exceed the current highest accuracy within 20 rounds, the model will automatically terminate early.

The correct rate, recall rate and F value commonly used in the field of natural language processing are used to evaluate the classification result of key viewpoint sentences and the results of overall sentiment classification.

4.2 Fine-Grained Sentiment Analysis Experiment and Result Analysis Based on Key Sentences

Fine-grained sentiment analysis comparison experiments are set up as follows:

(1) RNN. The word vector processed by Word2vec is used as the input of the RNN model, and the output is predicted by the Softmax layer. This experiment is the benchmark model for subsequent experiments.
(2) Bi-LSTM. The word vector processed by Word2vec is used as the input of the Bi-LSTM model, and the output is predicted by the Softmax layer.
(3) Bi-GRU. The word vector processed by Word2vec is used as the input of the Bi-GRU model, and the output is predicted by the Softmax layer.

The related experimental results are shown as in Table 2 and Fig. 3.

From the experimental results, it can be observed that:

(a) In the case of same input and output, the training effect of the RNN model is worse on the test set than those of the Bi-LSTM and Bi-GRU models. It is because the latter can capture the forward and backward bidirectional semantic dependence relationships.

Table 2. Comparison of fine-grained sentiment analysis results

		RNN	Bi-LSTM	Bi-GRU
Overall	P	0.8878	0.8968	0.9026
	R	0.8881	0.8988	0.9034
	F1	0.8877	0.8974	0.9029
Positive	P	0.9039	0.9242	0.9097
	R	0.886	0.9203	0.9287
	F1	0.8948	0.9223	0.9191
Neural	P	0.7976	0.6964	0.7258
	R	0.7283	0.5821	0.6716
	F1	0.7614	0.6341	0.6977
Negative	P	0.8854	0.8974	0.9186
	R	0.9137	0.9193	0.9099
	F1	0.8994	0.9082	0.9142

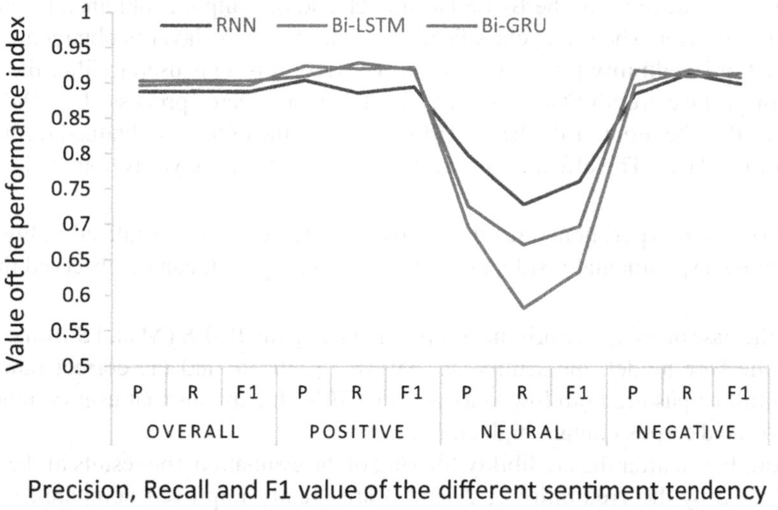

Fig. 3. Graphic illusion of the comparison of fine-grained sentiment analysis results.

(b) For the three basic models in Fig. 3, the emotional prediction results of positive and negative key viewpoints are better than that of the neutral viewpoints because users may have different degrees of deviation when publishing neutral opinions.

(c) From the results in TBALE II, the accuracy and recall rate of the GRU model is higher than that of the LSTM model, and the actual training speed of the GRU model is also faster. It is because the GRU has fewer parameters inside the model than the LSTM model, and the GRU is more suitable for shorter text. Because of its semantic learning ability, we choose Bi-GRU for the fine-grained sentiment analysis of viewpoint sentences.

4.3 Comment Sentiment Analysis Experiments and Result Analysis Based on Credibility Filtering

This section mainly discusses the impact of each credibility screening on the overall sentiment analysis of film reviews.

The experiments are set up as follows:

(1) Original material. The word vector processed by Word2vec is used as the input of the Bi-LSTM model, and the output is obtained through the Softmax layer. The result of the emotional layer classification of the text layer is obtained.
(2) The first credibility formula (18th line of Algorithm 1) is used to filter the corpus with positive credibility (credibility 1). The word vector processed by Word2vec is used as the input of the Bi-LSTM model, and the output is obtained from the Softmax layer. The effect classification result of the text layer is obtained.
(3) The second credibility formula (19th line of Algorithm 1) is used to filter the corpus with positive credibility (credibility 2). The word vector processed by Word2vec is used as the input of the Bi-LSTM model, and the output is obtained through the Softmax layer. The effect classification result of the text layer is obtained.
(4) The third credibility formula (20th line of Algorithm 1) is used to filter the corpus with positive credibility (credibility 3). The word vector processed by Word2vec is used as the input of the Bi-LSTM model, and the output is obtained through the Softmax layer. The effect classification result of the text layer is obtained.

The obtained experimental results are listed in Table 3, and visualized in Fig. 4. From the experimental results in TBALE III and Fig. 4, it can be observed that:

(a) In the case of using the original corpus and using the Bi-LSTM and Softmax layers as the base model, the sentiment analysis recall rate and the correct rate of the overall corpus are both low, only at about 70%. It explains that user comments in unfiltered corpus contain high emotional bias.
(b) From Fig. 4, after the credibility filtering of the evaluation, the results of the corpus selected by the credibility 1, 2, and 3 are greatly improved compared with the original corpus. It shows that when users comment the films, there is indeed a situation where the score is inconsistent with the expression emotion, which verifies the validity of our proposed credibility algorithm.
(c) From the comparison, the corpus filtered by credibility 2, 3, the accuracy and recall rate of the sentiment analysis of the text layer are better than the corpus filtered by credibility 1. It shows that the amount of emotional information of the corpus filtered by credibility 2 and 3 is higher, and the actual score of the user is closer to the emotional tendency of the user's comment. The weight of the viewpoint sentence is considered in the calculation of the credibility 2. The location of the opinion sentence and the length-weight of the viewpoint sentence are considered in

Table 3. Experimental results on the different corpus with different credibility filtering

		Origin	Credibility1	Credibility2	Credibility3
Overall	P	0.6983	0.757	0.7845	0.7892
	R	0.6959	0.7551	0.7837	0.7878
	F1	0.6951	0.7553	0.7833	0.7876
Positive	P	0.7259	0.8056	0.7662	0.7578
	R	0.6242	0.7389	0.7516	0.7771
	F1	0.6712	0.7708	0.7588	0.7673
Neutral	P	0.697	0.733	0.8087	0.8212
	R	0.7113	0.7784	0.7629	0.7577
	F1	0.7041	0.755	0.7851	0.7882
Negative	P	0.6688	0.7357	0.7712	0.78
	R	0.7554	0.741	0.8489	0.8417
	F1	0.7095	0.7384	0.8082	0.8097

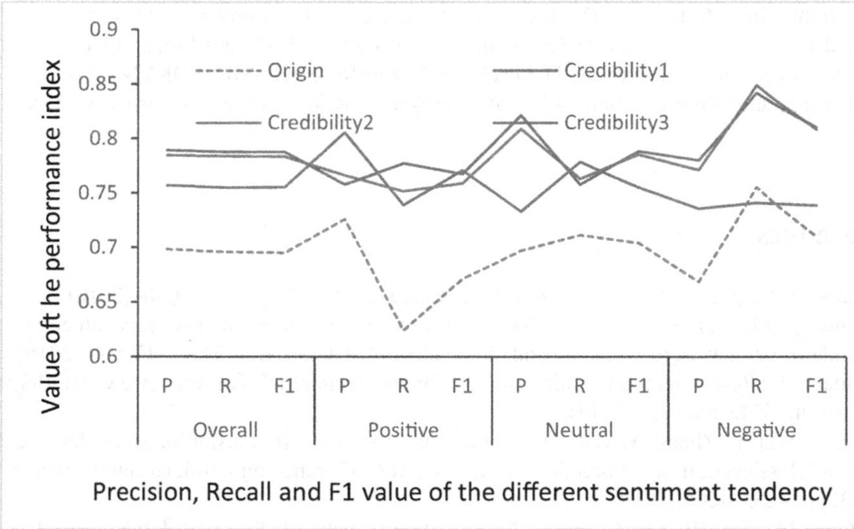

Fig. 4. Graphic illusion of the comparison of user review credibility experiment results.

the algorithm of credibility 3. The result proves that the position and the length of a viewpoint sentence are related to the user's expression of emotion: The longer the length of the viewpoint sentence, the higher the weight of the emotional expression in the text sentence; the viewpoint sentence appearing at the beginning or end of the text has a higher weight than the emotional expression of the viewpoint sentence in a sentence.

5 Conclusions and Future Work

In this paper, a Bi-GRU model is used to perform fine-grained sentiment analysis on key-viewpoint sentences. Based on its result, a film review's credibility algorithm is proposed, where a Bi-LSTM is used to analyze the sentiment of more credible reviews. Corpus screened by the credibility algorithm achieves higher accuracy than the original corpus, which verifies the validity of the credibility filtering algorithm. By considering the fine-grained sentiment analysis tendency, the position of the viewpoint sentence in the text, and the length of the view-point sentence, the credibility produced can be used to filter comments that do not match users rating for sentimental tendency. Furthermore, although the corpus in the experiments is constructed based on the Douban website Dataset, method proposed in this paper is also suitable for the other kinds of film review dataset. The requirement of the dataset is there are sentiment tendencies in the comment sentences and with a grade for the overall sentiment tendency. Therefore, the proposed method has the property of adaptability. However, the situation when users' comments are not related to the movie content is not recognized. Our future research work will focus on the relevance of the evaluation content and the content of the movie, and identify those false user comments.

Acknowledgment. This work is supported by National Natural Science Foundation of China under Grants No. 61671070, 61972364. National Science Key Lab Fund project 6142006190301, National Language Committee of China under Grants ZDI135-53, and Project of Developing University Intension for Improving the Level of Scientific Research–No.2019KYNH226, Qin Xin Talents Cultivation Program, Beijing Information Science & Technology University No. QXTCP B20190.

References

1. Zhao, Y.Y., Qin, B., Liu, T.: Sentiment analysis. J. Softw. **21**(8), 1834–1848 (2010)
2. Chuang, J.L., Xu, S.Z., Hu, Q.F.: Market research report on movie box office, an empirical study of internet word-of-mouth and movie box office. Chin. Bus. Theor. **17**, 6–7 (2016)
3. Huang, S.: Research on fine-grained opinion mining technologies of web review text. Beijing Institute of Technology (2014)
4. Li, Z., Wei, Y., Zhang, Y., et al.: Exploiting coarse-to-fine task transfer for aspect-level sentiment classification. In: Proceedings of the AAAI Conference on Artificial Intelligence, vol. 33, pp. 4253–4260 (2019)
5. Yang, M., Yin, W., Qu, Q., et al.: Neural attentive network for cross-domain aspect-level sentiment classification. IEEE Transactions on Affective Computing, 1–11 (2019)
6. Pontiki, M., Galanis, D., Pavlopoulos, J., et al.: Semeval-2014 task4: aspect based sentiment analysis. In: Proceedings of the 8th International Workshop on Semantic Evaluation, pp. 27–35 (2014)
7. Rosenthal, S., Farra, N., Nakov, P.: SemEval-2017 task4: sentiment analysis in twitter. In: Proceedings of SemEval (2017)
8. Guo, C., Wang, Z.Y.: Auto-construct of sentiment ontology tree for fine-grained opinion mining. J. Chin. Inform. Process. **27**(5), 75–83 (2014)
9. Liu, L., Wang, Y.H., Hang, W.: Fine-grained sentiment analysis oriented to product comment. J. Comput. Appl. **35**(12), 3481–3486 (2015)

10. Tang, X.B., Lan, Y.T.: Sentiment analysis of microblog product reviews based on feature ontology. Libr. Inform. Serv. **60**(16), 121–127 + 136 (2016)
11. Tang, D.Y., Qin, B., et al.: Effective LSTMs for target-dependent sentiment classification. In: Proceedings of the 26th International Conference on Computational Linguistics, pp. 3298–3307 (2016)
12. Ruder, S., Ghaffari, P., Breslin, J.G.: A hierarchical model of reviews for aspect-based sentiment analysis. In: EMNLP, pp. 999–1005 (2016)
13. Ma, D.H., Li, S.J., Zhang, X.D., Wang, H.F.: Interactive attention networks for aspect-level sentiment classification. In: IJCAI, pp. 4068–4074 (2017)
14. Xue, W., Li, T.: Aspect Based Sentiment Analysis with Gated Convolutional Networks (2018). arXiv preprint, arXiv:1805.07043
15. Guan, P.F., Li, B.A., Lv, X.Q.: Attention enhanced bi-directional LSTM for sentiment analysis. J. Chin. Inform. Process. **33**(02), 105–111 (2019)
16. Zhang, S.Q., Lv, X.Q.: Movie short-text reviews sentiment analysis based on multi-feature fusion. In: International Conference Proceeding Series. ACM (2018)
17. Paul, A.: Who gives a tweet evaluating microblog content value. In: Proceedings of the ACM 2012 Conference on Computer Supported Cooperative, pp. 471–474 (2012)
18. Sheng, Y.: Research on micro-blog information quality evaluation based on content—with sina micro-blog as an example. Inform. Sci. **31**(05), 51–54 + 66 (2013)
19. Hochreiter, S., Schmidhuber, J.: Long short-term memory. Neural Comput. **9**(8), 1735–1780 (1997)

Essence Computation Oriented Multi-semantic Analysis Crossing Multi-modal DIKW Graphs

Shijing Hu[1,2], Yucong Duan[1](✉), and Mengmeng Song[3]

[1] School of Computer Science and Cyberspace Security, Hainan University, Haikou, China
duanyucong@hotmail.com
[2] Yuanpei College, Peking University, Beijing, China
[3] School of Tourism, Hainan University, Haikou, China

Abstract. Based on our previous proposal of an existence level computation and reasoning approach called Existence Computation, which extends traditional identification of objects towards an existence level clarification, and a semantic expression mechanism called Relationship Defined Everything of Semantics which adopts the ideology that entities are defined or determined by more essential relationships or structures, we proposed in this work that although resource or content processing are embodied in every modal of the target content, the essence of the requirement of processing instead of the projected modal of the content is vital as the source of processing towards expected result. From the perspective of processing of multiple modal resources or content in the background of insufficient single modal resource to achieve progress in decision making and improve precision of computation with higher efficiency, we propose an approach called Essence Computation and Reasoning, originating in existence computation level and relationship defined everything of semantics models, to synchronize crossing modal processing towards minimizing the uncertainties through merging cross modals and inter modal or meso-scale capabilities of available data, information, knowledge and wisdom (DIKW). We show the application of essence computation and reasoning of modal transformation and deduplication analysis crossing multiple semantics modals of data modal, information modal and knowledge modal in an encoding and decoding case.

Keywords: DIKW graphs · Multi-semantic · Cross-modal transformation · Privacy protection · Edge computing · Essence computation

Supported by Natural Science Foundation of China Project (No. 61662021 and No. 72062015), Hainan Provincial Natural Science Foundation Project No. 620RC561, Hainan Education Department Project No. Hnky2019-13 and Hainan University Educational Reform Research Project No. HDJWJG03.

H. Gao et al. (Eds.): CollaborateCom 2020, LNICST 349, pp. 320–339, 2021.
https://doi.org/10.1007/978-3-030-67537-0_20

1 Introduction

The current data collection, information processing, and knowledge fusion processing mechanisms are relatively lagging behind the accelerated generation of multi-modal, multi-scale and even mesoscale content processing in the information world. It is difficult to maintain the cognitive processing capacity of social individuals and the ability to improve the dynamic balance of content growth with the help of current information processing facilities. Facing the challenge of cognitive overload in the context of multi-level, multi-dimensional, multi-modal, multi-scale social information network massive content, natural annotation of big data and multi-source association to obtain open domain knowledge, the current semantics with natural language as the core understanding and semantic fusion research is mostly based on the natural language semantics and context analysis and retrieval of multiple knowledge based on scene knowledge graphs, affair knowledge graphs, temporal knowledge graphs, and reinforcement learning. This type of analysis method faces arduous challenges in terms of understandable modeling theory and interpretable semantic representation technology when dealing with mixed subjective and objective multi-modal content with both logic and artistic expression. In the context of language evolution consistent with the human-computer understanding of cross-modal interaction intentions in the environment and society, existing solutions are oriented to event-oriented and context-driven concepts and semantic migration of concept combinations, small sample common sense semantic capture. The general cognitive intelligence model that is adjusted across the semantic scope of the abstract level and the knowledge and other modalities lacks robust and adaptable natural language semantic cognitive understanding basic theoretical innovation support.

Based on the previous accumulation of cross-modal, cross-scale and even mesoscale cognitive semantics research work, we start from the solution of the uncertainty presented by artificial intelligence processing problems, facing the complex content objects and complex processing processes of the information discipline, focusing on subjectivity and the semantic essence expression and analysis of the artificial intelligence interpretability problem in the mesoscale category between the objective structure association and dynamic transformation. For example, by exploring the multi-modal, multi-scale, fuzzy, and even mesoscale concept semantic human-computer understanding and communication expression model as the prerequisite for the standardized processing of the essence of the problem [7,8,10], the development of cognitive semantic clarification, measurement and value orientation based on concept semantic traceability can be adapted to semantic optimization technology to build an artificial intelligence interpretable theoretical system based on the determination of essential semantics that integrates consciousness science, cognitive linguistics, information science, philosophy, etc. The method provides a theoretical basis for a new generation of interpretable precision intelligence based on complexity.

The semantic content is modeled based on the DIKW graphs [5,6,11]. TR_{DIKW} (typed resources) can be divided into D_{DIKW}(data resources), I_{DIKW}(information resources) and K_{DIKW}(knowledge resources). D_{DIKW} are discrete elements obtained by direct observation. They have no meaning without context, are not associated with a specific human purpose, and express the attribute content of a single entity. I_{DIKW} record human behavior and are used to mine, analyze, and express the interactive relationship between two entities. The entity can be another person or an objective thing. I_{DIKW} are related to a specific purpose of human beings, and the relationship between two entities can be inferred through the purpose. K_{DIKW} are derived from D_{DIKW} and I_{DIKW} through structured formalization. K_{DIKW} summarized the entity relationship abstractly on the basis of I_{DIKW}.

Multi-semantics refers to the understanding that the semantic content has multiple different purposes, that is, the TR_{DIKW} in the semantic content can be derived from various TR_{DIKW} to obtain I_{DIKW} with different purposes. Multi-semantic is caused by missing of TR_{DIKW} in semantic content, such as lack of D_{DIKW} or I_{DIKW}, which leads to a wider understanding of the content with different purposes, or by redundancy of TR_{DIKW} in semantic content. Redundant D_{DIKW} or I_{DIKW} will produce understandings of different purposes when combined with different TR_{DIKW}. Section 3 and Sect. 4 discuss these two reasons respectively. D_{DIKW} and I_{DIKW} can be transformed across modalities after combining the corresponding K_{DIKW}. In the process of solving multi-semantic, it is necessary to complete the transformation from the original TR_{DIKW} in the semantic content to the new TR_{DIKW}. The transformed objects are divided into D_{DIKW} and I_{DIKW}. Section 5 discusses the transformation methods for those two transformed objects. In addition to semantics, inadvertent behaviors of users in daily life will generate a lot of user behavior content, such as motion content and sound content. After modeling the user behavior content based on the DIKW graphs, the D_{DIKW}, I_{DIKW}, and K_{DIKW} contained in the user behavior content can be encoded and decoded to transform new TR_{DIKW} across modalities. Section 6 and Sect. 7 discuss the encoding and decoding modes of user content.

2 Related Work

Knowledge Graph identification [13] and Graph based intelligence analysis [3] has been increasingly recognized as both big challenges and great opportunities. In the links of Knowledge Graph. Xiao (2016) [14] proposes a two level hierarchical generative approach for semantics representation of Knowledge Graph through extracting aspects and assigns categories. Xu (2016) [15] presents an approach of representing Knowledge Graph with a combination of structural and textual encoding. Chen (2009) [2] presents visualization approaches based on distinguishing among Data, Information and Knowledge. Some existing works [1] has elaborated on the challenge of recovering implicit information extraction through state machine and abductive inference, etc. McSherry (2009) [12] present a strategy of information extraction through queries while taking care of the data privacy concerns in the background of sequential and parallel compositions. We present a formalization of the basic concepts of Data, Information and Knowledge and propose to protect privacy resources in a three-tier architecture consisting of Data Graph, Information Graph and Knowledge Graph in a value driven or cost-effective manner [4]. We model multi-semantic content and implement the elimination of multi-semantic in terms of transformation of type in the formalized DIKW architecture.

3 Solve Multi-semantic Caused by Missing TR_{DIKW}

The missing TR_{DIKW} in semantic content has led to a reduction of the limitation of the scope of content understanding. By increasing the restriction on content understanding, the scope of content understanding can be narrowed, so that only one of the derived I_{DIKW} for different purposes is retained, thereby solving multi-semantics. Modeling the semantic content based on the data graph, information graph and knowledge graph. The missing TR_{DIKW} in semantic content can be divided into the missing of I_{DIKW} and the missing of D_{DIKW}.

3.1 Solve Multi-semantic Caused by Missing I_{DIKW}

Semantic content "In the summer night, user A stays in the study room". can correspond to the following D_{DIKW} and I_{DIKW}.

$$\begin{aligned} D_{0_1} &= (A|T_{PLACE}(INS(Studyroom))) \\ D_{0_2} &= (T_{TIME}(Night)) \\ D_{0_3} &= (T_{SEASON}(Summer)) \\ I_0 &= R_{CORRESPOND}(D_{0_1},\ R_{COMBINE}(D_{0_2},\ D_{0_3})) \end{aligned} \tag{1}$$

Since the semantic content lacks the I_{DIKW} of "what does user A do in the study room", there will be ambiguities in understanding the content. For example, combining the data resource D_{0_1} "user A stays in the study room" and the knowledge resource K_1 "the study room is a place for learning", it can be deduced that the purpose of user A staying in the study room is to learn. Combining the data resource D_{0_2} "night" and the knowledge resource K_2 "people usually sleep at night", it can be deduced that the user A may be sleeping in the study room. These two derivation methods are correct in the absence of other relevant resources, but they produce I_{DIKW} with different purposes, leading to multi-semantic.

$$\begin{aligned} K_1 &= R_{IN}(T_{ACT}(Study),\ T_{PLACE}(Studyroom)) \\ K_2 &= R_{AT}(T_{ACT}(Sleep),\ T_{TIME}(Night)) \end{aligned} \tag{2}$$

$$\begin{aligned} D_{0_1} + K_1 \rightarrow I_{new_1} &= R_{PURPOSE}(A,\ T_{ACT}(Study)) \\ D_{0_2} + K_2 \rightarrow I_{new_2} &= R_{PURPOSE}(A,\ T_{ACT}(Sleep)) \\ K_{new_1} &= R_{CONFLICT}(I_{new_1}, I_{new_2}) \end{aligned} \tag{3}$$

The scope of content understanding can be narrowed by adding relevant D_{DIKW} or I_{DIKW} to solve multi-semantic.

Solve by Adding D_{DIKW}. Obtain relevant data resources D_1 "The air condition in the bedroom is broken". D_2 "The air condition in the study room is good". Combining data resource D_{0_3} "Summer" and the knowledge resource

K_3 "Summer is hot", it can be deduced that the temperature in the bedroom is high and the temperature in the study room is low. Those D_{DIKW} increase the restrictions on the study room environment, thereby narrowing the scope of content understanding to temperature-related fields. Combined with the knowledge resource K_4 "People like to sleep in a cool place", it can be deduced that the temperature in the study room is low and suitable for sleeping, which supports the I_{DIKW} "User A stays in the study room is to sleep", thus solve the multi-semantic.

$$\begin{aligned}
&D_1 = (T_{FACILITY}(INS(AIR_CONDITION_{Bedroom}))|\ T_{CONDITION}(Broken))\\
&D_2 = (T_{FACILITY}(INS(AIR_CONDITION_{Studyroom}))|\ T_{CONDITION}(Normal))\\
&K_3 = R_{IS}(T_{SEASON}(Summer),\ T_{TEMP}(High))\\
&K_4 = R_{LIKE}(T_{PERSON},\ R_{IN}(T_{ACT}(Sleep),\ R_{IS}(T_{PLACE},\ T_{TEMP}(Low))))
\end{aligned} \tag{4}$$

$$\begin{aligned}
D_{0_3} + D_1 + K_3 \rightarrow D_{new_1} &= (T_{PLACE}(INS(Bedroom))|\ T_{TEMP}(High))\\
D_{0_3} + D_2 + K_3 \rightarrow D_{new_2} &= (T_{PLACE}(INS(Studyroom))|\ T_{TEMP}(Low))\\
D_{new_1} + D_{new_2} + K_4 \rightarrow I_{new_3} &= R_{LIKE}(A,\ R_{IN}(T_{ACT}(Sleep),\\
&\qquad T_{PLACE}(INS(Studyroom)))\\
K_{new_2} &= R_{SUPPORT}(I_{new_3},\ I_{new_2})
\end{aligned} \tag{5}$$

Algorithm. (1) Combine known data resources D_0 and information resources I_0 with related K_{DIKW} to derive new information resources I_{new_1} and I_{new_2} with different purposes.

(2) Search related data resources $D_{related}$ in the data graph.

(3) Combine $D_{related}$ with related I_{DIKW} and K_{DIKW} to further derive new information resource I_{new_3} that can narrow the scope of understanding.

(4) Determine the relationship between I_{new_3} and I_{new_1} with I_{new_2}. Retain I_{DIKW} supported by I_{new_3}, and delete other I_{DIKW}.

(5) Set the remaining I_{DIKW} as the final result to solve multi-semantic.

Solve by Adding I_{DIKW}. Obtain related information resource I_1 "User A does not like learning". Combining D_{0_2} "night" and knowledge resource K_5 "Those who like to learn may study at night", it can be deduced that user A is unlikely to study in the study room at this time. This I_{DIKW} excludes the I_{DIKW} "User A stays in the study room is to learn" out of the scope of understanding of the content, and the only I_{DIKW} remaining "User A stays in the study room is to sleep" is the final result, thus solve the multi-semantic.

$$\begin{aligned}
&I_1 = !R_{LIKE}(A,\ T_{ACT}(Study))\\
&K_5 = R_{AT}(R_{DO}(T_{PERSON}(R_{LIKE}(T_{PERSON},\ T_{ACT}(Study))),\ T_{ACT}(Study)),\\
&\quad T_{TIME}(Night))
\end{aligned} \tag{6}$$

$$\begin{aligned}
D_{0_2} + I_1 + K_5 \rightarrow I_{new_4} &= !R_{DO}(A,\ T_{ACT}(Study))\\
K_{new_3} &= R_{OPPOSE}(I_{new_4},\ I_{new_1})
\end{aligned} \tag{7}$$

Algorithm. (1) Combine known data resources D_0 and information resources I_0 with related K_{DIKW} to derive new information resources I_{new_1} and I_{new_2} with different purposes.

(2) Search related information resources $I_{related}$ in the information graph.

(3) Combine $I_{related}$ with related I_{DIKW} and K_{DIKW} to further derive new information resource I_{new_3} that can narrow the scope of understanding.

(4) Determine the relationship between I_{new_3} and I_{new_1} with I_{new_2}. Retain I_{DIKW} supported by I_{new_3}, and delete other I_{DIKW}.

(5) Set the remaining I_{DIKW} as the final result to solve multi-semantic.

3.2 Solve Multi-semantic Caused by Missing D_{DIKW}

Semantic content "User A's seniority is greater than user B's." can correspond to the following D_{DIKW} and I_{DIKW}.

$$\begin{aligned} D_{0_1} &= (A|T_{SENIORITY}) \\ D_{0_2} &= (B|T_{SENIORITY}) \\ I_0 &= R_{GREATER\_THAN}(D_{0_1},\ D_{0_2}) \end{aligned} \tag{8}$$

Because the semantic content lacks D_{DIKW} related to "user A's age and user B's age", there are different purpose of understanding the I_{DIKW} of "the relationship between user A and user B's age". Although there is a knowledge resource K_1: "higher generation may be older", with this K_{DIKW}, new I_{DIKW} "user A may be older than user B" can be deduced. However, there are many examples of people has high seniority with young age, so it is still impossible to exclude the I_{DIKW} "User A may be younger than user B".

$$\begin{aligned} K_1 = R_{GREATER\_THAN}(&T_{AGE}(T_{PERSON}(T_{SENIORITY}(High))), \\ &T_{AGE}(T_{PERSON}(T_{SENIORITY}(Low))) \end{aligned} \tag{9}$$

$$\begin{aligned} I_0 + K_1 \rightarrow I_{new_1} &= R_{PROBABLY\_GREATER\_THAN}(T_{AGE}(A),\ T_{AGE}(B)) \\ I_{new_2} &= R_{PROBABLY\_LESS\_THAN}(T_{AGE}(A),\ T_{AGE}(B)) \end{aligned} \tag{10}$$

It is also possible to reduce the scope of understanding by adding relevant D_{DIKW} or I_{DIKW}, thereby solving multi-semantic.

Solve by Adding D_{DIKW}. Obtain related data resources D_1 "User A is mature in mind". D_2 "User B is innocent". Combining D_1 and D_2, the I_{DIKW} "User A is more mature than user B" is deduced. Those D_{DIKW} increase the restriction on the mentally mature relationship between user A and user B when determining "the age of user A and user B", and further narrow the scope of understanding of the content. Thus the previously deduced I_{DIKW} "User A is older than user B" is supported.

$$\begin{aligned} D_1 &= (A|T_{MIND}(Mature)) \\ D_2 &= (B|T_{MIND}(Naieve)) \end{aligned} \tag{11}$$

$$\begin{aligned} D_1 + D_2 \rightarrow I_{new_3} &= R_{MATURE\_THAN}(A,\ B) \\ I_{new_1} + I_{new_3} \rightarrow I_{new_0} &= R_{GREATER\_THAN}(T_{AGE}(A),\ T_{AGE}(B)) \end{aligned} \tag{12}$$

Algorithm. (1) Combine known data resources D_0 and information resources I_0 with related K_{DIKW} to derive new information resources I_{new_1} and I_{new_2} with different purposes.

(2) Search related data resources $D_{related}$ in the data graph.

(3) Combine $D_{related}$ with related I_{DIKW} and K_{DIKW} to further derive new information resource I_{new_3} that can narrow the scope of understanding.

(4) Determine the relationship between I_{new_3} and I_{new_1} with I_{new_2}. Retain I_{DIKW} supported by I_{new_3}, and delete other I_{DIKW}.

(5) Set the remaining I_{DIKW} as the final result to solve multi-semantic.

Solve by Adding I_{DIKW}. Obtain relevant information resources I_1 "User B respects user A very much". Knowledge resource K_3 "People with low status respect people with high status". Combining I_1 and K_3, I_DIKW "User A has a higher status than user B" is deduced. Those D_{DIKW} increase the restriction on the status relationship between user A and user B when judging "the age of user A and user B", and further narrow the scope of understanding of the content. Thus the previously deduced I_{DIKW} "User A is older than user B" is supported.

$$\begin{aligned} &I_1 = R_{RESPECT}(B,\ A) \\ &K_3 = R_{RESPECT}(T_{PERSON}(T_{STATUS}(Low)),\ T_{PERSON}(T_{STATUS}(High))) \end{aligned} \tag{13}$$

$$\begin{aligned} I_1 + K_3 \rightarrow I_{new_3} &= R_{GREATER\_THAN}(T_{STATUS}(A),\ T_{STATUS}(B)) \\ I_{new_1} + I_{new_3} \rightarrow I_{new_0} &= R_{GREATER\_THAN}(T_{AGE}(A),\ T_{AGE}(B)) \end{aligned} \tag{14}$$

Algorithm. (1) Combine known data resources D_0 and information resources I_0 with related K_{DIKW} to derive new information resources I_{new_1} and I_{new_2} with different purposes.

(2) Search related information resources $I_{related}$ in the information graph.

(3) Combine $I_{related}$ with related I_{DIKW} and K_{DIKW} to further derive new information resource I_{new_3} that can narrow the scope of understanding.

(4) Determine the relationship between I_{new_3} and I_{new_1} with I_{new_2}. Retain I_{DIKW} supported by I_{new_3}, and delete other I_{DIKW}.

(5) Set the remaining I_{DIKW} as the final result to solve multi-semantic.

4 Solve Multi-semantic Caused by Redundant TR_{DIKW}

Redundancy in semantic content refers to the understanding that the D_{DIKW} or I_{DIKW} in the semantic content have multiple different purposes on the same issue. Model the semantic content based on DIKW graphs. The redundant of TR_{DIKW} in semantic content can be divided into I_{DIKW} redundancy and D_{DIKW} redundancy.

4.1 Solve Multi-semantic Caused by Redundant I_{DIKW}

Semantic content "User A likes to play basketball, user A hates sports." can correspond to the following I_{DIKW}.

$$\begin{aligned} I_{0_1} &= R_{LIKE}(A,\ T_{ACT}(Basketball)) \\ I_{0_2} &= R_{HATE}(A,\ T_{ACT}(Sports)) \end{aligned} \tag{15}$$

Knowledge resources K_1 "Playing basketball is a type of sport". K_2 "The relationship "hates" and the relationship "likes" contradicts". From I_{0_2} and K_1, user A hates sports, and playing basketball belongs to a type of sports, new information resource I_{new_1} "User A hates playing basketball can be deduced". With K_2 Knows that I_{0_1} contradicts I_{new_1}, therefore, for the question of "User A's attitude towards playing basketball", I_{0_1} and I_{0_2} have different understandings, representing the redundancy of I_{DIKW} in the semantic content.

$$\begin{aligned} K_1 &= R_{BELONG_TO}(T_{ACT}(Basketball),\ T_{ACT}(Sports)) \\ K_2 &= R_{OPPOSE}(T_{RELATION}(Like), T_{RELATION}(Hate)) \end{aligned} \tag{16}$$

$$\begin{aligned} I_{0_2} + K_1 \rightarrow I_{new_1} &= R_{HATE}(A,\ T_{ACT}(Basketball)) \\ I_{0_1} + I_{new_1} + K_2 \rightarrow K_{new_1} &= R_{CONFLICT}(I_{0_1}, I_{new_1}) \end{aligned} \tag{17}$$

From the above derivation, it can be seen that the redundant information resources I_{0_1} and I_{0_2} are contradictory, so there must be an error in one of them. It can help to determine the correctness of redundant I_{DIKW} by adding relevant D_{DIKW} or I_{DIKW}, thereby solving multi-semantic.

Solve by Adding D_{DIKW}. Obtain the spatial data resource D_1 related to user A "basketball court". There are relevant knowledge resources K_3 "The main activity in the basketball court is playing basketball". K_4 "People who often play basketball like to play basketball". Combining D_1 and K_3, user A often appears in the basketball court, so user A often plays basketball. Combined with K_4, user A often plays basketball, and people who play basketball frequently are likely to like basketball, indicating that user A is likely to like basketball and supports I_{0_1}. When the information resource I_{0_1} has supporting D_{DIKW} but the information resource I_{0_2} does not have supporting D_{DIKW}, it tends to determine that I_{0_1} is correct and I_{0_2} is wrong.

$$\begin{aligned} D_1 &= (A|T_{PLACE}(INS(BasketballCourt)) \\ K_3 &= R_{IN}(T_{ACT}(Basketball),\ T_{PLACE}(BasketballCourt)) \\ K_4 &= R_{LIKE}(T_{PERSON}(R_{DO}(person,\ T_{ACT}(Basketball))),\ T_{ACT}(Basketball)) \end{aligned} \tag{18}$$

$$\begin{aligned} D_1 + K_3 \rightarrow I_{new_2} &= R_{DO}(A|T_{ACT}(Basketball)) \\ I_{new_2} + K_4 \rightarrow I_{new_3} &= R_{LIKE}(A,\ T_{ACT}(Basketball)) \\ K_{new_2} &= R_{SUPPORT}(I_{new_3}, I_{0_1}) \end{aligned} \tag{19}$$

Algorithm. (1) Find the conflicting information resources I_{O_1} and I_{O_2}.
(2) Search related data resources $D_{related}$ in the data graph.
(3) Combine $D_{related}$ with related I_{DIKW} and K_{DIKW} to further derives new information resource I_{new} that can help judge the truth.
(4) Determine the relationship between I_{new} and I_{O_1} with I_{O_2}. Keep the result supported by I_{new}, and delete the other result.
(5) Set the result supported by I_{new} as the final result to solve multi-semantic.

Solve by Adding I_{DIKW}. Obtain related information resource I_1 "User A is a member of school's basketball team". There are relevant knowledge resources K_5 "The members of the basketball school team often play basketball". Combining I_1 and K_5, user A is a member of school's basketball team, so user A often plays basketball. Combined with K_4, user A often plays basketball, and people who often play basketball are likely to like to play basketball, indicating that user A is likely to like playing basketball and supports the information resource I_{O_1}. When the information resource I_{O_1} has supporting I_{DIKW} but the information resource I_{O_2} does not have supporting I_{DIKW}, it tends to determine that I_{O_1} is correct and I_{O_2} is wrong.

$$\begin{aligned} &I_1 = R_{IS\_A\_MEMBER\_OF}(A,\ T_{GROUP}(INS(BasketballTeam)) \\ &K_5 = R_{DO}(T_{PERSON}(R_{IN}(person,\ T_{GROUP}(BasketballTeam)),\ T_{ACT}(Basketball)) \end{aligned} \tag{20}$$

$$\begin{aligned} I_1 + K_5 \rightarrow I_{new_4} &= R_{DO}(A,\ T_{ACT}(Basketball)) \\ I_{new_4} + K_4 \rightarrow I_{new_5} &= R_{LIKE}(A,\ T_{ACT}(Basketball)) \\ K_{new_3} &= R_{SUPPORT}(I_{new_5}, I_{O_1}) \end{aligned} \tag{21}$$

Algorithm. (1) Find conflicting information resources I_{O_1} and I_{O_2}.
(2) Search related information resources $I_{related}$ in the information graph.
(3) Combine $I_{related}$ with related I_{DIKW} and K_{DIKW} to further derives new information resource I_{new} that can help judge the truth.
(4) Determine the relationship between I_{new} and I_{O_1} with I_{O_2}. Keep the result supported by I_{new}, and delete the other result.
(5) Set the result supported by I_{new} as the final result to solve multi-semantic.

4.2 Solve Multi-semantic Caused by Redundant D_{DIKW}

Semantic content contains data resources D_{O_1} "today's temperature is 30°". D_{O_2} "today's temperature is 20°".

$$\begin{aligned} D_{O_1} &= (T_{TEMP}(30)) \\ D_{O_2} &= (T_{TEMP}(20)) \end{aligned} \tag{22}$$

In response to the issue of "today's temperature", the data resources D_{O_1} and D_{O_2} are contradictory, indicating that upon the redundant data resources D_{O_1} and D_{O_2} there must be an error in one of them. It is possible to add relevant

D_{DIKW} or I_{DIKW} to help judge the correctness of redundant D_{DIKW}, thereby solving multi-semantic.

Solve by Adding D_{DIKW}. Obtain data resources D_1 "summer". D_2 "Hainan". Knowledge resource K_1 "Hainan has higher temperatures in summer". Combining D_1, D_2 and K_1, it can be deduced that today's temperature should be high. Thus, data resource D_{0_1} is supported. When the data resource D_{0_1} has supporting D_{DIKW} but the data resource D_{0_2} does not have supporting D_{DIKW}, it tends to determine that D_{0_1} is correct but D_{0_2} is wrong, thereby solving the multi-semantic.

$$\begin{aligned} &D_1 = (T_{SEASON}(Summer)) \\ &D_2 = (T_{PLACE}(Hainan)) \\ &K_1 = R_{IS}(R_{IN}(T_{PLACE}(Hainan),\ T_{SEASON}(Summer)),\ T_{TEMP}(High)) \end{aligned} \tag{23}$$

$$\begin{aligned} D_1 + D_2 + K_1 \rightarrow D_{new_1} &= (T_{TEMP}(High)) \\ K_{new_1} &= R_{SUPPORT}(D_{new_1},\ D_{0_1}) \end{aligned} \tag{24}$$

Algorithm. (1) Find conflicting data resources D_{0_1} and D_{0_2}.

(2) Search related data resources $D_{related}$ in the data graph.

(3) Combine $D_{related}$ with related I_{DIKW} and K_{DIKW}, further derives new data resource D_{new} that can help judge the truth.

(4) Determine the relationship between D_{new} and D_{0_1} with D_{0_2}. Keep the result supported by D_{new}, and delete the other result.

(5) Set the result supported by D_{new} as the final result to solve multi-semantic.

Solve by Adding I_{DIKW}. Obtain information resource I_1 "Data resource D_{0_1} comes from the Bureau of Meteorology". Information resource I_2 "Data resource D_{0_2} comes from the Internet". Knowledge resources K_2 "Data from professional institutions is more reliable than data from the Internet". Combining information resources I_1, I_2 and knowledge resources K_2, it can be deduced that data resources D_{0_1} are more reliable than data resources D_{0_2}. Thus, it can be judged that D_{0_1} is correct but D_{0_2} is wrong, thereby solving the multi-semantic.

$$\begin{aligned} &I_1 = R_{FROM}(D_{0_1},\ T_{INSTITUTE}(INS(MeteorologicalBureau)) \\ &I_2 = R_{FROM}(D_{0_2},\ T_{INTERNET}(INS(Website)) \\ &K_2 = R_{RELIABLE\_THAN}(T_{DATA}(R_{FROM}(data,\ T_{INSTITUTE})), \\ &\quad T_{DATA}(R_{FROM}(data,\ T_{INTERNET}))) \end{aligned} \tag{25}$$

$$\begin{aligned} I_1 + I_2 + K_2 \rightarrow I_{new_1} &= R_{RELIABLE\_THAN}(D_{0_1},\ D_{0_2}) \\ K_{new_2} &= R_{SUPPORT}(I_{new_1},\ D_{0_1}) \end{aligned} \tag{26}$$

Algorithm. (1) Find conflicting data resources D_{O_1} and D_{O_2}.

(2) Search related information resources $I_{related}$ in the information graph.

(3) Combine $I_{related}$ with related I_{DIKW} and K_{DIKW}, further derives new information resource I_{new} that can help judge the truth.

(4) Determine the relationship between I_{new} and D_{O_1} with D_{O_2}. Keep the result supported by I_{new}, and delete the other result.

(5) Set the result supported by D_{new} as the final result to solve multi-semantic.

5 Cross-modal Transformation of TR_{DIKW}

Whether in the detection of multi-semantic phenomena, or in the process of solving the multi-semantics and increasing the related TR_{DIKW}, it is necessary to complete the cross-modal transformation from the original TR_{DIKW} to the new TR_{DIKW}. Therefore, this section focuses on the cross-modal transformation of TR_{DIKW}. The target TR_{DIKW} of transformation can be divided into D_{DIKW} and I_{DIKW}.

5.1 Transform TR_{DIKW} to Purposed D_{DIKW}

The target TR_{DIKW} of transformation is D_{DIKW} "user A's occupation".

$$D_0 = (A|T_{OCCUPATION}(INS(Student)) \tag{27}$$

There are three ways to derive D_0, by combining D_{DIKW} with K_{DIKW}, by combining I_{DIKW} with K_{DIKW}, and by combining D_{DIKW} and I_{DIKW} with K_{DIKW}.

Transform D_{DIKW} Combined with K_{DIKW} to Purposed D_{DIKW}. Obtain related data resource D_1 "user A is 10 years old". There is relevant knowledge resources K_1 "people younger than 15 should go to school". Combining D_1 and K_1, user A is 10 years old, and his age is less than 15 years old, I_{DIKW} "user A should go to school" is deduced. So the target D_{DIKW} of "user A's occupation is student" can be further deduced.

$$\begin{aligned} &D_1 = (A|T_{AGE}(10)) \\ &K_1 = R_{SHOULD}(T_{PERSON}(R_{LESS\ THAN}(T_{AGE},\ 15)),\ T_{ACT}(Education)) \end{aligned} \tag{28}$$

$$\begin{aligned} D_1 + K_1 \rightarrow I_{new_1} &= R_{SHOULD}(A,\ T_{ACT}(Education)) \\ I_{new_1} \rightarrow I_0 &= R_{IS}(A,\ T_{OCCUPATION}(INS(Student)) \\ I_0 \rightarrow D_0 &= (A|T_{OCCUPATION}(INS(Student)) \end{aligned} \tag{29}$$

Transform I_{DIKW} Combined with K_{DIKW} to Purposed D_{DIKW}. Obtain relevant information resources I_1 "User A often goes to school". I_2 "User A does not have a teacher qualification certificate". Knowledge resources K_2 "students and teachers need to go to school frequently". K_3 "teachers have teacher qualification certificates". Combining I_1 and K_2, since user A often goes to school, it can be known that user A is a student or teacher. Combining I_2 and K_3, since user A does not have a teacher qualification certificate, so I_{DIKW} "user A is not a teacher" is deduced. In the case that user A is a student or teacher and user A is not a teacher, the target D_{DIKW} of "user A's occupation is a student" can be further deduced.

$$\begin{aligned}
&I_1 = R_{GO_TO}(A,\ T_{PLACE}(INS(School)))\\
&I_2 = !R_{OWN}(A,\ T_{LICENCE}(INS(TeacherCertification)))\\
&K_2 = R_{GO_TO}(T_{OCCUPATION}(Student)\ AND\ T_{OCCUPATION}(Teacher),\\
&\quad T_{PLACE}(School))\\
&K_3 = R_{OWN}(T_{OCCUPATION}(Teacher),\ T_{LICENCE}(INS(TeacherCertification))
\end{aligned} \tag{30}$$

$$\begin{aligned}
I_1 + K_2 \rightarrow I_{new_2} &= R_{IS}(A,\ T_{OCCUPATION}(INS(Student)\ OR\\
&\quad T_{OCCUPATION}(INS(Teacher))\\
I_2 + K_3 \rightarrow I_{new_3} &= !R_{IS}(A,\ T_{OCCUPATION}(INS(Teacher))\\
I_{new_2} + I_{new_3} \rightarrow I_0 &= R_{IS}(A,\ T_{OCCUPATION}(INS(Student))\\
I_0 \rightarrow D_0 &= (A|T_{OCCUPATION}(INS(Student))
\end{aligned} \tag{31}$$

Transform D_{DIKW} and I_{DIKW} Combined with K_{DIKW} to Purposed D_{DIKW}. Obtain relevant data resource D_1 "User A is 10 years old this year". Information resource I_1 "User A often goes to school". Knowledge resources K_2 "students and teachers need to go to school frequently". K_4 "teachers' age is generally greater than 20". Combining I_1 and K_2, due to user A often goes to school, it can be known that user A is a student or teacher. Combining D_1 and K_2, since user A is 10 years old this year, and the teacher's age is generally greater than 20 years old, I_{DIKW} "user A is not a teacher" is deduced. In the case that user A is a student or teacher, and user A is not a teacher, the target D_{DIKW} "user A's occupation is a student" can be further deduced.

$$\begin{aligned}
&D_1 = (A|T_{AGE}(10))\\
&I_1 = R_{GO_TO}(A,\ T_{PLACE}(INS(School)))\\
&K_2 = R_{GO_TO}(T_{OCCUPATION}(Student)\ AND\ T_{OCCUPATION}(Teacher),\\
&\quad T_{PLACE}(School))\\
&K_4 = R_{GREATER_THAN}(T_{AGE}(T_{OCCUPATION}(Teacher)),\ 20)
\end{aligned} \tag{32}$$

$$\begin{aligned}
I_1 + K_2 \rightarrow I_{new_2} &= R_{IS}(A,\ T_{OCCUPATION}(INS(Student)\ OR\\
&\quad T_{OCCUPATION}(INS(Teacher))\\
D_1 + K_4 \rightarrow I_{new_3} &= !R_{IS}(A,\ T_{OCCUPATION}(INS(Teacher))\\
I_{new_2} + I_{new_3} \rightarrow I_0 &= R_{IS}(A,\ T_{OCCUPATION}(INS(Student))\\
I_0 \rightarrow D_0 &= (A|T_{OCCUPATION}(INS(Student))
\end{aligned} \tag{33}$$

5.2 Transform TR_{DIKW} to Purposed I_{DIKW}

The target TR_{DIKW} of transformation is I_{DIKW} "User A likes to play soccer":

$$I_0 = R_{LIKE}(A,\ T_{ACT}(INS(PlaySoccer)) \tag{34}$$

There are three ways to derive I_0, deriving from D_{DIKW} with K_{DIKW}, deriving from I_{DIKW} with K_{DIKW}, and deriving from D_{DIKW} and I_{DIKW} with K_{DIKW}.

Transform D_{DIKW} Combined with K_{DIKW} to Purposed I_{DIKW}. Obtain user A's related spatial data resource D_1 "soccer field". Knowledge resources K_1 "The main activity in the soccer field is playing soccer" K_2 "People who often play soccer like to play soccer". Combining D_1 and K_1, since user A often appears in the soccer field, so user A often plays soccer. Combined with K_2, since user A often plays soccer, and people who often play soccer are likely to like to play soccer, which can further derive the target I_{DIKW} "User A likes to play soccer".

$$\begin{aligned} &D_1 = (A|T_{PLACE}(INS(SoccerCourt)) \\ &K_1 = R_{IN}(T_{ACT}(Soccer),\ T_{PLACE}(SoccerCourt)) \\ &K_2 = R_{LIKE}(T_{PERSON}(R_{DO}(person,\ T_{ACT}(Soccer)))T_{ACT}(Soccer)) \end{aligned} \tag{35}$$

$$\begin{aligned} D_1 + K_1 \rightarrow I_{new_1} &= R_{DO}(A|T_{ACT}(INS(Soccer))) \\ I_{new_1} + K_2 \rightarrow I_{0_{im}} &= R_{LIKE}(A,\ T_{ACT}(INS(Soccer)) \end{aligned} \tag{36}$$

Transform I_{DIKW} Combined with K_{DIKW} to Purposed I_{DIKW}. Obtain information resource I_1 "User A is a member of school's soccer team". Knowledge resources K_2 "People who often play soccer like to play soccer". K_3 "Members of school's soccer team often play soccer". Combining I_1 and K_3, since user A is a member of school's soccer team, so user A often plays soccer. Combined with K_2, since user A often plays soccer, and people who often play soccer are likely to like to play soccer, which can further derive the target I_{DIKW} "User A likes to play soccer".

$$\begin{aligned} &I_1 = R_{IS\_A\_MEMBER\_OF}(A,\ T_{GROUP}(INS(SoccerTeam)) \\ &K_2 = R_{LIKE}(T_{PERSON}(R_{DO}(person,\ T_{ACT}(Soccer))),\ T_{ACT}(Soccer)) \\ &K_3 = R_{DO}(T_{PERSON}(R_{IS\_A\_MEMBER\_OF}(person,\ T_{GROUP}(SoccerTeam)), \\ &\quad T_{ACT}(Soccer)) \end{aligned} \tag{37}$$

$$\begin{aligned} I_1 + K_3 \rightarrow I_{new_1} &= R_{DO}(A,\ T_{ACT}(INS(Soccer))) \\ I_{new_1} + K_2 \rightarrow I_{0_{im}} &= R_{LIKE}(A,\ T_{ACT}(INS(Soccer)) \end{aligned} \tag{38}$$

Transform D_{DIKW} and I_DIKW Combined with K_{DIKW} to Purposed I_{DIKW}. Obtain user A's related reading data resource D_2 "soccer news", and information resource I_2 "user A likes sports". Knowledge resources K_4 "People who often watch soccer news are interested in soccer sports events". K_5 "Sports include playing soccer, basketball and so on". Combining D_2 and K_4, since user A often reads soccer news, so user A is interested in soccer matches. Because user A's interest in soccer may only lie in watching soccer matches, the I_{DIKW} "user A is interested in soccer matches" cannot directly infer that user A likes to play soccer. Combining I_2 and K_5, since user A likes sports, and sports include playing soccer. Because user A may be more interested in playing basketball and other sports, this I_{DIKW} is not enough to directly infer that user A likes playing soccer. However, since it was previously deduced that user A is interested in soccer matches, combined with the I_{DIKW} "user A likes sports", the target I_{DIKW} "user A likes playing soccer" can be derived.

$$\begin{aligned} D_2 &= (A|T_{NEWS}(Soccer)) \\ I_2 &= R_{LIKE}(A|T_{ACT}(INS(SportsActivity))) \\ K_4 &= R_{INTERESTED\_IN}(T_{PERSON}(R_{READ}(person,\ T_{NEWS}(Soccer))), \\ &\quad T_{SPORTS}(Soccer)) \\ K_5 &= R_{INCLUDE}(T_{ACT}(SportsActivity),\ T_{ACT}(Soccer,\ Basketball,\ ...)) \end{aligned} \tag{39}$$

$$\begin{aligned} D_2 + K_4 &\rightarrow I_{new_2} = R_{INTERESTED\_IN}(A,\ T_{SPORTS}(Soccer)) \\ I_2 + I_{new_2} + K_5 &\rightarrow I_0 = R_{LIKE}(A,\ T_{ACT}(INS(Soccer)) \end{aligned} \tag{40}$$

6 Encoding Mode

The user behavior content that is easy to be observed mainly includes motion content and sound content. This section mainly discusses modeling and encoding modes for these two types of user behavior contents.

6.1 Encoding of Motion Content

Motion content is not limited to the user's overall movement, but also includes the movement of a single part of the user, such as hands, feet, head, etc., and the combined movement of multiple parts. There are two commonly used methods of observing and capturing the motion content. One is to directly record the motion content of various parts of the human body through a wearable device. The other is to collect motion image data through a camera, and then recognize the motion content. The motion content can be divided into D_{DIKW} and I_{DIKW}.

Encode to D_{DIKW}. Motion content can contain a variety of D_{DIKW}, which can be divided into hand movement data resource D_{hand}, foot movement data resource D_{feet}, head movement data resourceD_{head}, body movement Data resource D_{body} and so on. According to the type of D_{DIKW}, it can be divided into scalar data resource D_{scalar} and vector data resource D_{vector}.

Encode to Scalar D_{DIKW}. Scalar D_{DIKW} include but are not limited to distance $D_{distance}$, speed D_{speed}, acceleration $D_{acceleration}$, and so on.

$$\begin{aligned} D_{distance} &= DISTANCE_{observerd} \\ D_{speed} &= SPEED_{observerd} \end{aligned} \tag{41}$$

Scalar D_{DIKW} can be transformed into each other. For example, speed can be obtained by dividing distance by time, acceleration can be obtained by dividing the change in speed by time, and so on.

$$\begin{aligned} D_{speed} &= \frac{D_{distance}}{D_{time}} \\ D_{acceleration} &= \frac{D_{speed_{new}} - D_{speed_{old}}}{D_{time}} \end{aligned} \tag{42}$$

Encode to Vector D_{DIKW}. Vector D_{DIKW} include but are not limited to the direction of motion $D_{direction}$, the location of motion $D_{location}$, and so on.

$$\begin{aligned} D_{direction} &= DIRECTION_{observed} \\ D_{location} &= LOCATION_{observed} \end{aligned} \tag{43}$$

Vector D_{DIKW} can also be transformed into each other, for example, the direction of movement can be obtained from the change of position.

$$D_{direction} = D_{loaction_{new}} - D_{location_{old}} \tag{44}$$

Encode to I_{DIKW}. Motion content also contains a variety of I_{DIKW}. I_{DIKW} are obtained by combining the detected D_{DIKW} with purpose. These I_{DIKW} can be generally recognized movement information such as smile, finger snapping, clapping, etc., or movement information defined by the user, such as a specific gesture or a certain dance. For example, I_{smile} "smile" is an I_{DIKW} expressed by mouth movement. When it is detected that the movement of the corner of the mouth is obliquely upward, and the distance of the movement is about 1 cm, it can be determined that the corner of the user's mouth is raised to express the I_{DIKW} "smile".

$$\begin{aligned} D_{direction} &= (lip|T_{DIRECTION}(angle\_upward)) \\ D_{distance} &= (lip|T_{DISTANCE}(1cm)) \\ I_{smile} &= R_{COMBINE}(D_{direction}, D_{distance}) \end{aligned} \tag{45}$$

6.2 Encodeing of Sound Content

Sound content is not only limited to meaningful speech such as dialogue, but also includes sounds that have no practical meaning but such as pure music singing, which contains different kinds of pitches, timbres and volumes. The observation and capture of sound content mainly rely on microphones and other equipment to directly record the audio content. The sound content can be divided into D_{DIKW} and I_{DIKW}.

Encode to D_{DIKW}. The sound content can contain a variety of D_{DIKW}. According to the characteristics of the sound, it can be divided into pitch data resource D_{pitch} corresponding to the frequency of the audio, timbre data resource D_{timbre} corresponding to the waveform of the audio, volume data resource

D_{volume} corresponds to the loudness of audio and so on. The collection of sound D_{DIKW} can be divided into continuous form and discrete form.

The continuous form means that the observed value is directly assigned to the sound D_{DIKW} without any processing.

$$\begin{aligned} D_{pitch}(continuity) &= PITCH_{observed} \\ D_{volume}(continuity) &= VOLUME_{observed} \\ &\ldots \end{aligned} \tag{46}$$

Discrete form refers to setting a certain threshold k. When the observed value exceeds the threshold, the D_{DIKW} is set to a certain value, otherwise it is set to another value.

$$D_{pitch}(dispersed) = \begin{cases} PITCH_1 & PITCH_{observe} > PITCH_k \\ PITCH_2 & PITCH_{observe} \leq PITCH_k \end{cases}$$

$$D_{volume}(dispersed) = \begin{cases} VOLUME_1 & VOLUME_{observe} > VOLUME_k \\ VOLUME_2 & VOLUME_{observe} \leq VOLUME_k \end{cases} \tag{47}$$

Encode to I_{DIKW}. The sound content also contains a variety of I_{DIKW}. Regardless of the specific semantics of the voice, different information resources can also be expressed from the characteristics of the voice. For example, if it is recognized that the pitch of thc voice is higher than usual, the volume is higher usual, and the tone is very different from usual. It can be determined that the user's mood has fluctuated greatly, and the voice may express the I_{DIKW} "anger".

$$\begin{aligned} D_{pitch} &= (User, T_{PITCH}(high)) \\ D_{volume} &= (User, T_{VOLUME}(high)) \\ D_{timbre} &= (User, T_{TIMBRE}(different)) \\ I_{angry} &= R_{COMBINE}(D_{timbre}, R_{COMBINE}(D_{pitch}, D_{volume})) \end{aligned} \tag{48}$$

7 Decoding Mode

After the process of encoding, the D_{DIKW} and I_{DIKW} obtained may not be used directly. It is necessary to further decode the encoded D_{DIKW} and I_{DIKW} to obtain the target D_{DIKW} and I_{DIKW}. This section discusses the decoding modes of transforming the encoded TR_{DIKE} into target D_{DIKW} and I_{DIKW}.

7.1 Decode to D_{DIKW}

If the target TR_{DIKW} of decoding is D_{DIKW}, it can be obtained from the encoded D_{DIKW} or I_{DIKW}.

Transform Encoded D_{DIKW} to Target D_{DIKW}. In the case where the target D_{DIKW} is obtained from the encoded D_{DIKW}, the target D_{DIKW} and the encoded D_{DIKW} may be in the same modality, or in different modalities.

Transformation of D_{DIKW} of Same Modality. If the modalities of both D_{DIKW} are the same, it means that the type and dimension of the D_{DIKW} are same. Then the transformation process can be regarded as a homomorphic mapping f, mapping the encoded D_{DIKW} to the target D_{DIKW}. For example, assuming that the encoded data resource D_{raw} and the target data resource $D_{purpose}$ are one-dimensional continuous parameters. The value range of D_{raw} is 0 to 10000. The value of $D_{purpose}$ ranges from 0 to 100. Then, by scaling D_{raw} by one percent, the target data resource $D_{purpose}$ can be transformed.

$$\begin{aligned} D_{raw} &= (T_{num}) \\ D_{purpose} &= (T_{num}) \\ f &: T_{num} \rightarrow T_{num} \\ f(x) &= \frac{x}{100} \\ f(D_{raw}) &= D_{purpose} \end{aligned} \tag{49}$$

Transformation of D_{DIKW} of Different Modalities. For the transformation between D_{DIKW} of different modalities, the transformation process can be regarded as first performing cross-modal inference g on the encoded data resource D_{raw} to obtain the D_{DIKW} which is in same modal with target data resource $D_{purpose}$. Then performing homomorphic mapping f to obtain the target D_{DIKW}. The different modalities can be specifically divided into different data types and different data dimensions.

For different data types situation, type conversion is required. In the case that the encoded data resource D_{raw} is a numerical type and the target data resource $D_{purpose}$ is a logical type, a threshold value k can be set. And the D_{raw}, whose value is greater than k, is set to $true$, while the D_{raw}, whose value is less than or equal to k, is set to $false$. Then, the transformation from numerical data to logical data can be completed.

$$\begin{aligned} D_{raw} &= (T_{num}) \\ D_{purpose} &= (T_{logic}) \\ g &: T_{num} \rightarrow T_{logic} \\ g(x) &= \begin{cases} true & x > k \\ false & x \leq k \end{cases} \\ f &: T_{logic} \rightarrow T_{logic} \\ f(g(D_{raw})) &= D_{purpose} \end{aligned} \tag{50}$$

For different dimensions situation, dimensional compression or expansion is required. If the encoded data resource D_{raw} is three-dimensional data, and the target data resource $D_{purpose}$ is two-dimensional data. Then by design a mapping from three-dimensional space to two-dimensional space g, the dimensional compression from D_{raw} to $D_{purpose}$ can be completed.

$$\begin{aligned}
D_{raw} &= (T_{num}(dimension = 3)) \\
D_{purpose} &= (T_{num}(dimension = 2)) \\
g &: R^3 \rightarrow R^2 \\
g(x, y, z) &= (g_1(x, y, z), g_2(x, y, z)) \\
g_1(x, y, z) &= x + y + z \\
g_2(x, y, z) &= x * y * z \\
f &: T_{num} \rightarrow T_{num} \\
f(g(D_{raw})) &= D_{purpose}
\end{aligned} \tag{51}$$

Transform Encoded I_{DIKW} to Target D_{DIKW}. The encoded I_{DIKW} can also be transformed into target D_{DIKW}. Target D_{DIKW} can be divided into logical D_{DIKW} and numerical D_{DIKW}.

Transform Encoded I_{DIKW} to Logical D_{DIKW}. If the target D_{DIKW} is a logical D_{DIKW}, a specific I_{DIKW} can be associated with a specific logical expression. For example, the I_{DIKW} "laugh" represents *true*, and the I_{DIKW} "crying" represents *false*. Through this association between I_{DIKW} and logical expressions, the transformation from I_{DIKW} to target logical D_{DIKW} can be completed.

$$\begin{aligned}
f &: I \rightarrow D_{logic} \\
f(I) &= \begin{cases} true & I == I_{smile} \\ false & I == I_{cry} \end{cases} \\
f(I_{raw}) &= D_{purpose}
\end{aligned} \tag{52}$$

Transform Encoded I_{DIKW} to Numerical D_{DIKW}. If the target D_{DIKW} is a numerical D_{DIKW}, a specific I_{DIKW} can be associated with a specific value. For example, the I_{DIKW} "clap" represents "10", and clap twice represents "20". Through this association between I_{DIKW} and numerical values, the transformation from I_{DIKW} to target logical D_{DIKW} can be completed.

$$\begin{aligned}
f &: I \rightarrow D_{num} \\
f(I) &= 10 \quad (I = I_{clap}) \\
f(I_{raw}) &= D_{purpose}
\end{aligned} \tag{53}$$

7.2 Decode to I_{DIKW}

If the target of decoding is I_{DIKW}, it can only be obtained by transforming the encoded I_{DIKW} to target I_{DIKW}. For the transformation between I_{DIKW}, it is necessary to establish an association between the encoded I_{DIKW} and the target I_{DIKW}. For example, by establishing a connection between the I_{DIKW} "snap finger" and the target I_{DIKW} "complete a certain operation", the transformation from encoded I_{DIKW} to the target I_{DIKW} can be completed.

$$\begin{aligned}
f &: I \rightarrow I \\
f(I) &= I_{specified} \quad (I = I_{snap}) \\
f(I_{raw}) &= I_{purpose}
\end{aligned} \tag{54}$$

8 Conclusion

We propose to extend mesoscience practice to the field of information technology, conceptualize the mesoscale category, and carry out cross-modal and cross-scale semantic definition, modeling, analysis and measurement design towards identification and dealing with the complexity of objective uncertainty and subjective uncertainty in both conceptual category and cognitive category and the interaction category of objective and subjective. From the existence computation and reasoning paradigm and relationship defined everything of semantics expression mechanism, we model the essential semantic content based on the previous modeling approaches centering data graph, information graph, and knowledge graph, and analyze the reasons for the occurrence of multi-semantic situations. The reasons for the multi-semantic situation can be divided into two categories: the missing of typed resources and the redundancy of typed resources. Each category can be specifically subdivided into missing data resources, missing information resources, redundant data resources, and redundant information resources. For each specific situation, we give a variety of solutions, such as adding data resources or adding information resources, and design corresponding algorithms for each solution. We have also conducted in-depth exploration on the cross-modal transformation of type resources in the process of multi-semantic analysis, and discussed multiple transformation methods for target resources.

Relying on the cognitive semantic tracing and conceptualization strategies at the relevant existential level, the cross-modal and cross-scale core concept cognitive semantics of the DIKW model was systematically conceptualized and analyzed, and the construction of the DIKW systematic meta-model system was practiced. The category defines the formal semantics of DIKW core concepts in a way that multiple core concepts are conceptualized at the same time. It is proposed to expand the knowledge graph into Data Graph, Information Graph, Knowledge Graph and Wisdom Graph based on DIKW, which will help to promote the recognition of multi-modal mixed content by knowledge graph. Research on representation, modeling and optimization processing.

References

1. Appelt, D.E., Hobbs, J.R., Bear, J., Israel, D.J., Tyson, M.: A finite-state processor for information extraction from real-world text. In: PIJCAI (13). Chambéry, France, 28 August–3 September 1993, pp. 1172–1178 (1993)
2. Chen, M., et al.: Data, information, and knowledge in visualization. IEEE Comput. Graph. Appl. **29**, 12–19 (2009)
3. Coffman, T., Greenblatt, S., Marcus, S.: Graph-based technologies for intelligence analysis. Commun. ACM **47**, 45–47 (2004)
4. Duan, Y., Shao, L., Yang, X., Sun, X., Zhou, Z., Yu, L.: Data, information, and knowledge-driven manipulation between strategical planning and technical implementation for wireless sensor network construction. Int. J. Distrib. Sens. Netw. **13**(11) (2017)

5. Duan, Y., Sun, X., Che, H., Cao, C., Li, Z., Yang, X.: Modeling data, information and knowledge for security protection of hybrid IoT and edge resources. IEEE Access **7**, 99161–99176 (2019)
6. Duan, Y., Lu, Z., Zhou, Z., Sun, X., Wu, J.: Data privacy protection for edge computing of smart city in a DIKW architecture. Eng. Appl. Artif. Intell. **81**, 323–335 (2019)
7. Duan, Y.: Towards a periodic table of conceptualization and formalization on state, style, structure, pattern, framework, architecture, service and so on. In: SNPD 2019, pp. 133–138 (2019)
8. Duan, Y.: Existence computation: revelation on entity vs. relationship for relationship defined everything of semantics. In: SNPD 2019, pp. 139–144 (2019)
9. Duan, Y., Shao, L., Hu, G.: Specifying knowledge graph with data graph, information graph, knowledge graph, and wisdom graph. Int. J. Softw. Innov. **6**(2), 10–25 (2018)
10. Duan, Y., Cruz, C.: Formalizing semantic of natural language through conceptualization from existence. In: IJIMT, p. 6 (2011)
11. Duan, Y., Shao, L., Hu, G., Zhou, Z., Zou, Q., Lin, Z.: Specifying architecture of knowledge graph with data graph, information graph, knowledge graph and wisdom graph.. In: IEEE 15th (SERA), London 2017, pp. 327–332 (2017)
12. McSherry, F.: Privacy integrated queries: an extensible platform for privacy preserving data analysis. In: Proceedings of the ACM SIGMOD 2009, Providence, Rhode Island, USA, 29 June 29–2 July 2009, pp. 19–30 (2009)
13. Pujara, J., Miao, H., Getoor, L., Cohen, W.: Knowledge graph identification. In: Alani, H., et al. (eds.) ISWC 2013. LNCS, vol. 8218, pp. 542–557. Springer, Heidelberg (2013). https://doi.org/10.1007/978-3-642-41335-3_34
14. Xiao, H., Huang, M., Zhu, X.: A generative model for interpretable knowledge graph embedding (2016). http://arxiv.org/abs/1608.07685
15. Xu, J., Chen, K., Qiu, X., Huang, X.: Knowledge graph representation with jointly structural and textual encoding (2016). http://arxiv.org/abs/1611.08661

Distributed Reinforcement Learning with States Feature Encoding and States Stacking in Continuous Action Space

Tianqi Xu[1], Dianxi Shi[1,2(✉)], Zhiyuan Wang[1], Xucan Chen[1(✉)], and Yaowen Zhang[1]

[1] Artificial Intelligence Research Center (AIRC), National Innovation Institute of Defense Technology (NIIDT), Beijing 100071, China
dxshi@nudt.edu.cn, xcchen18@139.com

[2] Tianjin Artificial Intelligence Innovation Center (TAIIC), Tianjin 300457, China

Abstract. The practical application of reinforcement learning agents is often bottlenecked by the duration of training time. To accelerate training, practitioners often turn to distributed reinforcement learning architectures to parallelize and accelerate the training process. This work, we utilize the distributed reinforcement learning architecture to deal with continuous control tasks. The Importance Weighted Actor-Learner Architectures (IMPALA) decouples the acting and learning process to reduce queuing time. IMPALA attains higher scores on the new DMLab-30 set and the Atari-57 set because of its high performance, good scalability, and high efficiency. We extend IMPALA on the continuous control tasks with three changes. We encoder states into low dimensional data to establish an action distribution function that the agents have the ability to exploit and explore. A queue buffer is used to store a mini-batch data and discard them after training. In order to make the agent take appropriate action in the continuous control environment, we stack the past three steps states that attempt to make the robot moves smoothly. Finally, experiments are carried out on Mujoco tasks. The results show that our work is better than other distributed reinforcement learning algorithms.

Keywords: Distributed reinforcement learning · Importance sampling · Continuous control task · Scalable agent

1 Introduction

Deep reinforcement learning methods have recently achieved great success in a wide variety of domains through trial and error learning [3,9,11,13,14,19].

This work was supported in part by the Key Program of Tianjin Science and Technology Development Plan under Grant No. 18ZXZNGX00120 and in part by the China Postdoctoral Science Foundation under Grant No. 2018M643900.

H. Gao et al. (Eds.): CollaborateCom 2020, LNICST 349, pp. 340–353, 2021.
https://doi.org/10.1007/978-3-030-67537-0_21

While, one of the most important factors restricting the effect of reinforcement learning is the low efficiency of reinforcement learning sampling from the environment, especially in the complex terrain environment, simulation and mechanical control environment. Taking the StarCraft II environment [18] as an example, it consumes 3 to 4 s to start a StarCraft environment or reset a StarCraft environment. The maximum acceleration is about 16 times faster without turning on StarCraft graphics rendering. This works out to about five to ten seconds for training an episode, whereas training an intensive learning model would require tens of thousands or even hundreds of thousands of episodes, which is unacceptable.

In reinforcement learning scenarios [15], there is no static tag information available in advance, usually it's necessary to introduce the actual target system to obtain feedback information, which can be used to judge losses and gains, and then complete the closed-loop feedback of the entire training process [7]. A typical step is to collect feedback information and returns by observing the state of a specific target system. These information are used to adjust the parameters and train the model. According to the new training results, the output can be used to adjust the behavior of the target system and output to the target system. Then it affects the state change of the target system and completes the closed loop. In this way, the ultimate goal is to maximize total returns.

This work, we consider encoder the environment states, outputting the distribution of potential variables through a generation model [8], assuming that the distribution of potential variables is a Gaussian distribution, that the distribution of actions under the corresponding environmental state and updating the model through the probability and the return value of the action. In the environment, only through the data of the states, it can be fully quantified as the corresponding potential variables, so as to select the corresponding action a of the current states. By generating the distribution of potential variables, we can deal with high-dimensional input and continuous action space. We utilize a Scalable Distributed Deep-RL with Importance Weighted Actor-Learner Architectures [6]. The main task of the learner is updating the parameters of each neural network by obtaining the trajectory generated by actors to calculate stochastic gradient descent. Because the process of neural network training can be parallel, the learner uses GPU to calculate. The actor obtains the latest neural network parameters from the learner on a regular basis, and each actor starts up several simulation environments to use the latest policy synchronized by the learner and transmits the acquired trajectories back to the learner to update its neural network parameter. In order to make the agent take appropriate action in the continuous control environment, we stack the past three steps states that attempt to make the robot moves smoothly.

2 Related Work

DeepMind proposed asynchronous methods for deep reinforcement learning (A3C) [10]. It greatly improves the sampling efficiency by learning strategies simultaneously and asynchronously updating parameters through multiple actors. However, the A3C is generally inefficient in GPU utilization. The average server will have one or several GPUs, but the number of CPUs may reach dozens or even hundreds. Since each distributed Actor in the A3C algorithm needs to learn parameters, and it is obviously unrealistic to equip each actor with a GPU device, the A3C algorithm can only use CPU for training.

To take full advantage of the GPU, GA3C [1] proposed a hybrid CPU/GPU version of A3C. During training process of A3C, we need to copy networks for each parallel agent to collect samples and calculate the cumulative gradient. When the number of parallel agents is large, it consumes memory. GA3C algorithm module is mainly divided into agent, predictor, and trainer. Like A3C, the agent collects trajectories but does not need to copy a model separately. It only needs to add the current state as a request to the prediction queue before each action is selected. After n steps of the action are executed, the total returns of each step is calculated backward. The n(s_t, a_t, R, s_{t+1}) obtained are added to the training queue. Predictor will queue the request samples in the prediction queue as mini-batch and fill them into the GPU's network model, and return the actions predicted by the model to their respective agents. In order to reduce the delay, multi-threaded parallel multiple predictors can be used. Trainer uses the sample of training queue as mini-batch to fill in the network model of GPU to train and update the model. Similarly, in order to reduce the delay, multiple trainers can be used in parallel with multiple threads.

On the basis of the A3C algorithm, the Importance Weighted Actor-Learner Architectures (IMPALA) [6], a framework of large-scale reinforcement learning and training, which has high performance, good scalability, and high efficiency. Under the framework of large-scale computing, each Actor asynchronously samples, leading to the dislocation of sampled data and the implemented policy, instead of the complete on-policy sampling. In this case, the paper uses V-trace technology to eliminate the error of asynchronously sampled data.

However, these algorithms are limited to problems with a finite number of discrete actions. In control tasks, commonly seen in the robotics domain, continuous action spaces are the norm. For algorithms such as Deep Q Network [11] and A3C, the policy is only implicitly defined in terms of its value function, with actions selected by maximizing this function. In the continuous control domain this would require either a costly optimization step or discretization of the action space. While discretization is perhaps the most straightforward solution, this can prove a particularly poor approximation in high- dimensional settings or those that require finer grained control. Instead, a more principled approach is to parameterize the policy explicitly and directly optimize the long term value of following this policy.

3 Background

The problem of discounted infinite-horizon RL in Markov Decision Processes (MDP) is tough to solve. See [12] where the goal is to find a policy π that maximizes the expected sum of future discounted rewards: $V^\pi(x) = \mathbb{E}_\pi[\sum_{t\geq 0} \gamma^t r_t]$,

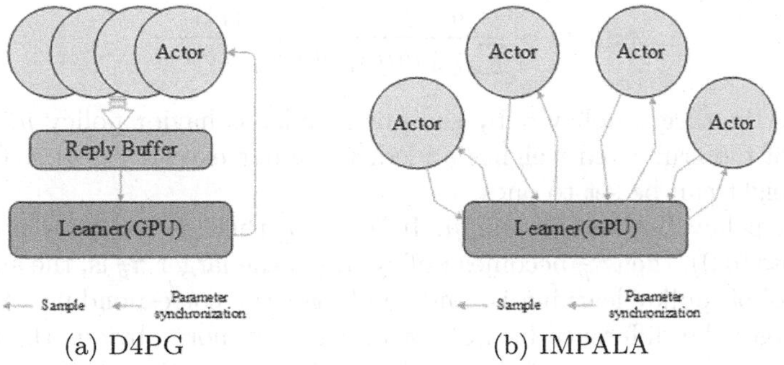

Fig. 1. Architecture schemes for D4PG and IMPALA. D4PG aggregates actors batch into a large training batch and the learner performs mini-batch SGD. IMPALA actors asynchronously generate data.

the $\gamma \in [0, 1)$ is the discount factor, $r_t = r(x_t, a_t)$ is the reward at time t, x_t is the state at time t (initial state $x_0 = x$), $a_t \sim \pi(\cdot|x_t)$ is the action generated by following some policy π. The goal of off-policy reinforcement learning algorithm is to learn the value function V^π of the target policy π using the trajectories generated by the behavior policy μ, which is usually different from the target policy.

3.1 Importance Weighted Actor-Learner Architectures

IMPALA decouples acting and learning by having the actor threads send actions Fig. 1, observations, and values while the learner thread computes and applies the gradients from a queue of actors experience. This will maximize GPU utilization and allow for increased sample throughput, resulting in high training speed on simpler environments such as Pong. With the increase of the number of learners, the worker policy begins to deviate from the learner policy, resulting in an obsolete policy gradient. The purpose of IMPALA is to get a better estimation of the current state value function $v_{(x)}$ following (1) based on the trajectories $(x_t, a_t, r_t)_{t=s}^{t=s+n}$ and the current state value function network.

$$v_s = V^\pi(x) + \sum_{t=s}^{s+n+1} \gamma^{t-s} (\prod_{i=s}^{t-1} c_i) \rho_t (r_t + \gamma V^\pi(x_t + 1) - V(x_t)) \qquad (1)$$

Where V^π is the value network, π is the target network of learner thread, μ is the behavior network of actor thread. ρ_t and c_i are defined as follows: $\rho_t = \min(\overline{\rho}, \frac{\pi(a_t|x_t)}{\mu(a_t|x_t)})$, $c_i = \min(\overline{c}, \frac{\pi(a_t|x_t)}{\mu(a_t|x_t)})$. They are truncated importance sampling (IS) weights.

ρ_t and c_i, as the the (truncated) IS weights, play different roles. ρ_t defines the fixed point of this update rule. Every time, updating state value function $V_\theta(x)$ toward $v(x)$, the state value function converged to is a value function between V^π and V^μ called $V_{\pi_{\overline{\rho}}}$. $\pi_{\overline{\rho}}$ is defined as

$$\pi_{\overline{\rho}}(a|x) = \frac{min(\overline{\rho}\mu(a|x), \pi(a|x))}{\sum_{b\in A} min(\overline{\rho}\mu(b|x), \pi(b|x))} \tag{2}$$

Update the target policy π by sampling under behavior policy μ. But with the limit of the truncated weights added, it cannot exceed $\overline{\rho}$ and $\overline{c}$. Generally, the two weight can be set to ones.

$\pi_{\overline{\rho}}$ is a policy between π and μ. If $\overline{\rho} = \infty$, then $\pi_{\overline{\rho}}$ becomes policy π, if $\overline{\rho} \to 0$ (close to 0), then $\pi_{\overline{\rho}}$ becomes policy μ. (so the larger $\pi_{\overline{\rho}}$ is, the smaller the deviation of off-policy learning is, and the larger the corresponding variance is.

The larger the difference between π and μ, the more obvious the off-policy is, the greater the variance of the results is. The truncation method is used to control the variance. $\overline{\rho}$ affects what value function to converge to, while $\overline{c}$ affects the speed of convergence to this value function.

It should be noted that v_s in the case of on-policy updating (i.e. $\pi = \mu$), and let $c_i = 1$, and $\rho_t = 1$, will become the following formula

$$v_s = V(x) + \sum_{t=s}^{s+n+1} \gamma^{t-s}(r_t + \gamma V(x_t+1) - V(x_t)) = \sum_{t=s}^{s+n+1} \gamma^{t-s} r_t + \gamma^n V(x_{s+n}) \tag{3}$$

V-trace targets can be calculated in an iterative way (4).

$$v_s - V(x) = \delta_t V + \gamma c_s(v_{s+1} - V(x_{s+1})) \tag{4}$$

3.2 Distributed Distributional DDPG [2]

The method adopted by Distributed Distributional DDPG (D4PG) is from the improvement of DDPG (Deep Deterministic Policy Gradient) algorithm, and contains some enhancement. These extensions include a distributional critic update, the use of distributed parallel actors, N-step returns, and prioritization of the experience replay.

First, and perhaps most crucially, it considers the inclusion of a distributional critic. The return of distributional update is a random variable Z_π, such that $Q_\pi(s,a) = \mathbb{E}Z_\pi(s,a)$. The distributional Bellman operator can be defined as

$$(\mathcal{T}_\pi Z)(s,a) = r(s,a) + \gamma\mathbb{E}[Z(s',\pi(s'))|s,a] \tag{5}$$

In order to use this function within the context of the actor-critic architecture introduced above, They parameterize this distribution and define a loss and write the loss as

$$L(\omega) = \mathbb{E}_\rho[d((T)_{\pi_{\theta'}} Z_{\omega'}(s,a), Z_\omega(s,a)] \tag{6}$$

Next, they utilize a modification to the DDPG update which utilizes N-step returns when estimating the TD error. Finally, they modify the standard training procedure in order to distribute the process of gathering experience.

4 IMPASS Algorithm

Importance Weighted Actor-Learner Architectures with States Stacking (IMPASS algorithm) extends IMPALA algorithm on the continuous control tasks with three changes. We encoder observations into low dimensional data to establish an action distribution function so that the agent has the ability to exploit and explore. A queue buffer is used to store a mini-batch data and discard them after training. In order to make the agent take appropriate action in the continuous control environment, we stack three steps past observations that attempt to make the robot move smoothly. Finally, experiments are carried out on Mujoco tasks. The results show that our work is better than the existing distributed reinforcement learning algorithms.

Algorithm 1. IMPASS

Input: batch size M, number of actors N, sample queue length K, learning rate schedule r, clipped important sample weight $\overline{\rho}$ and $\overline{c}$

Initialize network weights θ, ω at random
Initialize batch buffer m of size M
Launch K actors and replicate network weights θ, ω to each actor
Extract data from K to m
while True **do**
  for $t = 1 \to T$ **do**
    Get T trajectories $(S_{i:i+T}, a_{i:i+T}, \mu_{i:i+T}, \sigma_{i:i+T}, r_{i:i+T})$ from m
    Establish action distribution with μ and σ and calculate action probability
    Calculate probability of action a with target network
    Calculate the value function loss of critic network
$$\mathbb{E}[\frac{\pi(a_t|S_t)}{\mu(a_t|S_t)} \nabla \log \pi Q^{\mu}(S_t, a_t)]$$
    Calculate the policy gradient loss of actor network
$$\nabla_{\omega} \log \pi_{\omega}(a_t|S_t)(r_t + \gamma v_{t+1} - V_{\theta}(S_t))$$
    Calculate the entropy loss
$$-\nabla_{\omega} \sum_a \pi_{\omega}(a|S_t) \log \pi_{\omega}(a|S_t)$$
    Minimize the total loss
    Update network parameters $\theta \leftarrow \theta'$, $\omega \leftarrow \omega'$
    if $t = 0 \bmod t_{params}$ **then**
      Replicate network weights to the actors
    end if
  end for
end while

Actor-i starts synchronously with the Learner

repeat
  Samples action a from action distribution function
  Executes action a, gains reward r
  Transfers (S, a, μ, σ, r) to K
until learner finishes

4.1 States Feature Encoding

We consider to establish the action distribution function from the observations generated by actors, so adopt the method of feature encoding to get the potential variable distribution of input data. The probability distribution of observations is learned by a neural network, referring to the working principle of the Variational Auto-Encoder [8]. This kind of network is called auto-encoder. The probability distribution obtained is the action distribution of the agent in the current state. We assume that the probability distribution is Gaussian distribution.

In practical problems, the action space of many tasks is continuous, such as robot control. If there is a ready-made position controller, the expected output of the policy can be a position, which is naturally continuous. This is also applicable to speed control. In the continuous action space, it's unrealistic to calculate the probability of every action, because of infinite actions. Consequently, the problem can be solved to obtain a probability distribution of action in certain action spaces. At this point, we need to parameterize the policy, which represents the distribution of operations. The action probability distribution of continuous action space is usually a normal distribution, and its probability density function PDF is defined as

$$p(x) \doteq \frac{1}{\sigma\sqrt{2\pi}} \exp -\frac{(x-\mu)^2}{2\sigma^2} \tag{7}$$

A policy can be parameterized by means of mean and variance as

$$\pi(a|s,\theta) \doteq \frac{1}{\sigma(s,\theta)\sqrt{2\pi}} \exp -\frac{(a-\mu(s,\theta)^2}{2\sigma(s,\theta)^2} \tag{8}$$

This policy has two parameters $\theta = [\theta_\mu, \theta_\sigma]^T$, the parameters of the mean and variance, respectively. This two parameters are fitted by stochastic gradient descent. The selection of activation function in the final output layer must be a non-negative mapping, such as a soft-plus activation function.

After generating the probability density function of actions, actors can sample actions from the distributional function like stochastic policy as show in Fig. 3, which is a normal distribution. The mean and variance are all fitted by neural networks. Through this process, the model of mean and variance is optimized, but the operation of "sampling" is not differentiable and the result of sampling is differentiable. That sampling an action a from the distribution $p(x)$ (i.e. $N(\mu, \sigma^2)$) is equivalent to sampling an action a from the distribution $N(0, 1)$, then let $a = \mu + \epsilon * \sigma$.

We change the sampling from $N(\mu, \sigma^2)$ to $N(0, 1)$, and then get the result of sampling from $N(\mu, \sigma^2)$ by parameter transformation. Therefore, the operation of "sampling" does not need to participate in gradient descent. Instead, the result of sampling is involved, so that the whole model can be trained.

4.2 States Stacking

The agent chooses the appropriate action through the current environment, and obtains the reward value of the environmental feedback. Meanwhile, we usually expect the robot to move more smoothly in continuous control tasks. Having an agent retrospect the environmental observations it has just experienced may help it choose more appropriate actions. So we retain the observations of the past three steps in the state of each step as show in Fig. 2 that learned from DRQN's [4] approach to handling Partially Observable Markov Decision Process (POMDP). So the new environmental observations can be expressed as

$$S_t = S(s_{t-3}, s_{t-2}, s_{t-1}, s_t) \tag{9}$$

Next step, the new environment state s_{t+1} is joined in and the oldest data s_{t-3} will be discarded. There is no historical state to refer to when each environment is reset, so we choose to make four copies of the first step state as the environment initial state stack S_1.

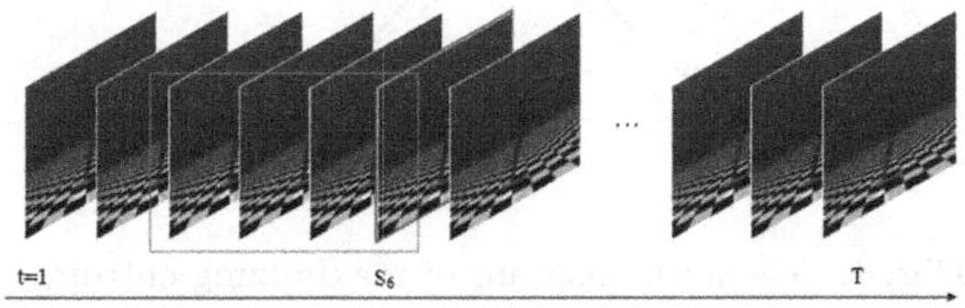

Fig. 2. The architecture of states stack

IMPASS uses a queue buffer to line the samples up that transmitted from different actors. The central learners constantly extract data from the queue to aggregate a batch of data for stochastic gradient descent.

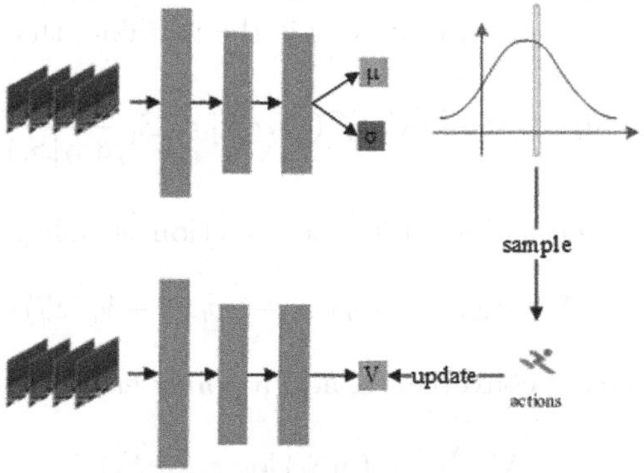

Fig. 3. The architecture of the network of actors and critics. The left figure shows the critic network, which has two layers of forward neural network. The right graph is the actor network, which encodes the observed values through two-layer network and outputs the mean and variance of Gaussian distribution.

4.3 Actor-Critic Method

Consider a parametric representation V_θ, of the value function and the current policy π_ω. Trajectories have been generated by actors following some behavior policy μ. The V-trace targets vs are defined by 1. At time t, the parameter θ of the value network is updated to target v_s by gradient descent on L2 loss. When updating critic network, the gradient direction is

$$(v_s - V_\theta(S_t))\nabla_\theta V_\theta(S_t) \tag{10}$$

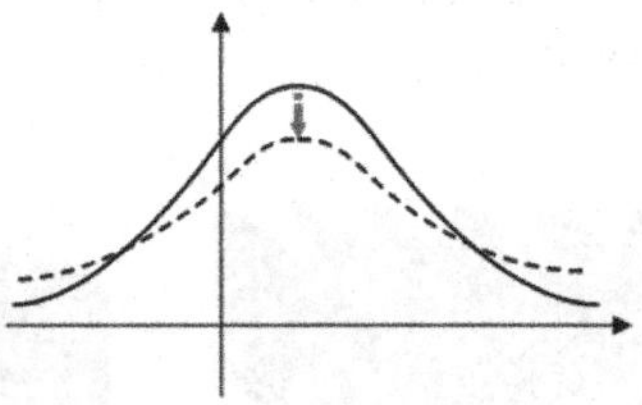

Fig. 4. Schematic diagram of maximizing entropy.

The actor network in Fig. 3 establishes the probability density function of the action distribution by fitting the means and variances of the environment state and obtains the action a of the current state S_t by sampling from the distribution.

The actor network will be updated in the direction of the off-policy learning (i.e. $\mathbb{E}_\mu[\log \mu(a_t|S_t)Q^\mu(S_t, a_t)]$). The goal of our algorithm is to learn target policy π, not behavior policy μ, so we need to replace $\mu \rightarrow (\frac{\mu}{\pi})\pi$ with the coefficient in the bracket, and the π after is the variable, that is

$$\mathbb{E}_\mu[log\mu(a_t|S_t)Q^\mu(S_t, a_t)] = \mathbb{E}[\frac{1}{\frac{\mu}{\pi}\pi}\nabla_\pi Q^\pi(S_t, a_t)] = \mathbb{E}[\frac{\pi(a_t|S_t)}{\mu(a_t|S_t)}\nabla \log \pi Q^\mu(S_t, a_t)] \tag{11}$$

Then, update the actor network in the direction of policy gradient

$$\rho_s = \nabla_\omega \log \pi_\omega(a_s|S_t)(r_t + \gamma v_{t+1} - V_\theta(S_t)) \tag{12}$$

To prevent premature convergence, add a policy entropy penalty to the gradient of the actor

$$-\nabla_\omega \sum_a \pi_\omega(a|S_t) \log \pi_\omega(a|S_t) \tag{13}$$

Before sampling from the distribution, we minimize the entropy multiplied by a coefficient. As show in Fig. 4, the sampling of actions will have more possibilities, which will increase the ability of exploration and avoid falling into local optimum. By adding the three gradients and converting them with appropriate coefficients (which are the hyper parameters of the algorithm), the global update is obtained.

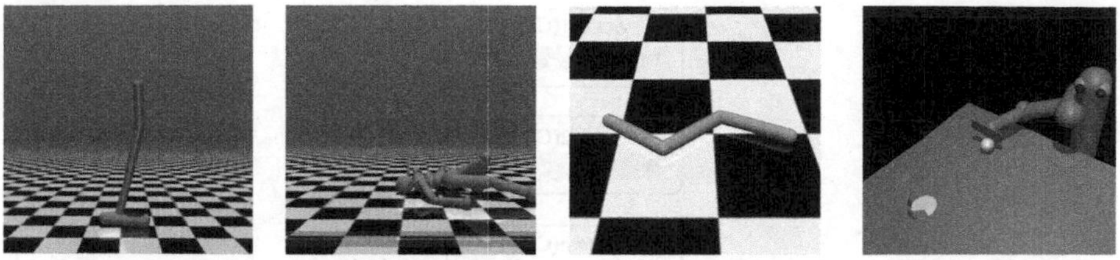

(a) Hopper (b) Humanoid (c) Swimmer (d) Striker

Fig. 5. Experiment environments of Mujoco

5 Experiment

This section will show the performance of the algorithm IMPASS in the continuous action space. Therefore, the training is carried out in each task and the experimental data is recorded regularly.

We first consider evaluating the performance of some simple physical control tasks using a set of benchmark tasks developed by the Mujoco physical simulator [16,17]. Run 1000 steps for each task, and provide instant and intensive rewards according to specific tasks $r_t = [0, 1]$ or sparse reward $r_t = 0, 1$. For each domain, the input presented to the agent consists of reasonable low-dimensional observations, many of which are composed of physical state, joint angle, etc. These observations range from 6 to 60 dimensions, but note that the difficulty of the task is not directly related to its dimensions. For example, Acrobot is one of the lowest dimensional tasks in this suite. Because of its controllability, it is more difficult to learn than other high-dimensional tasks.

We have carried out our experiments on the robot simulation suite Mujoco. Here we show the performance of our algorithm in four simulation scenarios. These four scenarios are Hopper, Humanoid (stand-up), Swimmer and Striker Fig. 5. The task in the Hopper scenario is to make a two dimensional one legged robot hop forward as fast as possible.Humanoid Make a three dimensional biped robot stand as fast as possible. This task involves a 3-link swimming robot in a viscous fluid, where the goal is to make it swim forward as fast as possible, by actuating the two joints. The origins of task can be traced back to Remi Coulom's thesis [5]. The purpose of the Striker task is to control the robot arm and strike a white ball into the goal with a white fence. In the initial state of the environment, the posture of the robot arm and the coordinates of the object are fixed values, and the coordinates of the goal are determined randomly. The task-achievement condition is that the XY coordinates of the object must fall within the goal range.

5.1 Learner-Actor Architecture

We use a scalable learner-actor architecture with multiple distributed actors to train the network. In a model like this, each actor interacts with the environment using a copy of the policy network parameters. The actor will periodically

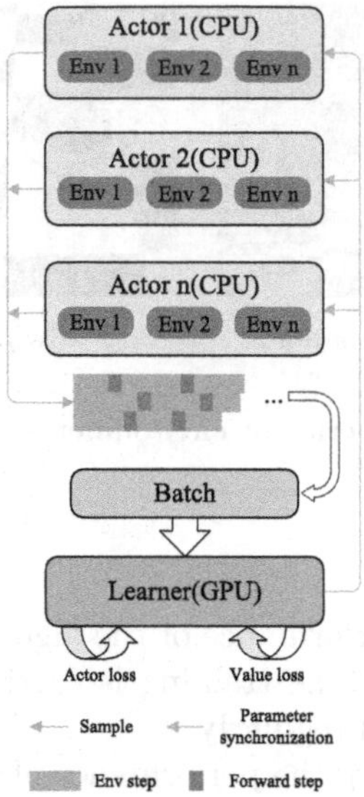

Fig. 6. The whole training process. Learner do back propagation on GPU. Each actor uses a CPU for environment inference, and each actor can run multiple simulation environments meanwhile.

transmits back the sampled data to the central parameter server, without affecting its pause of exploration, and the learners located in the central server will use these tracks to update network. Actors in this framework are not used to calculate policy network gradient. They just collect experiences and pass them on to the central learner. The learner uses a GPU device for stochastic gradient descent Fig. 6. Therefore, in such a architecture, the actor and the learner are work independent. In order to take advantage of the scale advantages of modern computing systems, IMPALA can support a single learner machine or multiple synchronous learner machines. Separating learning and acting in this way can also improve the throughput of the whole system because unlike batch A2C architecture, actors no longer need to wait for learning steps. This helps us train IMPALA in an interesting environment without having to face the difference in frame rendering time or task restart time delay.

Algorithm pseudocode for the IMPASS algorithm which includes all the above-mentioned modifications can be found in Algorithm 1.

5.2 Reinforcement Learning Framework

In order to improve the performance of the algorithm, we deploy the algorithm to a parallel reinforcement learning framework. PARL (Paddlepaddle Reinforcement Learning) is a high performance, flexible reinforcement learning framework. The framework supports large-scale parallel computing, and supports the training of multi-GPUs reinforcement learning model.

PARL provides a compact API for distributed training, allowing users to transfer the code into a parallelized version by simply adding a decorator. Call the decorated function to initialize the parallel communication. The instance obtained in this way has the same function as the original class. Because these classes are running on other computing resources, executing these functions no longer consumes the current thread computing resources. Real actors are running at the cpu cluster, while the learner is running at the local gpu with several remote actors.

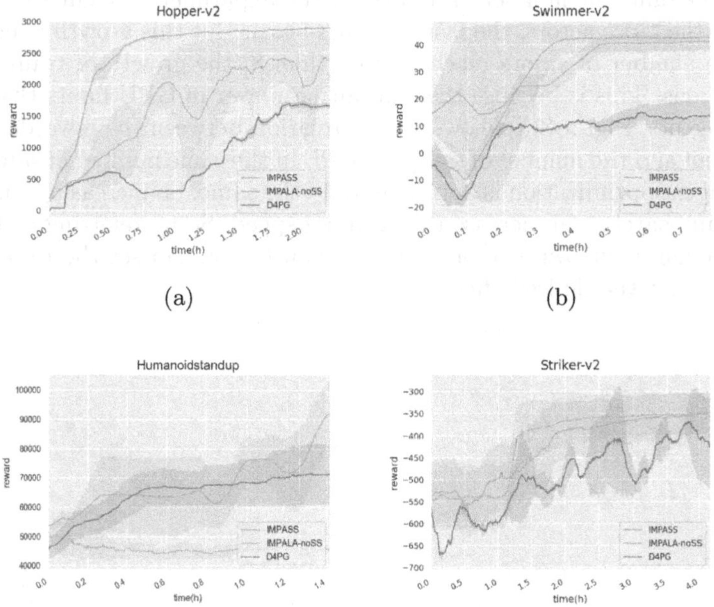

(a) (b)

(c) (d)

Fig. 7. Performance on different continuous control tasks: hopper, swimmer, striker and HumanStandup. The performance of IMPASS, IMPALA-noSS (IMPALA with states feature encoding and without states stacking) and D4PG. In order to show the fairness of the results, all experiments were run on the same computing platform, and the number of actor was 8.

6 Results

As shown in the Fig. 7, the performance of our algorithm in four Mujoco tasks. We use 8 actors and a single learner, and each actor runs 16 simulators in parallel. Our experiment is carried out on a computer that has one CPU with 16 cores and an Nvidia 1050Ti GPU. At the same time, our algorithm is compared with

D4PG, which is a distributed reinforcement learning algorithm too. In order to ensure the fairness of the experiment, this group of comparison experiments is also run on the same hardware device, and 8 actors are assigned. It can be seen that our agents are better than D4PG in most environments, and they can get a higher reward value. Even in the hopper environment where the performance advantage is not obvious, our agent is the first to reach the convergence state.

The learning speed of the central learner is closely related to the scale of the actor. Theoretically, the faster the sample return rate, the faster the learner can learn. But the computing performance of the GPU is limited. We also tested the effect of the different number of actors on the experimental results (Fig. 8). As expected, the more actors, the faster agents learn. But this growth is not linear. When the number of actors reaches more than 8, the growth of training speed is not obvious. This is because the computing power of GPU limits the training speed of agents. Figure 8(a) shows the correlation between the rewards obtained by the agent and the number of samples used. In the same number of samples, the speed of agent optimization is approximately the same. So increasing sample size and training batch, can increase the training speed. So by referring to Fig. 8(a), and combining your own computer performance, you can set the right number of actors to get the desired effect.

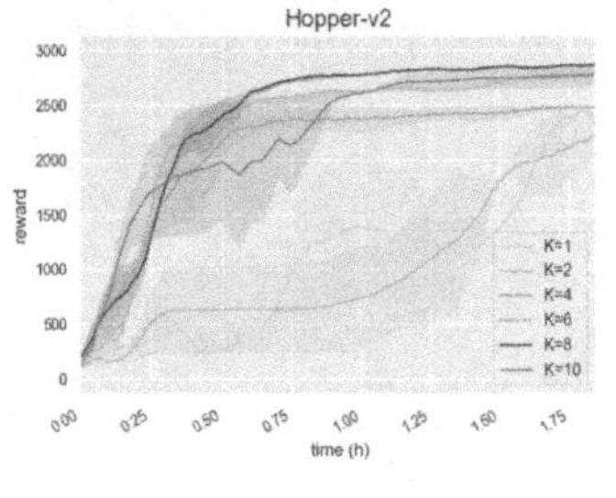

(a)

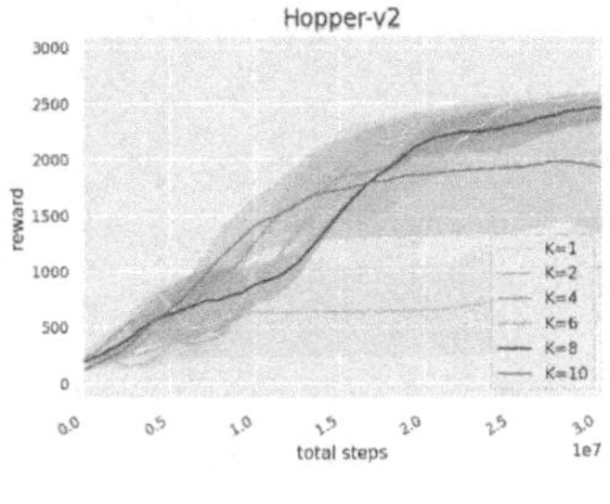

(b)

Fig. 8. Effect of different number of actors on experiments for both sample and time efficiency.

7 Conclusion

We have extended the highly scalable distributed agent, IMPALA, with a method of states feature encoding and states stacking. With its scalable distributed architecture, IMPASS can make efficient use of available computing device at small and large scale. It successfully shows that it has better exploration ability and more flexible expansibility in high-dimensional input space and continuous action space.

Through the error correction of v-trace, the asynchronous environment inference and network parameter synchronization of actor and learner are realized, which makes IMPASS have higher training speed and is superior to D4PG, which is also a distributed reinforcement learning algorithm optimized on the continuous action space. However, it also brings a problem that IMPASS has low sample utilization efficiency. At the same time, it gets a higher average reward value and with abundant environment frames, D4PG can gain higher average rewards with less sample capacity.

References

1. Babaeizadeh, M., Frosio, I., Tyree, S., Clemons, J., Kautz, J.: Reinforcement learning through asynchronous advantage actor-critic on a GPU. arXiv Learning (2016)
2. Barthmaron, G., et al.: Distributed distributional deterministic policy gradients. arXiv Learning (2018)
3. Barthmaron, G., et al.: Distributional policy gradients (2018)
4. Chen, C., Ying, V., Laird, D.: Deep q-learning with recurrent neural networks. Stanford Cs229 Course Report 4, 3 (2016)
5. Coulom, R.: Reinforcement learning using neural networks, with applications to motor control. Ph.D. thesis, Institut National Polytechnique de Grenoble-INPG (2002)
6. Espeholt, L., et al.: IMPALA: scalable distributed deep-RL with importance weighted actor-learner architectures. arXiv Learning (2018)
7. Kaelbling, L.P., Littman, M.L., Moore, A.W.: Reinforcement learning: a survey. J. Artif. Intell. Res. **4**(1), 237–285 (1996)
8. Kipf, T., Welling, M.: Variational graph auto-encoders. arXiv Machine Learning (2016)
9. Lillicrap, T., et al.: Continuous control with deep reinforcement learning (2016)
10. Mnih, V., et al.: Asynchronous methods for deep reinforcement learning, pp. 1928–1937 (2016)
11. Mnih, V., et al.: Human-level control through deep reinforcement learning. Nature **518**(7540), 529–533 (2015)
12. Puterman, M.L.: Markov Decision Processes: Discrete Stochastic Dynamic Programming. Wiley, Hoboken (2014)
13. Silver, D., et al.: Mastering the game of go with deep neural networks and tree search. Nature **529**(7587), 484–489 (2016)
14. Silver, D., et al.: Mastering the game of go without human knowledge. Nature **550**(7676), 354–359 (2017)
15. Sutton, R.S., Barto, A.G.: Reinforcement Learning: An Introduction (1999)
16. Tassa, Y., et al.: DeepMind control suite. arXiv e-prints arXiv:1801.00690 (2018)
17. Todorov, E., Erez, T., Tassa, Y.: MuJoCo: a physics engine for model-based control, pp. 5026–5033 (2012)
18. Vinyals, O., et al.: StarCraft II: a new challenge for reinforcement learning. arXiv Learning (2017)
19. Zoph, B., Vasudevan, V.K., Shlens, J., Le, Q.V.: Learning transferable architectures for scalable image recognition, pp. 8697–8710 (2018)

AI Application and Optimization

An Efficient Approach for Parameters Learning of Bayesian Network with Multiple Latent Variables Using Neural Networks and P-EM

Kaiyu Song, Kun Yue(✉), Xinran Wu, and Jia Hao

School of Information Science and Engineering, Yunnan University, Kunming 650500, China
kyue@ynu.edu.cn

Abstract. Bayesian network with multiple latent variables (BNML) is used to model realistic problems with unobservable features, such as diagnosing diseases and preference modeling. However, EM based parameter learning for BNML is challenging if there is a large amount of intermediate results due to missing values in the training dataset. To address this issue, we propose the clustering and P-EM based method to improve the performance of parameter learning. First, an innovative layer of neural network is defined based on Recurrent Neural Network (RNN) by incorporating the structural information of BNML into the Mixture of Generative Adversarial Network (MGAN), which can reduce the number of parameters by enabling clustering in an unsupervised manner. We then propose a Parabolic acceleration of the EM (P-EM) algorithm to improve the efficiency of convergence of parameter learning. In our method, the geometry knowledge is adopted to obtain an approximation of the paramcters. Experimental results show the efficiency and effectiveness of our proposed methods.

Keywords: Bayesian network · Latent variable · Generate adversarial network · Recurrent neural network · Parameter learning · Expectation maximization

1 Introduction

With the rapid development of Web2.0, unprecedented amount of user behavioral data can be collected and analyzed to facilitate the development of more realistic applications in economic and medical systems [6, 23]. In general, the features implied in user behavioral data are multi-dimensional. For instance, a user rates the movie, "*Titanic*", with a 5 star rating as he/she prefers *English* movies and *love* stories, shown as Fig. 1. Specifically, 75% of *female* users consider the *genre* first, and out of them, 88% of the users might prefer *love* stories. As it is seen in this example, multi-dimensional user preferences are often unobservable. Furthermore, various associations with uncertainties exist among users' properties, the attributes of the movies, the observed ratings, as well as other latent preferences on various other aspects of the movies. To enable the analysis of realistic applications, it is desirable to construct a human-like model to achieve interpretable inferences, in which the unobservable features and uncertain associations are represented.

H. Gao et al. (Eds.): CollaborateCom 2020, LNICST 349, pp. 357–372, 2021.
https://doi.org/10.1007/978-3-030-67537-0_22

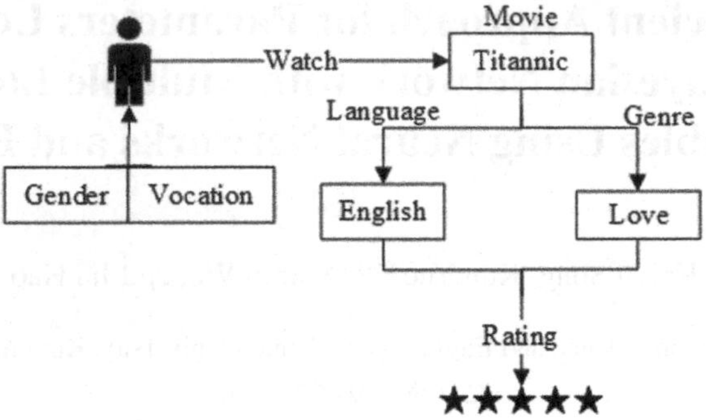

Fig. 1. An example of movie rating.

Probabilistic graphical model (PGM) is an effective framework to describe the dependency relationships among variables, and further quantify the above mentioned relationships. As an important PGM, Bayesian Network (BN) can be used to represent and infer the dependency relationships by a directed acyclic graph (DAG), combining with conditional probability tables (CPTs) [12]. In this paper, we adopt a BN with multiple latent variables, abbreviated as BNML, to model the unobserved features with uncertainties from user behavioral data. Oriented to preference modeling in movie rating as in Fig. 1, the corresponding BNML is illustrated in Fig. 2, where the latent variable is shown with dashed borders.

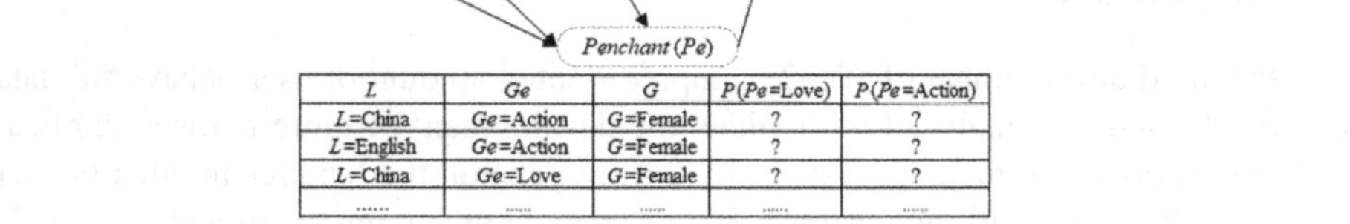

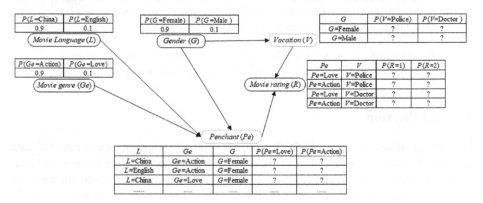

Fig. 2. An example of BNML.

The learning of BNML includes parameter learning and structure learning [12]. Generally, the structural expectation maximization (SEM) algorithm [22] is used for BNML learning. However, parameter learning in the SEM is highly complex due to the presence of missing values of latent variables. Expectation maximization (EM) [1] can be also adopted to fulfill parameter learning, where all possible combinations of the values of latent variables are generated for each incomplete sample. Then, the conditional probabilities of each complete sample are iteratively computed to obtain the optimal parameters. The complexity of parameter learning is $\Theta(pc^s)$, where p and s are the number of iterations and that of latent variables respectively. c is a constant number greater than 1, related to the number of parameters. Therefore, EM based parameter learning is also inefficient due to the large amount of intermediate results.

The execution time of parameter learning can be exponentially reduced by decreasing the number of latent variables, which is, however, not applicable to most cases of realistic problems. Therefore, it is essential to further investigate the method of learning parameters of BNML. To address this issue, we propose the technique by reducing the number of parameters to speed up the convergence of parameter learning.

We first propose a clustering based method to process the original dataset, which can significantly reduce the number of parameters. In this paper, Mixture of GAN (MGAN) [30] is adopted to implement clustering with an embedding process by generating a new distribution for each class in an unsupervised manner. Moreover, Recurrent Neural Network [15] (RNN) is the special neural network using the recurrent mechanism to pass the information from prior step, which could be used to process the information with structural characteristics. In this paper, RNN based MGAN (RMGAN) is proposed to cluster the observed features into several classes, by which the original samples are replaced by the outputs of RMGAN to reduce the number of parameters. Specifically, MGAN uses GAN to train the Gaussian mixture model [30] to fulfill clustering, in which a dense layer with the softmax activation function [29] is incorporated to generate the labels associated to the samples. By using this approach, we can build multiple generators and discriminators to use labels and approximate the hyperparameter of distribution for each class.

Then, we propose the RNN layer to refine the dense layer. In the proposed RNN layer, the values from the prior RNN cell and the RNN cells associated with the next RNN cell according to the DAG of BNML could be considered. This enables one-by-one correspondence of each feature to the RNN cells. Thus, the values of the prior and related cells can be used in the recurrent mechanism if there is an edge in the DAG of BNML between the two cells. By replacing the dense layer in MGAN with our proposed RNN layer, RMGAN can use the structural information of BNML to fulfill clustering and avoid eliminating useful information as much as possible. This establishes the relationship between neural network and BNML.

To further guarantee the efficiency of parameter learning, we use parabolic acceleration of the EM algorithm (P-EM) [4] to refine the EM based parameter learning. P-EM adopts an innovative geometrical method to increase the convergence of EM algorithm by estimating the approximated value of the maximum likelihood function [12]. Thus, a parameter learning method is proposed to reduce the number of iterations by using the updating function of P-EM to replace that of the EM based method.

Generally, the contributions of this paper are as follows:

- We propose RMGAN by rebuilding the recurrent mechanism in the RNN based on MGAN. Thus, clustering could be achieved and the information in the DAG of BNML could be incorporated. Further, the outputs of RMGAN could be used to reduce the number of parameters.
- We present an improved method for parameter learning based on P-EM to reduce the number of iterations and improve the efficiency of convergence.

- We conduct experiments on Alarm, MovieLens, and Cardiovascular datasets. Experimental results show that our proposed methods improve the efficiency of parameter learning by reducing the number of parameters and iterations.

2 Related Work

BN is widely used in real applications, such as medical analysis [6] and recommendation systems [23]. However, it is not easy to calculate the parameters of BN in presence of missing values. Several methods were proposed to perform parameter learning based on incomplete data, including EM, Robust Bayesian estimation, Monte-Carlo and Gaussian approximations [18].

To overcome the efficiency bottleneck in parameter learning, some optimization techniques were proposed, among which the concept of clique tree was incorporated to avoid repeated computations using shared computing [27]. Quasi Newton [11] was also used to make EM algorithm more efficient by adopting Quasi Newton function to generate an approximation instead of an accurate estimation. The Damped Anderson acceleration method [10] was further proposed by accelerating the iterations. Constraints were used to reduce the search space during the learning of parameters and structure [7, 8]. MapReduce was adopted to improved parameter learning by accelerating EM [2, 26]. However, few previous works were used for BNML, due to the large amount of intermediate results generated by missing values after adding multiple latent variables.

Dimensionality reduction achieved by embedding has been widely adopted as the effective means to improve the efficiency of machine learning algorithms such as categorization in computer vision [30]. In particular, several unsupervised methods can be used to implement clustering to finish categorization, e.g., k-means [20] and GAN [9]. Recently, ClusterGAN [21] was presented to fulfill clustering based on the potential feature space. Few previous works were used to process structural information, since it is difficult to generate structural information for most plications. In this paper, we are to implement clustering by incorporating the structural information from DAG.

3 RMGAN for Clustering

3.1 Preparation

Definition 1. BNML is a BN with more than one variable, denoted as $B(G, \theta)$, where $G = (V, E)$ is the DAG of B. V, O and L are the sets of all variables, observed variables and latent variables, respectively, $V = O \cup L$. E is the set of directed edges. θ is the set of probability parameters consisting of CPTs, and

- $\pi(V_i)$ denotes the set of parent nodes for the ith variables, V_i in G.
- θ_{ihk} denotes the parameter of V_i, where the value of V_i is equal to h and the value of its parent nodes take the kth combination.

Inputs. To avoid eliminating useful information as much as possible, the inputs of RMGAN include the features of some observed variables and the relationships in G

among these observed variables. First, some observed variables are chosen as the inputs of RMGAN. Then, we build the relationships among the chosen variables. For instance, it is reasonable to build a relationship between *gender* and *vocation*.

Then, we choose the adjacent matrix, denoted as G_{mm}, to save these relationships in G and then feed G_{mm} into RMGAN, where G_{mm}^{ij} denotes the values of E_{ij}, and E_{ij} denotes the relationship from V_i and V_j (V_i, $V_j \in V$, $i \neq j$). G_{mm}^{ij} is set to 1 if there is a relationship between V_i and V_j, otherwise, G_{mm}^{ij} is equal to 0. In addition, we keep the relationships among the chosen variables unchangeable in the structure part of SEM, such that the change of G in SEM does not affect RMGAN.

3.2 Building the Structure of RMGAN

Our clustering model built by RMGAN is based on MGAN [30]. The structure of MGAN includes several generators, discriminators and a dense layer. We propose a novel RNN layer, referred to as DRNN, to refine the dense layer. The generator and discriminator are the same as those in MGAN. The hypothesis of the whole DRNN is defined as $H(x) = f_D(Flatten(f_R(X,\ G_{mm})))$, where $f_R(X,\ G_{mm})$, *Flatten* and f_D are the hypothesis of RNN, flatten layer and dense layer, respectively.

RNN Layer. We propose the new component, called state gate, between two RNN cells to control the values generated by the recurrent mechanism. Specifically, we generate the number of RNN cells with the same number of the chosen variables in G_{mm}. This achieves one-by-one matching between features of the chosen variables and the RNN cells. For instance, the feature of the nth variable is fed to the nth RNN cells. In $E_{ij} = 1$ for $i = n$ and $j = n - 1$, the value of the $n - 1^{\text{th}}$ RNN cell is considered during recurrent. This can be achieved as follows:

$$\begin{cases} f(x^n) = f(x^{n-1}), n = 2 \\ f(x^n) = f(f^{n-1}(x^{n-1}) + S(x^n)), 2 < n < a \end{cases} \tag{1}$$

where $f(x)$ is the hypothesis of each RNN cell, $S(x)$ is the function of the state gate, and a is the number of observed variables in O. Details of building $f(x)$ refer to [15]. Matrix $\mathbf{y}$ is adopted to record the result of the RNN cell in the state gate, which is updated after each RNN cell transmits the result to the state gate as given in Eq. 2.

$$y = [f(x^1), f(x^2), \ldots\ldots f(x^n)] \tag{2}$$

If the RNN cell in the nth time step sends the result $f(x^n)$ to the state gate, the value of $\mathbf{y}[n]$ is updated by $f(x^n)$ as in Eq. 1. To calculate the values in the recurrent mechanism, we define the new function $S(x)$ in Eq. 3.

$$S(x^n) = \langle G_{mm}[n], y\rangle + f(x^{n-1}) \tag{3}$$

where $G_{mm}[n]$ denotes the values of the nth column of G_{mm}. y is the values in $\mathbf{y}$ and the inner product is dot product [28].

Therefore, if G_{mm}^{ij} between the current and other cells is equal to 1, where $i = n$ and $j = n - 1$, the values of hypothesis of related cells and prior cells are both considered during calculating the hypothesis of the current cell.

Flatten Layer. The flatten layer collects the features from RNN and sends them to the dense layer with softmax activation.

Dense Layer. The dense layer is similar to that in MGAN, composed of several fully connected layers. We set the activation of the output layer in the dense layer as softmax [29].

Therefore, DRNN can calculate the relationships in G shown as Fig. 3, where R_i is the ith RNN cell, T is the state gate, F is the flatten layer and M is the output layer of the dense layer that sets the activation function as softmax. We then put DRNN into MGAN and use the generators as well as discriminators to fulfill clustering. The training strategy of RMGAN is the same as the original training strategy in MGAN [30]. In addition, the outputs of RMGAN are the parameters of distributions and labels.

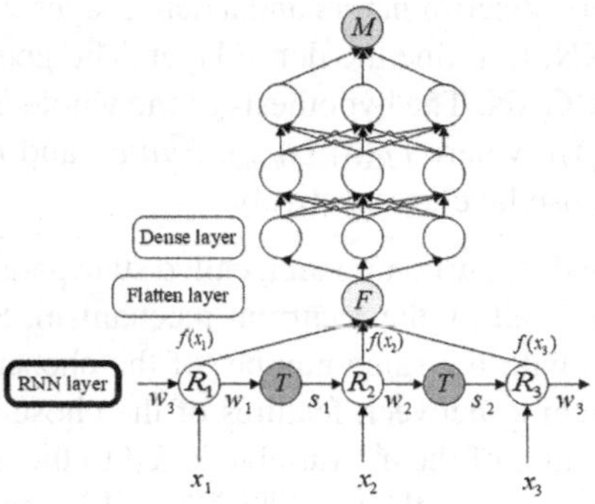

Fig. 3. Structure of the DRNN.

Finally, to reduce the number of combinations among variables, we rebuild the dataset by the outputs of RMGAN, used to generate new values for all samples and replace the original values. In other words, RMGAN generates new distributions $Para^i_{new}$ by using $Distri^*_i$ for the ith class, and $Para^i_{new}$ is the parameter of new distribution for the chosen variables. Then, $Para^i_{new}$ is used to replace the original sample values belonging to the ith class in the dataset. This reduces the number of combinations among the clustered variables in the rebuilt dataset to the number of classes, which is less than the original number of combinations. Correspondingly, the number of parameters in the CPTs is reduced [12]. Thus, the value of c is also accordingly reduced, since c is related to the number of parameters in the CPTs.

4 P-EM Based Parameter Learning

4.1 Preparation

Inputs. The inputs of PPL contain the DAG of BNML and the rebuilt dataset, denoted as $D = (D_1, D_2, ..., D_z)$ and $D_z = (O_1, O_2, ..., O_a)$. Note that replacement by generating D reduces the number of combinations in V, and further cleans the preliminary data cleaning, since the data is clustered into the same class to eliminate unnecessary information in the original dataset. Therefore, the efficiency of parameter learning is further improved by using the outputs of RMGAN.

EM Based Parameter Learning. EM based parameter learning (EPL) uses the EM based fragment updating method (EMFU) [12]. The whole process of EPL is based on SEM in which the DAG structure, G, is complete, since SEM can provide a complete G during parameter learning part. Therefore, G'' denotes as the complete G generated by SEM and contains all information of E among V. EMFU is composed of E-step, and M-step. In the E-step, EMFU extends D by all possible values of L to generate the extended dataset $D^e = (D_1^e, D_2^e, \ldots, D_z^e)$, where $D_z^e = (O_1 \ldots O_a, L_1 \ldots L_s)$. Then, the set of weights is denoted as W and calculated as follows:

$$m_{ihk} = \sum_{l=1}^{z} P(V_i = h, \pi(V_i) = k|(O_l, L_l)) \tag{4}$$

In the initial E-step, the set of initial parameters θ^0 is also generated randomly. In the M-step, θ_{ihk} is calculated based on D^e and the weights by Eq. 5.

$$LL(\theta_{ihk}) = \begin{cases} \frac{m_{ihk}}{\sum_{h=1}^{r_i} m_{ihk}}, & if \sum_{h=1}^{r_i} m_{ihk} > 0 \\ \frac{1}{r_i}, & else \end{cases} \tag{5}$$

where r_i is the cardinality of V_i. The EPL then turns to the E-step to repeat the above process until θ satisfies the terminating condition [12].

4.2 P-EM Based Parameter Learning

Our proposed PPL is based on EMFU by using the function in Eq. 6 based on Bézier curve [4] to calculate the approximate values of maximum likelihood estimation.

Note that PPL needs 3 seeds to calculate the approximate values. In the E-step of the proposed PPL, the weights are calculated based on θ^0 by E-step in EPL. In the M-step of the proposed PPL, θ^1, θ^2 and θ^3 are also calculated based on the weights using EPL to generate the required seeds, denoted by P^0, P^1 and P^2 respectively, where $P^0 = (\theta^0, LL(\theta^0)) = (\theta^0, \theta^1)$, $P^1 = (\theta^1, LL(\theta^1)) = (\theta^1, \theta^2)$ and $P^2 = (\theta^2, LL(\theta^2)) = (\theta^2, \theta^3)$. Then, $P^{new} = (\theta^{new}, LL(\theta^{new}))$ can be obtained using Eq. 6 and Eq. 7.

$$P^{new} = (1-t)^2 * P^0 + 2 * t * (1-t) * P^1 + t^2 * P^2 \tag{6}$$

where

$$t = 1 + u^{\delta} g \tag{7}$$

where δ is the number of iterations of Eq. 6, u and g are two constants. Then, the new value of P^0 is replaced by P^{new} to generate the new seeds P^1 and P^2 by the same way to repeat the above process until δ is equal to the maximum number of iterations. As the terminating condition, we consider Euclidean distance [13] between $LL(\theta_{ihk}^{\text{new}}))$ and $LL(\theta_{ihk}^2))$ to reduce the number of parameters to be updated in each iteration.

$$\left\| LL(\theta_{ihk}^{new}) - LL(\theta_{ihk}^2) \right\|_2 < \sigma \tag{8}$$

where σ is the threshold of terminating condition, which is the same as that of EMFU, and the E-step of PPL is the same as the EMFU. We use vectorization to optimize the updating process of our proposed M-step. This enables all parameters to be calculated at the same time. First, the seeds are converted into the vectors and stored in a $3 \times \xi$ matrix, denoted as *Vec*, where ξ is the number of parameters in θ corresponding to θ_{ihk}. Then, Eq. 9 can be obtained by incorporating the vectorization into Eq. 6.

$$P^{new} = (1 - t)^2 * Vec[0] + 2 * t * (1 - t) * Vec[1] + t^2 * Vec[2] \tag{9}$$

Note that θ_{ihk} will not be updated if Eq. 8 is satisfied. Our proposed M-step of PPL is given in Algorithm 1. Iterator, denoted as *iter*, is built to iterate the values for all parameters saved in P^{new}, P^2 and *Vec*. By using PPL, the number of iterations required for convergence is reduced comparing to EMFU. Moreover, reducing the iterating times during the M-step can further improve the fitting of parameters and SEM. The value of p in the complexity of P-EM is less than that in the EM based parameter learning. Therefore, the total number of iterations in SEM is reduced.

Algorithm 1 M-step of PPL based on vectorization

Input:

$G^{''}$, DAG of BNML generated by SEM
D^e, the extended D in E-step
W, the weights based on θ^0
u and g, two constants in Eq. 7
σ, threshold of the terminating condition
maxiter, the maximum number of iterations

Output: θ, the set of parameters of CPTs in BNML

Steps:

$\delta \leftarrow 0$; //number of iterations of Eq. 6
generating θ randomly;
generating seeds P^0, P^1 and P^2 based on W, $G^{''}$ and D^e by Eq. 5;
$Vec[0] \leftarrow P^0$; $Vec[1] \leftarrow P^1$; $Vec[2] \leftarrow P^2$; //vectorization
$\theta \leftarrow Vec[2]$; $\xi \leftarrow \theta.length$; //number of parameters in θ
$flag \leftarrow [0, \ldots, 0]_{1\times N}$ //label updated
while $\xi > 0$ **do** //number of parameters need to be updated
    $t \leftarrow 1+u^{\delta}g$; //by Eq. 7
    $P^{new} \leftarrow (1-t)^2 * Vec[0] + 2*t*(1-t)*Vec[1] + t^2 * Vec[2]$; //by Eq. 9
    $iter \leftarrow 0$; $iter.length \leftarrow \xi - 1$;//generate iterator for all θ_{ihk} in θ
    for each $LL(\theta_{iter}^{new})$ and $LL(\theta_{iter}^{2})$ in P^{new} and P^2 **do**
        if $\| LL(\theta_{iter}^{new}) - LL(\theta_{iter}^{2}) \|_2 < \sigma$ **and** $flag_{iiter} = 0$ **then**
            $\theta_{iiter} \leftarrow \theta_{iiter}^{new}$; $flag_{iter} \leftarrow 1$; $\xi \leftarrow \xi - 1$; $iter \leftarrow iter+1$; //$iter < iter.length$
        end if
    end for
    $\theta^0 \leftarrow \theta^{new}$;
    generating seeds P^0, P^1 and P^2 based on θ^0, $G^{''}$, and D^e by Eq. 5;
    $Vec[0] \leftarrow P^0$; $Vec[1] \leftarrow P^1$; $Vec[2] \leftarrow P^2$; $\delta \leftarrow \delta + 1$;
    if $\delta \geq maxiter$ **do**
        $iter \leftarrow 0$;
        for each $flag_{iter}$ in *flag* **do**
            updating θ_{iiter} by $Vec[2][iter]$; $iter \leftarrow iter+1$;
        end for
        return θ;
    end if
end while

5 Experiments

5.1 Experimental Setup

Datasets. We adopted 3 publicly available datasets to evaluate the proposed RMGAN and P-EM accelerate methods, including Alarm[1], MovieLens[2], and Cardiovascular[3].

- **Alarm** contains 37 feature. The size of this dataset was set as 500, 1000 and 2000. We used this dataset to evaluate the training speed of our proposed method with many observed and latent variables in the DAG.
- **MovieLens** contains 7 features. The size of this dataset was set as 10000, 20000 and 40000. We used this dataset to evaluate the feasibility of proposed method for illustrating the validity of the BNML.
- **Cardiovascular** contains 12 features. The size of this dataset was set as 10000, 20000 and 40000. We used this dataset to evaluate the performance of BNML that uses P-EM combining with RMGAN to fulfill parameter learning in different tasks comparing with MovieLens.

Parameter Settings. For all datasets, the parameters of Adam optimizer [14] were fixed to 5e−5. In the improved updating P-EM, the threshold, u, g and the maximum number of iterations were set to 0.01, 1.5, 0.2, and 50, respectively.

Environment. Our experiments were carried out on a PC with RTX 2070 GPU, e1230 v5 CPU and 8G RAM. All codes were developed in Python 3.6. The neural network framework was tensorflow2.0.

Competing Models. The classical EM was used as the baseline to measure the efficiency. PNN [24] was used to measure the feasibility of the neural network model and BNML combining with our training methods. The SVD matrix factorization (MF) [16] was also used to measure the feasibility of the BNML combining with our proposed RMGAN and P-EM.

Evaluation Metrics. The execution time was used to evaluate the efficiency of parameter learning based on RMGAN and P-EM. RMSE, MSE and MAE were adopted to measure the accuracy of prediction made by BNML trained by several methods and P-EM, defined as $RMSE = \sqrt[2]{\sum_{i=1}^{\pi} (y_i - y_i')^2/\pi}$, $MSE = \sum_{i=1}^{\pi} (y_i - y_i')^2/\pi$, and $MAE = \sum_{i=1}^{\pi} |y_i - y_i'|/\pi$ respectively, where y_i is the true value of the ith sample, π is the number of the samples, and y_i' is the predicted value.

[1] https://rdrr.io/cran/bnlearn/man/alarm.html.

[2] https://grouplens.org/datasets/movielens/1m/.

[3] https://www.kaggle.com/sulianova/cardiovascular-disease-dataset/data.

5.2 Experimental Results

Efficiency. To evaluate the efficiency of RMGAN combining with P-EM based parameter learning, we compared the execution time of the parameter learning using different methods. The comparisons were made for different sizes of Alarm dataset and different number of latent variables shown in Figs. 4, 5 and 6.

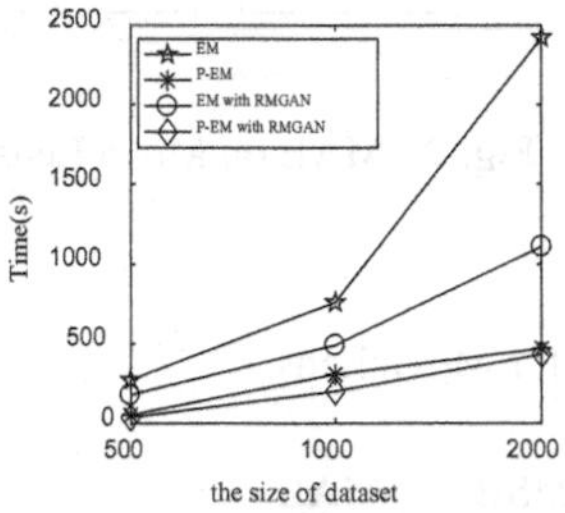

Fig. 4. Execution time on different sized datasets with 4 latent variables.

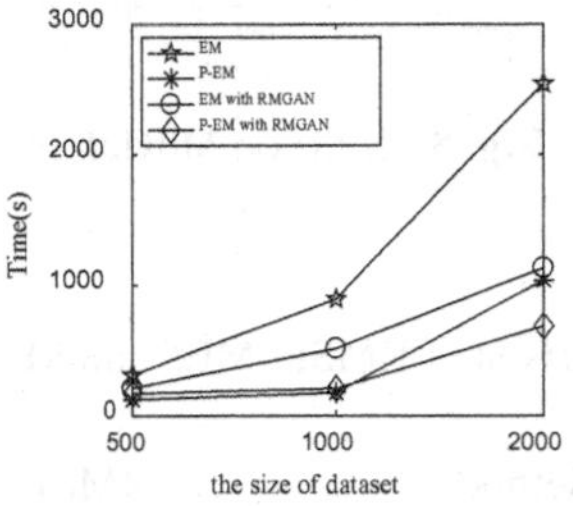

Fig. 5. Execution time on different sized datasets with 5 latent variables.

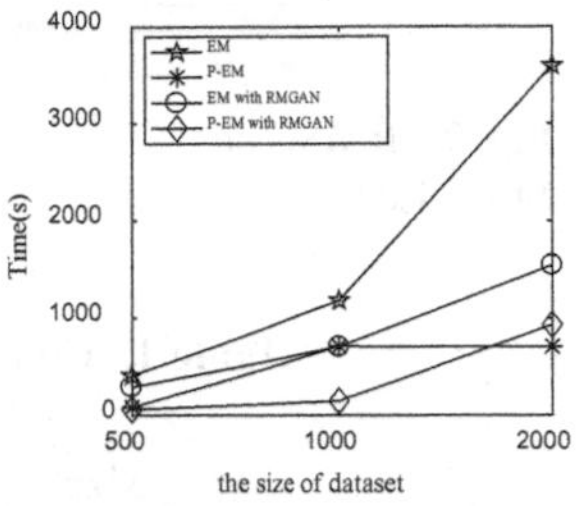

Fig. 6. Execution time on different sized datasets with 6 latent variables.

We chose 7 observed features and used RMGAN to fulfill the clustering and replacement. It can be seen from Fig. 4 that EM combining with RMGAN and P-EM reduce the execution time by 31.73% and 73.8% on average comparing with EM. It can also be seen in Figs. 5 and 6 that EM combining with RMGAN reduces almost 30% of execution time on average with 5 and 6 latent variables. This demonstrates that the proposed clustering method based on RMGAN improves the efficiency of parameter learning in BNML and P-EM. Moreover, we also made comparisons between P-EM combining with RMGAN and P-EM. As shown in Figs. 4, 5 and 6, P-EM combining with RMGAN reduces 74.79% of execution time on average compared with the EM in all conditions. It is shown that P-EM combining with RMGAN outperforms P-EM.

Feasibility. To evaluate the feasibility of BNML, we compared RMSE, MSE and MAE by different models on different sized MovieLens datasets, shown as Figs. 7, 8 and 9 and Table 1. The models include BNML using different methods to fulsill parameter learning, PNN and MF, in which the methods of parameter learning of BNML include P-EM with RMGAN, P-EM, EM, and EM with RMGAN. We then set the number of latent variables as 2 and chose the 2 observed features. 20% of the available dataset was considered as the test set and the remaining was used as training set. It can be seen from Table 1 that EM combining with RMGAN to finish parameter learning, on average, improves RMSE by 2.98%, MSE by 5.70%, and MAE by 3.39% respectively. This means the validity of clustering and the efficiency of RMGAN. As shown in Figs. 7, 8 and 9, the RMSE, MSE, and MAE based on BNML using P-EM to finish parameter learning are the same as those based on BNML using EM, which shows that P-EM achieves the same optimal values as EM.

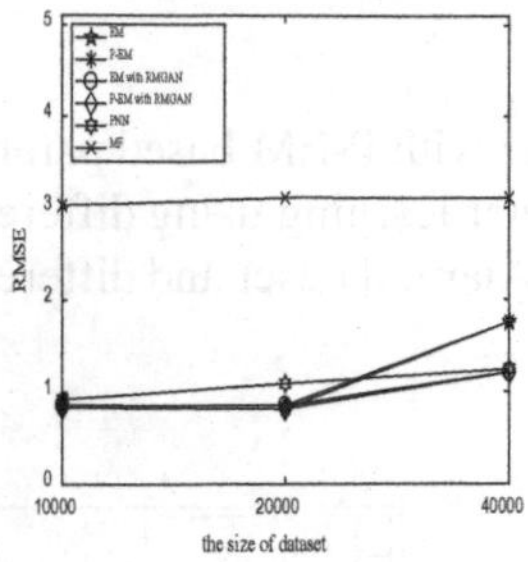

Fig. 7. RMSE on MovieLens.

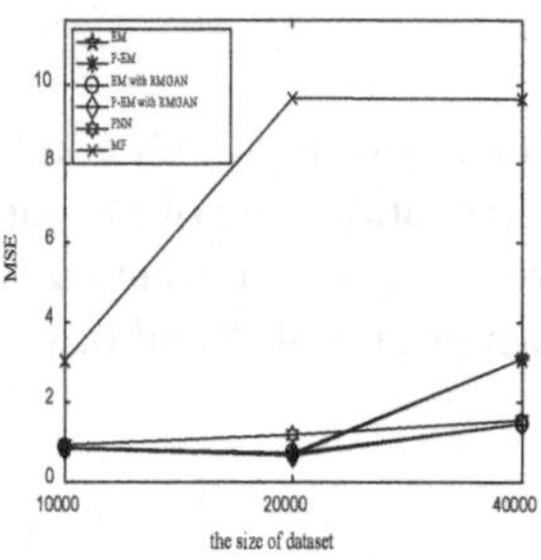

Fig. 8. MSE on MovieLens.

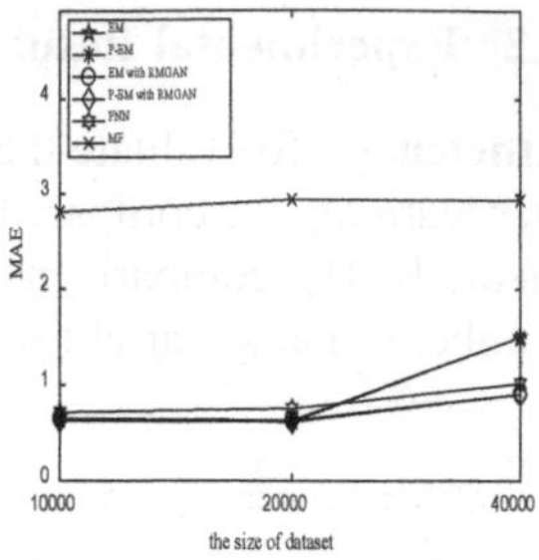

Fig. 9. MAE on MovieLens.

Table 1. Comparison of RMSE, MES and MAE for MovieLens.

	Size	Method	RMSE	MSE	MAE
BNML	10000	EM	0.85848	0.85848	0.64900
		EM with RMGAN	0.85848	0.85848	0.64900
		P-EM	**0.83366**	**0.83366**	**0.62700**
		P-EM with RMGAN	**0.83366**	**0.83366**	**0.62700**
	20000	EM	0.84897	0.72075	0.63075
		EM with RMGAN	**0.81319**	**0.72075**	**0.63075**
		P-EM	0.84897	0.66127	0.61489
		P-EM with RMGAN	**0.81319**	**0.66127**	**0.61489**
	40000	EM	1.75741	3.08850	1.50025
		EM with RMGAN	**1.21017**	**1.46450**	**0.90513**
		P-EM	1.75741	3.08850	1.50025
		P-EM with RMGAN	**1.21017**	**1.46450**	**0.90513**
PNN	10000	None	0.92168	0.92168	0.71250
	20000		1.08630	1.18000	0.75265
	40000		1.24358	1.54650	1.01600
MF	10000	None	3.02869	3.02869	2.81000
	20000		3.10913	9.66670	2.93820
	40000		3.10547	9.64397	2.93747

Additionally, we made comparisons between P-EM combining with RMGAN and P-EM, shown in Figs. 7, 8 and 9. The RMSE, MSE, and MAE for BNML using P-EM combining with RMGAN are lower than those for BNML using P-EM. This shows that P-EM combining with RMGAN outperforms P-EM in parameter learning. It can also be seen from Table 1 that BNML by our proposed method improves the performance of rating prediction by 15% and 80% comparing with PNN and MF respectively. Therefore, these results suggest that BNML is an effective model for realistic applications such as rating prediction.

To further illustrate the feasibility of BNML, we compared the RMSE, MSE and MAE based on different models on different sized Cardiovascular datasets and presented the results in Figs. 10, 11 and 12 and Table 2. The models include BNML using different methods to fulfill parameter learning, and PNN, in which the methods of parameter learning of BNML are the same as the experiment on MovieLens.

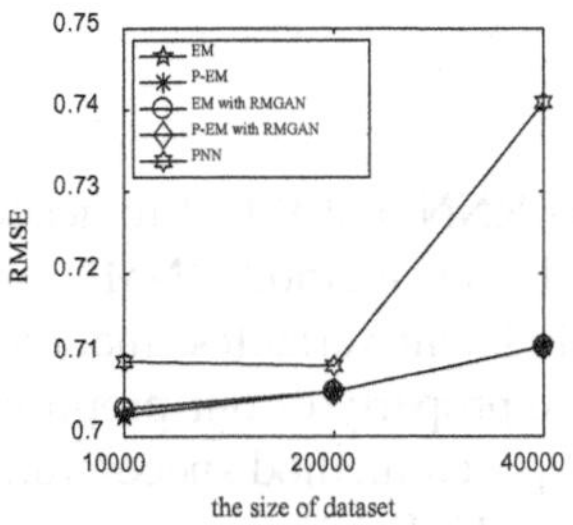

Fig. 10. RMSE on Cardiovascular.

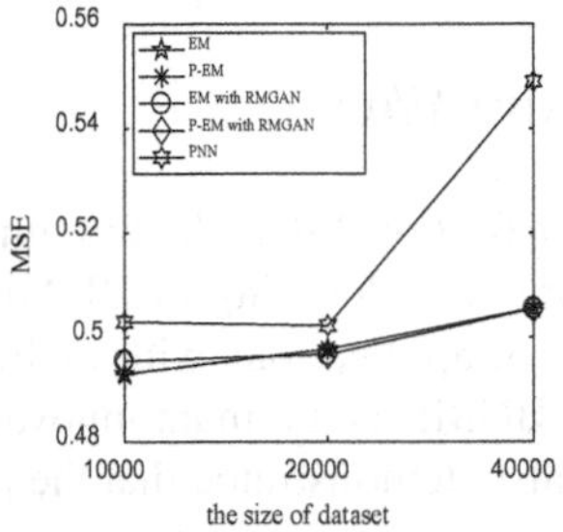

Fig. 11. MSE on Cardiovascular.

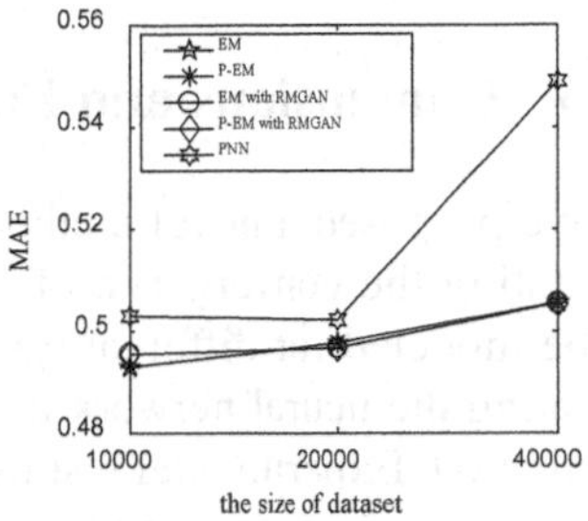

Fig. 12. MAE on Cardiovascular.

Table 2. Comparison of RMSE, MSE and MAE in Cardiovascular.

	Size	Method	RMSE	MSE	MAE
BNML	10000	**EM**	**0.70265**	**0.49300**	**0.49300**
		EM with RMGAN	0.70343	0.49550	0.49550
		P-EM	**0.70265**	**0.49300**	**0.49300**
		P-EM with RMGAN	0.70343	0.49550	0.49550
	20000	EM	0.70551	0.49775	0.49775
		EM with RMGAN	**0.70480**	**0.49675**	**0.49675**
		P-EM	0.70551	0.49775	0.49775
		P-EM with RMGAN	**0.70480**	**0.49675**	**0.49675**
	40000	**EM**	**0.71107**	**0.50562**	**0.50562**
		EM with RMGAN	**0.71107**	**0.50562**	**0.50562**
		P-EM	**0.71107**	**0.50562**	**0.50562**
		P-EM with RMGAN	**0.71107**	**0.50562**	**0.50562**
PNN	10000	None	0.70922	0.50300	0.50300
	20000		0.70869	0.50225	0.50225
	40000		0.74094	0.54900	0.54900

We first set one latent variable and chose 3 observed features to fulfill clustering. Since the features in MovieLens are discrete, we dropped the continuous features in

Cardiovascular. It can be seen from Table 2 that the performance of BNML using P-EM and RMGAN is, on average, improved by 0.6% in terms of RMSE, MSE and MAE. It can also be seen that our proposed method of parameter learning is effective on different types of datasets. Moreover, the BNML learnt by all methods in RMSE, MSE and MAE in Table 2 is, on average, improved by 0.7% comparing with PNN. By increasing the size of the dataset in Figs. 10, 11 and 12, BNML using different methods to finish parameter learning can achieve better RMSE, MSE and MAE comparing with PNN. This suggests that BNML is also feasible in Cardiovascular.

6 Conclusions and Future Work

We proposed a novel method based on MGAN, combining RNN and P-EM to accelerating the convergence of parameter learning of BNML. In our method BNML can be modeled for different types of applications with multiple latent variables. Incorporating the neural network into BNML is the main innovative property of our proposed method. Experimental results also demonstrated that the proposed method successfully achieves clustering and further improves the performance of BNML. There is however a disadvantage in selecting the clustering features. In our future research endeavors, we will explore a method based on GAN to address the feature selection problem. Another possible dimension for extending this work is to adopt Graph Neural Network to finish embedding instead of clustering.

Acknowledgments. This paper was supported by the National Natural Science Foundation of China (U1802271), the Science Foundation for Distinguished Young Scholars of Yunnan Province (2019FJ011), the Cultivation Project of Donglu Scholar of Yunnan University, and the Foundation for Undergraduates of Educational Department of Yunnan Province (2020Y0010).

References

1. Dempster, A.P., Laird, N.M., Rubin, D.B.: Maximum likelihood from incomplete data via the EM algorithm. J. R. Stat. Soc. Ser. B (Methodol.) **39**(1), 1–38 (1977)
2. Basak, A., Brinster, I., Ma, X., Mengshoel, O.J.: Accelerating Bayesian network parameter learning using Hadoop and MapReduce. In: 1st International Workshop on Big Data, Streams and Heterogeneous Source Mining: Algorithms, Systems, Programming Models and Applications, pp. 101–108. ACM, Beijing (2012)
3. Becker, A., Geiger, D.: A sufficiently fast algorithm for finding close to optimal clique trees. Artif. Intell. **125**(1–2), 3–17 (2001)
4. Berlinet, A., Roland, C.: Parabolic acceleration of the EM algorithm. Stat. Comput. **19**(1), 35–47 (2009)
5. Binder, J., Koller, D., Russell, S., Kanazawa, K.: Adaptive probabilistic networks with hidden variables. Mach. Learn. **29**(2–3), 213–244 (1997)
6. Bradley, A., Van Der Meer, R.: Upfront surgery versus neoadjuvant therapy for resectable pancreatic cancer: systematic review and Bayesian network meta-analysis. Sci. Rep. **9**(1), 1–7 (2019)
7. De Campos, C.P., Ji, Q.: Improving Bayesian network parameter learning using constraints. In: 19th International Conference on Pattern Recognition, pp. 1–4. IEEE, Tampa (2008)

8. Feelders, A., Van der Gaag, L.C.: Learning Bayesian network parameters under order constraints. Int. J. Approx. Reason. **42**(1–2), 37–53 (2006)
9. Goodfellow, I., et al.: Generative adversarial nets. In: Advances in Neural Information Processing Systems, pp. 2672–2680. MIT Press, Montreal (2014)
10. Henderson, N.C., Varadhan, R.: Damped Anderson acceleration with restarts and monotonicity control for accelerating EM and EM-like algorithms. J. Comput. Graph. Stat. **28**(4), 834–846 (2019)
11. Jamshidian, M., Jennrich, R.I.: Acceleration of the EM algorithm by using quasi-Newton methods. J. R. Stat. Soc. Ser. B (Stat. Methodol.) **59**(3), 569–587 (1997)
12. Jensen, F.V.: An Introduction to Bayesian Networks. UCL Press, London (1996)
13. Ji, Z., Xia, Q., Meng, G.: A review of parameter learning methods in Bayesian network. In: Huang, D.-S., Han, K. (eds.) ICIC 2015. LNCS (LNAI), vol. 9227, pp. 3–12. Springer, Cham (2015). https://doi.org/10.1007/978-3-319-22053-6_1
14. Kingma, D.P., Ba, J.: Adam: a method for stochastic optimization. In: 3rd International Conference on Learning Representations, pp. 7–9. ICLR, San Diego (2015)
15. Klapper-Rybicka, M., Schraudolph, N.N., Schmidhuber, J.: Unsupervised learning in LSTM recurrent neural networks. In: Dorffner, G., Bischof, H., Hornik, K. (eds.) ICANN 2001. LNCS, vol. 2130, pp. 684–691. Springer, Heidelberg (2001). https://doi.org/10.1007/3-540-44668-0_95
16. Koren, Y., Bell, R., Volinsky, C.: Matrix factorization techniques for recommender systems. Computer **42**(8), 30–37 (2009)
17. Langseth, H., Bangsø, O.: Parameter learning in object-oriented Bayesian networks. Ann. Math. Artif. Intell. **32**(1–4), 221–243 (2001)
18. Lauritzen, S.L.: The EM algorithm for graphical association models with missing data. Comput. Stat. Data Anal. **19**(2), 191–201 (1995)
19. Liao, W., Ji, Q.: Learning Bayesian network parameters under incomplete data with domain knowledge. Pattern Recogn. **42**(11), 3046–3056 (2009)
20. Likas, A., Vlassis, N., Verbeek, J.J.: The global k-means clustering algorithm. Pattern Recogn. **36**(2), 451–461 (2003)
21. Mukherjee, S., Asnani, H., Lin, E., Kannan, S.: Clustergan: latent space clustering in generative adversarial networks. In: 33th AAAI Conference on Artificial Intelligence, pp. 4610–4617. AAAI Press, Honolulu (2019)
22. Friedman, N.: The Bayesian structural EM algorithm. In: 14th Conference on Uncertainty in Artificial Intelligence (UAI 1998), pp. 129–138. Morgan Kaufmann Publishers Inc., San Francisco (1998)
23. Ono, C., Kurokawa, M., Motomura, Y., Asoh, H.: A context-aware movie preference model using a Bayesian network for recommendation and promotion. In: Conati, C., McCoy, K., Paliouras, G. (eds.) UM 2007. LNCS (LNAI), vol. 4511, pp. 247–257. Springer, Heidelberg (2007). https://doi.org/10.1007/978-3-540-73078-1_28
24. Qu, Y., et al.: Product-based neural networks for user response prediction. In: 16th International Conference on Data Mining, pp. 1149–1154. IEEE, Barcelona (2016)
25. Reed, E., Mengshoel, O.J.: Bayesian network parameter learning using EM with parameter sharing. In: 11th UAI Bayesian Modeling Applications Workshop co-located with the 30th Conference on Uncertainty in Artificial Intelligence, pp. 48–59. CEUR-WS.org, Quebec (2014)
26. Reed, E.B., Mengshoel, O.J.: Scaling Bayesian network parameter learning with expectation maximization using mapreduce. Proc. Big Learn. Algorithms Syst. Tools **3**, 1 (2012)
27. Shafer, G.R., Shenoy, P.P.: Probability propagation. Ann. Math. Artif. Intell. **2**(1–4), 327–351 (1990)
28. Wikipedia Contributors Inner product space. https://en.wikipedia.org/wiki/Inner_product_s pace. Accessed 07 Feb 2020

29. Wikipedia Contributors Softmax function. https://en.wikipedia.org/wiki/Softmax_function. Accessed 16 Aug 2020
30. Yu, Y., Zhou, W.J.: Mixture of GANs for clustering. In: 27th International Joint Conference on Artificial Intelligence, pp. 3047–3053. IJCAI.org, Stockholm (2018)

DECS: Collaborative Edge-Edge Data Storage Service for Edge Computing

Fuxiao Zhou and Haopeng Chen(✉)

Shanghai Jiao Tong University, Shanghai, China
{zhoufuxiao,chen-hp}@sjtu.edu.cn

Abstract. With the development of IoT and 5G, data are generated by the numerous smart end devices at each moment. Simultaneously, as the improvement of the hardware's performance, computing and storage are partly transferred to the edge of the Internet. However, the core cloud and massive data centers are still responsible for management and coordination. In more and more local-area and small-scale scenarios such as a parking lot, an office building, or a college campus, these scenarios also need the edge nodes to offload computing and storage tasks. Moreover, in order to decrease costs and be lightweight, these scenarios need to decouple with the core cloud partly. In this paper, we proposed a collaborative edge-edge data storage service called DECS for edge computing in local-area scenarios. DECS can make the edge nodes collaborate with others. Such as trade-off to pick the most appropriate edge node to offload storage or computing tasks. DECS can also replicate data or generate forwarding rules in advance by predicting data's popularity proactively.

In this paper, we evaluated DECS at two real scenarios compared with state-of-the-art research. The experiment results proved that DECS was more suitable for the local-area edge cluster. Which lowered the access latency, saved the total bandwidth, and improved the resource utilization of the whole edge cluster.

Keywords: Collaborative storage · Storage as a service · Edge computing · Small-scale · Local area · Resource management

1 Introduction

With the development of 5G and the Internet of Things (IoT), the amount of data is more substantial. Meanwhile, the devices between the cloud and the end devices (IoT devices, mobile phones, or laptops) are getting more powerful, such as base stations, gateway servers, or routers. Fog computing has been researched widely in recent years to solve the problem of how to effectively utilize and manage resources of the devices between the core cloud to the end devices [19].

The traditional Internet logic architecture model has been approximately divided into three layers [19], including the cloud layer, which is composed of

H. Gao et al. (Eds.): CollaborateCom 2020, LNICST 349, pp. 373–391, 2021.
https://doi.org/10.1007/978-3-030-67537-0_23

large data centers, such as Amazon's AWS. The fog layer consists of base stations, RANs, gateway servers, or routers. Moreover, the end layer contains IoT devices such as sensors, monitors, or mobile phones, laptops.

Generally speaking, the scenarios of edge computing are wide-range geographical such as metropolis computing, inter-metropolis computing, or smart city, and which can be seen in Fig. 1(a). In these scenarios, the core cloud and edge devices provide computing and storage resources for connected vehicles, health care, and smart delivery. However, in some small-scale and local area scenarios such as a college campus, a parking lot, a hospital, an industrial park even an office building. Which also need edge computing and multi-access edge computing (MEC) to provide computing and storage resources. These resources on edge nodes can process offloaded tasks from IoT devices or mobile phones.

For example, the scenario in an office building, which contains many IoT devices such as video surveillance monitors distributed on every floor, and many mobile phones used by many people worked in this building, which can be seen in Fig. 1(b). Each video monitor or sensor can offload tasks to some closer edge node, such as a router. Moreover, the task may be a realtime computing task such as video violations identification. The edge node which executes the identification task will analyze the video data and utilizes machine learning to identify the wrong behavior such as smoking and stealing. On the other hand, the users who have mobile phone also can offload tasks to a closer edge node to storage or read some data, and the task can be a computing task too.

Traditional distributed databases, file systems such as cassandras [4], or HDFS [8] can not be a solution directly to the local area and small-scale scenarios mentioned above because that these schemas do not consider the features of edge computing. Due to the edge nodes have limited resources, and different edge node's performance may have considerable differences with others. Moreover, the data stored inside the edge node are heterogeneous (may be structured or unstructured). Besides, the storage involved in edge computing considerate the distance and geographically location. Therefore, in this paper, we abstracted the storage as a service and proposed a collaborative edge-edge service. The storage service can be provided to end devices to write or read data. Meanwhile, this service can be an essential service to support other computing services or be a sub-service in more extensive services.

In this paper, we proposed a collaborative edge-edge data storage service for edge computing in the local area and small-scale scenarios called DECS. DECS only runs on the edge cluster, which is closer to end devices. DECS contains some modules such as Data Storage Module, Metadata Module, Data Processing Module, Trade-off Module, Prediction Module.

This paper mainly made these contributions as follows:

(1) This paper abstracted the storage as a service and proposed a collaborative edge-edge data storage service called DECS to utilize the resources in the edge cluster effectively.
(2) This paper designed and implemented the sub-modules of DECS, including the Metadata Module, Data Storage Module, Data Processing Module, Trade-off Module, and Prediction Module.

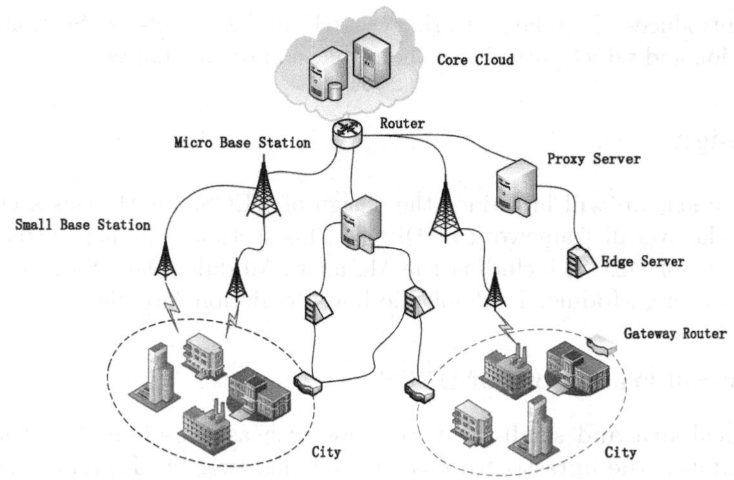

(a) Three-Layer Edge Architecture In Wide Area

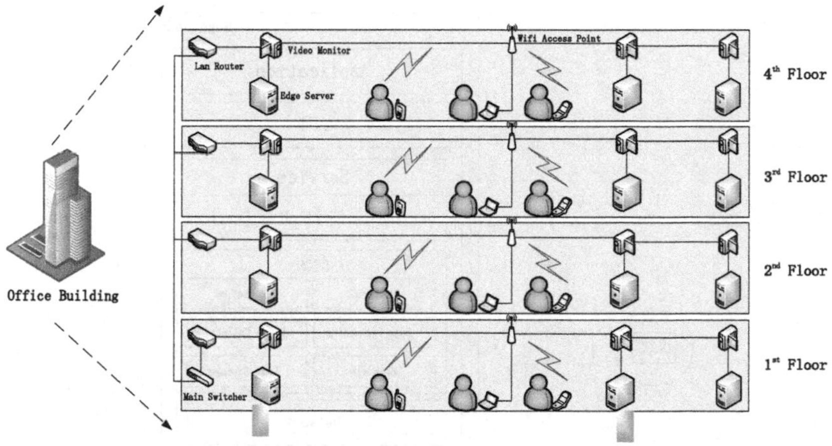

(b) An Example of Local Area Scenario: An Office Building

Fig. 1. Wide area and local area scenarios for edge computing

(3) This paper did experiments and evaluated DECS at two real scenarios compared with state-of-art research. This paper gave explanations and analysis of the experiment results.

The rest of this paper is organized as follows: Sect. 2 introduces the design of DECS, including the overall framework and every sub-module. Section 3 introduces the implementation and evaluation of DECS and its sub-modules. Section 4 mainly introduces the related work, research, and state-of-art. Section 5 is the final section and which introduces the conclusion of this paper.

2 Design

In this section, we will introduce the design of DECS. Firstly this section will describe the overall framework of DECS. This section will then introduce the sub-modules of DECS, including the Metadata Module, Data Storage Module, Data Processing Module, Trade-off Module, Prediction Module.

2.1 Overall Framework of DECS

In the local area and small-scale scenarios such as a building, a park, and a college campus, the infrastructure is relatively lightweight. Therefore the DECS is applicable to edge cluster, which is closer to end devices. DECS is organized as a micro-service running in every edge node of the edge cluster.

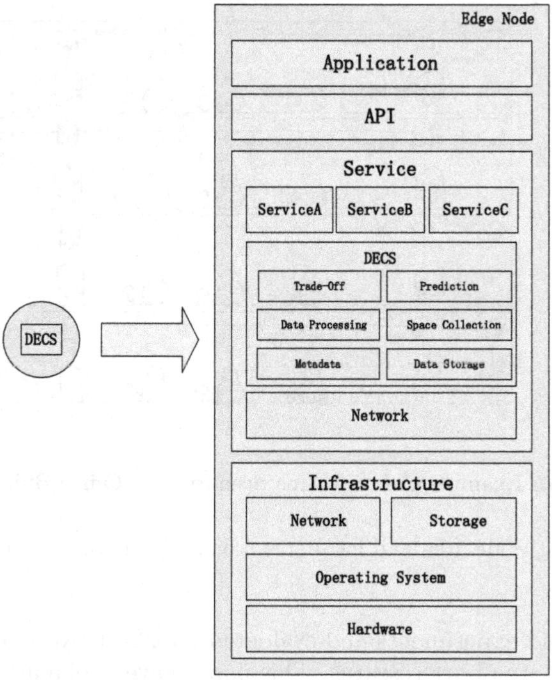

Fig. 2. Inner architecture of edge node and DECS

Edge nodes connect with each other by network cable, and DECS is run on every edge node. End devices are split into two types: the first one is IoT devices, which connects edge nodes by network cable usually. IoT devices can offload real-time computing tasks to the appropriate edge node, typically such as target recognition. The second type is mobile phone or laptop carried by users, and it is a MEC architecture, multi-users can access the same edge node. The users are movable to every position of the specific local area and small-scale scenario. Mobile phones can access the edge node by wifi or 4G/5G network, and the laptops can access the edge cluster by wifi or network cable.

DECS contains some sub-modules which can be seen in Fig. 2. There are the metadata module, data storage module, data processing module, prediction module, trade-off module, and space collection module. DECS is in the service layer, and there are other services in this layer. Part of them are underlying fundamental services such as network. Moreover, others are top-level and specific services such as some computing services offloaded from users, or inherent services attached to some applications. DECS can be seen as an underlying service because it integrates the fundamental storage as a sub-service. On the other hand, DECS contains some top-level services such as data processing sub-service.

Table 1. Sub-modules of DECS instruction

Module Name	Description
Metadata Module	This module is responsible for managing all the metadata and status of this edge cluster and synchronized meta-information between nodes by the Gossip protocol
Data Storage Module	This module is an underlying module to process the write and read operation received from the storage task sent by end devices
Data Processing Module	This module can execute some computing tasks, especially the data processing task in real-time or asynchronously, and only return the result
Trade-off Module	This module is responsible for weighing the cost against latency or resource utilization in the task offloading or data storage in the whole cluster
Prediction Module	The role of this module is to predict the appropriate place for task offloading and predict the possible access point to migrate or replicate data proactively in order to lower the latency of the whole cluster

The concise introduction of sub-modules in DECS can be seen in Table 1, the rest of this section will introduce every sub-module in detail.

2.2 Metadata Module in DECS

Metadata Module is a core sub-module in DECS. Each edge node shares its status, such as stored data meta-information, access information, machine status information, and resource utilization information to the whole cluster. When an end device accesses some edge node to get data, the edge node will first check its metadata to get whether the wanted data in this node. If not, this node will refer to metadata to get which node has the wanted data and forward the request to the target node.

Table 2. Definition and pattern of metadata

Name	Pattern
liveNodes	[nodeIp,*]
availableSpace	[{nodeIp:space},*]
nodeHasWhatFile	[{nodeIp:[file,*]},*]
nodeCpuInfo	[{nodeIp:cpuInfo},*]
whoStoreFiles	[{devIp:[file,*]},*]
hotNodeForSpecificDev	[{devIp:[nodeIp:[period:acc,*],*]},*]
forwardingRule	[{devIp:[nodeIp,*]},*]
fileStoredWhere	[{file:[nodeIP,*]},*]
fileAccessFromNodeIpEachPeriod	[{file:{NodeIp:[{period:acc},*],*}},*]
filePopularityFromNodeIpLastPeriod	[{file:[{nodeIp:pop},*]},*]

In DECS, we defined some types of metadata to represent each edge node's status, the shared information of the whole edge cluster, and the metadata of stored data. The specific definition and representation of metadata can be seen in Table 2. Such as the *liveNodes* represents the number of live nodes of the edge cluster. Other metadata will be explained in the remainder of this paper when used.

2.3 Data Storage Module in DECS

The data storage module is fundamental and essential in DECS. This module can be accessed individually or supporting the upper sub-module such as the data processing module. The data storage module provides two operations for users as basic operations, which are **write operation** and **read operation**, each operation contains the call to the trade-off module and prediction module.

The process of store data to the edge cluster by DECS can be seen in Algorithm 1, the type of data can be divided into two types as **temp data** and **regular data**. The difference between temp data and regular data is their generated source. Due to the data from IoT devices will be provided to the real-time task usually. After the task completed, the data will be discarded in most cases and only store the computed result. To save limited space in edge nodes, we will

Algorithm 1. The process of write operation.

Require:

End device's IP, ip_f
The type of data wanted to be written, t
Default access the close edge node, ip_c
The file name which can uniquely identify the file, $name$
The binary stream of file wanted to be written, $data$;

Ensure:

Whether the data is written successfully, $result$.

1: $liveNodes \leftarrow GetLiveNodes()$
2: $threshold \leftarrow GetStorageThresh()$
3: $closeNode \leftarrow GetCloseNodeInfo(ip_c)$
4: $exist \leftarrow CheckExist(name)$
5: **if** $exist == True$ **then**
6:     **return** $False$
7: **end if**
8: $UpdateMetadata(data, name, ip_f, t)$
9: $forwardRule \leftarrow QueryMeta(ip_f)$
10: **if** $forwardRule$!= $Null$ **then**
11:     $forwardNodes = GetFwdNodes(forwardRule)$
12:     $ip_t \leftarrow PickNodeW(forwardNodes, data, ip_c, t)$
13:     $UpdateMetadata(data, name, ip_t, t)$
14:     **return** $Store(ip_t, data, name, t)$
15: **end if**
16: $ip_t \leftarrow PickNodeW(liveNodes, data, ip_c, t)$
17: $UpdateMetadata(data, name, ip_t, t)$
18: **return** $Store(ip_t, data, name, t)$

mark this kind of data as temp data, which only lives a short time. The regular data is generated from mobile phones or laptops of users, and this kind of data will be stored for a long time unless some extreme cases.

In the process of storing data, this module checks whether this data existed by metadata *fileStoredWhere* firstly at line 4. Then this module will updata the metadata (*availableSpace*, *whoStoreFiles*). After the write operation, this module updates the metadata once again at the end at line 17, which updates the *fileStoreWhere* after tradeoff to record where the file is written.

At the beginning of each time period, the prediction module will predict the storage pattern for the devices that frequently store data in the previous periods. The predicted results will be recorded in the metadata *forwardingRule*, and the storage pattern can help devices to write data to some nodes that most likely to be accessed by other end devices. During the process of the write operation, the storage module will check the *forwardingRule* in metadata at line 11 to proactively forward this storage request to the target node in advance.

The process of write operation calls the trade-off module many times, such as *PickNodeW* at line 12, which needs to pick the most appropriate node to write

Algorithm 2. The process of read operation.

Require:

The type of data wanted to be read, t
Default access the close edge node, ip_c
The file name which can uniquely identify the file, $name$;

Ensure:

The binary stream of file wanted to be read, $result$.

1: $liveNodes \leftarrow GetLiveNodes()$
2: $existClose \leftarrow CheckExistClose(name, closeNode)$
3: **if** $existClose == True$ **then**
4:     $ip_t \leftarrow ip_c$
5: **else**
6:     $existNodes \leftarrow FindExistNode(liveNodes, name)$
7:     **if** $existNodes.lenth == 0$ **then**
8:         **return** $Null$
9:     **else**
10:         $ip_t \leftarrow PickNodeR(existNodes, name)$
11:     **end if**
12: **end if**
13: $updataMetadata(name, ip_t, t)$
14: **if** $t == Temp$ **then**
15:     **return** $ReadAndDelete(name, ip_t, t)$
16: **else**
17:     **return** $Read(name, ip_t, t)$
18: **end if**

data from the candidate nodes. It is necessary to filter the candidate nodes first to exclude the node that has no space to store the data.

The read operation process is relatively simple, which can be seen in Algorithm 2. The module checks whether the accessed edge node has the wanted data firstly. If not, this module will query the metadata *fileStoreWhere* to get the list of edge nodes that contain the wanted data at line 6. Regular data may exist in more than one node, and the reason for this case is that it is popular for this cluster. The prediction module proactively replicates this data to the other nodes in advance. However, the temp data only has one. When the wanted data has more than one replicas, this module will call the *PickNodeR* of the trade-off module at line 10 to determine to pick an appropriate node to read. If the desired data are temp data, these data will be removed at line 15 from the edge cluster to keep space enough.

2.4 Data Processing Module in DECS

The data processing module is a critical top-level sub-module in DECS. Its role is executing the computing tasks that are offloaded from the IoT devices or the users. The data processing task processed in this module can be split into two types, namely **with-data task** and **without-data task**, the difference between them is whether the task with or without accompanying data.

The differences can be seen in Table 3. If a requested task with data means this data processing task needs to process the data attached, the accompanying data will be written as temp data. The without-data task will process the existing data, and this kind of task will not modify the raw data.

Table 3. Differences between the two types of tasks

	With-Data Task	Without-Data Task
Request With Data	Yes	No
Modify Data	May modify carried data	No
End Target	IoT devices or users	Users
Data Type	Temp data	Regular data
Main Resource Needed	Computing Power	Data Resources

The process of processing with-data task can be seen in Algorithm 3. This module needs the supports of the data storage module to write or read temp data. This module will gather real-time CPU idle utilization of the candidate nodes and computes the real-time computing power for each. Then call the function *PickNodeP* in the trade-off module at line 10 to pick the best node to process the with-data task.

Algorithm 3. The process of processing with-data task.

Require:
The data processing task, *task*
The accompanying data with this task, *data*
The data storage module, *module*;
Ensure:
The computed result of this task, *result*.

```
1:  canStoreNodes ← module.GetCanStoreNodes(data)
2:  for node in canStoreNodes do
3:      cpuInfo ← GetCpuInfoFromMeta(node.GetIp)
4:      cpuIdle ← GetCpuIdle(node.GetIp())
5:      idleComputePower = cpuIdle · cpuInfo.GetFreq()
6:      canComputeNodes.Add(node, idleComputePower)
7:  end for
8:  nodes ← canComputeNodes ∩ canStoreNodes
9:  if nodes.lenth != 0 then
10:     bestNode ← PickNodeP(nodes)
11:     module.storeTemp(data, bestNode)
12:     return Process(bestNode)
13: end if
```

The process of processing without-data task is similar to the process of processing the with-data task. The difference between them is that the latter firstly traverses all nodes which can store the temp data and compute their computing power to pick the most appropriate node. However, the former only traverses the nodes that have the desired raw data. The rest of them are the same.

2.5 Trade-Off Module in DECS

The trade-off module is an auxiliary module in DECS, and the role of it is decision-making. This module needs to decide where to write or read data and where to offload computing tasks according to the distance, bandwidth, computing power, or free space. This module is called by the data storage module and data processing module, which can be seen in the algorithms mentioned above. The following will introduce the specific details of each function of this module.

The function *PickNodeW* at line 12 of Algorithm 1 means picking the most appropriate node to write data, the specific process is formalized as follows:

In Eq. (1), the l means the length (MB) of data, the s means the real-time network speed (MB/s) from the accessed node to the node$_i$, and the t_i means the estimated time (s) to transfer the data from the accessed edge node to the node$_i$.

$$t_i = \frac{l}{s} \tag{1}$$

This module trades off the most appropriate node to write data by comparing the score of s_i generated from Eq. (2). In the equation, the r_i means the real-time ratio of free space in node$_i$, and $r_i \epsilon [0, 1]$. The n_i means the predicted access amount after normalizing, and $n_i \epsilon [0, 1]$. The T_c is a constant, which means the transferring time (s) between the end devices and the accessed edge node. The α and β are the hyperparameter, which can be adjusted to change the weight, and $\alpha, \beta \epsilon [0, 1]$. The edge node with the highest score will be chosen as the target node.

$$s_i = \frac{\alpha \cdot r_i + (1-\alpha) \cdot n_i}{\beta \cdot (1 + \frac{t_i}{T_c})} = \frac{T_c \cdot [\alpha \cdot r_i + (1-\alpha) \cdot n_i]}{\beta \cdot (T_c + t_i)} \tag{2}$$

The function *PickNodeR* at line 10 of Algorithm 2 means picking the most appropriate node to read data, the role of this function is load balancing, and the specific process is formalized as follows:

$$score_i = \frac{\alpha \cdot (1-p_i)}{\beta \cdot (1 + \frac{t_i}{T_c})} = \frac{T_c \cdot [\alpha \cdot (1-p_i)]}{\beta \cdot (T_c + t_i)} \tag{3}$$

In Eq. (3), the p_i is the predicted access amount after normalizing which the access will request node$_i$,and $p_i \epsilon [0, 1]$, the higher the predicted access amount, the lower the score of node$_i$, which can balance loads. Other parameters have the same meaning as the previous equation, and the edge node with the highest score will be chosen as the target node.

The function *PickNodeP* at line 10 of Algorithm 3 means picking the most appropriate node to offload the data processing task, and the specific process is formalized as follows:

$$s_i = \frac{\alpha \cdot r_i + (1-\alpha) \cdot c_i}{\beta \cdot (1 + \frac{t_i}{T_c})} = \frac{T_c \cdot [\alpha \cdot r_i + (1-\alpha) \cdot c_i]}{\beta \cdot (T_c + t_i)} \tag{4}$$

The Eq. (4) is similar as the Eq. (2) in form, however, the meaning is different. In Eq. (4), the c_i means the free CPU computing power after normalizing. But in Eq. (3), the n_i means the predicted access amount after normalizing, and the $c_i \epsilon [0, 1]$.

2.6 Prediction Module in DECS

The prediction module can proactively replicate hot data in advance and predict the location that most likely to be accessed according to the user's storage pattern to write data by machine learning. This module runs at the beginning of each time period periodically and automatically, which is not called by any other module. The following will introduce the specific details of this module.

At the beginning of each time period, this module will check the metadata to acquire the access information of the files in this edge cluster of the last time period. The length of one period may be one day or be custom set. Each write operation (not temp data) from end device will be recorded in the metadata such as *whoStoredFiles*, and each read operation (not temp data) will increase the access amount record in metadata such as *fileAccessFromNodeIpEachPeriod*.

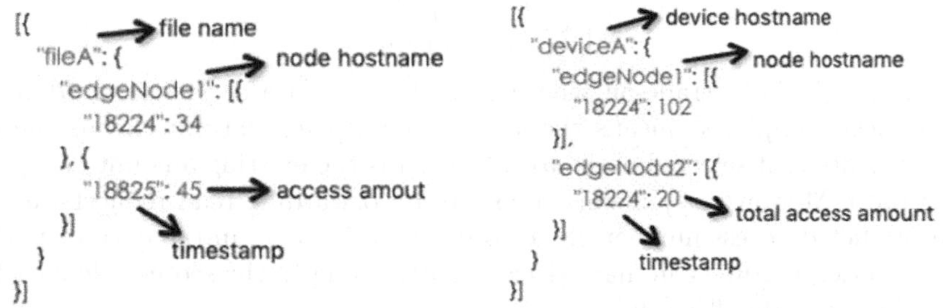

(a) *fileAccessFromNodeIpEachPeriod* (b) *hotNodeForSpecificDec*

Fig. 3. Examples of metadata helping to predict

The example of the metadata *fileAccessFromNodeIpEachPeriod* can be seen in Fig. 3(a). Which means the requests from end devices to access *fileA* hit the edge node *edgenode1* 34 times in *time period 18224* and 45 times in *time period 18225*. Another example is the metadata *hotNodeForSpecificDec* can be seen in Fig. 3(b). Which means the requests to access files stored by *deviceA* hit edge node *edgenode1* 102 times totally in *time period 18224*, and hit *edgenode2* 20 times in *time period 18224*. These two metadata can help to predict the access amount in the current time period, and the predicted result can decide the metadata *forwardingRule* to forward the storage request during each write operation and decide whether replicate the data to a hot node.

At the beginning of each time period, when this module predicts the most appropriate node to form the *forwardingRule*. This module gets a total access amount list for each period in each edge node for the files stored by the specific device, and the results are organized as a triple <*deviceIp, edgeNodeIp, accessRecordEachPeriod*>. If the length of *accessRecordEachPeriod* is too short or the last access amount of *accessRecordEachPeriod* is too lower, this item will be excluded and not be predicted, or else, this model will train an LSTM model using the list *AccessRecordEachPeriod* as training data to predict the access amount of node *edgeNodeIp* for the device *deviceIp*. If the predicted access amount greater than the threshold, this item will be recorded in the metadata *forwardingRule*. When the same device stores file again, the request will be forwarded to the hot edge node, and the data will be stored in the selected edge node. The thresholds mentioned above are the hyperparameters and can be adjusted according to the actual situation.

When this module predicts whether to replicate hot data, this module will acquire the metadata *fileAccessFromNodeIpEachPeriod* to get the access amount for each file at each edge node, and the result also organized as a tripe <*fileName, edgeNodeIp, AccessRecordEachPeriod*>. The rest of process of predicting is the same as above, this model will train an LSTM model for the eligible list *accessRecordEachPeriod* and predict the access amount in node *edgeNodeIp*, if the access amount more significant than the threshold, then will be decided whether replicating data to this hot edge node.

$$\begin{cases} s_r = \frac{r_i \cdot T_c}{e^{n+1} \cdot (T_c + t_i)} \\ s_f = \frac{T_c}{p_i \cdot e \cdot (T_c + t_i)} \end{cases} \tag{5}$$

Equation (5) is to trade-off whether replicating data or just forwarding the read requests simply, s_r means the score of replicating data, r_i means the free space utilization of selected node $node_i$, and n is the existing amount of replicas of this data. Moreover, s_f means the score of forwarding read requests, and p_i is the predicted access number of this data in node $node_i$ in the current period. The prediction module can make decisions according to the scores calculated by Eq. (5) in the trade-off module.

3 Implementation and Evaluation

In this section, we will introduce the implementation and performance evaluation of proposed DECS, including the experiment parameters, experimental data, and experiment results analysis.

3.1 Implementation

We implemented the collaborative edge-edge data storage service DECS based on the *Spring Cloud* micro-service framework, and each DECS run as a service in each edge node. Moreover, each edge node was running on the Ubuntu operating system. We limit the CPU cores, memory size, bandwidth, and disk size differently at each edge node to simulate the practical heterogeneity. In each edge node, we utilized the Redis to store the metadata, and the Redis runs at cluster mode can synchronize information with others using the gossip protocol. DECS provided its service using the RESTful API, and end devices such as IoT devices or smartphones can access these RESTful API by HTTP communication.

3.2 Experimental Parameters and Data

In this experiment, we used fifteen computers with the CPU Intel(R) Xeon(R) E5-2620 v3 with main frequency 2.40 GHz, and this CPU had eight cores. The computer has 32G memory and 2 TB disk space. These fifteen computers are acting as the edge devices and run our data services DECS and other services. In this experiment, to store the heterogeneous data, we stored the data as file, and this paper only considered the data's size, ignoring the data's content and semantics.

All experiment data were generated from the end devices, the end devices, including the video surveillance devices such as *Hikvision DS-2CD3T45(D)-I3/I5/I8*, smartphones such as *iphone 11*, *Mi 8*, and laptops such as *Macbook Pro*. The data generated by the end devices can be many types such as image, text video, music, or others.

Table 4. Edge node types and resource limit

	Small node	Medium node	Large node
Amount	7	5	3
CPU cores	2	4	8
Memory (GB)	8	16	32
Disk size (GB)	100	500	1000

We divided the edge nodes into three categories according to the performance, namely small, medium, and large nodes, and the detailed parameters can be seen in Table 4.

Table 5. Experiment parameters

Parameter	Value
Experimental time (hours)	30
Size of each time period (hour)	1
Epoch of LSTM	100
Batch size of LSTM	1
Step size of LSTM	5
Layers of LSTM	2
Neurons of input layer	16
Neurons of output layer	1

The Table 5 are the detailed experiment parameters. This experiment was running lasting 30 h, and each time period's size was set to one hour. Other parameters were the hyperparameters of LSTM.

3.3 Experimental Scenarios and Applications

- **Path Fitting Application:** We used this application as an experimental scenario at a parking lot. In this scenario, end devices were IoT devices, mostly such as video surveillance or sensors. This application needed computing and storage resources.
- **Edge Dropbox Application:** We used this application as an experimental scenario at our university campus. In this scenario, end devices were mobile phones or laptops used by users. This applicable needed the storage resource.

These two applications ran at the edge devices mentioned above.

3.4 Baseline

In this experiment, we compared DECS with two baselines:

- **No Service:** The first one is that no data service runs on the edge node, which means each edge node does not collaborate to store.
- **ElfStore:** The second one is ElfStore, a resilient data storage service proposed by Sumit Kumar Monga et al. for federated edge and fog [15]. ElfStore also needs collaboration between edge nodes to store data.

3.5 Performance Evaluation

Firstly, we evaluated the data storage module and prediction module in DECS. For the write operation, the experiment result can be seen in Fig. 4. The left figure represents the ratio of forwarded write requests over time. The right figure represents read latency to access the data written by the same end device over time. The result shows that the ratios of forwarded write requests of DECS and ElfStore were increased, and the average read latency of DECS and ElfStore was decreased over time.

These results proved that DECS could build forwarding rules accurately for this end device when the same end device frequently writes data. Which can record this end devices' storage pattern, and forward this end device's write request to the edge node which is most likely to be accessed later. The results proved that DECS forwarded writing operations proactively in advance to decrease the read latency of the edge cluster effectively.

Then we evaluated the read operation in DECS, and the results can be seen in Fig. 5. These two subfigures represented the read latency of the path fitting application and the edge dropbox application. Each subfigure can be split into two parts, namely before replicating and after replicating. The experiment result proved that the prediction module in DECS could replicate the hot data to the

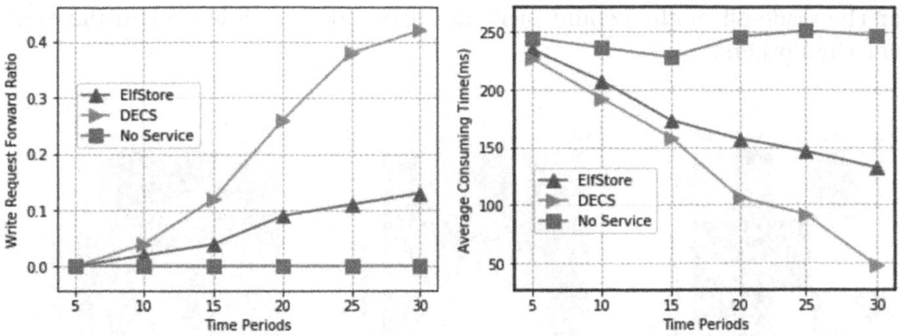

Fig. 4. Read latency for the same writer's data

target edge node in advance, proactively referencing the predicted access amount. In the path fitting application, most of the data were temp data, these temp data was not be replicated, so the improvement of reading performance was not evident. In the edge dropbox application, most of the data were regular data, and the read latency for hot data decreased evidently for DECS after replicating or migrating hot data.

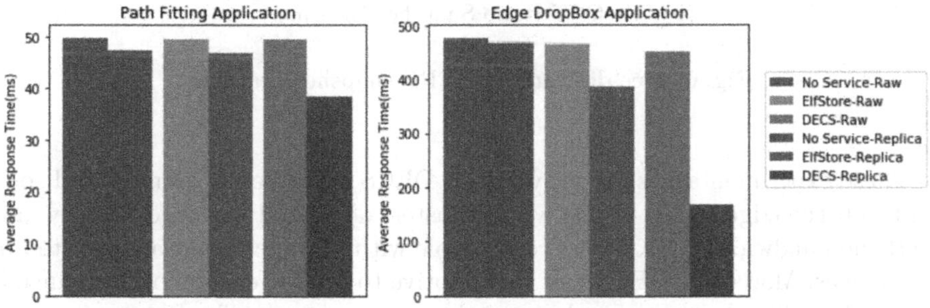

Fig. 5. Read latency before and after replicating or migrating data

Then we evaluated the trade-off module and data processing module in DECS. The experiment results can be seen in Fig. 6. We collected the free disk space ratio and idle CPU ratio snapshot of all the fifteen edge devices in this experiment.

The Fig. 6(a) represents the free disk ratio collected after all time periods of the experiment. This result proved that the trade-off module could pick edge node which with more free disk size to store data. The Fig. 6(b) shows the idle CPU ratio snapshot of all the fifteen edge nodes at a random time. We can see that when the data processing module process the offloaded tasks, the trade-off module can pick the freest edge node to execute the task. This result proved that the trade-off module could pick the edge node that idle CPU utilization is lower than others.

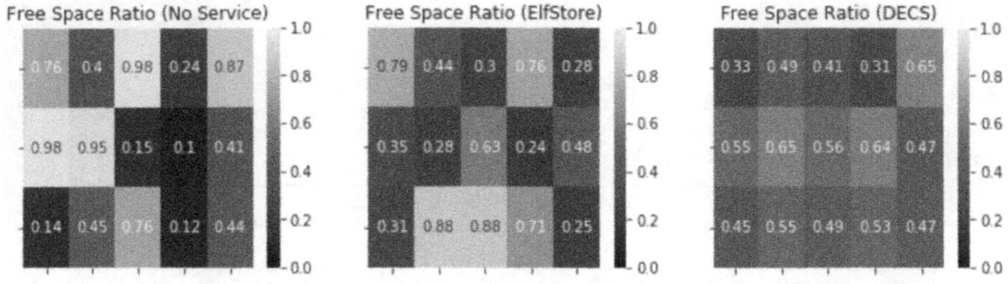

(a) Free Disk Space Ratio Heatmap

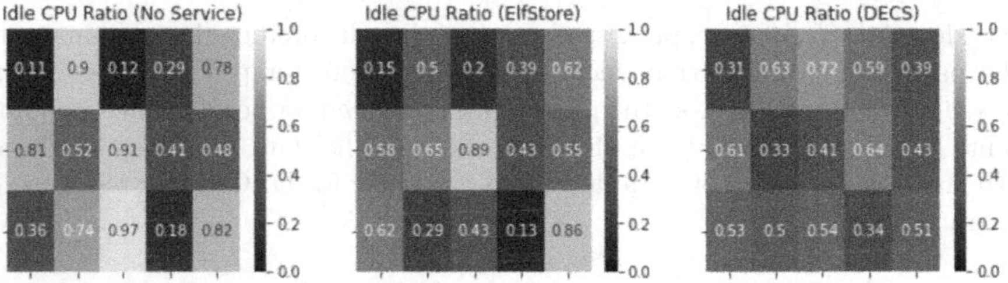

(b) Idle CPU Ratio Snapshot Heatmap

Fig. 6. Free disk and idle CPU snapshot heatmap

These experimental results proved that DECS could better manage and coordinate all the edge nodes of the whole cluster, decreased the read latency, and saved the bandwidth. DECS was compatible with all types of data generated by end devices. Moreover, DECS can be adaptive to heterogeneous performance of edge nodes, offload storage, and computing tasks more evenly.

4 Related Work

Some current researches such as [1–3, 9] proposed approaches to solve the resources management, peers discovering, or self-adaptation problems on the decentralized edge nodes without core cloud. Study [1] proposed a resource management framework in the decentralized edge cluster decoupled from the cloud to deploy latency-sensitive IoT applications. Moreover, research [9] proposed a decentralized storage system, which utilized the proposed peer living loop algorithm to find other live nodes. What is more, the study [3] proposed a self-adaptive computing framework at the edge to manage uncertainties at runtime and satisfy the requirements of a decentralized data-intensive system. Furthermore, study [2] proposed an approach for coordinating resources and computations in edge clusters, which can tolerate unreliability by self-adaptation to device failure, mobility, and withdrawal.

Other researches implemented the decentralized schemas based on the blockchain technology such as [5,6,17,20], and these studies concentrated on the security and collaboration of edge nodes. Study [6] analyzed the advantages and disadvantages of decentralized cloud storage solutions using blockchain technology. Moreover, research [5] proposed a multiagent and collaborative governance approach of decentralized edge micro clouds with blockchain-based distributed ledgers to improve the efficiency in decision-making. This paper [20] proposed a blockchain-based decentralized architecture called Microthings Chain, which supporting cross-domain data sharing safely. The study [17] proposed that using blockchain technology as a platform hierarchical and distributed control systems, and utilizing the edge nodes as the executive level for the actual process.

Some current researches focused on providing services to end devices based on edge nodes or collaborated with the cloud, or the service framework on edge, such as [11,14,16,18]. Study [14] proposed the deployment of edge-computing in 5G NFV environment and services-based architecture. Research [18] proposed an algorithm called OREO to jointly optimizes dynamic service caching and task offloading to solve service heterogeneity and decentralized coordination challenges. Moreover, another paper [11] used W-DAG to model the data-intensive service or business and make resource allocation between the cloud data center and edge nodes to improve QoS. The study [16] proposed a cloud-edge collaboration framework for IoT data analytics that can provide data analytics services.

There were studies concentrated on the storage at the edge. Which includes collaborated with the cloud, forecasted for pre-store, self-adaptive according to specific situations, such as [7,10,12,13]. The study [10] proposed a single namespace and resource-aware federation file system called EDGESTORE for edge nodes, which can aggregate storage namespace and federate resources of edge nodes to enable high resource-sharing in the federation. The study [12] proposed an effective model for edge-side collaborative storage in data-intensive edge computing, which can forward the offload requests to the with-data edge nodes. The study [7] proposed an edge storage method based on multi-view learning, which predicted the resources that need to be pre-stored based on multi-feature linear discriminant analysis, including their real-time performance and user's location. The study [13] presented a three-layer architecture model for data storage management on edge, including an adaptive algorithm that dynamically finds a trade-off between providing high forecast accuracy necessary for efficient real-time decisions.

5 Conclusion

With the development of IoT and 5G and the improvement of the hardware's performance, edge nodes of the Internet also have the power to store data or process tasks. However, we found that many local-area and small-scale scenarios also need the computing or storage resources of the edge nodes, and for lower costs, these scenarios usually partly decoupled with the core cloud.

In this paper, we proposed a collaborative edge-edge data storage service for edge computing in local-area scenarios called DECS. DECS runs at each edge node, which not only provides the fundamental data storage sub-service but also manages and coordinates the storage and computing resource of the whole cluster by the trade-off module and the prediction module. Especially, the latter can replicate data or forward requests in advance by predicting the access amount proactively. We implemented the proposed service and evaluated the performance compared with state-of-art research. The results proved that DECS was more suitable for these scenarios, lowered the access latency, and saved total bandwidth by the prediction beforehand. Moreover, the trade-off module improved the resource utilization of the whole edge cluster.

Acknowledgment. This paper is supported by Project 213.

References

1. Avasalcai, C., Tsigkanos, C., Dustdar, S.: Decentralized resource auctioning for latency-sensitive edge computing, pp. 72–76, July 2019. https://doi.org/10.1109/EDGE.2019.00027
2. Casadei, R., Viroli, M.: Coordinating computation at the edge: a decentralized, self-organizing, spatial approach, pp. 60–67, June 2019. https://doi.org/10.1109/FMEC.2019.8795355
3. D'Angelo, M.: Decentralized self-adaptive computing at the edge, pp. 144–148, May 2018
4. Decandia, G., et al.: Dynamo: Amazon's highly available key-value store. **41**(6), 205–220 (2007)
5. Freitag, F.: On the collaborative governance of decentralized edge microclouds with blockchain-based distributed ledgers, pp. 709–712, December 2018. https://doi.org/10.1109/WI.2018.000-7
6. Gabriel, T., Cornel-Cristian, A., Arhip-Calin, M., Zamfirescu, A.: Cloud storage. A comparison between centralized solutions versus decentralized cloud storage solutions using blockchain technology, pp. 1–5, September 2019. https://doi.org/10.1109/UPEC.2019.8893440
7. Gao, Q., Gao, L., Xue, T., Zhu, X., Zhao, X., Cao, R.: Multi-view learning based edge storage management strategy, pp. 366–371, August 2018. https://doi.org/10.1109/CBD.2018.00072
8. Ghemawat, S., Gobioff, H., Leung, S.T.: The Google file system. In: Proceedings of the 19th ACM Symposium on Operating Systems Principles, Bolton Landing, NY, pp. 20–43 (2003)
9. Gheorghe, A., Crecana, C., Negru, C., Pop, F., Dobre, C.: Decentralized storage system for edge computing, pp. 41–49, June 2019. https://doi.org/10.1109/ISPDC.2019.00009
10. Khan, A., Muhammad, A., Kim, Y., Park, S., Tak, B.: EDGESTORE: a single namespace and resource-aware federation file system for edge servers, pp. 101–108, July 2018. https://doi.org/10.1109/EDGE.2018.00021
11. Li, X., Lian, Z., Qin, X., Abawajyz, J.: Delay-aware resource allocation for data analysis in cloud-edge system, pp. 816–823, December 2018. https://doi.org/10.1109/BDCloud.2018.00122

12. Li, Y., Luo, J., Jin, J., Xiong, R., Dong, F.: An effective model for edge-side collaborative storage in data-intensive edge computing, pp. 92–97, May 2018. https://doi.org/10.1109/CSCWD.2018.8465306
13. Lujic, I., Maio, V.D., Brandic, I.: Efficient edge storage management based on near real-time forecasts, pp. 21–30, May 2017. https://doi.org/10.1109/ICFEC.2017.9
14. Lv, H., Chen, D., Wang, Y.: Deployment of edge-computing in 5G NFV environment and future service-based architecture, pp. 811–816, December 2018. https://doi.org/10.1109/CompComm.2018.8780937
15. Monga, S.K., Ramachandra, S.K., Simmhan, Y.: ElfStore: a resilient data storage service for federated edge and fog resources, pp. 336–345 (2019)
16. Moon, J., Cho, S., Kum, S., Lee, S.: Cloud-edge collaboration framework for IoT data analytics, pp. 1414–1416, October 2018. https://doi.org/10.1109/ICTC.2018.8539664
17. Stanciu, A.: Blockchain based distributed control system for edge computing, pp. 667–671, May 2017. https://doi.org/10.1109/CSCS.2017.102
18. Xu, J., Chen, L., Zhou, P.: Joint service caching and task offloading for mobile edge computing in dense networks, pp. 207–215, April 2018. https://doi.org/10.1109/INFOCOM.2018.8485977
19. Yousefpour, A., et al.: All one needs to know about fog computing and related edge computing paradigms: a complete survey (2018)
20. Zheng, J., Dong, X., Zhang, T., Chen, J., Tong, W., Yang, X.: MicrothingsChain: edge computing and decentralized IoT architecture based on blockchain for cross-domain data sharing, pp. 350–355, October 2018. https://doi.org/10.1109/NANA.2018.8648780

DCT: A Deep Collaborative Filtering Approach Based on Content-Text Fused for Recommender Systems

Zhiqiao Zhang(✉), Junhao Wen, and Jianing Zhou

School of Big Data and Software Engineering, Chongqing University, Chongqing, China
{zhangzhiqiao,jhwen,jnzhou}@cqu.edu.com

Abstract. Recommender systems commonly make recommendations by means of user-item interaction ratings. One of the basic methodologies of recommendation is collaborative filtering, which exploits users' and items' latent space features to make predictions of personalized ranking list for an individual user. However, most of existing collaborative filtering approaches only employ explicit interaction ratings data to predict user preferences, and neglect the necessary of exploiting implicit feedback data and auxiliary information in promoting the performance recommendation. In this paper, we raise a novel model of recommendation on the basis of neural networks architecture. Concretely, the model exploits both of interaction data and content text information as input and adopts two parallel neural networks to learn the latent feature representations of users and items for a better performance. We utilize three kinds of real-world data to make extensive evaluations on the model. The experimental results reveal that the method we proposed dramatically outperforms the state-of-the-art methodologies and achieves expressively improvement in performance.

Keywords: Recommender systems · Deep learning · Textual information

1 Introduction

With the development of technology, numerous applications emerge in large numbers resulting in the expansion size in data which has led to the problem such as 'information overload'. Recommender systems (RSs) are an effective means of solving this phenomenon that can find the valuable information to meet users' demands quickly and efficiently. It provides unique user with a personalized recommendation service. The fundamental essence of RS is prediction which utilizes the information of users' preferences on items to recommend [1, 2]. In addition, RSs are everywhere which have been adopted into numerous fields, such as information retrieval, e-commerce and so on.

The methodologies of RSs major include collaborative filtering (CF), content-based recommendation, and hybrid recommendation. CF is the most widely applied approach, which leverages the similarities among items to recommend for users. In other words, if users have bought some products in the past, the strategy of collaborative filtering infers

H. Gao et al. (Eds.): CollaborateCom 2020, LNICST 349, pp. 392–405, 2021.
https://doi.org/10.1007/978-3-030-67537-0_24

that they would be interested in similar stuffs in the future [3]. Matrix factorization (MF) is one of the most commonly used methodologies of collaborative filtering due to its simplicity and scalability. MF employs latent feature factors to denote to items and users respectively, which will be adopted to obtain predicted ratings of users on a set of items.

To make predictions of personalized ranked lists on items for recommendation, matric factorization captures users' and items' low-dimensional space through ratings interaction matrix. Whereas, the method only using the explicit ratings data is insufficient to make a good top-n recommendation [4–6]. Many existing methods take the implicit feedback into consideration for a better performance [7–9]. Such as collaborative denoising autoencoder (CADE) [10] and neural collaborative filtering (NCF) [11]. Nevertheless, both of them only employ implicit feedback and neglect the explicit ratings.

In the meanwhile, the growing of the number over of stuffs and users in applications will sharpen sparseness of rating data of users on items, which will impact the performance of recommendation. Consequently, several methods take account of ratings and auxiliary information (e.g. social relationship, item description documents, etc.). Some previous methodologies (e.g. stacked denoising auto-encoder (SDAE) and latent dirichlet allocation (LDA)) adopt item description documents to enhance the performance [12, 13]. Nevertheless, those hybrid recommendation methodologies commonly only use bag-of-words model to capture content information, neglecting the semantic features and deep latent features, which will affect the performance in predictions.

To dispose the above-mentioned phenomenon, we exploit two parallel deep neural networks to deal with interaction data and auxiliary information at the same time for getting a better performance. Moreover, owing to its ability of powerful representation learning, deep learning has obtained significant progress in image identification, speech recognition, nature language processing and so on. Especially, applying deep neural networks into recommender systems facilitates deeper extract the users' and items' latent representations [14–16].

In our work, we address a novel collaborative filtering method on the basis of neural networks. The model leverages both of interaction data and auxiliary information to make recommender prediction through two parallel networks. In details, we capture items' and users' deep latent feature representations from the interaction ratings data by utilizing the multi-layer neural network. In the meanwhile, we convert content text into latent feature vectors through another parallel neural network layers for enhancing the final item representation. Moreover, we take full advantage of explicit data and implicit feedback information to optimize the model, which could be better to learn the latent features of items and users. As for the experiments, we compared performance variation the model proposed with other five methodologies on three kinds of real-world datasets. In addition, we make extensively evaluations on hyper-parameters of the model. Experimental results signify that the approach we raised is effective.

The rest part organization of this paper is as bellow. First, Sect. 2 has a brief review of related work. Next, Sect. 3 states problems and structure of the model concretely. Then, Sect. 4 analyzes the experiments over three kinds of data. Finally, Sect. 5 outlines the contributions and future work.

2 Related Work

2.1 Collaborative Filtering

It can be perceived as a process of word-of-mouth for the strategy of collaborative filtering (CF). It makes recommendations for users by making use of the rating data and other auxiliary information. Some models of collaborative filtering are quite remarkable.

Wang [17] combined LDA with probabilistic matrix factorization (PMF) in order to optimize the method performance of CF. Hu [18] made full use of implicit feedback data to identify the unique properties. They proposed that the interaction information of users on items can be regarded as indicant of positive and negative preference in reference to disparate confidence levels. Lee [19] adopted variational autoencoders (VAE) to extent the model of CF, in the meanwhile, they leveraged auxiliary information to intensify items' and users' representations. Liang [20] also applied VAE to the model of collaborative filtering, which utilized implicit feedback. Based on that, then they devised a generative model with multinomial likelihood.

The methods of CF try to take advantage of the auxiliary information or learn unsupervised topic models for promoting the performance. However, by reason of data sparse, the low-dimensional representations learned from the model are not perfectly efficient. Meanwhile, these methods only exploit bag-of-words models and neglect the semantic features of content text, which will lead to some limitations. Whereas, deep learning can be utilized to address this issue.

2.2 Deep Learning

Deep Learning lets computational models with multi-processing-layers extract representations of data with abstraction multi-levels [21]. It has a significant attention in object detection, natural language processing (NLP), information retrieval and so on. Furthermore, deep learning has made great progress in recommender systems. And the recommendation models using the strategy of deep learning adopt deep neural networks, then employ user-item interaction information as input for obtaining items' and users' latent representations respectively for which can be used to make personalized prediction.

Xue [22] presented to combine the explicit ratings with implicit feedback, and capture users' and items' low-dimensional representations by utilizing deep neural networks. He [11] raised a common neural collaborative filtering architecture on the basis of deep learning, modeling with linearity and non-linearity of multi-layers. Zheng [23] exploited comments to obtain latent feature space of items and users respectively through parallel neural networks. Then, they jointed items' features and users' preference of learned from model to predict ratings.

These recommendation methods have proved that it is effective to obtain items' and users' latent representations through the content text and deep neural network respectively. Nonetheless, to our knowledge, there is no previous methods that obtain the latent representations of items, users and content text at same time by adopting the deep neural networks, and take both of explicit and implicit feedback into consideration as well.

3 Proposed Model

First, we state the problem of our model in this section. Then, we analyze the model addressed in details. Finally, we put forward the objective function to optimize the method.

3.1 Problem Statement

Assuming that there are M users $U = \{u_1, \ldots, u_M\}$, N items $V = \{v_1, \ldots, v_N\}$ and each item contains content text. On account of users interacting with items, make $R \in \mathbf{R}^{M \times N}$ denotes to the interaction matrix, where R_{ij} indicates the rating attained from user i interacted with item j. If R_{ij} is unknown, we mark it *unk*.

Commonly, existing two kinds of methodologies to devise interaction matrix $Y \in \mathbf{R}^{M \times N}$ of users on items based on the implicit feedback of R, where Y_{ij} indicates the $(i, j)_{th}$ entry of Y. Several recommendation methods employ Eq. (1) to model matrix Y, which regard the observed ratings as 1 and neglect the significance of the value of ratings. While some others construct user-item matrix Y by adopting Eq. (2) that reserves the value of ratings which can be applied to show the degree of preference of user u_i on item v_j. Based on this, we employ Eq. (2) to model interaction matrix Y. Meanwhile, we mark zero if ratings of users on items are unknown, which can be taken as implicit feedback in our work.

$$Y_{ij} = \begin{cases} 0 & if \ R_{ij} = unk \\ 1 & otherwise \end{cases} \tag{1}$$

$$Y_{ij} = \begin{cases} 0 & if \ R_{ij} = unk \\ R_{ij} & otherwise \end{cases} \tag{2}$$

Recommender systems are typically deemed to calculate the unknown entry of matrix Y, which means ratings prediction. Model-based collaborative filtering methods commonly make use of latent factor model (LFM) that exploits the dot product of user' and items' latent vectors for predicting.

$$\hat{Y}_{ij} = F^{LFM}(u_i, v_j | \Theta) \tag{3}$$

Where u_i and v_j respectively indicate the i_{th} user and j_{th} item. $\hat{Y}_{ij}$ is the predicted value of rating by adopting u_i and v_j. And F denotes to the function of mapping the parameters into predicted values. In our work, we follow the LFM and extent it.

Besides, on account of the recommending tasks which are related to content text. In our work, we exploit a sequence $X_j = (w_1, \cdots, w_l)$ with l word tokens to denote to the j_{th} item content text, where each word token is in the vocabulary containing W words.

3.2 Proposed Model

Facing the problem of data sparseness, many existing collaborative filtering methodologies try to adopt auxiliary information (e.g. social relationship, item description documents, etc.) for promoting the performance of recommendation. Whereas, most of previous models do not take the relevance of rating matrix and content text into consideration. They only employ one kind of information data as input of the model to predict. In addition, some traditional methods use topic model through unsupervised learning to pre-train text as features, which is insufficient to obtain the deep latent features of textual.

In our model, we exploit both of interaction ratings data and content text for promoting the performance of recommendation. Concretely, the model adopts two parallel neural networks to deal with interaction dataset and content text information at the same time. Figure 1 demonstrates our model's structure in details. We obtain the deep latent feature representations of items and users from interaction ratings matrix by utilizing the multi-layer neural network. Meanwhile, we convert content text into item latent representations through the embedding-layer and neural network layers. Employing the representation of embedding-based is helpful for learning the latent text features owing to word embedding can be pre-trained on large corpus through the unsupervised way. In addition, neural networks can better understand the deep latent features of content text. We adopt linear strategy to fusion the two kinds of item representations extracted from two parallel neural networks to enhance the final item representation. Moreover, we utilize users' and items' representations to predict the ratings.

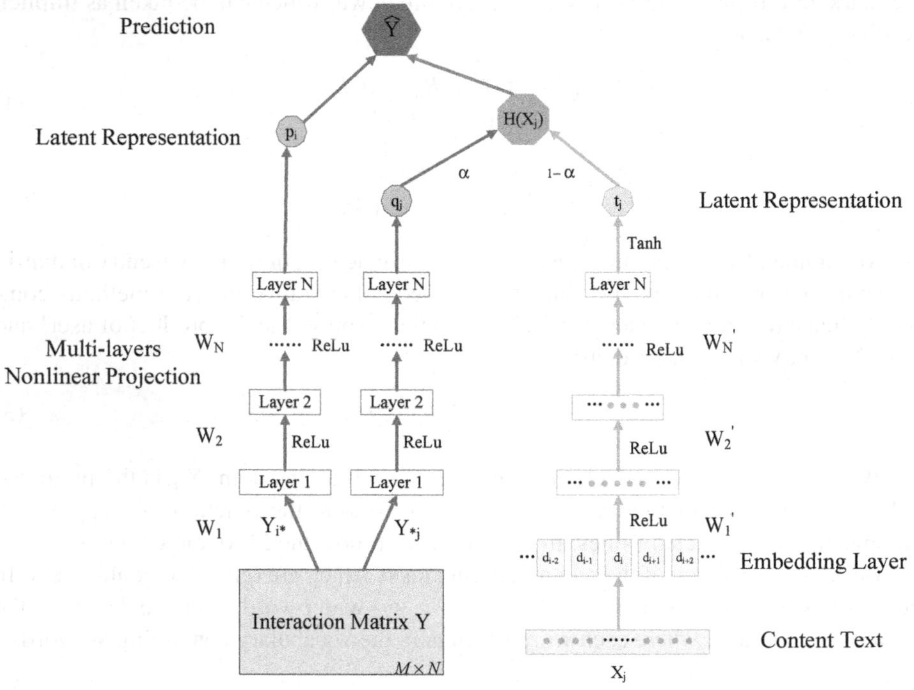

Fig. 1. The architecture of the proposed model.

Furthermore, we adopt Eq. (2) to model the interaction matrix Y, where Y_{i*} denotes to the high dimension vector of user u_i which demonstrates the ratings of i_{th} user over all of items and Y_{*j} is the high dimension vector of item v_j which means ratings of j_{th} item graded by all users. The q_j and p_i respectively indicate the latent representations of item v_j and user u_i. Whereas, we do not directly take the q_j as the item's final representation. The model fuses the latent vector q_j and text latent feature vector t_j which extracted from interaction matrix and item content text respectively. Therefore, the item's final latent representation becomes as Eq. (4).

$$H(X_j) = \alpha q_j + (1 - \alpha)t_j \tag{4}$$

Where α denotes to be a hyper-parameter. And t_j is the content text latent feature vector. Concretely, at first, the content text of item v_j is deemed to be a sequence $X_j = (w_1, \cdots, w_l)$ with l words. In addition, we leverage the word embedding model to process the item content text and adopt a lookup table where each word can be expressed by a single vector in the vocabulary. Then, we joint vectors of words in text sequence X_j to get a numeric text matrix $D_j^{\rho \times l}$ through the embedding layer.

$$D_j = \begin{bmatrix} & | & | & | & \\ \cdots & d_{i-1} & d_i & d_{i+1} & \cdots \\ & | & | & | & \end{bmatrix}$$

Where ρ indicates embedding dimension size of each word vector d_i and l signifies the length of content text. At last, we obtain text feature vector t_j through the deep neural network. In our model, we adopt tanh as activation function on the output-layer in text neural network for getting better performances.

$$tanh(x) = \frac{e^x - e^{-x}}{e^x + e^{-x}} \tag{5}$$

Besides, in the neural network layers, we employ b_i and W_i respectively indicate the i_{th} bias and weight matrix of the middle-hidden-layers l_i, i $\in [1, N]$. In the meanwhile, in order to prevent the vanishing gradient, we adopt ReLU as activation function.

$$\text{ReLU}(x) = max(0, x) \tag{6}$$

Moreover, in our work, we follow and extent LFM to predict ratings which is formulated as Eq. (7). And the model employs H(·) instead of q_i as the final item latent representation on account of integrating item content text feature factors into the model. H(·) can be referred as items' universal feature extractor, which can be extended in various way.

$$\hat{Y}_{ij} = p_j^T H(X_j) \tag{7}$$

3.3 Loss Function

It is another key component to devise an advisable objective function for optimizing the method in our work, which is on the basis of unobserved implicit feedback information and observed ratings. The most common of objective function is indicated as Eq. (8).

$$L = \sum\nolimits_{y \in Y} l(y, \hat{y}) + \Omega(\theta) \tag{8}$$

Where $\Omega(\theta)$ denotes to the regularizer and $l(\cdot)$ is the loss function. Most of existing recommendation methodologies common use squared loss as loss function. However, the squared loss could not do well with implicit feedback data on account of the value of implicit feedback in Y_{ij} is a binarized 1 or 0 indicating that if users interact with items or not. He [11] proposed a binary cross entropy loss function with implicit feedback as follows.

$$L = \sum_{(i,j)\in Y} Y_{ij}\, log\hat{Y}_{ij} + (1 - Y_{ij})\, log\left(1 - \hat{Y}_{ij}\right) \tag{9}$$

To integrate the explicit ratings into loss function, extending Eq. (9) to get a new normalized cross entropy loss as below.

$$L = -\sum_{(i,j)\in Y} \frac{Y_{ij}}{max(R)}\, log\hat{Y}_{ij} + \left(1 - \frac{\hat{Y}_{ij}}{max(R)}\right) log\left(1 - \hat{Y}_{ij}\right) \tag{10}$$

Where $\max(R)$ is the maximum in all ratings, which is utilized for normalization. Based on this, diverse value of Y_{ij} has different impacts on loss. In our work, we take the normalized cross entropy loss with implicit and explicit feedback as an objective function. However, for the predicted value of $\hat{Y}_{ij}$ can be negative, we exploit Eq. (11) to convert the value of predictions, where σ is an extremely low number and the value is $1.0e^{-8}$.

$$\hat{Y}'_{ij} = max\left(\sigma, \hat{Y}_{ij}\right) \tag{11}$$

4 Experiments

We make extensive evaluations of performances on the model proposed in this section. Firstly, we briefly review the settings of experiments in details, covering implementation details, evaluation criterion and datasets. Secondly, we compare performances of the model with various settings according to the extensive experiments, which contains thc number of neural network layers, integrating ratio, negative sampling ratio and so on.

4.1 Experimental Settings

Datasets. To indicate efficacy of model presented for personalized ranking recommendation, we exploit three kinds of real-world datasets to conduct experiments. One is MovieLens 100 K (ML-100 K), another is MovieLens 1 M (ML-1 M), and a third is Amazon. Those three kinds of datasets contain the explicit ratings which are on a scale of 1 to 5. The data of Amazon covers the comments of items, which can be deemed as content text. Nonetheless, MovieLens does not have the description of items, we captured the content text related to the items from IMDB. In the meanwhile, since the data of Amazon is highly sparse. We filtrated the Amazon data and reserved the users which have at least 5 interactions with the items. In addition, we removed the items which do not have the relevant content text information on each data. Table 1 illustrates the characters of three kinds of datasets specifically.

Table 1. Statistics of three real-world datasets.

Datasets	Users	Items	Ratings	Rating density
ML-100 K	943	1526	95349	6.626%
ML-1 M	6040	3544	993482	4.641%
Amazon	3374	23963	38648	0.048%

Evaluation. We exploited leave-one-out valuation used in [24–26] to make appraisal of performances on each model. And we split each dataset by holding-out users' recent interaction as test sets and reserving residual dataset as training sets. However, ranking all items for each user would be too time-consuming. In view of this, we adopted the approach in [11] and [22] to sample 100 items that have no interaction with users at random, which we can rank rest items according to the prediction among these items. We utilized the metrics of normalized discounted cumulative gain (NDCG) and hit ratio (HR) to make evaluations on models' performances, where NDCG indicates the ranking ability and quality, HR intuitively reveals that if the item of test is appeared on ranked lists.

Implementation Details. We exploited python and tensorflow to implement the method proposed. In the experiments, mini-batch Adam [27] was used to optimize the model. And each mini-batch's size was set to be 128. Learning rate was set to be 0.0001. In the meanwhile, in our work, to initialize word vectors through the pre-trained word embedding, we used the CBOW [28] which is based on Gensim, and the size of word dimension was set as 200. As for hyper-parameters, we set the value of integrating ratio α as 0.7.

4.2 Performance Comparison

To demonstrate the improvement of our approach on the performance of recommendation, we made an intuitive and full comparison on the performance variance of the model raised and some other the-state-of-art approaches as below.

1. **BPR** [25]. Bayesian personalized ranking is the one of classical methods in ranking for recommendation. The method used pairwise ranking loss to optimize the matrix factorization model. We took the BPR as a baseline method for comparison.
2. **CTR** [18]. Collaborative topic regression combines probabilistic matrix factorization with LDA, and employs both of ratings and documents context to recommend. In the experiments, we took CTR as the baseline methodology for comparing.
3. **NeuMF** [11]. NeuMF employs cross entropy loss to optimize the model. It combines generalized matrix factorization with multi-layer perceptron to get a better performance in item recommendation. However, it only utilizes the ratings data to predict.

4. **ConvMF** [29]. Convolutional matrix factorization employs convolutional neural networks to map content text into probabilistic matrix factorization. Nonetheless, different from our method, it only leverages neural networks to deal with content text. As for the interaction information, it still takes the traditional strategy of PMF. We adjusted the hyper-parameters of this model in accord with it.
5. **DMF** [22]. Deep matrix factorization employs deep neural networks to get users' and items' latent feature factors. However, vary from our model, it only takes interaction ratings information into consideration. This is a state-of-the-art deep neural networks methodology for recommendation, we adjusted the hyper-parameters of this model in accord with it.
6. **DCT**: The model we proposed, which is on the basis of neural networks architecture. We exploit both of interactions matrix of user on item and content text as input for a better performance in rating prediction. Meanwhile, we utilize two parallel neural networks to learn the latent features respectively at the same time.

Table 2 summarizes the comparative results that indicates all models' performances on three kinds of datasets respectively by making use of metrics of NDCG and HR. In details, as is showed in the experiments, contrasted with DMF and NeuMF, which only take use of interaction ratings through neural networks, our method has a significantly improvement on the real-world datasets. It proves that integrate the auxiliary content text into model is helpful in promoting the performance in recommendation and solving the problem of data sparse. Meanwhile, compared with ConvMF, which adopts review document and do not consider the deep latent features of items and users in interaction ratings data, our method also has a breakthrough on three kinds of data, and the results reveal that employ two parallel deep neural networks to train interaction ratings and content text at same time is valid, which can improve the performance.

Table 2. Comparisons over different methodologies.

Datasets	Metrics	Method					
		BPR	CRT	NeuMF	ConvMF	DMF	DCT
ML-100 K	HR	0.534	0.626	0.672	0.701	0.648	**0.896**
	NDCG	0.397	0.401	0.409	0.592	0.415	**0.667**
ML-1 M	HR	0.446	0.598	0.705	0.688	0.712	**0.864**
	NDCG	0.383	0.421	0.457	0.579	0.464	**0.627**
Amazon	HR	0.314	0.487	0.569	0.589	0.575	**0.638**
	NDCG	0.202	0.385	0.408	0.421	0.410	**0.454**

4.3 Sensitivity to Hyper-parameters

Depth of Neural Network Layers. Extracting the latent representations of items and users from the interaction data through the multi-layer neural network plays an important part in our model. Consequently, an extensive evaluation was conducted on the performance of our model with disparate in numbers of hidden-layers. Figure 2 states the impact on the performance with different in numbers of layers on each iteration. However, we only show the first 15 iterations on the metrics of HR and NDCG on account of the space limitation. As shown in the figure, compared with 1-layer and 3-layers structure, the method with 2 hidden-layers gets best performance on ML-1 M and ML-100 K. Nevertheless, on the dataset of Amazon, our model with 1-hidden-layer performs best on the metric of HR during the iteration, while the model with 2 and 3 hidden-layers have a similar performance on the metric of NDCG instead. Even so, deep layers seem to be a little helpful for prediction accuracy. Maybe due to the reason of our model extracts the auxiliary text latent factor to enhance item representation through a parallel neural network, which does have a good effect on the performance.

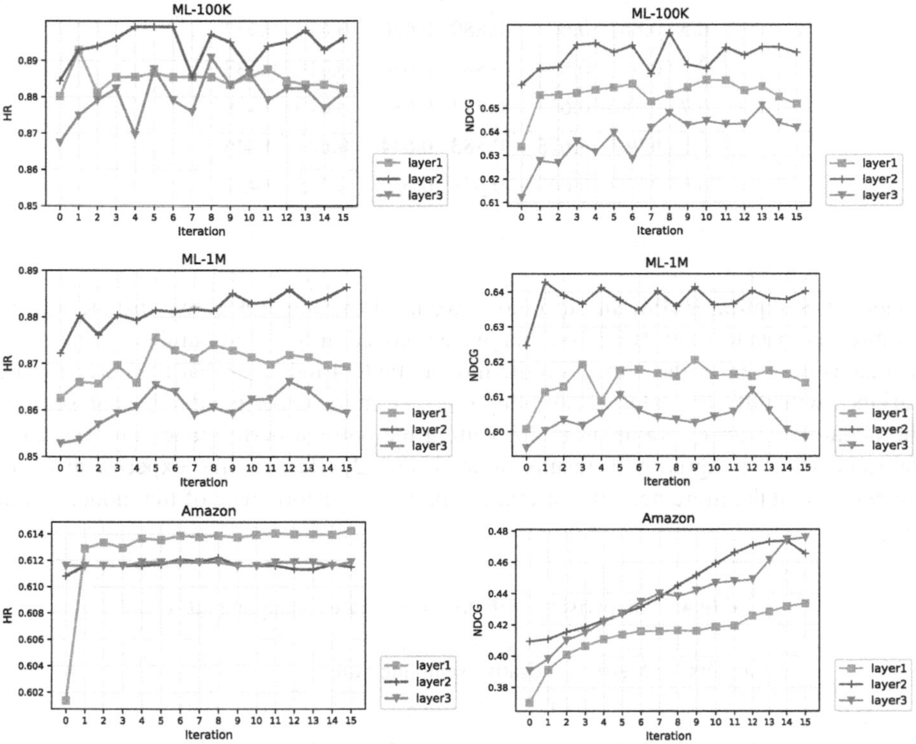

Fig. 2. Experimental results with different layers

Integrating Ratio. In our model, two kinds of item latent vectors are fused into one final item representation through the hyper-parameter α referred to the Sect. 3.2. Table 3 demonstrates the performance variance with disparate values of α on both of three kinds of datasets. As is shown in the table, the value of α around the number of 0.2 and 0.8 achieve the best performance. And it can be proved that integrate the representations of content text into the model and utilize two parallel neural networks to learn the representations from interaction data and content text at the same time seems useful for improving the performance in recommendation.

Table 3. Comparisons with different integrating ratio

α	ML-100 K		ML-1 M		Amazon	
	HR	NDCG	HR	NDCG	HR	NDCG
0.1	0.899	0.669	0.876	0.632	0.612	0.432
0.2	**0.905**	**0.672**	**0.885**	**0.640**	**0.623**	**0.451**
0.3	0.886	0.663	0.879	0.637	0.617	0.438
0.4	0.899	0.676	0.866	0.626	0.612	0.436
0.5	0.883	0.669	0.880	0.636	0.613	0.434
0.6	0.893	0.674	0.882	0.634	0.616	0.433
0.7	0.886	0.663	0.874	0.630	0.619	0.438
0.8	**0.901**	**0.675**	**0.883**	**0.644**	**0.621**	**0.446**
0.9	0.894	0.666	0.878	0.636	0.615	0.437

Negative Sampling Ratio. In our model, we need to sample negative instances from unobserved data to train. Based on this, we employed different in numbers of negative instances to observer the impact on the model. Table 4 indicates that the performance variance with different value of sampling ratio. From the experimental, we can see that the negative instance's value around 5 getting the optimal performance on both three datasets, which is agree with results of work in [22]. Besides, the experiment result expresses that the more negative instances, the better performance of the model would be.

Table 4. Comparisons with different negative sampling ratio.

Datasets	Metrics	Negative sampling ratio					
		1	2	5	7	9	10
ML-100 K	HR	0.889	0.890	**0.896**	0.892	0.892	0.882
	NDCG	0.637	0.655	**0.664**	0.666	0.663	0.657
ML-1 M	HR	0.874	0.878	**0.887**	0.885	0.886	0.883
	NDCG	0.629	0.635	**0.639**	0.637	0.634	0.636
Amazon	HR	0.609	0.612	**0.627**	0.623	0.611	0.613
	NDCG	0.426	0.437	**0.452**	0.432	0.423	0.433

Final Latent Space Factors Latent factors are another sensitive parameter in the method. We employ the model of 2 hidden-layers structure to conduct this experiment. Table 5 demonstrates the performance with disparate in numbers of factors on a scale of 8 to 128. As can be clearly seen from results, the model utilizing 128 latent factors gets the best performance over three kinds of datasets. In general, the representations with more factors can be more effective in performance for sparse datasets.

Table 5. Comparisons with different factors

Factors	ML-100 K		ML-1 M		Amazon	
	HR	NDCG	HR	NDCG	HR	NDCG
8	0.871	0.623	0.808	0.542	0.575	0.409
16	0.889	0.643	0.851	0.601	0.583	0.418
32	0.892	0.669	0.871	0.623	0.604	0.426
64	0.904	0.676	0.882	0.638	0.609	0.435
128	0.905	0.682	0.889	0.648	0.617	0.445

5 Conclusion and Future Work

In this paper, we present a new approach of recommendation, which integrates content text as auxiliary information into neural networks architecture to enhance items' representations. Meanwhile, this method exploits both of the explicit ratings and implicit feedback information to better understand the latent features of items and users. Compared with other approaches, the model we addressed utilizing two parallel neural networks to deal with interaction data and text information at the same time is proved to be valid and achieves a dramatically improvement performance in item recommendation.

As for the future work, we attempt to joint other auxiliary data (e.g. user description documents, social network, etc.) into the model for getting a better performance. In addition, we will try to devise some other methodologies to capture text latent factors by adopting some other natural language processing methods which can learn the semantic context better.

Acknowledgement. This research was supported in part by the National Key Research and Development Program of China under grant 2018YFF0214700.

References

1. Lu, J., Wu, D., Mao, M., et al.: Recommender system application developments: a survey. Decis. Support Syst. **74**, 12–32 (2015)

2. Bobadilla, J., Ortega, F., Hernando, A., et al.: Recommender systems survey. Knowl.-Based Syst. **46**, 109–132 (2013)
3. Linden, G., Smith, B., York, J.: Amazon.com recommendations: item-to-item collaborative filtering. IEEE Internet Comput. **7**(1), 76–80 (2003)
4. Ashley-Dejo, E., Ngwira, S., Zuva, T.: A survey of context-aware recommender system and services. In: International Conference on Computing, Communication and Security (ICCCS), pp. 1–6, IEEE (2015)
5. Bi, Z., Zhou, S., Yang, X., Zhou, P., Wu, J.: An approach for item recommendation using deep neural network combined with the Bayesian personalized ranking. In: Wang, X., Gao, H., Iqbal, M., Min, G. (eds.) CollaborateCom 2019. LNICST, vol. 292, pp. 151–165. Springer, Cham (2019). https://doi.org/10.1007/978-3-030-30146-0_11
6. Abdulkarem, H.F., Abozaid, G.Y., Soliman, M.I.: Context-aware recommender system frameworks, techniques, and applications: a survey. In: International Conference on Innovative Trends in Computer Engineering (ITCE), pp. 180–185. IEEE (2019)
7. He, R., Julian, M.: VBPR: visual Bayesian personalized ranking from implicit feedback. In: 30th AAAI Conference on Artificial Intelligence, pp. 144–150 (2016)
8. Purushotham, S., Liu, Y., Kuo, C.C.J.: Collaborative topic regression with social matrix factorization for recommendation systems. In: 29th International Conference on International Conference on Machine Learning (ICML), pp. 691–698 (2012)
9. Li, H., Diao, X., Cao, X., et al.: Collaborative filtering recommendation based on all-weighted matrix factorization and fast optimization. IEEE Access **6**, 25248–25260 (2018)
10. Wu, Y., DuBois, C., Zheng, A.X., et al.: Collaborative denoising auto-encoders for top-N recommender systems. In: 9th ACM International Conference on Web search and Data Mining (WSDM), pp. 153–162 (2016)
11. He, X., Liao, L., Zhang, H., et al.: Neural collaborative filtering. In: 26th International Conference on World Wide Web (WWW), pp. 173–182 (2017)
12. Ling, G., Lyu, M.R., King, L.: Ratings meet reviews, a combined approach to recommend. In: 8th ACM conference on Recommender systems (RecSys), pp. 105–112 (2014)
13. McAuley, J., Leskovec, J.: Hidden factors and hidden topics: understanding rating dimensions with review text. In: 7th ACM Conference on Recommender Systems, pp. 165–172 (2013)
14. He, X., Du, X., Wang, X., et al.: Outer product based neural collaborative filtering. In: IJCAI Conference, pp. 2227–2233 (2018)
15. Cheng, H., Koc, L., Harmsen, J, et al.: Wide & deep learning for recommender systems. In: 1st Workshop on Deep Learning for Recommender Systems, pp. 7–10 (2016)
16. de Souza Pereira Moreira, G., Ferreira, F., da Cunha, A.M.: News session-based recommendations using deep neural networks. In: 3rd Workshop on Deep Learning for Recommender Systems, pp. 15–23 (2018)
17. Wang, C., Blei, D.M.: Collaborative topic modeling for recommending scientific articles. In: 17th ACM SIGKDD International Conference on Knowledge Discovery and Data Mining, pp. 448–456 (2011)
18. Hu, Y., Koren, Y., Volinsky, C.: Collaborative filtering for implicit feedback datasets. In: 8th IEEE International Conference on Data Mining, pp. 263–272 (2008)
19. Lee, W., Song, K., Moon, I.C.: Augmented variational autoencoders for collaborative filtering with auxiliary information. In: ACM on Conference on Information and Knowledge Management, pp. 1139–1148 (2017)
20. Liang, D., Krishnan, R.G., Hoffman, M.D., et al.: Variational autoencoders for collaborative filtering. In: Word Wide Web Conference, pp. 689–698 (2018)
21. LeCun, Y., Bengio, Y., Hinton, G.: Deep learning. Nature **521**, 436 (2015)
22. Xue, H.J., Dai, X., Zhang, J., et al.: Deep matrix factorization models for recommender systems. In: IJCAI Conference, pp. 3203–3209 (2017)

23. Zheng, L., Noroozi, V., Yu, P.S.: Joint deep modeling of users and items using reviews for recommendation. In: 10th ACM International Conference on Web Search and Data Mining, pp. 425–434 (2017)
24. Bayer, I., He, X., Kanagal, B., et al.: A generic coordinate descent framework for learning from implicit feedback. In: 26th International Conference on World Wide Web, pp. 1342–1350 (2017)
25. He, X., Zhang, H., Kan, M.Y., et al.: Fast matrix factorization for online recommendation with implicit feedback. In: 39th International ACM SIGIR Conference on Research and Development in Information Retrieval, pp. 549–558 (2016)
26. Rendle, S., Freudenthaler, C., Gantner, Z., et al.: BPR: Bayesian personalized ranking from implicit feedback. In: Conference on Uncertainty in Artificial Intelligence, pp. 452–461 (2012)
27. Kingma, D.P., Ba, J.: Adam: A method for stochastic optimization. unpublished (2014)
28. Mikolov, T., Sutskever, I., Chen, K., et al.: Distributed representations of words and phrases and their compositionality. In: Neural Information Processing Systems, pp. 3111–3119 (2013)
29. Kim, D., Park, C., Oh, J., et al.: Convolutional matrix factorization for document context-aware recommendation. In: 10th ACM Conference on Recommender Systems, pp. 233–240 (2016)

Real-Time Self-defense Approach Based on Customized Netlink Connection for Industrial Linux-Based Devices

Ming Wan[1], Jiawei Li[1], and Jiangyuan Yao[2](✉)

[1] School of Information, Liaoning University, Shenyang, China
wanming@lnu.edu.cn, lijiawei_lnu@hotmail.com
[2] School of Computer Science and Cyperspace Security, Hainan University, Hainan, China
yaojy@hainanu.edu.cn

Abstract. With the deep integration of IT (Information Technology) and OT (Operational Technology), various Linux operating systems have been successfully applied in critical industrial devices, such as Linux-based IIoT (Industrial Internet of Things) controllers or gateways, and the vulnerabilities of these systems may become a new breakthrough for the organized and high-intensity attacks. In order to prevent malwares from corrupting or disabling industrial Linux-based devices, this paper proposes a novel real-time self-defense approach, which can be easily developed without redesigning the basic software and hardware platform. By establishing the customized Netlink connection between kernel mode and user mode, this approach can monitor all application processes, and block each new malicious application process, which cannot conform to the trusted white-listing rules. All experimental results show that the proposed approach has a comparative advantage to effectively detect and prevent the malware-related attacks, and provides a self-defense function for industrial Linux-based devices, which meets their availability due to the millisecond resolution.

Keywords: Self-defense · Customized Netlink · Application process · Industrial Linux-based devices

1 Introduction

In recent years, the development of information-enabled industry has promoted the large-scale integration and interconnection of current industrial control systems, and the original self-imposed isolation of various industrial components has been broken [1]. As a result, the trend of increasing vulnerabilities appears as the nonlinear superposition, and today's industrial control systems are suffering from serious security threats due to the organized and high-intensity attacks [2–4]. Especially, advanced persistent threats have become the most popular and dangerous attack pattern, which infiltrates and destroys industrial infrastructures by exploiting various hidden zero-day vulnerabilities [5]. Although the researchers are giving full concentrations and efforts to excavate the

H. Gao et al. (Eds.): CollaborateCom 2020, LNICST 349, pp. 406–420, 2021.
https://doi.org/10.1007/978-3-030-67537-0_25

shortcomings of industrial hardware and software platforms, some hidden vulnerabilities still exist and are difficult to perform the patch updating [6, 7]. The main reasons can be summarized as follows: firstly, most of industrial vulnerabilities derive from the specialized function features which can be fully utilized by a professional adversary, and the traditional IT vulnerability mining technique is difficult to apply in wide range; secondly, although the number of industrial vulnerabilities is limited, their hiding ability becomes more disconcerting as their number decreasing; thirdly, the patch updating presents an obvious hysteretic nature, and it is a daunting task to repair these vulnerabilities because industrial control systems must ensure the uninterrupted running in a long term.

In order to achieve the deep integration of IT (Information Technology) and OT (Operational Technology), various Linux operating systems have been widely applied in critical industrial devices, for example, an open-source industrial IoT gateway is designed by exploiting the micro-service of Docker, whose software solutions are mainly based on Linux [8]; a software CNC (Computerized Numerical Control) controller which runs the Linux real-time operating system is developed to implement the EPL (Ethernet Powerlink) stack as a real-time kernel module [9]; SIMATIC WinCC which is widely used in today's industrial control systems can also provide the cross-platform support for Linux [10]. As a result, the vulnerabilities of Linux operating system become a subject worthy of special attention, and some potential security threats which aim at damaging or infecting the Linux operating system are coming to the surface. Typically, Mirai and Brickerbot malwares may pose serious downtime risks to Linux-based IIoT (Industrial Internet of Things) devices, because they can make an electronic device completely useless in a DDoS (Distributed Denial-of-Service) or PDoS (Permanent Denial-of-Service) attack [11, 12]. Although the physical layer authentication or secure transmission can provide an important opportunity to enhance IoT security [13, 14], the operating system-level security threats may still become a controversial issue. In order to deal with these security threats and challenges, some security defense technologies have been designed to ensure the availability of critical industrial devices, and these technologies can be roughly divided into two categories: indirect defense and direct defense. Furthermore, the indirect defense employs some external defense approaches to safeguard critical industrial devices, and one representative example is industrial firewall [15]. However, the indirect defense belongs to the traditional passive defense technologies, which cannot reply the targeted and multi-variability behaviors of industrial attacks. Differently, the direct defense strengthens the self-protection ability of critical industrial devices, and its defense mechanism is similar to the autoimmunity. As an example, trusted computing is a feasible approach which can provide the system integrity verification and identity authentication [16–18], but its main drawback is that it need redesign the software and hardware platform to add the trusted computing module.

In this paper, we propose a novel real-time self-defense approach for the Linux operating system, and this approach can be conveniently embedded without requiring the redesign of basic software and hardware platform. More specifically, this approach can monitor all application processes in kernel mode by establishing the customized Netlink connection, and check each new application process in user mode by comparing with the trusted white-listing rules. If one incoming application process is inconsistent with all white-listing rules, this process will be killed and an accompanying alarm will be generated. In order to verify its defense effect, we perform the experimental verification and compared discussion by simulating the Modus/TCP runtime environment. Moreover, the experimental results show that our self-defense approach has an obvious advantage to effectively detect and prevent the malware-related attacks, and obtains the millisecond resolution which meets the availability of industrial Linux-based devices.

The main objectives and contributions of this paper are summarized as follows: firstly, we present the basic design principle of self-defense approach based on customized Netlink connection, and analyze its three significant aspects, which can directly relate to the benefits against the redesign of basic software and hardware platform; secondly, we depict the detailed execution procedures of self-defense approach based on customized Netlink connection, which can expound the simplicity and feasibility of self-defense process implementation; thirdly, we perform a lot of qualitative and quantitative experiments to verify the fine effectiveness of self-defense approach, and we hope that our approach can bring a feasible opportunity for the cyber security defense of industrial Linux-based devices, especially Linux-based IIoT devices.

The rest of this paper is organized as follows: Sect. 2 outlines the detailed design of self-defense approach based on customized Netlink connection, including basic design principle of self-defense approach and main execution procedures of white-listing monitoring and defense. Section 3 presents our experiments and discussions, including functional testing, comparative analysis and file protection verification. Finally, Sect. 4 provides some concluding remarks regarding this research.

2 Self-defense Approach Based on Customized Netlink Connection

2.1 Basic Design Principle

In general, the application process in the Linux operating system can be created by the following two ways: one is the indirect creation by generating an additional process copy, and the other is the direct creation by using the system call *execve()*. Actually, the additional process copy can also invoke the function *execve()* in its final phase. When a new user software boots, the kernel will be motivated to gain control of CPU, and the interface *system_call()* will be used to find the corresponding system call service routine according to the Linux system call table *sys_call_table*. In this case, we can further modify *sys_call_table*, and change the *execve()* address to the user-defined *hook_execve()* address, which is an important function installed in kernel mode. Before executing the native system call *execve()*, we can monitor and block the new application process. Figure 1 shows the basic design principle of our approach, which contains three significant aspects:

1) Netlink connection: the Netlink connection is an excellent method to achieve inter-process communications between kernel mode and user mode, and it can be established by some mature APIs (Application Program Interfaces) [19]. Moreover, the customized Netlink connection can support asynchronous communications from kernel mode to user mode, and send new process events to user mode without interfering with the stable runtime of other application processes.

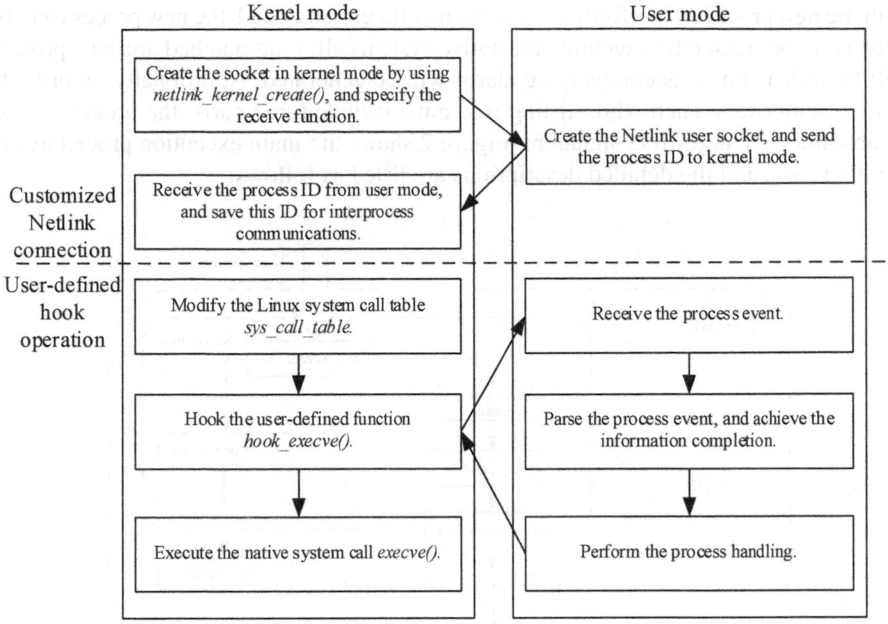

Fig. 1. Basic design principle of self-defense approach based on customized Netlink connection.

2) Hook operation: the hook operation can intercept the normal execution flow of one application, and insert some other operations or the terminated instruction. According to the Linux system call table *sys_call_table*, we can design a user-defined function *hook_execve()* before the system call *execve()*, and analyze the new process creation of executable file.

3) LKM (Loadable Kernel Module) mechanism: LKM mechanism [20], which improves the scalability and maintainability of Linux kernel, can provide an external interface for dynamic loading and unloading of modules. In our approach, the Netlink connection and hook operation are designed as LKMs, which smoothly run as two additional parts of Linux kernel.

The main advantages of our approach can be summarized as follows: firstly, the customized Netlink connection can establish a communication path between kernel mode and user mode, and it can enhance the real-time performance to receive new process events from kernel mode; secondly, the process handling is only performed in user mode, and the fewer operations in kernel mode cannot affect the system stability; finally, due to the frequent process creation, our approach can effectively use the simultaneous multi-threading technology in user mode to perform the process event parsing, information completion and process handling.

2.2 White-Listing Monitoring and Defense

Based on the above design principle, we further set the trusted white-listing rules which contain all regular application processes after the system boots, and compare these rules with the new process to perform the process handling: if matched, the new process will be allowed to be created by executing the native system call; if unmatched, the new process will be killed and an accompanying alarm will be generated. Additionally, in order to ensure uniqueness, each white-listing rule must include three parts: the process name, its absolute path and MD5 signature. Figure 2 shows the main execution procedures of our approach, and the detailed descriptions are listed as follows:

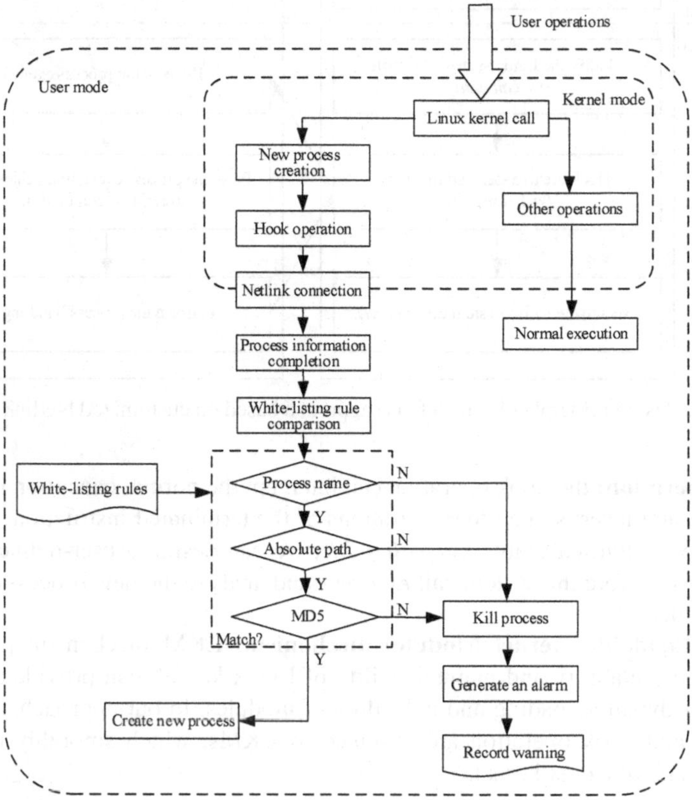

Fig. 2. Main execution procedures of white-listing monitoring and defense.

Step 1: Load LKMs, including the customized Netlink connection module and user-defined hook operation module. On the one hand, this step can establish the communication between kernel mode and user mode by inserting the customized Netlink connection module; on the other hand, this step can modify the process creation procedure, and add the user-defined hook operation module.

Step 2: Establish the customized Netlink connection. In user mode, the socket for the communication between kernel mode and user mode is created, and the corresponding information can be sent to kernel mode.

Step 3: Monitor each new process creation. In kernel mode, when the Linux system call for one new process creation is monitored, this process event (including the process

ID and its absolute path) is sent to user mode, and the information completion (including the process name and MD5 signature) is achieved in user mode.

Step 4: Check the new process. In user mode, the new process is checked by comparing with each white-listing rule, including the process name, its absolute path and MD5 signature.

Step 5: Perform the process handling. According the compared result, this step judges the legitimacy of new process: if matched, the new process will be allowed to be created by executing the native system call; if unmatched, the new process and its child processes will be killed.

Step 6: Generate an alarm. For the abnormal process creation, an alarm is generated in user mode, and the warning information can be recorded to the log, such as the process name, its absolute path, MD5 signature and the alarm time.

3 Experimental Verification and Discussion

3.1 Functional Testing and Verification

Various soft PLCs (Programmable Logic Controllers) have been developed to complete industrial process control and automated production, for example, the OpenPLC can run on the Linux operating system to emulate a traditional PLC, which can use Modbus/TCP to implement the data acquisition and control management in industrial control communications [21]. In order to facilitate the functional verification, we install the Modbus/TCP master/slave software on the Linux operating system to simulate the Modus/TCP runtime environment. Under normal circumstances, we suppose the Modbus/TCP master executes a read operation whose purpose is to read the coil status of Modbus/TCP slave by using the function code "01" [22].

In order to verify the defense effect, we construct a malicious script "hack-kill.sh" to simulate a devastating attack, whose main purpose is to destroy the Modbus/TCP slave process. More specifically, this common attack can threaten the Modbus/TCP application, and the attack principle can be briefly described as follows: by searching through all running processes, this malicious script "hack-kill.sh" can locate the Modbus/TCP slave process, and forcibly kill it to destroy the normal industrial control activities. In this experiment, we respectively perform the same attack with and without our self-defense approach, and the experimental results are shown in Fig. 3. From this figure we can see that, when the system is not protected by our self-defense approach, Fig. 3(a) shows that the malicious script "hack-kill.sh" can be unknowingly executed, and the Modbus/TCP slave process is killed. More specifically, the malicious script finds the Modbus/TCP slave process ID 4690, and forcibly kills it to destroy proper functioning. Differently, when the system is zoned and protected by our self-defense approach, Fig. 3(b) shows that the malicious script "hack-kill.sh" can be successfully prevented without disturbing the availability of Modbus/TCP application. That is, before the malicious script performs destructive actions, our self-defense approach can check and kill the incoming process ID 8541 generated by the malicious script. From the compared results we can conclude that, our self-defense approach can effectively detect and restrict the script-related attacks, which use the native Linux script running mechanism to destroy various industrial applications.

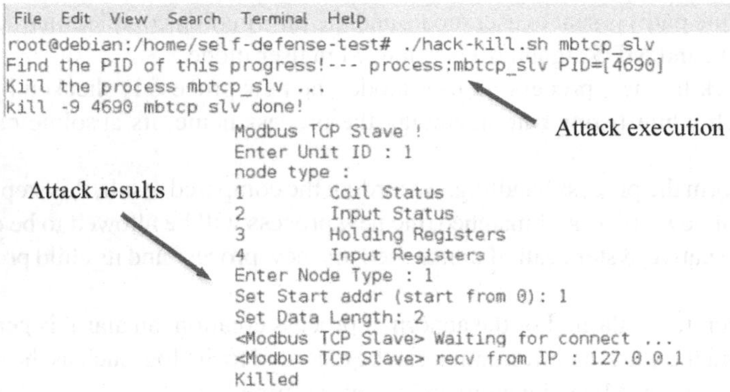
File Edit View Search Terminal Help
root@debian:/home/self-defense-test# ./hack-kill.sh mbtcp_slv
Find the PID of this progress!--- process:mbtcp_slv PID=[4690]
Kill the process mbtcp_slv ...
kill -9 4690 mbtcp slv done!
Attack execution
Modbus TCP Slave !
Enter Unit ID : 1
node type :
Attack results
1 Coil Status
2 Input Status
3 Holding Registers
4 Input Registers
Enter Node Type : 1
Set Start addr (start from 0): 1
Set Data Length: 2
<Modbus TCP Slave> Waiting for connect ...
<Modbus TCP Slave> recv from IP : 127.0.0.1
Killed

(a) Attack execution and results without our self-defense approach

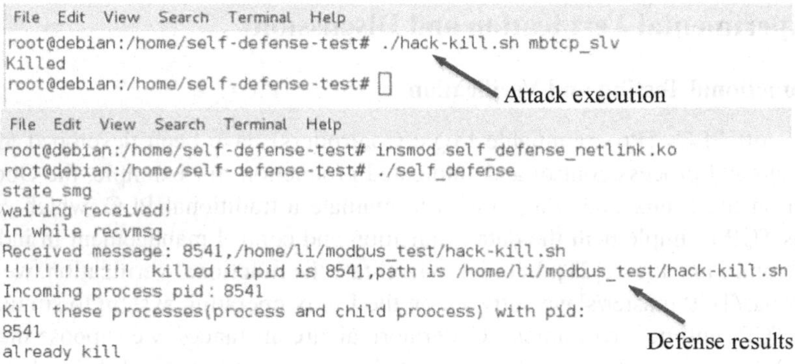
File Edit View Search Terminal Help
root@debian:/home/self-defense-test# ./hack-kill.sh mbtcp_slv
Killed
root@debian:/home/self-defense-test#
Attack execution
File Edit View Search Terminal Help
root@debian:/home/self-defense-test# insmod self_defense_netlink.ko
root@debian:/home/self-defense-test# ./self_defense
state_smg
waiting received!
In while recvmsg
Received message: 8541,/home/li/modbus_test/hack-kill.sh
!!!!!!!!!!!! killed it,pid is 8541,path is /home/li/modbus_test/hack-kill.sh
Incoming process pid: 8541
Kill these processes(process and child proocess) with pid:
8541
already kill
Defense results

(b) Attack execution and defense results with our self-defense approach

Fig. 3. Compared results by executing the malicious script "hack-kill.sh" with and without our self-defense approach.

Different from the malicious script, many malware-related attacks always execute the malicious codes which are designed by using some application development languages. In order to verify the corresponding defense effect, we develop one malware "hack-falsify" by using the classic Linux C program. Moreover, this malware can simulate a disturbing attack, whose fundamental purpose is to intentionally tamper with some critical file contents after searching for the specified target path and file. Similarly, we also perform the same attack with and without our self-defense approach, and the experimental results are shown in Fig. 4. In this experiment, the malware "hack-falsify" attempts to rewrite some critical parameter values in the Modbus/TCP log file "modbus.log", which may have a significant impact on the auditing of technological process. From this figure we can see that, when the system is not protected by our self-defense approach, Fig. 4(a) shows that the malware "hack-falsify" can be easily executed, and some critical parameter values in Lines 2 and 3 have been secretly modified, for example, the parameter *rtid* has been modified to "hacked". Conversely, when the system is zoned and protected by our self-defense approach, Fig. 4(b) shows that the malware "hack-falsify" can be successfully prevented, and the incoming process ID 6467 is killed before the malware performs disturbing actions. To sum up, the compared results offer further proof that our self-defense approach can effectively detect and restrict some malware-related attacks, which develop some malicious applications to disturb various industrial activities.

File Edit View Search Terminal Help
root@debian:/home/self-defense-test# ./hack-falsify /home modbus.log
searching....
load_file:/home/li/modbus_test/modbus.log
Attack execution
File Edit View Search Terminal Help
root@debian:/home/self-defense-test# cat /home/li/modbus_test/modbus.log
<Modbus TCP Master> Enter Unit ID : <Modbus TCP Master> 2020-03-24 19:56:58,rtid = 1, mbus.fc = 1
{<Modbus TCP Master> 1999-99-99 99:99:99,rtid = <hacked!!!!!!!!>, mbus.fc = 1225556},
{<Modbus TCP Master> 2020-04-11 21:39:46,rtid = <hacked!!!!!!!!>, mbus.fc = 1225556},
<Modbus TCP Master> 2020-03-24 19:57:01,rtid = 2, mbus.fc = 1
<Modbus TCP Master> 2020-03-24 19:57:04,rtid = 3, mbus.fc = 1
<Modbus TCP Master> 2020-03-24 19:57:07,rtid = 4, mbus.fc = 1
<Modbus TCP Master> 2020-03-24 19:57:10,rtid = 5, mbus.fc = 1
Attack results

(a) Attack execution and results without our self-defense approach

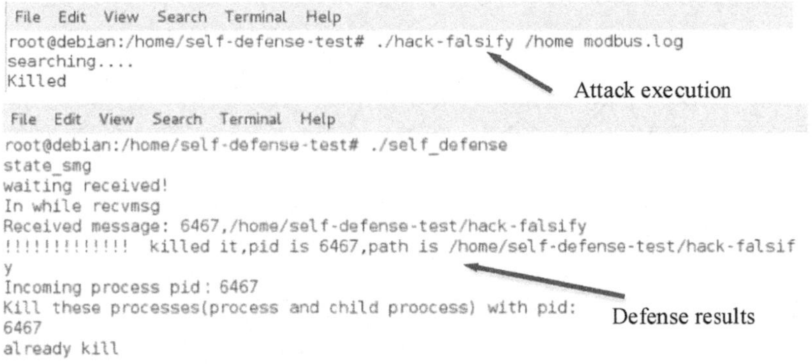
File Edit View Search Terminal Help
root@debian:/home/self-defense-test# ./hack-falsify /home modbus.log
searching....
Killed
Attack execution
File Edit View Search Terminal Help
root@debian:/home/self-defense-test# ./self_defense
state_smg
waiting received!
In while recvmsg
Received message: 6467,/home/self-defense-test/hack-falsify
!!!!!!!!!!!!! killed it,pid is 6467,path is /home/self-defense-test/hack-falsify
Incoming process pid: 6467
Kill these processes(process and child proocess) with pid: 6467
already kill
Defense results

(b) Attack execution and defense results with our self-defense approach

Fig. 4. Compared results by executing the malware "hack-falsify" with and without our self-defense approach.

3.2 Performance Comparison

In order to illustrate the advantage of high efficiency, we select the consuming time to handle different process events as an important performance indicator, because an excellent defense approach can take short CPU time to achieve the noticeable protection. Moreover, we compare with a user-mode defense approach, whose main design principle can be summarized as follows: by using the native Netlink connector built in the Linux operating system, this approach can get the new process ID after triggering the system call *execve()*, and search the file system */proc* to obtain the absolute path of this new process. Additionally, this process event can be checked and handled by comparing with its own white-listing rules. For a comprehensive analysis, we consider the following two situations: one is the consuming time comparison for one trusted process event, which can be regarded as a normal application conforming to the white-listing rules; the other is the consuming time comparison for one malicious process event, which can be treated as an abnormal application deviating from the white-listing rules.

The consuming time comparison for one trusted process event is shown in Fig. 5, and different curves depict the consuming time changes of two approaches under 25 experiments, respectively. From this figure we can see that, the consuming time of our self-defense approach is significantly lower than the one of user-mode defense approach. More precisely, the average consuming time of user-mode defense approach reaches about 0.83 ms, and the average consuming time of our self-defense approach is only about 0.69 ms. That is to say, our self-defense approach to handle one trusted process event can improve the efficiency by 17.45%.

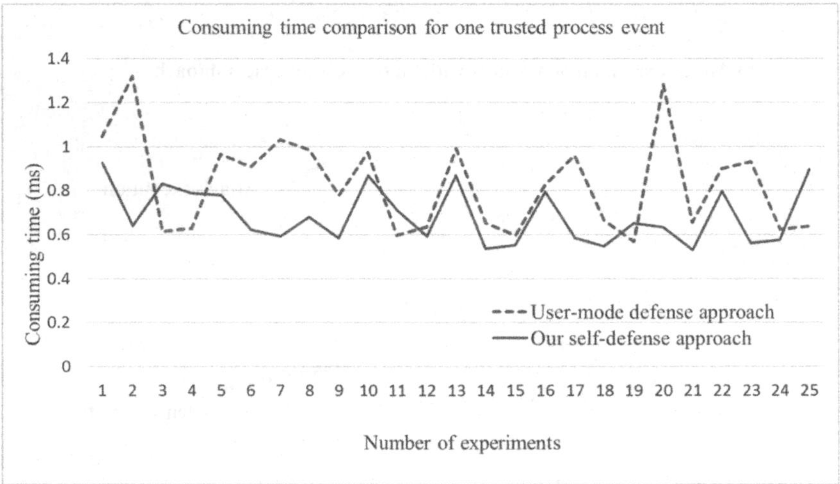

Fig. 5. Consuming time comparison for one trusted process event.

Similarly, Fig. 6 shows the consuming time comparison for one malicious process event under 25 experiments, and our self-defense approach also has a distinct advantage to prevent an abnormal application. Especially, the average consuming time of user-mode defense approach reaches about 1.77 ms, and the average consuming time of our self-defense approach is only about 1.63 ms. Namely, the increased efficiency of our self-defense approach to handle one malicious process event can reach 7.92%. From the above experimental results we can conclude that our self-defense approach can obtain the millisecond resolution which meets the availability of industrial Linux-based devices, and have better real-time performance to protect Linux operating systems.

In extreme cases, when monitoring each new process creation, our self-defense approach can obtain more accurate process event, whose application process may be created and released instantaneously. For example, Fig. 7 shows the different absolute

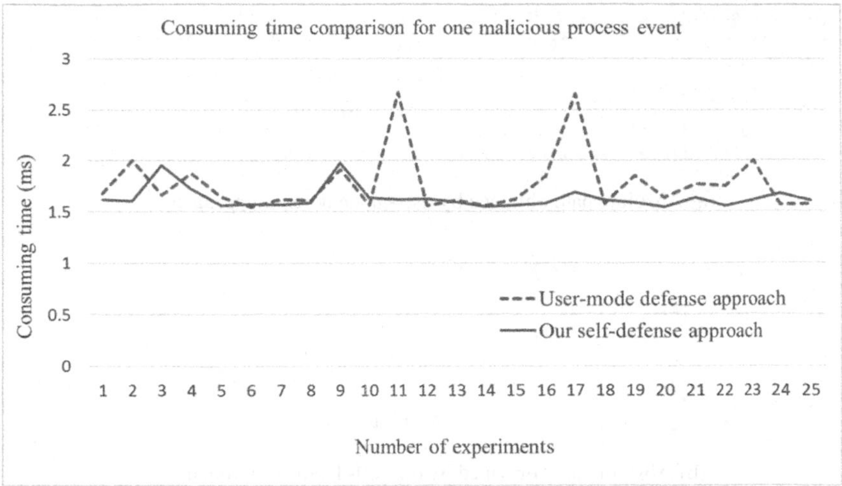

Fig. 6. Consuming time comparison for one malicious process event.

paths obtained by two approaches under the same process ID. From this figure we can find that, the same process ID is 14551, and our self-defense approach can obtain its full absolute path. On the contrary, the user-mode defense approach only shows "proc is not exist", namely, this approach cannot obtain any path information of process ID 14551. The main reason for this case is that before the user-mode defense approach searches the file system */proc* to obtain the absolute path, the application process has been released and the corresponding process information in the file system */proc* has been deleted. Therefore, under the same defense function, our self-defense approach can have a great advantage in information acquisition of new process event, which may provide more detailed warning information for a significant post-audit.

Compared with the user-mode defense approach, our self-defense approach not only has better real-time performance to satisfy industrial high availability requirements, but also forms a relatively complete process event to support security vulnerability analysis and auditing. In a word, our self-defense approach can provide more effective solutions to protect industrial Linux-based devices.

3.3 File Protection Design and Verification

Based on the hook operation mechanism, we can further develop a file protection function, which can further prevent an illegal access to some sensitive files. By using the hook operation, this file protection function can intercept the system calls to shared library functions, and insert some new executing codes which achieve the file access control. In the dynamic link libraries of Linux operating system, the environment variable *LD_PRELOAD* can affect the runtime linkers of various application programs, and preferentially allow to load one dynamic link library before running the application program. In other words, we can design a new library function to replace the original system call *read()*, and this new library function can check the protection status of one sensitive

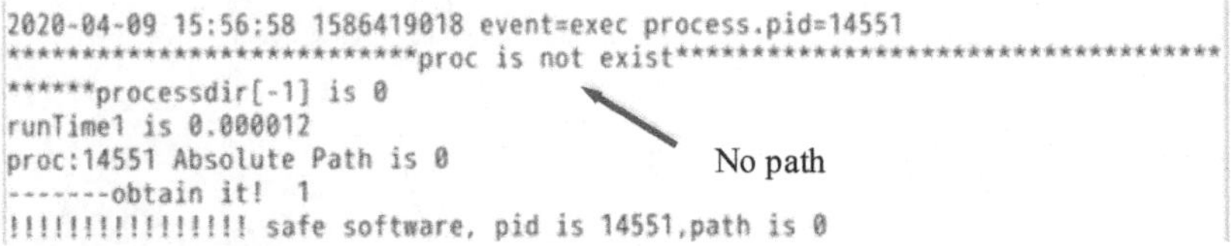
2020-04-09 15:56:58 1586419018 event=exec process.pid=14551
****************************proc is not exist************************************
******processdir[-1] is 0
runTime1 is 0.000012
proc:14551 Absolute Path is 0
-------obtain it! 1
!!!!!!!!!!!!!!!! safe software, pid is 14551,path is 0
No path

(a) Absolute path obtained by user-mode defense approach

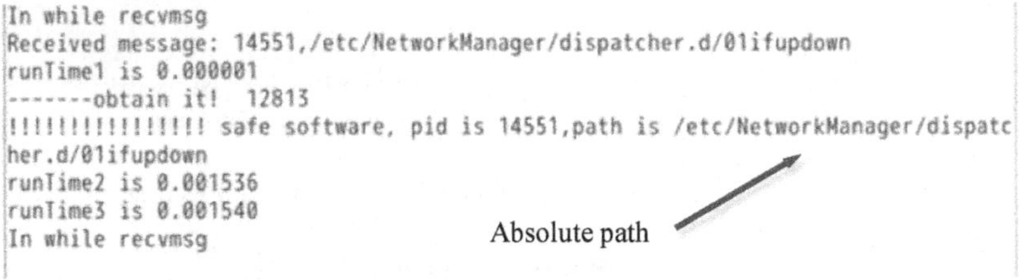
In while recvmsg
Received message: 14551,/etc/NetworkManager/dispatcher.d/01ifupdown
runTime1 is 0.000001
-------obtain it! 12813
!!!!!!!!!!!!!!!! safe software, pid is 14551,path is /etc/NetworkManager/dispatc
her.d/01ifupdown
runTime2 is 0.001536
runTime3 is 0.001540
In while recvmsg
Absolute path

(b) Absolute path obtained by our self-defense approach

Fig. 7. Different absolute paths obtained by two approaches under the same process ID.

file to authorize or deny its access. More specifically, if one sensitive file is set to the protected file, this new library function will deny its access request, and generate an accompanying alarm. Actually, the most important advantage of file protection function is to prevent unauthorized disclosure of information when an adversary tries to penetrate and view some sensitive files by using some common Linux commands. Figure 8 shows the main execution procedures of file protection function, and the detailed descriptions are listed as follows:

Step 1: Load new library function in the dynamic link libraries. Based on the environment variable *LD_PRELOAD*, this new library function can own the highest priority. When one file is accessed, the hook operation of file protection can be executed before triggering the original system call *read()*.
Step 2: Monitor each new file access. If one new file access is monitored, the file protection function can be triggered.
Step 3: Check and handle the new file access. By comparing with the pre-determined absolute path which relates to the protected file, this step can judge the legitimacy of new file access: if matched, this file needs to be protected, and the corresponding file access can be denied; if unmatched, the file contents can be allowed to view.
Step 4: Generate an alarm. For the abnormal file access, an alarm is generated in user mode, and the warning information can be recorded to the log.

In order to verify the file protection effect, we suppose that one adversary has successfully penetrated some critical industrial Linux-based devices, and wants to steal confidential information by viewing some sensitive files. In general, two different methods can be easily carried out to achieve this goal: one is to view these sensitive files by using some common Linux commands, such as "cat" and "tail", and the other is to view these sensitive files by applying some customized malwares, for example, we

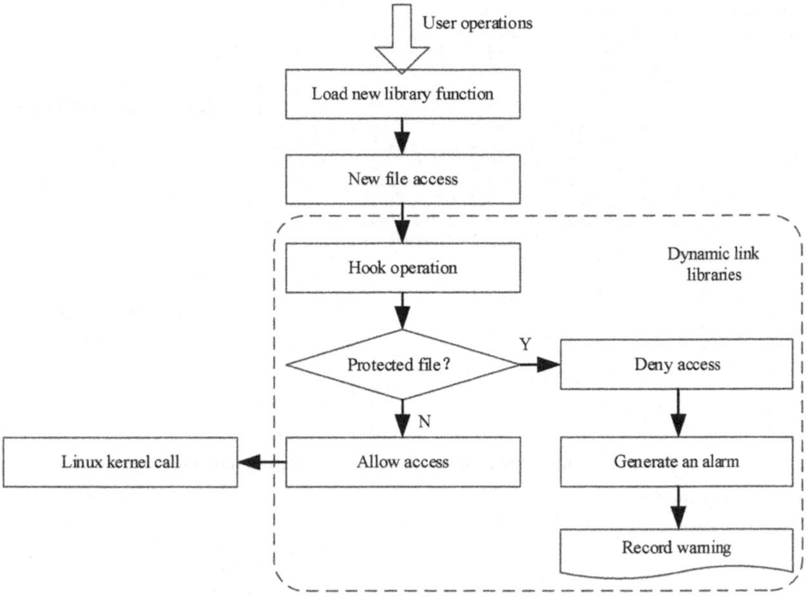

Fig. 8. Main execution procedures of file protection function.

develop a malware "hack-read" to realize this function, and this malware is designed by using the classic Linux C program. In our experiment, we also perform the same attack with and without our file protection function, and the experimental results are shown in Fig. 9. Moreover, we suppose that the file "modbus_secret_data" is a sensitive log which stores some key Modbus/TCP communication data, and the malware "hack-read" can directly search to view this file. From this figure we can see that, when the file "modbus_secret_data" is not protected by our file protection function, Fig. 9(a) shows that both the common Linux command "cat" and the malware "hack-read" can be easily executed to view all sensitive log records. Differently, when the file "modbus_secret_data" is protected by our file protection function, Fig. 9(b) shows that both the common Linux commands and the malware "hack-read" cannot be executed to view any sensitive log record, and the terminal can generate an alarm "it's an illegal access". Additionally, it is worth mentioning that three different Linux commands ("cat", "tail" and "tac") can be successfully prevented by our file protection function in Fig. 9(b). Therefore, the above compared results convincingly demonstrate that our file protection function can effectively protect sensitive information from leaving the confines of normal industrial production.

File Edit View Search Terminal Help
root@debian:/home/self-defense-test# cat modbus_secret_data
2020-06-14 16:04:20 1471334660 192.168.1.103 192.168.1.3 1 10 1
2020-06-14 16:04:21 1471334661 192.168.1.102 192.168.1.2 1 5 1
2020-06-14 16:04:21 1471334661 192.168.1.104 192.168.1.4 1 8 0
2020-06-14 16:04:22 1471334662 192.168.1.2 192.168.1.102 5 5 1
2020-06-14 16:04:22 1471334662 192.168.1.2 192.168.1.102 5 5 0
2020-06-14 16:04:22 1471334662 192.168.1.103 192.168.1.3 1 10 1
2020-06-14 16:04:22 1471334662 192.168.1.2 192.168.1.102 6 100 250
2020-06-14 16:04:22 1471334662 192.168.1.2 192.168.1.102 6 100 0
2020-06-14 16:04:22 1471334662 192.168.1.102 192.168.1.2 3 100 0
2020-06-14 16:04:22 1471334662 192.168.1.2 192.168.1.102 6 200 500
root@debian:/home/self-defense-test# ./hack-read
2020-06-14 16:04:20 1471334660 192.168.1.103 192.168.1.3 1 10 1
2020-06-14 16:04:21 1471334661 192.168.1.102 192.168.1.2 1 5 1
2020-06-14 16:04:21 1471334661 192.168.1.104 192.168.1.4 1 8 0
2020-06-14 16:04:22 1471334662 192.168.1.2 192.168.1.102 5 5 1
2020-06-14 16:04:22 1471334662 192.168.1.2 192.168.1.102 5 5 0
2020-06-14 16:04:22 1471334662 192.168.1.103 192.168.1.3 1 10 1
2020-06-14 16:04:22 1471334662 192.168.1.2 192.168.1.102 6 100 250
2020-06-14 16:04:22 1471334662 192.168.1.2 192.168.1.102 6 100 0
2020-06-14 16:04:22 1471334662 192.168.1.102 192.168.1.2 3 100 0
2020-06-14 16:04:22 1471334662 192.168.1.2 192.168.1.102 6 200 500
"cat" attack and results
"hack-read" attack and results
(a) Attack execution and results without file protection function
File Edit View Search Terminal Help
root@debian:/home/self-defense-test# export LD_PRELOAD="/home/self-defense-test/self-defense-file.so"
root@debian:/home/self-defense-test# cat modbus_secret_data
it's an illegal access
root@debian:/home/self-defense-test# tail modbus_secret_data
it's an illegal access
root@debian:/home/self-defense-test# tac modbus_secret_data
it's an illegal access
"cat" attack and defense results
File Edit View Search Terminal Help
root@debian:/home/self-defense-test# export LD_PRELOAD="/home/self-defense-test/self-defense-file.so"
root@debian:/home/self-defense-test# ./hack-read
it's an illegal access
"hack-read" attack and defense results

(b) Attack execution and defense results with file protection function

Fig. 9. Compared results with and without our file protection function.

4 Conclusion

Various Linux operating systems have been widely applied in critical industrial devices (especially Linux-based IIoT controllers or gateways), whose vulnerabilities can be viciously exploited by some potential security threats. In order to protect industrial Linux-based devices, this paper proposes a real-time self-defense approach based on the customized Netlink connection, which can be easily embedded without requiring the redesign of basic software and hardware platform. On the one hand, this approach can establish the interprocess communication between kernel mode and user mode to enhance the real-time performance; on the other hand, this approach can handle new process events in user mode without affecting the system stability. Additionally, this approach can be further developed to realize the file protection function. All experimental results show that the proposed self-defense approach can effectively prevent the malware-related attacks, and have better efficiency due to the millisecond handling latency. In a word, the proposed self-defense approach can successfully protect industrial Linux-based devices without disturbing their availability.

Acknowledgements. This work is supported by the Program of Hainan Association for Science and Technology Plans to Youth R & D Innovation (Grant No. QCXM201910), the Natural Science Foundation of Liaoning Province (Grant No. 2019-MS-149), the National Natural Science Foundation of China (Grant No. 61802092), and the Scientific Research Setup Fund of Hainan University (Grant No. KYQD (ZR) 1837).

References

1. Cheminod, M., Durante, L., Valenzano, A.: Review of security issues in industrial networks. IEEE Trans. Ind. Inform. **9**(1), 277–293 (2013)
2. Lyu, X., Ding, Y., Yang, S.: Safety and security risk assessment in cyber-physical systems. IET Cyber-Phys. Syst.: Theory Appl. **4**(3), 221–232 (2019)
3. Wu, G., Sun, J.: Optimal switching integrity attacks on sensors in industrial control systems. J. Syst. Sci. Complex. **32**(1), 1290–1305 (2019)
4. Adepu, S., Kandasamy, N.K., Zhou, J., Mathur, A.: Attacks on smart grid: power supply interruption and malicious power generation. Int. J. Inf. Secur. **19**(2), 189–211 (2019). https://doi.org/10.1007/s10207-019-00452-z
5. Yuan, H., Xia, Y., Zhang, J., Yang, H., Mahmoud, M.: Stackelberg-game-based defense analysis against advanced persistent threats on cloud control system. IEEE Trans. Ind. Inform. **6**(3), 1571–1580 (2020)
6. Pogliani, M., Quarta, D., Polino, M., Vittone, M., Maggi, F., Zanero, S.: Security of controlled manufacturing systems in the connected factory: the case of industrial robots. J. Comput. Virol. Hacking Tech. **15**(3), 161–175 (2019). https://doi.org/10.1007/s11416-019-00329-8
7. Wan, M., Shang, W., Zeng, P.: Double behavior characteristics for one-class classification anomaly detection in networked control systems. IEEE Trans. Inf. Forensics Secur. **12**(12), 3011–3023 (2017)
8. Nguyen-Hoang, P., Vo-Tan, P.: Development an open-source industrial IoT gateway. In: 2019 19th International Symposium on Communications and Information Technologies (ISCIT), pp. 201–204. IEEE, Ho Chi Minh City (2019)
9. Erwinski, K., Paprocki, M., Grzesiak, M., Karwowski, K., Wawrzak, A.: Application of ethernet powerlink for communication in a Linux RTAI open CNC system. IEEE Trans. Ind. Electron. **60**(2), 628–636 (2013)
10. Tufail, H., Anwar, M., Qasim, I., Azam, F.: Towards the selection of optimum alarms system in leading industry automation software. In: 2019 8th International Conference on Industrial Technology and Management (ICITM), pp. 241–246. IEEE, Cambridge (2019)
11. Su, J., Vasconcellos, D., Prasad, S., Sgandurra, D., Feng, Y., Sakurai, K.: Lightweight classification of IoT malware based on image recognition. In: 2018 IEEE 42nd Annual Computer Software and Applications Conference (COMPSAC), pp. 664–669. IEEE, Tokyo (2018)
12. Kolias, C., Kambourakis, G., Stavrou, A., Voas, J.: DDoS in the IoT: Mirai and other Botnets. Computer **50**(7), 80–84 (2017)
13. Zhang, N., et al.: Physical layer authentication for internet of things via WFRFT-based Gaussian gag embedding. IEEE Internet Things J. **7**, 9001–9010 (2020)
14. Zhang, N., Wu, R., Yuan, S., Yuan, C., Chen, D.: RAV: relay aided vectorized secure transmission in physical layer security for internet of things under active attacks. IEEE Internet Things J. **6**(5), 8496–8506 (2019)
15. Lee, S., Lee, S., Yoo, H., Kwon, S., Shon, T.: Design and implementation of cybersecurity testbed for industrial IoT systems. J. Supercomput. **74**(9), 4506–4520 (2017). https://doi.org/10.1007/s11227-017-2219-z

16. Yuan, J., Li, X.: A reliable and lightweight trust computing mechanism for IoT edge devices based on multi-source feedback information fusion. IEEE Access **6**, 23626–23638 (2018)
17. Maene, P., Götzfried, J., Clercq, R., Müller, T., Freiling, F., Verbauwhede, I.: Hardware-based trusted computing architectures for isolation and attestation. IEEE Trans. Comput. **67**(3), 361–374 (2018)
18. Ashraf, N., Masood, A., Abbas, H., Latif, R., Shafqat, N.: Analytical study of hardware-rooted security standards and their implementation techniques in mobile. Telecommun. Syst. **74**(3), 379–403 (2020). https://doi.org/10.1007/s11235-020-00656-y
19. Jia, J., Liu, G., Han, D., Wang, J.: A novel packets transmission scheme based on software defined open wireless platform. IEEE Access **6**, 17093–17118 (2018)
20. Zarrabi, A., Samsudin, K., Adnan, W.A.W.: Linux support for fast transparent general purpose checkpoint/restart of multithreaded processes in loadable kernel module. J. Grid Comput. **11**, 187–210 (2013)
21. Alves, T., Buratto, M., Souza, F., Rodrigues, T.: OpenPLC: an open source alternative to automation. In: IEEE Global Humanitarian Technology Conference (GHTC), pp. 585–589. IEEE, San Jose (2014)
22. Wan, M., Shang, W., Kong, L., Zeng, P.: Content-based deep communication control for networked control system. Telecommun. Syst. **65**(1), 155–168 (2016). https://doi.org/10.1007/s11235-016-0223-x

API Misuse Detection Based on Stacked LSTM

Shuyin OuYang[1], Fan Ge[1], Li Kuang[1](✉), and Yuyu Yin[2]

[1] School of Computer Science and Engineering, Central South University, Changsha 410075, HN, China
kuangli@csu.edu.cn

[2] Hangzhou Dianzi University, Hangzhou 310027, ZJ, China

Abstract. In modern software engineering, API (Application Programming Interface) is widely used to develop applications rapidly by reusing data structure, frameworks, class libs, and etc. However, due to the considerable number of interfaces, lack of documents and timely maintenance and updates, APIs are often used in a wrong way. Therefore, it has become an important problem to detect API misuse in an automatic way. Many existing automatic API detecting methods do not make full use of APIs' potential semantic information and independent integrity of each API. In this paper, we employ Stacked LSTM to learn the API usage specification to detect the API misuse defects. Specifically, first, we obtain ACSG (API Call Syntax Graph) through the static analysis of source code. And then, based on ACSG, we generate API sequences, and transform the sequences into <precious API sequence, next API> for training. Third, in order to represent the APIs in a semantic way, we apply word2vec as a pre-training model to embed features of each API. Though the stacked LSTM model, we regard embedding precious API sequence as the input to model the API use specifications and discover the potential API misuse defects by judging whether the next API is in the output (API probability list) or not. We design experiments to evaluate the effectiveness our method with Java Cryptography APIs and their used code, and the results show the advancement of our proposed method.

Keywords: API misuse detection · Static analysis · Pre-training model · Semantic representation · LSTM

1 Introduction

Nowadays, API (Application Programming Interface) plays an important role in software developing. The developers can save a lot of time to access libraries and frameworks via APIs. According to the ProgrammableWeb's reports, the number of APIs recently included in its website has exceeded 23,403. Since 2015, an average of more than 2,000 APIs has been added per year. As for the first six month of 2019, the API directory has seen over 1,320 new APIs added, which has an increase of over 30% over the previous years. More and more enterprises choose reusing existing data structure, frameworks and class libs to complete the rapid development of new software projects.

H. Gao et al. (Eds.): CollaborateCom 2020, LNICST 349, pp. 421–438, 2021.
https://doi.org/10.1007/978-3-030-67537-0_26

Although API reduces projects' development effort by accessing underlying services, it brings the challenges for ensuring the correctness of programs. API misuse often occurs in the projects, which includes using redundant APIs, using wrong APIs, missing key APIs, ignoring the handling of exceptions that may be thrown in some APIs, which may cause defects or even serious security problems. API misuse is an inevitable problem while developers are using APIs to develop projects. Generally speaking, API misuse is caused by the following reasons: (1) There are lots of APIs and usually an application involves various API interactions. For example, Java Encache API is used to create a cache between distributed nodes, and the interaction between distributed nodes requires the support of Java RMI (remote call) API. (2) The quality of API documentation is not good enough, and the low-level documentation is the main obstacle to API learning and using. (3) API need to be continuously upgraded and maintained, but the corresponding API documentation is not updated synchronously in time.

To address the issues of API misuse detection, the existing methods can be grouped as two kinds. The first kind is utilizing API's semantic information. Inspired by Natural Language Processing, software engineering starts to take "naturalness" of code into consideration. Many studies apply the powerful models in Natural Language Processing area into code analysis, code completion and bug detection. They discover the potential semantic features of underlying codes and take advantage of these features to make contributions for research on source code. For example, some studies finished their work based on n-gram statistical language model to solve code completion problem [26–28]. However, due to the fact that the syntax of the code is the rule of communicating with machines, there are also some unique attributes among code structures. Thus, the second kind is discovering code's grammatical structure. There are many studies to learn the usage specifications of code through structure and pattern mining [6–12, 19–25]. Applications of above methods are common in software repositories mining, documentations, API patterns, summarization and anomaly detection. However, there are following challenges existed in these methods: While analyzing source code structure, people often tend to mine the connection relation between APIs and use pattern matching to find misuse of code structure. The API's semantic information is often ignored.

In this paper, we employ Stacked LSTM to learn the API usage specification by feeding API sequences with semantic representation, so as to detect the API misuse defects. First, we perform a static analysis of Java code files which aims to obtain ACSG (API Call Syntax Graph) of each file. Then, we design a mining algorithm to get API sequences on ACSGs and progressively transform the sequences generated into previous API sequences and corresponding next API, which are regarded as the input and output of our model proposed later respectively. In order to utilize the properties of the API itself, we apply word2vec to acquire the semantic representation of each API by mapping the relation between APIs and API sequences into the relation between words and sentences. Next, we learn API usage specification by Stacked LSTM, which generates the next API probability list as output. We determine API misuse by checking whether the actual API is contained in the prediction list or not. Finally, we verify our method by compared with the existing method. Two experiments are conducted, using the precision, recall rate, and F1 score as the criteria. The results show that our proposed model is the best among the experimental models.

This paper is organized as follows. Section 2 introduces the related works about Service adaption and API misuse detection. Section 3 presents the overall process and its implementations. Section 4 introduces the design and analysis of the experiment designed with two schemes of model. Finally, in Sect. 5 we summarize and prospect this paper.

2 Related Work

In this section we review some of related work in the area of API misuse detection. We simply divide the existing studies into two parts, namely language models and specification mining and defect detection.

Utilizing Semantic Information. According to "The naturalness hypothesis" [1], software corpuses have similar statistical properties to natural language corpuses. Codes in software can be regarded as sequences of token, Thus, the model applied in Natural Language Processing can be also used for solving problems in API misuse detection area.

Recently, language models have been used successfully in tasks including code completion fault localization [2, 4] and code conventions checking [3]. Pu et al. [13] treated the program statement as a sequence of tokens, and they propose purely syntactic learning, through which the candidate missing or fault program statement can be generated as one token at a time. Via modifying and training the seq2seq neural network model, they achieve the goal of finding and recovering the defects in MOOCs. Raychev et al. [15] designed a scalable analysis to extract sequence of method calls. They abstract the sentence with holes from the partial program with hole at event-level, where event is the semantics representation of source codes. In order to discover the completions for the code misuses, they employ statistical language model to find the highest possible bug-occurred sentence. Ray et al. [5] proposed a cache language model to extend the traditional language models by applying an addiction cache to capture the regularities in local context. Then, they apply the bug-finder to detect the lines of code where are more likely to detect defects. Tu et al. [18] proposed "source code is localized", which shows that source code tends to take semantic features in local contexts. In order to achieve the goal of code suggest, they recommend the next token based on the current context. The cache language model they applied are capable of this task.

We can see from these studies above that it is very useful to applying Natural Language Processing models to API misuse detection. On the other hand, the "localness" in source code, which refers to the complexity of code structures, still is the chronic and main challenge for API misuse detection when applying language models.

Exploring Code's Structure. Code can be considered as a kind of language to communicate with each other through a specific compiler environment. Specification mining and defect detection aims to find a finite set of patterns from source code, which consists of part of human-interpretable behaviors. And it can present the mined patterns to software engineers without annotation or supervision.

Many algorithms and techniques have been proposed for programming rule mining and misuse detection. [6–12, 14, 19–25] Oh et al. [6] presented a method for building an adaptive static analyzer. Via Bayesian optimization, they learn a good sophisticated parameterized strategy for discovering the specifications from the real-world C programs code structures. In order to solve the problem of assessing final students' code, Piech et al. [7] introduced a neural network to model students' programs as linear maps of their code structures. Then, the feedback algorithm makes use of these maps to find the misuses for Stanford University's CS1 course's students' code assignments. Wang et al. [12] designed Bugram for bug detection. Based on the assumption on specification that the API call token is only related to the n tokens before it, the occurrence probability of the API call token sequence appearing in the software project is calculated. Though linking the occurrence of the token sequence with API misuse, Bugram achieves the goal of automatic defect detection. Wang et al. [25] set up the recurrent neural network to learning API use specification. Their study makes a context-based prediction on the API code, and finds out the potential API misuse by comparing the prediction results with the actual code.

The existing specification-based code defects detection methods show great performance on analyzing structural information on real-world codes. Therefore, learning API usage specification is helpful for API misuse detection.

3 Method

We design a new API-misuse detect process which is shown in Fig. 1, that makes full use of the strengths of previous study of API-misuse and figures out the shortcomings summarized above. As shown in Fig. 1, the specific steps are listed as follows:

- Static Analysis. We design API Call Syntax Graph (ACSG), a presentation of API usage which can capture the order between API calls and data interactions which can distinguish misuses from correct usages.
- Data Generation. We design a new API call sequences mining algorithm, which can generate all the API call sequences into <precious API sequence, next API>. Through learning the API usage specification, we apply Word2Vec as a pre-training model to achieve representation for each API which can make use of API semantic features among API sequences.
- Model Training and Prediction. We design Stacked LSTM model for next API call prediction on the basis of previous API call sequences.

3.1 Static Analysis

Abstract syntax tree (AST) is a power tool to map the Java code into a Tree data structure. We use Javaparser to parse out the structure of the source Java code. Amann et al. [32] found the AUG (API-Usage Graph) as a presentation of API usages, which is a direct, connected multi graph with named nodes and edges. However, due to the fact that AUG contains too much details about API calls, so we simplify the AUG and propose our

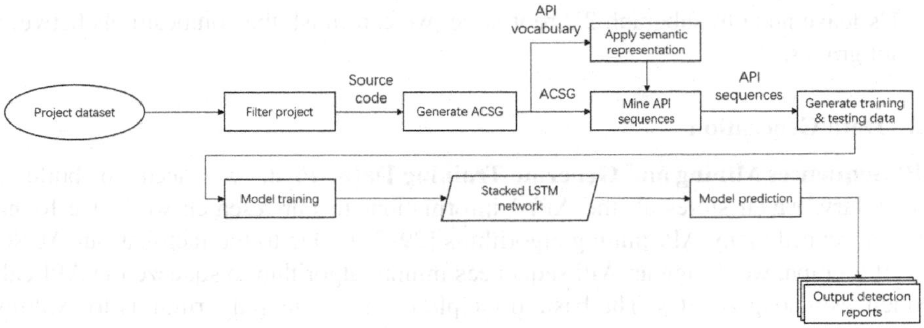

Fig. 1. The overview of our method

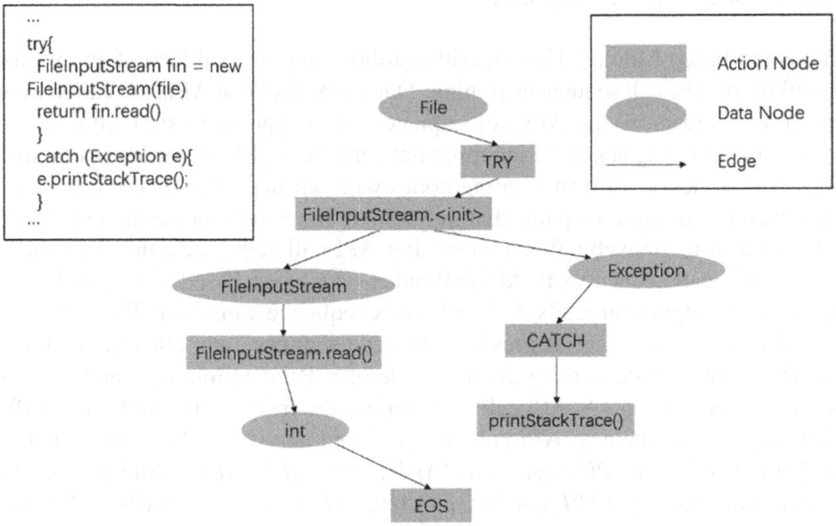

Fig. 2. Example for ACSG

API Call Syntax Graph (ACSG) for further API sequences mining. Figure 2 shows an example of ACSG in this article.

The following definitions are involved in the construction of ACSG:

a. Node: Node represents a node in ACSG, which includes Action Node and Data Node. Action Node contains API calls, method calls and control statements. Data Node presents the objects and values that appears in the source code. The Node with 0 in-degree can be regarded as root node, and the Node with 0 out-degree can be regarded as leave node.
b. Edge: Edge represent the edge which connects the parent node to the child node. In this paper, we use Edge to show the order relation between adjacent nodes.
c. Graph: Graph represents the ACSG. Graph can be composed by subgraphs. And the root of Graph represents the start of ACSG. By adding the edge from the subgraph

1's leave node to subgraph 2's root node, we can finish the compositions between subgraphs.

3.2 Data Generation

API Sequences Mining and Generate Training Data. First, we need to build a vocabulary, which stores all the APIs' information. In our research work, we found and referenced many API mining algorithms [29–31]. Due to the fact that our ACSG is still a graph, we design an API sequences mining algorithm to squeeze the API call sequences among ACSGs. The basic principle of this mining algorithm is to exploits the API sequences which start with in-degree of 0 nodes and end with out-degree of 0 nodes. Table 1 shows our API sequences mining algorithm:

The algorithm follows 2 key ideas:

- Exhaustion-based Mining: This algorithm follows the general idea of an exhaustion algorithm for API call sequences mining. Due to the fact that APIs are called in order from top to bottom, so the API call sequences are supposed to start with one of the initial nodes. Aiming at generating more data, it mines API call sequences by starting nodes with in-degree of 0 and ending nodes with out-degree of 0. The key idea here is that traverse all possible paths that may satisfy the condition mentioned above.
- Generate data recursively: When squeezing API call sequences, the algorithm first converts API call sequences to API call index sequence *s* based on vocabulary. Subsequently, the algorithm reads API call index sequence *s* in order. The training data is divided into 2 parts: one is previous API call sequence, the other is the next API call. The algorithm starts from the first node of *s*. By determining whether the next node exists by reading *s* 's API call, it generates the training data in form <API call index sequence *si*, the next API call *ci + 1*>, and persists to the local training data file. For example, an API sequence is $[API_1, API_2, API_3, API_4]$. And the final training data generated is $[<[API_1], API_2>, <[API_1, API_2], API_3>, <[API_1, API_2, API_3], API_4>]$.

After generating the API call sequences and counting the frequency of API calls in the generated API sequences, a vocabulary with API call indexes corresponding to the API calls is established and persisted to the local vocabulary file.

API Embedding. Word2vec is a kind of word embedding algorithm that provides state-of-the-art outcomes on various linguistic tasks. Due to Distributional Hypothesis [11], which proposes that words that appear in the same context tend to have close relations. As a pre-training model, Word2vec can represent words as d-dimensional vectors, so that words which have close relations with others can have similar vector representation [33, 34]. We treat the API call sequences generated above as sentence in the text, such that adjacent APIs can be semantically embedded into similar vectors. API embeddings is a solution to the problem of numerically representing APIs.

Let $\varphi = \{a_i : i = 1, 2, \ldots, V\}$ be the ordered set of APIs, where V is the size of the API vocabulary and a_i is the i-th API. Usually, the APIs in the API sequence are represented as one-hot vectors,

$$\beta_1 = (1, 0, \ldots, 0), \ldots, \beta_V = (0, 0, \ldots, 1) \quad (1)$$

Table 1. API sequences mining algorithm

```
def API_sequences_mine(acsg: ACSG)
sequences = ∅
sequences = find_all_sequence(acsg)
training_data = generate_training_data(sequences)
return training_data

def find_all_sequence(acsg: ACSG)
sequences = ∅
Start = {nodes are with 0 in-degree in A}
End = {nodes are with 0 out-degree in A}
for sequence_in_acsg in all_sequences_in_acsg: // Ex-
haustion-based sequences mining
  if sequence's start node in Start && sequence's end
node in End:
    sequences = sequences∪sequence_in_acsg
return sequences

def generate_training_data(sequences)
training_data = ∅
for sequence in sequences:   //Generate training data
recursively
  sequence -> <previous API sequence, next API>
  training_data = training_data∪sequence
return training_data
```

where β_i is the hot-vector representation of the API a_i. In this work, we define $A(a_i) \rightarrow \alpha_i$, where α_i is the embedded d-dimensional representation of the API a_i. The way we construct A is through Skip-gram model [16], which could predict a target word given a set of words called context words. We define context APIs as the APIs that appears in the context of target API. The context APIs of an API are defined as the set of APIs are at a distance less than or equal to c from each occurrence of the target API in the API call sequences, where c is a constant which is defined by us. For example, if one wants to let the neural network to show the representational vector of target API a_t, where the context APIs are $\{a_{t-c}, a_{t-c+1}, \ldots, a_{t+c-1}, a_{t+c}\}$. The neural network can be showed as in Fig. 3, where the input layer is the target API a_t, the projection layer can predict the context APIs and finally return the *Context*(t) in the form if a vector to represent API a_t.

Through word2vec, we can put all the API sequences as input for training, and get a model of semantic representation to the API by analyzing the semantic information of the context APIs. Via this model, we can represent representation of each API that appears in the vocabulary file.

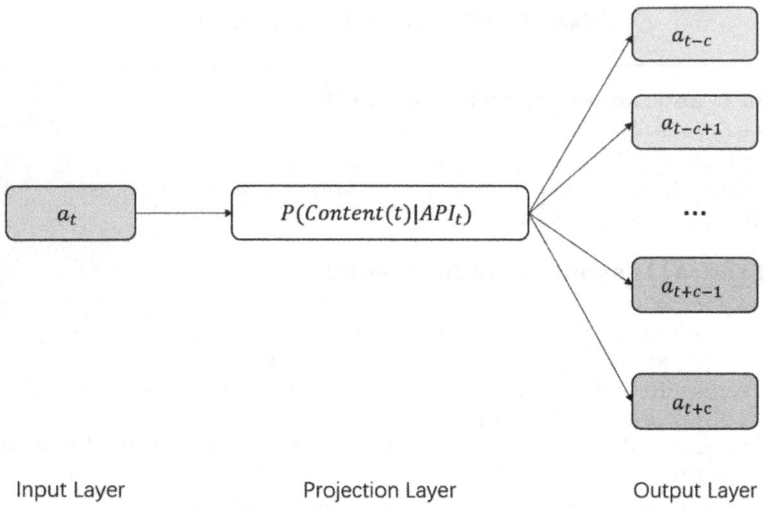

Fig. 3. Word2Vec model

3.3 Model Training and Prediction

During this step, based on TensorFlow framework, we use Python3 to build a deep learning model aiming to learn the training data obtained from the original code fragments. After achieving the model training, we use the model to predict the next API calls located after a certain API call sequence. By compared the predicted API call with target API call, we can judge whether the target API call is suitable or not, so this stage is mainly divided into two parts: model training and model prediction.

Since being regarded as a research hotspot in the field of artificial intelligence, Neural networks have already been used in regression and classification issues. Due to the fact that Recurrent Neural Networks (RNNs) have the structure that can memorize the previous information and apply it to the calculation of the current output, RNNs have the capacity of modeling sequential data. As one of the improved algorithms of RNN, it is worth noting that long short-term memory (LSTM) is designed to solve the long-term dependencies problem for modeling sequential data [16, 17, 35], such as time series, which shows great potential in modeling API call prediction.

Review on LSTM. In addition to the input layer and output layer, the standard deep LSTM model consists of a number of hidden layers which include LSTM layers and fully connected layers. As a basic layer of LSTM architecture, LSTM layers have a group of LSTM cells which can map the output sequences depending on the given input sequence represented by (x_1, ..., x_T). Each LSTM cell (shown as Fig. 4) owns its independent weights and biases, which is similar to ANN's neural node. Different from other structures, LSTM cell structure can delete or add information to the cell state through a unit called gate. Every LSTM cell's internal information we call it cell state. There are three gates in LSTM cell, including a forget gate, an input gate and an output gate. The forget gate determine what information needs to be discarded in the cell state. The input gate updates the cell state by the flow of input activation. The output gate controls the flow of output activation into the next LSTM cell. Here comes the denotation of LSTM operations. In the LSTM network layer, the forget gate shows as f_t, the input gate shows as i_t, the output gate shows as o_t, the cell state shows as c_t, and the hidden layer output shows as h_t. And at the previous time step, we represent the cell memory as c_{t-1}, and the hidden layer output as h_{t-1}. The specific operation in LSTM cell are described as follows:

$$f_t = \sigma\left(W_f \cdot \left[h_{t-1}, x_t\right] + b_f\right) \tag{2}$$

$$i_t = \sigma\left(W_i \cdot \left[h_{t-1}, x_t\right] + b_i\right) \tag{3}$$

$$o_t = \sigma\left(W_o \cdot \left[h_{t-1}, x_t\right] + b_o\right) \tag{4}$$

$$\tilde{C}_t = \tanh(W_C \cdot \left[h_{t-1}, x_t\right] + b_C) \tag{5}$$

$$C_t = f_t * C_{t-1} + i_t * \tilde{C}_t \tag{6}$$

$$h_t = o_t * \tanh(C_t) \tag{7}$$

where W_α (here $\alpha = \{f, i, o, C\}$) represents the weight matrices corresponding to different gates (e.g. forget gate, input gate, output gate or tanh layer as shown in Fig. 4), and b_α denotes the corresponding bias vector. In Fig. 4, $\tilde{C}_t$ represent the candidate information of cell state created by the tanh layer; σ is the logistic sigmoid unit; tanh is the hyperbolic tangent unit.

When it comes to the workflow of this LSTM cell, the input of this structure contains 3 parts: C_{t-1}, h_{t-1}, x_t, while the output includes 2 parts: C_t, h_t(y_t equals to h_t). The relationship between x_t and y_t is that x_t equals to the last layer's y_t. Back propagation of LSTM is utilized to update the weights during the training process. Subsequently, the LSTM last output will be used to predict a specific API call followed by a fully connected layer. Eventually, the next API call can be obtained by the previous API call sequences.

The standard deep LSTM network used for sequential data modeling is composed of plenty pf LSTM layers and fully connected layers (dense layers). The Fully Connected layers are located between LSTM layers and output layers, which are used to connected these two layers. Fully Connected Layers have full connections with previous layer's activation nodes. The embedding layer also can be used in this structure, which can transfer sparse matrix into higher-dimensional dense matrix. In addition, dropout layer which is not shown in Fig. 5, can be added after each layer to prevent overfitting of models. The direct effect of dropout layer is to reduce the number of intermediate features, thereby reducing redundancy, which means increasing the orthogonality between each feature of each layer.

The Full Deep LSTM Network (F-LSTM). The construction of this model is based on the deep learning framework TensorFlow 1.15.0 as the implementation framework. It is noted that, to implement this LSTM architecture, the main calculation of this model

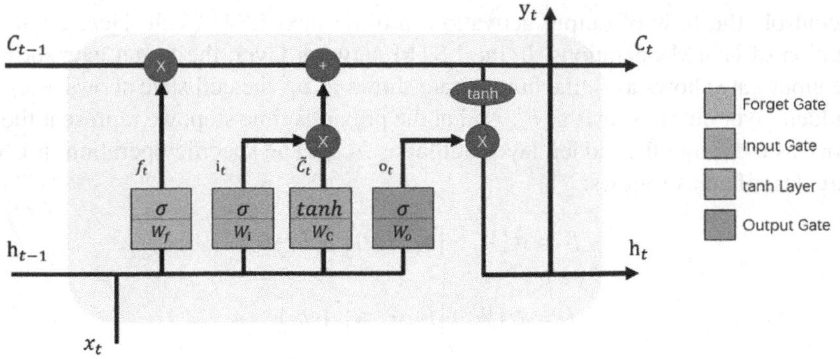

Fig. 4. LSTM cell

is shown as follows: (1) Through the embedding layer, we use the word2vec model to generate the corresponding vector to represent the API's semantic information. (2) The input vector converted by the embedding layer passes through the dropout layer, through which it discards part of the information, aiming to increase the model's robustness. (3) Send the processed data to the LSTM processing unit provided by TensorFlow for training. (4) The output of the model needs to go through the dense layer, which can convert the output into desired format, and we define the output of dense layer as logits. Subsequently, logits undergo softmax processing and cross-entropy calculation to obtain the probability distribution and loss value of API calls respectively. Besides, TensorFlow will optimize the parameters based on the given optimization function.

In this content of API prediction modeling, let us denote the input front API sequences as $x = \{x_1, x_2, \ldots, x_n\}^T \in R^{n \times p}$ and the output API as y, where p represent the number of input features and y is represented as one-hot vector. x is vector whose matrices whose rows stand for time steps and columns for features (the dimension of API). The full deep LSTM network takes the target API's preceding API sequence (x) as input and map it to the output of the model, target predicted API (y), where each cell within a LSTM layer is connected with its two neighbors via $\{t-1, t, t+1\}$. At each time step, all input features at the current time step are fed into the deep LSTM network. The process keeps sending input features to the network via repeated LSTM cells across the entire temporal space, which could build a chain-like structure to keep the long-short term time dependencies. Figure 5 illustrates the basic structure of F-LSTM which aims to model the target API (y) given its preceding API sequence (x).

In conclusion, the F-LSTM network models the corresponding input & output relation along the temporal space. But a main disadvantage of F-LSTM is that it has a massive demand for training efforts especially for long API sequences that require a large amount of computation memory. To solve this problem, we modify and propose the Stacked Deep LSTM network.

The Stacked Deep LSTM Network (S-LSTM). We present the S-LSTM model which takes the stacked input of a certain number of time steps, $\{X_1, X_2, \ldots, X_t\}^T$, to predict the output y. Assuming that we have only one feature, we can illustrate the concept of S-LSTM as Fig. 6. Via embedding layer, the original sequences of input X is

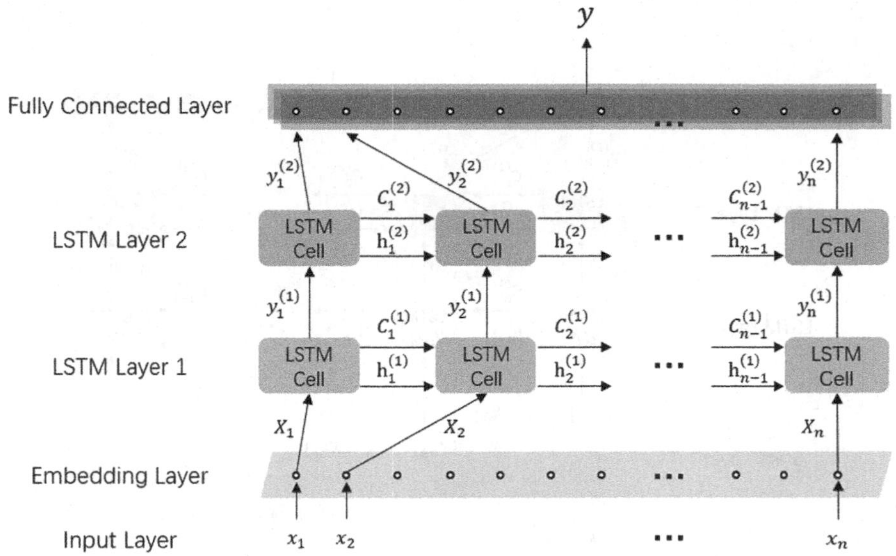

Fig. 5. The structure of full deep LSTM

divided into multiple stacks, which forms the new input $\tilde{X} = \left\{\tilde{X}_1, \tilde{X}_2, \ldots, \tilde{X}_s\right\}$, where s is the number of the stacks. Therefore, each stack we created can be considered as one new time step and reduce the temporal space the model costs. Note that $sw \leq n$ (if $sw < n$, we apply zero-padding to meet the disparity). Every input stack consists of a fixed length of original API sequence, which is considered as the new input features to fed the LSTM cells.

S-LSTM not only reduces the temporal dimension but also represent the semantic meaning of each API in the previous API sequence. It is worth noting that, w is an empirical number and depends on the specific embedding result as well as the dominant mode of the dynamic model response. In addition, S-LSTM is observed to provide better training and prediction performance as compared to LSTM-f and other models as showed in Sect. 4.

3.4 Model Prediction

In this stage, on the basis of the model trained in deep learning model training, we use the previous sequence of API calls for prediction as input, predict the probability distribution of the API call at the next position, and sort the probability distribution to obtain the API call probability list at the next position. This API call probability list will be used as an important part of defect detection.

4 Experiment

This section presents the implementation of applying LSTM model to predict the target API. In our work, the experimental evaluation is divided into 2 parts: (1) The model

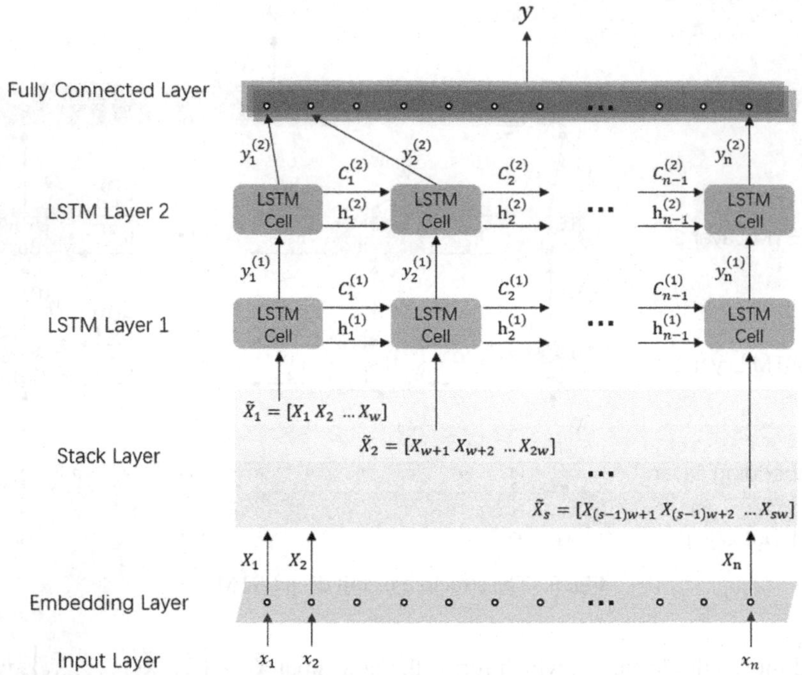

Fig. 6. The structure of stacked deep LSTM

training experiment. The purpose of this part is to optimize the accuracy and reliability of the prediction of the API call sequence through model. (2) The code defect detection experiment. The purpose of this part is to detect the model's effectiveness and usability in API misuse defect detection.

4.1 Experimental Object

Java cryptography extension (JCE) is a package provided by JDK, which [9] can provide the implementation of cryptographic primitive, including block ciphers and message authenticate codes (MACs). Java cryptography APIs provided by JCE are under the package named javax.crypto. According to related research [9], by separating the implementation details for the users, developer can use it conveniently to achieve the encryption and decryption functions. Also, these APIs provide multiple modes and configuration setting options. However, to use and combine these APIs correctly can be challenges for developers. In the following parts, we will focus on the experimental object of Java cryptography APIs to illustrate the ability of our models in defect detection.

4.2 Model Training Experiment

The Dataset of this article was obtained from Git, a free and open source distributed version control system, through which we could manage no matter what kinds of projects.

GitHub is a large-scale open source code hosting platform and version control system based on Git. The open source code on GitHub is a massive data source [20, 21] for mining projects which contain Java cryptography APIs. Then, we generate our original data by using "javax.crypto" as the key word from the project we search via GitHub.

In order to achieve comparative experiments, we use the parameter configuration proposed in Wang's work [8]. In our work, we compared the training effects of three different models, which include Bugram (3-gram and 4-gram are selected), the deep learning LSTM model proposed by Wang's work [8] (we will call it D-LSTM later), F-LSTM and S-LSTM. The essential difference between our models and D-LSTM is whether use word2vec in the padding layer or not. The specific parameter configuration is HIDDEN_SIZE = 250, NUM_LAYER = 2, LR = 0.002, NUM_EPOCH = 20. As shown in Fig. 7, we show the loss of 3 different models, where the horizontal axis is the number of epochs, and the vertical axis is the loss value.

In this configuration, the model's effect is shown in Fig. 8. The horizontal axis is the number of epochs, and the vertical axis is the accuracy.

Obviously, we can tell that F-LSTM and S-LSTM perform well in both accuracy and loss value. We can intuitively see from the Fig. that the adding Word2Vec has a great influence on the training of the model. Although model training capabilities shown by the F-LSTM and S-LSTM are very similar, S-LSTM still has about 1% better in accuracy and 0.05 lower in loss value than F-LSTM. Finally, the D-LSTM achieve 80.3% accuracy and 0.772 loss value; the F-LSTM achieve 83.2% accuracy and 0. 622 loss value; the S-LSTM achieve 84.2% accuracy and 0.567 loss value.

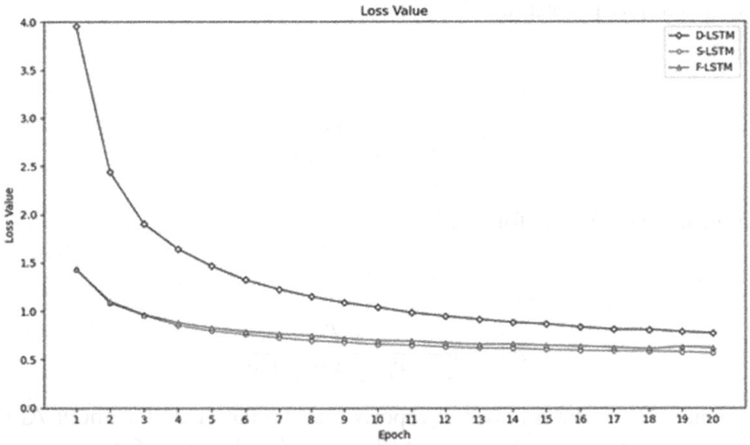

Fig. 7. Loss value of models

4.3 Code Defect Detection Experiment

Based on the models we have proposed above, the code defect detection experiment is designed to show the ability of code defect detection on specific projects. First, we define the evaluation standard for this experiment as follows:

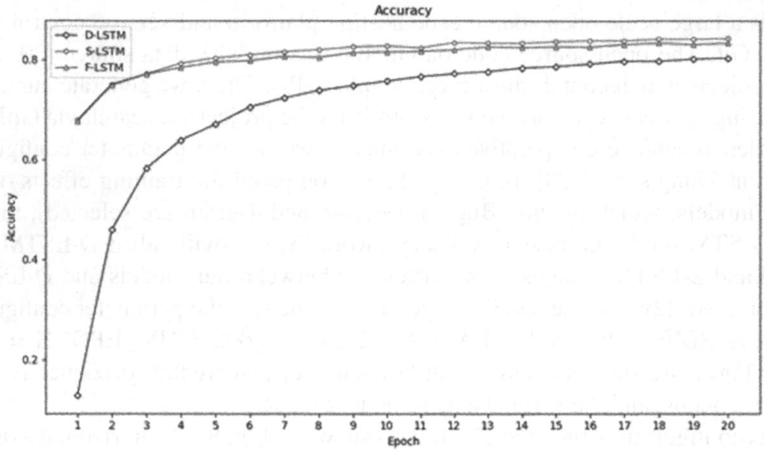

Fig. 8. Accuracy of models

1. We define the precision as follow:

$$Precision = \frac{TP}{TP + FP} \tag{8}$$

2. We define the recall as follow:

$$Recall = \frac{TP}{TP + FN} \tag{9}$$

3. We define the F1 score as follow:

$$F1 = \frac{2 \times TP}{2 \times TP + FP + FN} \tag{10}$$

where we define *TP* as the number of reports that the target API without API misuse can be correctly predicted in the target API call probability list; *FP* as the number of reports that the target API with API misuse can be correctly predicted in the target API call probability list; *FN* as the number of reports that the target API with API misuse cannot be correctly predicted in the target API call probability list. Due to the fact that the score of precision and recall cannot reflect the result of experiment comprehensively, we decide to use F1 score which is the harmonic mean of precision and recall, as the final standard for the evaluation of our experimental results.

In our experiment, according to the actual API misuse of Java Cryptography API in our survey results, we select 8 API misuse codes which are collected from high-quality projects from open code platform, such as GitHub and SourceForge. And we use these codes as the test set for our models' evaluation. The code defect detection experimental results are shown as Figs. 9, 10, 11. The lines in these three figures represent different experimental models respectively, and the x-axis represents the value of the acceptable threshold (top-k). Figure 9 shows the F1 score value of each model; Fig. 10 shows the precision value of each model; Fig. 11 shows the recall rate value of each model.

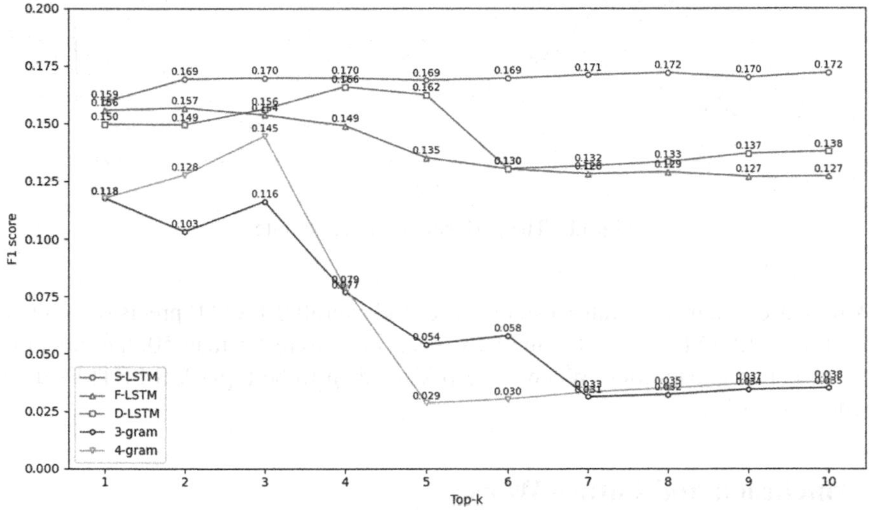

Fig. 9. The performance of F1 score

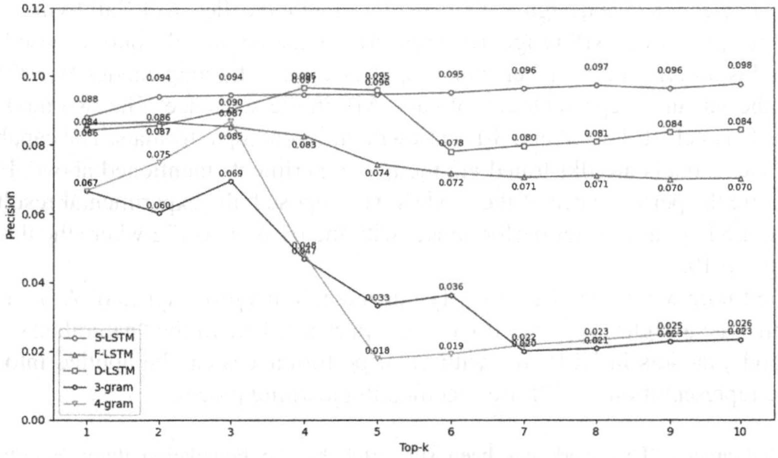

Fig. 10. The performance of precision

It can be seen from the results that S-LSTM' precision is always higher than F-LSTM's. The recall rates of both models are very low when they are at top-1, but as the

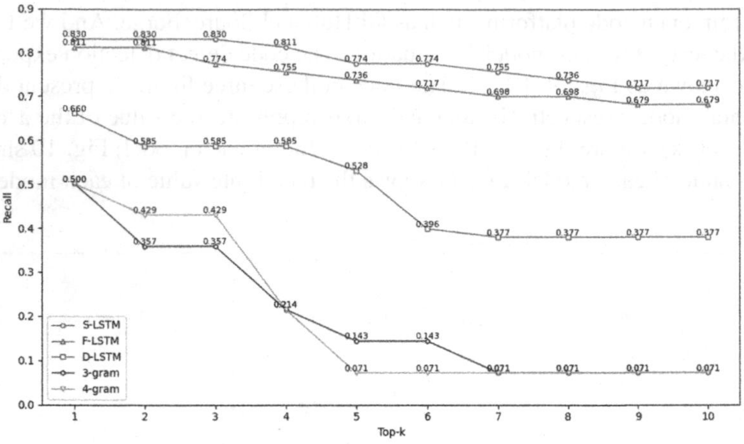

Fig. 11. The performance of recall rate

top-k increases, their recall rates also increase, and overall S-LSTM' precision is always higher than F-LSTM's. Since the model's recall rate is greater than 50, the model has defect detection capabilities. When the top-k is taken to be top-10, S-LSTM's defect detection works best.

5 Conclusion and Future Work

This paper proposes a deep learning approach, based on a long short-term memory (LSTM) recurrent neural network, for API usage specification learning and API misuse defect detection. We design a static analysis method called API Call Syntax Graph (ACSG) for presenting API usage. Different from existing models, our proposed architecture adds an embedding layer which applies the pre-training model Word2Vec to achieve the semantic representation of each API in the sequence. The two models are developed, namely, F-LSTM and S-LSTM with different input formats. The capabilities of these two models are illustrated via the two experiments mentioned above. Finally, we compare the performance of the models we proposed, the experimental results turn out that S-LSTM has a better performance, with the F1 score 0.172 where the threshold is among top-10.

In the future work, we plan to verify our method in various types of APIs in order to demonstrate a higher applicability of our approach. Due to the "naturalness" of the source code, models in NLP area with great performances can be applied into either semantic representation or API use specification learning process.

Acknowledgement. This work has been supported by the Foundation item: National Key R&D Program of China (2018YFB1402800); National Natural Science Foundation of China (61772560).

References

1. Allamanis, M., et al.: A survey of machine learning for big code and naturalness. ACM Comput. Surv. (CSUR) **51**(4), 1–37 (2018). Author, F., Author, S.: Title of a proceedings paper. In: Editor, F., Editor, S. (eds.) CONFERENCE 2016, LNCS, vol. 9999, pp. 1–13. Springer, Heidelberg (2016)
2. Nguyen, S., et al.: Combining program analysis and statistical language model for code statement completion. In: 2019 34th IEEE/ACM International Conference on Automated Software Engineering (ASE). IEEE (2019)
3. Allamanis, M., et al.: Learning natural coding conventions. In: Proceedings of the 22nd ACM SIGSOFT International Symposium on Foundations of Software Engineering (2014)
4. White, C., Vendome, M., Linares-Vásquez, M., Poshyvanyk, D.: Toward deep learning software repositories. In: MSR 2015, pp. 334–345 (2015)
5. Ray, B., Hellendoorn, V., Godhane, S., Tu, Z., Bacchelli, A., Devanbu, P.: On the naturalness of buggy code. In: Proceedings of the International Conference on Software Engineering (ICSE) (2016)
6. Oh, H., Yang, H., Yi, K.: Learning a strategy for adapting a program analysis via Bayesian optimisation. In: Proceedings of the Conference on Object-Oriented Programming, Systems, Languages & Applications (OOPSLA) (2015)
7. Piech, C., Huang, J., Nguyen, A., Phulsuksombati, M., Sahami, M., Guibas, L.J.: Learning program embeddings to propagate feedback on student code. In: Proceedings of the International Conference on Machine Learning (ICML) (2015)
8. Pradel, M., Sen, K.: Deep learning to find bugs (2017)
9. Proksch, S., Lerch, J., Mezini, M.: Intelligent code completion with Bayesian networks. ACM Trans. Softw. Eng. Methodol. (TOSEM) **25**, 1–31 (2015)
10. Rabinovich, M., Stern, M., Klein, D.: Abstract syntax networks for code generation and semantic parsing. In: Proceedings of the Annual Meeting of the Association for Computational Linguistics (ACL) (2017)
11. Raychev, V., Vechev, M., Krause, A.: Predicting program properties from "big code". In: Proceedings of the Symposium on Principles of Programming Languages (POPL) (2015)
12. Wang, S., Chollak, D., Movshovitz-Attias, D., Tan, L.: Bugram: bug detection with n-gram language models. In: Proceedings of the International Conference on Automated Software Engineering (ASE) (2016a)
13. Pu, Y., Narasimhan, K., Solar-Lezama, A., Barzilay, R.: sk_p: a neural program corrector for MOOCs. In: Proceedings of the Conference on Systems, Programming, Languages and Applications: Software for Humanity (SPLASH) (2016)
14. Xue, Y., Wang, J., Liu, Y., Xiao, H., Sun, J., Chandramohan, M.: Detection and classification of malicious javascript via attack behavior modelling. In: ISSTA 2015, pp. 48–59 (2015)
15. Raychev, V., Vechev, M., Yahav, E.: Code completion with statistical language models. In: Proceedings of the Symposium on Programming Language Design and Implementation (PLDI) (2014)
16. Xie, K., Wen, Y.: LSTM-MA: a LSTM method with multi-modality and adjacency constraint for brain image segmentation. In: 2019 IEEE International Conference on Image Processing (ICIP). IEEE (2019)
17. Xue, H., Huynh, D.Q., Reynolds, M.: SS-LSTM: a hierarchical LSTM model for pedestrian trajectory prediction. In: 2018 IEEE Winter Conference on Applications of Computer Vision (WACV). IEEE (2018)
18. Tu, Z., Su, Z., Devanbu, P.: On the localness of software. In: Proceedings of the International Symposium on Foundations of Software Engineering (FSE) (2014)

19. Kersten, M., Murphy, G.C.: Using task context to improve programmer productivity. In: FSE, pp. 1–11. ACM (2006)
20. Kuhn, A., Ducasse, S., Gírba, T.: Semantic clustering: identifying topics in source code. Inf. Softw. Technol. **49**(3), 230–243 (2007)
21. Allamanis, M., Barr, E.T., Bird, C., Sutton, C.: Suggesting accurate method and class names. In: Proceedings of the Joint Meeting of the European Software Engineering Conference and the Symposium on the Foundations of Software Engineering (ESEC/FSE) (2015)
22. Allamanis, M., Tarlow, D., Gordon, A., Wei, Y.: Bimodal modelling of source code and natural language. In: Proceedings of the International Conference on Machine Learning (ICML) (2015)
23. Allamanis, M., Peng, H., Sutton, C.: A convolutional attention network for extreme summarization of source code. In: Proceedings of the International Conference on Machine Learning (ICML) (2016)
24. Allamanis, M., Brockschmidt, M., Khademi, M.: Learning to represent programs with graphs. In: Proceedings of the International Conference on Learning Representations (ICLR) (2018)
25. Wang, X., Chen, C., Zhao, Y.F., Peng, X., Zhao, W.Y.: API misuse bug detection based on deep learning. Ruan Jian Xue Bao/J. Softw. **30**(5), 1342–1358 (2019). (in Chinese). http://www.jos.org.cn/1000-9825/5722.htm
26. Hindle, A., Barr, E.T., Su, Z., Gabel, M., Devanbu, P.: On the naturalness of software. In: Proceedings of the International Conference on Software Engineering (ICSE) (2012)
27. Nguyen, A.T., Nguyen, T.N.: Graph-based statistical language model for code. In: Proceedings of the International Conference on Software Engineering (ICSE) (2015)
28. Nguyen, T.T., Nguyen, A.T., Nguyen, H.A., Nguyen, T.N.: A statistical semantic language model for source code. In: Proceedings of the Joint Meeting of the European Software Engineering Conference and the Symposium on the Foundations of Software Engineering (ESEC/FSE) (2013)
29. Fowkes, J., Sutton, C.: Parameter-free probabilistic API mining across GitHub. In: Proceedings of the 2016 24th ACM SIGSOFT International Symposium on Foundations of Software Engineering (2016)
30. Nguyen, P.T., et al.: FOCUS: a recommender system for mining API function calls and usage patterns. In: 2019 IEEE/ACM 41st International Conference on Software Engineering (ICSE). IEEE (2019)
31. Chen, C., et al.: Mining likely analogical apis across third-party libraries via large-scale unsupervised API semantics embedding. IEEE Trans. Softw. Eng. (2019)
32. Sven, A., Nguyen, H.A., Nadi, S., et al.: Investigating next steps in static API-misuse detection. In: 2019 IEEE/ACM 16th International Conference on Mining Software Repositories (MSR), pp. 265–275. IEEE (2019)
33. Lilleberg, J., Zhu, Y., Zhang, Y.: Support vector machines and word2vec for text classification with semantic features. In: 2015 IEEE 14th International Conference on Cognitive Informatics & Cognitive Computing (ICCI* CC). IEEE (2015)
34. Zhang, D., et al.: Chinese comments sentiment classification based on word2vec and SVMperf. Expert Syst. Appl. **42**(4), 1857–1863 (2015)
35. Zhang, R., et al.: Deep long short-term memory networks for nonlinear structural seismic response prediction. Comput. Struct. **220**, 55–68 (2019)

Edge Computing and CollaborateNet

Location-Aware Edge Service Migration for Mobile User Reallocation in Crowded Scenes

Xuan Xiao[1], Yin Li[2], Yunni Xia[1(✉)], Yong Ma[3], Chunxu Jiang[4], and Xingli Zhong[5]

[1] College of Computer Science, Chongqing University, Chongqing 400044, China
xiayunni@hotmail.com
[2] Institute of Software Application Technology, Guangzhou and Chinese Academy of Sciences, Guangzhou 511000, China
[3] School of Computer and Information Engineering, Jiangxi Normal University, Nanchang 330022, China
[4] Chongqing Key Laboratory of Smart Electronics Reliability Technology, Chongqing 401331, China
[5] CISDI R&D Co. Ltd., Chongqing 401122, China

Abstract. The mobile edge computing (MEC) paradgim is evolving as an increasingly popular means for developing and deploying smart-city-oriented applications. MEC servers can receive a great deal of requests from equipments of highly mobile users, especially in crowded scenes, e.g., city's central business district (CBD) and school areas. It thus remains a great challenge for appropriate scheduling and managing strategies to avoid hotspots, guarantee load-fairness among MEC servers, and maintain high resource utilization at the same time. To address this challenge, we propose a coalitional-game-based and location-aware approach to MEC Service migration for mobile user reallocation in crowded scenes. Our proposed method includes multiple steps: 1) dividing MEC servers into multiple coalitions according to their inter-euclidean distance by using a modified k-means clustering method; 2) discovering hotspots in every coalition area and scheduling services based on their corresponding cooperations; 3) migrating services to appropriate edge servers to achieve load-fairness among coalition members by using a migration budget mechanism; 4) transferring workloads to nearby coalitions by backbone network in case of workloads beyond the limit. Experimental results based on a real-world mobile trajectory dataset for crowded scenes, and an urban-edge-server-position dataset demonstrate that our method outperforms existing approaches in terms of load-fairness, migration times, and energy consumption of migrations.

Keywords: Service migration · Load fairness · Coalitional game · Crowded scenes · Hotspot discovery · Mobile trajectory · User reallocation · Backbone network

H. Gao et al. (Eds.): CollaborateCom 2020, LNICST 349, pp. 441–457, 2021.
https://doi.org/10.1007/978-3-030-67537-0_27

1 Introduction

Mobile edge computing (MEC) refers to the concept of processing data at mobile edge network. The edge is similar to a distributed cloud with proximity close to end users. It is usually built upon small-scale data centers close to the data sources to guarantee low latency, high reliability, and scalability [8,11]. As shown in Fig. 1, services in edge nodes are situated close to users and an edge server (ES) can only contact the users that fall into its coverage area. Due to user mobility and the constraint of the user-server proximity, in practice, server-side services need to be migrated from their original nodes to new ones to maintain the user-server proximity. In reality, population in motion in crowded areas, e.g., supermarkets, can lead to unbalanced distribution of highly mobile users [20], and thus edge servers can receive quite different amounts of requests from users even when they are located at nearby areas [21]. For example, a subway station usually attracts a lot of users while a park or a bookstore near the subway usually attracts much fewer. Consequently, edge servers deployed in the station can show higher load than those in the park or bookstore. A smart strategy that is capable of handling user mobility and migrating tasks from highly-loaded servers to ones with low load is thus in high need [2]. An ES should deal with load peaks and very spiky patterns s. Moreover, MEC servers are generally equipped with lightweight computing components and limited storage. This exacerbates the challenge to avoid hotspot effects on ESs, and load-fairness among MEC servers, as well as appropriate utilization rates of edge servers [1,2]. Thus, user reallocation from highly-loaded ES to lowly-loaded ES should be performed by service migration [3,5]. To overcome these limitations, we propose a novel coalitional game-theoretic approach to location-aware MEC service migration (CGL-SM) for crowded scenes. It includes a coalition formation strategy and mechanisms for load-balancing. The coalitions are formed by using a modified k-means algorithm, where its payoff is proportional to load-fairness of the ESs in the corresponding coalition. Load-balancing is maintained through service migrations among ESs in the coalition and avoiding hotspots by

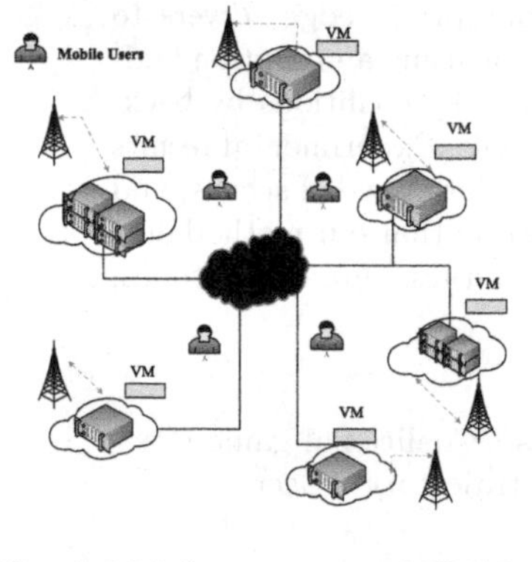

Fig. 1. Mobile users upon MEC System

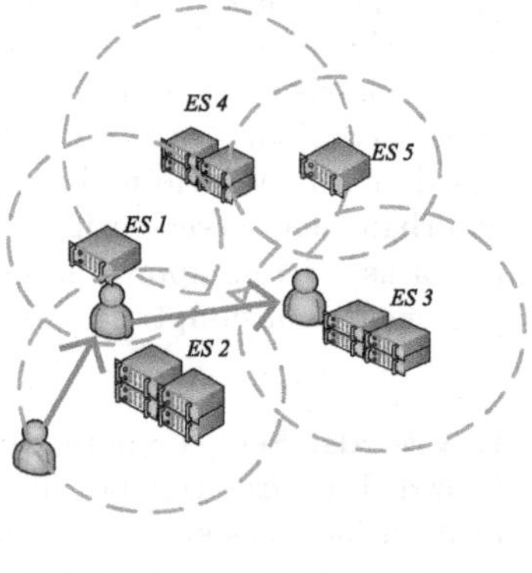

Fig. 2. Mobile User's Service Migration

using an ES workloads detection mechanism. To validate our proposed method, we carry out a case study based on a well-known edge users allocation dataset (EUA dataset) [25] and a crowded-scene mobile user trajectory dataset [29]. We show that our method beats its competitors in terms of: 1) workload-fairness of the involved ESs in highly-crowded and highly-overloaded situations; 2) number of service migrations performed; 3) energy consumption of migrations.

2 Related Work

Most existing works in this direction focus on the edge user allocation problem in the MEC environment [7,15,19,27], or location prediction of mobile users for MEC systems [17,21,22]. Some work [3] has researched crowded scenes from the perspective of computer graphics. However, how to migrate services upon MEC infrastructures in crowded scenes is less considered and studied. Robicquet *et al.* [28,29] provided a mobile trajectory dataset in crowded scenes by drone in the Stanford campus, as shown in Fig. 4 and 5. Some work has researched hotspot discovery issues. For example, Anchuri *et al.* [23] have researched hotspot discovery problem in service-oriented architecture, however, it is not in a specific scenario. Huang *et al.* [14] have researched the roadside hotspot in edge computing based on internet of vehicle from the perspective of protecting security from attacks, and their method is based on a Stackelberg game approach. This is the most similar research that is also focusing hotspot issues in edge, however, their proposed method is mainly for the purpose of avoiding internet attacks and thus service migration is not considered.

In a crowded scene, a single MEC server is inadequate and a group of MEC servers is usually required for cooperative tasks [12]. Generally, user requests and mobile resource allocation technology could be applied in this scenario [23,30]. A number of works considered focus on D2D communications to appropriately allocate user requests with multiple MEC servers. For example, He *et al.* [13] considered using D2D communication for task offloading and resource management in a multi-user and distributed mobile edge cloud resource environment. However, in a highly mobile environment, D2D is not a mainstream solution due to the fact it tends to lose connectivity stability when load is high [4].

Recently, service migration is regarded as a highly effective means for load balancing, and task offloading with mobile users. As can be seen in the example illustrated in Fig. 2, a mobile user moves freely and it is assumed generally that their moving area is a circle (note that this assumption is widely used in related works [19,27]). By radio access network, edge servers could cooperate to execute tasks collectively. For example, Pang *et al.* [7] developed a loosely coupled fog radio access network model leveraging low-end infrastructures such as small cells' power to achieve ultralow latency by exploiting the joint-edge-computing and near-range communications techniques. Some incooperative-game-based model is usually used for user reallocation or resource reallocation. He *et al.* [19] proposed a game-theoretic approach that formulates the edge users allocation problem as a potential game. Decreasing System-cost is taken as the metric for service migrations among edge servers, and the less system-cost is better. Wang *et al.* [15] formulated the service migration problem as a Markov decision process. Their formulation captured general cost models and provided a mathematical framework to design optimal service migration policies. Locations of mobile users in service migration are also researched frequently. Wu *et al.* [20] promoted a user-centric location prediction approach by leveraging users' social information. They considered that each user is with private location information that is not shared to others, and proposed a factor-graph-learning model that takes into account not only user's social and network information, but also inter-user correlation information. Tan *et al.* [21] proposed a location-aware load prediction which deals with user mobility by correlating load fluctuations of edge datacenters in physical proximity. Yin *et al.* [22] presented a decision-support framework for provisioning edge servers for online services providers.

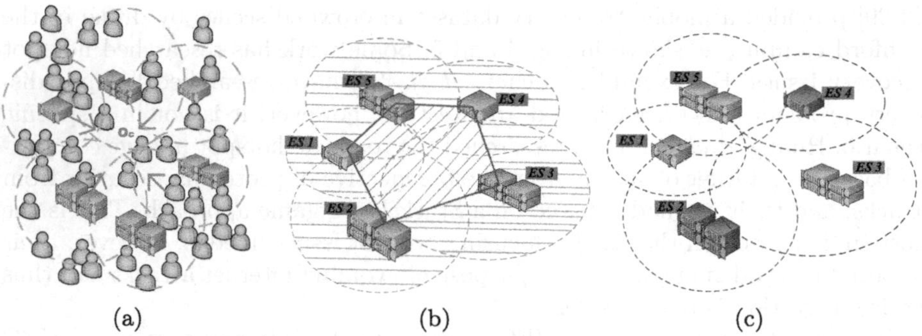

Fig. 3. Crowded Scene and Hotspot in an ES Coalition Area (Color figure online)

3 System Model

3.1 Crowd Model Scenario

If a large number of users gather toward the same destination, such as a subway station, a school gate or a plaza, then a crowded area O_c emerges. Figure 3(a) shows an illustrative example of the gathering pattern, where highly mobile users' paths are bidirectional and users move periodically towards and away from O_c. Figure 3(b) shows the ES coalition graph with such patterns, where crowded areas are marked with blue lines and hotspot areas are marked with red lines. As various existing works [20,28] did, we consider that the coverage area of the MEC ESs is larger than the hotspot area. Figure 3(c) demonstrates that ES2 and ES4 are marked red and they are affected by crowds of users. Therefore, they should transfer some workloads to other ESs in the same coalition. If all ESs in the same coalition have no remaining capacity, then the workload is transferred to a nearby coalition by backbone network [24].

Fig. 4. Deathcircle Scene in Stanford

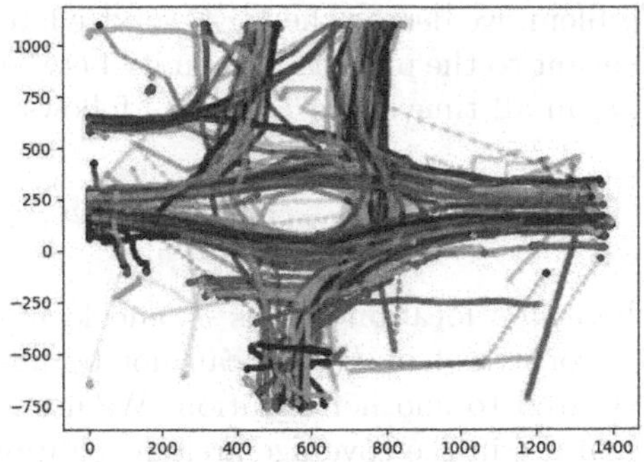

Fig. 5. Trajectories in Deathcircle

3.2 Service Migration Model

Originally, an application user u_i can be allocated to an ES e_j only if it is covered by e_j, i.e., $COV(e_j)$. It is usually covered by many ESs, however, it will choose the nearest one originally as shown in (1):

$$\begin{aligned} &u_1 \in COV(e_{j1}),\ COV(e_{j2}),\ COV(e_{j3}) \\ &u_i \longrightarrow e_{j1},\ s.t.\ d_i^{j1} < d_i^{j2} < d_i^{j3} \end{aligned} \tag{1}$$

where d_i^{j1} is the distance between e_{j1} and u_i. s_i could be migrated to the ES in another coalition when u_i is at the boundary of any two coalitions. $traj$ is the set of trajectories of all users, which is also the input of *Algorithm 2* below. Each user's movement is decided by its corresponding mobile trajectory as given in (2). $traj_i$ denotes the moving trajectory of u_i's location: y_i^t, and it can be denoted by:

$$\begin{aligned} traj &= \{traj_i \mid u_i \in U\} \\ traj_i &= \{\ y_i^t = <LO_i^t, LA_i^t> \mid t \in T\} \end{aligned} \tag{2}$$

where LO_i^t and LA_i^t denote the longitude and latitude values at time point t, U is the set of users, T denotes all time slices, N_s the number of all services. According to [20], we use the following model to estimate service migration overhead. We consider that every service have a budget B_m. A unit migration budget, B_i, is consumed, whenever s_i is migrated. Migration stops when the corresponding budget is used out. The remaining budget of all services: B_r, is thus:

$$\begin{aligned} &\mathrm{B}_m^i = B_m - \mathrm{n}^i * B_i \quad s.t.\ B_i \propto w_i,\ B_m^i > 0 \\ &B_r = N_s * B_m - \sum_{i=1}^{N_s} * n^i * B_i \end{aligned} \tag{3}$$

where B_m^i is the remained budget of s_i, w_i denotes the workload of s_i, B_i is a unit budget cost related to w_i, n^i is the total migration times of s_i. Thus, the service with a larger workload should be moved for a fewer times. In our problem, we denote the MEC service latency at time point t as $l(p_i^t, y_i^t)$, which is relevant to the physical proximity between u_i and the matched ES. Then latency of s_i in all time slices T, i.e., L_i follows:

$$L_i = \sum_{t=1}^{T} l(p^t, y_i^t), \; s.t. \; B_m^i > 0 \tag{4}$$

where u_i's location at t is y_i^t and location of service s_i in ES at t is p^t. With the constraint of B_m^i, s_i can not be migrated within the coalition before it is migrated to another coalition. We use M_i to denote the maximum delay of s_i when it's in the coverage area of coalition c. It follows:

$$L_i < M_i \tag{5}$$

If u_i steps into another coalition's area, s_i gets a new budget. If the user is still in the coalition area, but out of the coverage area of the original ES, the service is migrated to another ES as well. The overall energy consumption for service migration, i.e., E_S can be obtained as:

$$E_S = \sum_{i=1}^{N_s} n^i * E_m * w_i \tag{6}$$

where E_m is the migration energy for a unit workload.

3.3 Capacity and Workload Model

Each edge server e_j has a capacity of F^j, and F_i^j denotes the service workload of u_i placed on server e_j. It should satisfy that The aggregate workload of each resource type incurred by all allocated users must not exceed the capacity of their assigned server in (6). The total workloads generated by all users allocated to an edge server must not exceed its remaining capacity as shown in (6). Assume that if services are placed on e_j, they should satisfy the function $PL(e_j)$ in (6), which indicates the set of services placed on e_j.

$$\begin{aligned} &\mathrm{Q(e_j)} = \{u_i \mid \mathrm{u}_i \in e_j\} \\ &\mathrm{PL(e_j)} = \{\mathrm{s}_i \mid s_i \in e_j\} \\ &F^j > \sum_{u_i \in Q(e_j)} F_i^j \end{aligned} \tag{7}$$

F^j is the capacity value of an edge server e_j, $Q(e_j)$ is a function to indicate the set of users whose services are placed on e_j. The total workloads of users in $Q(e_j)$ mustn't exceed F^j. Assume that W_j is the workload of e_j, and it equals all the services' workloads on e_j. R_j is the resource utlization rate of e_j, and it can be computed as:

$$R_j = \frac{W_j}{F^j} \quad s.t. \; W_j \neq 0, R_j < z_j \tag{8}$$

where z_j is the maximum utilization rate of e_j, and R_j should not exceed z_j.

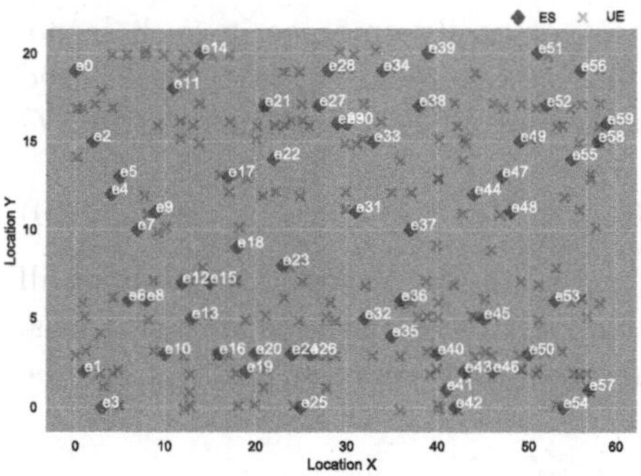

Fig. 6. Map of ESs and Users (Color figure online)

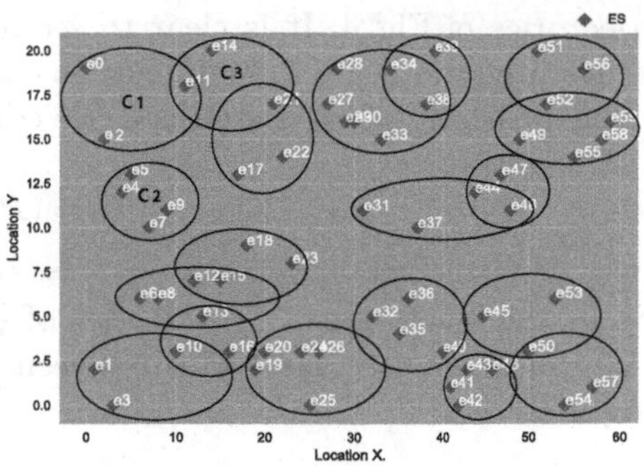

Fig. 7. Coalitions by Modified K-means

4 Coalitional Game for Load-Balancing

4.1 Framework Overview

The overall framework includes 3 major steps:1) using k-means to divide the coalitions as in pseudocode lines 1–11 in *Algorithm 1*; 2) discovering the hotspot happened in the area covered by ESs in the coalition, and then reallocate the workloads inside the coalition members; 3) transferring workloads to members of nearby coalitions from the overloaded ESs by backbone network [24] in pseudocode lines 21–23 in *Algorithm 2*. As shown in Fig. 7, c_1 could transfer workloads to nearby coalitions, such as: c_2, c_3. Coalitions, which are next to the overloaded ES, i.e., overloaded ES2 and ES4 in Fig. 3(c), should be choosen in prior.

4.2 Location-Aware Coalition Formation

Algorithm 1: Coalition Formation Based on k-means

Input: P, H, k, J
Output: C

1 Build a coordinate system for the ESs in map, and confirm their x, y coordinates
2 Compute the ES distances set E according to (10)
3 **for** *every h in H* **do**
4 | Form a coalition according to k-means
5 **end**
6 Finally, it groups e_1,e_2...e_N as k coalitions
7 **if** e_{j1} *is close to another coalition* **then**
8 | Add e_{j1} to that coalition, renew C
9 **end**
10 **if** *for any coalition* c*, its* n_c *is larger than* J **then**
11 | Divide this coalition by k-means
12 **end**

Figure 4 illustrates crowds of people gathering in a circular area in Stanford Campus [28,29]. Edge server is in this area as well. Figure 5 illustrates the users' trajectories of Fig. 4. It is clear to see a hotspot easily emerges in such scene. Service migrations are in high need to counter the emergence of hotspots caused by the gathering of massive users. We consider that all ESs construct a graph G as shown in (9).

$$G = (P, E) \tag{9}$$

where P is the set of vertices as $P = \{e_1, e_2, e_3 ... e_N\}$. N is the number of all ESs in the whole area. E is the set of euclid distances between any two ESs. The euclid distance $Dist(a, b)$ between any two edge servers: e_{j1}, e_{j2} can be computed as:

$$\begin{aligned} &E = \{Dist(j1, j2) \mid e_{j1}, e_{j2} \in P\} \\ &Dist(j1, j2) = \sqrt{(ELO_{j1} - ELO_{j2})^2 + (ELA_{j1} - ELA_{j2})^2} \end{aligned} \tag{10}$$

ELO_{j1}, ELA_{j1} are the longitude and latitude of e_{j1}, and the same to e_{j2}. According to E, close ESs are divided into a group as a coalition. To partition the ESs in a large area into several small units, the locations of ESs are the key factor. We use the k-means method to divide the ESs and employ the locations of ESs as the deciding factor for partition. As shown in Algorithm 1, the partition takes hotspot areas, i.e., H, $H = \{h_1, h_2, h_3 ... h_k\}$ of human crowds as the input, where k is both the number of hotspot areas and also the number of ES coalitions. The partition algorithm takes H, P as the inputs as well and generates the set of coalitions as C. n_c is the number of ESs in a single coalition c and it's bounded by J. J is flexible, as it varies according to the size of the hotspot area. If the hotspot area is wide, it needs more ESs to cover. Coalitions may share the borders as shown in pseudocode lines 7–8 of *Algorithm 1*. k is decided by the number of hotspot areas. d is the number of iteration, and thus the time complexity of pseudocode lines 1–6 is $O(k * N * d)$. For the modified operations for each coalition in pseudocode lines 7–11, if the total modification times is g, then the final time complexity is $O(k * N * d + g)$. As shown in Fig. 6, 100 base stations (BSs) colocated with ESs are presented, and users are distributed around them. Users are marked in blue, and ESs in red. To make the coalitions fine-grained, the k-means-based clustering analysis process is iterated until n_c fits the population distribution. As shown in Fig. 7, all BSs are divided into a number of coalitions according to Algorithm 1.

4.3 Coalitional Game Model for Workload Allocation

A coalitional game Γ consists of two essential elements [10]: 1) a set of players $N = \{1, 2 ...\}$, in this paper (ESs are modelled as players); 2) a characteristic value ν that specifies the value created by different subsets of the players. i.e., the payoff of a coalition c. Here maximizing the payoff $\nu(c)$ means maximizing the coalition's load fairness.

$$\Gamma = (N, \nu) \tag{11}$$

Every edge server is modelled as a player. Players are assumed to be rational to join a coalition c. As a participant, each ES wants to keep a moderate utility.

Service migration is the strategy for them to adjust the workloads distribution among ESs in the coalition.

$$c = \{e_j \mid e_j \in \Gamma\} \tag{12}$$

The coalition's load variance $D(c)$ should be bounded by:

$$D(c) = \frac{1}{n-1}\sum_{j=1}^{n}(W_j - \bar{W}_c)^2 \tag{13}$$

For each coalition, it should keep a low level of variance $D(c)$ to avoid imbalance of workloads.

$$\begin{aligned} &Min\ D(c) \\ &s.t.\ F^j > \sum_{i=1}^{n} F_i^j \end{aligned} \tag{14}$$

where $\bar{W}_c$ is the average workload of all ESs in a coalition c. Std is the standard deviation based on $D(c)$. As the constraint, workloads of the services placed on the jth ES should not exceed its capacity F^j. Here, we stipulate that the payoff of a coalition is $\nu(c)$, and it can be obtained as:

$$\begin{aligned} &\nu(c) = \frac{1}{Std} \\ &s.t.\ d_{i,j} = 1,\ 0 < R_j < z_j \end{aligned} \tag{15}$$

where $d_{i,j}$ is a boolean variable to indicate whether the ith service is placed on the jth server. Edge servers in a coalition communicate with each other and migrate the services among the coalition members to balance the workloads on them. The resulting optimization object is thus to maximize ν with constraint of B_m^i:

$$\begin{aligned} &Max\ \ \nu(c) \\ &s.t.\ \nu \neq 0,\ B_m^i > 0 \end{aligned} \tag{16}$$

A coalition $c = \{e_1, e_2...e_{n_c}\}$ includes edge servers grouped by *Algorithm 1*. Users' locations change with time, and they may form a hotspot at any time slice. Pseudocodes line 1–9 in *Algorithm 2* describe three criteria for judging whether a hotspot in the ES's coverage area is discovered or an ES is qualified to be a source ES, namely: 1) any ES e_j in the coalition is over utilized, namely, the utilization rate of it is higher than its threshold value R^h, which is set according to the edge server itself; 2) any ES receives the most workloads among all ESs in the coalition; 3) any ES's workload exceeds the average workload of ESs in the coalition. Based on these criteria, an ES could be regarded as a source ES. It is constrained that any service must be migrated to another ES which covers its user. Pseudocodes in line of 9–17 in *Algorithm 2* illustrate the service migration operation. As a consequence, the subsequent payoff of the coalition, i.e., $\nu^{'}(c)$, should exceed $\nu(c)$. $\nu^{'}(c)$ is the payoff of the coalition after one service migration is performed.

$$\begin{aligned} &\forall\, e_{j1}\ e_{j2} \in c,\ u_i \in e_{j1},\ W_{j1} > W_{j2} \\ &\textbf{If } \nu(c) < \nu'(c), \textbf{ after service migration} \\ &\textbf{Then do } u_i \xrightarrow{move} e_{j2} \\ &s.t.\ traj_i \in Cov(e_{j1}, e_{j2}), B_m^i > 0 \end{aligned} \tag{17}$$

In *Algorithm 2*, $Cov(e_{j1}, e_{j2})$ denotes the coverage area of e_{j1} and e_{j2}. In a coalition, to match the destination ES and the source one, let $ord(j)$ denote the ascending order of the j_{th} ES according to the workload. A highly-loaded ES is matched with a lowly-loaded ES with $ord(n_c - ord(j))$. If the ES with $ord(n_c - ord(j))$ does not cover the user, then consider the ES with $ord(n_c - ord(j) - 1)$ or $ord(n_c - ord(j) + 1)$. The match operation is in pseudocode line 11 in *Algorithm 2.* m_s denotes all service migrations performed during the whole process. Fairness improvement is calculated according to the decrease of the standard deviation of the workloads distributed on ESs in a coalition after the migration is conducted, i.e., ΔStd.

Algorithm 2: CGL-SM
(Service Migration in a coalition c)

Input: U, S, c, $traj$, B_m
Output: Updated match of U, S, c, B_m^i
1 Step 1: Hotspot or source ES detection
2 **for** *all ESs in c* **do**
3   **if** *1. any ES's utilization rate exceeds its highest threshold value* R^h *;*
4   *2. any ES in c receives the most workloads ;*
5   *3. any ES receives workload that exceeds the average workload in c.* **then**
6   Take this ES as a source ES according to the criterion: 1, 2, 3.
7   **end**
8 **end**
9 Step 2: Perform service migrations
10 **for** *all ESs in c* **do**
11   **if** *a source ES* e_{j1} *and* e_{j2} *can be matched* **then**
12   **for** *services on* e_{j1} *and in Cov(*e_{j1}*,* e_{j2}*)* **do**
13   **if** *any* s_i *can be migrated based on (17) and its* $B_m^i > 0$ **then**
14   *migrate* s_i *from* e_{j1} *to* e_{j2}
15   **end**
16   *else* **if** *no services are qualified to migrate* **then**
17   *Stop Service Migration*
18   **end**
19   **end**
20   **end**
21   else **if** *ESs in c could not handle workloads from the hotspot* **then**
22   Transfer services to neighbor coalitions or central cloud by backbone network
23   **end**
24 **end**

4.4 Complexity Analysis

The complexity of *Algorithm 2* can be examined as several steps. n_c is the number of ESs in a coalition. The time complexity of the match step is $O(n_c/2)$. The process of service migration depends on the number of service migrations. If the maximum number of service migrations performed in a round of match operation is m, then the time complexity of it is $O(m)$. Finally, the overall time complexity is $O(m * n_c/2)$.

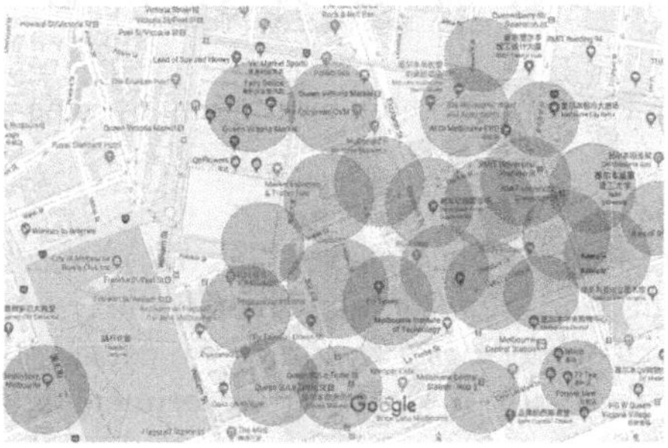

Fig. 8. Experimental Area in Melbourne

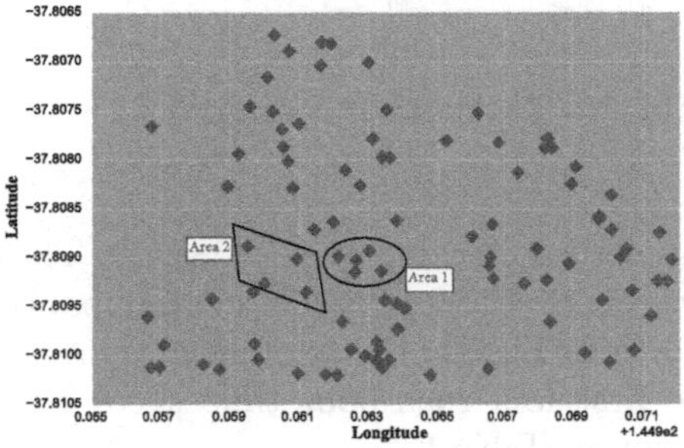

Fig. 9. ESs in Some Areas of Melbourne

5 Experiment Setting and Evaluation

5.1 Benchmark Policies

We compare our CGL-SM with existing user allocation algorithms and a no coalition formed algorithm: 1) EUAGame [19], an incooperative game-based approach applied in for user allocation in MEC; 2) Greedy [27], a proximity-priority-based migration method; 3) No-Cos, a non-coalitional variant of CGL-SM for showing the performance gain by the coalitional model.

5.2 Experimental Settings

In Fig. 8, base stations (BS) are distributed based on Google Map [26]. We depict the locations of BS in part of Melbourne from dataset [25] as shown in Fig. 9. According to the modified k-means algorithm, we select two typical crowded areas based on ES coalitions, i.e., *area 1* and *area 2* as illustrated in Fig. 9. The total area is around 0.06 km^2. Here J is set as 7 according to the distribution of these BS. R^h of any ES is set to 80%. The workload capacity of each ES is set to 1800–2000 according to the service workloads as shown in Fig. 10, which is based on a public workload dataset CoMon [6]. Based on workloads, B_m is set to 30. E_m is set to 2 mJ. To evaluate the effectiveness of our approach, we conducted experiments on a real-world crowded-scene mobile user trajectory

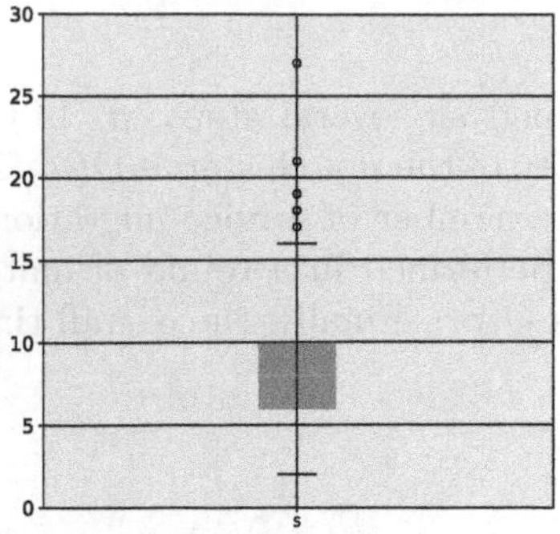

Fig. 10. Workloads of user services

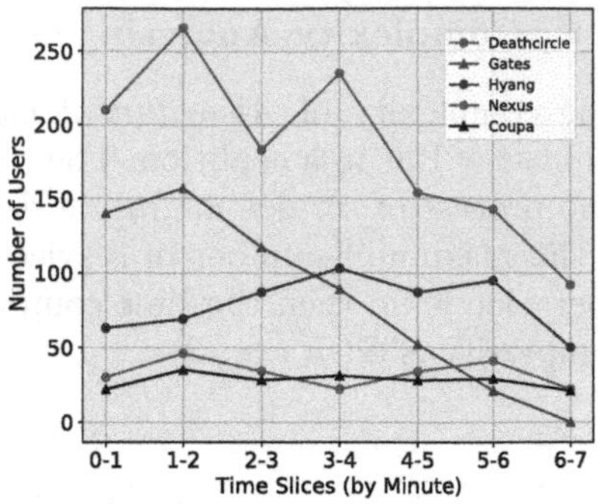

Fig. 11. The number of users by time

dataset for crowded scenes [28], i.e., The Stanford Drone dataset [29]. We choose five scenes: Gates, Deathcircle, Coupa, Hyang, and Nexus as shown in Table 1. As shown in Fig. 11, the number of users in each scene varies by time. We consider the first four rush-hour time slices $T = \{T_1, T_2, T_3, T_4\}$, in our experiment as shown in Table 2.

Table 1. Scenes in two areas

Scenarios	*Area 1*	*Area 2*	Total users	Time (min)
Scene 1	Deathcircle	Deathcircle	871	6:56
Scene 2	Gates	Gates	409	5:03
Scene 3	Hyang	null	326	6:19
Scene 4	Nexus	Nexus	129	6:22
Scene 5	Coupa	Coupa	117	6:39

Table 2. Four time slices

Time Slices	T_1	T_2	T_3	T_4
Period (min)	00:00–1:00	1:00–2:00	2:00–3:00	3:00–4:00

5.3 Experiment Evaluation and Analysis

We evaluated all approaches by: 1) m_s, B_r and E_S, which evaluate quality of service migrations; 2) ΔStd, which evaluates the load fairness; 3) the frequency of using backbone network by the impact of N_S.

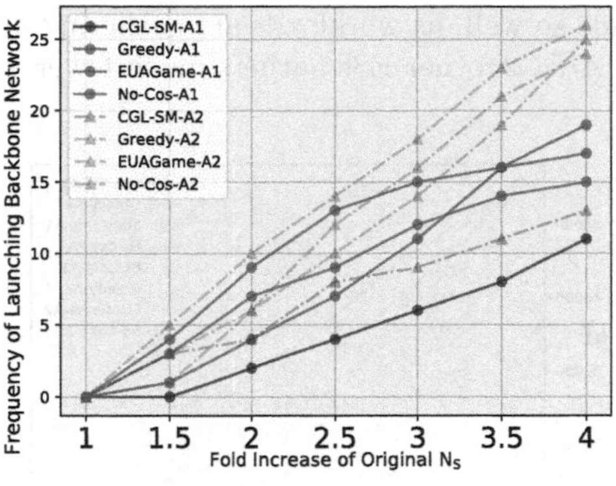

Fig. 12. Using backbone network by fold change of N_S

Impact of N_S on Using Backbone Network: When the number of services rises rapidly, the ESs in a coalition are allowed to transfer the workloads to other ESs by using backbone network. As shown in Fig. 12 (suffix *A*1 and *A*2 denote *Area 1* and *Area 2*), the occurrence rate of using backbone network of CGL-SM is clearly lower than those of its competitors due to the fact that CGL-SM achieves a better load distribution among coalition members.

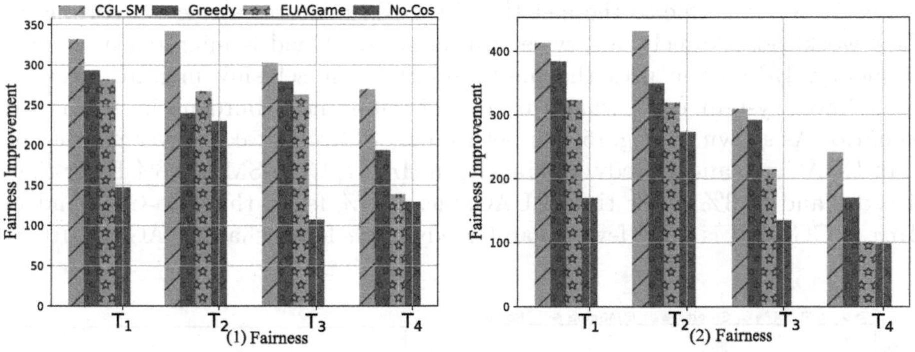

Fig. 13. Δ Std in *Area 1*, *Area 2*

Load Fairness ($\boldsymbol{\Delta Std}$): As shown in Fig. 13, all four migration methods improve the fairness of workload among all ESs, but our approach achieves the highest amount of fairness improvement at all time slices in both areas. The average advantages of CGL-SM over EUAGame, Greedy, No-Cos in *Area 1* are 16.9%, 8.7%, 35.6% in terms of ΔStd, and 24.3%,7.6%, 45.4% in *Area 2*, respectively. As can be seen, the advantage of our proposed method is achieved due to the fact that it chooses ESs appropriately while EUAGame tends to choose the ES with lowest load and Greedy tends to choose the nearest one. No-Cos does not balance loads so well, as workloads are gathering at the boundary ESs between *Area 1* and *Area 2*, other ESs gathers much fewer workloads.

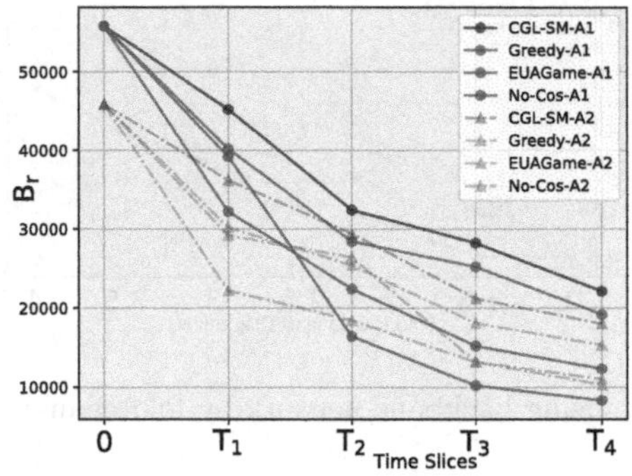

Fig. 14. B_r in two areas at different time slices

Service Migrations (B_r, m_s, E_S): As shown in Fig. 14 (suffix *A*1 and *A*2 denote *Area 1* and *Area 2*), B_r in *Area 1* and *Area 2* are depicted. CGL-SM also costs least budget in both cases (in *Area 1*, CGL-SM is 8.3%, 10.6%, 19.3% less than Greedy, EUAGame, No-Cos, and in *Area 2*, it is 5.1%,12.8%, 13.5%, respectively). It is due to the fact that CGL-SM migrates services according to their workloads. Namely, a service with larger workload is migrated in a lower frequency. EUAGame uses the most budget, as it selfishly migrates services for a lower system cost, which inversely creates more burdens in a crowded scenario. As shown in Fig. 15, for both areas, CGL-SM takes fewer migrations than EUAGame and Greedy. Averagely, in *Area 1*, CGL-SM is 5.6% fewer than Greedy, and 16.3% fewer than EUAGame, 11.5% fewer than No-Cos. And in *Area 2*, CGL-SM is 6.1% fewer than Greedy, 9.8% fewer than EUAGame,10.5%

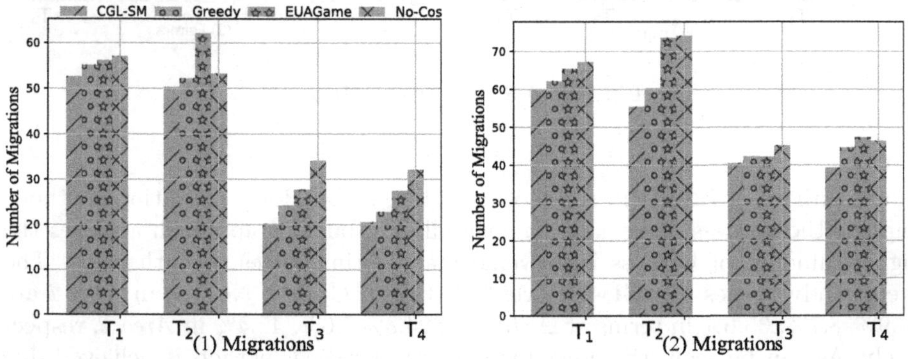

Fig. 15. m_s in *Area 1*, *Area 2*

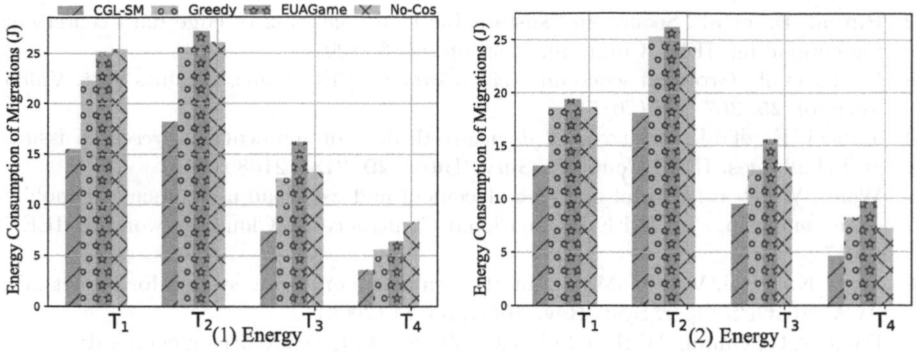

Fig. 16. E_S in *Area 1*, *Area 2*

fewer than No-Cos. As shown in Fig. 16, the energy consumption of migrations in two areas are depicted. We see that CGL-SM has the least energy consumption for migrations compared to other approaches (in *Area 1*, CGL-SM is 15.6%, 28.3%, 21.5% less than Greedy, EUAGame, No-Cos, and in *Area 2*, it is 13.1%, 25.8%, 22.3%, respectively). The advantage of our approach is achieved due to the fact that: 1) the objective model of CGL-SM aims at decreasing the variance of ES workloads in a coalition. Meanwhile, the mobility of users and the user-server proximity constraint are appropriately exploited in balancing workloads of ESs; 2) as CGL-SM performs fewer migrations, then it consumes less migration-overhead. 3) some baseline algorithm (e.g. EUAGame) aims to decrease system cost by using more service migrations, which is not effective in a crowded scenario.

6 Conclusion

In this paper, we propose a location-aware MEC service migration approach for mobile user reallocation in crowded scenes. The proposed method leverages a modified-k-means-based strategy for the formation of coalitions, a workload-based method for the detection of hotspots, and a load-fairness-based strategy for ES workload allocation. Additionally, coalitions are connected by backbone network in case that massive workloads of services could not be sustained by a single coalition. A case study based on real-world datasets of edge-user distribution and trajectory traces demonstrates that our proposed method beats its peers in terms of load-fairness, the number of service migrations required, and energy consumption of migrations. In future work, we will explore a hybrid approach, namely, D2D and edge-cloud mode.

References

1. Niu, X., et al.: Workload allocation mechanism for minimum service delay in edge computing-based power Internet of Things. IEEE Access **7**, 83771–83784 (2019)

2. Puthal, D., et al.: Secure and sustainable load balancing of edge data centers in fog computing. IEEE Commun. Mag. **56**, 60–65 (2018)
3. Li, T., et al.: Crowded scene analysis: a survey. IEEE Trans. Circuits Syst. Video Technol. **25**, 367–386 (2015)
4. Jameel, F., et al.: A survey of device-to-device communications: research issues and challenges. IEEE Commun. Surv. Tutor. **20**, 2133–2168 (2018)
5. Wang, W., et al.: Virtual machine placement and workload assignment for mobile edge computing. In: IEEE International Conference on Cloud Networking. IEEE (2017)
6. Park, K.S., Pai, V.S.: CoMon: a mostly-scalable monitoring system for PlanetLab. ACM SIGOPS Oper. Syst. Rev. **40**(1), 65–74 (2006)
7. Pang, A.C., Chung, W.H., Chiu, T.C., Zhang, J.: Latency-driven cooperative task computing in multi-user fog-radio access networks. In: Proceedings of the IEEE 37th International Conference on Distributed Computing Systems (ICDCS), pp. 615–624, June 2017
8. Mohiuddin, I., Almogren, A.: Workload aware VM consolidation method in edge/cloud computing for IoT applications. J. Parallel Distrib. Comput. **123**, 204–214 (2019)
9. Myerson, R.B.: Game Theory: Analysis of Conflict. Harvard Univ. Press, Cambridge (1991)
10. Saad, W., Han, Z., Debbah, M., Hjorungnes, A., Basar, T.: Coalitional game theory for communication networks: a tutorial. IEEE Signal Process. Mag. **26**(5), 77–97 (2009)
11. He, Y., Ren, J., Yu, G., Cai, Y.: D2D communications meet mobile edge computing for enhanced computation capacity in cellular networks. IEEE Trans. Wirel. Commun. **18**, 1750–1763 (2019)
12. Mehmood, Y., et al.: Internet-of-Things-based smart cities: recent advances and challenges. IEEE Commun. Mag. **55**(9), 16–24 (2017)
13. Baccour, E., Erbad, A., Mohamed, A., Guizani, M.: CE-D2D: dual framework chunks caching and offloading in collaborative edge networks with D2D communication. In: 2019 15th International Wireless Communications and Mobile Computing Conference, IWCMC 2019, pp. 1550–1556. (Institute of Electrical and Electronics Engineers Inc.) (2019)
14. Huang, X., Yu, R., Pan, M., Shu, L.: Secure roadside unit hotspot against eavesdropping based traffic analysis in edge computing based internet of vehicles. IEEE Access **6**, 62371–62383 (2018)
15. Wang, S., et al.: Dynamic service migration in mobile edge computing based on Markov decision process. IEEE/ACM Trans. Netw. **27**, 1272–1288 (2019)
16. Liu, L., Liu, X., Zeng, S.: Research on virtual machines migration strategy based on mobile user mobility in mobile edge computing. J. Chongqing Univ. Posts Telecommun. (Nat. Sci. Ed.) **31**(2) (2019)
17. He, H., Qiao, Y., Gao, S., Yang, J., Guo, J.: Prediction of user mobility pattern on a network traffic analysis platform. In: International Workshop on Mobility in the Evolving Internet Architecture, pp. 39–44 (2015)
18. Guo, Q., Huo, R., Meng, H.: Research on reinforcement learning-based dynamic power management for edge data center. In: 2018 IEEE 9th International Conference on Software Engineering and Service Science (ICSESS) (2018)
19. He, Q., et al.: A game-theoretical approach for user allocation in edge computing environment. IEEE Trans. Parallel Distrib. Syst. **31**, 515–529 (2020)

20. Wu, Q., Chen, X., Zhou, Z., Chen, L.: Mobile social data learning for user-centric location prediction with application in mobile edge service migration. IEEE Internet of Things J. **6**, 7737–7747 (2019)
21. Le Tan, C.N., et al.: Location-aware load prediction in Edge Data Centers. In: 2017 Second International Conference on Fog and Mobile Edge Computing (FMEC) (2017)
22. Yin, H., et al.: Edge provisioning with flexible server placement. IEEE Trans. Parallel Distrib. Syst. **28**(4) (2017)
23. Anchuri, P., Sumbaly, R., Shah, S.: Hotspot detection in a service-oriented architecture. In: Proceedings of the 2014 ACM International Conference on Information and Knowledge Management, CIKM 2014, Shanghai, China, pp. 1749–1758 (2014)
24. Zhao, H., Deng, S., Zhang, C., Du, W., He, Q., Yin, J.: A mobility-aware cross-edge computation offloading framework for partitionable applications. In: 2019 IEEE International Conference on Web Services (ICWS), Milan, Italy, pp. 193–200 (2019)
25. He, Q.: Swinbine University of Technology EUA Dataset. https://sites.google.com/site/heqiang/eua-repository
26. Google Map. https://www.google.com/maps/@-37.8081158,144.9622256,17z?hl=zh-CN
27. Peng, Q., et al.: Mobility-aware and migration-enabled online edge user allocation in mobile edge computing. In: Proceedings - 2019 IEEE International Conference on Web Services, ICWS 2019, Part of the 2019 IEEE World Congress on Services, pp. 91–98. Institute of Electrical and Electronics Engineers Inc. (2019)
28. Robicquet, A., Sadeghian, A., Alahi, A., Savarese, S.: Learning social etiquette: human trajectory understanding in crowded scenes. In: Leibe, B., Matas, J., Sebe, N., Welling, M. (eds.) ECCV 2016. LNCS, vol. 9912, pp. 549–565. Springer, Cham (2016). https://doi.org/10.1007/978-3-319-46484-8_33
29. Stanford Drone Dataset (2018). http://cvgl.stanford.edu/projects/uav_data/. Accessed 26 Aug 2018
30. Xia, X., Chen, F., He, Q.: Cost-effective app data distribution in edge computing. IEEE Trans. Parallel Distrib. Syst. **32**(1), 31–44 (2020). https://doi.org/10.1109/TPDS.2020.3010521

Bidding Strategy Based on Adaptive Differential Evolution Algorithm for Dynamic Pricing IaaS Instances

Dawei Kong[1], Guangze Liu[2], Li Pan[1], and Shijun Liu[1](✉)

[1] School of Software, Shandong University, Jinan, China
wei2white@163.com, {panli,lsj}@sdu.edu.cn
[2] Pennsylvania State University, State College, PA, USA
gbl5120@psu.edu

Abstract. In recent years, with the development of cloud computing technology and the improvement of infrastructure performance, cloud computing has developed rapidly. In order to meet the diverse needs of users and to maximize the revenue of cloud computing service providers, cloud providers have launched auction-type instances like Amazon Spot instances in the AWS cloud. For dynamic pricing cloud instances, how to select appropriate instance or instance group among multiple instances and make reasonable bids to optimize its own costs is a great challenge. This paper models the dynamic pricing instance pricing and multi-instance combination problem as a constrained optimization problem. Then we introduce the basic differential algorithm and proposes an adaptive differential evolution algorithm to optimize the combination of price bidding based on the optimal cost and the use of instances. Finally, we use real dynamic pricing instance price data released by the Amazon cloud to verify the optimization strategy. The experimental results show that the adaptive differential evolution algorithm has a better optimization effect on short-term task requirements and long-term task requirements.

Keywords: Dynamic pricing model · Iterative algorithm · Amazon cloud · Spot instance

1 Introduction

This computing resource sharing pool is called "cloud". Cloud computing brings together many computing resources. In other words, the power of computing can be regarded as a commodity. Cloud service types can be divided into three categories, infrastructure as a service (IaaS), and platform as a service (PaaS) and software as a service (SaaS). IaaS is one of the main service categories. It provides virtual computing resources, such as virtual machines, storage, networks, and operating systems, to individuals or organizations of cloud computing providers. Encapsulating basic resources such as hardware devices into services for users to use, customers cannot manage or control the cloud infrastructure,

H. Gao et al. (Eds.): CollaborateCom 2020, LNICST 349, pp. 458–478, 2021.
https://doi.org/10.1007/978-3-030-67537-0_28

but can control the operating system, storage, and applications deployed by themselves, and can also partially control the network components used. For example, Amazon's EC2 [1] and Google Compute Engine [2].

In recent years, with the development of cloud computing technology and the improvement of infrastructure performance, cloud computing has developed rapidly. Not only in the professional research field related to cloud computing but also more and more companies and individuals are beginning to deploy their businesses and applications to the cloud. Users can quickly build large and complex network applications, which greatly reduces equipment costs by using cloud computing platforms. Cloud platform providers virtualize computing resources through virtualization technology. The user uses the instances provided by the cloud platform to pay for the use of resources in the cloud platform. The payment methods of cloud computing services mainly include on-demand payment and scheduled payment. On-demand instance type is which users pay corresponding fees based on the instance type and usage time. Scheduled instances are which users pay some fees in advance and then receive a discount on the price when using the instance. In addition, in order to meet the user's diversity of instance requirements and to maximize the benefits of cloud computing service providers, cloud providers have launched auction-type instances. Amazon Cloud has launched Spot instances in the AWS cloud. Compared with the price of on-demand instances, Spot instances receive an ultra-low discount on usage fees [3]. Unlike the fixed pricing of on-demand instances, Spot instances use dynamic pricing, and the dynamic price of Spot instances is called the hourly price of spot instances. The spot price of each instance type in each available region is set by Amazon EC2 and gradually adjusted according to the long-term supply and demand of Spot instances. As long as there is the available capacity and the maximum price per hour requested by the user exceeds the Spot price, each Spot instance will run [4]. Amazon Web Service (AWS) prices Spot instances based on the application of Spot instances and the resource situation of EC2 instances. However, Amazon has not announced the detailed pricing strategy of spot instances. When users use bidding instances, they cannot accurately predict the price changes of the instances in the future [5].

In addition to cloud computing providers providing cloud computing resources, how to define service charges is also a key issue. The rationality of the pricing method can greatly influence the choice of users. Both Amazon's EC2 and Google's GCE have defined detailed charging methods. In order to meet the diverse needs of users, in order to improve the utilization of system resources and maximize the revenue of cloud platforms, cloud providers are also constantly enriching the billing methods of cloud platforms. Spot instances with dynamic pricing.

When users apply for the use of Spot instances, they need to bid on the instances, and users need to bid based on their experience. This undoubtedly increases the difficulty of users using Spot instances. In order to let users referring to the auction price, Amazon released historical data on the price of each type of spot instance in the first 90 days. Therefore, it is possible to predict future price changes by investigating 90-day price data, which helps to formulate reasonable bidding strategies. Amazon adjusted its pricing strategy for Spot on November 17, 2018. Prior to this, the prices of all types of spot instances fluctuated greatly. Now Amazon Cloud has shifted from a short-term resource application and pricing strategy for remaining resources to a pricing strategy based on long-term benefit. Therefore, the current prices of many spot instances have stabilized. After the price of Spot instances has stabilized, it is more convenient for users to make bids and formulate reasonable use strategies. It is meaningless to predict future prices for stable and constant price data. Through the research on the pricing historical data of Spot instances, it is found that in certain available areas, for certain types of instances, the price of Spot instances still has great volatility. Therefore, when users apply for instance resources, they often need to apply for multiple instances at the same time to meet the needs of the task. In this dynamic pricing instance model, how to combine resources to achieve the maximum profit bidding strategy is a very important problem. Resource optimization and scheduling problem belong to NP problems and NP problems are difficult to solve or time cost. Therefore, in our problem, we seek the approximate optimal solution of the problem using heuristic differential evolutionary algorithms. In this paper, we optimize mutation methods and adaptive parameters of the basic differential evolution algorithms. Experiment results show that our adaptive algorithm has a better result.

2 Related Work

Cloud computing has developed into a key information technology and system model. Cloud computing-related technologies have made great progress, and the scale of cloud computing has grown significantly, becoming a popular direction in the research field [6]. For research in the field of cloud computing, cloud pricing models and resource optimization have always been a hot research topic [7].

The Spot instance launched by Amazon cloud has very obvious characteristics in its bid payment method and dynamic price. Therefore, many scholars and research institutes have published a lot of research on Amazon's Spot instance pricing strategy, price prediction, and maximizing returns [8]. The physical resources of cloud computing providers often have a lot of idle resources, which causes a lot of waste of resources. In order to minimize the economic losses caused by idle resources, cloud providers need to incentivize users to use these idle resources [9]. Kaminski B and Szufel P analyzed the factors that control the price of Amazon EC2 spot instances and proposed an adaptive bidding strategy that can digest the cost of use. It also proves that the bidding close to the spot price and the dynamic switching between instances is an effective and easy to implement strategy [10].

Efficiently providing resources is a challenging problem in the cloud computing environment because it is dynamic and needs to support heterogeneous applications. Although VM (Virtual Machine) technology allows multiple workloads to run concurrently and use shared infrastructure, it still cannot guarantee application performance. Therefore, current cloud data center providers either do not provide any performance guarantees or prefer static VM allocation rather than dynamic allocation, which leads to inefficient utilization of resources. In addition, due to the execution of different types of applications (such as HPC and web), workloads may have different QOS (Quality of Service) requirements, which makes resource provision more difficult. Early work either focused on a single type of SLA (service level agreement) or on the resource usage pattern of applications (such as web applications), resulting in inefficient use of data center resources. In this article, we deal with the problem of resource allocation in a data center that runs different types of application workloads, especially non-interactive and cross-operating system applications. Vivek H. Bharad and Hitesh A. Bheda proposed admission control and scheduling mechanism that maximizes resource utilization and profits while ensuring that user QOS requirements specified in the SLA are not met. In order to better provide and utilize data center resources, it is important to understand the different types of SLAs and the applicable penalty and workload combinations. Compared with static server integration, the proposed mechanism provides great improvements and reduces SLA violations [11].

The reference factor for user bidding is often learned from historical price data or bidding strategies. Some studies use the historical price data of bidding instances to record the change of the price of each instance from p_i to p_j as the state transition, as well as obtain the probability of the state transition from p_i to p_j statistically according to the historical data. Since each instance corresponds to the values of Ni prices, a Ni * Ni price probability transfer matrix [8] is obtained, which the problem is modeled as a Markov chain model and to be solved [12].

Many scholars have studied the optimization of parallel tasks [13, 14]. Differential Evolution Algorithm (DE) was first proposed by Storn and Price in 1995. It is mainly used to solve real number optimization problems [15]. This algorithm is a type of group-based adaptive global optimization algorithm, which is a type of evolutionary algorithm. Because of its simple structure, easy implementation, fast convergence, and strong robustness, it is widely used. In recent years, differential evolution algorithms have been used in constrained optimization calculation [16], clustering optimization calculation [17], and nonlinear optimization control [18]. Like the genetic algorithm [19], the differential evolution algorithm is also an optimization algorithm based on modern intelligence theory, which refers to the optimization of constrained optimization calculations and the direction of optimization search through the group intelligence generated by cooperation and competition among individuals within the group.

3 Dynamic Pricing IaaS Instances

In this section, we introduce dynamic pricing instances and the method of applying dynamic pricing IaaS cloud instances. Cloud platform users usually purchase IaaS cloud services in the form of VM, according to the payment of related fees, to run the user's tasks. Each cloud platform service provider platform (such as Amazon EC2) has varied types of VM (virtual machine) with multiple physical resource configurations. Such as c4.xlarge, c4.2large, and c4.4xlarge. Generally speaking, high-performance VM instances usually have a higher price than others. When users buy IaaS instances from the cloud platform, they can choose instance type (such as c4.xlarge), operating system (such as Unix or Windows), and other optional items according to the actual situation. In the model of this paper, users need to apply for instance resources based on IaaS, and each user needs to apply for a set of instances in order to complete the work in time. In fixed-price cloud instances, users only need to consider the pricing of the current instance and their budget to select the instance. For dynamic pricing instances, users need to give pricing and make strategic choices. As an example, in the Amazon Spot instance scenario, the price of a Spot instance is affected by many factors. In addition to the performance of the instance, it is also affected by the number of users and the application status of the instance. Many times, we can see that better-performing instances have a lower price than lower-performing instances. In this case, the user needs to adjust the resource usage strategy to maximize the user's revenue.

3.1 Fixed Pricing Cloud Instance Model

The price of each fixed pricing cloud instances model is fixed and it does not change over time. So, users only need to consider the budget and the performance requirements of the instances. As shown in Fig. 1.

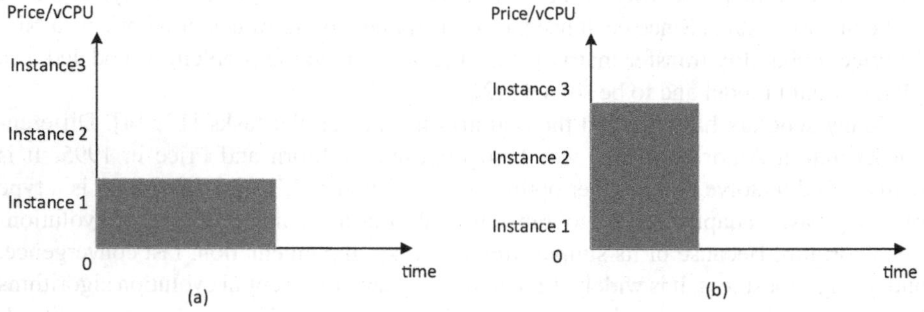

Fig. 1. Fixed pricing instances

In Fig. 1, Instance 1, Instance 2, and Instance 3 are three different performance instance types. C_i (capacity) is used to represent the performance of different instances, and Pi represents the price of different instances. Now there are $C_{\text{Instance 3}} > C_{\text{Instance 2}} > C_{\text{Instance 1}}$, $P_{\text{Instance 3}} > P_{\text{Instance 2}} > P_{\text{Instance 1}}$. The X-axis of the coordinate axis represents the time the task runs after the user selects an instance, and the Y-axis represents the price of different performance instances. The rectangular part of the figure represents the cost that the user needs to pay after selecting an instance. It can be seen that when the user submits the same task, regardless of the task due date, select a different instance type, and the actual fee paid Also different. In the fixed-price model, the pricing of each instance is fixed. The user selects an instance according to the budget and pays the actual fee according to the used price. This model is widely adopted by cloud providers due to its ease of use. For example, Amazon's on-demand instances use fixed pricing instance models. The advantage of this model is that the price of each instance is fixed, and the performance is stable. The user selects and pays a predictable fee according to the actual situation so that the budget can be well controlled while completing the task.

3.2 Cloud Instance Model for Dynamic Prices

The instance model based on bidding requires users to adopt competitive bidding methods to bid, and users need to specify the highest acceptable bid for a certain instance of the application. Under normal circumstances, when the user's bid is higher than the cloud service provider's bid, the application instance gets run. If the user's bid is lower than the cloud service provider's pricing (the pricing is immediate and constantly changing), the bidding instance fails as shown in Fig. 2.

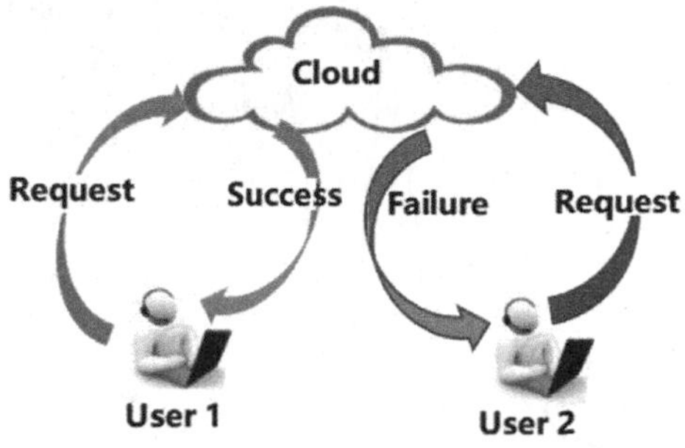

Fig. 2. Applying the bidding instance

The price of the bidding instance is specified by the cloud service provider. The cloud service provider will adjust the instance price according to the number of applications for the bidding instance, idle instance resource, and some other conditions. In this auction model, the price of cloud instances is usually dynamic. Most cloud service providers have launched bidding instances, such as Google Cloud's bidding instance, Alibaba Cloud's bidding instance, and Amazon's Spot instance.

3.3 Features of Amazon Spot Instances

Amazon's bidding instance is named Spot instance. Spot instance is Amazon's idle computing resources of Amazon EC2. Since the Spot instance is based on the bidding model, the price of the Spot instance fluctuates dynamically. Figure 3 shows the price data with the instance type is r3.8xlarge, the region is us-east-1, and the operating system is Linux.

As shown in Fig. 3, Amazon Cloud set the price of Spot instances in real-time. Since Amazon Cloud has not announced a detailed price strategy, it can only speculate that the price of Spot instances is related to the remaining capacity of Spot instances and the demand for Spot instances. In short, Amazon updates the price of Spot instances every 5 min. To use a Spot instance, the user needs to submit a Spot instance request, specify the type of instance, the region, and the highest price that the Spot instance can accept per hour. Users can propose bid prices based on historical price analysis of instance provided by Amazon. We use Amazon EC2 API and the AWS management console to obtain the historical price data of each Spot instance. When the user's highest bid price exceeds the current Spot instance's official pricing, the user's Spot instance application is accepted, and Spot instance is allocated to the user. The instance will continue to run until the user chooses to release it or Spot instance's official pricing is higher than the user's bidding price. According to the resource allocation strategy or revenue, Amazon Cloud will also have a certain probability of recovering the Spot instance applied by

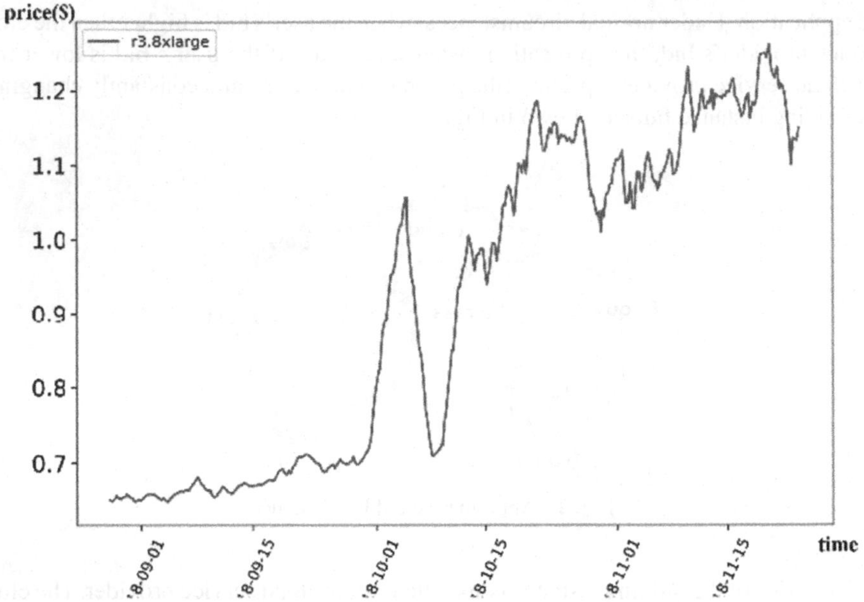

Fig. 3. A dynamic pricing instance

the user. In the dynamic pricing model of Spot instances, users cannot simply make the optimal resource application plan as easily as using fixed pricing instances. Users need to give a reasonable bidding strategy to adjust their workload. It is difficult to find the best answer to optimizing strategies.

4 Dynamic Bidding Modeling

4.1 Dynamic Bidding Modeling

Users apply for a dynamic price instance and give a bid price before using the dynamic pricing instances. If users bidding price is higher than the cloud provider's pricing, the instances are successfully allocated and executed. While these instances are running for user tasks, there are still many situations that they will be terminated by the cloud management system. Such as Amazon Spot instance, when the Spot instance capacity is insufficient, or the price rises above the bid, the applied instance will be terminated by the cloud management system. Amazon Cloud has not announced a specific price strategy, therefore, for users, there may be unexpected instances that may be terminated. The possibility of this kind of termination case that approximates the random nature greatly increases the difficulty of the relevant research on the auction case. Therefore, the research problem in this paper does not consider the situation where the cloud provider has been re-managed due to internal policy adjustments. This article can only consider that when the price of the dynamic instance fluctuates, the instance is terminated when the price rises above the user's highest accepted bid. When the price drops, if the cloud provider's pricing is lower than our bidding instance, it will still run. As shown in Fig. 4.

In Fig. 4, we can see how the price changes in the dynamic price instance and how the user's bid affects the bidding instance application. The blue line in the figure represents the price change value of the dynamic pricing instance in the future time period, the green line represents the user's bid price, and the abscissa is the application time of the instance, that is, the running time of the instance. Amazon's Spot instance billing has a feature. When Amazon EC2 actively terminates the user's Spot instance, the user has the opportunity to waive all or part of the fee. This is also a big difference between Spot instances and on-demand instances. Amazon EC2 has launched the finest example of billing per second. The new billing method of billing per second further helps users reduce the cost of using Spot instances. Spot instances that meet the per-second billing conditions: During the first hour of launch, if the instance is terminated by Amazon EC2 for price reasons, no fees will be incurred; if an interruption occurs after more than 1 h, it will be calculated in seconds and make it more accurate to the actual usage time; if the user actively terminates the Spot instance, even if it is less than 1 h, it will be charged according to the actual usage time in seconds. Spot instances billed by the hour: During the first hour of launch, if the instance is terminated by Amazon EC2 due to price, there will be no cost; if the Spot instance price exceeds the user's bid, the Spot instance will be in the instance hour In the middle of the interruption, the user does not need to pay for the interrupted part of less than 1 h; if the user actively terminates the Spot instance in the middle of the instance hour, he needs to pay for the hour [20].

In Fig. 4(a), the user's bid is much higher than the highest price of the instance in 10 h. It can be intuitively known that this is not an optimal bidding strategy. But the advantage of this bidding strategy is that the user can keep the bid instance applied to keep running unless the cloud management system forcibly withdraws the bid instance. In Fig. 4(b), the user's bid for the first 3 h is higher than the price of the cloud bidding instance, so the instance is executed. During the period of 3 h to 5 h, the bidding instance of the user becomes inexecutable because the price of the instance rises above the user's highest bid.

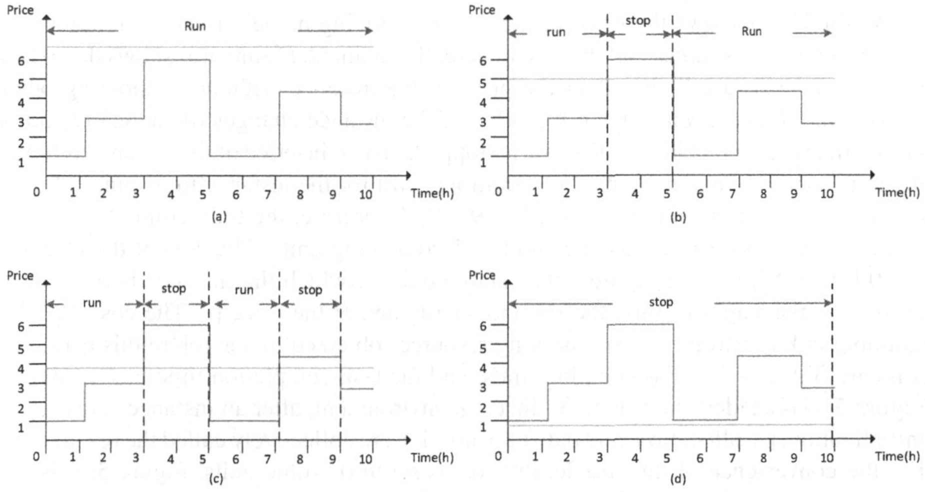

Fig. 4. The bidding price of a dynamic pricing instance (Color figure online)

Between 5 h and 10 h, the price drop instance of the bidding instance becomes feasible. The bidding strategies proposed by different cloud providers are slightly different here. After the user's bidding instance is terminated, some cloud service providers will restart the user's bidding instance when the price reaches the requirements, while some cloud service providers will directly terminate the user. The instance is running and takes back the bidding instance. The situation in Fig. 4(c) is similar to that in Fig. 4(b). The difference is that the maximum bid price of the user becomes lower, and the highest bid price is lower than the bid instance price during the application period. In Fig. 4(d), since the user's bid is always lower than the actual price of the bidding instance, the user has never been used by the instance, and the auction process fails.

For dynamic pricing instances, users need to consider not only the combination of rented instance types but also auction bidding. The auction price largely determines the time and economic benefits of task completion. Generally, when the user gives a higher bid, more computing resources can be obtained, so the task can be completed in a shorter time. However, it also needs to pay more usage fees. On the contrary, if the bidding price of the user is too low, only fewer computing resources can be obtained, and the task completion time will become longer.

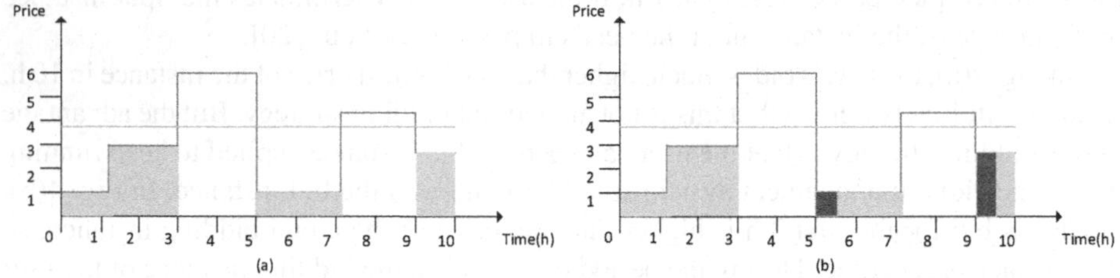

Fig. 5. The restart cost of dynamic pricing instances

While Fig. 5 shows the price of the dynamic pricing model, users bid relationship with computing resources and the cost of actually obtained. Figure 5(a) shows that when other users' bids are higher than the price of the instance without considering other factors, the instance runs. When the price of the instance changes dynamically and is higher than the user's bid, the instance is stopped. When the price of the instance is lower than the user's bid again, there is no additional cost for the instance to re-run. In Fig. 5 (a), it is executed at time [0, 3], [5, 7], [9, 10]. Therefore, the total computing power obtained by the user is an instance unit of 5 computing units. The cost of the instance in 10 h $p = \sum pi * ti$. P_i represents a stable price at which the instance is run, and t_i represents the length of time the instance is running at the price p_i. The cost P of the example in Fig. 5(a) is 9, the computing resources obtained by the job in this period of time are $3 + 2 + 1 = 6$ computing units, and the task completion time is $T = W/6C$. Figure 5(a) is an ideal environment. In a real environment, after an instance is restarted, initialization, job allocation, and data transmission are collectively called the restart cost. For the convenience of this article, the cost is set to 0.5-time units. Figure 5(b) is the case when the cost calculating is restarted. At this time, the cost P of the instance is still 9, the actual computing resources obtained are $3 + 1.5 + 0.5 = 5$ computing units, and the task completion time is T = W/5C. Figure 5 considers bidding for a single instance. If bidding for multiple instances is considered at the same time, the problem becomes more complicated.

The solid color lines in Fig. 6 are actual price data of three type instances in 10 h. The yellow line indicates the instance with three VCPU units and is named as instance 3. The red line indicates the instance with two VCPU units and is named as instance 2. The blue line indicates the instance with one VCPU units and is named as instance 1. C_i represents the *i*-th computing power of the instance. The dotted lines represent the bid price for three instances under the current price bidding strategy. In Fig. 6(a) and Fig. 6(b), the computing resources obtained computing resources of 30 units in 10 h. The cost in Fig. 6(a) is 41.5, and the cost in Fig. 6(b) is 45.5. According to different bidding strategies, the three instances are bid to obtain the same computing resources, and the same workload can be processed within 10 h. The costs of users under the two bidding strategies are indeed very different. In Figs. 6(c) and 6 (d), the computing resource for user's tasks is 30 computing units. According to the difference between two bidding strategies of (c) (d), the bidding strategy in Fig. 6(c) takes 5 h to complete the task while the bidding strategy in Fig. 6(d) takes 7 h to complete the task. So, it is clear that different bidding strategies will affect the cost and completion time of the task.

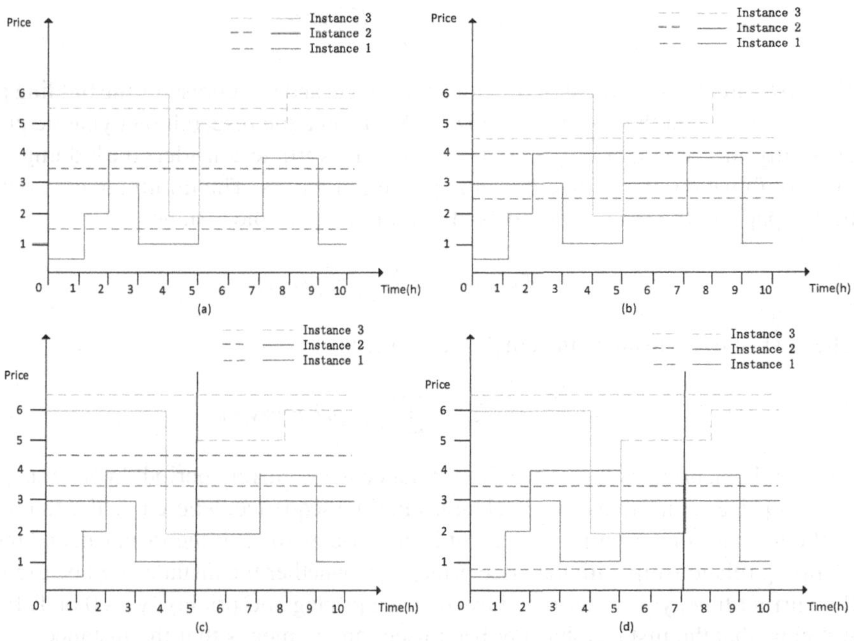

Fig. 6. Comparison of different bidding strategies (Color figure online)

4.2 Dynamic Bidding Model

Users choose IaaS instances to run their jobs. In the research questions of this article, they believe that users are pursuing the economic benefits and quality of tasks. Therefore, users will try to choose a better use strategy to maximize their economic benefits. It is assumed here that the tasks submitted by users are easily segmental, parallelizable, and computationally intensive, such as graphics rendering, video encoding, and machine learning. The total workload of the task is expressed as Workload, simply written as W. W represents the time required for each job to run and complete on an instance of unit performance. Assuming that the performance of the instance of the unit CPU is C, ideally, the completion time of the task using a single CPU is T = W/C. When a user rents multiple instances to form a combination of instances, the parallel CPU calculation unit is n. For example, if the user rents a single CPU instance and a 2CPU instance at the same time, the parallel CPU unit is 3. According to the above assumptions about tasks, tasks are splittable and easy to parallel, so the actual running time of the task is $T_{real} = W/(nC)$, then $T_{real} \leq T$. As you can see, users can rent more instances to complete tasks faster. However, it also needs to pay more for instance rental costs. Therefore, it is necessary to balance the time and economic benefits of renting instances. The optimization of user costs can be expressed as:

$$u^* = \underset{u \in U}{argmin}\, Expenses \tag{1}$$

Expenses are the user's cost of renting an instance, and u* represents the bidding price of the instance when the cost is minimized. At present, the research on dynamic pricing is generally an example of a bidding strategy [8]. This article considers the bidding problem of a combination of multiple dynamic pricing instances. The minimum cost-optimized in this paper is the sum of the costs of combining multiple instances.

$$Expenses = \sum\nolimits_1^n Expenses_i$$

The cost of each instance in actual usage time is

$$Expenses_i = \sum\nolimits_1^T f(price_i)*y_i*t_i$$

Among them, $price_i$ is the price of the instance in the current period. In general, $price_i$ is a set of price values that contain changes in time sequence, where t_i is the running time of the instance in a running cycle, that is, the pricing time of the instance, and there is a mapping relationship with the price $price_i$. y_i is whether the instance will be used within the current time, y_i is determined by instance pricing and pricing, y_i is 0 or 1. If y_i is 0, it means that the instance has not been used, and y_i means that the instance is running and starts charging.

$$Workload = \sum\nolimits_1^n \sum\nolimits_1^T \mathrm{Workload}_{ij},$$

We use *res* to represent whether the instance is in a restarted state or no after being stopped. *res* is 0 to indicate that the instance is newly started. *res* is 1 to indicate that the instance is to be restarted. It takes *restart_cost* hours to restore the job site. Therefore, no job was run within *restart_cost* hours. The optimization problem equation is:

$$b^* = \underset{u \in U}{argmin} \sum\nolimits_1^n \sum\nolimits_1^T f\left(price_{ij}\right) * y_{ij} * t_{ij} \tag{2}$$

Subject to

$$Workload = \sum\nolimits_1^n \sum\nolimits_1^T \text{Capacity}_{ij} * (t_{ij} - \text{restart\_cost} * res_{ij}) * y_{ij}$$

In the research problem of this paper, users submit resource requests for Workload calculation and bid on a group of instances through bidding, that is, $b = [b_1, b_2, ..., b_n]$, where n is the type of instance. $b*$ is the optimal solution to be sought. The restrictions indicate that the workload requested by the user must be satisfied.

5 Iterative Optimization Algorithm for Dynamic Pricing

How users can bid to minimize revenue is the issue to be studied in this section. In our problem, we do not seek the optimal solution of the strategy but obtain the approximate optimal solution of the problem through a heuristic algorithm. Under the condition of meeting the user's needs, obtaining the approximate optimal solution can greatly improve the efficiency of the system and reduce the user's expense. We mainly use maximum price bidding algorithms, differential evolutionary algorithms, and adaptive differential evolution algorithm. The maximum price bidding algorithms are intuitive and easy to understand. Differential evolutionary is to imitate the natural evolution of the solution of our target constant optimization, know to meet the conditions of the problem solution. Our adaptive differential evolution algorithm optimizes the parameters of thc basic evolution algorithm to meet the complex nonlinear data problems. We will explain the algorithms below.

5.1 The Highest Bidding Auction Strategy

First, we introduce a heuristic bidding strategy. The heuristic algorithm is easy to implement and has achieved good results in work. It is called the maximum price bidding strategy. When using the highest price auction strategy, users consider the possible price range of the bidding instance and always bid the highest price. In the maximum price bidding strategy, because users always bid at the highest possible price, the bidding instance will never be terminated by the cloud service provider. In other words, the maximum price bidding strategy can ensure the consistent execution of any job. Therefore, the time to complete the task is often minimized. Secondly, assuming that there is an optimal strategy in the bidding strategy, if the strategy of bidding at the highest price is not much different from the actual benefit of this optimal strategy, then the bidding at the maximum price is also similar to the optimized bidding strategy. This is more dependent on the price data distribution of the bidding instance, for example, the price change range of the bidding instance is relatively small.

When using the bidding strategy with the maximum price, as shown in Fig. 7, the user doesn't need to consider the change of instance price. They only need to know the maximum value of the bidding instance's price within a specified time and bid at this maximum value. In the maximum price bidding model, since the instance keeps running once applied, the job can generally be completed in the shortest time.

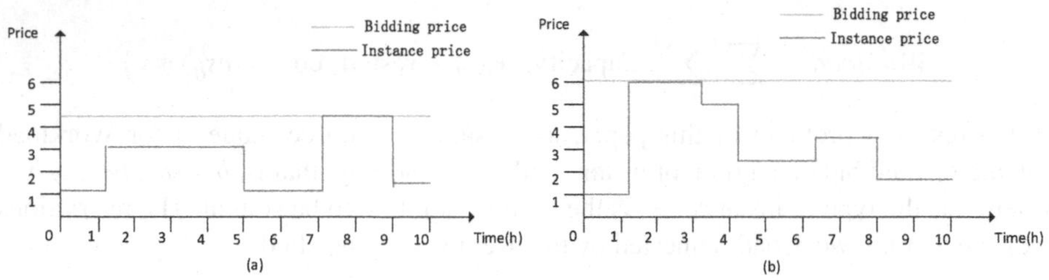

Fig. 7. The highest bidding auction strategy

The cost the user needs to pay is

$$Cost = \sum_{i=1}^{n} Price_i * t_i$$

$Price_i$ in the formula represents the stable bid price for a period of time, and t_i represents the length of time that the price of the bid instance remains at $Price_i$. If the running time of the task is T, then there is

$$T = \sum_{i=1}^{n} t_i.$$

5.2 Heuristic Global Optimization Algorithm Differential Evolution Algorithm

The basic idea of the algorithm is to start with a randomly generated initial population and generate a new individual by summing the vector difference of any two individuals in the population with the third individual, and then combine the new individual with the corresponding individual in the contemporary population In comparison, if the fitness of the new individual is better than the fitness of the current individual, the old individual will be replaced with the new individual in the next generation, otherwise, the old individual will be preserved. Through continuous evolution, retain good individuals, eliminate inferior individuals, and guide the search towards the optimal solution. The algorithm steps are shown in Fig. 8.

Step 1: Initialize. N_P population individuals need to be randomly and uniformly generated in the solution space. The first thing to determine is the dimension D of each individual. X_{Low} and X_{High} represent the upper and lower limit vectors of the D-dimensional space, respectively. Generate the ith individual of the population: $x_i^0 = X_{Low} + \text{random}(0, 1) * (X_{High}\text{-}X_{Low})$, $i \in [1, 2, \ldots, N_P]$. The initial population of N_P individuals $X^0 = [x_1^0, x_2^0, \ldots, x_{NP}^0]$ determine the variation factor F and crossover CR,

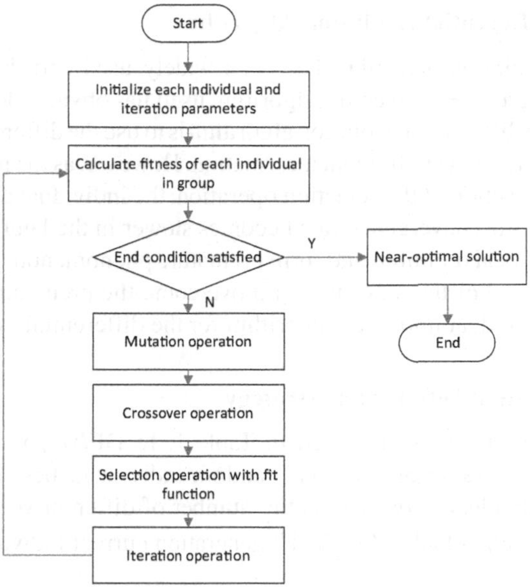

Fig. 8. Differential evolution procession

determine the maximum algebra G_{max}, the number of individuals N_P is generally greater than 4.

Step 2: For individuals in the Gth x_i^G $(i = 1, 2, \ldots, Np)$, calculate the fitness value PE (x_i^G) of each individual, and calculate the best individual with the best fitness.

Step 3: Mutation. For the ith individual in the Gth belt, three different individuals in the Gth band except with x_i^G which is $x_{r1}^G, x_{r2}^G, x_{r3}^G$ are selected, r1, r2, r3, $i \in [0, N_P]$ and r1 $\neq$ r2 $\neq$ r3 $\neq i$, Generate new mutant individuals $v_i^{G+1} = x_{r1}^G + F\ (x_{r2}^G\text{-}x_{r3}^G)$, generate mutant population $v^{G+1} = [v_1^{G+1}, v_2^{G+1}, \ldots, v_{NP}^{G+1}]$. F $\in$ [0, 2].

Step 4: Intersection. Cross the original individual x_i^G and the mutated individual v_i^{G+1} to get the cross individual u_i^{G+1}, CR $\in$ [0, 1] is the cross probability, r = random (0, 1) is the average distribution from 0 to 1. Cross operation can maintain the diversity of the population.

$$u_{ij}^{G+1} = \begin{cases} v_{ij}^{G+1}, & r_{ij} \leq CR \\ x_{ij}^G, & else \end{cases} \tag{3}$$

j $\in$ [0, D], it represents a dimension of each individual. Generally speaking, the greater the value of CR, the faster the convergence rate, but the convergence rate will decrease after exceeding a certain value, and the CR value will tend to premature.

Step 5: Selection is to determine whether the mutated individual u_i^{G+1} will become a new individual in the next generation. According to the greedy strategy, choose the x_i^{G+1} with better fitness of x_i^G and u_i^{G+1} as the G + 1 generation.

Step 6: Repeat steps 2–5 until the end condition of the iteration is reached.

5.3 Improved Differential Evolution Algorithm

The differential evolution algorithm has been widely used and developed due to its simplicity and efficiency. However, the algorithm itself has obvious deficiencies. The key step variation of the differential evolution algorithm is to use the difference information of the population to correct the individual information. However, as the number of iterations increases, under the action of the selection operation, the individual differences continue to decrease, so that the convergence rate becomes slower in the later iterations, and it is easy to fall into the local optimum and form premature phenomenon. In order to improve the convergence speed of the algorithm and overcome the premature algorithm, many scholars have proposed an improved algorithm for the differential evolution algorithm.

5.3.1 Improvement of Differential Strategy

Several differential strategies are listed in Table 1. In DE/x/y, x represents the base individual vector, the value of x is rand random individual, best optimal individual, current optimal individual. y represents the number of different vectors. x_{best}^{G} is the ith generation optimal individual, x_i^G is the ith generation current individual.

Table 1. Differential strategies

Differential strategy	Expression
DE/rand/1	$v_i^{G+1} = x_{r1}^G + F * \left(x_{r2}^G - x_{r3}^G\right)$
DE/rand/2	$v_i^{G+1} = x_{r1}^G + F * \left(x_{r2}^G + x_{r3}^G - x_{r4}^G - x_{r5}^G\right)$
DE/best/1	$v_i^{G+1} = x_{best}^G + F * \left(x_{r2}^G - x_{r3}^G\right)$
DE/best/2	$v_i^{G+1} = x_{r1}^G + F * \left(x_{r2}^G + x_{r3}^G - x_{r4}^G - x_{r5}^G\right)$
DE/randtobest/1	$v_i^{G+1} = x_i^G + F * \left(x_{best}^G - x_i^G\right) + F * \left(x_{r1}^G - x_{r2}^G\right)$

DE/rand/1/ and DE/best/2/ are currently the most widely used and most successfully applied differential strategies. DE/rand/1/ is conducive to maintaining the diversity of groups, and the DE/best/2/ strategy is more concentrated on the convergence speed of the algorithm. In our improved differential strategy, this article uses a combination strategy of DE/rand/1/ and DE/best/2. The parameter λ_{rand} represents the probability of the algorithm adopting the DE/rand/1 differential strategy,$\lambda_{\mathrm{rand}} \in [0, 1]$. At the beginning of the iteration, the algorithm uses more DE/rand/1, which can maintain the diversity of races and prevent falling into the local optimal solution. In the later stage of the iteration, DE/best/2 is used more so that the algorithm can converge to the optimal solution faster. The differential strategy is as follows:

$$v_i^{G+1} = \begin{cases} x_{r1}^G + F * \left(x_{r2}^G - x_{r3}^G\right), & random[0, 1] < \lambda_{rand} \\ x_{r1}^G + F * \left(x_{r2}^G + x_{r3}^G - x_{r4}^G - x_{r5}^G\right), & else \end{cases}$$

5.3.2 Adaptive Mutation Factor

The mutation factor F is an important parameter of the differential evolution algorithm that controls the mutation of individuals in the population. Generally, the greater the F, the greater the impact on individual variation, which is conducive to maintaining the diversity of the population. However, the larger the F value, the greater the fluctuations in the mutant individuals, which will reduce the search efficiency and the accuracy of finding the global optimal solution. When the F value is smaller, the impact on individual variation is smaller, which can serve the purpose of searching for the optimal solution locally, but it is easy to make the algorithm premature. We improve the linear method presented in paper [21] by adopting a linear method of adjusting the variation factor F. Through experiments, it is known that too small or too large F generally improves the algorithm less. When the value of F is 0.4 to 1, it tends to have better results. Introduce parameters here $\xi = random\left[\left(\frac{G_{max}-G_i}{G_{max}}\right)^2, \frac{G_{max}-G_i}{G_{max}}\right]$. L represents the length of the interval that changes randomly, and the changing trend of L is $0 \rightarrow 1/4 \rightarrow 0$. The adaptive mutation factor in this paper is expressed as follows:

$$F_x = \xi * F_{max} + (1 - \xi) * F_{min} \tag{5}$$

$F_i = \max(F_x, F_{min})$ In this article, there is a tendency for the mutation to become smaller gradually so that in the early stage of the iteration, there is a high probability that *Fmax*, and in the later stage of the iteration, the value of F has a greater probability of approaching *Fmin*. The differential evolution algorithm can maintain the population diversity in the early stage, which prevents prematurely, and may accurately locate the local optimal solution in the later stage. Not only maintains the trend of F from large to small but also makes the variation factor have certain randomness to better maintain population diversity and better find the global optimal solution.

5.3.3 Adaptive Crossover Factor

The crossover factor *CR* is another important parameter of the differential evolution algorithm. The crossover factor has an important influence on the convergence speed of the algorithm. The larger the value of *CR*, the greater the contribution of the mutant individual to the iteration, which is beneficial to the local search of the algorithm and accelerates the convergence rate of the algorithm; the smaller the value of *CR*, the greater the contribution of the current individual to the iteration, which is conducive to maintaining population diversity. Beneficial to the global search of the algorithm. A good search algorithm should maintain the diversity of the population in the early stage of the iteration and strengthen the convergence speed of the local search in the later stage. Therefore, this article uses the adaptive crossover factor to express as follows:

$$CR_x = (1 - \xi) * CR_{max} + \xi * CR_{min} \tag{6}$$

$CR_i = \max(CR_x, CR_{min})$ The difference factor in this paper has a greater probability to obtain a smaller value at the beginning of the iteration, which can well maintain the diversity of races, and has a greater probability to obtain a larger value in the later period, which allows the algorithm to quickly It converges to the local optimal solution, while the adaptive parameters maintain certain randomness.

6 Experimental Evaluation

In this section, this article downloads the actual Spot price historical data from the Amazon cloud official website platform. The experiment selects 12 types of instances and optimizes the bidding strategy for 12 instances through the adaptive parameter differential evolution algorithm.

6.1 Data Description

Downloading the actual spot historical price data from Amazon, the price data has a period of nearly 3 months from 2018.9.28 to 2018.11.23. Amazon cloud instances are classified into general-purpose instances according to optimization and usage, computing-optimized instances, memory-optimized instances, accelerated computing instances, storage-optimized instances, etc. Generally speaking, users need the same kind of instances for one job application Types of. The experiments in this paper are set to be computationally intensive. Therefore, the experiments in this paper chose the instances of computational optimization as the research object in the experiment and selected 12 types of instances of computational optimization as the types of instances of our bidding strategy optimization. The computing-optimized instance operating system is Linux/ Unix, and the usable area is us-east-1a. Table 2 lists the names and detailed parameters of the 12 computationally optimized instances of our experiment.

Table 2. Instances configuration comparison

Instance type	Number of VCPU	Memory	Instance type	Number of VCPU	Memory
c5.large	2	4	c5d.large	2	4
c5.xlarge	4	8	c5d.xlarge	4	8
c5.2xlarge	8	16	c5d.2xlarge	8	16
c5.4xlarge	16	32	c5d.4xlarge	16	32
c5.9xlarge	36	72	c5d.9xlarge	36	72
c5.18xlarge	72	144	c5d.18xlarge	72	144

Figure 9 shows the price data for 12 instances. Due to the large difference in prices of different instances, to better show the fluctuation of the output price of the instance, we show the prices of the instances in four sub-graphs. It can be seen from the figure that the price of the instance fluctuates greatly.

6.2 Settings of Experimental Parameter

This article conducts strategy optimization experiments on the real data set of Amazon Cloud Spot instances. The experiment uses a differential evolution algorithm to optimize the bidding strategy of the dynamically priced IaaS instance. The parameters of the differential evolution algorithm are the minimum value of the F mutation factor 0.4, the

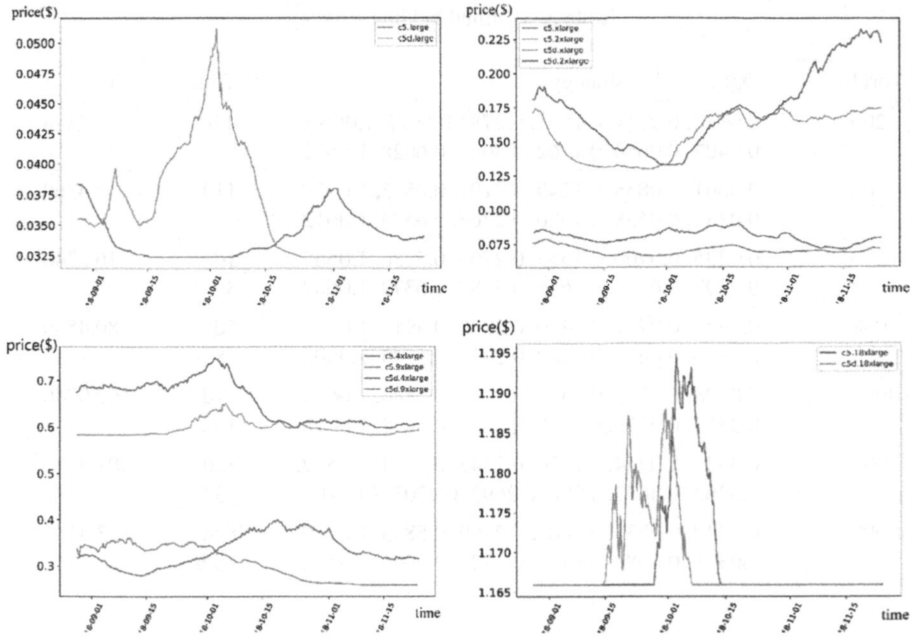

Fig. 9. Comparison of instances price data

maximum value 1, the minimum value of the *CR* crossover factor 0.3, the maximum value 0.9, the data dimension is the number of instances, and the number of populations 15. The number of iterations is 200. First, the workload is used as the task's limiting condition, and the vector value of the individual in the population is used to calculate the time required to meet the current workload, and then the time is calculated to obtain the fitness function value of the optimization problem. In the experiment, a variety of Workload values were entered, and our optimization method was tested under different conditions of demand, and the optimal bidding strategy under each workload, and the task completion time and total cost under the current optimal bidding strategy were obtained.

6.3 Experiment Results and Analysis

Table 3 shows the optimization results of the bidding strategy through the adaptive differential evolution algorithm. H represents hour and D represents the day. The workload is the demand of the user task, that is, the running time of the CPU with the unit performance required by the task. The strategy combination is the optimal bidding combination found by the differential evolution algorithm under each workload. The completion time represents the time required to complete the current workload under the current strategy, and finally the total cost. Through the differential evolution algorithm, the minimum cost value of the task whose workload satisfies is found.

Table 4 compares the performance of the algorithm. In the strategy optimization for handling different workloads, the adaptive differential evolution algorithm has a

Table 3. Optimal bidding strategy

Workload	Optimization strategy	Time	Cost
200	0.0334, 0.0714, 0.1567, 0.2783,0.5677, 1.0666, 0.0407, 0.0640, 0.1762, 0.3069, 0.6028, 1.0622	4 h	3.5848
500	0.0301, 0.0858, 0.1249, 0.3197, 0.0585, 1.0727, 0.0382, 0.0638, 0.1686, 0.2763, 0.6509, 1.0612	11 h	8.4360
1000	0.0335, 0.0700, 0.1553, 0.2766, 0.5836, 1.0567, 0.0297, 0.0686, 0.1656, 0.3188, 0.6379, 1.0684	1d 3d	16.3240
5000	0.0343, 0.0797, 0.1490, 0.2832, 0.5917, 1.0478, 0.0310, 0.0665, 0.1160, 0.2434, 0.5945, 1.0500	5d 18 h	80.4540
10000	0.0308, 0.0632, 0.1347, 0.2758, 0.5840, 1.0649, 0.0313, 0.0676, 0.1640, 0.2557, 0.5472, 1.0678	11d 13 h	162.0740
30000	0.0335, 0.0744, 0.1570, 0.2543, 0.5904, 1.0572, 0.0362, 0.0677, 0.988, 0.2692, 0.4303, 1.0510	32d 13 h	491.9586
50000	0.0340, 0.0733, 0.1487, 0.2499, 0.5863, 1.0610, 0.0301, 0.0589, 0.1533, 0.2724, 0.5946, 1.0562	82d 18 h	823.4141

very good performance improvement. Meanwhile, when the workload increases, the preferential ratio maintains a relatively stable price, which shows that the adaptive differential evolution algorithm in this paper has a good optimization effect on short-term task requirements and long-term task requirements. The adaptive differential evolution algorithm has an average cost optimization of 12.48% compared to the maximum bidding strategy, and an average cost optimization ratio of 3.53% over the basic differential evolution algorithm.

Table 4. Comparison of optimization results

	Maximum bid	Differential evolution	Adaptive differential evolution	Better than the biggest bid	Special differential evolution discount
200	4.8370	3.7116	3.5848	25.98%	3.41%
500	9.6741	8.6415	8.1620	15.60%	5.54%
1000	19.3486	17.4566	16.3420	15.63%	6.38%
5000	90.9054	82.6310	80.4540	11.50%	2.63%
10000	179.4283	165.6529	162.0740	9.36%	2.16%
30000	529.1734	504.3787	491.9586	4.88%	2.46%
50000	880.0985	841.2941	823.4141	4.41%	2.13%

7 Conclusion

This paper introduces the dynamic pricing cloud service instance types and introduces the characteristics of cloud dynamic pricing instance types. Then explained the analysis of the bidding strategy of the dynamic pricing cloud service instance and compared it with the fixed price instance. Then, the optimization strategy of the dynamically priced cloud instance is modeled, and the expression for optimizing the user's job cost is obtained. Then it introduces the maximum bidding strategy and the bidding strategy based on the differential evolution algorithm of the global optimization algorithm. In the problem of this paper, the model can simultaneously optimize the bidding and usage strategies of multiple instances. Finally, download the real price history data of the Amazon cloud dynamic instance Spot instance, and calculate the performance parameters of the optimized instance in reference 12, and use the adaptive differential evolution algorithm to perform dynamic pricing instances when users use the combination of instances. Strategy optimization to achieve the purpose of reducing costs. The experimental results show that the adaptive differential evolution algorithm has a good optimization effect for both short-term task requirements and long-term task requirements. The adaptive differential evolution algorithm has an average cost optimization of 12.48% compared to the maximum bidding strategy, and an average cost optimization ratio of 3.53% over the basic differential evolution algorithm.

In this paper, the types of jobs are limited and only consider parallel tasks. In a real environment, users' tasks are complex and diverse. So, how the bidding strategy is suitable for various task types is the problem that needs to be considered.

Acknowledgement. The authors would like to acknowledge the support provided by the National Key R&D Program of China (2018YFB1404501), and the Young Scholars Program of Shandong University.

References

1. Amazon Cloud EC2 (2020). https://aws.amazon.com/ec2/
2. Google Compute Engine (2020). https://cloud.google.com/compute/
3. Yang, S., Pan, L., Wang, Q., et al.: Subscription or pay-as-you-go: optimally purchasing IaaS instances in public clouds. In: International Conference on Web Services, pp. 219–226 (2018)
4. Amazon Cloud Spot instance (2020). https://docs.aws.amazon.com/AWSEC2/latest/UserGuide/using-spot-instances.html
5. Wolski, R., Brevik, J.: Providing statistical reliability guarantees in the AWS spot tier. In: High Performance Computing Symposium (2016)
6. Vecchiola, C., Pandey, S., Buyya, R., et al.: High-performance cloud computing: a view of scientific applications. In: International Symposium on Pervasive Systems, Algorithms, and Networks, pp. 4–16 (2009)
7. Karunakaran, S., Krishnaswamy, V., Sundarraj, R.P.: Decisions, models and opportunities in cloud computing economics: a review of research on pricing and markets. In: Australian Symposium on Service Research and Innovation (2014)
8. Tang, S., Yuan, J., Li, X., et al.: Towards Optimal bidding strategy for Amazon EC2 cloud spot instance. In: International Conference on Cloud Computing, pp. 91–98 (2012)

9. Benyehuda, O.A., Benyehuda, M., Schuster, A., et al.: Deconstructing Amazon EC2 spot instance pricing. Electron. Commer. **1**(3), 1–20 (2013)
10. Kaminski, B., Szufel, P.: On optimization of simulation execution on Amazon EC2 spot market. Simul. Model. Pract. Theory **58**, 172–187 (2015)
11. Bharad, V.H., Bheda, H.A.: SLA-based virtual machine management for heterogeneous workloads in a cloud datacenter. Int. J. Comput. Appl. (0975 – 8887) **112**(16) (2015)
12. Zhu, J., Hong, J., Hughes, John G.: Using Markov chains for link prediction in adaptive web sites. In: Bustard, D., Liu, W., Sterritt, R. (eds.) Soft-Ware 2002. LNCS, vol. 2311, pp. 60–73. Springer, Heidelberg (2002). https://doi.org/10.1007/3-540-46019-5_5
13. Zheng, B., Pan, L., Yuan, D., Liu, S., Shi, Y., Wang, L.: A truthful mechanism for optimally purchasing IaaS instances and scheduling parallel jobs in service clouds. In: Pahl, C., Vukovic, M., Yin, J., Yu, Q. (eds.) ICSOC 2018. LNCS, vol. 11236, pp. 651–659. Springer, Cham (2018). https://doi.org/10.1007/978-3-030-03596-9_47
14. Shao, Q., Liu, S., Pan, L., Yang, C., Niu, T.: A market-oriented heuristic algorithm for scheduling parallel applications in big data service platform. In: 2018 IEEE 42nd Annual Computer Software and Applications Conference (COMPSAC), Tokyo, Japan, pp. 677–686 (2018)
15. Storn, R., Price, K.: Differential evolution – a simple and efficient heuristic for global optimization over continuous spaces. J. Glob. Optim. **11**(4), 341–359 (1997)
16. Kim, H., Chong, J., Park, K., et al.: Differential evolution strategy for constrained global optimization and application to practical engineering problems. In: IEEE Conference on Electromagnetic Field Computation, vol. 43, no. 4, pp. 1565–1568 (2006)
17. Omran, M.G., Engelbrecht, A.P.: Self-adaptive differential evolution methods for unsupervised image classification. In: IEEE Conference on Cybernetics and Intelligent Systems, pp. 1–6 (2006)
18. Zhang, R., Ding, J.: Non-linear optimal control of manufacturing system based on modified differential evolution. In: IMACS Multiconference on Computational Engineering in Systems Applications. IEEE (2006)
19. Onwubolu, G.C., Babu, B.V.: New Optimization Techniques in Engineering. Springer, Berlin Heidelberg (2004). https://doi.org/10.1007/978-3-540-39930-8
20. Amazon spot instance detail. https://aws.amazon.com/cn/blogs/china/amazon-ec2-spot-instance-detail/
21. Liang-Hong, W., Hui-Nan, W.: Adaptive quadratic differential evolution algorithm. Control Decis. **21**(8), 898–902 (2006)

SBiNE: Signed Bipartite Network Embedding

Youwen Zhang[1], Wei Li[1], Dengcheng Yan[1(✉)], Yiwen Zhang[1], and Qiang He[2]

[1] School of Computer Science and Engineering, Anhui University, Heifei 230601, China
wen070864@gmail.com, {91019,yanzhou,zhangyiwen}@ahu.edu.cn

[2] School of Software and Electrical Engineering, Swinburne University of Technology, Melbourne, VIC 3122, Australia
qhe@swin.edu.au

Abstract. This work develops a representation learning method for signed bipartite networks. Recent years, embedding nodes of a given network into a low dimensional space has attracted much interest due to it can be widely applied in link prediction, clustering, and anomalous detection. Most existing network embedding methods mainly focus on homogeneous networks with only positive edges and single node type. However, negative edges are more valuable than positive edges in certain analysis tasks. Even though the work on signed network representation learning distinguishes between positive and negative edges, it does not consider the difference in node types. Moreover, bipartite network representation learning which considers two types of vertices do not tell link signs. In order to solve this problem, we further consider the link sign on the basis of the bipartite network to conduct signed bipartite network analysis. In this paper, we propose a simple deep learning framework SBiNE, short for signed bipartite network embedding, which both preserves the first-order (i.e., observed links) and second-order proximity (i.e., unobserved links but have similar sign context), and then by optimizing the objective function, experiments on three datasets show that our proposed framework SBiNE is competitive in link sign prediction task.

Keywords: Signed bipartite networks · Network embedding · Link sign prediction

1 Introduction

Not all networks in real-world are the same, and in fact, relationship between entities can be expressed via biological networks, social networks and communication networks. Most previous work have primarily focus on signed networks with single node type or bipartite networks with only positive links. However, the network structure of signed bipartite networks is often overlooked. A signed

H. Gao et al. (Eds.): CollaborateCom 2020, LNICST 349, pp. 479–492, 2021.
https://doi.org/10.1007/978-3-030-67537-0_29

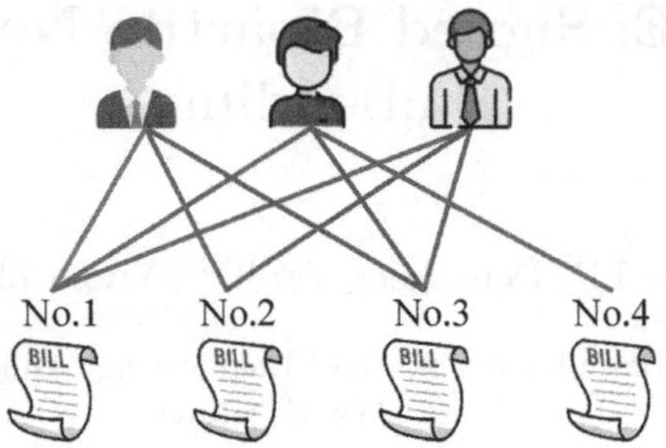

Fig. 1. Congresspersons vote "Yea" or "Nay" for the bills

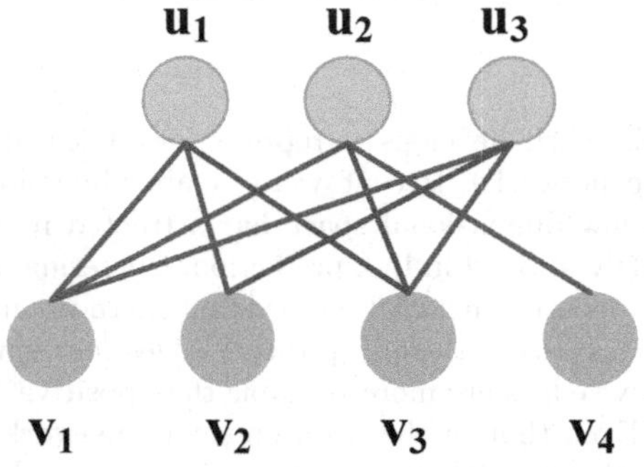

Fig. 2. The signed bipartite network corresponding to Fig. 1

bipartite network is a special network with two sets of nodes, meanwhile, edges exist only between different types of nodes. In addition, there are no edges between nodes of the same type. In fact, signed bipartite networks are involved in many domains of our lives. For example, in the United States Congress, the voting records are modeled as a signed bipartite social network (i.e., contains both positive and negative connections) between the representatives and the bills [1]. In several analytical tasks, negative links have proven to be significantly useful in improving positive link prediction [2] and recommendation performance [3].

In order to perform predictive analysis on the signed bipartite network, this requires us first to learn low-dimensional dense representations for nodes. One common form of data mining is network embedding which transforms network information into low-dimensional dense vectors, while preserving the network topology and using it as an input to existing machine learning algorithms, processing tasks such as node classification [4], link prediction [5], recommendation [6–9] and visualization [10].

Most existing signed network representation learning are based on homogeneous networks [11], which are not applicable to signed bipartite networks when having two node types. Our work investigates how to effectively learn the node embeddings which can well preserve the original network structure. Motivated by LINE [12], we introduce second-order proximity to signed bipartite networks to measure the similarity between nodes of the same type since there is no link between them. The first-order proximity is measured by the embedding of observed edges which will be learned by deep learning. For better understanding, we take Fig. 2 as an example. From the perspective of topology structure, LINE [12] and BiNE [13] will consider that user u_1, u_2 and u_3 are similar because they have multiple common neighbours. But if we take the sign between congressperson and bill into consideration, although satisfying the second-order proximity in structure, u_1 and u_3 actually are of big difference because of their different sign context. The major contributions of our method are listed as follows:

- Based on bipartite networks, we further consider the link sign difference to perform signed bipartite network representation learning.
- To measure the similarity between nodes in signed bipartite network, we design a suitable objective function and then propose a simple deep learning framework SBiNE for signed bipartite network representation learning, which learns low-dimensional vectors for nodes via optimizing the objective function.
- We conduct experiments on three real-world datasets. Experimental results demonstrate the effectiveness of the proposed framework SBiNE.

The rest of this paper is organized as follows. We introduce some related works in Sect. 2. In Sect. 3, the motivation is supplied for this article. In Sect. 4, we introduce the proposed framework SBiNE with the details about the signed bipartite network embedding objective function. In Sect. 5, we perform empirical evaluations with discussion. Finally, we present the conclusion in Sect. 6.

2 Related Work

Network representation learning or network embedding aims to learn the low-dimensional vector representation of a given network. Initially, network representation learning algorithms were mainly based on matrix feature vectors. Spectral analysis [14] algorithm obtains k-dimensional node representations by calculating the top-k eigenvectors of the Laplacian matrix. The well-known algorithm Deepwalk [15] is inspired by the word representation vector model Skip-gram in NLP. It treats the sequence of nodes as sentences and learns the node embedding from the sequence of random walks.

LINE [12] improves Deepwalk [15] and preserves first-order and second-order proximity. Moreover, it is suitable for large-scale directed, undirected, and weighted networks. GraRep [16] introduces higher-order proximity between nodes on this basis. The bipartite network embedding BiNE [13] proposed a novel optimization framework by accounting for both the explicit relations and implicit relations in learning the vertex representations. Likewise, heterogeneous information network embedding aims to embed multiple types of nodes into a low-dimensional vector space. RHINE [17] utilized the structural characteristic of Affiliation Relation(ARs) and Interaction Relation(IRs) and then proposed a novel Structure Relation-aware Heterogeneous Information Network Embedding model. However, the aforementioned methods concerning homogenous network, bipartite network, and heterogeneous network cannot directly be applied to signed networks because they don't consider negative links. Signed network embedding which considers negative links like SiNE [11] introduced structural balance theory into signed social networks and optimized the objective function based on deep learning to automatically learn signed network embedding. Other signed network representation learning, like SNEA [18], both considers the network structure and node attributes, which makes the link sign prediction task a significant improvement. Although these signed network embedding methods consider the link sign, the node type is still single so they also cannot be applied to signed bipartite network embedding analysis. Thus, we focus on the problem of learning low-dimensional vectors for nodes in the signed bipartite networks by utilizing the power of deep learning. More specifically, we design a simple framework SBiNE which optimizes a multi-layer perceptron based objective function to learn signed bipartite network embedding automatically.

3 Motivation

In this section, we briefly illustrate the motivation of our research according to Fig. 1 and Fig. 2. Specifically, Sect. 3.1 discusses the necessity of studying signed bipartite networks, and Sect. 3.2 explains why first-order and second-order proximity should be introduced into signed bipartite networks.

3.1 Why Should Signed Bipartite Networks Be Researched?

The relationship between entities can be represented by networks, and different networks have different characteristics. For example, for a homogeneous network of social relationships between blog authors on the BlogCatalog website, node2vec [19] designed a biased random walk to effectively explore various node neighbors. For a bipartite network containing authors and publishers in DBLP. BiNE [13] learns the representation vector of nodes from two perspectives of explicit and implicit relations. The two-layer relationship between users and items in e-commerce viewing and purchasing constitutes a multi-dimensional bipartite network, hence MINES [20] aims to learn the representation of nodes in each dimension of the network structure. Heterogeneous information network of entities, words and categories in Wikipedia, Zhao et al. [21] build the co-occurrence matrices between same and different types of nodes, and use coordinate matrix factorization to jointly learn the representations of entities, words and categories from all matrices. The links of trust and distrust between people in Epinions form a signed social network, and SiNE [11] uses triangular structure balance theory to design the objective function and optimize it.

The previous network representation learning work has ignored a common form of the signed network—signed bipartite network, which inherits the advantages of signed networks with negative edges that indicate distrust or dislike, and there are two sets of nodes. It should be noted that links only exist between nodes of different types.

3.2 Why Should the First-Order and Second-Order Proximity Be Introduced?

We take Fig. 2 as an example to clearly illustrate these two proximities in signed bipartite networks. Purple nodes are the representatives in the United States Congress and orange nodes are the bills they have voted on. Blue positive link or red negative link denotes a congressperson voted "Yea" or "Nay" for the bill respectively. Naturally, the observed links in Fig. 2 preserve the first-order explicit relationship between the node sets of U and V. u_1 and v_1 node pair is such the first-order proximity relationship. Apparently, those observed links are not sufficient for preserving the global network structures. However, the second-order proximity of the signed bipartite network is quite different from the previous unsigned homogenous network. Node v_1, v_2 and v_3 are the common neighbors of u_1 and u_3. From the perspective of LINE [12], it assumes that u_1 and u_3 have several neighbor nodes in common, so they are supposed to be similar. However, if we consider the link sign context, the result will be different. Node v_1 is the common bill that u_1 and u_3 vote, but these two persons show different voting preferences on v_2 and v_3. Hence, although u_1 and u_3 have similar neighbors, we assume that they are still dissimilar. We apply the proven equation [22] to fit the problem of second-order implicit relationship between the node sets of U or V.

In study [22], it proved that LINE(2nd) is actually factoring two different matrices separately. For each directed edge (v_i, v_j), it defines v_j as the "context" of v_i. V denote a matrix whose i-th column is the vertex embedding $\mathbf{v}_i$ and U denote a matrix whose j-th column is the "context" embedding $\mathbf{u}_j$. It figured out that LINE(2nd) is factoring a matrix $M^{(2)} = V^T U$. According to the objective function in LINE(2nd), it characterizes the matrix $M^{(2)}$ that LINE(2nd) is actually factoring:

$$M_{ij}^{(2)} = PMI(v_i, v_j) - logk \tag{1}$$

4 Signed Bipartite Network Embedding

We describe the framework SBiNE to preserve the explicit relationship with the positive or negative sign between two sets of nodes and the implicit relationship to measure the same set of nodes with similar sign context since there is no link between them. Experiments show that neither of these two conditions can be missed. Considering the first-order proximity relationship between representatives and bills is far more not enough, the similarity between congresspersons also does important. Node vectors are used to denote the features of congresspersons and bills, and edge embeddings represent the "Yea" or "Nay" relationship between a congressperson and a bill. First, we summarize the major notations used throughout this paper. Next, the objective function designed for signed bipartite networks will be introduced and then we will explain the details of the simple framework SBiNE.

4.1 Problem Formulation

Notations. Given a signed bipartite network G = (U, V, E) where U = $\{u_1, u_2, \ldots u_m\}$ and V = $\{v_1, v_2, \ldots v_n\}$ is the set of the m, n nodes in G respectively. E denotes the set of observed edges between the nodes in the set U and V. Each edge $e_{ij} \in E$ is associated with a sign $\mathbf{A}_{ij}$. $\mathbf{A}$ is the $m \times n$ adjacency matrix of G. We use $\mathbf{A}_{ij} = 1$ or -1 to represent a positive or negative link between u_i and v_j , and $\mathbf{A}_{ij}$=0 denotes no link.

Problem Definition. The purpose of designing the framework SBiNE is to effectively learn the representation of the nodes in the signed bipartite network, so as to use the learned vectors of two sets of nodes (i.e., $\mathbf{X}^{M \times d}$ and $\mathbf{Y}^{N \times d}$) as input to the link sign prediction task. Each row of $\mathbf{X}$ is an embedding vector with d features and the same goes for $\mathbf{Y}$. In the embedding Euclidean space, the observed explicit links in the signed bipartite graph and the implicit relationships between the set of U or V should be properly preserved. The input and output of SBiNE can be defined as follows:

Input: A signed bipartite network G = (U, V, E) and its adjacency matrix $\mathbf{A}$.

Output: D-dimensional embedding vector for each node u_i ,v_j.

4.2 First-Order and Second-Order Proximity in Signed Bipartite Network

Some recent studies on signed networks have shown that negative and positive edges have significantly different properties. Therefore, we cannot directly apply the method of unsigned network representation learning to the signed network. In addition, there are two set of nodes in the signed bipartite network. Therefore, we design a novel applicable objective function for the special network of signed bipartite networks. We first clarify the proximity that exists in the signed bipartite network, which is also our motivation for constructing the objective function.

Condition 1: second-order proximity between nodes of the same type. It measures the similarity between u_i and u_k, and is used to distinguish nodes of the same type that have common neighbors but quite different sign patterns.

Condition 2: first-order proximity according to the observed links composed of node sets of U and V. To measure the relation between u_i and v_j, we learn the edge vector $\mathbf{E}_{ij}$ for them.

Previous work [18] show that the second-order proximity actually factorizes the Pointwise Mutual Information (PMI) [23] matrix of each node pair with a constant shift. MSE is chosen as the loss function, so the objective function of the second-order implicit relationship can be written as:

$$O_1 : \arg\min \sum_{e_{ij} \in E} \frac{1}{|E|} \left(f\left(\mathbf{x}_i\right) \cdot f\left(\mathbf{y}_j\right) - \mathrm{PMI}(i,j) + \log k\right)^2 \tag{2}$$

Since the signed bipartite network is an undirected network, each edge can be treated as two directed edges with opposite directions. Therefore, we convert the expression of PMI into:

$$PMI(i,j) = \log \frac{\sum_{p=1}^{m} \deg\left(u_p\right) + \sum_{q=1}^{n} \deg\left(v_q\right)}{\deg\left(u_i\right) \deg\left(v_j\right)} \tag{3}$$

Where $\deg(u_i)$ and $\deg(v_j)$ are the degrees of nodes u_i and v_j, respectively. In [22], to avoid time-consuming, k defines the negative samples. We empirically set $k = 15$ in this work. The $\mathbf{x}_i$ and $\mathbf{y}_j$ in expression (1) are the d-dimensional embeddings of nodes u_i and v_j, respectively, and will be continuously learned during the training process of the framework SBiNE. The mapping function f performs dimensionality reduction operations on $\mathbf{x}_i$ and $\mathbf{y}_j$ respectively, and after obtaining new feature vectors, the dot product is performed between the node pairs. Details about the function f will be discussed in the following subsection.

For condition 2, given that there are two kind of edges in the signed bipartite network: positive edges and negative edges, we also assume that the sign of the edges follows the Bernoulli distribution as in work [24], either -1, otherwise 1. The feature of the edge is the relationship between the learning nodes u_i and v_j. The edge representation vector $\mathbf{E}_{ij}$ adopts the element-wise product of two node embedding $\mathbf{x}_i$ and $\mathbf{y}_j$ after the dimensionality reduction of the mapping function f as input. The objective function of condition 2 is shown in Eq. (4).

$$O_2 : \arg\min \sum_{e_{ij} \in E} \frac{1}{|E|} \log g\left(\mathbf{E}_{ij}\right)^{\frac{1+\mathbf{A}_{ij}}{2}} \left(1 - g\left(\mathbf{E}_{ij}\right)\right)^{\frac{1-\mathbf{A}_{ij}}{2}} \tag{4}$$

Considering the learned node embeddings are supposed to satisfy above two proximities, we can finally obtain the objective function as follows:

$$\begin{aligned} O : \arg\min \sum_{e_{ij} \in E} \frac{\beta}{|E|} \left(f\left(\mathbf{x}_i\right) \cdot f\left(\mathbf{y}_j\right) - PMI(i,j) + \log k\right)^2 \\ - (1-\beta) \log g\left(\mathbf{E}_{ij}\right)^{\frac{1+\mathbf{A}_{ij}}{2}} \left(1 - g\left(\mathbf{E}_{ij}\right)\right)^{\frac{1-\mathbf{A}_{ij}}{2}} \end{aligned} \tag{5}$$

The mapping function g models the relationship between the edge representation vector space and the link sign. Details will be explained in the framework SBiNE. The hyper-parameter $\beta \in [0,1]$ is used to control the influence of these two conditions during the process of learning node representations (Fig. 3).

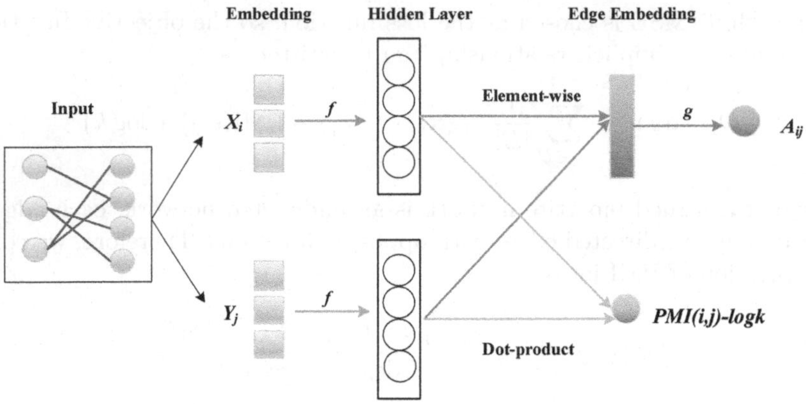

Fig. 3. The architecture of the proposed deep learning framework SBiNE

4.3 The Architecture of SBiNE

Since we have the objective function, our following task is to learn well-represented vectors for the nodes in the signed bipartite network, as well as suitable nonlinear mapping functions f and g. Deep learning technology provides strong technical support for nonlinear representation learning [25]. We use multi-layer perceptron to model the nonlinear mapping functions f and g. By optimizing the objective function (5), the node embeddings and the mapping functions can be learned simultaneously via back-propagation, since the node embedding is integrated into the neurals.

The input to the framework is a signed bipartite network G with two sets of initial node embeddings $\mathbf{X}_i$ and $\mathbf{Y}_j$. For better understanding, we first give an explanation of the hidden layer. The outputs of the hidden layer of the two neural networks are given as:

$$\begin{aligned} f\left(X_i\right) &= \tanh\left(W^{11}X_i + b^{11}\right) \\ f\left(Y_i\right) &= \tanh\left(W^{12}Y_i + b^{12}\right) \end{aligned} \tag{6}$$

Where tanh is the hyperbolic tangent function, other activation functions like sigmoid, ReLU are also effective. $\mathbf{W}^{11}$ and $\mathbf{W}^{12}$ are the weights of the hidden layer and b^{11}, b^{12} are the bias. The primary function of the hidden layer is to reduce the dimension of the node embeddings on the basis of retaining effective features. The first step is to achieve the second-order proximity in the signed bipartite network. With regard to the node pair (u_i ,v_j), after a dot product is operated on $f(\mathbf{x}_i)$ and $f(\mathbf{y}_j)$. We hope its value as close as possible to $PMI(i,j){-}logk$ through back propagation. The loss function here uses the mean square error MSE. The second step is to achieve the first-order proximity between the node pair (u_i,v_j) in signed bipartite network. We use edge embedding to represent this first-order proximity relationship which will be well learned via deep learning. We perform element-wise operation on $f(\mathbf{x}_i)$ and $f(\mathbf{y}_j)$, and then get their edge embedding $\mathbf{E}_{ij}$. $\mathbf{E}_{ij}$ will be the input of the new single neural with sigmoid nonlinearity. The output of the neural is:

$$g(\mathbf{x}_i, \mathbf{y}_i) = \text{sigmoid}(\mathbf{E}_{ij}) = \text{sigmoid}\left(f(\mathbf{x}_i) \odot f(\mathbf{y}_j)\right) \tag{7}$$

5 Experimental Results

We empirically evaluate our proposed signed bipartite network embedding framework SBiNE. By applying the method to three real-world datasets, we seek to answer the following two questions. First, does network representation learning for signed bipartite network provide an improvemance for predicting link signs? Second, how does dimension d of the learned vectors affect the performance of the model? For better understanding, we also conduct parameter sensitivity analysis.

5.1 Datasets

The three datasets used in the experiment are all from [18]. Bonanza is similar to e-commerce sites like Amazon or eBay, where every user can buy and sell a variety of goods. A buyer can rate the product purchased from a seller when a transaction is finished. To distinguish between the buyer and the seller, we use the node sets U and V to represent them. U.S.Senate and U.S.House datasets represent the role call votes combined from 1st to 10th United States Congress. We represent the senators or representatives by the set U and the bills that were voted by V. The positive link indicates that the senate support a bill, the negative link is the opposite. Table 1 has detailed information about these three datasets. Apparently, it can be seen that Bonanza is an extremely unbalanced dataset with only 2.02% negative links.

Table 1. Statistics of the three datasets

Dataset	m	n	Positive links	Negative links	$\lvert E\rvert$
Bonanza	7,919	1,973	97.98%	2.02%	36.543
U.S.Senate	1,056	145	98.88%	44.69%	27.083
U.S.House	1,281	515	53.96%	46.04%	114.378

5.2 Experimental Settings

Next, we first discuss the settings used for above three datasets on task of link sign prediction. For all three datasets, 80% of all the links are randomly selected as training set , so the remaining 20% used to test the model performance. To the best of our knowledge, only [26] is devoted to proposing balance theory in signed bipartite networks to predict link signs. Therefore, this work is the first study to carry out representation learning research on the signed bipartite networks. In order to compare with [26], we also use both F1 and AUC for evaluation.

5.3 Comparison Results

We seek to find out whether the signed bipartite network representation learning can improve the mining task of predicting the link sign on the three datasets. The task of link sign prediction is to predict the sign of the given links. In Tables 2 and 3, we can see the comparison results of AUC and F1. Because of the high imbalance of the Bonanza dataset, we will discuss the results of link sign prediction of this dataset separately. The definitions of the related six baseline methods [26] are as follows:

- **Degree Based Supervised Classifier (SCd):** By extracting features from each node, Scd trains the model by constructing the training dataset composed of links of positive and negative semantic links. With the trained model, it can predict arbitrary link sign.
- **Signed Catterpillars Based Supervised Classifier (SCsc):** Compared with Scd, this method extract features from characteristics of neighbor nodes based on balance theory.
- **Matrix Factorization (MF):** A traditional matrix factorization method to predict the unknown link sign by optimizing the node feature vectors via biadjacency matrix.
- **Matrix Factorization with Balance Theory (MFwBT):** It calculates if using balance theory could suggest a positive or negative link for nodes that do not exist in the network.
- **Lazy Random Walk (LazyRW):** LazyRW which is used to as a comparison against the SBRW performs a random walk to the neighbor node of the current node in a probabilistic way.
- **Signed Bipartite Random Walk (SBRW):** This method performs random walk on the unipartite signed network constructed by an adjacency matrix.

In the latter two datasets, our SBiNE model achieved the best performance on both AUC and F1 metrics for predicting the link sign. To be specifically, AUC and F1 are higher than the highest values of AUC and F1 of the other six methods, respectively.

Table 2. Results of predicting the link sign in terms of AUC and F1 metrics

Metric	Algorithm	Bonanza	U.S.Senate	U.S.House
AUC, F1	SCd	(0.553,0.959)	(0.638,0.654)	(0.625,0.635)
	SCsc	(**0.664**,0.674)	(0.812,0.823)	(0.827,0.837)
	MF	(0.593,0.903)	(0.792,0.812)	(0.831,0.846)
	MFwBT	(0.608,0.905)	(0.814,0.827)	(0.834,0.848)
	LazyRW	(0.547,**0.979**)	(0.808,0.821)	(0.815,0.827)
	SBRW	(0.582,0.949)	(0.836,0.849)	(0.846,0.858)
	SBiNE	(0.626,0.954)	**(0.915,0.857)**	**(0.934,0.869)**

In the extremely unbalanced dataset of Bonanza, we observed that among all the six comparison methods except SCsc and MFwBT, F1 are all greater than 0.9, but AUC is less than 0.6. Obviously, this classification results have bad performance. To further analyze the problem of highly imbalance in the Bonanza dataset, we made three sets of experiments and compared them with the six baseline methods. By adding different weights to the positive and negative sample categories in the loss function, we obtain three sets of results. When AUC and F1 are relatively balanced, AUC in SBiNE is second only to SCsc, and F1 is second only to SCd and LazyRW. However, both AUC and F1 performs better than MFwBT. If F1 is higher, the pair of (AUC, F1) is also better than LazyRW with the highest F1 in the six methods. According to the third set of results, we try to increase the AUC value by training as much as possible, then the results obtained are also higher than the SCsc with the highest AUC in the six methods.

Table 3. Results of link sign prediction with regard to Bonanza dataset

Algorithm	SCd	SCsc	MF	MFwBT	LazyRW	SBRW	SBiNE		
AUC	0.553	0.664	0.593	0.608	0.547	0.582	0.626	0.555	**0.668**
F1	0.959	0.674	0.903	0.905	0.979	0.949	0.954	**0.981**	0.886

5.4 Impact of Dimension d

In order to analyze the impact of dimension d on SBiNE, we set $k = 15$ as a fixed value, and $\beta = 0.05$. Meanwhile, we vary d as {8, 16, 32, 64, 100, 128, 200}. Bonanza is an extremely unbalanced dataset with only 2.02% of negative links, so we conducted this experiment on the two other datasets. From (a) (b), we observe that if d is too small, for example, when the dimension is 8, the F1 of the two datasets is very low, then the learned node feature vectors cannot adequately represent the structural characteristics of the original node. When $d \in [8, 32]$, the value of F1 and AUC of both datasets are improving. The U.S. Senate dataset slowly decreases after the AUC reaches its peak when the dimension is 32, while F1 slowly increases and then decreases. The AUC and F1 of the U.S. House dataset rapidly decreased and then increased after reaching the peak, and then slowly declined. When the dimension d is too large, it tends to overfit. Considering the moderate size of the dataset, we finally set the dimension of the vector to a reasonable value of 32.

5.5 Parameter Sensitivity Analysis

Since k has little effect on the results in the link sign prediction, we empirically set k to 15. So in this section, we performed hyper-parameter analysis on all three datasets to observe the effect of β on the model performance. The hyper-parameter $\beta \in [0, 1]$ is mainly affected by two conditions that control the proximity of nodes in the signed bipartite network. When $\beta = 1$, SBiNE only uses the first condition, that is, the second-order proximity of the nodes in the bipartite network; when $\beta = 0$, SBiNE only uses the second condition; when $\beta \in (0, 1)$, the method combines two conditions at the same time. The influence of β on the link sign prediction task is shown in Table 4 above. We can see from the Table 4 that when $\beta < 1$, F1 and AUC are better than those when $\beta = 1$. It shows that we cannot just consider the second-order proximity between nodes of the same type. When $\beta = 0.05$, SBiNE considers two conditions at the same time. Meanwhile, the classification effect of link sign prediction performs best. It indicates that second-proximity is important but not as important as first-proximity (Fig. 4).

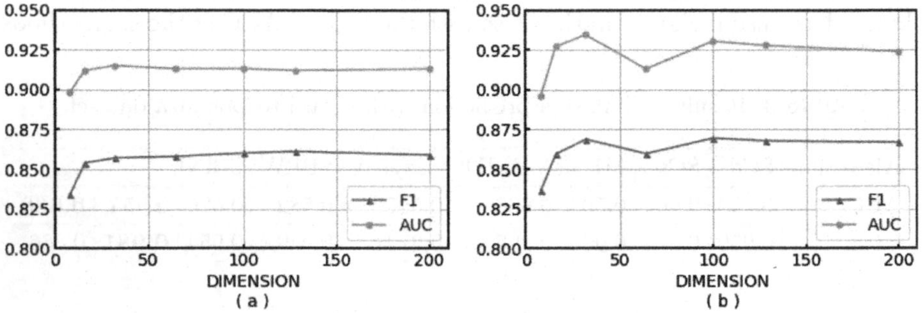

Fig. 4. The impact of embedding dimension d on SBiNE for link sign prediction

Table 4. The sensitivity of SBiNE to β in link sign prediction

Dataset	Metric	β				
		0	0.05	0.1	0.5	1
Bonanza	AUC	0.948	**0.954**	0.716	0.624	0.630
	F1	0.595	0.626	0.660	0.693	**0.755**
U.S.Senate	AUC	0.913	**0.915**	0.912	0.911	0.510
	F1	0.855	**0.857**	0.854	0.854	0.711
U.S.House	AUC	0.933	**0.934**	0.933	0.856	0.507
	F1	0.866	**0.869**	0.867	0.797	0.691

6 Conclusion

In this paper, we are committed to applying network representation learning to a special network called signed bipartite network. Furthermore, we introduce a new object function suitable for signed bipartite network embedding and propose a simple framework SBiNE to optimize it. According to the analysis of link sign prediction on the above three signed bipartite networks, we get the following two conclusions. First, the learned low-dimensional node vectors can effectively preserve the original structural topology. Through the construction of our framework SBiNE, the Hadamard product of the learned node embeddings can significantly distinguish the positive and negative links compared to [26]. Structural balance theory [27] in signed social networks has been extensively applied to improve the performance across measuring [28] and mining applications [29]. However, we can't find the trace of the triangle in signed bipartite networks due to the existence of heterogeneous nodes. In our future work, we decide to adopt balance in signed bipartite networks [26] for signed bipartite network embedding analysis.

Acknowledgement. This work was supported by the National Key Research and Development Program of China (No. 2019YFB1704101), the National Natural Science Foundation of China (no. 61872002U1936220) and the Natural Science Foundation of Anhui Province of China (no. 1808085MF197).

References

1. Derr, T., Tang, J.: Congressional vote analysis using signed networks. In: 2018 IEEE International Conference on Data Mining Workshops (ICDMW), pp. 1501–1502. IEEE (2018)
2. Leskovec, J., Huttenlocher, D., Kleinberg, J.: Predicting positive and negative links in online social networks. In: Proceedings of the 19th International Conference on World Wide Web, pp. 641–650 (2010)
3. Ma, H., Lyu, M.R., King, I.: Learning to recommend with trust and distrust relationships. In: Proceedings of the Third ACM Conference on Recommender Systems, pp. 189–196 (2009)
4. Bhagat, S., Cormode, G., Muthukrishnan, S.: Node classification in social networks. In: Aggarwal, C. (ed.) Social Network Data Analytics, pp. 115–148. Springer, Boston (2011). https://doi.org/10.1007/978-1-4419-8462-3_5
5. Liben-Nowell, D., Kleinberg, J.: The link-prediction problem for social networks. J. Am. Soc. Inform. Sci. Technol. **58**(7), 1019–1031 (2007)
6. Zhang, Y., et al.: Covering-based web service quality prediction via neighborhood-aware matrix factorization. IEEE Trans. Serv. Comput. (2019)
7. Zhang, Y., Cui, G., Deng, S., Chen, F., Wang, Y., He, Q.: Efficient query of quality correlation for service composition. IEEE Trans. Serv. Comput. (2018)
8. Zhang, Y., Yin, C., Wu, Q., He, Q., Zhu, H.: Location-aware deep collaborative filtering for service recommendation. IEEE Trans. Syst. Man Cybern.: Syst. (2019)
9. Zhang, Y., Zhou, Y., Wang, F., Sun, Z., He, Q.: Service recommendation based on quotient space granularity analysis and covering algorithm on spark. Knowl.-Based Syst. **147**, 25–35 (2018)
10. Maaten, L., Hinton, G.: Visualizing data using t-SNE. J. Mach. Learn. Res. **9**(Nov), 2579–2605 (2008)
11. Wang, S., Tang, J., Aggarwal, C., Chang, Y., Liu, H.: Signed network embedding in social media. In: Proceedings of the 2017 SIAM International Conference on Data Mining, pp. 327–335. SIAM (2017)
12. Tang, J., Qu, M., Wang, M., Zhang, M., Yan, J., Mei, Q.: Line: Large-scale information network embedding. In: Proceedings of the 24th International Conference on World Wide Web, pp. 1067–1077 (2015)

13. Gao, M., Chen, L., He, X., Zhou, A.: Bine: Bipartite network embedding. In: The 41st International ACM SIGIR Conference on Research & Development in Information Retrieval, pp. 715–724 (2018)
14. Belkin, M., Niyogi, P.: Laplacian eigenmaps and spectral techniques for embedding and clustering. In: Advances in Neural Information Processing Systems, pp. 585–591 (2002)
15. Perozzi, B., Al-Rfou, R., Skiena, S.: Deepwalk: online learning of social representations. In: Proceedings of the 20th ACM SIGKDD International Conference on Knowledge Discovery and Data Mining, pp. 701–710 (2014)
16. Cao, S., Lu, W., Xu, Q.: GraRep: learning graph representations with global structural information. In: Proceedings of the 24th ACM International on Conference on Information and Knowledge Management, pp. 891–900 (2015)
17. Lu, Y., Shi, C., Hu, L., Liu, Z.: Relation structure-aware heterogeneous information network embedding. In: Proceedings of the AAAI Conference on Artificial Intelligence, vol. 33, pp. 4456–4463 (2019)
18. Wang, S., Aggarwal, C., Tang, J., Liu, H.: Attributed signed network embedding. In: Proceedings of the 2017 ACM on Conference on Information and Knowledge Management, pp. 137–146 (2017)
19. Grover, A., Leskovec, J.: node2vec: Scalable feature learning for networks. In: Proceedings of the 22nd ACM SIGKDD international Conference on Knowledge Discovery and Data Mining, pp. 855–864 (2016)
20. Ma, Y., Ren, Z., Jiang, Z., Tang, J., Yin, D.: Multi-dimensional network embedding with hierarchical structure. In: Proceedings of the Eleventh ACM International Conference on Web Search and Data Mining, pp. 387–395 (2018)
21. Zhao, Y., Liu, Z., Sun, M.: Representation learning for measuring entity relatedness with rich information. In: Twenty-Fourth International Joint Conference on Artificial Intelligence (2015)
22. Wang, Q., Wang, Z., Ye, X.: Equivalence between line and matrix factorization. arXiv preprint arXiv:1707.05926 (2017)
23. Church, K.W., Hanks, P.: Word association norms, mutual information, and lexicography. Comput. Linguist. **16**(1), 22–29 (1990)
24. Song, W., Wang, S., Yang, B., Lu, Y., Zhao, X., Liu, X.: Learning node and edge embeddings for signed networks. Neurocomputing **319**, 42–54 (2018)
25. Bengio, Y., Courville, A., Vincent, P.: Representation learning: a review and new perspectives. IEEE Trans. Pattern Anal. Mach. Intell. **35**(8), 1798–1828 (2013)
26. Derr, T., Johnson, C., Chang, Y., Tang, J.: Balance in signed bipartite networks. In: Proceedings of the 28th ACM International Conference on Information and Knowledge Management, pp. 1221–1230 (2019)
27. Cygan, M., Pilipczuk, M., Pilipczuk, M., Wojtaszczyk, J.O.: Sitting closer to friends than enemies, revisited. In: Rovan, B., Sassone, V., Widmayer, P. (eds.) MFCS 2012. LNCS, vol. 7464, pp. 296–307. Springer, Heidelberg (2012). https://doi.org/10.1007/978-3-642-32589-2_28
28. Derr, T., Aggarwal, C., Tang, J.: Signed network modeling based on structural balance theory. In: Proceedings of the 27th ACM International Conference on Information and Knowledge Management, pp. 557–566 (2018)
29. Chiang, K.Y., Natarajan, N., Tewari, A., Dhillon, I.S.: Exploiting longer cycles for link prediction in signed networks. In: Proceedings of the 20th ACM International Conference on Information and Knowledge Management, pp. 1157–1162 (2011)

Usable and Secure Pairing Based on Handshake for Wrist-Worn Smart Devices on Different Users

Xiaohan Huang[1], Guichuan Zhao[1], Qi Jiang[1,2(✉)], Xindi Ma[1], Youliang Tian[3], and Jianfeng Ma[1]

[1] School of Cyber Engineering, Xidian University, Xi'an 710071, China
609117168@qq.com, 1078161458@qq.com, {jiangqixdu, xdma}@xidian.edu.cn, jfma@mail.xidian.edu.cn

[2] Peng Cheng Laboratory, Network Communication Research Centre, Shenzhen 518055, China

[3] College of Computer Science and Technology, Guizhou University, Guiyang, China
youliangtian@163.com

Abstract. Wrist-worn smart devices are being used to share various sensitive personal information in various fields such as social, medical, sports, etc. Secure pairing establishing a trusted channel between the involved devices is a prerequisite to ensure data transmission security. Handshake has been proposed to realize secure pairing between devices worn by different users without pre-shared knowledge, the participation of third parties and complex user interactions. However, existing schemes cannot meet the practical requirement in terms of time delay and security. In this paper, we proposed a feasible handshake based secure pairing scheme, which utilizes the handshake acceleration data. Specifically, we quantify the features of acceleration data through random threshold values, to shorten the handshake time required for guaranteeing the length of the negotiated key. Besides, we propose an optimal feature selection algorithm that improves the success rate and security of the system. What's more, security analysis indicates that our solution can resist man-in-the-middle attacks. Experiments are performed on our scheme, which show that the proposed scheme is robust and secure. Users only need to take a few seconds to perform simple operations, and the devices can automatically pair securely.

Keywords: Handshake · Secure pairing · Smart wearable devices · Acceleration · Key negotiation

1 Introduction

Nowadays, wrist-worn smart devices are ubiquitous in our lives, the embedding of sensors (e.g., accelerometer, gyroscope and heartbeat detector) enable them widely used in health monitoring, activity recognition, and personal assistance. In some social occasions, an increasing number of people use their wrist-worn smart devices to sharing various personal information, such as business cards, music, and personal pictures, etc.

H. Gao et al. (Eds.): CollaborateCom 2020, LNICST 349, pp. 493–510, 2021.
https://doi.org/10.1007/978-3-030-67537-0_30

A typical application scenario of employing wrist-worn devices for information sharing is shown in Fig. 1. Users wearing the devices are considered to be in secure domain, and they share data through wireless channels [1], such as Bluetooth and Wi-Fi.

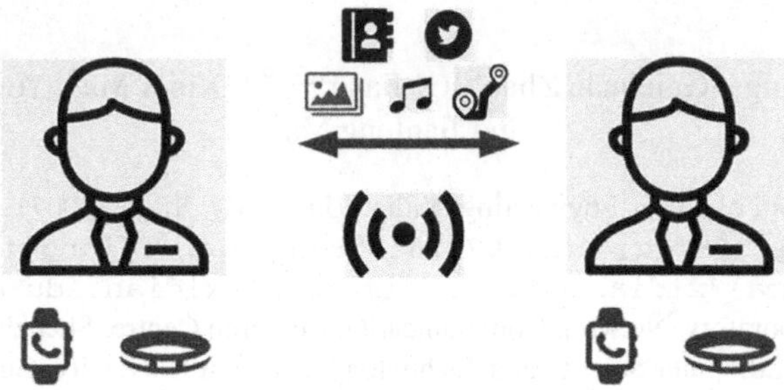

Fig. 1. The scenario of information sharing.

Unfortunately, the wireless channels for information sharing is inherently open due to its public nature, which could cause various types of attacks such as eavesdropping [2], tampering, and man-in-the-middle (MITM) attack [3–5]. Therefore, to share personal information securely, secure paring between two devices that have never known each other before is an urgent requirement, based on which a secure communication channel can be established to transmit sensitive information [6]. However, it faces the following challenges to fulfill secure pairing in this setting. First, two pairing devices on different users do not have any prior security context or a common point of trust, making the authentication between them difficult to achieve. Second, as for the users, almost zero effort should be required to complete secure pairing [1], since complicate user operations may cause users without security consciousness to skip the secure pairing procedure. Third, the energy consumption and time delay should be minimized to be fit the resource-constrained wrist-worn smart devices.

A number of secure pairing solutions have been proposed [7–12], while few of existing solutions caters to the scenario shown in Fig. 1, which is the concern of this paper. Although several works tried to address this issue [13–16], none of them satisfies both security and usability required by practical use, making them still have some distance from an ideal secure pairing scheme for wrist-worn devices on different users.

Handshake, a common form of physical contact between human beings, is one of the most promising solution to achieve secure pairing in this scenario. Handshake-based schemes usually takes handshake acceleration as the common input for secure pairing, however, existing methods cannot meet both security and usability requirements. For example, the scheme in [17] and [18] both have good usability, while their auxiliary data used for the key negotiating are not secure enough, which may reveal the secret key information. The scheme in [19] guarantees the security, while its key generation rate is low, requiring longer user handshake time, leading to poor usability.

Therefore, in this paper, we propose a novel handshake-based secure pairing scheme which achieves both the security and usability. Specifically, we obtain optimized features from the accelerometer data through feature extraction and feature selection, from which we extract the similar witness and negotiate a symmetric key through fuzzy commitment, implementing secure pairing between the devices. In summary, the main contributions of this paper are outlined as follows.

- We propose a secure pairing scheme based on handshake acceleration, which enables secure pairing between wrist-worn smart wearable devices equipped by different individuals.
- We quantify the acceleration features through random threshold values, which increases the rate of witness generation thanks to making full use of the information contained in the acceleration data.
- We sort the acceleration features based on Euclidean distance and propose the optimal feature selection algorithm, which makes the witness generated by the two devices has more similarity. In addition, to ensure that the auxiliary data will not leak the acceleration information, we introduce the chaff features in the optimal feature selection algorithm.
- We evaluate the performance and security of our scheme by experiments and theoretical analysis. The results show that the proposed wrist-worn smart devices pairing scheme is robust and is able to resist both the passive attacks and active attacks. The key generation rate can reach 87bit/s, it only takes about three seconds to perform secure pairing.

The remainder of this paper is organized as follows: Sect. 2 reviews the related work. Section 3 introduces the required preliminary. Section 4 provides the details of our proposed scheme. In Sect. 5, the optimal feature selection algorithm is detailed. We evaluate the system by evaluating its performance in Sect. 6 and analyzing its security in Sect. 7, followed by the conclusion in Sect. 8.

2 Related Work

Most wearable devices today transmit data via short-range wireless communication technologies such as Bluetooth, Wi-Fi, etc., through which the data is easily leaked and tampered with [20]. Security is a matter of concern for information transmissions between them [21]. A simple method for pairing two devices is to have the users to enter the same password on both devices. However, it is shown that the passwords chosen by users are generally easy to guess. Another method is that, a random number is generated and displayed on the output interface of one device, then it is typed by the user on another device to be paired with [22]. This method is vulnerable to shoulder attacks and lacks user friendliness [23].

To avoid above problems, similar data matching based secure paring schemes have been explored, in which sensors embedded in devices are used to collect common context generated by users [24, 25]. Since shaking is a common behavior which can generate the same context between pairing devices, shaking based device pairing schemes [10, 11] have been proposed, in which the acceleration generated by shaking the devices simultaneously is employed to pair the devices. Specifically, the acceleration signals produced by the shaking patterns are pseudo-random, unique, and difficult to reproduce. Smart-Its

Friends [11] is the first shaking based scheme, which simply uses context matching to pair devices. However, it is vulnerable to MITM, as the context is transmitted in plaintext. ShaVe [10], which extends Smart-Its Friends, combines the session key generated by DH exchange protocol with the accelerometer data collected by shaking to form an acceleration time series, which are compared by both sides to fulfill secure pairing. However, the scheme entails much computational cost due to the encryption operations involved, and is still subject to MITM. Hence, secure pairing based on common context comparisons is not immune to attacks. To address this issue, shaking motion based secure pairing schemes [8, 10] have been proposed, which can not only accomplish device pairing, but also ensure secure transmission subsequently. ShaCK [10] and [8] extract features from the three-dimensional acceleration signal generated by the movement pattern of synchronous shaking devices, which are used to generate symmetric keys. However, they have security issues and are not suitable for the scenario shown in Fig. 1.

Recently, inspired by shaking based secure pairing, in [17–19], the device motion pattern generated by the handshake are used to establish a secure channel between devices, which is attractive for secure pairing between wrist-worn devices worn by different users. Regrettably, existing schemes still cannot satisfy both security and usability required by practical use. In [17], it is proposed to detect the handshake action in real time by extracting acceleration, gyroscope from multiple sensors, which is of high computation cost. Besides, check bits of the Hamming code and parity digit are used as auxiliary data to reconcile a session key, which will cause the leakage of the key information, reducing the security of the scheme. In [18], principal component analysis (PCA) is introduced to reduce the dimension of raw acceleration data. However, its key generation rate is low, since the processing of ambiguous bits in this procedure increases the pairing time. In [19], the perturbation vector based fuzzy cryptography (PVFC) is proposed to reduce the computing overhead. However, the method in [19] has low usability due to the long time delay incurred by the feature collection procedure.

3 Preliminaries

In this section, we introduce the important preliminaries of the proposed scheme.

3.1 Euclidean Distance and Hamming Distance

We first introduce two different types of distance used in this paper, Euclidean Distance and Hamming Distance.

Euclidean Distance refers to the straight-line distance between two points in Euclidean space. In this paper, we denote the Euclidean distance between two points x and y as $\Delta(x, y)$, which can be calculated by formula (1), where $|\cdot|$ represents the absolute value.

$$\Delta(x, y) = |x - y| \tag{1}$$

Hamming distance is a concept that represents the number of different bits corresponding to two bit strings of the same length. In this paper, we use $dis(str1, str2)$ to represent the Hamming distance between two bit strings $str1$, $str2$.

3.2 Fuzzy Commitment

Fuzzy commitment [26, 27], the combination of error correcting codes and cryptography, is a useful primitive for biometric authentication. In this paper, we utilize the Bose-Chaudhuri-Hocquenghem (BCH) code [28].

We denote the function of fuzzy commitment as $fc(\cdot)$, as shown in formula (2), where k is a randomly selected key, $h(\cdot)$ represents the anti-collision hash function.

$$fc(w, k) = (h(k), X) \tag{2}$$

X can be obtained by formula (3),

$$X = (c \oplus w) \tag{3}$$

where $\oplus$ represents XOR operation, and c is obtained by BCH encoding of k.

To decommit, c' is obtained from w' and X first (as shown in formula (4)), then k' can be recovered from c' by BCH decoding. Finally, the values of $h(k)$ and $h(k')$ are compared to confirm whether k is successfully restored.

$$c' = (X \oplus w') \tag{4}$$

If $dis(w, w')$ is lower than the error tolerance of the BCH code, then k can be recovered by w'.

4 Handshake Based Secure Pairing

In this section, we first give the overall system processes of the protocol, and then introduce the details of our handshake based secure pairing scheme. The devices on two sides are denoted as A and B.

The protocol includes four phases: data preprocessing, feature selection and reconciliation, witness generation and key binding, and key verification. In the data preprocessing phase, the acceleration data collected by the device is processed to reduce its dimension. In the feature selection and reconciliation phase, the optimal feature selection algorithm is employed to select the reliable features which are used to generate witness. In the stage of witness generation and key binding, the witness is generated by the extraction algorithm, then is bound with k to obtain auxiliary data. In the key verification phase, the device recovers the original key through the auxiliary data, then determines whether the pairing is successful by comparing the hash value of the key.

4.1 Data Preprocessing

In order to detect the start of the handshake accurately, users are required to press a button on the wrist-worn devices to indicate that it is about to start handshaking and devices pairing. Besides, 1–2 s of quiescence is needed before the handshake. It is worthy to note that these additional interaction can greatly simplify the detection of the start of the handshake and avoid other unnecessary monitoring process. Moreover, the required interactions are very simple and in line with the user's usage habits. After the data is collected, dimensionality reduction and synchronization are performed to improve the probability of successful pairing.

In our scheme, the three-dimension accelerometer data generated by the handshake is used. Since the positions of the two devices worn on users' wrists vary from person to person, the collected three-dimension acceleration time series lack spatial alignment and cannot be compared directly. To this end, the root mean square of the X-axis, Y-axis, and Z-axis of accelerometer data (see Fig. 2(a)) is calculate to reduce the three dimension data to one dimension as shown in Fig. 2(b). We refer to the data obtained after processing as acceleration magnitude.

Furthermore, the accelerometer data, which are collected by devices independently, needs to be synchronized. As shown in Fig. 2(a), the handshake acceleration exhibits periodicity. Fig. 2(b) shows the acceleration magnitude, from which we can obviously find that the magnitude of acceleration fluctuates significantly. As it is verified that the time when the two sides generated data with almost zero acceleration magnitude is very close in [19], we take the intersection of the axis $x = 0$ and the acceleration magnitude curve of the first complete period (i.e., the acceleration magnitude in red arrow in the Fig. 2(b)) as the starting point. In this paper, the acceleration magnitude time series after synchronization is used as the feature sequence f.

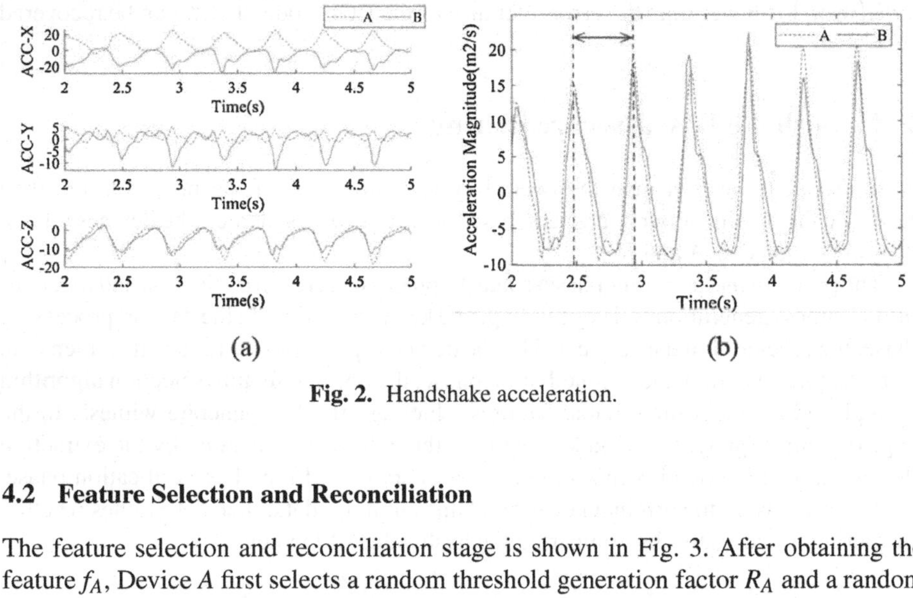

Fig. 2. Handshake acceleration.

4.2 Feature Selection and Reconciliation

The feature selection and reconciliation stage is shown in Fig. 3. After obtaining the feature f_A, Device A first selects a random threshold generation factor R_A and a random key k_A. Then, R_A and its identity ID_A are sent to device B. Then both sides start to agree on the optimal features, which is detailed in Sect. 5. Device A uses R_A and R_B respectively to calculate its own optimal feature indexes $I_{R_A}^A$ and $I_{R_B}^A$, and sends them to device B. After receiving the $I_{R_A}^B$, $I_{R_B}^B$ sent by device B, device A compares the preselected indexes I_R generated by both devices to obtain $I_R{}^*$, that is $I_{R_A}^* \leftarrow I_{R_A}^A \cap I_{R_A}^B$, $I_{R_B}^* \leftarrow I_{R_B}^A \cap I_{R_B}^B$. $x \cap y$ represents the intersection of x and y.

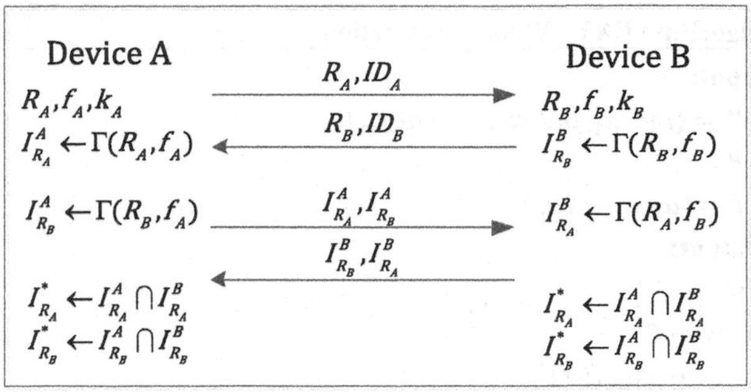

Fig. 3. Feature selection and reconciliation.

4.3 Witness Generation and Key Binding

The witness generation and key binding stage is given as follows, as shown in Fig. 4.

1. Device A calculates the witness w_{R_A}, w'_{R_B} respectively based on the preselected feature indexes $I^*_{R_A}$, $I^*_{R_B}$ obtained in the reconciliation phase, where $w_{R_A} \leftarrow EXT(I^*_{R_A}, R_A, f_A)$, $w'_{R_B} \leftarrow EXT(I^*_{R_B}, R_B, f_A)$. The witness generation algorithm EXT is shown in Fig. 5. At the beginning, w is set as an empty bit string $\varnothing$., and the fixed random number generation algorithm *rand* with the seed R is executed to generate random threshold sequence value $th = \{th_1, th_2, \ldots, th_n\}$. Next, the witness is gained by comparing the random threshold value with the feature sequence. If the feature value α_{index_i} is larger than the random threshold value th_{index_i}, a bit 1 is generated; otherwise, a bit 0 is generated, where $index_i$ is an element in I^*_{RA} or I^*_{RB} and the α_{index_i} is the value in f_A, | means bit concatenation. Finally, the witness w is compressed into a string of length *len*.

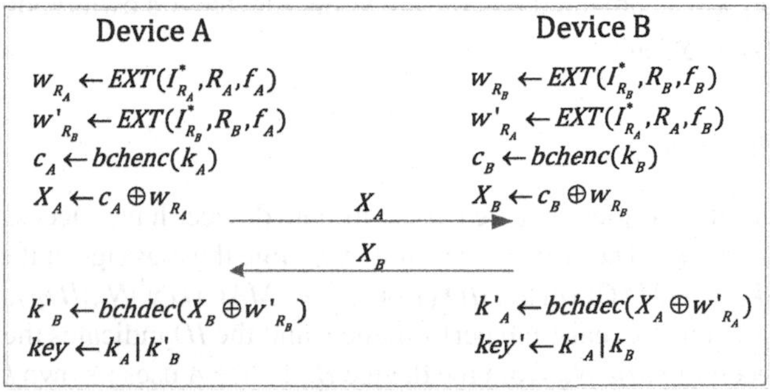

Fig. 4. Witness generation and key binding.

Algorithm EXT Witness Generation

Input:

$I^* = \{index_1, index_2, \ldots, index_m\}$

R

$f = \{\alpha_1, \alpha_2, \ldots, \alpha_n\}$

Output:

w

$w \leftarrow \emptyset$

$th \leftarrow rand(R)$

for $i = 0$ to m do

  if $th_{index\ i} - \alpha_{index\ i} \geq 0$ then

    $w \leftarrow w|1$

  else

    $w \leftarrow w|0$

  end if

end for

$w \leftarrow w[1{:}\,len] \oplus w[len + 1{:}\,length(w)|1^{2*len - length(w)}]$

Fig. 5. Witness generation algorithm.

2. Device A calculates c_A by encoding the random key k_A with the BCH code, that is, $c_A \leftarrow bchenc(k_A)$, where $bchenc(\cdot)$ represents the BCH encoding function. The length of the information bits of the BCH code is equal to the length of k_A.
3. Device A generates auxiliary data X_A by binding the w_A with c_A, i.e., $X_A \leftarrow c_A \oplus w_A$, then sends the X_A to device B.
4. After receiving the X_B sent by device B, device A use w'_{R_B} generated in (1) to recover the random key k_B selected by the device B, that is $k'_B \leftarrow bchdec(X_B \oplus w'_{R_B})$, where $bchdec(\cdot)$ is the BCH decoding function.
5. The final key key is obtained by concatenating Device A's own random key k_A with the recovered key k'_B.

4.4 Key Verification

In the final phase, it is verified that the two legitimate devices have successfully extracted the same secret key key. The two devices first exchange the message authentication code (MAC), i.e., $M_A \leftarrow MAC(key|N_A|ID_A)$, $M_B \leftarrow MAC(key'|N_B|ID_B)$, where N is a random number to ensure real-time performance, and the ID indicates the identity of the device. After receiving M_B, N_B from the device B, device A uses its own key to calculate the $M'_B \leftarrow MAC(key|N_B|ID_B)$, then verifies whether $M'_B = M_B$. If the equation holds, the pairing is successful.

5 Optimal Feature Selection

After data processing (as described in Sect. 4.1), the noise in the acceleration data will affect the success rate of device pairing. Therefore, we propose the optimal feature selection algorithm, which can select reliable features from the feature sequence f for device pairing. In our algorithm, first the generated raw features are sorted, then the optimal features are selected from the sorted features, and finally the index I_R of the optimal features are sent as the auxiliary data to the pairing device.

5.1 Sort Based on Euclidean Distance

We find that the larger $\Delta(\alpha_i, th_i)$ is, the more robust the bit generated from the feature through experiments. Hence, sorting the Euclidean distance $\Delta(\alpha_i, th_i)$ of each feature in ascending order and then selecting those later features (features with larger Euclidean distance) is a feasible method to obtain reliable features. However, just sorting by $\Delta(\alpha_i, th_i)$ to select optimal features is not enough, since the sorting reliability can be still affected by the slight delay of one of the devices. In order to reduce the bias caused by delay, we propose an algorithm called minimum distance within the window. Specifically, we call $window_i = [\alpha_{i-\delta}, \ldots, \alpha_i, \ldots, \alpha_{i+\delta}]$ as a window of feature α_i with the size δ, and $min(\Delta(window_i, th_i))$ is the smallest α_i of $\Delta(\alpha_i, th_i)$ corresponding to $window_i$. We do not directly take $\Delta(\alpha_i, th_i)$ as the sorting criteria, but introduce the values δ before and after α_i to reduce the misleading caused by the slight lag, and use $min(\Delta(window_i, th_i))$ as the standard to sort.

5.2 Secure Feature Selection

To filter out the noisy features that are most likely to influence the success rate of device pairing, we propose the optimal feature selection algorithm Γ (see Fig. 6), where n is defined as the total number of original features, and q is defined as the number of optimal features selected by Γ, the corresponding optimal feature sequence is defined as Q.

We take the index i of each element in f and the corresponding $min(\Delta(window_i, th_i))$ as a key-value tuple $(i, min(\Delta(window_i, th_i)))$, and put them into a list, that is $add(list, (i, min(\Delta(window_i, th_i))))$, then the $list$ is sorted by $sort(\cdot)$ algorithm. As $min(\Delta(window_i, th_i))$ becomes smaller, the probability of the offset occurring between the α_i of devices will increase. We filter the feature sequence and select the last q key-value tuples in the sorted $list$ as the feature sequence Q for generating w. However, the direct exchange of the indexes i of the last q key-value tuples will expose the information of f, so we protect f by deleting some tuples in Q randomly, and we call these deleted tuples as chaff features. The algorithm $genchaff(\cdot)$ selects chaff features randomly from a sequence. In order to ensure that the minimum number of optimal features indexes selected by the devices on both sides is not less than len, that is, $Count(U_A \cap U_B) > len$, where $Count(U_A)$ and $Count(U_B)$ represent the number of features finally selected by users A and B, which is calculated by $q - cn$. Then $Count(U_A \cap U_B) > len$ can be simplified as:

$$Count(U_A)/Count(U_B) > \frac{n + len}{2} \tag{5}$$

and the value range of cn can be obtained:

$$cn < q - \frac{\mathrm{n} + \mathrm{len}}{2} \tag{6}$$

Finally, we arrange the key values of the tuples in Q in a sequential manner as the preselected feature indexes I_R.

Algorithm Γ Optimal feature selection

Input:

R

$f = \{\alpha_1, \alpha_2, \ldots, \alpha_n\}$

Output:

I_R

$th \leftarrow rand(R)$

$list \leftarrow \emptyset$

for $i = 0$ to n do

  $add(list, (i, min(\Delta(window_i, th_i))))$

end

$sort(list)$

$Q \leftarrow list(n - q + 1{:}n)$

$chaff \leftarrow genchaff(Q)$

$Q \leftarrow Q \backslash chaff$

$I_R = getkeys(Q)$

Fig. 6. Optimal feature selection algorithm.

6 Performance Evaluation

6.1 Experiment Setup

In our experiments, the iPhones (iPhone6-CPU: A8 1.4 GHz, OS: iOS 11.0 and iPhone8-CPU: A11 2.74 GHz, OS: iOS 11.0) is tied to the volunteers' wrist to simulate the wrist-worn devices. We recruited 16 volunteers, containing 8 males and 8 females. Their ages are ranging from 22 to 25. In the following experiments, the handshake data is generated by volunteers shaking hands in pairs (one volunteer will shake with the remaining 15 volunteers), each handshake including 16 ups and downs and is repeated 5 times to ensure the accuracy of the data, the accelerometer sampling frequency is set to be 200 Hz. It is worth noting that the pairing of devices through a short-time handshake is user-friendliness. In our scheme, a handshake time of 3 to 4 s is enough to complete the pairing, which is convenient to user.

6.2 Witness Similarity

The success rate of secure pairing is directly related to the similarity of the witness w and w', which are generated from the optimal features. In this section, we first evaluate the sorting based on the minimum distance within the window, and then experimentally analyze the effect of window size δ and q on the similarity of the witness. In our scheme, we sort the features in ascending order based on $min(\Delta(window_i, th_i))$.

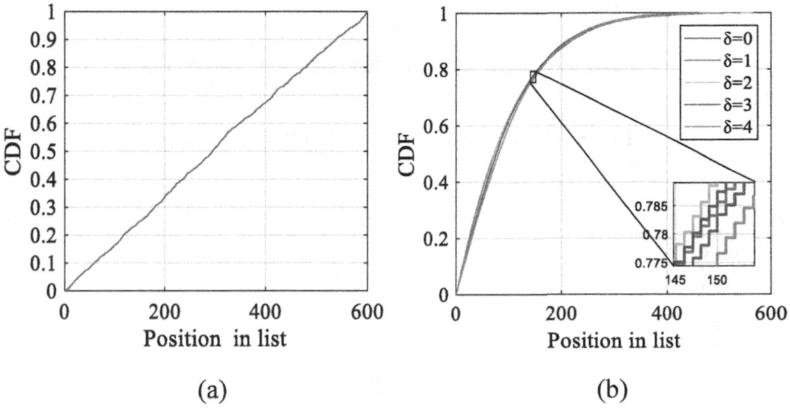

Fig. 7. CDF of indexes of the offset features.

Definition 1: If th_i and its corresponding feature values $\alpha_{A,i}$, $\alpha_{B,i}$ satisfy $(th_i - \alpha_{A,i}) * (th_i - \alpha_{B,i}) \leq 0$, we deem the index i of the th_i as the index of the offset feature.

The first 3 s of the synchronized acceleration magnitude is used as original feature data in our experiments, from which 600 acceleration magnitude features will be generated. The distribution of the indexes of the offset features is shown in Fig. 7. Fig. 7(a) shows the cumulative distribution function (CDF) of indexes of the offset features of the unsorted feature sequence, and Fig. 7(b) shows that of the sorted feature sequence. As can be seen, the indexes of the offset features are uniformly distributed before sorted, are mostly distributed at the front of the sequence after sorted. The smaller the index i is, the faster the CDF grows, and 80% of the indexes of the offset feature are ranging from 0 to 150. Besides, it can be seen from the partially enlarged diagram of Fig. 7(b) that in the range around 150, the CDF performance is better when $\delta = 2$.

In optimal feature selection algorithm, the window size δ and q are two important parameters which affect our feature selection directly, thus affecting the similarity of the witness w. We evaluate the similarity of the witness by changing the two important parameters, and the similarity of the two witness is measured by Hamming distance. Fig. 8 shows the variation of the average value of the $dis(w, w')$ under different q and δ, from which we can see that as the number of selected feature q decreases, $dis(w, w')$ decreases exponentially. When q is less than 320, $dis(w, w')$ approaches 0, while the feature number q is selected to be less than 500, $dis(w, w')$ becomes smaller and the decreasing tendency slows down, which is less than 7. In addition, when q is taken around 500, the $dis(w, w')$ is the smallest when $\delta = 2$.

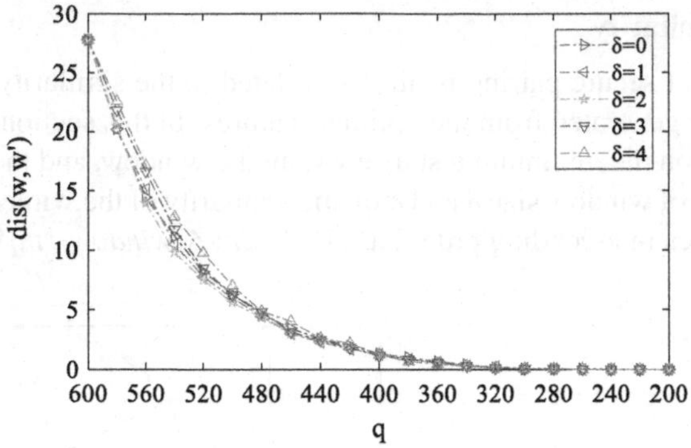

Fig. 8. Average $dis(w, w')$ with different q and δ.

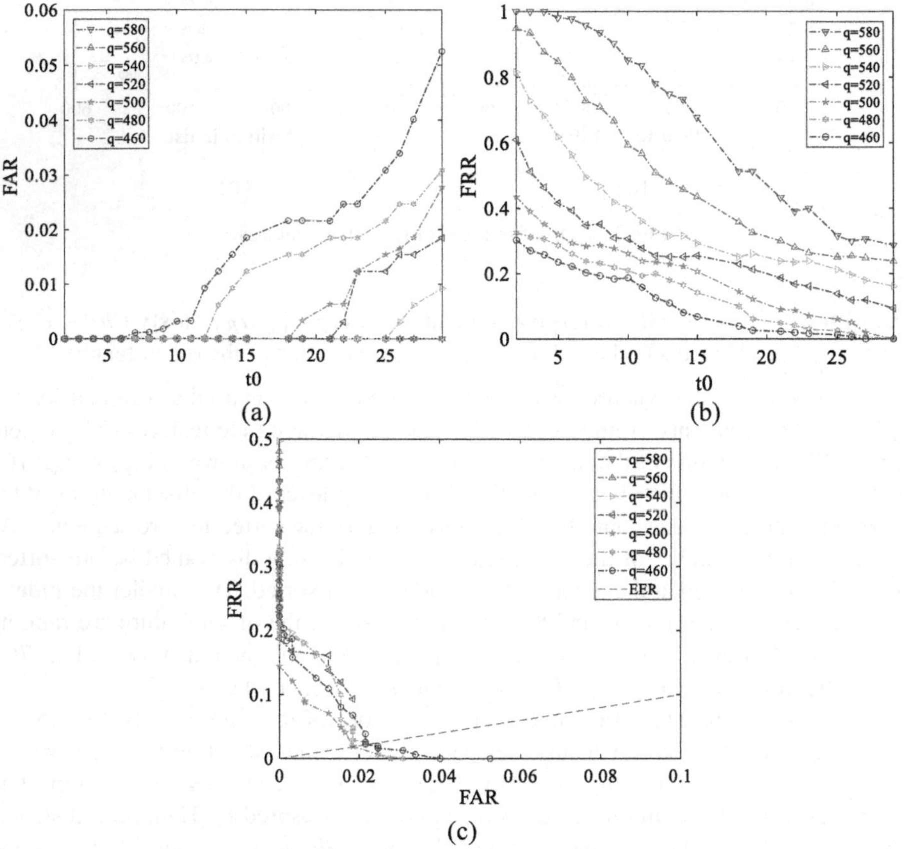

Fig. 9. FRR, FAR, EER of the scheme with different q and $t0$.

6.3 System Accuracy

System accuracy is the ability of the system to pair two devices correctly. It requires that only devices worn by users who shake hands with each other can be paired and generate the same symmetric key.

In our scheme, there are two main factor impacting system accuracy. One is the optimal feature selection algorithm, the second is the parameter of the BCH code. Since the feature selection algorithm directly affects the similarity of witness, and the similarity of witness is related to the accuracy of the BCH code, we evaluate the joint impact of q and different BCH codes of the system accuracy. We denote a BCH code as $(ln, lk, t0)$ where ln represents the total length of the BCH code, lk represents the length of the information bits, and $t0$ represents the capability of noise tolerance.

The false rejection rate (FRR) and false acceptance rate (FAR) are used to evaluate the system accuracy. To ensure that the key can provide high security and filter out as much noise data as possible, we set the ln of a BCH code as 255.

In the experiment, the main concern is the variation of FRR and FAR and their correlation when different BCH codes are selected. We can see from Fig. 9(a) that the higher the value $t0$ is, the lower the FRR is. Conversely, from Fig. 9(b), we can see that the higher the value $t0$ is, the higher the FAR is. Noted that if the system accuracy is needed to be ensured, FAR and FRR must be reduced at the same time. As shown in Fig. 9(c), we can find an equilibrium point, i.e., equal error probability (EER) when $q = 500$.

6.4 Key Generation Rate

In this section, we evaluate the key generation rate and compare it with the state-of-the-art of acceleration-based pairing schemes, as shown in Table 1. In our scheme, the BCH code we choose is (255, 18, 131) and q is set as 100, as our scheme has better accuracy in this case. As can be seen in Table 1, our key generation rate is higher than other schemes, this shows that our solution quickly complete device pairing and session key negotiation while ensuring accuracy, which is more usable than other acceleration-based schemes.

Table 1. Key generation rate

Scheme	Generation rate/s
Walkie-talkie [12]	26/s
Inter-Pulse-Interval [9]	2 ~ 16/s
Bandana [7]	1 ~ 2/s
Shake to Communicate [19]	16 ~ 32/s
Shake-n-Shack [18]	25 ~ 48/s
SDP via Handshake [17]	80/s
Our proposed scheme	87/s

7 Security Analysis

In this paper, we consider two types of attackers, passive attackers and active attackers. Specifically, active attackers are divided into two classed: one is a MITM attacker who attempts to pair with a legitimate device by intercepting and tampering with the messages; another one is to obtain useful information by observing the handshake between two legitimate users. Passive attackers try to crack the pairing key by eavesdropping and monitoring the information transmitted in the open channel.

In order to evaluate the impact of the two attackers on the security of our system, we constructed experimental data sets for them respectively. The active attacker data set is generated by two illegal volunteers simulating the handshake of legitimate volunteers, and the passive attacker data set is the auxiliary data transmitted between legal devices during the pairing process.

7.1 Witness Security

To evaluate the security of w, we apply Shannon entropy to measure the randomness of w. Since w is a bit string of 0 and 1, the entropy of each bit is between 0 and 1. Fig. 10 shows the CDF of the average entropy of each bit in w. It can be seen that the average entropy of each bit is close to 1. For each set of experiments, the average Shannon entropy is greater than 0.92 and the largest one is up to 0.982. For the 255-bit witness, the entropy contained in w can be calculated by the following formula (7), which is 250.41. Therefore, we demonstrate that the generated w is secure.

$$\mathrm{E} = \sum\nolimits_{i=0}^{m} e_i \tag{7}$$

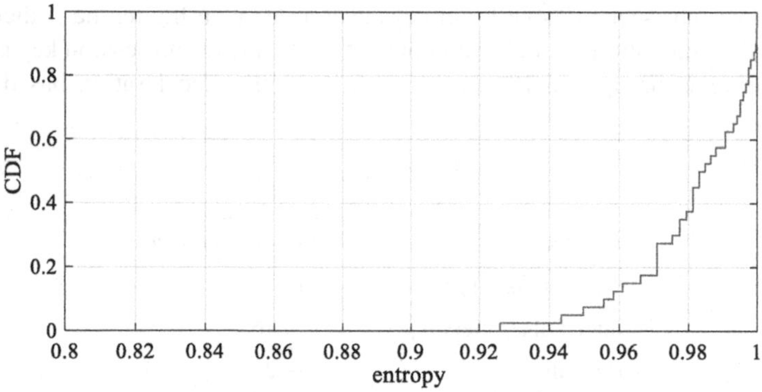

Fig. 10. CDF of the entropy of w.

7.2 Auxiliary Data Security

It is required that the transmitted auxiliary data I_R and X cannot reveal the information of the feature and the negotiated key, otherwise, it will be used by passive attackers.

In our scheme, we randomly delete a certain number of chaff features from the optimal q features, which makes it difficult for the attackers to distinguish the indexes of features that are close to the random threshold values. To guess the acceleration data from I_R, the attackers must identify the hidden chaff features from the missing indexes. The probability of finding all cn chaff features $P_{chafftuple}(cn)$ is obtained as follows:

$$P_{chafftuple}(cn) = 1/C_{n-q+cn}^{cn} \tag{8}$$

To find all the chaff features, the attacker needs at most $1/P_{chafftuple}(cn)$ attempts, so we define the security $S_{chaff}(cn)$ guaranteed by the chaff tuple as shown in formula (9).

$$S_{chaff}(cn) = -log_2 P_{chafftuple}(cn) \tag{9}$$

Take the BCH length *len* as 255 as an example. Fig. 11 shows the security that the chaff features can guarantee under different number of optimal features q. It can be seen that when $q = 520$, it reaches the peak. The security $S_{chaff}(cn)$ can reach up to 120 bits when q is in the range of 480 to 560. Therefore, our auxiliary data I_R is secure.

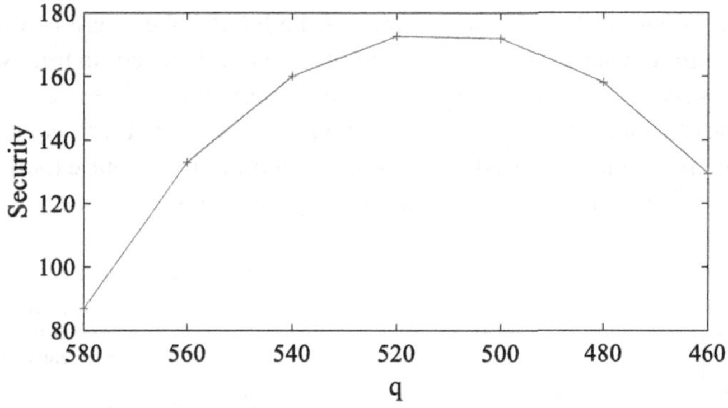

Fig. 11. Security guaranteed by chaff tuple.

The robustness of the auxiliary data $X = w \oplus bchenc(k)$ relies on the witness w and the random key k. The randomness of w is proved by experiments in Sect. 7.1, and the key k is randomly generated, so the $bchenc(k)$ obtained by BCH encoding is random. Therefore, the auxiliary data X is also secure.

7.3 Key Security

The session key *key* is concatenated by k_A,k_B, which are randomly selected by two devices, so its security depends on its length. As shown in Fig. 12, the length of the session key changes with the variation of noise tolerant ability $t0$ when the BCH code length *len* is 255. It can be seen that when the noise tolerance ability $t0$ is lower than 30, the length of the session key *key* is greater than 126. If the attacker wants to guess the negotiated *key* by brute force, it will take up to 2^{126} attempts.

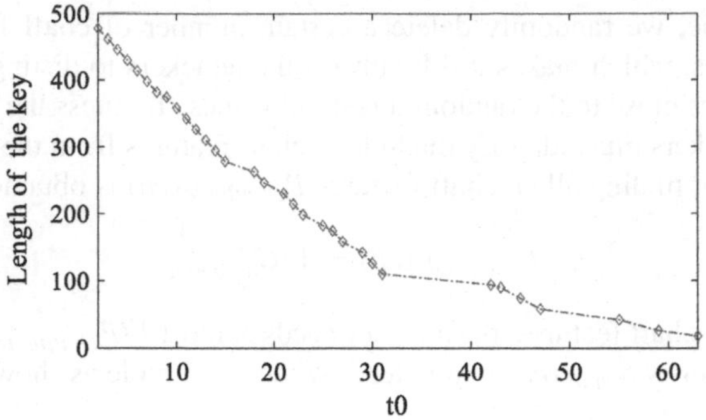

Fig. 12. Length of the session key *key* under different t0.

7.4 Mimicking Attack

The mimicking attacker observes the handshake between two legitimate users, and mimics it in real time in an attempt to obtain similar acceleration data collected by a legitimate device. We did imitation attack and obtained accelerometer value by imitating others handshake. As shown in Fig. 13, when mimics the handshake, there is a time lag from seeing the legitimate user's handshake to the react. Therefore, compared with a random handshake, it is more difficult for an attacker to generate a similar acceleration magnitude time series to pair with the legitimate device. Moreover, there is no success case that the mimicking attacker paired with a legitimated device in our experiments, which shows that our solution is suitable for practical applications.

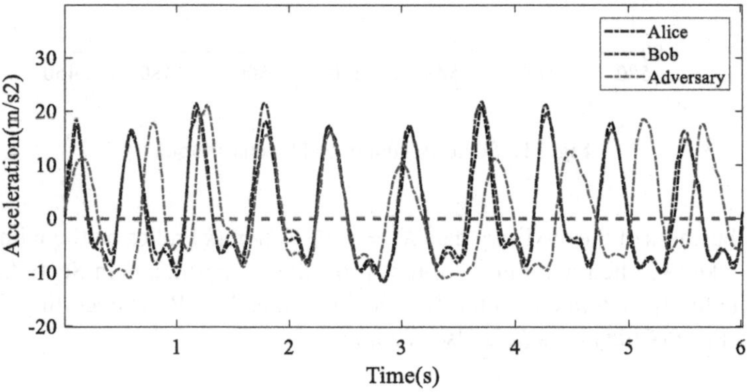

Fig.13. The acceleration magnitude gained by the mimicking attacker and the legitimate user.

8 Conclusion

In this paper, we proposed a robust and user-friendly secure pairing scheme for wrist-worn smart devices based on handshake acceleration, which is of high key generation rate. In our scheme, we used the three-dimension inertial acceleration sensor on the off-the-shelf wrist-worn smart devices to record handshake patterns. We used the random threshold-based witness generation algorithm to improve the key agreement rate. To increase the success rate and ensure the security of the scheme, optimal feature selection algorithm, and BCH based fuzzy commitment are used. In the future work, we will improve our scheme by reducing system overhead and strengthening system security.

Acknowledgement. This work is supported by the National Natural Science Foundation of China (61672413, U1708262, 61872449, 61772548, 61902290, 61772008), the project "The Verification Platform of Multi-tier Coverage Communication Network for oceans (LZC0020)", Scientific Research Program Funded by the Education Department of Shaanxi Province (Program No. 20JY016), the Fundamental Research Funds for the Central Universities, the Innovation Fund of Xidian University, Natural Science Foundation of Shaanxi Province (2019JM-109), Key Research and Development Program of Shaanxi (2019ZDLGY12–04, 2020ZDLGY09–06), China Postdoctoral Science Foundation (2018M640962).

References

1. Fomichev, M., Álvarez, F., Steinmetzer, D., Gardner-Stephen, P., Hollick, M.: Survey and systematization of secure device pairing. IEEE Commun. Surv. Tutorial **20**, 517–550 (2018)
2. Zhang, N., Wu, R., Yuan, S., Yuan, C., Chen, D.: RAV: relay aided vectorized secure transmission in physical layer security for internet of things under active attacks. IEEE Internet Things J. 8496–8506 (2019)
3. Mirzadeh, S., Cruickshank, H., Tafazolli, R.: Secure device pairing: A survey. IEEE Commun. Surv. Tutorials **16**, 17–40 (2014)
4. Zhang, N., Cheng, N., Lu, N., Zhang, X., Mark, J., Shen, X.: Partner selection and incentive mechanism for physical layer security. IEEE Trans. Wireless Commun. **8**, 4265–4276 (2015)
5. Chen, D., Zhang, N., Cheng, N.: Physical layer based message authentication with secure channel codes. IEEE Trans, Dependable Secure Comput (2018)
6. Jiang, Q., Zhang, N., Ni, J., Ma, J., Ma, X.: Unified biometric privacy preserving three-factor authentication and key agreement for cloud-assisted autonomous vehicles. IEEE Trans. Veh, Technol (2020)
7. Schürmann, D., Brüsch, A., Sigg, S., Wolf, L.: BANDANA – body area network device-to-device authentication using natural gait (2017)
8. Groza, B., Mayrhofer, R.: SAPHE - simple accelerometer based wireless pairing with heuristic trees. In: Proceedings of the 10th International Conference on Advances in Mobile Computing & Multimedia (2012)
9. Sun, Y., Wong, C., Yang, G.Z., Lo, B.: Secure key generation using gait features for body sensor networks (2017)
10. Mayrhofer, R., Gellersen, H.: Shake well before use: Intuitive and secure pairing of mobile devices. IEEE T. Mob. Comput. **8**, 792–806 (2009)
11. Holmquist, L.E., Mattern, F., Schiele, B., Alahuhta, P., Beigl, M., Gellersen, H.-W.: Smart-its friends: a technique for users to easily establish connections between smart artefacts. In: Abowd, G.D., Brumitt, B., Shafer, S. (eds.) UbiComp 2001. LNCS, vol. 2201, pp. 116–122. Springer, Heidelberg (2001). https://doi.org/10.1007/3-540-45427-6_10
12. Xu, W., Revadigar, G., Luo, C., Bergmann, N., Hu, W.: Walkie-Talkie: motion-assisted automatic key generation for secure on-body device communication. In: 2016 15th ACM/IEEE International Conference on Information Processing in Sensor Networks (IPSN) 2016 15th

ACM/IEEE International Conference on Information Processing in Sensor Networks (IPSN), p. 12 (2016)
13. Liu, H., Wang, Y., Yang, J., Chen, Y.: Fast and practical secret key extraction by exploiting channel response (2013)
14. Liu, H., Yang, J., Wang, Y., Chen, Y., Koksal, C.E.: Group secret key generation via received signal strength: Protocols, achievable rates, and implementation. IEEE T. Mob. Comput. **13**, 2820–2835 (2014)
15. Schürmann, D., Sigg, S.: Secure communication based on ambient audio. IEEE T. Mob. Comput. **12**, 358–370 (2013)
16. Yue, Q., Srinivasan, K., Arora, A.: Shape matters, not the size: a new approach to extract secrets from channel. In: ACM Workshop on Hot Topics in Wireless (2014)
17. Guo, Z., Gao, X., Ma, Q.: Secure device pairing via handshake detection. Tsinghua Science and Technology (2018)
18. Shen, Y., Yang, F., Du, B., Xu, W., Wen, H.: Shake-n-Shack: enabling secure data exchange between smart wearables via handshakes. In: IEEE International Conference on Pervasive Computing and Communications (2018)
19. Jiang, Q., Huang, X., Zhang, N., Zhang, K., Ma, X., Ma, J.: Shake to communicate: Secure handshake acceleration-based pairing mechanism for wrist worn devices. IEEE Internet Things **6**, 5618–5630 (2019)
20. Zhang, N., Lu, N., Cheng, N., Mark, J., Shen, X.: Cooperative spectrum access towards secure information transfer for CRNs. IEEE J. Sel. Areas Commun. **2013**, 2453–2464 (2013)
21. Zhang, N.,et al.: Physical layer authentication for internet of things via WFRFT-based Gaussian tag embedding. IEEE Internet Things J. (2020)
22. Christian, G., Kaisa, N.: Manual authentication for wireless devices. RSA Cryptobytes **7**(1), 29–37 (2004)
23. Chong, M.K., Mayrhofer, R., Gellersen, H.: A survey of user interaction for spontaneous device association. ACM Comput. Surv. **47**, 1–40 (2014)
24. Chong, M.K., Gellersen, H.: Classification of spontaneous device association from a usability perspective (2010)
25. Ming, K.C., Gellersen, H.: Usability classification for spontaneous device association. Pers. Ubiquitous Comput. **16**, 77–89 (2012)
26. Juels, A., Wattenberg, M.: A fuzzy commitment scheme (1999)
27. Jiang, Q., Chen, Z., Ma, J., Ma, X., Shen, J., Wu, D.: Optimized fuzzy commitment based key agreement protocol for wireless body area network. IEEE Trans. Emerg. Topics Comput. (2019)
28. Bose, R.C., Raychaudhuri, D.K.: On a class of error correcting binary group codes (1960)

Classification and Recommendation

SC-GAT: Web Services Classification Based on Graph Attention Network

Mi Peng[1], Buqing Cao[1(✉)], Junjie Chen[1], Jianxun Liu[1], and Bing Li[2]

[1] School of Computer Science and Engineering & Hunan Provincial Key Laboratory for Services Computing and Novel Software Technology, Hunan University of Science and Technology, Xiangtan, China
pengm12138@qq.com, buqingcao@gmail.com, hnust_cjj@163.com, ljx529@gmail.com

[2] School of Computer, Wuhan University, Wuhan, China
bingli@whu.edu.cn

Abstract. The classification of Web services with high similarity is conducive to the promotion for service management and service discovery. With the increasing number of Web services, how to accurately and efficiently classify the Web services becomes an urgent and challenging task. Although the existing methods achieve significant results in the task for service classification via integrating the structure information of service network with the content features of service node, it fails to discriminate the importance of neighbor services in the service network on the service node needed to be classified. To solve this problem, we propose a Web services classification method based on graph attention network. Firstly, according to the composition and shared annotation relationship of Web services, it applies the description documents, tags of Web services and the call relationship between mashups and services to build a service relationship network. Then, the attention coefficient of service nodes in the network is calculated by the self-attention mechanism, and different service nodes in the neighborhood are assigned different weights to classify Web services. Through the graph attention network, the content features of Web service can be well integrated with its structure information. Also, the learned attention weight is more interpretable. The experimental results on the real dataset of ProgrammableWeb platform show that the precision, recall and macro-F1 of the proposed method are greatly improved compared to those of GCN, Node2vec, DeepWalk and Line.

Keywords: Web service · Attention mechanism · Graph attention network · Service classification

1 Introduction

As the main technology of implementing SOA architecture, Web services can perform distributed computing and integrate data effectively. Web API is a typical Web service, which refers to the API function interface on the network to meet the various needs of

H. Gao et al. (Eds.): CollaborateCom 2020, LNICST 349, pp. 513–529, 2021.
https://doi.org/10.1007/978-3-030-67537-0_31

developers and users. Nowadays, Web APIs have become a core resource of Web and mobile applications. According to recent statistics of the ProgrammableWeb platform, its number has exceeded 20000, and more mashup developers use Web APIs for service composition. Therefore, how to quickly and accurately select the desired Web API from a large number of services has become a challenging issue for developers. Service classification can greatly reduce the time and space cost of Web service search and discovery [1, 2].

At present, many scholars have studied the classification of Web APIs. Most of them [3–5] are based on functional attributes to classify services, such as mining WSDL documents to build feature vectors, using TF-IDF, cosine similarity and other similarity measurement methods to calculate the similarity between vectors [6]. In addition, some researchers use LDA topic model or its extended model [7–9] to mine the hidden topic information of services, and use the topic vector to represent the description documents of services, and then carry out similarity calculation and service clustering [8, 9]. However, due to the short length of the WSDL document and the sparse features, these methods that only consider service content information are difficult to achieve good results [10]. Besides its content information, Web services have direct or indirect relationships with other information (such as tags, mashups, etc.), which can describe the functional attribute characteristics of Web services from multiple perspectives [11]. Therefore, some methods are put forward to exploit auxiliary relations such as tags into the process of service classification, and obtain the positive correlation between services by analyzing the tag relations between services with similar functions [6, 8]. At the same time, many mashup developers combine services with different functions to form some new applications (called Mashups), from which we can mine the negative correlation between different services to enhance the service classification. Therefore, this paper analyzes the relationship between service nodes in the relationship network, and uses the composition relationship and shared annotation relationship between Web services to form a complex service relationship network. In recent years, some related work has also carried out the network construction but basically did not consider the important weight of nodes in the service relationship network. Graph convolution network (GCN) [12] integrates the structure information of the graph with the features of the node to achieve good results in the task for node classification. However, GCN explicitly assigns a non-parametric weight to the adjacent nodes when combining with the adjacent nodes for feature aggregation, which is tough to assign different learning weights to different neighbor nodes.

In recent years, the attention mechanism [13] has been widely adopted in deep learning methods, which can use the self-attention mechanism to calculate different attention coefficients of neighbor nodes. And then, the linear combination of features and attention coefficients can better aggregate features according to the contribution of neighbor nodes. Therefore, some researchers propose graph attention network (GAT) [14], which applies self-attention mechanism to graph neural network. Through the parameterization of the weights between the nodes, the neural network is used for training and learning to assign more weights to important nodes. Inspired by this work, we introduce graph attention network into service classification, and propose a novel Web service classification method based on graph attention network, called SC-GAT. This method takes the description document and service relationship network as input, then outputs the embedding features of Web service. We apply the learned low dimension vector to represent the service node, which can characterize not only the structure of service network, but also the attribute information of the service node, which to some extent solves the problem of feature sparsity in the previous methods. Finally, we adopt activation function to predict the classification of Web services.

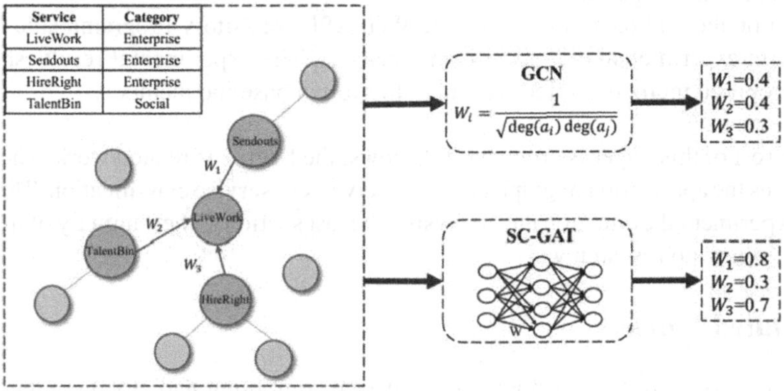

Fig. 1. The scene of service classification (take service "Livework" and its neighbor services "Sendouts", "Hireright" and "Talentbin" for example. Among them, a_i is the service, W_i is the weight of neighbor service).

Figure 1 shows an example of service classification method based on GCN and SC-GAT. Some methods, such as GCN, which consider the use of structural information to represent multi-node aggregation, usually calculate the preference weight of different neighbor nodes through the calculation of node degree. For example, the importance weights of neighbor services "Sendouts", "Hireright" and "Talentbin" calculated by the formula to the central service "Livework" are 0.4, 0.3 and 0.4 respectively, while it is obvious that the service "Sendouts" and "Hireright" are consistent with the category of "Livework", but they have not received good attention. The service "Talentbin" has a different category from "livework", but it gets similar weight to "Sendouts". This method not only needs to obtain the structural information of the service in advance, but this simple weight calculation method can not get the importance weight of the neighbor node well. In this paper, the proposed method SC-GAT parameterizes the weights of neighboring services, uses neural networks to train them according to the features of the nodes, and continuously adjusts the parameters during the training process. Finally, the services "Sendouts" and "Hireright", which are more similar to the central service "Livework", learned the importance weights of up to 0.8 and 0.7 respectively. The final learned parameters can well represent the importance of different neighbor nodes and consequently facilitate Web services classification.

In summary, the contributions of this paper are as follows:

- To our best knowledge, this is the first work to introduce graph attention network into Web service classification, which is significant to weigh the importance of neighbor services on the service node needed to be classified.
- We propose a novel Web service classification based on graph attention network, by exploiting the graph attention mechanism, which uses node features of the service to learn their parametric weights and mines the influence of Web service information on its structure calculation, avoiding costly matrix calculation or relying on the structure of the pre-known graph.
- Based on the real dataset on an online Web APIs repository ProgrammableWeb, we perform experimental comparison and analysis. The experimental results show that the proposed method is effective and outperforms baseline methods.

The rest of this paper is arranged as follows: the Sect. 2 is related work. The Sect. 3 introduces the application of graph attention network to service classification. The Sect. 4 is the experimental evaluation and analysis. The last section is the summary of this paper and the follow-up research work.

2 Related Work

In recent years, with the development of service computing and cloud computing, the discovery and mining of Web services has become a hot research direction. Research shows that efficient Web service classification can effectively improve the performance of Web service discovery [15]. At present, there are two main Web services classification methods, i.e., service classification based on functional semantics [16] and service classification based on QoS (quality of service) [17–19]. However, due to the short timeliness and difficult to obtain QoS information of services, it is rarely used in the Web services classification.

Web Services classification based on functional semantics divides services with high similarity into the same category by calculating the functional similarity of Web services. Chen et al. [6] proposed a Web services clustering method, called as wtcluster, which uses WSDL documents and tags to cluster Web services. Elgazzar et al. [20] designed a new WSDL documents mining technology, and successfully cluster it into similar Web service groups. This method extracts the key information of WSDL documents and clusters them according to service similarity. However, the limited information and sparse features of the service documents limit the clustering performance to some extent. In addition, some researchers exploit the auxiliary information of services into their service clustering process. For example, Wu et al. [21] presented a hybrid Web service tag recommendation strategy WSTRec considering tag co-occurrence, tag mining technology and semantic relevance measurement. Shi et al. [16] devised a probability topic model MR-LDA considering multiple Web service relationships, which can model the relationship between the combination of Web services and the relationship between Web services sharing tags. Cao et al. [22] put forward a Web service classification method based on the topic attention mechanism by combining the local hidden state vector with the global LDA topic vector. The above methods take into account service relationships into the training process of the model and improve service classification performance. However, the hidden relationships between Web services are not fully mined.

Actually, the composition relationship and shared annotation relationship of Web services form a complex service relationship network, and the research on network or graph data has attracted extensive attention. In the service relationship network, we regard the service itself as a node, and the edge is built through the mutual relationship between services. Nowadays, the network representation learning technology [23] can better learn the embedding vector of structure graph for the task of node classification. Therefore, inspired by the research of Yao [23], we propose a Web service classification method based on graph attention network. By weighing the importance of neighborhood service nodes, we use self-attention strategy to calculate the hidden representation of each service node in the service network.

3 Methods

The framework of the proposed method is shown in Fig. 2. It includes four parts: (1) data preprocessing; (2) service similarity computation; (3) graph attention network model construction; (4) service classification. In the data preprocessing part, we crawl Web services from the network and obtains meta information, such as service description text, tags and mashup after preprocessing. Then, API feature matrix and similarity matrix are obtained by calculating service meta information. Next, we construct the graph attention network model, input the API feature matrix and similarity matrix into the graph attention network model, and calculate the network embedding through self-attention mechanism. Finally, the softmax function is used to predict the categories of Web services.

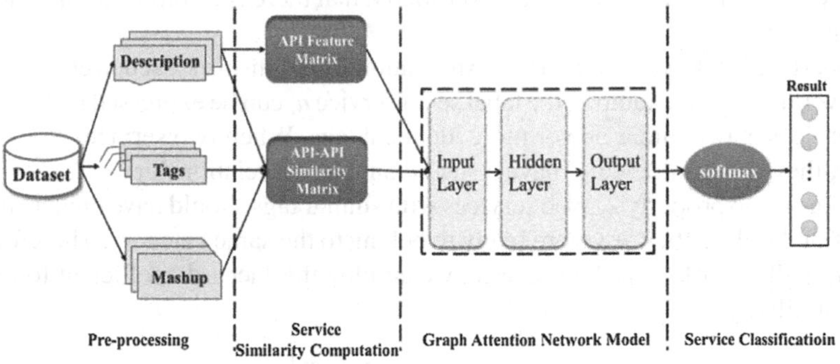

Fig. 2. The framework of service classification

3.1 Pre-processing

After crawling Web services from the network, we perform preprocessing to extract useful information of each Web service description document. The specific operations are as follows:

Remove Punctuation. Lowercase all the words in the description document to facilitate subsequent word stemming. At the same time, the punctuation in the text is removed, leaving only meaningful words.

Word Segmentation. Filter out the functional words in the language, remove the stop words in a sentence, and then list the remaining words. Each element is the word after removing the stop words.

Word Stemming. We implement word stemming for description document. For example, the root of create and created is create, and these words have the same meaning.

3.2 Web Service Similarity Computation

We define service relation network as $G = (V, E)$, which consists of a set of service nodes $V = \{a_i : 1 \le i \le N\}$ and a set of edge $E = \{(a_i, a_j) : 1 \le i, j \le N\}$. Among them, N is the number of Web services, and the edge (a_i, a_j) between nodes is determined by comparing the similarity and threshold of service nodes. When the calculated similarity is higher than a certain threshold, there is one edge between the two services.

In this paper, we use the word2vec [25] word embedding algorithm to learn the potential semantic vector of Web service description document. We define the text similarity between service a_i and service a_j as $Des(a_i, a_j)$, and combine the relationship between service and tag, mashup to determine the node similarity. According to our survey [24], Web services have the relationship of function composition and label common annotation, which is defined as follows:

Property 1 (Web Service Composition Relationship): a Web service may be called once or more by a mashup developer. If two services a_i and a_j at least one time are called by a user at the same time, it is considered that there is a composition relationship between a_i and a_j.

Property 2 (Web Service Sharing Annotation Relationship): each Web service contains several tags. For example, the label set of service a_i can be expressed as T_i. Different Web services may contain one or more identical tags. When two services share at least one tag, they are considered to have a shared annotation relationship.

According to property 2, Web services with similar tags should have higher similarity in function level, so they are more likely to belong to the same category. Therefore, after extracting all tags of each Web service, we employ the Jaccard coefficient to calculate the tag similarity:

$$Jac(a_i, a_j) = \frac{|T_i \cap T_j|}{|T_i \cup T_j|} \tag{1}$$

Where T_i, T_j are the tags owned by service a_i and a_j. $|T_i \cap T_j|$ represents the number of common tags owned by two services, $|T_i \cup T_j|$ represents the union of the number of tags owned by two services respectively.

The total similarity matrix is obtained by the following formula (2):

$$Sim(a_i, a_j) = Neg(a_i, a_j)(\alpha Des(a_i, a_j) + (1 - \alpha)Jac(a_i, a_j)) \tag{2}$$

Where α is the user's preference weight for the service description document similarity, and $(1 - \alpha)$ is the weight for tag similarity. $Neg(a_i, a_j)$ describes the correlation between

Web services. This correlation is obtained from the property 1. If two Web services have been called by the same Mashup, it is determined that the two Web services have different functions and should be divided into different categories, and the value is 0. On the contrary, if the two Web services have not been called by the same Mashup, the value is 1.

Consequently, we can take the total similarity $Sim(a_i, a_j)$ as the element value of the matrix to build the API-API similarity matrix.

3.3 GAT Model

After performing pre-processing and service similarity computation, we adopt GAT to learn the network embedding of Web services. The specific framework is shown in Fig. 3, including four parts, i.e., input layer, linear layer, attention layer and output layer.

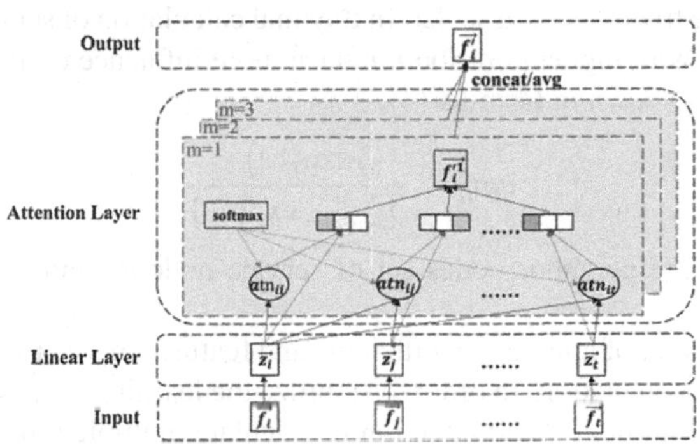

Fig. 3. The framework of GAT (take 3-heads attention as an example)

Input Layer. We define the initial node features of services as $\mathrm{F} = \left\{\overrightarrow{f_1}, \overrightarrow{f_2}, \ldots, \overrightarrow{f_N}\right\}$, and the feature vector $\overrightarrow{f_i}$ of each service a_i is a one-hot coding vector $\{0, 1\}^L$, where L is the number of features in each node. It is worth noting that this initial feature vector is discrete and high-dimensional, and its dimension depends on the number of feature words extracted by word2vec. Our goal is to optimize it and learn the continuous and low-dimensional embedding vector.

Linear Layer. In order to make the features enough representable, we need at least one learnable linear process to transform the input features into higher-level features. Therefore, we apply a shared linear transformation parameterized by the weight matrix to each node through the linear layer $\overrightarrow{z_i} = W\overrightarrow{f_i}$, $W \in \mathbb{R}^{\mathrm{L}' \times \mathrm{L}}$. Where L' is the dimension of target output feature, and the weight matrix represents the relationship between input feature and output feature.

Attention Layer. Generally speaking, the relationship between the service in the network space with a certain range will be closer. We use masked attention to apply graph structure to the model, that is, only the first-order neighbor nodes of each service node are calculated. The attention mechanism is realized by a single-layer feedforward neural network and parameterized by a weight vector $\vec{\theta} \in \mathbb{R}^{2L'}$. We concatenate the feature of the central service and the feature of neighbor service as the same query and key, so that we can learn the word dependency within the concatenating vector and automatically capture the correlation between the central node and the neighbor node. At the same time, the activation function *Leakyrelu* is applied to retain the coding of positive signal and small negative signal.

$$c_{ij} = \text{LeakyReLU}(\vec{\theta}^T [W\vec{f_i} \,||\, W\vec{f_j}\,]) \tag{3}$$

In order to make the coefficients between different services easy to compare, we utilize softmax to normalize the contribution of all neighbor services, and get the contribution of neighbor service a_j to the new feature of center service a_i. This attention mechanism is shared among all node pairs atn : $\mathbb{R}^{N'} \times \mathbb{R}^{N'} \to \mathbb{R}$, that is to say, the attention coefficient we finally get is a scalar. In the final calculation of service node features, neighbor services with higher contribution have more influence on the classification of current service. As shown in formula (4):

$$atn_{ij} = \frac{\exp(c_{ij})}{\sum_{h \in Ne_i} \exp(c_{ij})} \tag{4}$$

Where Ne_i is the neighbor nodes set of service node a_i, and $||$ is the connection operation.

The weighted sum of attention coefficients and features of several neighbor services is the result of one calculation. In order to stabilize the learning process of self-attention mechanism, we adopt multi-head attention to calculate multiple representations of services several times. For the aggregation of multiple representations, there are different ways, such as cascading, weighted average, and so on. In the first attention level, we can simply get the embedding vector of the final service node by concatenating without additional parameter learning. The specific calculation way is shown in formula (5):

$$\vec{f_i'} = \Big\|_{m=1}^{M} (\sigma\left(\sum_{j \in Ne_i} atn_{ij}^m W^m \vec{f_j}\right)) \tag{5}$$

Where atn_{ij}^m is the attention coefficient standardized of the m-th calculation for service a_j to service a_i. $\sigma(\cdot)$ is the non-linear activation function. And $\|$ is the connection operation, connecting the multiple results calculated.

Output Layer. It is worth noting that the average value of several attention coefficients is used to derive the final feature representation of network embedding, which is shown in formula (6):

$$\vec{F_i} = \sigma\left(\frac{1}{M} \sum\nolimits_{m=1}^{M} \sum\nolimits_{j \in N_i} attn_{ij} \vec{f_j'}\right) \tag{6}$$

Where M is the number of attention heads, that is, the number of functions used to calculate the attention coefficient in the multi-head attention mechanism. m represents the m-th function to calculate the attention coefficient.

3.4 Service Classification

We input the embedding features of the service from the GAT model into a full connection layer, and output the probability distribution of all the candidate Web service categories by using the softmax function. Softmax transforms the output value of multi classification into a relative probability, which indicates the probability that the node belongs to a certain category. The calculation way of it is shown in formula (7), where K is the number of candidate Web service categories:

$$\text{softmax}\left(\overrightarrow{F_i'}\right) = \frac{\exp\left(\overrightarrow{F_i'}\right)}{\sum_{k=1}^{K} \exp\left(\overrightarrow{F_k'}\right)} \tag{7}$$

In the process of model training, we employ the cross-entropy loss function to learn the parameters of the model, which is usually chosen as the objective function of the multi classification problem. Cross-entropy describes the distance between the actual output probability and the expected output probability. The smaller the value is, the closer the two probability distributions are. The specific calculation function of it is shown in formula (8):

$$Loss = -\sum_{k=1}^{K} y_k \log \frac{1}{p_k} \tag{8}$$

Where K is the number of service categories. y_k is the indicator variable of service nodes, which is obtained by the one-hot coding. If the service node is consistent with the corresponding category k, $y_k = 1$, otherwise $y_k = 0$; p_k is the prediction probability for the service belonging to the category k.

4 Experiment

4.1 Dataset Description and Experimental Setup

We crawled 17,783 Web APIs from ProgrammableWeb platform as the dataset of service classification. For each Web API, its information includes name, description text, category, tags and other information. Because the experimental dataset is too large, the top 10, 15, 20, 25 and 30 categories with the largest number of Web APIs are selected as the experimental dataset. The distribution of the top 30 categories with the largest number is shown in Table 1. During training, we randomly reorganized the experimental data, and then 60% of the dataset is selected as the training set, 20% as the verification set and 20% as the test set. Adam [26] method is used as the optimizer of the model. The learning rate is equal to 0.005, the batch size is 1, the number of attention heads is 8, and service similarity threshold is set to 0.8.

4.2 Baselines

DeepWalk [27]: Deepwalk uses random walk to get the local information of the network, and learns the vector representation of the service nodes according to the local information. Then it also applies random walk to attain the vertex sequence as the corpus, adopt the word embedding model for training, and learn the one-dimensional feature representation of each service node.

Table 1. Top 30 categories order by number

Category	Number	Category	Number	Category	Number
Tools	850	Telephony	338	Games	240
Financial	758	Reference	308	Photos	228
Messaging	601	Security	305	Music	221
eCommerce	546	Search	301	Stocks	200
Payments	526	Email	291	Cloud	195
Social	501	Video	289	Data	187
Enterprise	472	Travel	284	Bitcoin	173
Mapping	437	Education	275	Other	165
Government	369	Transportation	259	Project Management	165
Science	368	Advertising	254	Weather	164

Node2Vec [28]: Node2vec is an improved method based on DeepWalk. It obtains the corresponding sequence of each service point by using a specific walk way, defines two parameters to balance the influence of BFS and DFS, and considers the local and global information of the service graph structure.

Line [29]: Line uses the first-order similarity model and the second-order similarity model between the training vertices, and employs the edge sampling method to measure the degree of close connection between the service nodes, so as to obtain the similar service nodes for classification..

GCN [12]: GCN takes the standardized graph structure and service node features as the input, extracts the spatial features of the topological graph by using the eigenvalues and eigenvectors of the Laplacian matrix of the graph, and finally uses softmax to score and predict the category of service.

4.3 Evaluation Metrics

$MacroF1$ is chose as the evaluation metric to measure the performance of the proposed method and baselines. By calculating the recall (Rec_i) and precision (Pre_i) of each service category, we get the average recall (Rec_{ma}) and average precision (Pre_{ma}) of all N service categories. Among them, the value of recall is the proportion of correctly classified Web APIs in all Web APIs of this service category, and the value of precision is the proportion of Web APIs of this service category in the final service classification result. $MacroF1$ is the harmonic average value of recall and precision. They can be defined as below:

$$Rec_{ma} = \frac{\sum_i^N Rec_i}{N} \tag{9}$$

$$Pre_{ma} = \frac{\sum_{i}^{N} Prec_i}{N} \tag{10}$$

$$F1_{ma} = \frac{2 \times Rec_{ma} Pre_{ma}}{Rec_{ma} + Pre_{ma}} \tag{11}$$

4.4 Experimental Result

The experiment respectively tests the top 10, 15, 20, 25 and 30 service categories with the greatest number of Web APIs, and the experimental results are shown in Table 2 and Fig. 4. It can be seen from the experimental results that the proposed method is superior to all four baseline methods in terms of precision, recall and Macro-F1.

More concretely, with the increase of the number of categories, the experimental performances of all methods gradually decline. The reason is that the increasing of the content information contained in service category and the corresponding selected features makes service classification more difficult. At the same time, the decreasing, uneven number of Web APIs in some categories also leads to the declining of service classification performance.

As for DeepWalk, Node2Vec and Line, they mainly obtain the structure information and classify the services into different categories according to the functional network graph of Web APIs. While SC-GAT not only contains the structural information, but also employs the content information of services to determine the importance of different neighbor nodes. And the classification result of SC-GAT is obviously better than the three methods using only a single structural information.

It can be found that when the number of service categories is small, such as 10, the classification result of GCN is similar to that of SC-GAT. However, when the number of service categories increases to 20 or 30, the classification result of SC-GAT superiors to that of GCN. The reason is that the weight calculation between neighbor nodes is single in GCN. With the increase of the number of service categories, the structure of graph becomes rich and complex, which raises the difficulty of service classification. In the SC-GAT, it utilizes the self-attention mechanism to mine the characteristics of the nodes themselves and calculate the contribution of the neighbor nodes, which is completely dependent on the features of nodes and can be independent of the graph structure. Therefore, the representation ability of SC-GAT is improved and service classification is more accurate.

In addition, we make use of t-sne to visualize the classification results in our proposed method and the other four baseline methods. The top-10 categories with the largest number of services are adopted in the data. When the number of categories increases, the number of services contained in some categories decreases, and the division between categories has certain fuzziness, which is not conducive to the display of visual effect. We use the node's color to represent the category of the service node, and calculate the embedding of nodes by t-sne to get the location of nodes. As can be seen from Fig. 5, the service classification results of SC-GAT are basically better than the other four baseline methods. For the Line, DeepWalk and Node2Vec models, most of the service nodes are mixed together, which can not get a satisfactory result of classification. In the SC-GAT, the nodes of most service categories are very close to each other.

Table 2. F1-measure comparison of different serivce classification methods

Methods	Categories				
	10	15	20	25	30
Line	0.54371	0.48640	0.45376	0.45828	0.42714
Node2Vec	0.62957	0.59856	0.56300	0.56880	0.55107
DeepWalk	0.64052	0.60186	0.57076	0.56802	0.54734
GCN	0.82075	0.72173	0.67771	0.60425	0.55630
SC-GAT	**0.82716**	**0.73412**	**0.71343**	**0.70690**	**0.65837**

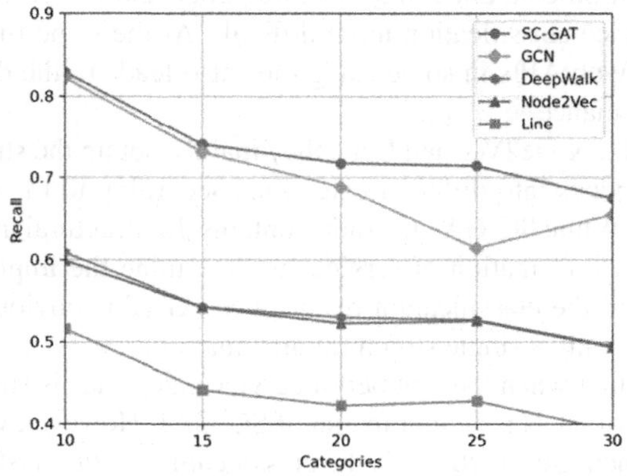

(b) Recall

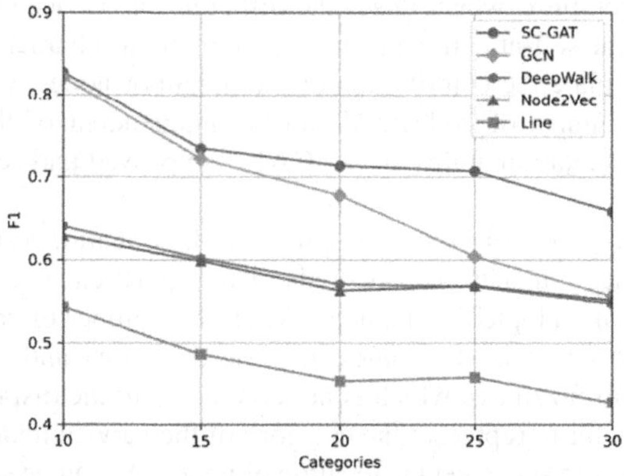

(c) F1

Fig. 4. Performance comparison of different service classification methods

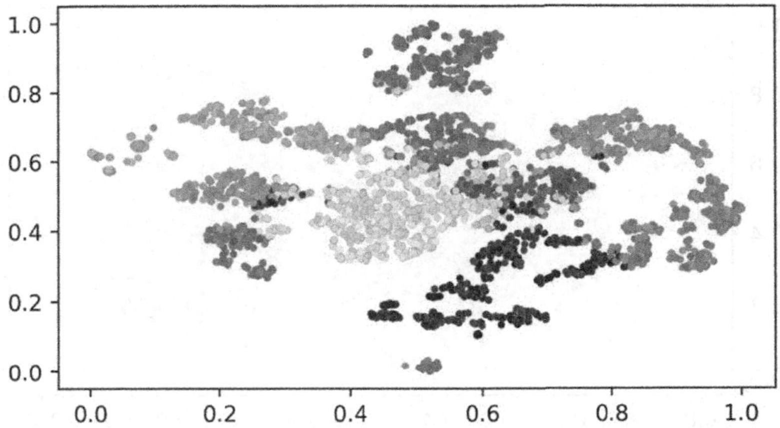

(a) SC-GAT

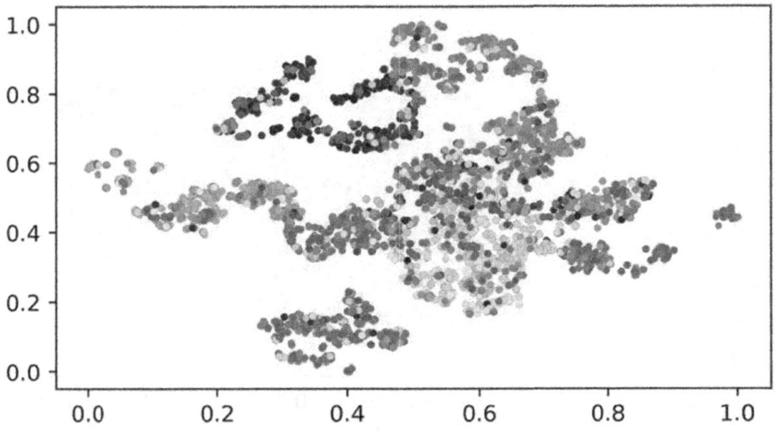

(b) GCN

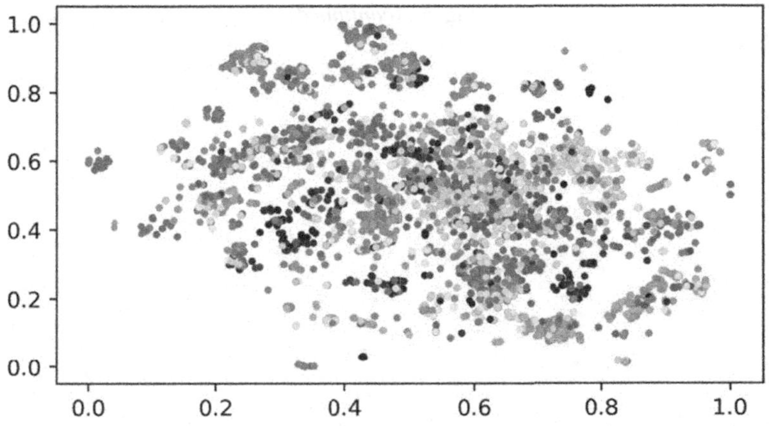

(c) DeepWalk

Fig. 5. Service classification visualization for different methods when the number of service categories is 10

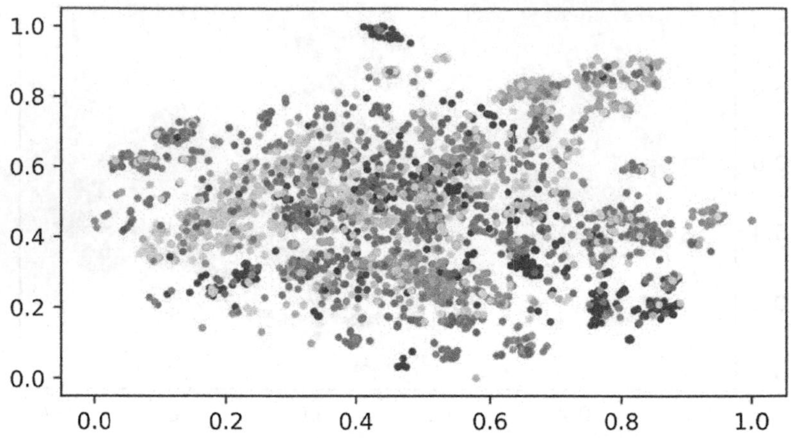

(d) Node2Vec

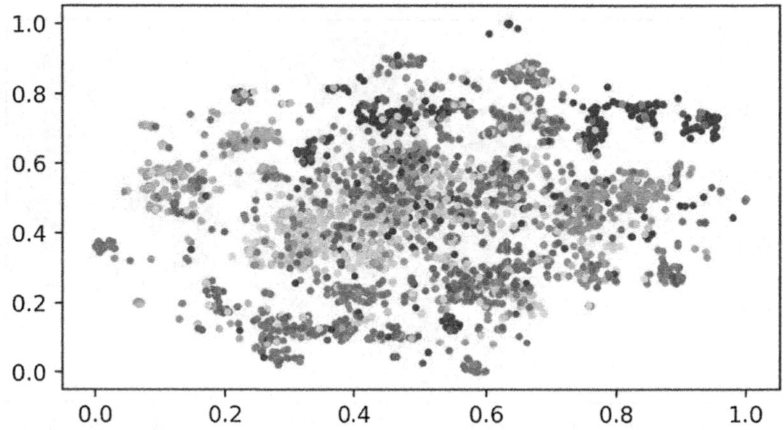

(e) Line

Fig. 5. (*continued*)

4.5 Conclusion and Future Work

In this paper, we propose a Web service classification method based on graph attention network. It adopts self-attention mechanism to calculate the characteristics of service nodes to identify different contribution degrees for different service nodes in the neighborhoods, and determine the service category according to the influence of the first-order neighbor nodes. The experimental results on real datasets verify the effectiveness of the proposed method. In the future work, we will consider to extend the SC-GAT model to include edge features, in order to further improve the performance of service classification. At the same time, due to the heterogeneity of Web services information, we consider introducing heterogeneous information network in future work, and recommend services according to its rich node types and semantic information.

Acknowledgement. Our work is supported by the National Natural Science Foundation of China (No. 61873316, 61872139, 61832014 and 61702181), the National Key R&D Program of China (2018YFB1402800, 2017YFB1400602), and the Natural Science Foundation of Hunan Province (No. 2018JJ3190, 2018JJ2136). Buqing Cao is the corresponding author of this paper.

References

1. Zhou, Z., Sellami, M., Gaaloul, W., Barhamgi, M., Defude, B.: Data providing services clustering and management for facilitating service discovery and replacement. IEEE Trans. Autom. Sci. Eng. **10**(4), 1131–1146 (2013)
2. Skoutas, D., Sacharidis, D., Simitsis, A., Sellis, T.: Ranking and clustering web services using multicriteria dominance relationships. IEEE Trans. Serv. Comput. **3**(3), 163–177 (2010)
3. Zhang, M., Liu, X., Zhang, R., Sun, H.: A web service recommendation approach based on QoS prediction using fuzzy clustering. In: 2012 IEEE Ninth International Conference on Services Computing, pp. 138–145 (2012)
4. Wang, H., Yang, Z., Yu, Q.: Online reliability prediction via long short term memory for service-oriented systems. In: 2017 IEEE International Conference on Web Services, pp. 81–88 (2017)
5. Xia, B., Fan, Y., Tan, W., Huang, K., Zhang, J., Wu, C.: Category-aware API clustering and distributed recommendation for automatic mashup creation. IEEE Trans. Serv. Comput. **8**(5), 674–687 (2015)
6. Chen, L., et al.: WTCluster: utilizing tags for web services clustering. In: Kappel, G., Maamar, Z., Motahari-Nezhad, H.R. (eds.) ICSOC 2011. LNCS, vol. 7084, pp. 204–218. Springer, Heidelberg (2011). https://doi.org/10.1007/978-3-642-25535-9_14
7. Min, S.H.I., Jian-Xun, L.I.U., Dong, Z., Bu-Qing, C.A.O., Yi-Ping, W.E.N.: Multi-relational topic model-based approach for web services clustering. Chin. J. Comput. **42**(4), 820–836 (2019)
8. Chen, L., Wang, Y., Yu, Q., Zheng, Z., Wu, J.: WT-LDA: user tagging augmented LDA for web service clustering. In: Basu, S., Pautasso, C., Zhang, L., Fu, X. (eds.) ICSOC 2013. LNCS, vol. 8274, pp. 162–176. Springer, Heidelberg (2013). https://doi.org/10.1007/978-3-642-45005-1_12
9. Shi, M., Liu, J., Zhou, D., Tang, M., Cao, B.: WE-LDA: a word embeddings augmented LDA model for web services clustering. In: 2017 IEEE International Conference on Web Services, pp. 9–16 (2017)

10. Cao, B., et al.: Mashup service clustering based on an integration of service content and network via exploiting a two-level topic model. In: 2016 IEEE International Conference on Web Services, pp. 212–219 (2016)
11. Cao, B., Liu, X.F., Rahman, M.M., Li, B., Liu, J., Tang, M.: Integrated content and network-based service clustering and web APIs recommendation for mashup development. IEEE Trans. Serv. Comput. **13**(1), 99–113 (2020)
12. Kipf, T.N., Welling, M.: Semi-supervised classification with graph convolutional networks. In: 2017 International Conference on Learning Representations (2017)
13. Peng, C., Sun, Z., Bing, L., Wei, Y.: Recurrent attention network on memory for aspect sentiment analysis. In: Proceedings of the 2017 Conference on Empirical Methods in Natural Language Processing, pp. 452–461 (2017)
14. Veličković, P., Cucurull, G., Casanova, A., Romero, A., Lio, P., Bengio, Y.: Graph attention networks. arXiv preprint arXiv:1710.10903 (2017)
15. Li, H., Liu, J., Cao, B., Tang, M., Liu, X., Li, B.: Integrating tag, topic, co-occurrence, and popularity to recommend web APIs for mashup creation. In: 2017 IEEE International Conference on Services Computing, pp. 84–91 (2017)
16. Shi, M., Liu, J., Zhou, D., Tang, M., Xie, F., Zhang, T.: A probabilistic topic model for mashup tag recommendation. In: 2016 IEEE International Conference on Web Services, pp. 444–451 (2016)
17. Xiong, W., Wu, Z., Li, B., Gu, Q.: A Learning approach to QoS prediction via multi-dimensional context. In: 2017 IEEE International Conference on Web Services, pp. 164–171 (2017)
18. Zhang, Y., Zheng, Z., Lyu, M.R.: WSPred: a time-aware personalized QoS prediction framework for web services. In: 2011 IEEE 22nd International Symposium on Software Reliability Engineering, pp. 210–219 (2011)
19. Kang, G., Liu, J., Tang, M., Liu, X., Cao, B., Xu, Y.: AWSR: active web service recommendation based on usage history. In: 2012 IEEE 19th International Conference on Web Services, pp. 186–193 (2012)
20. Elgazzar, K., Hassan, A.E., Martin, P.: Clustering WSDL documents to bootstrap the discovery of web services. In: 2010 IEEE International Conference on Web Services, pp. 147–154 (2010)
21. Wu, J., Chen, L., Zheng, Z., Lyu, M.R., Wu, Z.: Clustering web services to facilitate service discovery. Knowl. Inf. Syst. **38**(1), 207–229 (2013). https://doi.org/10.1007/s10115-013-0623-0
22. Cao, Y., Liu, J., Cao, B., Shi, M., Wen, Y., Peng, Z.: Web services classification with topical attention based Bi-LSTM. In: Wang, X., Gao, H., Iqbal, M., Min, G. (eds.) CollaborateCom 2019. LNICST, vol. 292, pp. 394–407. Springer, Cham (2019). https://doi.org/10.1007/978-3-030-30146-0_27
23. Yao, L., Mao, C., Luo, Y.: Graph convolutional networks for text classification. In: Proceedings of the AAAI Conference on Artificial Intelligence, vol. 33, pp. 7370–7377 (2019)
24. Shi, M., Liu, J., Cao, B., Wen, Y., Zhang, X.: A prior knowledge based approach to improving accuracy of web services clustering. In: 2018 IEEE International Conference on Services Computing, pp. 1–8 (2018)
25. Mikolov, T., Chen, K., Corrado, G., Dean, J.: Efficient estimation of word representations in vector space. arXiv preprint arXiv:1301.3781 (2013)
26. Kingma, D., Ba, J.: Adam: a method for stochastic optimization. In: 2015 International Conference on Learning Representations (2015)
27. Perozzi, B., Al-Rfou, R., Skiena, S.: DeepWalk: online learning of social representations. In: Proceedings of the 20th ACM SIGKDD International Conference on Knowledge Discovery and Data Mining, pp. 701–710. Association for Computing Machinery, New York (2014)

28. Grover, A., Leskovec, J.: Node2vec: scalable feature learning for networks. In: Proceedings of the 22nd ACM SIGKDD International Conference on Knowledge Discovery and Data Mining, pp. 855–864. Association for Computing Machinery, New York (2016)
29. Tang, J., Qu, M., Wang, M., Zhang, M., Yan, J., Mei, Q.: LINE: large-scale information network embedding. In: Proceedings of the 24th International Conference on World Wide Web, pp. 1067–1077. International World Wide Web Conferences Steering Committee, Republic and Canton of Geneva, CHE (2015)

A Novel Multidimensional Comments and Co-preference Aware Recommendation Algorithm

Yanmei Zhang(✉), Nana Song, Xiaoyi Tang, and Huaihu Cao

Central University of Finance and Economics, Beijing 100081, China
zhangym@cufe.edu.com

Abstract. A recommendation system creates a personalized experience for each customer, which helps companies boost the average order value and the amount of revenue generated from each customer. In a typical recommendation system, comments typify the group wisdom of users, which can reflect their feelings toward the product in multiple dimensions. Co-preference mirrors common preference of a group of users. By mining the multidimensional comments and co-preference relationship comprehensively, it is justifiable to recommend products that both have a good reputation and conform to users' interests. However, the existing related methods have two problems. Firstly, there is lack of further consideration on how to fully utilize comments of products from multiple dimensions for recommendation. Secondly, how to mine co-preference relationship and combine it with multidimensional comments for recommendation is seldom considered. Therefore, a novel recommendation algorithm is proposed, which mines the comments from multiple dimensions and then converges it with co-preference relationship for recommendation. Experiments conducted on two real-world datasets reveal that our proposed method improves the accuracy in terms of MAE and RMSE, compared with state-of-the-art algorithms.

Keywords: E-commerce · Personalized recommendation · Comment mining · Natural language processing · Co-preference relationship

1 Introduction

Recommendation systems in the e-commerce platform, which work as intelligent agents to assist people to locate what they need, are become more and more indispensable. The data that are relatively easy to acquire are ratings [1, 2] and comments [3–5], but the limitations of both are also obvious. First, product comments and ratings are arbitrary, and the actual ratings for products are generally higher [3]. For example, a seller may gain a good reputation because of a particularly high rating of one of the products; therefore, users believe that other products sold by the seller are also good, but the actual situation may differ. Second, ratings and comments are relatively simple and cannot represent users' real concerns about the product. Third, some e-commerce platforms

H. Gao et al. (Eds.): CollaborateCom 2020, LNICST 349, pp. 530–549, 2021.
https://doi.org/10.1007/978-3-030-67537-0_32

(e.g., Taobao) give high ratings by default to users who forget to give ratings, which do not represent the real experiences and feelings of users. Therefore, how to effectively utilize the available information to recommend high-quality products that meets users' demand is the key for recommendation systems.

At present, the related work of recommendation algorithms is mainly divided into two categories: one is the recommendation algorithm based on comments, which digs out the users' concerns from the comments and adds them to the preference model to achieve product recommendation that meets the characteristics of different users. There are still problems such as insufficient reflection of user interest differences. The other is the recommendation algorithm based on co-preference (common preference) relationship, which adds the co-preference factors into the recommendation model and uses the behavioral preference information of the co-preference users to achieve the recommendation for the target users. However, there are still personalized recommendations such as the single emotion of the users in the co-preference relationship.

In response to the above problems, a novel recommendation algorithm named MCCA (Multidimensional Comments and Co-preference Aware) algorithm framework is proposed. From the perspective of two kinds of information, based on the rating matrix, multi-dimensional comments and Co-preference relationships are integrated into the recommendation model to achieve the purpose of accurately positioning users' consumption tendency. In order to further reflect the user's emotional feelings for different products, we have also improved the rating prediction model, which also considers the user's historical ratings, product ratings based on comments, and ratings of co-preference user groups. It has been verified that the algorithm proposed in this paper has improved the prediction effect.

The main contributions of our proposed algorithm are as follows:

First, the MCCA recommendation algorithm is proposed. It mines users' comments from multiple dimensions and co-preference relationship so as to get effectively recommendation results.

Second, experiments were conducted on two real-world datasets of different sizes, and the results showed that our proposed algorithm could achieve more accurate prediction results. Compared with the state-of-the-art algorithms, the average effect improved by 7.2% in terms of mean absolute error (MAE) and root mean square error (RMSE).

2 Related Work

In general, the related work mainly includes two types: recommendation algorithms based on comments and those based on co-preference relationship.

2.1 Recommendation Algorithms Based on Comment Mining

As users usually focus more on comments on products, it is important to mine those comments. Research has been conducted in the field of natural language processing. For example, Turney [4] examined comments at the sentence level and proposed a simple unsupervised learning approach. Zhuang et al. [5] disposed comments at the word level and obtained more personalized user information through fine-grained mining. Studies at the word level include two basic tasks: domain word clustering and emotion analysis.

Domain word clustering refers to extraction from innumerable texts of words that belong to specific domains and automatically grouping them into semantic categories. For example, Zhang et al. [3] clustered words with LDA, and Luo et al. [6] used Word2Vec for clustering. However, LDA differs from Word2Vec [6] in that LDA is based on document information and learns semantic relevance, while Word2Vec is based on adjacent words information and learns semantic similarity.

In terms of emotion analysis, the two main methods are statistical and semantic. The statistical method implies determining the emotional tendency of words with statistical models. For example, Turney [4] established a database of positive and negative seed words. Semantic methods mainly use the existing knowledge base for analysis. For example, Li et al. [7] used WordNet to judge the emotional tendency. Other emotional dictionaries such as SentiWordNet [8] and HowNet [9] also exist.

With regard to recommendation based on comments, Zhang et al. [4] proposed CommTrust, which generated dependency relations of comments and clustered words with LDA to rank sellers. Wei et al. [10] extracted the tag information from the interaction between users and movies, constructed users' preferred theme model, combined users' ratings of movies with the tag, and used the improved SVD algorithm to make personalized recommendations for users. Kharrat et al. [11] proposed a recommendation algorithm based on the semantics of comments with SlopeOne and SVD. Liu [9] constructed an algorithm with collaborative filtering recommendation by considering users' interests and contextual constraints. Ma et al. [12] suggested that users' nonthemed content is helpful in showing the similarity of users' interests, proposed a framework to extract features, and introduced it into the algorithm based on matrix decomposition (MF).

Some studies merge the text information of comments with the collaborative filtering algorithm and obtain better recommendation effects. For example, McAuley et al. [13] believed that the themes of comments reflect the features of users and products, and proposed the Hidden Factors as Topics (HFT). They considered the dimensions of ratings and subjects of comments, and combined the latent factor model (LFM) and LDA. Zhang et al. [14] proposed the explicit factor model (EFM), which extracted the features of products and users through a periodic sentiment analysis of comments and merges them with the hybrid matrix decomposition framework. Zhang [15] focused on the word-level affective analysis of comments and emphasized its key role in cold start, interpretability of recommendation results, and generation of features of products and users. Bao [16] proposed a matrix decomposition model TopicMF that considers ratings and comments simultaneously. The topic is derived from the comments through non-negative matrix decomposition such that the topic distribution parameters are consistent with the corresponding potential user factors and product factors. Ling [17] suggested ratings meet reviews (RMR), which combines the theme model of comments with the mixed Gaussian function model based on ratings, and uses Gibbs sampling to learn the model parameters to improve the accuracy of the recommendation model.

However, recommendation algorithms based on comments make limited use of the multidimensional nature of comments. Moreover, one of the most important differences between comments and ratings is that comments reflect users' feeling about products from multiple dimensions such as quality, price, and logistics. Therefore, it is necessary to examine the comments from multiple dimensions and word levels for personalized recommendations.

2.2 Recommendation Algorithms Based on Co-preference Relationship

Users' preference are different in different cases, so scholars try to model users' preference to achieve more accurate recommendations. For example, Liu [18] designed a three-layer model to consider users' preference, examined the hidden preference layer between users and products, and proposed a model-based collaborative filtering algorithm. A few studies have added co-preference factors into preference modeling for users. For example, Chen et al. considered the co-preference relationship between users in the context of social network [19].

Golbeck [15] combined social networking and online scoring to isolate several features that demonstrated the co-preference characteristics of users. It was found that the co-preference relationship based on the overall similarity could predict user ratings more accurately by experimenting on filmTrust; Georgoulas and Vlachou et al. [20] introduced a user-centered similarity calculation method in the consideration of user's preference. Guo and Xu et al. [21] considered the recommendation of friends, the proposed framework can model user relationships and learn the strength of user relationships to better infer potential co-preference relationships between users.

Guo and Zhang et al. [22] proposed the TrustSVD algorithm that used matrix decomposition and considered the co-preference between users. They think that the co-preference relationship had a very important influence on the user's score. Co-preference users may have similar preference, so the influence of the co-preference user should be taken into account in the user's scoring prediction of the item; Deng et al. [23] used the deep learning for automatic feature selection, introduced the impact of social networks, and proposed a co-preference perception recommendation method based on matrix decomposition (DLMF).

At present, although the recommendation algorithms proposed in some related work do not involve direct co-preference relationships, it is related to the research of our work. Yang and Lei et al. [24] considered the trust factor in the process of matrix decomposition, and proposed the Trust MF model. In the model, trust relationship and trusted relationship were used for user' recommendation. Wei and He et al. [25] proposed the IRCD-CCS and IRCD-ICS models to solve the problem of cold start based on the SDAE model [26] that extracted information features and improved the SVD model [27] (timeSVD++); Dang and Ignat [21] proposed dTrust to avoid simple social recommendation methods using personal information; Golbeck's TidalTrust algorithm [16] provided a way to calculate the weight of users' direct neighbors in the network.

Although there are some related studies, few focuses on mining comments from multiple dimensions. Moreover, since the social network is difficult to obtain on e-commerce platforms, it is more feasible to mine the co-preference relationship hidden between users. As far as we know, there is few work integrating multidimensional comments and co-preference relationship to support recommendation.

3 MCCA Recommendation Method

Users' comments contain a significant amount of information, which mirror their experience and feelings about products from multiple dimensions, including quality, price, experience, and logistics. Therefore, it is necessary to comprehensively mine and utilize users' comments from multiple dimensions for recommendations. Moreover, co-preference relationship based on product interest among users should also be considered. Co-preference relationship represents the users' relationship with the similar interests and preference. In general, users are more willing to purchase the same products with their co-preference friends. If the comments and co-preference relationship can be considered comprehensively, the users' feelings and experiences regarding products can be revealed from multiple aspects, so as to make more accurate recommendations.

Therefore, the MCCA recommendation algorithm is proposed and the framework can be seen in Fig. 1. The upper and bottom rectangles with black borders are the offline and online processes, respectively. The arrows represent the algorithm flow, and the hollow arrow on the left represents the updating process of user-product rating matrix and comments. In the offline process, dependency relations in the form of (head, dependent) are mined from comments. Next, the "head" words are clustered and the rating of every cluster with "dependent" words is calculated. Finally, a comment rating of each product is generated based on the comment. In the online process, when users enter the system, the co-preference relationship set of the users is obtained bas ed on the Pearson similarity of their historical ratings. Finally, users' ratings for products are predicted in order to recommend for users.

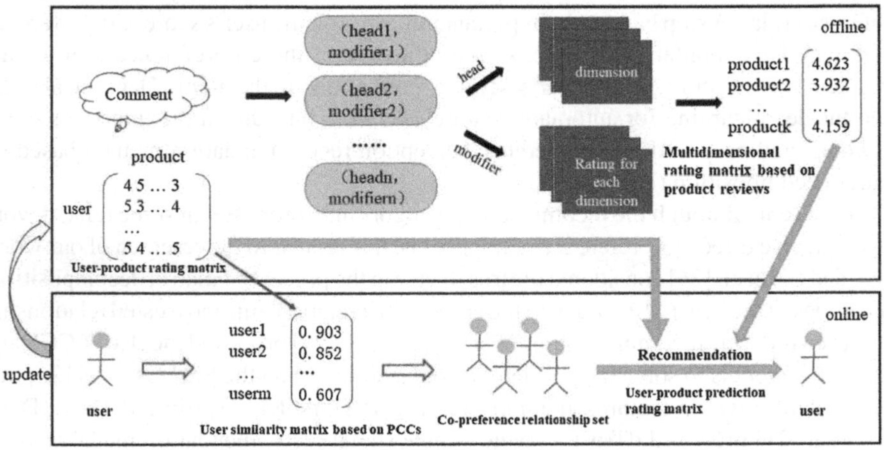

Fig. 1. The framework of MCCA recommendation algorithm

3.1 Comment Mining

Generation of Dependency Relations. Unlike ratings, a user's comments are usually subjective and unstructured; therefore, natural language processing techniques are used to mine and process comments. The Stanford Parser Syntax analysis is a two-step tool used to analyze English comment data. The first step is to use the syntactic analysis tool to address the commodity review corpus and excavate the syntactic dependencies in comments. The second step helps choose the dependency relations, which can reflect the modification of emotional words ("dependent" words) to subject words ("head" words) in the forms of (head, dependent). Figure 2 shows the structure and relationship pairs of a simple English comment sentence "I like it very much" by Stanford Parser.

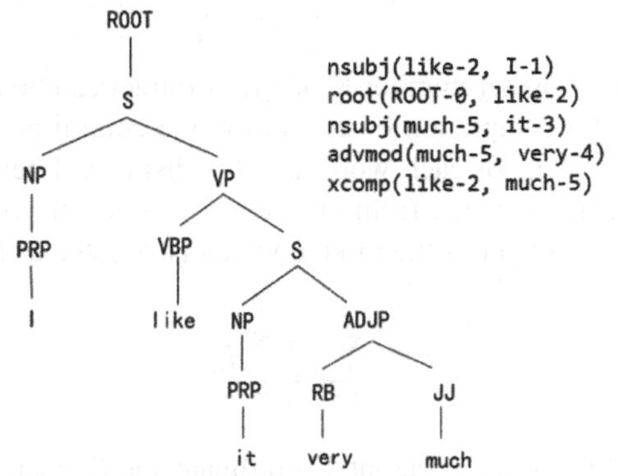

Fig. 2. Analysis result of Stanford Parser tool (*ROOT- the statement to process text; S- phrase combination; NP- noun phrase; VP- verb phrase; PRP- personal pronoun; VBP- link verb; ADJP- adjective phrase; RB - adverb; JJ- adjective; NN - noun; VB - verb; amod - adjective modification; advmod - adverb modification; nsubj - noun subject; acomp - adjective complement; xcomp - lack of subject clause complement*)

In the comments, "dependent" words are used to modify the "head" words, and there is a modified dependency between the "head" words and "dependent" words. The Stanford Parser syntactic analysis tool expresses the dependency between words in the sentence and generates different kinds of dependency pairs. Zhang et al. [5] suggested that the dependency relations in the forms of (head, dependent) can be obtained from four dependency relations: amod (NN, JJ), advmod (VB, RB), nsubj (JJ, NN), and acomp (VB, JJ).

Clustering of Dependency Relations. As users usually focus on different aspects of products, the subject words ("head" words) should be clustered after generating (head, dependent) relations to examine the comments multidimensionally. K-means is used to cluster subject words.

Distance in K-means. Given that the objects of K-means clustering are not numerical values but subject words, Word2Vec is used to represent each word as a vector corresponding to the hidden space. Word2Vec is a tool based on deep learning, which could convert words into vectors and calculate the similarity of the words by measuring the distance of their corresponding vectors. Table 1 describes the words and their similarity that are semantically similar to "man" calculated by the Word2Vec.

Table 1. Semantically similar words and similarity of "man"

Semantically Similar Word	Similarity
Spider	0.60792303
Men	0.5677488
Progs	0.53041357
Henchmen	0.527988

The Selection of K. The cluster number K of each product can be identified by the line graph method. After the completion of clustering, the central point M_i of K clusters, the corresponding cluster C_i of each word, and the distance d between two words are obtained. The sum of the distance from all words to the center point of the cluster is considered as the measurement of the model, which is denoted as D_k:

$$D_k = \sum_{i=1}^{K} \sum_{X \in C_i} d_{x,M_i} \tag{1}$$

Let different K values be the horizontal coordinate and D_k be the vertical coordinate as follows:

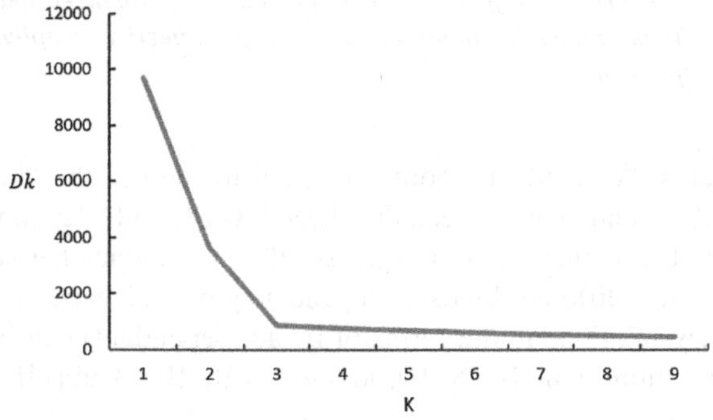

Fig. 3. Cluster number K and corresponding $\boldsymbol{D_k}$

Figure 3 illustrates that the bigger the K, the smaller the D_k, and K = 3 is an inflection point, which indicates D_k goes down very fast when K = 1 to 3 and then stabilizes, so this inflection point is the best K value.

Sentiment Analysis. Sentiment analysis is to examine users' emotional tendencies in comments that include positive, negative, and neutral. In the MCCA algorithm, SentiWordNet is used to assess emotion words, and positive, negative, and neutral emotion words correspond to +1, –1, and 0 ratings, respectively. Meanwhile, when negative relation is detected, the polarity of emotion words should be inverted.

Comment Ratings. Users usually concentrate on different aspects, and their interest can be found with the word frequency of dimensions. Let w_i denote the weight of each dimension, which is the ratio of the number of (modifier, head) dependency relations in the dimension to those in total. The recommendation index R is defined as follows:

$$\mathrm{R} = 5 * \sum_{i}^{m} w_i * (\frac{p_i - n_i}{p_i + n_i}), \tag{2}$$

where R denotes the comment rating of product, and m denotes dimension numbers of the product. In addition, w_i represents the weight of each dimension as above. p_i and n_i represent the number of positive and negative emotional words of a dimension, respectively.

3.2 Co-preference Relationship

Co-preference relationship is usually obtained through users' past ratings or interaction, and its value is [0, 1], where 0 indicates no co-preference and 1 indicates complete co-preference. An algorithm based on Pearson similarity is introduced to obtain co-preference relationship. The similarity between user u and user v is defined as follows:

$$S_{u,v} = \frac{\sum_i (r_{u,i} - \bar{r}_u)(r_{v,i} - \bar{r}_v)}{\sqrt{\sum_i (r_{u,i} - \bar{r}_u)^2}\sqrt{\sum_i (r_{v,i} - \bar{r}_v)^2}}, \tag{3}$$

where $r_{u,i}$ denotes the rating of user u to product i, and $\overline{r_u}$ denotes the average rating of user $u.r_{v,i}$ denotes the rating of user v to product i, and $\overline{r_v}$ denotes the average rating of user v.

Papagelis et al. [28] defined co-preference relationship degree with Pearson similarity. When the Pearson similarity between users is greater than a given threshold, the co-preference relationship degree is $S_{u,v}$; otherwise, the co-preference relationship degree is 0.

$$t_{u,v} = \begin{cases} s_{u,v}, ifs_{u,v} > \theta_s, |I_{u,v}| > \theta_I \\ 0, otherwise \end{cases}, \tag{4}$$

where $I_{u,v}$ represents a set of products that both user u and user v rated. θ_s is the threshold of Pearson similarity, which should be determined by experimenting, while θ_I is threshold of $I_{u,v}$ and the value of θ_I is 2.

3.3 Rating Prediction

The MCCA algorithm is proposed based on the TrustSVD algorithm. $\mu + \mathrm{b_j}$ represents the overall rating bias of product j in the TrustSVD algorithm, while it is replaced by c_j, which denotes the comment rating of product j in the MCCA algorithm. The proposed rating prediction model is as follows:

$$\hat{r}_{u,j} = b_u + c_j + q_j^T \left(p_u + |I_u|^{-\frac{1}{2}} \sum_{i \in I_u} y_i + |T_u|^{-\frac{1}{2}} \sum_{v \in T_u} w_v \right), \tag{5}$$

where q_j^T and p_u are the product feature matrix and the user feature matrix, respectively; c_j denotes comment rating of product j ; b_u denotes user bias; I_u is the collection of products rated by user u , and $q_j^T |I_u|^{-\frac{1}{2}} \sum_{i \in I_u} y_i$ represents the change in user factors based on implicit feedback. In addition, T_u denotes the set of co-preference relationship users of user u, and $q_j^T |T_u|^{-\frac{1}{2}} \sum_{v \in T_u} w_v$ denotes the influence of co-preference relationship users.

To avoid the issue of overfitting, the loss function of the model presented adopts the regularization strategy. The objective function is as follows:

$$\mathrm{L} = \frac{1}{2}\sum_{u}\sum_{j \in I_u}(\hat{r}_{u,j} - r_{u,j})^2 + \frac{\lambda_t^I}{2}\sum_{u}\sum_{v \in T_u'}(\hat{t}_{u,v} - t_{u,v}^I)^2 + \frac{\lambda}{2}\sum_{u}|I_u|^{-\frac{1}{2}}b_u^2 + \frac{\lambda}{2}\sum_{j}|U_j|^{-\frac{1}{2}}c_j^2$$
$$+ \sum_{u}\left(\frac{\lambda}{2}|I_u|^{-\frac{1}{2}} + \frac{\lambda_t}{2}|T_u|^{-\frac{1}{2}}\right)||p_u||_F^2 + \frac{\lambda}{2}\sum_{j}|U_j|^{-\frac{1}{2}}||q_j||_F^2 + \frac{\lambda}{2}\sum_{i}|U_i|^{-\frac{1}{2}}||y_i||_F^2$$
$$+ \frac{\lambda}{2}|T_v|^{-\frac{1}{2}}||w_v||_F^2$$

The model is solved by stochastic gradient descent method, and the Pseudo code is as follows:

Algorithm 1: The MCCA Algorithm

Input: user-product rating matrix R, co-preference relationship matrix T, comment rating matrix C, learning rate a, maximum number of iterations K, λ and λ_t

Output: predicted user-product rating matrix $\hat{R}$

1 Initialize P, Q, W, B. Let *Integer* = 0

2 Do

3 For each $r_{u,j}$ in R, each $e_{u,v}$ in T

4 Calculate the rating of user u to product j, $\hat{r}_{u,j}$

5 Calculate $e_{u,j} = \hat{r}_{u,j} - r_{u,j}$

6 Calculate $e_{u,v} = \hat{t}_{u,v} - t_{u,v}^I$

7 $b_u = b_u - a(e_{u,j} + \lambda|I_u|^{-\frac{1}{2}}b_u)$

8 $p_u = p_u - a(e_{u,j}q_j + \lambda_t e_{u,v}w_v + (\lambda|I_u|^{-\frac{1}{2}} + \lambda_t|T_u^I|^{-\frac{1}{2}})p_u)$

9 $q_j = q_j - a(e_{u,j}(p_u + |I_u|^{-\frac{1}{2}}\sum_{i \in I_u} y_i + |T_u^I|^{-\frac{1}{2}}\sum_{v \in T_u} w_v) + \lambda|U_j|^{-\frac{1}{2}}q_j)$

10 $\forall i \in I_u, y_i = y_i - a(e_{u,j}|I_u|^{-\frac{1}{2}}q_j + \lambda|U_i|^{-\frac{1}{2}}y_i)$

11 $\forall v \in T_u, w_v = w_v - a(e_{u,j}|T_u|^{-\frac{1}{2}}q_j + \lambda_t e_{u,v}p_u + \lambda|T_v^I|^{-\frac{1}{2}}w_v)$

12 End for

13 Integer = Integer + 1

14 While (Integer < K)

15 Calculate $\hat{R}$ with the latest parameters

16 Return $\hat{R}$

The steps of the MCCA algorithm are as follows: First, input user-product rating matrix R, co-preference relationship matrix T, comment rating matrix C, learning rate a, maximum number of iterations K, λ, and λ_t, and initialize user feature matrix P, product feature matrix Q, co-preference feature matrix W, rating bias matrix B, and iteration number *Integer* (line 1). Next, go into the iteration loop. In each iteration, calculate

$\hat{r}_{u,j}$, $e_{u,j}$, and $e_{u,v}$ with the parameters obtained from the previous iteration, while the initial values are used in the first iteration (lines 4–6). Next, the updated parameters are obtained with Stochastic Gradient Descent (lines 7–11). End the loop when *Integer* up to K. Finally, measure the predicted user-product rating matrix $\hat{R}$ with the latest parameters and return it (lines 15–16).

4 Experiments

In this section, several experiments are conducted to select parameters and compare MCCA to some classical algorithms and state-of-the-art algorithms. The experiments are aimed at addressing the following research questions:

RQ1: How to select the cluster number K of K-means?
RQ2: How to select the threshold of co-preference relationship?
RQ3: How much dose the ratios of training sets and test sets affects the algorithm performance?
RQ4: How does MCCA perform when it runs on two data sets of different sizes and sparsity?

4.1 Dataset

The experiments in the present study use a real-world dataset, that is, a comment dataset of Amazon audio equipment, which includes 1,428 users, 846 products, 10,250 ratings and comments, and Amazon book comment dataset, which includes 68,218 users, 31,191 products, 10,05,011 ratings and comments. Table 2 shows specific information of the datasets.

Table 2. Dataset statistics

Dataset	User	Product	Rating (comment)	Density
Audio equipment	1428	846	10250	0.85%
Book	68218	31191	1005011	0.05%

The co-preference relationship among users is calculated by Pearson similarity. The optimal threshold is determined by the grid search approach. In the experiment, different thresholds were taken, and different sets of co-preference users were obtained. Table 3 describes the number of co-preference users under different thresholds in the two data sets.

Table 3. Co-preference relationship statistics

Threshold	Co-preference (audio equipment)	Co-preference (book, $*10^5$)
0.1	5998	29.74
0.2	5686	26.28
0.3	5466	21.03
0.4	5027	19.72
0.5	4614	17.91
0.6	4285	14.69
0.7	3843	11.15
0.8	2709	8.42
0.9	1869	6.95
1.0	1050	5.38

4.2 Evaluation Metrics

In the experiments, common metrics, including MAE and RMSE, were used, which could reflect the recommendation accuracy.

1) MAE [6]: It is the average of the absolute error, which can reflect the prediction error by calculating the difference between predicted ratings and actual ratings. The smaller the value, the higher is the recommendation accuracy.

$$\mathrm{MAE} = \frac{\sum_{u,j} \left| \hat{r}_{u,j} - r_{u,j} \right|}{N}, \tag{7}$$

where $\hat{r}_{u,j}$ denotes the predicted rating of user u to product j, and $r_{u,j}$ denotes the actual rating of user u to product j, and N is the total number of ratings in the test set.

2) RMSE [6]: It is the standard deviation between predicted ratings and actual ratings, which measures more rigorously. The smaller the value, the higher is the recommendation accuracy.

$$\mathrm{RMSE} = \sqrt{\frac{\sum_{u,j} \left(\hat{r}_{u,j} - r_{u,j} \right)^2}{N}}, \tag{8}$$

where $\hat{r}_{u,j}$ denotes the predicted rating of user u to product j, and $r_{u,j}$ denotes the actual rating of user u to product j, and N is the total number of ratings in the test set.

4.3 Experiment Algorithms

The comparison algorithm selected in the present study includes a classical random algorithm and state-of-the-art algorithms closely related to MCCA, including CommTrust [3] (the algorithm considering comments), HFT [25] (the algorithm considering comments and ratings), and the algorithms based on singular value decomposition or matrix decomposition such as SVD++ [29], TrustSVD [21], and TrustMF [24]).

1) Random: The algorithm generates user ratings for products randomly, and it is a basic algorithm of the recommendation system.
2) CommTrust [3]: The algorithm fetches dependency relations of comments and clustered words with LDA in order to rank sellers, which represents the group wisdom of users.
3) HFT [13]: Hidden Factors as Topics (HFT) considers the dimensions of ratings and themes of comments together, and combines LFM and LDA. In the experiment, the parameters are set as follows: the learning rate is 0.005, and the number of iterations is 100.
4) SVD++ [29]: The SVD++ considers implicit feedback based on singular value decomposition (SVD) [29]. In the experiment, the parameters are set as follows: $\lambda = 0.1$, the learning rate is 0.005, and the number of iterations is 100.
5) TrustSVD [30]: This algorithm considered trust based on SVD++. In the experiment, the parameters are set as follows: $\lambda = 1.2$, $\lambda t = 0.9$, the learning rate is 0.005, and the number of iterations is 100.
6) TrustMF [23]: Based on matrix decomposition, this algorithm makes recommendations from the viewpoint of trust and being trusted. In the experiment, the parameters are set as follows: $\lambda = 0.001$,$\lambda t = 1$, the learning rate is 0.005, and the number of iterations is 100.

4.4 Experiment Result Analysis

1) Experiment 1: This experiment answers RQ1. For each product i, the best cluster number K_i can be determined by the line graph method. However, for repeatability of the algorithm, a uniform cluster number K should be selected. In this experiment, the best cluster numbers of 100 products K_1~K_{100} are generated first to determine the range of the uniform cluster number K, shown in Fig. 4.

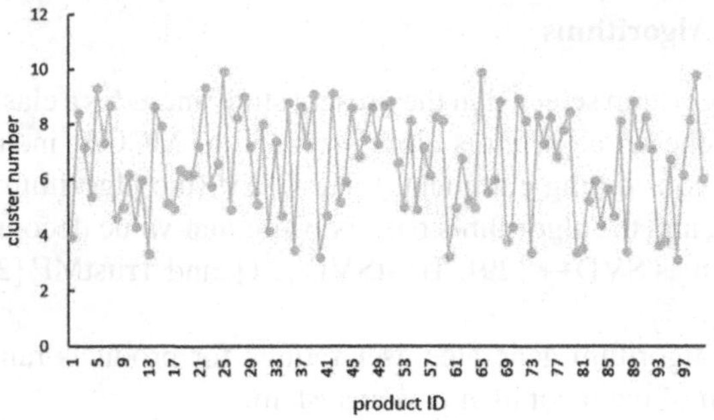

Fig. 4. Distribution of best cluster numbers of 100 products

Figure 4 shows that the range of the best cluster numbers of 100 products is between 3 and 10. Different K values in the range are taken in the following experiments. We choose 80% of the data as a training set and the remaining 20% as a test set. The results are fundamentally the same in other ratios of training set to test set. The parameters in the experiment are set as follows: $\lambda = 1.2$, $\lambda t = 0.9$, the learning rate is 0.005, and the number of iterations is 100.

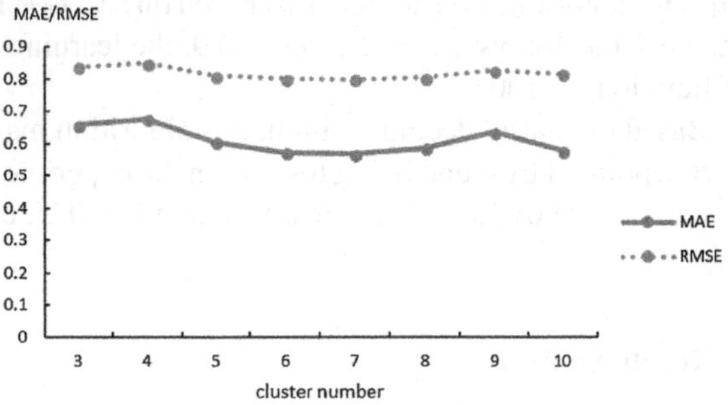

Fig. 5. MAE and RMSE of different cluster numbers

Figure 5 shows that the distribution of MAE and RMSE in the number of clusters is approximately the same. The MAE and RMSE of MCCA are minimum when K = 7, which implies that K = 7 is the optimal uniform cluster number and should be used in subsequent experiments.

2) Experiment 2: This experiment answers RQ2. Different similarity thresholds are set as 0.7, 0.8, 0.9, and 1. If the Pearson similarity between a target user and other users is larger than the threshold, co-preference relationship will be constructed between them. The MAE and RMSE of MCCA with co-preference relationship in different similarity thresholds are shown in Tables 4 and 5. The settings of ratio of training set to test set and parameters are the same as in Experiment 1.

Table 4. MAE and RMSE in different thresholds (audio equipment)

Threshold	MAE	RMSE
0.1	0.571858	0.809445
0.2	0.571855	0.809465
0.3	0.571846	0.809469
0.4	0.571806	0.809264
0.5	0.571798	0.809257
0.6	0.571802	0.809272
0.7	0.571787	0.809215
0.8	0.571703	0.809068
0.9	**0.571629**	**0.808931**
1.0	0.571706	0.809076

Table 5. MAE and RMSE in different thresholds (book)

Threshold	MAE	RMSE
0.1	0.773954	0.982867
0.2	0.773876	0.982693
0.3	0.773838	0.982625
0.4	0.773812	0.982603
0.5	0.773769	0.982481
0.6	0.773733	0.982441
0.7	0.773791	0.982502
0.8	0.773702	0.982387
0.9	**0.773647**	**0.982277**
1	0.773723	0.982406

Tables 4 and 5 show that the MAE and RMSE are minimum when the threshold is 0.9, which implies that the experiment performance is best when Pearson similarity of at least 0.9 is considered as co-preference relationship.

3) Experiment 3: This experiment answers RQ3. In the experiment, different ratios of training sets and test sets were selected for the experiment. In the algorithm considering co-preference, co-preference relationship was used. A value of 0.9 with better performance was selected as the similarity threshold. In the selection of clustering number in MCCA algorithm, K = 7 is selected as the optimal clustering number. The performance of MCCA algorithm in terms of MAE and RMSE under the two data sets is shown in Figs. 6 and 7.

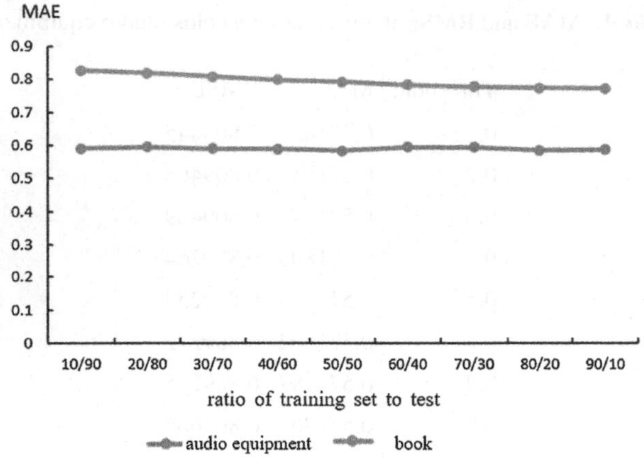

Fig. 6. MAE of different ratio of training set to test

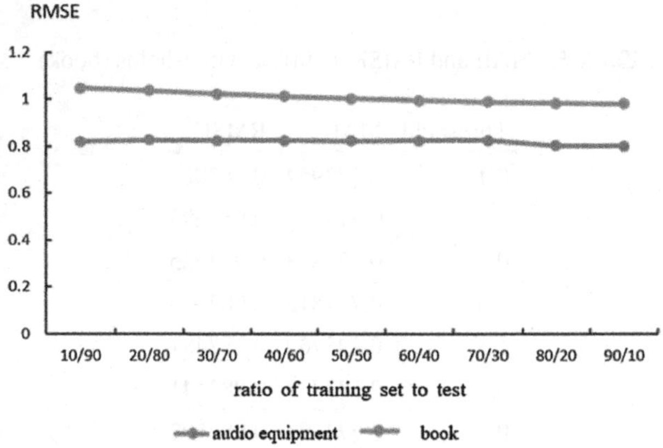

Fig. 7. RMSE of different ratio of training set to test

4) Experiment 4: This experiment answers RQ4. The performance of the CommTrustSVD algorithm is compared with classical Random algorithm as well as state-of-the-art algorithms, including CommTrust, HFT, SVD++, TrustSVD, and TrustMF by calculating the MAE and RMSE. We set K = 7 and threshold of Pearson similarity as 0.9; other settings parameters are the same as in Experiment 1. The results are shown in Figs. 8, 9, 10 and 11.

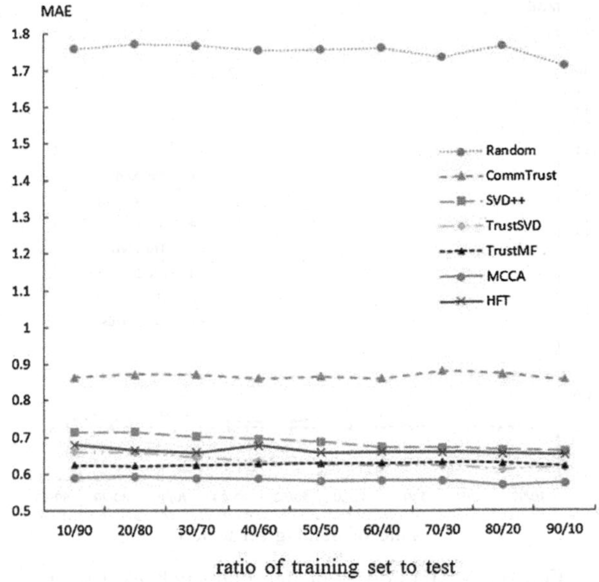

Fig. 8. MAE of different algorithms in audio equipment dataset

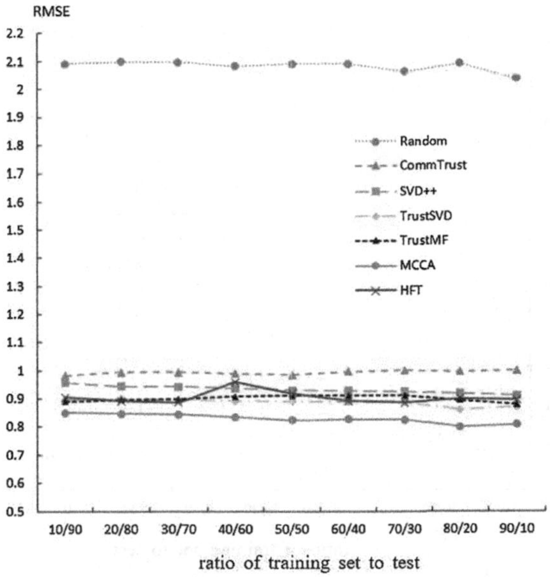

Fig. 9. RMSE of different algorithms in audio equipment dataset

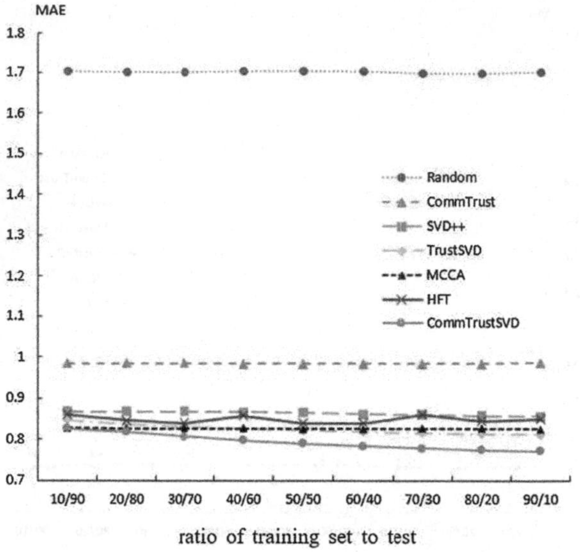

Fig. 10. MAE of different algorithms in book dataset

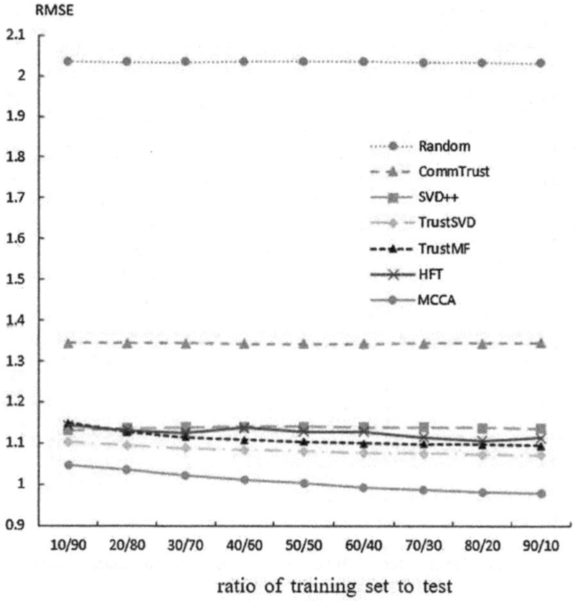

Fig. 11. RMSE of different algorithms in book dataset

Figures 8, 9, 10 and 11 show that in terms of all ratios of training set to test of the two datasets, the proposed algorithm MCCA slightly improved in terms of MAE and RMSE compared with these baseline algorithms. For comparison algorithms, HFT, TrustSVD, and TrustMF are better than Random, CommTrust, and SVD++.

Random is based on neither products nor users' interest, so the effect is the worst. CommTrust is a ranking algorithm that only considers comments and not users' preference. Although its performance is better than the Random algorithm, it does not reflect users' personalization, so the recommendation effect is not satisfactory. The traditional SVD++ algorithm adds implicit feedback information; however, compared with TrustSVD, TrustMF, and MCCA, it does not consider users' co-preference or other information.

HFT considers comments based on collaborative filtering, and the prediction effect is better than SVD++. Notably, in the case of sparse data, the effect of HFT is better than SVD++. This indicates that adding comments into the matrix decomposition algorithm can fairly alleviate data sparsity. However, under the present experimental environment, the performance of HFT algorithm is not as good as that of the algorithm with co-preference, such as TrustSVD, TrustMF, and MCCA.

In terms of MAE, TrustSVD performs better than TrustMF in datasets with higher proportion of training set, while TrustMF performs better than TrustSVD in datasets with lower proportion of training set. It indicates that in the present experimental environment, TrustSVD is suitable for datasets with dense data, while TrustMF is suitable for datasets with sparse data. However, the results of MAE and RMSE are not always consistent. In terms of the ratio of each training set and test set, the RMSE of TrustSVD is lower than that of TrustMF. Therefore, in terms of RMSE, the performance of TrustSVD is better than that of TrustMF. Both these algorithms consider the co-preference relationship among users, but since the dataset does not obtain the explicit co-preference relationship, it is replaced by the co-preference user set with the similarity threshold of 0.9. The results of the experiment show that the effect is favorable.

In conclusion, compared with the baseline algorithms, the MCCA shows certain improvement in terms of MAE and RMSE. Compared with TrustSVD and TrustMF, the experimental effect improved by nearly 7.2%. Compared with HFT, considering both comment and co-preference, the experimental effect improved by nearly 12.5%. MCCA uses the multidimensional comment information of products to fairly alleviate data sparsity. Moreover, compared with TrustSVD, MCCA directly replaces the average rating and item deviation in TrustSVD with the comment score that reflects the characteristics of products, which reduce the errors in the iterative calculation process, improve the accuracy of recommendation, and fairly alleviate the data sparsity.

5 Conclusion

The MCCA algorithm considered comments and co-preference relationship is proposed. Rather than using ratings alone and solving the "good reputation" problem, our proposed method uses comments from multiple dimensions, meanwhile considering users' co-preference relationship by Pearson similarity relations, and finally using the MCCA algorithm to predict user ratings. To some extent, the proposed algorithm can solve the problem of lack of objectivity resulting from overall high ratings, and it can fairly alleviate the data sparsity. Experiments conducted on two real data sets show that the proposed algorithm can make a more accurate rating prediction.

However, natural language processing on word level has limitations. For example, the dependency relations (price, low) indicates that the product is cheap and fine, but "low" may be considered a negative word. Therefore, future research might analyze comments at the sentence level, which may use machine learning for specific context training.

References

1. Ahmed, A.: Modeling trust-aware recommendations with temporal dynamics in social networks. IEEE Access **8**, 149676–149705 (2020)
2. Li, Y.K.: A novel implicit trust recommendation approach for rating prediction. IEEE Access **8**, 98305–98315 (2020)
3. Zhang, X.: CommTrust: computing multi-dimensional trust by mining E-commerce feedback comments. IEEE Trans. Knowl. Data Eng. **26**, 1631–1643 (2014)
4. Turney, P.: Semantic orientation applied to unsupervised classification of reviews. In: Meeting on Association for Computational Linguistics, Philadelphia, pp. 417–424 (2002)
5. Zhuang, L.: Movie review mining and summarization. In: ACM International Conference on Information and Knowledge Management, Arlington, Virginia, pp. 43–50 (2006)
6. Luo, J.: Domain word clustering based on word2vec and semantic similarity. In: 33th Chinese Control Conference, Nanjing (2014)
7. Li, C.: Mining dynamics of research topics based on the combined LDA and WordNet. IEEE Access **7**, 6386–6399 (2019)
8. Vijjini, A.R.: A sentiwordnet strategy for curriculum learning in sentiment analysis. In: 25th International Conference on Applications of Natural Language to Information Systems, Germany, pp. 170–178 (2020)
9. Zhu, S., Li, Y., Shao, Y., Wang, L.: Building semantic dependency knowledge graph based on HowNet. In: Hong, J.-F., Zhang, Y., Liu, P. (eds.) CLSW 2019. LNCS (LNAI), vol. 11831, pp. 525–534. Springer, Cham (2020). https://doi.org/10.1007/978-3-030-38189-9_54
10. Wei, S.: A hybrid approach for movie recommendation via tags and ratings. Electron. Commer. Res. Appl. **18**, 83–94 (2016)
11. Kharrat, F.: Recommendation system based contextual analysis of Facebook comment. In: Computer Systems and Applications, pp. 1–6. IEEE, Agadir (2017)
12. Ma, W.: Your tweets reveal what you like: introducing cross-media content information into multi-domain recommendation. In: 27th International Joint Conference on Artificial Intelligence, Palo Alto, pp. 3484–3490 (2018)
13. Mcauley, J., Leskovec, J.: Hidden factors and hidden topics: Understanding rating dimensions with review text. In: 7th ACM Conference on Recommender Systems, Hong Kong, pp. 165–172 (2013)
14. Zhang, Y.: Explicit factor models for explainable recommendation based on phrase-level sentiment analysis. In: International ACM SIGIR Conference on Research & Development in Information Retrieval, Gold Coast, pp. 83–92 (2014)
15. Zhang, Y.: Incorporating phrase-level sentiment analysis on textual reviews for personalized recommendation. In: Eighth ACM International Conference on Web Search and Data Mining, Shanghai, pp. 435–440 (2015)
16. Bao, Y.: TopicMF: simultaneously exploiting ratings and reviews for recommendation. In: Twenty-Eighth AAAI Conference on Artificial Intelligence, Québec (2014)
17. Ling, G.: Ratings meet reviews, a combined approach to recommend. In: 8th ACM Conference on Recommender Systems, CA, pp. 105–112 (2014)
18. Ifrim, G.: Enhancing collaborative filtering by user interest expansion via personalized ranking. IEEE Trans. Syst. Man Cybern. Part B (Cybern.) **42**(1), 218–233 (2011)

19. Chen, T.: Recommendation algorithm based on trust in social network environment. J. Softw. **3**, 771–781 (2017)
20. Georgoulas, K.: User-centric similarity search. IEEE Trans. Knowl. Data Eng. **29**(1), 200–213 (2016)
21. Guo, D.: User relationship strength modeling for friend recommendation on Instagram. Neurocomputing **239**, 9–18 (2017)
22. Yang, B.: Social collaborative filtering by trust. IEEE Trans. Pattern Anal. Mach. Intell. **39**(8), 1633–1647 (2017)
23. Deng, S.: On deep learning for trust-aware recommendations in social networks. IEEE Trans. Neural Netw. Learn. Syst. **28**(5), 1164–1177 (2017)
24. Jamali, M.: A matrix factorization technique with trust propagation for recommendation in social networks. In: ACM Conference on Recommender Systems, pp. 135–142 (2010)
25. Wei, J.: Collaborative filtering and deep learning based recommendation system for cold start items. Expert Syst. Appl. **69**, 29–39 (2017)
26. Wang, H.: Collaborative deep learning for recommender systems. In: 21th ACM SIGKDD International Conference on Knowledge Discovery and Data Mining, Sydney, pp. 1235–1244 (2015)
27. Koren, Y.: Collaborative filtering with temporal dynamics. Commun. ACM. **53**(4), 89–97 (2010)
28. Liu, Z.H.: Recommendation Algorithm fusing implicit similarity of users and trust. In: 21st IEEE International Conference on High Performance Computing and Communications, Zhangjiajie, pp. 2084–2092 (2019)
29. Koren, Y.: Factorization meets the neighborhood: a multifaceted collaborative filtering model. In: ACM SIGKDD International Conference on Knowledge Discovery and Data Mining, Las Vegas, pp. 426–434 (2008)
30. Guo, G.: TrustSVD: collaborative filtering with both the explicit and implicit influence of user trust and of item ratings. In: AAAI Conference on Artificial Intelligence, Austin, pp. 123–129 (2015)

A Deep Recommendation Framework for Completely New Users in Mashup Creation

Yanmei Zhang[1](✉), Jinglin Su[1], and Shiping Chen[2]

[1] Information School, Central University of Finance and Economics, Beijing 100081, China
Zhangym@cufe.edu.cn
[2] Software and Computational Systems CSIRO Data61, Sydney, Australia
shiping.chen@data61.csiro.au

Abstract. When service business is in evolution from B2B to B2C model, a cold-start problem raises for service composition due to the completely new clients with no historical records. Therefore, it is of great importance to solve the cold-start problem brought by completely new users. In this paper, we propose a recommendation framework for completely new users in Mashup creation based on deep-learning technology. Firstly, this framework extracts the mapping relationship between Mashup description and APIs offline by the deep neural network. Then, when the completely new users have the Mashup demands online, the matching APIs are recommended for them by using the mapping relationship. The experimental results with real-world datasets show that our proposed model outperforms the state-of-the-art ones in term of both accuracy and recall rate. The accuracy of the proposed method is 1.34 times higher than that of the state-of-the-art methods, and the recall rate is 1.55 times higher than that of the state-of-the-art methods. Moreover, considering that the new user history invocational data is very sparse, the performance of the proposed method can be greatly improved on the denser dataset.

Keywords: Service recommendation · Recommendation framework · Completely new user · Deep neural network · Mashup

1 Introduction

Web Mashups are Web applications developed using the contents and services available online [1]. Compared with traditional "developer-centric" composition technologies, such as BPEI and WSCI, Mashup provides a flexible and easy-of-use way for service composition on Web [2]. As the largest and most active collection of Web APIs and Mashups, the Programmable Web consists of more than 20,000 APIs until Jan-2020 [3]. Both the B2C business model and massive service information bring new challenges to service recommendation in the process of Mashup creation. First, there are lots of new users without any service invocation records. Second, some service Mashup platforms (e.g., Programmable Web) store composite service information (e.g., which APIs a Mashup is composed of), instead of any history of user invocation so that which the

H. Gao et al. (Eds.): CollaborateCom 2020, LNICST 349, pp. 550–566, 2021.
https://doi.org/10.1007/978-3-030-67537-0_33

corresponding historical data are quite scarce. Without any historical service invocation raises new challenges for service composition and recommendation.

The types of service recommendation models are increasingly diverse now. Firstly, demands matching methods [4–7] usually use Natural Language Processing (NLP) technology for demand matching. However, it cannot accurately describe the complex relationship between users and services. Secondly, some methods [8–10] mainly deal with the original matrix and expand the original scoring matrix with implicit feedback. However, for completely new users, the implicit feedback information is insufficient to effectively expand the original matrix. The third type of methods [11–14] focuses on the relationship of user-service, which mainly improves the process of user-service relationship modeling. But inadequate initial information makes it hard to build accurate models for new users. How to describe the complex relationship and make the model more suitable for new users becomes an intractable issue.

This paper proposes a novel deep recommendation framework for completely new users. The framework is divided into two parts: (a) offline learning and (b) online recommendation. In the offline part, the mapping relationship is extracted between Mashup description and APIs by deep learning techniques. In the online part, the User-API matrix is predicted for recommendation by the user demands and the mapping relationship obtained from the offline part. The contributions of this paper are highlighted as follows:

1. We propose a novel deep recommendation framework for completely new users, which combines the technologies of user feature extraction, matrix decomposition and learning mapping relationship by the neural network. The framework is low coupling so that each part can be updated or replaced independently.
2. Large experiments on the real data sets have been conducted and the experimental results prove that the accuracy of the proposed method has been improved, which is 1.34 times higher than the state-of-the-art methods, and the recall rate has been improved, which is 1.55 times higher than the state-of-the-art methods.

The rest of this paper is organized as follows: Sect. 2 presents the related work. Section 3 describes the proposed framework. Section 4 analyzes the experimental results on real data set. The last part draws a conclusion.

2 Related Work

According to the recommendation algorithm framework, the related work is divided into the following three categories:

2.1 Methods Focusing on Requirements Matching

Texts play an important role in the recommended system. It is one of the major directions of method improvement to accurately mine the user demands by converting the natural language into machine language.

Traditional NLP models such as LDA(Latent Dirichlet Allocation) [15], Word2Vec [16] and Doc2Vec [17], eventually extract long-text natural language as a word vector. But some fixed phrases have unique meanings of the words themselves. Gu et al. [4] proposed a method to decompose text into discourse units rather than word vectors in NLP so that the original meaning can be maintained for analysis. Then the relationship matrix of services can be generated based on discourse analysis and LSI (Latent Semantic Index) model for recommendation. But this method shows unclear relationship definitions. Some scholars have noticed the traditional NLP has a mediocre interpretability record. Xiong et al. [5] extract the semantic relationship of text as a "verb-noun-verb" triplet and generates a corresponding "object-attribute-value" triplet semantic map for recommendation. Although this method has better interpretability, it may lead to many complex sentences that are difficult to be parsed in the demand description of completely new users. Lin et al. [6] used a non-negative matrix decomposition method TNMF based on word correlation matrix. However, this method cannot provide effective query based on word vectors, and the recommendations could not accurately perform on small data sets. LIAN Tao et al. [7] proposed a recommendation algorithm based on the mixed LDA, which transports the historical message to pseudo-document information for analyzing. The method extracts detailed features by LDA and predicts the blank terms in the matrix by neighbor fields. Although this method defined the complex relationship between users and projects not so good, it provided important ideas for this paper. LDA is a topic model proposed by Blei et al. [15], that is divided into different levels according to documents, topics and words. After continuous improvement, LDA has become one of the most classical text analysis models. So, we choose the LDA model to extract more accurate features of the text.

2.2 Methods Focusing on Matrix

Most cold-start recommendation methods focus on how to effectively use the original matrix containing information about users and services. Song et al. [8] supplement the user feature matrix based on the user's click to learn his product preference online, build an online product cluster tree, and improved the accuracy by continuously collecting feedback. However, the life of service recommendation API is too short to seek the optimal solution through long-term exploration. Most methods use other sources of information to make up for the lack of rating data. Barjasteh et al. [9] proposed to use the similarity between users and items to alleviate the problem of cold start. This method generates the evaluation submatrix by excluding cold-start users and items from the original evaluation matrix, and then processes the cold-start matrix according to the similar matrix. But this method needs to have enough information about cold-start, and the recovery effect of the 0-1 matrix in the problem is unknown. Other scholars like Zhou et al. [10] proposed an iterative optimization method based on functional matrix factorization (FMF). The method iterates between the construction of the decision tree and the extraction of potential configuration files and considers the regularization scheme of tree structure. But this method pays more attention to matrix factorization technology, ignores the content features, as an important data source. Rendle et al. [11] combined the BPR(Bayesian Personalized Ranking) with matrix decomposition technique and KNN for recommending. This method optimized the original matrix by increasing the probability of user preference and provided an important idea for 0-1 matrix decomposition in this paper.

2.3 Methods Focusing on Relationship

Data relationship mining is the main part of recommendation method. It is one of the key points of recommendation method, i.e. how to construct the mapping relationship between users and services.

Yuan et al. [12] proposed a method for pairing a cold start item with the original item. But this method involves less numerical operations and the accuracy of the method is mediocre. Obadic et al. [13] proposed a method based on DNN. The DNN learns the mapping relationship directly between the content of the item and the potential factors obtained by the UI matrix. He et al. [18] developed a general framework named NCF, short for Neural network based Collaborative Filtering, which can express and generalize matrix factorization. The method leverages a multi-layer perceptron to learn the user-item interaction function. Xiong et al. [19] proposed a hybrid method named DHSR which aims to capture underlying complex interactions between mashups and services according to their invocation history and the corresponding functionalities.

The above methods show that the neural network accurately captures the relationship between services and mashups, thus brings better performance in service recommendation. Based on this observation, we decide to use DNN to learn relational mapping to improve recommendation accuracy.

There is no uniform specification for service descriptions, which results in either no service description or unclear service description making it difficult for service recommend systems to deliver high quality of recommendation services. Hao et al. [14] proposed a method based on target reconstruction service description (TRSD) which can recommend service items that are more in line relevant application scenarios for users. But if the new user enters the service platform for the first time, the text of user demand may be blurred, and the effect of reconstructing the hidden valuable information in the text may not be ideal.

The common mashup data stored in the service recommendation platform is of high value. Gao et al. [20] proposed to model the user's historical preferences for mashups and APIs separately in a single framework. Shi et al. [21] proposed a mashup-API-Tag recommendation model, which shows the potential value of tags. Mbipom et al. [22] proposed that the method of query could also be improved based on academic terms. Although these algorithms have some shortcomings, they provide some valuable ideas for our framework design.

In conclusion, the methods focusing on requirements matching by NLP can better mine user demands. Focusing on the original matrix can effectively alleviate the cold-start problem, but this method could not accurately mine the demands of users. Focusing on mapping relationship can learn the relationship more accurately, but it is easy to fall into difficulties in sparse data. In a word, current work cannot effectively solve the cold-start problem, i.e. lack of user data, caused by completely new users.

3 The Deep Recommendation Framework

First of all, in the III-A part, we explain the framework proposed in this paper and propose a method based on clustering to solve the case of sparse data. Then, the last three parts describe each component of the LDB model presented in this article.

3.1 Overall Framework

The novel deep recommendation framework we propose combines user feature extraction, matrix factorization and neural network learning mapping. Based on the selection of specific application technology, we call this method LBD (LDA + BPR + DNN) for short (see Fig. 1)

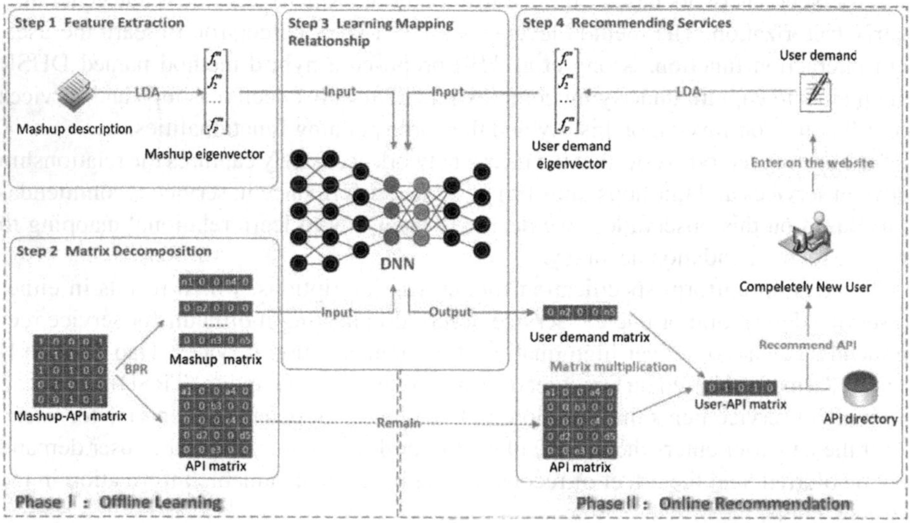

Fig. 1. LBD algorithm framework

LBD Method. As is shown in the figure, the method is divided into two parts: offline learning and online recommendation.

The offline part is shown in phase I. Mashup information is used to represent users' information in the context. The LDA model is used to process the limited service description text in the Mashup data and extract the Mashup description eigenvector. Then the Mashup-API invocation matrix is decomposed into the potential factor matrix of Mashup and API by using BPR (Bayesian personalized ranking). After getting data support from these two steps, the DNN was trained to learn the complex mapping of Mashup description eigenvectors and Mashup potential factor matrices offline.

The online part is as shown in phase II. When the new user inputs demand online which will be processed by the same LDA model to obtain user demand eigenvector. And the potential factor matrix of user demand is mapped by DNN, which has learned the complex mapping relationship. Finally, a new User-API matrix is obtained for recommendation by multiplying the potential factor matrix of user demand and the potential factor matrix of API obtained by matrix decomposition in the offline phase.

LBD-Clustering Method. To effectively overcome the problem of high sparsity in datasets, we propose an LBD-Clustering method based on the improved LBD. The function of clustering is to divide the samples in the dataset into disjoint subsets. According to topic clustering method [23], the original data is compressed to reduce data sparsity, and the clustering results are used as the new input validation effect of the method.

For sparse data, we cluster the Mashup entries in the original data. In the feature extraction stage, LDA can directly give the topics to which each service description belongs. According to formula (1), the corresponding vectors of all descriptions belonging to the same topics are added together to get the eigenvectors of the topic. n is the total number of words in the document collection. t presents the topic collection.

$$\mathrm{V}(t) = \sum\nolimits_{i=0}^{n} v(i)\{i \in t\} \tag{1}$$

3.2 Feature Extraction Based on Mashup Description

The service recommendation mainly relies on the service descriptions and user demand descriptions. Eigenvector extraction is the first step of the method to accurately define the relationship between user demand and service function. The extraction result has a great impact on the recommendation. Two important factors that determine the validity of extraction are data sources and vectorization. The following two aspects are introduced separately.

Data Sources Processing. Mashup classification and Mashup description are used in our method. The Mashup classification describes the application area and service type of the API. The Mashup description introduces the basic information of the service comprehensively such as the function and application scenario of the API and plays a major role in feature extraction. Since the completely new users only have natural language description of demand, we analyze the Mashup description as the text of user demand. When processing the Mashup description in a long text, we need to filter out the redundant information to improve the effect of feature extraction. As shown in Table 1, and we use the part-of-speech (POS) tag function to analyze part of speech, only retain vocabulary such as verbs, name forms, adjective adverbs and so on.

While processing the long text, we also obtained the Mashup classification tags. The classification tags are relatively independent, and there is no obvious relationship between the classification tags and the original long text of the Mashup description. We have integrated these tags with the long text as the service description of the API, and then extracted its eigenvector. In order to highlight the importance of the Mashup classification tags compared to general vocabularies, we have increased the number of times they appear in the text to improve their effect in feature extraction.

Text Vectorization for Demand Matching. The classification tags are relatively independent, and there is no obvious relationship between the classification tags and the

Table 1. Text preprocessing instructions.

Description	Example
Stop words	tool, solution, way, xml, platform, etc.
Frequency higher than 0.5	service, web, response, method, API, etc.
Frequency less than 0.01 and mistake	eventcategorie, myhurricane, phyloinformatic, hotukdeal, etc.
Meaningless word	Like prepositions, pronouns, articles etc.

original long text of Mashup description. Therefore, we use the LDA to extract the features of the Mashup description. The model of LDA is shown in Fig. 2 and the symbol convention is shown in Table 2. The feature extraction generates two files from the source document: the index table file and the vocabulary file. The Mashup description document is represented by multiple topics, which can not only solve the synonym problem but also effectively solve the problem of ambiguity.

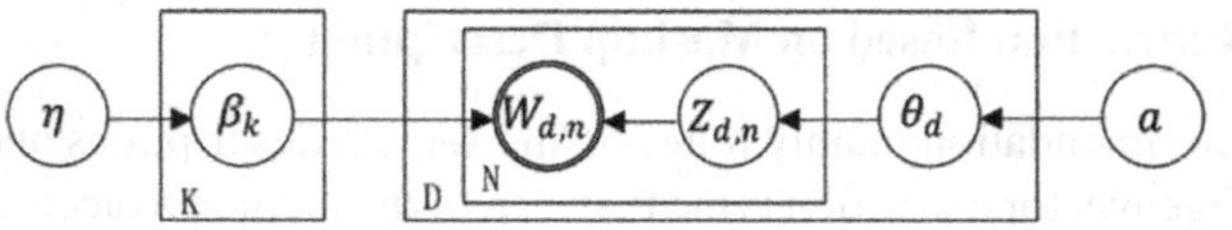

Fig. 2. LDA model diagram.

We extract text features according to the model shown in Fig. 2 α is the k-dimensional vector, which represents the proportion of the corresponding topic in the text. The text information is vectorized by using the formula (2) probability model:

$$p(\theta, z, w|\alpha, \beta) = p(\theta|\alpha)\prod\nolimits_{n=1}^{N} p(z_n|\theta)p(w_n|z_n, \beta) \tag{2}$$

β, calculated by the formula, is the v-dimension vector, which represents the probability value of the words in each topic. The entire β is combined as a k × V matrix, feature extraction of the text is made based on the matrix. The feature extraction is based on the description information of Mashup in the training set, we integrate the description text information and establish the LDA model. Through continuous training, different topic-k in the description information of Mashup is extracted clearly, and finally we can obtain the description eigenvector η of Mashup.

We take the eigenvector of the Mashup service description as the input of the deep neural network, and then take the corresponding Mashup potential factor matrix as the specified output to train the neural network model. Therefore, the next main task is how to choose the method to get the better Mashup potential factor matrix.

3.3 Matrix Decomposition Based on BPR

The common matrix decomposition methods in the recommended system include SVD, NMF, PMF [24] and SVD++ [25]. However, completely new users without service

Table 2. Symbolic representation of LDA model diagram.

Symbol	Explain	Symbol	Explain
α	Dirichlet super parameters of text-topic distribution	W	Words in text
η	Dirichlet super parameters of topic-word distribution	N	Total words in text
θ	Text-topic distribution	D	Total text
β	Topic-word distribution	K	Total topics
Z	Topic of words in text	V	Total words in dictionary

invocation history can only be recommended by analyzing the Mashup data against the description of user demand. What's more, the Mashup API invocation matrix is a 0-1 matrix instead of the scoring matrix. However, NMF and SVD++ are more focused on matrix decomposition with implicit feedback such as click and browse. They are more suitable for the scoring matrix so that they are ineffective for this problem. In the process of research, we find that BPR can effectively solve the decomposition problem of the 0–1 matrix by sorting reconstruction during matrix decomposition. So, we choose BPR as the main technology of matrix decomposition.

BPR (Bayesian Personalized Ranking) is a sort method based on matrix decomposition. It processes the score of users (display feedback "1", implicit feedback "0") into a set of pair pairs <i, j>, where i is the API with a score of "1" and j is the API with a score of "0". Assuming that a user has L "1" scores and D "0" scores, the user has L × D pair pairs, so that the data set is represented by triple <u, i, j> and the physical meaning is: for user u, API interface I has higher priority than interface j. The training results are two decomposed matrices W and H. K is a smaller dimension that needs to be defined by itself. For any user u, we can calculate its ranking score for API interface i according to formula (3) as follows:

$$\bar{x}_{ui} = w_u * h_i = \sum\nolimits_{f=1}^{k} w_{uf} h_{if} \tag{3}$$

From the ranking scores of user u for all APIs, we find the largest k-score, which is our real recommendation set top k API combination for user U. In the specific operation, we iterate the two submatrix W, h of the initial random according to the following formula (4, 5, 6) until the matrix converges to output the submatrix W, H:

$$w_{uf} = w_{uf} + \alpha\left(\sum\nolimits_{(u,i,j)\in D} \frac{1}{1 + e^{\bar{x}_{ui} - \bar{x}_{uj}}}\left(h_{if} - h_{jf}\right) + \lambda w_{uf}\right) \tag{4}$$

$$h_{if} = h_{if} + \alpha\left(\sum\nolimits_{(u,i,j)\in D} \frac{1}{1 + e^{\bar{x}_{ui} - \bar{x}_{uj}}} w_{uf} + \lambda h_{if}\right) \tag{5}$$

$$h_{if} = h_{if} + \alpha\left(\sum\nolimits_{(u,i,j)\in D} \frac{1}{1 + e^{\bar{x}_{ui} - \bar{x}_{uj}}} w_{uf} + \lambda h_{if}\right) \tag{6}$$

Where α is the gradient step, λ is the regularization parameter and K is the dimension of the decomposition matrix. We regard 1 in the 0–1 matrix as the items with the highest user preference ranking in the conventional U-I matrix, and the Mashup-API matrix is decomposed by using the derived formula. Experimental results show that due to the sparsity of data, the BPR model can effectively decompose the 0–1 matrix, thus improving the final effect of the method.

3.4 Learning Mapping Relationship by DNN

The DNN can be used to learn the mapping relationship between the eigenvector of Mashup description and the Mashup potential factor matrix, which is obtained by decomposing the Mashup-API matrix. Since the API potential factor matrix obtained by matrix decomposition is fixed, it is equivalent to learn the complex mapping relationship between Mashup description and API. The eigenvectors of service description extracted by LDA are simplified enough that DNN is suitable for most classification tasks and has excellent learning ability.

We decided to use the neural network with black box property to learn the relationship, and to accurately define its mapping relationship by fitting large-scale training data sets. The DNN is shown in Fig. 3.

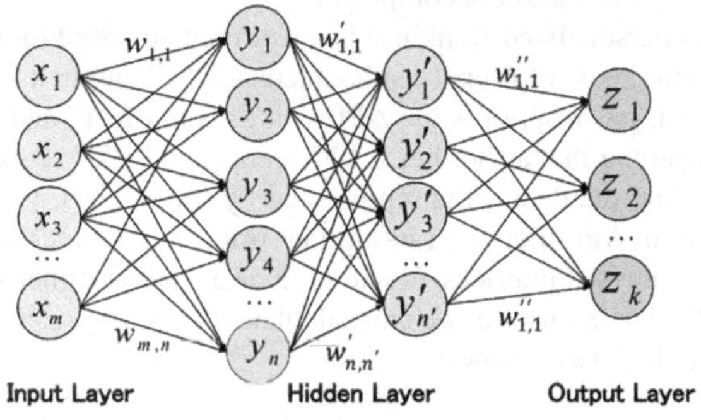

Fig. 3. DNN model.

The DNN is composed of many neurons, which are essentially the same as the perceptron. The neurons are trained by inputting x with weight W and activating function, that is, the linear relationship of multiple formulas (7) plus activating function δ (z), where B represents the threshold constant.

$$z = \sum w_i x_i + b \tag{7}$$

The weight w_i and the bias b are randomly initialized, then the training samples obtained in the earlier part are input into the neurons, the forward propagation has calculated the output. Then get the mean squared error by calculating between the target and the output. The error is propagated back to the hidden layers to update the parameters until the error to our expectation.

The number of layers in a neural network directly determines its performance ability. Although more layers may bring better effects, as the number of layers in a neural network deepens, the optimization function is more and more likely to fall into the local optimal solution, even deviate from the real global optimal solution. Sometimes, the performance of the deep network trained with limited data is not as good as that of the shallow network. Therefore, the layer number of DNN in our proposed model and the number of neurons in each layer will be the focus of the following discussion.

So far, the training stage of DNN has been completed, and the next step is to use the trained DNN to predict the services required by completely new users.

3.5 Prediction of Matching APIs for Completely New Users

After using LDA to extract the features of Mashup description, we can obtain the user demand eigenvector in the same vector space by using the same LDA to process the demanded text of a completely new user. Then, the Mashup demand eigenvector is input into the trained DNN, and the potential factor matrix of the user demand can be predicted. The latent factor matrix of Mashup demand is the same as the implicit space of API latent factor matrix obtained by BPR decomposition. The final User-API matrix can be obtained by multiplying the latent factor matrix of user demand and the latent factor matrix of API.

4 Experiment

The experiment will try to answer the following research questions:

RQ1: How to tune parameters to optimize the performance of the LBD framework?
RQ2: How does LBD perform in solving completely new user problems, compared with the state-of-the-art methods?
RQ3: How much does the performance of the LBD framework improve when the data set becomes denser?

The experiments are conducted on a 2.20 GHZ Core i7 processor and 16 GB RAM under Windows 10. All methods are implemented in Python Language.

4.1 Data Set Introduction

The real data set from Programmable Web is crawled for the experiment. The data set includes 6295 Mashup data where 1496 APIs are used. In addition to building the corresponding Mashup-API matrix through the data set, we also capture the corresponding description of Mashup, 4656 Mashup classification tags and five types of services. The statistical characteristics of the data are shown in Table 3.

4.2 Evaluation Metrics

To measure the recommended accuracy of the method we proposed, we adopt three classical metrics: precision, recall rate and F1-score. In the following comparison experiments, we used P, R and F1 respectively to represent them. The higher the calculation results of the three evaluation criteria, the better the recommendation effect of the method.

Table 3. Symbolic representation of LDA model diagram

Relations (X-Y)	Number of X	Number of Y	Number of (X-Y)
Mashup-API	6295	1496	13185
Mashup-Tag	6295	4656	6295
Mashup-Content	6295	6284	13185
Mashup-Type	6295	5	422

4.3 Baseline Approaches

To verify the superiority of our proposed method in solving completely new user problems, we select four state-of-the-art methods LDA-CF [7], BPR-KNN [11], NCF [18] and DHSR [19].

- BPR-KNN [11]: This method uses implicit feedback data with BPR to learn the k-nearest neighbor (KNN) model.
- LDA-CF [7]: This method uses LDA to analyze user text similarity and then makes recommendation based on collaborative filtering.
- NCF [18]: This method uses neural collaborative filtering(NCF) to learn the relationship between mashups and services, then recommends relevant service.
- DHSR [19]: This method uses a novel deep learning based hybrid approach to recommend Web service. The invocation interactions between mashups and services as well as their functionalities are seamlessly integrated into a deep neural network, which can be used to characterize the complex relations between mashups and services.

4.4 Experimental Results and Analysis

Parameter Tuning. *DNN structure optimization.* There are two main factors affecting the performance of the deep neural network model: the number of hidden layers, and the number of neurons contained in each hidden layer. So, we will try to find out the best DNN structure in this experiment. About the other parameters in the DNN models, we chose the same in each experiment. We choose the ReLU as the activating function and set the learning rate to 0.1.

To observe the influence of hidden layer number on model effect, we use neural network from 1 to 3 hidden layers to carry out experiments, and the results of neural network methods with different hidden layers are shown in Table 4.

Result: When the number of hidden layers is 2, the accuracy, recall rate and F1 value of the model are the highest. When the number of hidden layers increases to 3, the performance of the method decreases greatly.

Analysis: Because DNN has more hidden layers than the Perceptron, it can learn more complex non-linear relations. However, because the task of learning the relationship between Mashup description features and potential factors of Mashup is relatively simple for the neural network, two hidden layers are enough to learn the potential relationship, and more hidden layers will be used to fit the phenomenon.

Table 4. Performance of the number of hidden layers

Number of hidden layers	P	R	F1
1	0.481	0.352	0.406511
2	0.496	0.391	0.437285
3	0.463	0.325	0.381916

In addition, in view of the influence of the number of neurons in the hidden layer of the neural network on the effectiveness of the method, we have also carried out relevant parametric adjustment experiments. In order to facilitate the test, the number of neurons in the hidden layer is set to the same constant. On the premise of 2 hidden layers, the number of neurons in each hidden layer is set to 500, 1000 and 1500 respectively, and the performance is compared in Table 5.

Table 5. Performance of number of neurons

Number of neurons	P	R	F1
500	0.496	0.391	0.437285
1000	0.494	0.400	0.442058
1500	0.488	0.395	0.436602

Result: The accuracy decreases with the increase of the number of neurons in the hidden layer. When the number of neurons in the hidden layer is set to 1000, thc method has the highest recall rate and F1 value.

Analysis: Because the complexity of describing text information is general, when the number of neurons in the hidden layer increases to 1500, the feature extraction process will have overfitting phenomenon. After comprehensive consideration, we take the DNN with two hidden layers and 1000 neurons in each layer as the final choice of LBD.

Parameter Optimization in Prediction Part. In the final recommendation prediction part, we search for some users being similar to new users through pre-processing and then recommend APIs. We compare the selection of similar users and the number of recommended APIs.

We use the control variable to optimize K and N. First, we ensure that the number K of recommended APIs is fixed, change the number of similar users N to carry out the control experiment. Second, select the best N value as the final choice, and then carry out the second round of control experiment for different API recommendation number K. The results are shown in Table 6 and Table 7.

Table 6. Performance of different number of similar users

With K = 20, the number of N	P	R	F1
10	0.183	0.253	0.212
20	0.191	0.292	0.231
30	0.230	0.345	0.276

Table 7. Performance of different recommended number of API

With N = 30, the number of K	P	R	F1
5	0.215	0.286	0.245
10	0.236	0.314	0.269
15	0.254	0.356	0.296
20	0.230	0.345	0.276
25	0.192	0.329	0.242

Data Set Partition. The data set partitioning will affect the result of the method, so we conducted a comparative experiment of data sets partitioning on the basic of DNN structure optimization. The data set is divided into the training set and test set according to the ratio of 8:2, 7:3 and 6:4 respectively. The experimental results are shown in Table 8.

Table 8. Performance of data set partition

Training set: Test set	P	R	F1
8:2	0.462	0.360	0.404672
7:3	0.478	0.392	0.430749
6:4	0.441	0.351	0.390886

Result: According to the ratio of training set and testing machine, the recall rate R reaches the highest when the experimental data set is divided into 7:3, and the method effect decreases when the ratio is too high or too low.

Analysis: When the partition ratio of training set and testing machine is 8:2, the method appears over fitting phenomenon, which is caused by the increase of the sparsity of training data set with the increase of training set. When the partition ratio of training set and testing machine is 6:4, the method appears under fitting phenomenon, which is because the training data is small, which is not enough to support the learning task of neural network. Therefore, in this method, the data set is divided into training set and test set in the ratio of 7:3.

Compared Experiment. In order to verify the performance superiority of our proposed method in predicting the demands of the completely new users, we select four baseline methods for comparison.

In this part, we choose the K = 15, N = 30 and take the average value of the code running results of 20 times as the last result of each method. For the baseline methods, we have also carried out parameter tuning experiments to get the best performance. The comparison results of LBD and other methods are shown in Fig. 4.

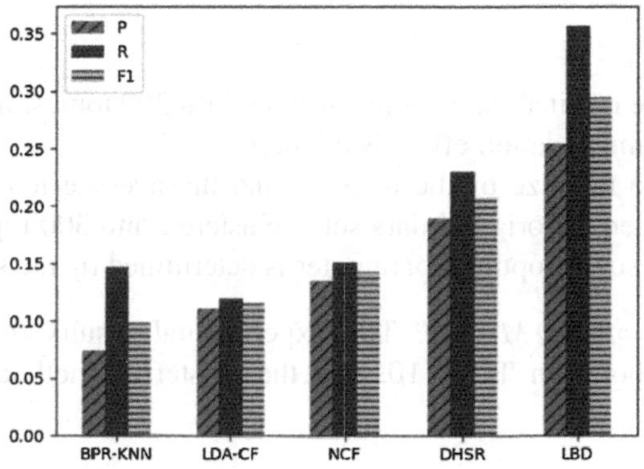

Fig. 4. Performance of different methods

Result: The performance of LBD is significantly improved compared with the other four methods in terms of precision and recall. The precision and recall rate of LBD is 1.34 times and 1.55 times respectively of best baseline method DHSR.

Analysis: BPR-KNN approach decomposes 0-1 matrix effectively and makes APIs recommendation using KNN method. LDA-CF method uses LDA model to analyze the similarity of service description text, and then recommends APIs based on collaborative filtering method. LDA-CF improves the performance better than the BPR-KNN, which proves the analysis of service description plays a key role for the content-based recommendation. However, these two methods fail to make use of the data comprehensively.

NCF approach captures the non-linear relationship between mashups and services based on a neural network framework. Compared with the BPR-KNN and LDA-CF, NCF can characterize the relations between mashups and services better thus it achieves better performance. DHSR approach adopts a hybrid deep learning based recommendation method. Though the DHSR shows that the hybrid recommend method has more advantages in service recommendation, our method LDB performs better than it.

Experiment on Data Sparsity. *Clustering of Data Sets.* In addition, the best clustering number of LBD-Clustering method is tested. We compared the Topic number selection method of clustering operations, and the results are shown in Table 9.

Table 9. Performance of different numbers of Topics

The number of Topic	P	R	F1
100	0.235	0.703	0.352
200	0.266	0.683	0.382
300	0.283	0.657	0.396
400	0.276	0.631	0.384
500	0.252	0.615	0.358

Result: When the original data set is clustered into 300 topics, the recall rate is the highest and the recommendation effect is the best.

Analysis: Due to the size of the data set and the average length of the description information, when the original data set is clustered into 300 topics, it has a better classification effect, so this optimal parameter is determined by the size of our data set.

Sparsity Experiment of LBD Method. The experimental results for data sparsity of LBD method are shown in Table 10, and the clustering method is referred to as LBD-Clustering.

Table 10. Performance of different method effects

Method	P	R	F1
LBD	0.254	0.356	0.295
LBD-clustering	0.283	0.657	0.396

Result: Compared with the original LBD method, LBD-Clustering has improved accuracy, recall rate and F1 score.

Analysis: From the comparison of the experimental results of LBD-Clustering and LBD, because LBD-Clustering carries out classification prediction based on the nature of the experimental data set, the recommendation effect has been greatly improved, which shows that the LBD method has certain expansion in classification prediction, it is also proved that with the increase of data density, the performance of our LBD method still has a lot of room to improve.

5 Conclusion

In order to solve the problem of insufficient information bring by the completely new users for recommendation, we propose a novel deep recommendation framework. At first, the deep neural network is used to extract the mapping relationship between Mashup description and APIs offline. Then, the mapping relationship between the Mashup description extracted in advance can predict matching APIs for new users through their requirements. This framework can completely rely on the text of new users' demands to recommend services. The experimental results on real-world datasets show that the proposed LBD framework has greater improvement in terms of accuracy and recall than the state-of-the-art algorithms.

Acknowledgment. This work was supported by the National Natural Science Foundation of China (Nos. 61602536, 61773415, 61672104).

References

1. Almarimi, N., Ouni, A., Bouktif, S., Mkaouer, M.W., Kula, R.G., Saied, M.A.: Web service API recommendation for automated mashup creation using multi-objective evolutionary search. Appl. Soft Comput. **85**, 105–830 (2019)
2. Liu, X., Hui, Y., Sun, W., et al.: Towards service composition based on mashup. In: IEEE Congress on Services. IEEE (2007)
3. Huang, K., Fan, Y., Tan, W.: An empirical study of programmable web: a network analysis on a service-mashup system. In: ICWS2012, pp. 552–559 (2012)
4. Gu, Q., Cao, J., Peng, Q.: Service package recommendation for mashup creation via mashup textual description mining. In: ICWS 2016, pp. 452–459 (2016)
5. Xiong, W., Wu, Z., Li, B., Gu, Q., Yuan, L., Hang, B.: Inferring service recommendation from natural language API descriptions. In: ICWS 2016, pp. 316–323 (2016)
6. Lin, C., Kalia, A.K., Xiao, J., Vukovic, M., Anerousis, N.: NL2API: a framework for bootstrapping service recommendation using natural language queries. In: ICWS 2018, pp. 235–242 (2018)
7. Lian, T., Ma, J., Wang, S., Cui, C.: LDA-CF: a mixture model for collaborative filtering. J. Chin. Inf. Process. **28**, 129–150 (2014)
8. Song, L., Tekin, C., Schaar, M.V.D.: Online learning in large-scale contextual recommender systems. IEEE Trans. Serv. Comput. **9**(3), 433–445 (2017)
9. Barjasteh, I., Forsati, R., Masrour, F., Esfahanian, A.-H., Radha, H.: Cold-start item and user recommendation with decoupled completion and transduction. In: Conference on Recommender Systems (2015)
10. Zhou, K., Yang, S.H., Zha, H.: Functional matrix factorizations for cold-start recommendation. In: Proceeding of the 34th International ACM SIGIR Conference on Research and Development in Information Retrieval, SIGIR 2011, Beijing, China, 25–29 July 2011. ACM (2011)
11. Rendle, S., Freudenthaler, C., Gantner, Z., Schmidt-Thieme, L.: BPR: Bayesian personalized ranking from implicit feedback. In: Conference on Uncertainty in Artificial Intelligence, pp. 452–461 (2009)
12. Yuan, J., Shalaby, W., Korayem, M., et al.: Solving cold-start problem in large-scale recommendation engines: a deep learning approach. In: IEEE International Conference on Big Data, pp. 1901–1910. IEEE (2017)
13. Obadić, I., Madjarov, G., Dimitrovski, I., Gjorgjevikj, D.: Addressing item-cold start problem in recommendation systems using model based approach and deep learning. In: Trajanov, D., Bakeva, V. (eds.) ICT Innovations 2017. CCIS, vol. 778, pp. 176–185. Springer, Cham (2017). https://doi.org/10.1007/978-3-319-67597-8_17
14. Hao, Y., Fan, Y., Tan, W., Zhang, J.: Service recommendation based on targeted reconstruction of service descriptions. In: ICWS 2017, pp. 285–292 (2017)

15. Blei, D.M., Ng, A.Y., Jordan, M.I., et al.: Latent dirichlet allocation. J. Mach. Learn. Res. **3**, 993–1022 (2003)
16. Mikolov, T., Chen, K., Corrado, G., Dean, J.: Efficient estimation of word representations in vector space. In: ICLR Workshop (2013)
17. Rong, X.: word2vec parameter learning explained. arXiv preprint arXiv:1411.2738 (2014)
18. He, X., Liao, L., Zhang, H., Nie, L., Hu, X., Chua, T.-S.: Neural collaborativefiltering. In: Proceedings of the 26th International Conference on World Wide Web, pp. 173–182 (2017)
19. Xiong, R., Wang, J., Zhang, N., Ma, Y.: Deep hybrid collaborative filtering for web service recommendation. Expert Syst. Appl. **110**, 191–205 (2018)
20. Gao, W., Chen, L., Wu, J., Bouguettaya, A.: Joint modeling users, services, mashups and topics for service recommendation. In: ICWS 2016, pp. 260–267 (2016)
21. Shi, M., Liu, J., Zhou, D., Tang, M., Xie, F., Zhang, T.: A probabilistic topic model for mashup tag recommendation. In: ICWS 2016, pp. 444–451 (2016)
22. Mbipom, B., Massie, S., Craw, S.: An E-learning recommender that helps learners find the right materials. In: AAAI 2018, pp. 7928–7933 (2018)
23. Zhang, M.: Forward-stagewise clustering: an algorithm for convex clustering. Pattern Recognit. Lett. **128**, 283–289 (2019)
24. Mnih, A., Salakhutdinov, R.: Probabilistic matrix factorization. In: Neural Information Processing Systems, pp. 1257–1264 (2007)
25. Koren, Y.: Factorization meets the neighborhood: a multifaceted collaborative filtering model. In: Knowledge Discovery and Data Mining, pp. 426–434 (2008)

Combining Feature Selection Methods with BERT: An In-depth Experimental Study of Long Text Classification

Kai Wang, Jiahui Huang, Yuqi Liu, Bin Cao(✉), and Jing Fan

Zhejiang University of Technology, Hangzhou, China
{wangkai,huangjh,liuyuqi,bincao,fanjing}@zjut.edu.cn

Abstract. With the introduction of BERT by Google, a large number of pre-training models have been proposed. Using pre-training models to solve text classification problems has become the mainstream. However, the complexity of BERT grows quadratically with the text length, hence BERT is not suitable for processing long text. Then the researchers proposed a new pre-training model XLNet to solve the long text classification problem. But XLNet requires more GPUs and longer fine-tuning time than BERT. To the best of our knowledge, no attempt has been done before combining traditional feature selection methods with BERT for long text classification. In this paper, we use the classic feature selection methods to shorten the long text and then use the shortened text as the input of BERT. Finally, we conduct extensive experiments on the public data set and the real-world data set from China Telecom. The experimental results prove that our methods are effective for helping BERT to process long text.

Keywords: Text classification · Long text · BERT · Feature selection

1 Introduction

Text classification has been widely studied and used in many real applications, such as e-mail filtering [7], news classification [1], complaint classification [26], etc. Traditional machine learning methods have been widely used in text classification, such as Logistics Regression (LR) [8,9], Support Vector Machine (SVM) [10,17] and Random Forest (RF) [3]. However, these methods need to construct feature engineering in advance and use One-Hot encoding to represent words. The disadvantage of this is that the relationship between words and the semantic connection between contexts cannot be obtained. With the development of deep learning and the powerful representation capabilities of word vectors, the use of neural networks to solve text classification problems has become the mainstream. The neural networks do not need to construct feature engineering and can capture the semantic information of the text, have achieved excellent classification results. Researchers have proposed many neural network models based on

H. Gao et al. (Eds.): CollaborateCom 2020, LNICST 349, pp. 567–582, 2021.
https://doi.org/10.1007/978-3-030-67537-0_34

CNN and RNN with good classification results, such as TextCNN [11], TextRNN [15,30], Bi-LSTM [31] and TextRCNN [14].

Subsequently, the pre-training model BERT [6] was proposed, and it has achieved great results in many NLP tasks. The network architecture of BERT is composed of multiple layers of Transformers [25], which are based on Multi-head self-attention [25]. Since the computational complexity and space complexity of Transformers are $O(n^2)$, where n is the length of the input sequence, Transformers require huge time and GPU memory to process long sequences, thus BERT requires huge GPU memory for long text. The input text length of BERT is limited to a fixed length L, generally $L = 128,\ 256,\ or\ 512$. However, in long text classification tasks, the length of many texts will exceed L. Take the real-world data of China Telecom as an example, as shown in Fig. 1. As can be seen from Fig. 1, when $L = 512$, only a small number of texts are longer than L. When $L = 256$, there are more than 10,000 texts with the text length exceeding L. When $L = 128$, more than half of the texts are longer than L. Limited by the GPU, L is usually 128 or 256 in practice, which means that BERT cannot obtain all the information of many texts.

Researchers have proposed many solutions to this shortcoming that Transformer cannot handle long sequences. Child et al. [4] proposed the Sparse Transformer, which reduces the complexity of Multi-head self-attention through factorization. Dai et al. [5] proposed Transformer-XL to cut long text into fragments and establish connections between fragments to solve the problem of long dependence. Rae et al. [21] proposed a Compression Transformer based on Transformer-XL. Then another pre-training model XLNet [29] which is based on Transformer-XL has been proposed and has become one of the most effective models. XLNet outperforms BERT in many NLP tasks and solves the problem that BERT cannot handle long texts. However, XLNet requires more GPUs and longer fine-tuning time than BERT [16].

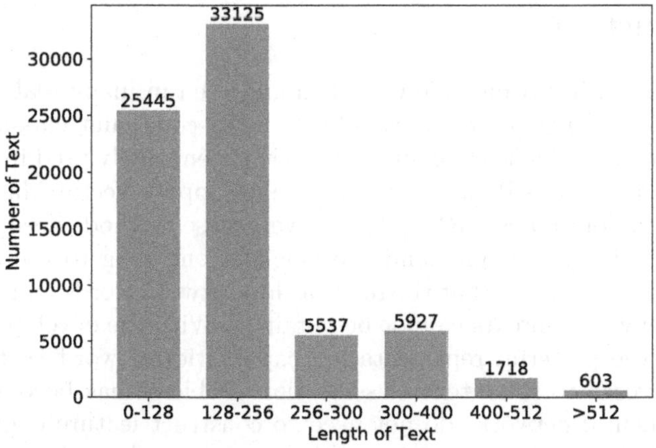

Fig. 1. Text length distribution of China Telecom text data (one month)

In this paper, we adopt six feature selection methods to improve the performance of BERT on long texts. The first four methods are traditional feature selection methods, including TF-IDF [22], TextRank [18], Mutual Information (MI) [13] and χ^2-Statistic (CHI) [23]. We use those four feature selection methods to shorten the length of the text while retaining the key information of the text, and then put the shortened text into the BERT. Considering that the key points of the text may appear at the end of the text, we put forward the next two simple truncation methods: Tail-only and Head+Tail. Tail-only means only to keep the last L tokens of the text, and Head+Tail means to keep the first $L/2$ tokens and the last $L/2$ tokens. Besides, we add XLNet as our baseline. Finally, we conduct extensive experiment on the public data set and real-world data set from China Telecom, experimental results show that our methods are effective.

The contributions of this paper can be concluded as follows:

- We put forward a very common problem in reality that BERT is not suitable for processing long text.
- We propose the method that combines feature selection methods and BERT to process long text. To the best of our knowledge, no attempt has been done before.
- We conduct extensive experiments based on public data set and real-world data set, experimental results prove the effectiveness of our methods.

The rest of this paper is organized as follows. In Sect. 2, we present the related work. Section 3 outlines the preliminaries. In Sect. 4, we describe our experimental results on real-world data set. Finally, we conclude our paper in Sect. 5.

2 Related Works

There are basically two ways to solve the problem that BERT cannot handle long text. The first is to split the input text into several short sentences, and use BERT to process each short sentence separately. The second method is to improve the Transformer to reduce the complexity of the Transformer so that BERT can handle long text.

Yang et al. [28] segmented the long text into several sentences, then put each sentence into the BERT, score each sentence, then score the long text according to the sentence score, and finally process the long text. Pappagari et al. [20] segmented the input text into several smaller chunks, and then put them into the model like BERT to obtain the representation of each chunk, and finally use either a recurrent LSTM network or another Transformer to perform the actual classification. However, this method cannot obtain the semantic information between the chunks.

Dai et al. [5] improved Transformer and proposed Transformer-XL. In order to get the ability to capture long-range dependencies, Transformer-XL proposed a segment-level recurrence mechanism and introduced a memory module, which can keep the connection between the segments. They also proposed a novel relative positional embedding scheme which outperforms the Transformer's original absolute positional system. Sukhbaatar et al. [24] improve the calculation efficiency of the Transformer, they proposed an adaptive attention span that allows the model to adaptively select the context length for processing, thereby reducing the running time and memory usage of the Transformer. Rae et al. [21] proposed the Compressive Transformer, a simple extension to the Transformer which maps past hidden activations (memories) to a smaller set of compressed representations (compressed memories). Kitaev et al. [12] proposed Reformer, they replaced dot-product attention by one that used locality-sensitive hashing, changing the complexity from $O(L^2)$ to $O(LlogL)$. Besides, they adopted reversible residual layers. Beltagy et al. [2] proposed Longformer. The attention mechanism of Longformer is a drop-in replacement for the standard self-attention and combines a local windowed attention with a task motivated global attention.

3 Preliminaries

In this section, we will briefly introduce our feature selection methods: TF-IDF, TextRank, Mutual Information (MI) and χ^2-Statistic (CHI), finally, we will introduce the BERT.

3.1 TF-IDF

TF-IDF (Term Frequency-Inverse Document Frequency) is used to measure the importance of words to sentences. TF refers to the number of times a given word appears in the texts, IDF means that the fewer texts that contain the word, the greater the IDF, indicating that the word has a good ability to distinguish categories. TF and IDF can be calculated with the following formula:

$$TF_w = \frac{n_w}{N}, \; IDF = log(\frac{D}{d_w + 1}) \tag{1}$$

where n_w means the number of the word w, N means the number of all the words, D means the number of all the texts, d_w means the number of texts which contain the word w. Finally, TF-IDF can be calculated with the following formula:

$$TF - IDF = TF * IDF \tag{2}$$

3.2 TextRank

TextRank builds a network through the adjacent relationship between words, and then uses PageRank [19] to iteratively calculate the rank value of each node, and sort the rank value to get the keyword. The PageRank formula is as follows:

$$PR(V_i) = (1 - d) + d * \sum_{j \in In(V_i)} \frac{1}{Out(V_j)} PR(V_j) \tag{3}$$

where $PR(V_i)$ represents the rank value of node V_i, $In(V_i)$ represents the set of predecessor nodes of node V_i, $Out(V_j)$ represents the set of successor nodes of node V_j, d is the damping factor for smoothing.

TextRank keyword extraction for sentences requires only word segmentation of the text, each word is a node. At the same time, assuming that each word is only related to n words in its vicinity, and an undirected edge corresponding to n words in its vicinity is connected, so that a graph is formed and the Rank value of each word can be calculated.

3.3 χ^2-Statistic

χ^2-Statistic (CHI) is a measure of the degree of association between the feature item t_i and category C_j, and it is assumed that the relationship between t_i and C_j conforms to the χ^2 distribution with first-order degrees of freedom. The higher the χ^2 statistic of the feature for a certain class, the greater the correlation, the more category information it carries.

Table 1. The relation between feature item and category

Feature item	Category	
	C_j	$\sim C_j$
t_i	A	B
$\sim t_i$	C	D

As shown in Table 1, Let N be the total number of texts, A represents the number of texts belonging to category C_j and containing t_i, B represents the number of texts not belonging to category C_j but containing t_i, C represents the number of texts belonging to C_j but not containing t_i, and D represents the number of texts that neither belong to C_j category nor contain t_i. The CHI value of t_i and C_j can be calculated according to the formula:

$$\chi^2(t_i, C_j) = \frac{N \times (A \times D - C \times B)^2}{(A + C) \times (B + D) \times (A + B) \times (C + D)} \tag{4}$$

For multi-class problems, suppose M is the number of all categories, we calculate the CHI value of t_i for each category separately, and then take the maximum value:

$$\chi^2_{MAX}(t_i) = max_{j=1}^{M} \chi^2(t_i, C_j) \tag{5}$$

3.4 Mutual Information

Mutual Information (MI) is similar to χ^2-Statistic, the greater the MI, the greater the degree of co-occurrence of feature t_i and category C_j. Similar to the definition of χ^2-Statistic, the mutual information between t_i and C_j can be calculated by the following formula:

$$I(t_i, C_j) = log(\frac{A \times N}{(A + C) \times (A + B)}) \tag{6}$$

Similar to the CHI value, Multi-class mutual information can be obtained by the following formula:

$$I_{MAX}(t_i) = max_{j=1}^{M}[P(C_j) \times I(t_i, C_j)] \tag{7}$$

3.5 BERT

BERT (Bidirectional Encoder Representations from Transformers), a pre-trained model whose goal is to use large-scale unlabeled training corpora to obtain a textual representation containing rich semantic information, and achieved good results in many NLP tasks. The main structure of BERT is Transformer. Transformer is formed by stacking several encoders and decoders, abandoning traditional CNN and RNN, and the entire network structure is composed of the attention mechanism. The input of BERT combines three feature embeddings: token embedding, position embedding, and segment embedding. Token embedding converts each word into a fixed-dimensional vector. Segment embeddings are used to distinguish two sentences. Position embedding encodes the position information of the word into a feature vector. BERT is a multi-task model, and its task is composed of two self-supervised tasks, namely MLM and NSP. MLM refers to masking some words randomly from the input expectation during training, and then predicting the word through the context. The task of NSP is to determine whether sentence B is the bottom of sentence A.

Although BERT has achieved good results in many NLP tasks, due to the defects of the Transformer, BERT has become inadequate when facing long texts.

4 Experimental Evaluation

In this section, we evaluate the performance of each method based on different fixed length L. We first shorten the text whose text length larger than L by different feature selection methods, and then use the shortened text as the input of BERT. Our feature selection methods including TF-IDF, TextRank, MI, CHI, Tail-only, Head+Tail. In addition, we add XLNet as our baseline. For the experimental results, our evaluation metrics including Accuracy, Precision, Recall, F1-score, Hamming-loss and ROC curve & AUC. In order to prove the effectiveness of our methods, we conduct experiments on both public data set and real-world data set. All experiments in our work are evaluated on a computer with Intel Core(TM) i9-9940X 3.30 GHz CPU and RTX 2080Ti graphics and Python 3.6.10. We conduct the experiment of XLNet on 3 GPUs and other experiments on 1 GPU.

4.1 Datasets

We employ two groups of data sets namely IFLYTEK[27] and DianXin.

IFLYTEK is a collection of application description text from the famous Chinese AI company iFLYTEK. IFLYTEK has a total of 190 categories, the training set has 12,133 texts, and the test set has 2599 texts. The text length distribution of IFLYTEK is shown in Fig. 2. Considering that 512 tokens can basically express the semantic information of the text, when L = 512, the experimental difference cannot be well reflected, so we conduct our experiments when $L = 256$ and $L = 128$ respectively.

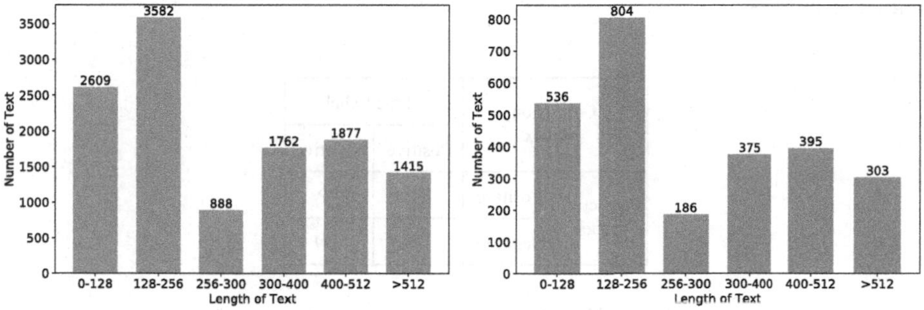

(a) Training set (b) Test set

Fig. 2. Text length distribution of IFLYTEK

DianXin is a collection of real-world complaint texts from China Telecom. DianXin is divided into training set and test set, a total of 580 categories. The training set has 156,834 texts and the test set has 38,890 texts. The text length distribution of DianXin is shown in Fig. 3. From the figure we can see that there are nearly 12% of the texts whose length larger than 256, and there are more than 50% of the texts whose length larger than 128. Therefore, we conduct experiments when $L = 128$ and $L = 256$ respectively to prove that our methods are helpful for BERT to process long texts.

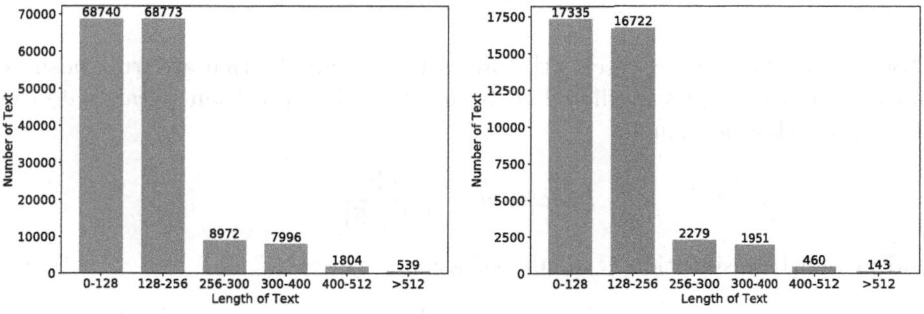

(a) Training set (b) Test set

Fig. 3. Text length distribution of DianXin

4.2 Evaluation Metrics

Confusion Matrix. Figure 4 is a confusion matrix of binary classification. TP (True Positive) indicates the number of samples that predict the positive class as a positive class. FN (False Negative) indicates the number of samples that predict the positive class as the negative class. FP (False Positive) indicates the number of samples that predict the negative class as the positive class. TN (True Negative) represents the number of samples that predict the negative class as a negative class.

In multi-classification, we use the *OVR* method, which means that when we perform classification tasks in a certain class, we treat all other classes as negative classes.

Confusion Matrix		True Label	
		Positive	Negative
Predictive Label	Positive	TP	FP
	Negative	FN	TN

Fig. 4. Confusion matrix

Accuracy. Accuracy is the overall evaluation of the model, and it is the most intuitive metric for evaluating the quality of the model. The accuracy is calculated in the binary classification as:

$$Accuracy = \frac{\mathrm{TP} + \mathrm{TN}}{\mathrm{TP} + \mathrm{TN} + \mathrm{FN} + \mathrm{FP}}$$

and in multi-classification is calculated as:

$$Accuracy = \frac{\sum \mathrm{TP}}{\mathrm{TP} + \mathrm{TN} + \mathrm{FN} + \mathrm{FP}}$$

Precision. Precision represents the proportion of samples that are truly positive among all the samples predicted to be positive. The precision is calculated in the binary classification as:

$$Precision = \frac{\mathrm{TP}}{\mathrm{TP} + \mathrm{FP}}$$

and in multi-classification is calculated as:

$$Precision = \frac{1}{n}\sum_{i=1}^{n} P_i$$

Recall. Recall is the percentage of predicted positive classes among all positive classes. The recall is calculated in the binary classification as:

$$Recall = \frac{\mathrm{TP}}{\mathrm{TP} + \mathrm{FN}}$$

and in multi-classification is calculated as:

$$Recall = \frac{1}{n}\sum_{i=1}^{n} R_i$$

F1-Score. Precision and recall are a pair of contradictory quantities. When precision is high, recall tends to be relatively low, and when recall is high, precision tends to be relatively low. So in order to better evaluate the performance of the classifier, F1-Score is generally used as the evaluation standard to measure the comprehensive performance of the classifier. F1-Score is calculated as:

$$F_1 = \frac{2 \cdot precision \cdot recall}{precision + recall}$$

Hamming-Loss. Hamming-loss is applicable to the problem of multi-classification. In short, it is to measure the loss between the predicted label and the real label, and the value is between 0 and 1. A loss of 0 indicates that the prediction result is exactly the same as the real result, and a loss of 1 indicates that the model is completely contrary to the result we want.

ROC Curve & AUC. ROC (Receiver Operating Characteristic) curve is a commonly used model evaluation method in binary classification. The abscissa of the ROC curve is the false positive rate (FPR), which is calculated as:

$$FPR = \frac{\mathrm{FP}}{\mathrm{FP} + \mathrm{TN}}$$

the ordinate of the ROC curve is the true positive rate (TPR), which is calculated as:

$$TPR = \frac{\mathrm{TP}}{\mathrm{TP} + \mathrm{FN}}$$

The greater the FPR, the more actual negative classes are predicted in the positive class, and the greater the TPR, the more actual positive classes in the positive class.

AUC (Area under Curve): The area under the Roc curve, between 0.1 and 1. As a numerical value, AUC can directly evaluate the quality of the classifier. The greater the value, the better the classifier.

4.3 Experimental Result of IFLYTEK

Due to the small scale of the IFLYTEK, we calculated the weighted-average precision, recall and F1-score. The experimental results when $L = 256$ are shown in Table 2. In terms of accuracy, recall and hamming-loss, the CHI+BERT method achieves the best results. On precision, CHI+BERT has almost the same result as BERT. On F1-sore, CHI+BERT also achieved results second only to XLNet. Overall, the CHI+BERT method outperforms BERT in four metrics, which means that the feature selection method is helpful for BERT to process long texts.

Table 2. Experimental results of IFLYTEK when $L = 256$

Experimental methods	Accuracy	Precision	Recall	F1-score	Hamming-loss
BERT	0.608	0.592	0.608	0.585	0.392
TF-IDF+BERT	0.595	0.565	0.594	0.570	0.405
TextRank+BERT	0.605	0.592	0.605	0.584	0.395
MI+BERT	0.600	0.573	0.600	0.576	0.400
CHI+BERT	**0.611**	0.589	**0.611**	**0.592**	**0.389**
Tail-only	0.586	0.567	0.586	0.562	0.414
Head+Tail	0.606	**0.593**	0.606	**0.586**	0.394
XLNet	0.598	**0.607**	0.598	**0.593**	0.402

Table 3. Experimental results of IFLYTEK when $L = 128$

Experimental methods	Accuracy	Precision	Recall	F1-score	Hamming-loss
BERT	0.601	0.587	0.601	0.582	0.399
TF-IDF+BERT	0.596	0.574	0.596	0.575	0.404
TextRank+BERT	0.594	0.578	0.594	0.574	0.406
MI+BERT	0.569	0.540	0.569	0.543	0.431
CHI+BERT	**0.606**	**0.587**	**0.606**	**0.585**	**0.394**
Tail-only	0.556	0.520	0.556	0.525	0.444
Head+Tail	**0.602**	0.581	**0.602**	0.580	**0.398**
XLNet	0.590	**0.598**	0.590	**0.588**	0.410

The experimental results when $L = 128$ are shown in Table 3. As can be seen from the table, the CHI+BERT method has achieved the best results in accuracy, recall and hamming-loss, and it is second only to XLNet and better than BERT in precision and F1-score. CHI+BERT method outperforms BERT in all metrics, which proves the effectiveness of feature selection.

In conclusion, On the IFLYTEK data set, the CHI+BERT method performs better than BERT, which to a certain extent shows that the combined feature selection methods help BERT to process long text.

4.4 Experimental Results of DianXin

In order to demonstrate the effectiveness of our method in reality, we implemented our experiments on the real-world data set DianXin. And we added the evaluation metric ROC & AUC. We calculate the macro-average precision, recall and F1-score.

The experimental results when $L = 256$ are shown in Table 4. It can be seen from Fig. 3 that when $L = 256$, about 88% of the texts are less than 256 in length, and only about 12% of the texts has feature selection operations. Therefore, the evolution metrics in Table 4 are very close. In terms of accuracy, the accuracy of CHI+BERT, Tail-only and XLNet surpasses BERT, and the gap between CHI+BERT and XLNet is very small. As for precision, CHI+BERT is the highest, next is TF-IDF+BERT. In terms of recall, XLNet has greatly improved, followed by TF-IDF+BERT. XLNet also achieved the best results in F1-score. As for hamming-loss, CHI+BERT and XLNet both have good performance.

The ROC curve and AUC value of each method are shown in Fig. 5. For the two ROC curves, the results of each method are not much different, the macro curve of TF-IDF+BERT is better, and the micro curve of TextRank is better.

Overall, CHI+BERT and XLNet are the best two methods, surpassing BERT in 5 evaluation metrics, and the accuracy of the two is almost the same. The result proves that our methods are effective.

Table 4. Experimental results of DianXin when $L = 256$

Experimental methods	Accuracy	Precision	Recall	F1-score	Hamming-loss
BERT	0.593	0.408	0.354	0.368	0.407
TF-IDF+BERT	0.590	**0.414**	**0.356**	**0.369**	0.410
TextRank+BERT	0.591	**0.409**	0.353	0.365	0.409
MI+BERT	0.593	0.407	0.352	0.366	0.407
CHI+BERT	**0.597**	**0.415**	**0.355**	**0.369**	**0.403**
Tail-only	**0.594**	0.403	**0.355**	0.366	**0.406**
Head+Tail	0.591	0.401	0.352	0.363	0.409
XLNet	**0.599**	**0.409**	**0.380**	**0.383**	**0.401**

The experimental results when $L = 128$ are shown in Table 5. It can be seen from Fig. 3 that when $L = 256$, about 40% of the texts are less than 256 in length, and about 60% of the texts has feature selection operations. It is obvious from Table 5 that the results of most methods are better than BERT. The five evaluation metrics of TF-IDF+BERT, CHI+BERT, Tail-only, Head+Tail and

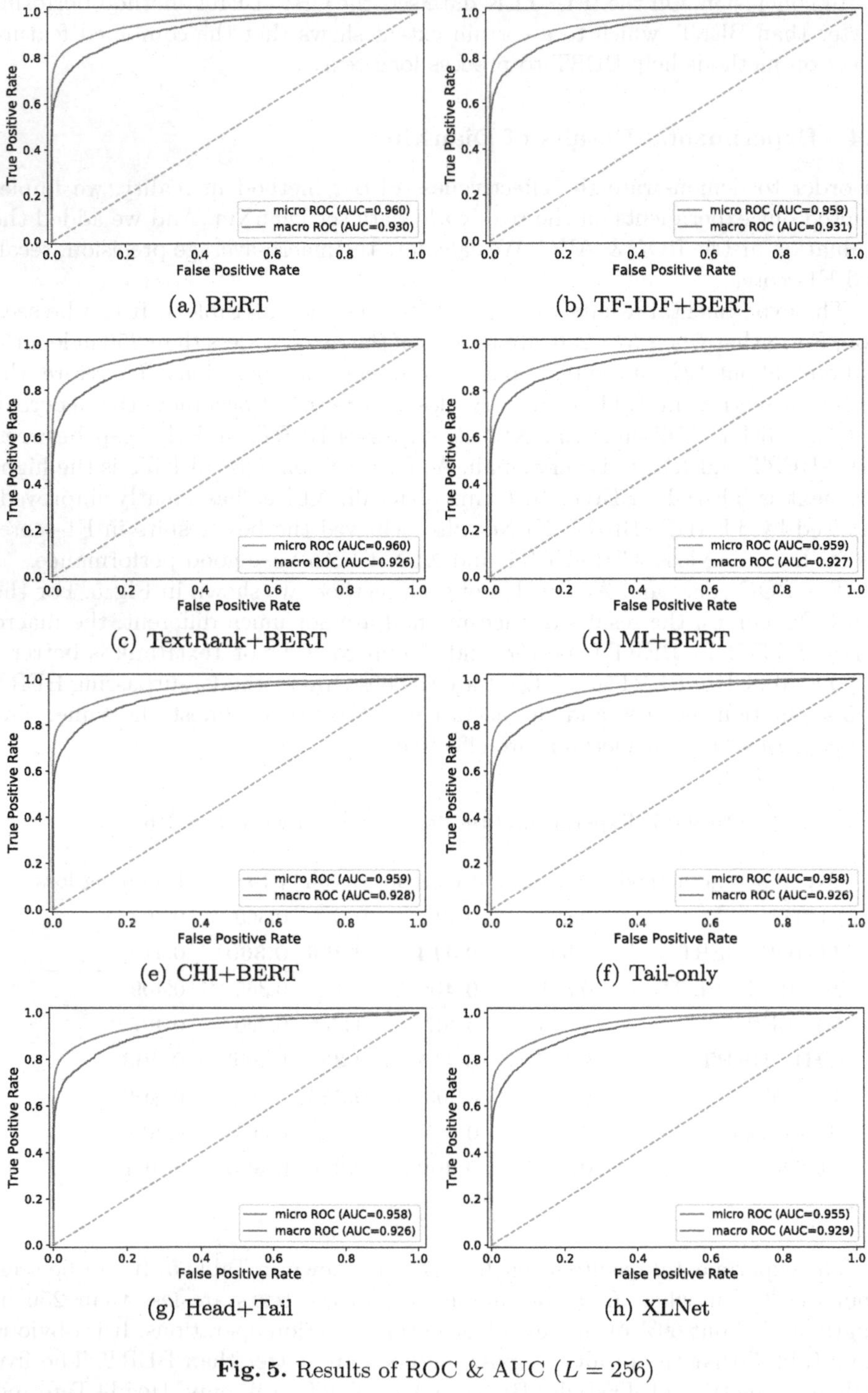

(a) BERT (b) TF-IDF+BERT (c) TextRank+BERT (d) MI+BERT (e) CHI+BERT (f) Tail-only (g) Head+Tail (h) XLNet

Fig. 5. Results of ROC & AUC ($L = 256$)

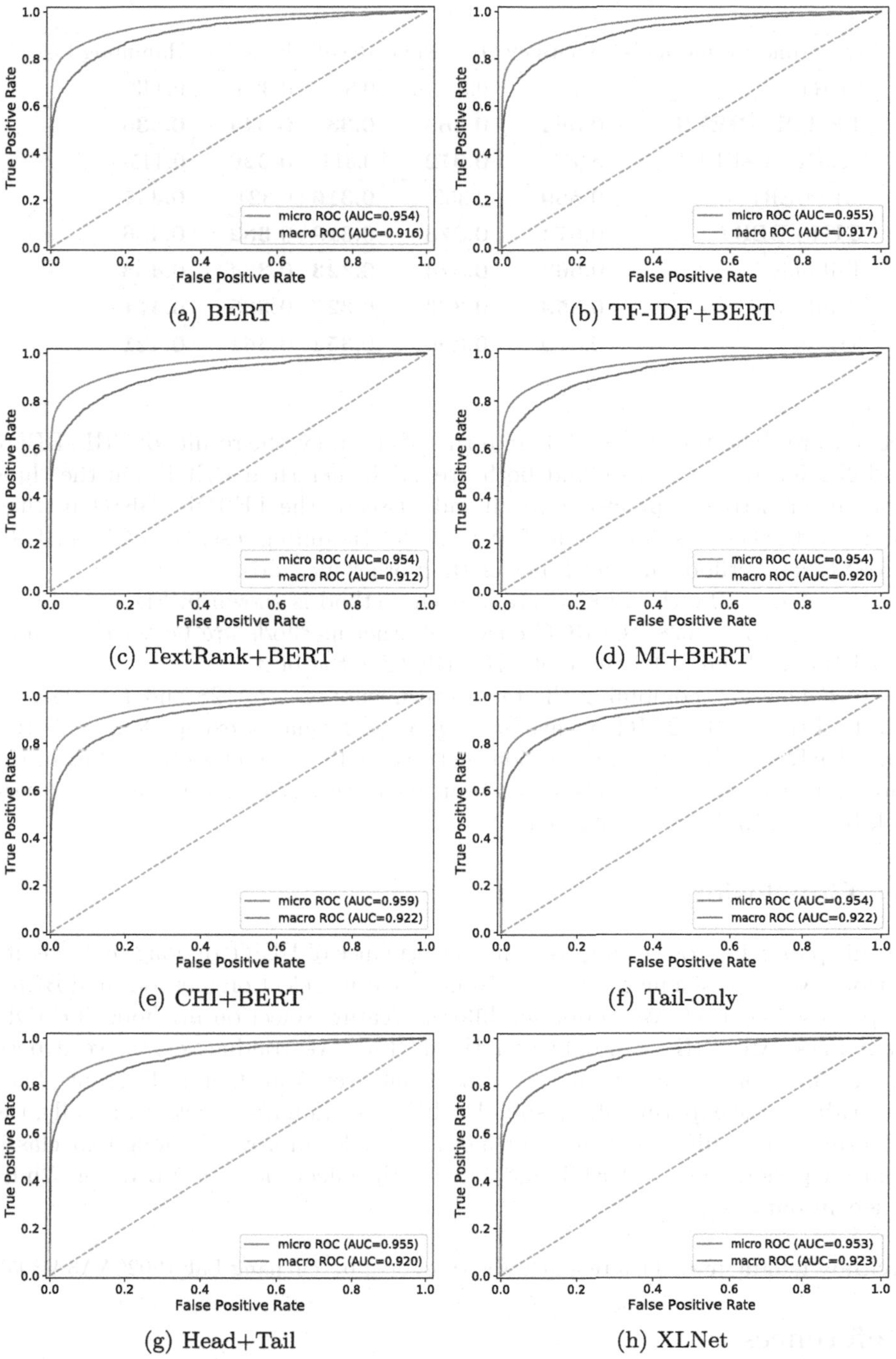

(a) BERT (b) TF-IDF+BERT (c) TextRank+BERT (d) MI+BERT (e) CHI+BERT (f) Tail-only (g) Head+Tail (h) XLNet

Fig. 6. Results of ROC & AUC ($L = 128$)

Table 5. Experimental results of DianXin when $L = 128$

Experimental methods	Accuracy	Precision	Recall	F1-score	Hamming-loss
BERT	0.557	0.363	0.315	0.325	0.443
TF-IDF+BERT	**0.564**	**0.385**	**0.334**	**0.344**	**0.436**
TextRank+BERT	0.555	**0.372**	0.311	**0.326**	0.445
MI+BERT	**0.559**	0.359	**0.316**	0.324	**0.441**
CHI+BERT	**0.574**	**0.373**	**0.319**	**0.332**	**0.426**
Tail-only	**0.562**	**0.374**	**0.323**	**0.334**	**0.438**
Head+Tail	**0.559**	**0.383**	**0.327**	**0.338**	**0.441**
XLNet	**0.579**	**0.390**	**0.354**	**0.361**	**0.421**

XLNet are all better than BERT. In terms of accuracy, the results of CHI+BERT and XLNet are very close, and both are 2% higher than BERT. On the three evaluation metrics of precision, recall and F1-score, the TF-IDF+BERT method achieved the best results except XLNet. As for Hamming-loss, both CHI+BERT and XLNet methods are much better than other methods.

When $L = 128$, the ROC curve of each method is shown in Fig. 6. Unlike $L = 256$, when $L = 128$, the ROC curves of other methods are better than those of BERT, and the ROC curve of CHI+BERT is the best.

In conclusion, combining the two experiments of $L = 128$ and $L = 256$, we can find that CHI+BERT always has a good performance except XLNet, better than BERT in almost all evaluation metrics, and is very close to XLNet in the most intuitive accuracy. This shows that our feature selection methods are very helpful for BERT processing long text.

5 Conclusion

In this paper, in order to improve the performance of BERT on long text classification. We propose a method that combines feature selection methods and BERT to process long text. We adopt six different feature selection methods: TF-IDF, TextRank, MI, CHI, Tail-only and Head+Tail. We conduct extensive experiments on public data set and real-world data set from China Telecom. Then we evaluate the experimental results by different evaluation metrics. Finally, the experimental results show that our methods could improve the long text classification performance of BERT and achieve the effect close to XLNet on China Telecom data set.

Acknowledgments. This research was sponsored by Zhejiang Lab (2020AA3AB05).

References

1. Ahmed, H., Traore, I., Saad, S.: Detecting opinion spams and fake news using text classification. Secur. Priv. **1**(1), e9 (2018)

2. Beltagy, I., Peters, M.E., Cohan, A.: Longformer: the long-document transformer. arXiv preprint arXiv:2004.05150 (2020)
3. Breiman, L.: Random forests. Mach. Learn. **45**(1), 5–32 (2001)
4. Child, R., Gray, S., Radford, A., Sutskever, I.: Generating long sequences with sparse transformers. arXiv preprint arXiv:1904.10509 (2019)
5. Dai, Z., Yang, Z., Yang, Y., Carbonell, J., Le, Q.V., Salakhutdinov, R.: Transformer-xl: attentive language models beyond a fixed-length context. arXiv preprint arXiv:1901.02860 (2019)
6. Devlin, J., Chang, M.W., Lee, K., Toutanova, K.: BERT: pre-training of deep bidirectional transformers for language understanding. arXiv preprint arXiv:1810.04805 (2018)
7. Diao, Y., Lu, H., Wu, D.: A comparative study of classification based personal e-mail filtering. In: Terano, T., Liu, H., Chen, A.L.P. (eds.) PAKDD 2000. LNCS (LNAI), vol. 1805, pp. 408–419. Springer, Heidelberg (2000). https://doi.org/10.1007/3-540-45571-X_48
8. Genkin, A., Lewis, D.D., Madigan, D.: Large-scale Bayesian logistic regression for text categorization. Technometrics **49**(3), 291–304 (2007)
9. Hosmer Jr., D.W., Lemeshow, S., Sturdivant, R.X.: Applied Logistic Regression, vol. 398. Wiley, Hoboken (2013)
10. Joachims, T.: Text categorization with support vector machines: learning with many relevant features. In: Nédellec, C., Rouveirol, C. (eds.) ECML 1998. LNCS, vol. 1398, pp. 137–142. Springer, Heidelberg (1998). https://doi.org/10.1007/BFb0026683
11. Kim, Y.: Convolutional neural networks for sentence classification. arXiv preprint arXiv:1408.5882 (2014)
12. Kitaev, N., Kaiser, Ł., Levskaya, A.: Reformer: the efficient transformer. arXiv preprint arXiv:2001.04451 (2020)
13. Kraskov, A., Stögbauer, H., Grassberger, P.: Estimating mutual information. Phys. Rev. E **69**(6), 066138 (2004)
14. Lai, S., Xu, L., Liu, K., Zhao, J.: Recurrent convolutional neural networks for text classification. In: Twenty-Ninth AAAI Conference on Artificial Intelligence (2015)
15. Liu, P., Qiu, X., Huang, X.: Recurrent neural network for text classification with multi-task learning. arXiv preprint arXiv:1605.05101 (2016)
16. Liu, X., Wangperawong, A.: Transfer learning robustness in multi-class categorization by fine-tuning pre-trained contextualized language models. arXiv preprint arXiv:1909.03564 (2019)
17. Manevitz, L.M., Yousef, M.: One-class SVMs for document classification. J. Mach. Learn. Res. **2**(Dec), 139–154 (2001)
18. Mihalcea, R., Tarau, P.: TextRank: bringing order into text. In: Proceedings of the 2004 Conference on Empirical Methods in Natural Language Processing, pp. 404–411 (2004)
19. Page, L., Brin, S., Motwani, R., Winograd, T.: The PageRank citation ranking: bringing order to the web. Tech. rep., Stanford InfoLab (1999)
20. Pappagari, R., Zelasko, P., Villalba, J., Carmiel, Y., Dehak, N.: Hierarchical transformers for long document classification. In: 2019 IEEE Automatic Speech Recognition and Understanding Workshop (ASRU), pp. 838–844. IEEE (2019)
21. Rae, J.W., Potapenko, A., Jayakumar, S.M., Lillicrap, T.P.: Compressive transformers for long-range sequence modelling. arXiv preprint arXiv:1911.05507 (2019)
22. Ramos, J., et al.: Using TF-IDF to determine word relevance in document queries. In: Proceedings of the First Instructional Conference on Machine Learning, vol. 242, pp. 133–142, New Jersey (2003)

23. Satorra, A., Bentler, P.M.: A scaled difference chi-square test statistic for moment structure analysis. Psychometrika **66**(4), 507–514 (2001)
24. Sukhbaatar, S., Grave, E., Bojanowski, P., Joulin, A.: Adaptive attention span in transformers. arXiv preprint arXiv:1905.07799 (2019)
25. Vaswani, A., et al.: Attention is all you need. In: Advances in Neural Information Processing Systems, pp. 5998–6008 (2017)
26. Wang, S., Wu, B., Wang, B., Tong, X.: Complaint classification using hybrid-attention GRU neural network. In: Yang, Q., Zhou, Z.-H., Gong, Z., Zhang, M.-L., Huang, S.-J. (eds.) PAKDD 2019. LNCS (LNAI), vol. 11439, pp. 251–262. Springer, Cham (2019). https://doi.org/10.1007/978-3-030-16148-4_20
27. Xu, L., et al.: Clue: a Chinese language understanding evaluation benchmark. arXiv preprint arXiv:2004.05986 (2020)
28. Yang, W., Zhang, H., Lin, J.: Simple applications of BERT for ad hoc document retrieval. arXiv preprint arXiv:1903.10972 (2019)
29. Yang, Z., Dai, Z., Yang, Y., Carbonell, J., Salakhutdinov, R.R., Le, Q.V.: XLNet: generalized autoregressive pretraining for language understanding. In: Advances in Neural Information Processing Systems, pp. 5753–5763 (2019)
30. Yang, Z., Yang, D., Dyer, C., He, X., Smola, A., Hovy, E.: Hierarchical attention networks for document classification. In: Proceedings of the 2016 Conference of the North American Chapter of the Association for Computational Linguistics: Human Language Technologies, pp. 1480–1489 (2016)
31. Zhou, P., Qi, Z., Zheng, S., Xu, J., Bao, H., Xu, B.: Text classification improved by integrating bidirectional LSTM with two-dimensional max pooling. arXiv preprint arXiv:1611.06639 (2016)

Defending Use-After-Free via Relationship Between Memory and Pointer

Guangquan Xu[1(✉)], Miao Li[1], Xiaotong Li[1], Kai Chen[2], Ran Wang[3], Wei Wang[4], Kaitai Liang[5], Qiang Tang[6], and Shaoying Liu[7]

[1] College of Intelligence and Computing, Tianjin University, Tianjin, China
{losin,lixiaotong}@tju.edu.cn, limiao_tju@163.com
[2] Institute of Information Engineering, Chinese Academy of Sciences, Beijing, China
chenkai@iie.ac.cn
[3] Security Center, JD.com, Beijing, China
wangran8088@gmail.com
[4] Beijing Jiaotong University, Beijing, China
wangwei1@bjtu.edu.cn
[5] Surrey Centre of Cyber Security, University of Surrey, Guildford, UK
k.liang@surrey.ac.uk
[6] New Jersey Institute of Technology, Newark, USA
qiang@njit.edu
[7] Hosei Univesity, Tokyo, Japan
sliu@hosei.ac.jp

Abstract. Existing approaches to defending Use-After-Free (UAF) exploits are usually done using static or dynamic analysis. However, both static and dynamic analysis suffer from intrinsic deficiencies. The existing static analysis is limited in handling loops, optimization of memory representation. The existing dynamic analysis, which is characterized by lacking the maintenance of pointer information, may lead to flaws that the relationships between pointers and memory cannot be precisely identified.

In this work, we propose a new method called UAF-GUARD without the above barriers, in the aim to defending against UAF exploits using fine-grained memory permission management. In particular, we design a key data structure to support the fine-grained memory permission management, which can maintain more information to capture the relationship between pointers and memory. Moreover, we design code instrumentation to enable UAF-GUARD to precisely locate the position of UAF vulnerabilities to further terminate malicious programs when anomalies are detected.

We implement UAF-GUARD on a 64-bit Linux system. We carry out experiments to compare UAF-GUARD with the main existing approaches. The experimental results demonstrate that UAF-GUARD is able to effectively and efficiently defend against three types of UAF exploits with acceptable space overhead and time overhead.

H. Gao et al. (Eds.): CollaborateCom 2020, LNICST 349, pp. 583–597, 2021.
https://doi.org/10.1007/978-3-030-67537-0_35

Keywords: Use-after-free vulnerability · Fine-grained memory permission management · Static instrumentation

1 Introduction

As a memory corruption flaw, Use-After-Free (UAF) vulnerability is defined by Common Weakness Enumeration (CWE) as an attack that "referencing memory after it has been freed can cause a program to crash, use unexpected values, or execute code" [1]. A single UAF vulnerability cannot be exploited in isolation, and it must be exploited along with other heap memory exploitation techniques (e.g., Heap Spray [2]). As a result, UAF vulnerability exploitation can often be discovered from most heap vulnerabilities. The exploit of UAF vulnerabilities may result in horrifying system corruption, such as arbitrary read, write-back, and malicious code execution. Arbitrary read may lead to information leakage, including the system information, user sensitive data, and process memory layout. Once attackers get the process information, they can bypass system security guards, e.g., Canary [3], PIE [4], ASLR [5], and further combine with other vulnerabilities to launch attacks to computer systems. Arbitrary write-back and code execution may make control flow be easily hijacked in such a way that attackers may execute preset instructions, potentially get shell and obtain all the permissions of a system.

There have been some proposed techniques in the literature to address the UAF vulnerability exploit problem.

Static Detection and Defense. The static analysis is defined to work as follows: a) converting binary programs into an uniform Intermediate Representation (IR), in which the conversion methods use small instruction sets, e.g., REIL IL [6], Bincoa [7] and BAP [8]); b) analyzing the code of IR with symbolic execution [9] and abstract interpretation [10]; c) building the Control Flow Graph (CFG) to describe the relationships inter and intra functions. Specifically, using the analysis results of the famous commercial reverse software IDA Pro [11] to avoid the repeated analysis for a given function; d) extracting the UAF vulnerabilities from the analysis results.

Traditional dynamic solutions, such as [12–14], choose to nullify the corresponding dangling pointer - a pointer which points to a freed memory - after freeing an object. However, it may easily imperil the original program logic and be extremely difficult to detect the vulnerability where the related pointer is fulfilled by calculating relative offsets.

Memory error detector, such as [15–17], can also be used to capture UAF vulnerabilities at runtime. By maintaining the allocated memory status, it may identify if a given memory is used after being freed. However, if attackers can control the heap memory allocation process, for instance, they make use of Heap Spray [2] to force the program to reallocate memory on a freed memory, this approach may be prone to be bypassed.

```
#include <stdlib.h>
#include <string.h>
    using namespace std;

    int main(){
        char *ptr = (char *) malloc(40);
        free(ptr);
        strcpy(ptr, "Use-After-Free");
        return 0;
    }
```

Fig. 1. An example of Type I UAF vulnerability.

```
#include <stdlib.h>
#include <string.h>
    using namespace std;

    int main(){
        char *ptr = (char *) malloc(40);
        char *ptr2 = ptr;
        _int64 val = (_int64) ptr;
        free(ptr);
        ptr = 0;
        strcpy(ptr2, "Use-After-Free");
        strcpy((char *)(val + 0x8), "Use-After-Free");
        return 0;
    }
```

Fig. 2. An example of Type II and III UAF vulnerabilities.

We make the following contributions:

- We propose an effective Use-After-Free (UAF) vulnerabilities detection and defense approach called UAF-GUARD. It can prevent pointer abuse and further defend the exploits of all types of UAF vulnerabilities (we define in this work) through checking the permission of the pointer-to-memory at runtime.
- We design a key data structures, which can maintain more information between pointer and memory, to concisely represent the relationship of them.
- We implement UAF-GUARD and evaluate its performance. The evaluation result shows that our UAF-GUARD approach can efficiently detect all the defined vulnerabilities with 5.7% space on average and 22% time overhead, respectively.

2 Background

In this section, we introduce the principle and exploitation of UAF vulnerabilities, and we define three types of UAF vulnerabilities to describe the characteristics of UAF and the difficulty of defending.

2.1 Use-After-Free Exploits

When an operating system allocates dynamically memory to a process through the `brk` or `mmap` system call (in Linux), the memory allocated to the process should be managed by the memory management mechanism in order to get rid of memory fragmentation. But we have to state that after being freed by these mechanisms, the allocated memory will yield UAF exploits if there is a way to access this memory.

2.2 Three Types of UAF Vulnerabilities

To understand the UAF vulnerability more clearly, we categorize the UAF vulnerabilities into three types as shown in Table 1 according to the source of exploited pointers and the difficulty of exploits.

Table 1. Three types of UAF vulnerability.

UAF vulnerability Type	Attack Means
Type I	original allocated pointer
Type II	fulfilled by pointer propagation
Type III	fulfilled by calculating relative offset

Type I allows attacker to directly leverage the pointer that is allocated with the corresponding memory space, to run read/write/execute operations. This type of vulnerability is due to the lack of carrying out the nullifying of the pointer in time, which results in dangling pointer. We state that Type I is the most common UAF vulnerability that may be easily exploited, and we show an example of the vulnerability of this type in Fig. 1.

In Type II, a pointer is obtained by pointer propagation (see `ptr2` in line 7 of Fig. 2). It has to rely on an existing pointer which points to a freed memory. For example, the pointer variable `ptr` is returned when memory space is allocated, but the value of the variable `ptr2` is made identical to `ptr` by assignment. The UAF vulnerabilities caused by `ptr2` should belong to Type II (please refer to an example in Fig. 2, at line 11).

Type III enables the pointer to be obtained by directly calculating offset, rather than by using the existing pointer which has already pointed to a freed memory. Type III vulnerability is typically exploited by combining multiple processing logic (please refer to Fig. 2, at line 12). Its exploitation may require the combination of other vulnerabilities, such as integer overflow and stack segment overflow.

3 UAF-GUARD Design

In this section, we present an overview of UAF-GUARD. To improve the performance of defending against the UAF vulnerabilities, we propose two key data structures before proceeding to the design of UAF-GUARD. We then introduce our technical details w.r.t. static instrumentation and runtime defending. Finally, we propose a target instruction filtering algorithm to optimize the detection overhead.

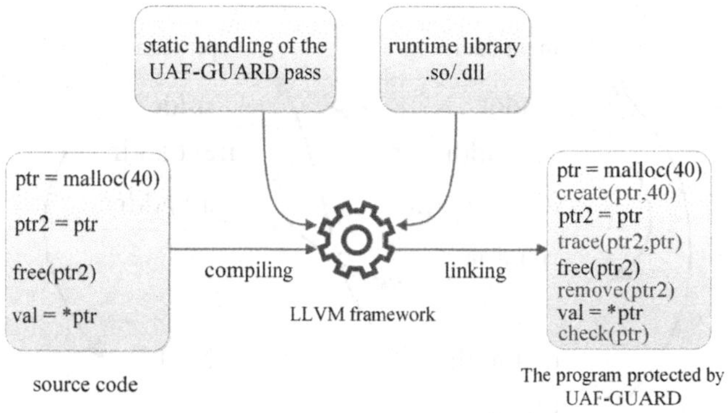

Fig. 3. The overall framework of UAF-GUARD.

3.1 Overview

To detect the UAF vulnerabilities, UAF-GUARD checks the permission of pointer to memory for each pointer operation at runtime and strictly limits the usage of pointer in user space.

As described in Sect. 2.1, a successful exploitation over a UAF vulnerability requires the following needs: a) the memory management mechanism frees the allocated memory, b) attackers read/write the freed memory or execute code, and c) proceed to subsequent exploitations, including reallocating the freed memory, information leakage and control flow hijack.

An overview of our design is illustrated in Fig. 3. The overview is based on LLVM [18] compiler framework/system. The "UAF-GUARD pass" in Fig. 3 is realized by the LLVM PASS (which is a module of code optimization in LLVM). UAF-GUARD participates in the compilation and linking process of the source code so that it can a) insert the instrumentation code in the compiled target program, b) link the compiled function library with the binary program during the linking process, and c) implement the instrumentation function to produce the protected binary program.

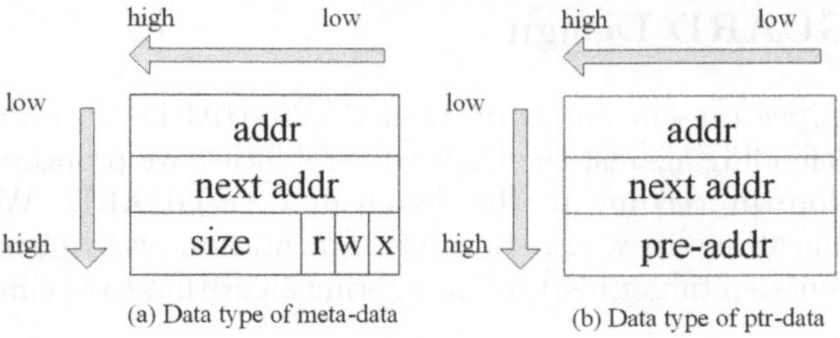

Fig. 4. Data types of meta-data and ptr-data.

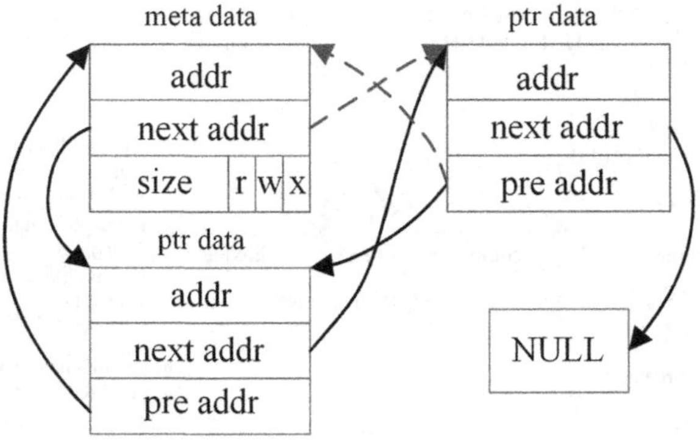

Fig. 5. Data structure of the relationship between memory and pointer.

3.2 Key Data Structures

To verify the permission of pointer-to-memory in a convenient way, we first design a key data structures: data structure for memory, pointer and their relationship.

Data Structure for Memory, Pointer and Their Relationship. We here design "special" data structures to record the information of memory, pointer and their relationship, which will be used for subsequent maintenance of the permission of pointer-to-memory.

Figure 4(a) shows the data type of the meta-data which describes the memory block. The meta-data includes three fields: `addr`, `next-addr`, and `size|rwx`. `addr` indicates the starting position of the data in memory chunk. `next-addr` represents the next data structure address in the linked list associated with the memory, which contains two types of address: a) the data structure address describing the pointer information (which is the permission of the pointer to the memory), and b) 0xFFFFFFFF (taking 32-bit system for example) indicating that there is no pointer pointing to the memory. The last field `size|rwx` consists of two parts, in which `size` is the size of the memory. Since the memory block allocated by `ptmalloc2` is 8 byte aligned in 32-bit operation system - i.e. the size of this field is a multiple of 8, and thus we can utilize the last three bits of this field (`rwx`) to store the permissions (i.e., read/write/execute) of memory. Such permission information will be checked in our later instrumentation realization (see Sect. 3.3).

Figure 4(b) illustrates the data type ptr-data which describes the pointer information. ptr-data has three fields: `addr`, `next-addr`, and `pre-addr`, in which `addr` indicates the address of pointer variable, `next-addr` shares the same definition as in meta-data, and `pre-addr` indicates the previous data structure address in the linked list associated with the pointer. The last field contains two types of address: a) the data structure type address describing the pointer information which has permission on the same memory, and b) the address of the memory meta-data which referred by the pointer, indicating that the pointer has the permission to the memory.

Figure 5 presents the data structure of the relationship between memory and pointers. Memory and pointer metadata are stored in a red-black tree structure. In addition, UAF-GUARD needs a doubly linked list for describing the permission of pointers to memory. This is done by utilizing the fields including pointer information in meta-data data type and ptr-data data type. The dotted line indicates the indirect pointing relationship. The operation in the doubly linked list is relatively simple through the pointer operations, including inserting and deleting elements.

The data types proposed in this section can be used to describe the information of memory and pointer at bit-level granularity, which means that our memory permission tracking provides a bit-level granularity. Traditional process memory permissions are coarse-grained, usually divided by "segment" or further by "section", which owns the consistent access permissions. Our UAF-GUARD divides memory permissions more properly. The minimum unit of memory permission is 64 bit. Bit-level granularity makes more precise and flexible control of memory permissions, and hence it performs better in preventing pointer abuse. In addition, the doubly linked list is sufficient to describe the many-to-one relationship between pointers and memory.

In real-world applications, the number of the pointers linked to the same memory is relatively small (since `m` is small), therefore the complexity of the traversal operation should be acceptable.

3.3 Technical Details

In this section, we introduce the technical details of our UAF-GUARD, including the mechanism of static instrumentation and the design of the runtime library for achieving the goal of runtime defenses.

Static Instrumentation. Static instrumentation for our UAF-GUARD is implemented in the LLVM compiler framework/system, therefore the source code can be compiled into an intermediate representation based on LLVM IR. The UAF-GUARD aims to achieve the following goals based on LLVM: a) generate the corresponding meta-data for each memory allocation, and b) for each pointer operations, propagate the associated memory information or updating the memory permissions of the pointer, so as to protect all the heap memory blocks.

In the compilation phase, we traverse each instruction in each function, parse all related instructions with the pointer operations using LLVM IR, and save the corresponding parameters and opcodes into the preset variables. We classify the parsed instructions obtained from previous step according to the LLVM IR instruction set, and insert the corresponding instrumentation functions according to each instruction.

The handling details for each type of instruction in LLVM IR are described as follows.

a) Memory allocation, reading and writing instructions.

- Memory allocation/free instructions. Since the `call` instruction sends request to the heap for memory space, it should be invoked. Taking C++ in Ubuntu Linux x86 system as an example, we use `call@malloc(k)` as the memory request function that returns is `(i8*)ptr`, where `k` is a positive integer. In the UAF-GUARD, we insert an instrumentation function `create(ptr, k)` to record the allocated memory block after this instruction. If the memory block is successfully allocated, the record is inserted into the data structure. Freeing the memory, we need to use `call@free(ptr)`, therefore, we should insert an instrumentation function `remove(ptr)` to delete the record of the memory block before the instruction. If the record is successfully deleted, the instruction will be executed.
- Memory reading instruction. Since the load instruction leverages the pointer to read process memory, which is equivalent to using the read permission of the pointer, we insert an instrumentation function `check(ptr)` before the load instruction. The load instruction usually uses `%val = load i32, i32* %ptr`. If `check(ptr)` doesn't throw an exception, the instruction will be executed (but not the other way around). We insert an instrumentation function `trace(ptr, val)` before the instruction if `val` is a pointer type. If `ptr` is a valid heap memory address after parsing, `val` will get the permission of the memory, which is called "permission propagation". Otherwise, if `ptr` is an invalid address, the instruction cannot be executed.
- Memory writing instruction. The write instruction usually uses `store i32 3, i32* %ptr`, before which we insert an instrumentation function `check(ptr)`. Since the instruction uses the pointer to write into the process memory, which is equivalent to making use of the write permission of the pointer. If `check(ptr)` doesn't throw an exception, the instruction will be executed, and otherwise, it will not be executed.

b) Arithmetic operation and bit operation instructions, including `add/fadd`, `sub/fsub`, `mul/fmul`, `udiv/sdiv/fdiv`, `urem/srem/frem`, `shl`, `lshr`, `ashr`, `and`, `or`, `xor`. UAF-GUARD focuses on the instructions associated with the heap memory pointer. When the arithmetic result includes pointer `ptr`, we insert an instrumentation function `trace(val, ptr)` before the instruction (if the instruction directly assigns the result to the pointer `ptr`). Namely, the right value `val` calculated by this instruction is assigned to the pointer variable `ptr`, so as to change the address pointed by the pointer. If the value is a valid address, the `ptr` will get the permission of the memory and be recorded into the UAF-GUARD data structure.

Runtime Defending. According to the preset instrumentation functions in the compilation process of the static instructions, we design a runtime library for UAF-GUARD to fulfil the task of runtime defending. There are four main instrumentation functions, known as `create()`, `trace()`, `check()`, and `remove()`.

`create()`. The instrumentation function `create(ptr, k)` is to build the permission relationship between pointer and memory. The `ptr` is a pointer storing the address of the memory that is allocated by the heap memory management mechanism, and `k` is the size of the heap memory space that requested by users. The design of `create()` is described as follows.

a) If the value of `ptr` is null, which means that `malloc()` has not successfully allocated the memory, then we stop building the permission relationship.
b) Otherwise, we build the meta-data of memory block and ptr-data of pointers according to the value of `ptr` and `k`, and establish the relationship of meta-data and ptr-data in doubly linked list. Moreover, we insert the meta-data and ptr-data into the enhanced red-black tree to accelerate the querying process. We use the memory address and pointer address as the keys of meta-data and ptr-data, respectively.

`check()`. The instrumentation function `check(ptr)` is to check the permission of pointer to memory, where the `ptr` is the pointer to be checked. The process of `check()` is described as follows.

a) Get the address and value of the pointer from `ptr`.
b) Search the information of the pointer in the structure. If the ptr-data of this pointer cannot be found, it indicates that the pointer is abused. Otherwise, if we get the information of the ptr-data, we can further trace the meta-data of the memory associated by the pointer in the doubly linked list, and calculate the scope of the memory block to determine if 1) the value of `ptr` is within the scope so as to prevent the pointer being obtained by assigning or calculating offsets; and 2) the pointer has the corresponding permissions in memory read, write, and execute. If the `ptr` can satisfy the above two conditions, the `ptr` will pass the detection.

`trace()`. The instrumentation function `trace(ptr1, ptr2)` is to transfer the memory permission between pointers, where the `ptr1` is a heap memory address or a pointer that points to a heap memory block, the `ptr2` is the pointer to be assigned, and `trace()` is an overloaded function to handle different parameters. The process of `trace()` works as follows.

a) If `ptr1` is a pointer, we determine if `ptr1` is legal by checking its address and value (that is, if the memory which `ptr1` points to is an allocated available memory).
b) If `ptr1` is a heap memory address, we determine if the address is a valid and allocated heap memory address.
c) If `ptr1` is illegal, the status information will be recorded and the program will throw an exception.
d) If `ptr1` is legal, we check if the information of `ptr2` has been built. If not, we build the ptr-data of `ptr2`, which will be inserted into the tree structure. Lastly, the permission transferring is accomplished after the `ptr2` is inserted into the doubly linked list of `ptr1`.

`remove()`. The instrumentation function `remove(ptr)` is to remove the permission of pointer to memory. The process of `remove()` works as follows.
a) If `ptr` is a heap memory address, we check if the address is valid. If not, it indicates that the memory to be freed is the pointer's unprivileged memory, unallocated memory or freed memory. Otherwise, we maintain the data structure to find the pointer information ptr-data of all associated pointers and meta-data of the corresponding memory block. After that, these elements from the tree structure are removed.
b) If `ptr` is pointer, removing the permission indicates the end of the pointer life cycle. We further decide if the `ptr` exists in the tree structure. If yes, delete them from the doubly linked list and the tree structure. Otherwise, it indicates that the pointer is not pointing to a memory block.

4 Experimental

To verify the effectiveness and efficiency of our design, we present a prototype implemented for our UAF-GUARD and compare it with the related ones. Our experiments include the security and performance tests. Note that the source code of our UAF-GUARD is publicly accessible: https://github.com/UAF-GUARD/sourcecode.

4.1 Experimental Environment

First, we evaluate the defending effectiveness of the UAF-GUARD over all types of UAF vulnerabilities, and further test the well-known CVE vulnerabilities in multiple versions of programs including Chrome, Wireshark, and OpenSSL. We further calculate the runtime overhead of the UAF-GUARD in the compiling phase. To bring simplicity and fairness in the comparison, we use the same pending programs across the related works, e.g., bzip2, and gcc. Finally, we analyze the runtime overhead of the UAF-GUARD based on the compilation results, and prove that UAF-GUARD can ensure the security of program at runtime.

All the experiments in this paper are accomplished on a PC with the configuration: eight-core Intel Core i7-6700HQ CPU @ 2.60 GHz, 16 GB RAM and 500 GB SSD, with 64-bit Ubuntu 16.04 (Linux Kernel 4.16).

Table 2. Comparison between UAF-GUARD and DANGNULL.

Vulnerabilities	Incidence	Position Of Vulnerabilities	Type	Detection Result	
				UAF-GUARD	DANGNULL
CVE-2010-2939	OpenSSL 1.0.0a, 0.9.8,0.9.7	0x800000000 22ba510	II	SIGSEGV	SIGSEGV
CVE-2016-4077	Wireshark 2.0.0-2.0.3	-	II	SIGSEGV	SIGSEGV
CVE-2013-2909	Google Chrome 30.0.1599.66	0x1bfc9901ece1	II	SIGSEGV	NORMAL
CVE-2013-2909	Google Chrome 30.0.1599.66	0x7f2f57260968	II	SIGSEGV	NORMAL
CVE-2013-2918	Google Chrome 30.0.1599.66	0x490341400000	III	SIGSEGV	TRIGGER
CVE-2013-2922	Google Chrome 30.0.1599.66	0x60b000006da4	II	SIGSEGV	NORMAL
CVE-2013-6625	Google Chrome 31.0.1650.48	0x897ccce6951	II	SIGSEGV	SIGSEGV
CVE-2012-5137	Google Chrome 23.0.1271.95	0x612000046c18	II	SIGSEGV	ASSERTION

4.2 Effectiveness of Detection

Our experiment uses the same vulnerabilities included in the three vulnerable Chromium versions in DANGNULL. To make the experimental results convincing, we leverage two mainstream software, known as OpenSSL and Wireshark, to exam their open vulnerabilities.

Table 2 is the comparison results between UAF-GUARD and DANGNULL. By "SIGSEGV", "TRIGGER", "NORMAL" and "ASSERTION" we mean "throw a exception", "trigger the vulnerability", "run in a normal way", and "the security assertion of Chrome", respectively. Compared to DANGNULL, UAF-GUARD is able to detect and terminate the execution of instruction in time. It is worth mentioning that DANGNULL does not throw SIGSEGV exception in some cases, such as CVE-2013-2909, and CVE-2013-2922. The reason behind is that the program can enter other execution paths through the branch structure after nullifying the pointer, and hence the program can re-execute normally. As a result, the position of the vulnerabilities cannot be reported as UAF-GUARD does. The ASSERTION result for CVE-2012-5137 is caused by the security assertions in Chrome, rather than by DANGNULL itself.

Through the above analysis based on real-world vulnerabilities, we can conclude that UAF-GUARD is highly competitive with the other runtime detection methods. Although DANGNULL is slightly more lightweight, the insufficient pointer information makes it suffer from the potential risk of bypassing.

4.3 Performance

In the performance tests we test the changes of the program at the static compiling and linking, which may include both the file size and the instrumentation functions, and we further test the extra overhead generated by UAF-GUARD at runtime.

Overhead at Compiling and Linking Process. We measure the changes of the programs by counting the difference between the file size before and after compiling and linking, and gathering statistics for the instrumentation functions. The experiment counts the related information at compiling-linking process for 16 programs, including 11 C programs and 5 C++ programs. We implement the test for UAF-GUARD based on the LLVM Compiler project, and the experimental results are shown in Table 3.

Table 3. The space overhead of UAF-GUARD in compiling and linking. Note *Before Compilation* represents the size of program before compiling, *Increase* indicates the increased size of program for detection after compiling and linking, *Percentage* shows the size of fixed dynamic library without calculation, and *Instrumentation Function* represents the number of instrumentation functions inserted in detection.

Name	Language	File Size				
		Before	Increase			
			DANGNULL	Percentage	UAF-GUARD	Percentage
gcc	C	8380KB	768KB	4.7%	618KB	2.7%
soplex	C++	4292KB	453KB	1.9%	428KB	0.9%
povray	C++	3383KB	513KB	4.2%	479KB	2.6%
h264ref	C	1225KB	420KB	4.1%	423KB	2.7%
gobmk	C	5594KB	416KB	0.8%	426KB	0.6%
chromium	C++	1858MB	10MB	0.5%	7M B	0.3%
sjeng	C	276KB	386KB	5.8%	396KB	2.2%
namd	C++	1182KB	382KB	1.1%	417KB	2.3%
hmmer	C	814KB	396KB	3.2%	420KB	3.7%
sphinx3	C	541KB	389KB	3.5%	422KB	5.9%
milc	C	351KB	386KB	4.6%	419KB	8.3%
astar	C++	195KB	378KB	4.1%	410KB	10.2%
bzip	C	172KB	378KB	4.7%	399KB	5.2%
mcf	C	53KB	376KB	11.3%	404KB	26.4%
libquantum	C	106KB	378KB	7.5%	400KB	9.4%
lbm	C	37KB	374KB	10.8%	393KB	8.1%
Average		117.7MB	1.1MB	4.55%	0.8MB	5.7%

Table 3 shows that for the small programs, the space overhead of UAF-GUARD is quite close to that of DANGNULL, but for the larger programs, the space overhead of the program is less than that of DANGNULL, with a decrease from 0.2% to 2% (the first 6 rows in Table 3).

Since the size of the different programs differs, the amount of instrumentation function should be different accordingly. This leads to a fact that the changes of the file size are also inconsistent. UAF-GUARD requires the support from the runtime library with a size of about 390KB. But we state that this extra overhead is necessary for recording the pointer status so that we can guarantee the precise detection and defense over the UAF vulnerability exploits.

Runtime Overhead. We design and implement a chrome plug-in for UAF-GUARD, which records the results of accessing the Alexa top 100 websites. The average increased time overhead is 22%. Due to space limit, we here only list the results for the top 8 (out of the 100) websites in Table 4. Each group of the results represents the average time taken by the website from the loading to finishing rendering of the web pages in the 1000 repeated experiments. In this way, we try to precisely reflect the user's experience after deploying UAF-GUARD.

Table 4. The rendering time of accessing web pages by the chromium. Note that *Requests* and *DOM Nodes* represent the number of requests and the number of DOM Nodes from loading to finishing rendering the pages, respectively. *Rendering Time* shows the rendering time after the chromium protected by UAF-GUARD and DANGNULL, in which the time is the mean of the results from one thousand experiments.

Website	Complexity of Page		Rendering Time				
	Requests	DOM Nodes	Original Time(s)	DANGNULL(s)	Increase(%)	UAF-GUARD(s)	Increase(%)
qq.com	92	2604	0.53	0.75	41.5	0.69	30.2
youtube.com	54	2397	3.11	4.13	32.8	3.90	25.4
baidu.com	21	142	0.21	0.25	19.0	0.25	19.0
taobao.com	80	1069	0.31	0.38	22.6	0.38	22.6
google.com	24	360	1.11	1.35	21.6	1.36	22.5
amazon.com	216	1508	2.28	2.66	16.7	2.67	17.1
gmail.com	52	240	1.82	2.23	22.5	2.24	23.1
twitter.com	18	668	3.45	3.81	10.4	3.97	15.1
Average	80	1124	1.73	1.95	21.7	1.94	21.9

In Table 4, the average time overheads of UAF-GUARD and DANGNULL are increased by 21.9% and 21.7% respectively, which shows that both methods share similar performance w.r.t. the average time overhead. The rendering time of a normal website is within about 2 s, and the extra time overhead is about 0.5 s, which is almost "imperceptible" to website clients. With the increase of DOM node number, our UAF-GUARD performs better than DANGNULL (please refer to the first 2 rows in Table 4).

5 Conclusion

In this work, we have proposed a novel approach called UAF-GUARD, based on a design over fine-grained memory permission management, with the aim to identify and locate UAF vulnerabilities. We have also conducted experiments in which we implement UAF-GUARD on a 64-bit Linux system and employ it to transform a program into a practical version. The experimental results demonstrate that our UAF-GUARD outperforms other existing approaches in terms of detection accuracy and overhead. It also shows that our UAF-GUARD is able to effectively and efficiently defend all the three types of UAF exploits.

In future work, we plan to design an automated UAF vulnerability discovery system for defending exploiting heap memory vulnerabilities with more description information of pointers.

References

1. CWE-416: Use After Free. https://cwe.mitre.org/data/definitions/416.html. Accessed 11 Oct 2019
2. Ratanaworabhan, P., Livshits, V.B., Zorn, B.G.: NOZZLE: a defense against heap-spraying code injection attacks. In: Proceedings of the 18th Conference on USENIX Security Symposium (SSYM 2009), Berkeley, CA, USA, pp. 169–186 (2009)
3. Canary (buffer overflow). http://www.cbi.umn.edu/securitywiki/CBI_Computer Security/MechanismCanary.html. Accessed 11 Oct 2019
4. Position Independent Executables (PIE). https://access.redhat.com/blogs/766093/posts/1975793. Accessed 11 Oct 2019
5. Address space layout randomization (ASLR). https://searchsecurity.techtarget.com/definition/address-space-layout-randomization-ASLR. Accessed 11 Oct 2019
6. Dullien, T., Porst, S.: REIL: a platform-independent intermediate representation of disassembled code for static code analysis. In: Proceedings of Cansecwest, Vancouver (2009)
7. Bardin, S., Herrmann, P., Leroux, J., Ly, O., Tabary, R., Vincent, A.: The BINCOA framework for binary code analysis. In: Gopalakrishnan, G., Qadeer, S. (eds.) CAV 2011. LNCS, vol. 6806, pp. 165–170. Springer, Heidelberg (2011). https://doi.org/10.1007/978-3-642-22110-1_13
8. Brumley, D., Jager, I., Avgerinos, T., Schwartz, E.J.: BAP: a binary analysis platform. In: Gopalakrishnan, G., Qadeer, S. (eds.) CAV 2011. LNCS, vol. 6806, pp. 463–469. Springer, Heidelberg (2011). https://doi.org/10.1007/978-3-642-22110-1_37
9. Ye, J., Zhang, C., Han, X.: POSTER: UAFChecker: scalable static detection of use-after-free vulnerabilities, New York, NY, USA, pp. 1529–1531 (2014)
10. Dolan-Gavitt, B., Hulin, P., et al.: LAVA: large-scale automated vulnerability addition. In: 2016 IEEE Symposium on Security and Privacy (SP), San Jose, CA, pp. 110–121 (2016)
11. Hex-Rays. https://www.hex-rays.com/. Accessed 11 Oct 2019
12. Lee, B., Song, C., Jand, Y., et al.: Preventing use-after-free with dangling pointers nullification. In: Symposium on Network and Distributed System Security (NDSS), San Diego, CA, USA, pp. 8–11 (2015)

13. Caballero, J., Grieco, G., Marron, M., Nappa, A.: Undangle: early detection of dangling pointers in use-after-free and double-free vulnerabilities. In: Proceedings of the 2012 International Symposium on Software Testing and Analysis, New York, NY, USA, pp. 133–143 (2012)
14. Kouwe, E., Nigade, V., Giuffrida, C.: DangSan: scalable use-after-free detection. In: Proceedings of the Twelfth European Conference on Computer Systems, New York, NY, USA, pp. 405–419 (2017)
15. Serebryany, K., Bruening, D., Potapenko, A., Vyukov, D.: AddressSanitizer: a fast address sanity checker. In: Proceedings of the 2012 USENIX Conference on Annual Technical Conference, Berkeley, CA, USA, p. 28 (2012)
16. Nethercote, N., Seward, J.: Valgrind: a framework for heavyweight dynamic binary instrumentation. In: Proceedings of the 28th ACM SIGPLAN Conference on Programming Language Design and Implementation, New York, NY, USA, pp. 89–100 (2007)
17. Nagarakatte, S., Zhao, J., Martin, Milo M.K., Zdancewic, S.: CETS: compiler enforced temporal safety for C. In: Proceedings of the 2010 International Symposium on Memory Management, New York, NY, USA, pp. 31–40 (2010)
18. The LLVM Compiler Infrastructure. https://llvm.org/. Accessed 11 Oct 2019

Internet of Things

Networked Multi-robot Collaboration in Cooperative–Competitive Scenarios Under Communication Interference

Yaowen Zhang[1], Dianxi Shi[1,2(✉)], Yunlong Wu[1,2(✉)], Yongjun Zhang[3], Liujing Wang[2], and Fujiang She[1]

[1] Artificial Intelligence Research Center (AIRC), National Innovation Institute of Defense Technology (NIIDT), Beijing 100071, China
{dxshi,ylwu1988}@nudt.edu.cn

[2] Tianjin Artificial Intelligence Innovation Center (TAIIC), Tianjin 300457, China

[3] National Innovation Institute of Defense Technology (NIIDT), Beijing 100071, China

Abstract. In this paper, we consider a scenario where a team of predator robots collaboratively survey an area for preventing the invasion from opponent robots. In this scenario, the predator robots can share the sensing information of the prey robots through wireless communication. In order to constrain the surveillance performance of the predator robots, besides the prey robots, some interfering robots are added to break the communication connectivity between the predator robots. This is a typical "cooperative–competitive" decision problem involving multiple optimization variables from electromagnetic and geographic domains, which makes it challenging to solve. For this problem, we first propose the perception and communication models of the robots. Then, with these models, we formulate the problem and adopt multi-agent reinforcement learning (MARL) to solve it. Furthermore, considering the long training-time cost of traditional MARL, we propose a scenario curriculum learning (SCL) training strategy, which can reduce the computation time and improve the performance by evolving the scenarios from simplicity to complexity. The effectiveness of the proposed method is verified by the analysis and simulation results. The results show that the SCL strategy can reduce the training time by 13%.

Keywords: Electromagnetic and geographic domains · Multi-agent reinforcement learning · Scenario curriculum learning

1 Introduction

With the rapid development of artificial intelligence and automation technology, robots are widely used in various fields such as industries, safety, military, and

Supported by the National Key Research and Development Program of China under Grand No. 2017YFB1001901, the Key Program of Tianjin Science and Technology Development Plan under Grant No. 18ZXZNGX00120 and the National Natural Science Foundation of China under Grant No. 61906212.

H. Gao et al. (Eds.): CollaborateCom 2020, LNICST 349, pp. 601–619, 2021.
https://doi.org/10.1007/978-3-030-67537-0_36

scientific research [1]. Compared with single-robot systems, multi-robot systems (MRS) can effectively improve the execution efficiency of tasks through collaboration and have greatly enhanced survivability and adaptability in complex environments [2]. Multi-robot collaboration often relies on information sharing and interaction between individuals through networking [3]. However, in a complex environment, more diversity and confrontation exist. In particular, in the case of active signal jamming, communication connectivity will face challenges. Therefore, how to ensure multi-robot coordination in complex environments is a challenge in the robotics field.

An intuitive way to model the behavior of an MRS is to predefine the action rules for each robot [4]. However, enumeration of the entire situations is difficult. Moreover, the action of each robot further results in a continuously changing environment [5]. Therefore, it's challenging to solve the decision problem of multi-robot for "cooperative–competitive" environment in "electromagnetic–geography" multi-domain. Because it need to be involved multiple optimization variables from electromagnetic and geographic domains from practicality.

In this paper, We first model the perception and communication of robots, then describe the problem with constraints as Markov decision process (MDP), and use Multi-Agent Reinforcement Learning (MARL) to solve it. The current MARL solution concept is based on centralized training and distributed execution mode. This mode can effectively improve the adaptability and convergence of the algorithm to complex environments by considering the action policies of all agents [6]. And it's widely used to solve decision problems in multi-agent systems. Some typical algorithms include DDPG [7], MADDPG [8], and COMA [9]. The advantage of the distributed learning mode for multi-agent system is that each agent in the system, having its own independent action policy, can optimize its own behavior by interacting with other agents. We analyze and model this multi-domain problem and aim to solve it using the MADDPG algorithm, because this algorithm has the advantages of distributed training and good expansibility.

In addition, each agent in MARL regards other agents as part of the environment and it observes and interacts with the environment to obtain reward, thereby the strategy of the multi-agent system could be updated [8,9]. With the increasing number of agents, the computational complexity and run-time of MARL increases exponentially [6]. In a multi-agent system, every agent aims to learn the best response to the behavior of others. If the other agents also adapt their strategy, the learning target moves, that is, the trained system strategy would be non-stationary [10], and efficiency will also be far away from expectations. In order to improve the stationary and efficiency of the algorithm, we propose a training method based on complex scenario curriculum learning (SCL).

However, current researches on MARL focus on relatively simple environments [4]. Many standardized environments such as the Arcade Learning Environment [11] are simple plane environment for the RL algorithm benchmark. Other platforms such as StarCraft Multiagent (a real-time strategy game) [12] and Hanabi (a multiplayer card game) [13] also involve single domains. To the best of our knowledge, this is the first time reinforcement learning is used to solve the problem of a MRS under "cooperative–competitive" in an "electromagnetic–geography" multi-domain environment. The following are the main contributions of this paper:

(1) We build the communication and perception models of the robots involving in the scenario. In the models, we further consider the communication interference effects from the interferencing robot.
(2) We formulate the "cooperative-competitive" problem which involves the cooperation between homogeneous robots and competition between heterogeneous robots.
(3) We propose a training method based on transfer learning for improving training efficiency and effectiveness. We adopt different scenarios to train the policies of the robots and can obtain a better cooperation strategy than the directly training.
(4) Numerical results demonstrate that the optimized collaboration model could be trained by DRL and we analyze the difference and advantage between two algorithms. We further prove the improvement of SCL in complex scenarios.

The rest of the paper is organized as follows. Section 2 presents the scenario modeling, and Sect. 3 presents the solution by implement MADDPG with depth-first search algorithm (DFS). In Sects. 4 and 5, we introduce SCL and describe the transfer process step by step. Section 6 describes the simulation experiments conducted to solve and verify method optimization. Finally, we conclude the study in Sect. 7.

2 Modeling

2.1 Problem Description

The scenario is shown in Fig. 1(a). A team of homogeneous predator robots are responsible for patrolling the sensitive area and maintaining communication connectivity among robots. The task of prey robots, acting as an opponent agent, is to break through the sensitive area from the intruding area to invade into the protected area. And heterogeneous robots, such as the predator robot and prey robot in the geospace, as well as the predator and interfering robot in the electromagnetic field, competitive to achieve own goals. The round ends when the prey robots are captured by the predator robots while the prey robots are in the protected area. Interfering robots (also opponent agents) disturb the communication links between predator robots.

Based on the above description, the problem can be modeled as an MDP [14] and defined as a tuple $\{S, A, R, P, \gamma\}$, where S be the state set for the MRS; A_i and O_i be the action and observation space of agent i. If an agent has limited perception ability and the environment state changes dynamically, the environment is partially observable for the agent. Therefore, we could define transition probabilities between states $P : S \times A_1 \times ... \times A_N \mapsto S$. The reward

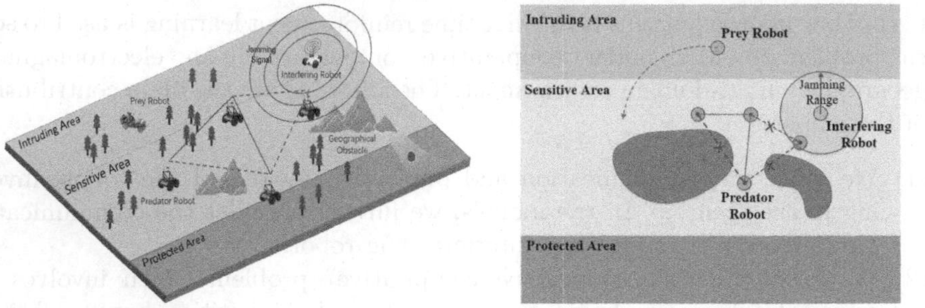

(a) Multi-agent in multi-domain scenario diagram. (b) The scenario modeling diagram.

Fig. 1. There are 4 homogeneous robots patrol the sensitive area collaboratively, and heterogeneous robots confrantation described in Fig. 1(a). We can model this scenario simply as Fig. 1(b).

function for an agent is $r_i : S \times A_i \mapsto \mathbb{R}$ and the observation is $\mathbf{o_i} : S \mapsto O_i$. The goal of agent i is to learn the optimal policy π_i^* that maximizes the expected return with a discount factor γ within a time range T:$R_i = \mathbb{E}_{s_i,a_i}\left[\sum_{t=0}^{T} \gamma^t r_i^t\right]$.

2.2 Robot Model

In this section, we first introduce the notation (Table 1) involved in our model, and then describe three types of robots in our problem: predator, prey, and interfering robots. The relationship between them is as follows: cooperative relationship between predator robots, meanwhile the prey and interfering robots also cooperate to confront predator robots. Competitive relationship between predator robots and opponent (prey and interference) robots. We consider the scenario is a two-dimensional planar environment. The position of robot i can be expressed as $p_i = [x_i, y_i], x_i, y_i \in \mathbb{R}$, and robot actions can be expressed as $a_i = [v_i, \theta]$ ($a_i \in A_i, v_i \in [V_{\max}, V_{\min}], \theta \in [0, 2\pi)$), where $V_{\max}$ and $V_{\min}$ are maximum and minimum speeds and θ is the orientation angle.

Table 1. Notation table.

Symbol	Explanation
$\mathbf{o}$	Observation vector
$\mathbf{u}$	Perception vector
$\mathbf{c}$	Communication vector
$\mathbf{r}_{i,c}$	Communication radius of predator robot i
U_{c}^i	Joint-communication scope of predator robot i
U_{s}^i	Uni-sensor scope of predator robot i
$r_i^{(k)}$	k−th reward of robot i

Predator Robots. The task of a predator robot is to capture preys within its own observable scope, and the perception vision can be shared between connective predators via communication. Therefore, we define their behavior to comprise collaboration, patrol, and active connection. Observation vector $\mathbf{o}_{i,t} \in O_{i,t}$ of predator i at time t consists of two parts: perception vector $\mathbf{u}_{i,t}$ and communication vector $\mathbf{c}_{i,t}$. $\mathbf{u}_{i,t}$ is a list of detected opponent robots, and $\mathbf{u}_{i,t} = \left[u_{i,t}^1, ..., u_{i,t}^j, ..., u_{i,t}^P\right]$, where P is the number of opponents. $\mathbf{c}_{i,t}$ is a list of connective teammates (predators), and $\mathbf{c}_{i,t} = \left[c_{i,t}^{P+1}, ..., c_{i,t}^{P+k}, ..., c_{i,t}^{P+I}\right]$, where I is the number of interconnected predators. $I + P \leq N - 1$ holds, where N is the total number of robots.

We assume the perception scope as a circle with radius r_s for $\mathbf{u}_{i,t}$, and distance between predator i and opponent j as the Euclidean distance $d_{i,j} = ||p_i - p_j||$. The element u_i^j in vector $\mathbf{u}_i$ could be defined as

$$u_i^j = d_{i,j}, d_{i,j} \leq r_\mathrm{s} \tag{1}$$

In addition, the communication scope is modeled as a circle with radius r_c for $\mathbf{c}_{i,t}$. In practice, communication scope is larger than that of perception, so we have $r_\mathrm{s} < r_\mathrm{c}$. The element c_i^{P+k} ($k \in [1, I]$) in vector $\mathbf{c}_i$ could be defined as

$$c_i^{P+k} = d_{i,k}, d_{i,k} \leq r_\mathrm{c} \tag{2}$$

In order to maintain connectivity under EMI signal, predators adopt two modes: conventional communication (lower power for energy saving) and strong communication (high-power and strong directivity, consumes more energy, and resists interference to a certain extent) [15]. The two modes can be modeled as circles with different radii. Interfering robot l has a circular interference area with a radius of r_o. For predator robots i and k, the communication radius of robot i is

$$r_{i,\mathrm{c}} = \begin{cases} r_\mathrm{h}, d_{i,k} \leq d_{i,l} \text{ or } d_{i,l} \geq r_\mathrm{o} \\ r_\mathrm{l}, d_{i,k} > d_{i,l} \text{ and } d_{i,l} < r_\mathrm{o} \end{cases} \tag{3}$$

Mode selection of the current predator robot according to the relationship of $d_{i,l}$ and $d_{i,k}$ (robot k is the nearest collaborative predator) [16]. When robot i is in the signal jamming area and $d_{i,k} \leq d_{i,l}$, the communication will be jammed due to EMI suppression by the interfering robot l. Then, the strong communication mode is on and the radius $r_{i,\mathrm{c}}$ is r_h. In other conditions, the conventional communication mode is maintained and radius $r_{i,\mathrm{c}}$ is r_l. For ensuring a certain energy consumption, the strong communication mode with high power consumption has smaller radius than the conventional mode, i.e., $r_\mathrm{h} < r_\mathrm{l}$.

The communication scope of robot i can be expressed as $\sigma_i = \{p_x | \forall d_{i,x} \leq r_{i,\mathrm{c}}\}$, where x is the index of discrete geospatial. An arbitrary predator k satisfying $p_k \in \sigma_i$ can be used as the mobile relay of robot i, and joint-communication scope of robot i can be expanded to

$$U_\mathrm{c}^i = \{\bigcup_j \sigma_j | \forall d_{i,j} \leq r_{i,\mathrm{c}}, p_j \in \sigma_i, \forall j \in [1, N]\} \tag{4}$$

By applying DFS [17], we can obtain the interconnected predator set G^i about predator i recursively. Based on networking group G^i, the uni-sensor scope can be expanded to

$$U_{\mathrm{s}}^i = \{p_x | \forall d_{j,x} \leq r_{i,\mathrm{s}}, j \in G^i\} \tag{5}$$

The predator accumulation reward $R_{i,t}$ is sum by four rewards of obstacle avoidance $R_i^{(0)}$, communication maintenance $R_i^{(1)}$, task $R_i^{(2)}$ and patrol $R_i^{(3)}$. We define obstacles as the set of location points occupied by an area on a two-dimensional plane $C_{\mathrm{obs}} \subset \mathbb{R}^2$. $R_{\mathrm{u}}^{(0)}$ is the obstacle detection reward and we have $R_i^{(0)} = R_{i,\mathrm{u}}^{(0)} + R_{i,\mathrm{w}}^{(0)}$; $R_{i,\mathrm{u}}^{(0)} = \sum_{t=0}^{T} \gamma^t r_{i,\mathrm{u}}^{(0)}$ is given by

$$r_{i,\mathrm{u}}^{(0)} = \begin{cases} -a, p_i \in C_{\mathrm{obs}} \\ 0, p_i \notin C_{\mathrm{obs}} \end{cases} \tag{6}$$

$R_{i,\mathrm{w}}^{(0)}$ is the reward for the joint communication group of predators repelling obstacles and $R_{i,\mathrm{w}}^{(0)} = \sum_{t=0}^{T} \gamma^t r_{i,\mathrm{w}}^{(0)}$ is given by

$$r_{i,\mathrm{w}}^{(0)} = \begin{cases} -b, C_{\mathrm{obs}} \cap U_{\mathrm{c}}^i \neq \varnothing \\ 0, C_{\mathrm{obs}} \cap U_{\mathrm{c}}^i = \varnothing \end{cases} \tag{7}$$

Where $a, b > 0$, p_i is position of robot i, and In the 2-D plane, the joint communication group of interconnected predators aims to avoid obstacles to maintain a good connectivity.

We define $R_i^{(1)} = \sum_{t=0}^{T} \gamma^t r_i^{(1)}$ in terms of the communication vector $\mathbf{c}_i$ as

$$r_i^{(1)} = \exp\left(k^{(1)} \max \mathbf{c}_i\right) - 1 \tag{8}$$

Where $k^{(1)} < 0$ is the scale factor. For disconnected predators, the current robot can actively attempt to reconnect and form a group again.

When a prey robot j's position satisfies $p_j \in \bigcup_{i=1}^{N} U_{\mathrm{s}}^i$, accumulation task reward $R_i^{(2)} = \sum_{t=0}^{T} \gamma^t r_i^{(2)}$ formulate in terms of the perception vector $\mathbf{u}_i$ as

$$r_i^{(2)} = k^{(2)} \min \mathbf{u}_i \tag{9}$$

Where $k^{(2)} < 0$ is the scale factor. The predators chase the prey that is within the perception scope.

For limited observable scope range, predators need patrol in the prescribed area if there has no prey robot appeared in the uni-sensor scope. We have $R_i^{(3)} = \sum_{t=0}^{T} \gamma^t r_i^{(3)}$ as

$$r_i^{(3)} = 1, p_i \in X_i \tag{10}$$

Prey Robot. We assumed that prey robots have global observation, so they could obtain every robot position in the environment. The observation vector of prey robot j is $\mathbf{o}_j = [d_{j,1}, ..., d_{j,k}, ..., d_{j,N}]$. Accumulation reward $R_{j,t}$ for prey robot j sum by three rewards of obstacle avoidance $R_j^{(0)}$, confrontation $R_j^{(1)}$ and task $R_j^{(2)}$. Here, obstacle avoidance reward $R_j^{(0)}$ is similar to that in the predator robot model.

Confrontation reward $R_j^{(1)} = \sum_{t=0}^{T} \gamma^t r_j^{(1)}$ is defined for keeping away from predators during intrusion. Relationship with $\mathbf{o}_i$ could be expressed through scale factor $\rho^{(1)} > 0$ as

$$r_j^{(1)} = \rho^{(1)} \min \mathbf{o}_i \tag{11}$$

We use function $g(x)$ represents the relationship between robot j position and task reward $R_j^{(2)} = \sum_{t=0}^{T} \gamma^t r_j^{(2)}$. The depth of prey j position is linearly increasing with the reward.

$$r_i^{(2)} = g^{(2)}(p_i) \tag{12}$$

Interfering Robot. The main task of interfering robots l is to interfere with as many predator robots as possible while they avoid obstacles (the reward for avoiding obstacles, $R_l^{(0)}$, is same as that of predators). Therefore, we have

$$R_l = R_l^{(0)} + \sum_{t=0}^{T} \gamma^t ||X^{(i)}||, X^{(i)} = \{i | \forall d_{l,i} \le r_o, i \in [1, N]\} \tag{13}$$

2.3 Problem Model

Based on the above constraints, we can model this problem for arbitrary robot i,

$$\begin{aligned} \max_{\pi_i} \quad & J_i = \mathbb{E}_{s_i, a_i}[R_{i,t}] \\ \text{s. t.} \quad & o_{i,t} \in O_{i,t}, \\ & a_{i,t} \in A_{i,t}, \\ & t \in [1, T]. \end{aligned} \tag{14}$$

The optimization goal J_i is the sum of cumulative rewards of the infinite horizon and R_i obtained by the interaction of agent with the environment. The optimization variable for this problem is each agent's action policy π_i. The action vector $a_{i,t}$ of agent i at time step t should be satisfied with the behavior policy according to the environment observation $o_{i,t}$ under the constraint of its action space A_i. Our objective is to determine an optimal behavior policy π_i that satisfies the constraints of multi-agent scenarios, so that the objective function can reach the expected maximum in an infinite time range.

3 Solution

3.1 Scenario Hypothesis

For the simulation of our model operation, the following constraints are considered: (1) delay or bandwidth in communication and the error of physical coordinates caused by sensors are ignored. Thus, once an agent is detected, its coordinates can be accessed immediately; (2) unlike [18], we do not use any differential dynamic model for agents; (3) any prey agent in the attack scope of predator agents is considered to be destroyed immediately.

3.2 System Model

We solve various scenarios with a specific actor–critic DRL algorithm framework. The entire system is shown in Fig. 2. The actor network is used to calculate the actions of the current agent according to the states observed from the environment. A critic network is used to evaluate the computation results of the actor network. The critic network has observation information of all agents, thus improving the performance of the mixed cooperative–competitive behavior. A replay buffer pool $\mathcal{D}$ is minibatch for collecting experience of tuples (S, A, R, S') from the environment. In the training process, the input of the actor network is its own observation value about the environment state, while the input of the critic network is not only its own observation value but also other agents' observation values. The critic network calculates the Q value of the state–action pair of the actor network, which is used to update the parameters of actor network.

In this way, using an actor–critic framework, each agent can receive information from other agents for training (i.e., centralized training) and perform actions through its own observation (i.e., decentralized execution). Therefore, each agent can optimize its own behavior policy through the information of other agents. For the trained model, each agent can use the action calculated by the actor network to interact with the environment. Even if an agent has only partial state information, it can still make appropriate decisions to perform actions.

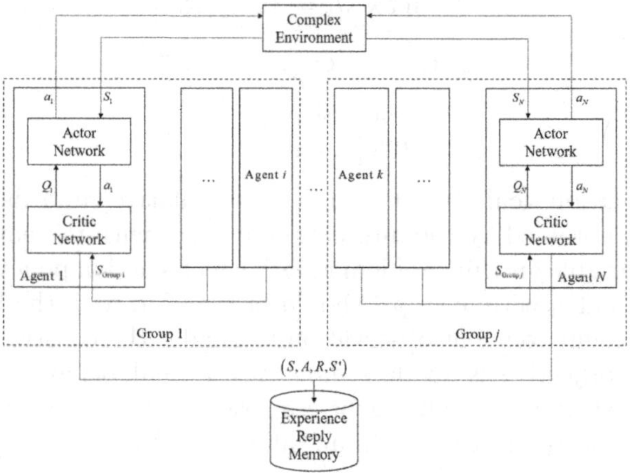

Fig. 2. Actor–critic framework of collaborative agent in networking group.

We assume a set of continuity strategies for all agents $\mu = \{\mu_1, \mu_2, \dots, \mu_N\}$. If each μ_i corresponds to a parameter vector θ_i in network, then the parameterization strategy mapped to the network can be expressed as $\theta = \{\theta_1, \theta_2, \dots, \theta_N\}$. Then, we obtain the gradient of the expected revenue of agent i as

$$\nabla_{\theta_i} J(\mu_i) = \mathbb{E}_{S,a \sim \mathcal{D}}[\nabla_{\theta_i} \mu_i(a_i|o_i) \nabla_{a_i} Q_i^{\mu}(S^{(k)}, a^{(k)})|_{a_i=\mu_i(o_i)}]. \tag{15}$$

Where Q_i^μ is the action value function of an agent. Under the condition of limited observation, agent i accesses the actions $a^{(k)} = \{a_i, ..., a_k\}$ of other agents (i.e., the agents form a group). At the same time, agent i obtains the status information $S^{(k)} = \{a_i, ..., a_k\}(k \leq N)$ of other agents in one group. Finally, the Q_i^μ value of agent i is output. The replay buffer pool $\mathcal{D}$ contains a tuple $(S, S', a_1, ..., a_N, r_1, ...r_N)$, which records the experience of all agents. Therefore, the value function Q_i^μ is updated as

$$\begin{aligned} \mathcal{L}(\theta_i) &= \mathbb{E}_{S,a,R,S'}\left[(Q_i^\mu(S, a_1, ..., a_N) - y)^2\right], \\ y &= r_i + \gamma Q_i^{\mu'}(S', a'_1, ..., a'_N)|_{a'=\mu'(o_j)} \end{aligned} \tag{16}$$

4 Scenario Curriculum Learning

In MADDPG, a complex environment not only leads to a huge amount of computation, but multi-agent also bring stability issues [10]. Curriculum learning (CL) is defined as a machine learning concept and is designed to improve the performance for transfer learning. In [19], CL was first combined with RL. One major direct application of CL in RL is to deal with complex tasks [20–22]. In CL, the goal is to improve the final asymptotic performance or decrease the computation time by generating a series of tasks. Tasks can be trained individually before progressing to learning on the final task [23].

However, most existing studies (such as those mentioned above) focus on single agents on CL. Although some existing approaches consider CL in a multi-agent system, they utilize CL in an extremely simple manner. Moreover, a single environment is a contrast to the environment considered in our article in that the number of agents and sparse rewards are constant. We propose a multi-agent CL named complex scenario curriculum learning (SCL) as shown in Fig. 3.

SCL solves the non-stationary and multi-agent training effect by starting from learning a simple multi-agent scenario and gradually increasing the number of agents and complication to finally learn the target task. Two kinds of transfer method are proposed across different order and training parameters, which can boost the performance of training on the win rate.

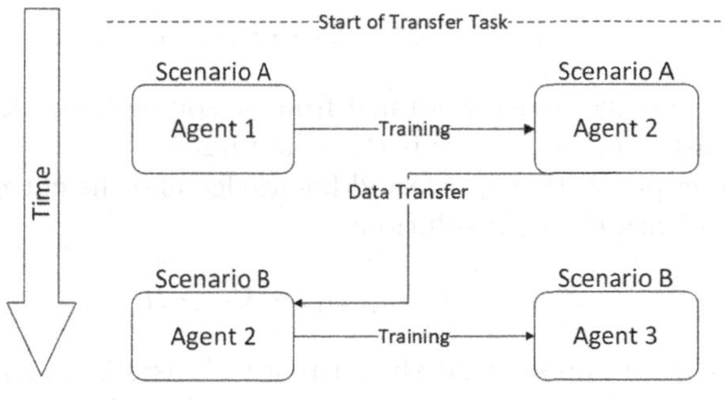

Fig. 3. Agent training in different scenarios.

Figure 3 shows the agent transfer process in different scenarios. First, the order of multi-source task transfer training is determined according to the scene complexity function ϕ . In the initial scene A, N agents are trained for t_1 iterations until approximate convergence, and then the strategy of agents is expressed as $\pi_{t_1}^1 = \{\pi_1^1, \pi_2^1, \ldots, \pi_N^1\}$. Then, the above strategy is taken into scene B as the initial condition. Afterwards, agents are trained for t_2 iterations until approximate convergence. Likewise, the model finally converges well in the final scenario. Its algorithm described as Algorithm 1.

Algorithm 1: Scenario Curriculum Learning

Input: Objective task M_t, scenario complexity m, originating task set $\mathcal{M}_s = \{M_0, M_1, \ldots\}$

Output: Solution of objective task M_t, $\mathcal{H}$

Generate curriculum tasks sequence $\mathbb{O} \leftarrow orderTasks(\mathcal{M}_s, m)$;

while $M_i \subseteq \mathbb{O}$ **do**

  $\mathcal{K}_i \rightarrow \mathcal{H}_i$;

  $\mathcal{H}_i \times \mathcal{K}_{i+1} \rightarrow \mathcal{K}_{transfer}$;

  $\mathcal{K}_{transfer} \times \mathcal{K}_{i+1} \rightarrow \mathcal{H}_{i+1}$;

return $\mathcal{H}_t$;

For single-source task transfer, namely, only a given task source, the agent can extend the prior knowledge learned from the source to the target task [24]. The process of RL modeling of each task is equivalent to an MDP process, so the task space can be represented by a set $\mathcal{M} = \{M_0, M_1, \ldots\}$.

The process of transfer can be expressed as: input the knowledge of the target task, and output the new solution $\mathcal{H}$ through training in the new scene.

$$\mathbb{A}_{learn} : \mathcal{K} \rightarrow \mathcal{H} \tag{17}$$

$\mathcal{K}$ represents the knowledge space of the source task as prior knowledge and $\mathcal{H}$ represents the solution space of the source task. In the knowledge transfer stage, according to the correlation between the source and the target task, appropriate knowledge is generated. This process can be expressed as follows:

$$\mathbb{A}_{transfer} : \mathcal{K}_s^n \times \mathcal{K}_t \rightarrow \mathcal{K}_{transfer} \tag{18}$$

Where $\mathcal{K}_s^n$ is the knowledge obtained from N source tasks, $\mathcal{K}_t$ is the knowledge of target tasks, and $\mathcal{K}_{transfer}$ is the final target.

In the learning phase, the transferred knowledge and the current task knowledge are used to learn the final solution:

$$\mathbb{A}_{t-learn} : \mathcal{K}_{transfer} \times \mathcal{K}_t \rightarrow \mathcal{H} \tag{19}$$

From Eq. (4.3), we can see that the current task uses $\mathcal{K}_{transfer}$ as additional knowledge when learning; therefore, the transfer algorithm and solution space dimension should be consistent with those of the target task. Thus, we use the policy π_t^0={π_1^0, π_2^0, ... ,π_N^0}.

which has been trained in source task M_0 after t iterations as the next task's initial solution. It is expressed as

$$\pi_s(p(S_t)) = \sigma[\pi_t^0(S_t)] \tag{20}$$

Where π_t^0 is the initial policy of the target task and π_s is the policy of the source task. During scenario transfer, the correspondence between the state space and action space of source task M_t^0 and target task M_s is $p : S_t \rightarrow S_s$ and $\sigma : A_t \rightarrow A_s$, respectively.

5 Scenarios

5.1 Scenario 1: Global Observation

In this section, we first consider a basic scenario involving two types of roles: the predator robot and the prey robot. Every robot has a global perception and communication scope. The confrontation scene is described in Fig. 4.

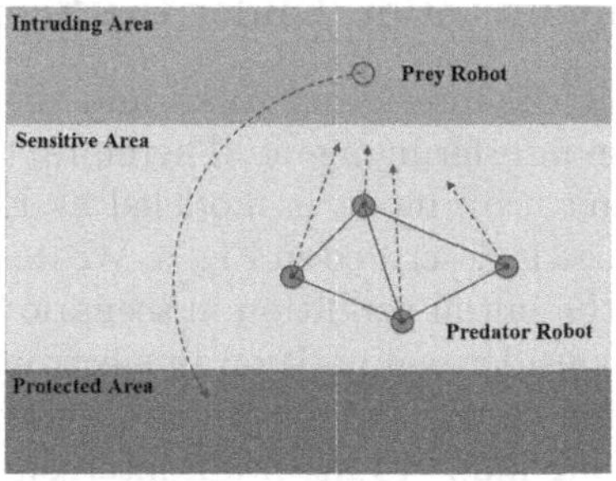

Fig. 4. Predators chase a prey robot synergistically.

In this part, we expect the predator robots to learn the action policy of collaborative encirclement and prey robots to learn the invitation policy by maximizing sum J in finite-horizon t_1 cumulative rewards as Eq. 2.14, and then reach Nash equilibrium in the competition between the two sides.

5.2 Scenario 2: Partial Observation

Based on Subsect. 5.1, we consider that a predator robot has the ability to have a local communication in Fig. 5(a) and partial perception in Fig. 5(b). Without considering communication interference, the constraints of sensing and connecting are, respectively, the same as Eq. (2.1) and Eq. (2.2) in Subsect. 2.2. And the extended areas in joint communication and the uni-sensor are expressed as Eq. (2.4) and Eq. (2.5), respectively. The reward functions of both sides are the same as Eq. (2.8)–Eq. (2.12).

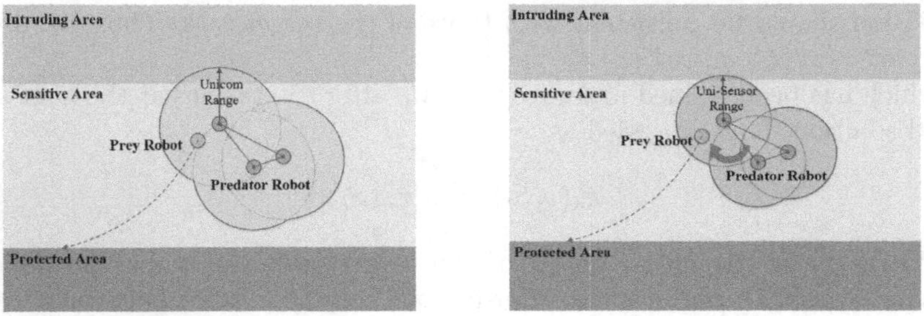

(a) Establish connection. (b) Extend sensor scope.

Fig. 5. Interconnective robots form a network group as Fig. 5(a) and extend uni-sensor scope to maximize perception range as Fig. 5(b).

The trained model described in Subsect. 5.1 is taken as the initial input for training the ability of maintaining the communication group based on the hunting strategy of predator robots. The predators in common group can chase the perceived prey robot collaboratively.

5.3 Scenario 3: Communication Under the Signal Jamming

In this part, we introduce the interfering robot and the suppression of the EMI signal transmitted by the interfering agent. Therefore, the radius of the conventional or strong communication mode is modeled as Eq. (2.3) in Subsect. 2.2. The process of this scenario is described in Fig. 6. We deem the scenario 2 trained model in t_2 iteration as the initial condition in scenario 3, and run at t_3 horizon.

In this scenario, we consider competition in electromagnetic domain. Predator which in the EMI area could patrol with conventional mode without joint-perception or switch strong mode to maintain interconnection.

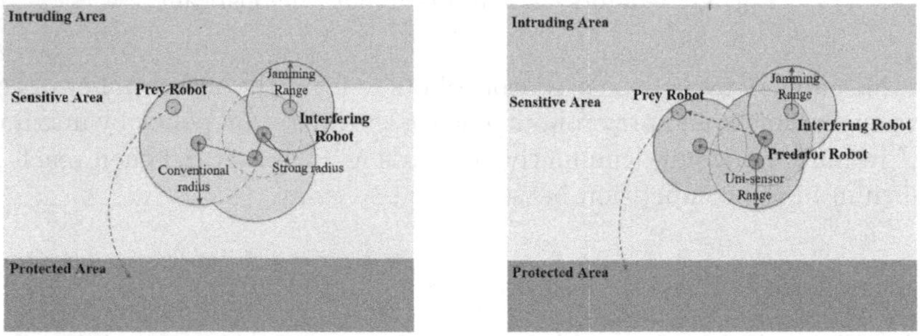

(a) Convert communication mode under signal interference. (b) Joint perception area by mobile relay robot.

Fig. 6. Strong communication mode helps maintain connection when predator robot position is in EMI signal area. Another predator can move closer to satisfied communication condition for maximum joint-perception scope.

5.4 Scenario 4: Collaboration Under the Interference

On the basis of the model trained in Subsect. 5.3 after iteration t_4, we continue to train the behavior of robots in the scene. The modeling detail is consistent with that in Sect. 2 as shown in Fig. 1(b).

The obstacles not only could hinder robots movement but also interrupt communication link. In the geographic field, both robots take action in confrontation on the premise of avoiding obstacles. In the electromagnetic field, competition happens between the two sides with "interference–counteract interference".

6 Simulation

We performed simulations to verify the effectiveness of the models proposed in Sect. 3 and optimization in Sect. 4. The workspace is a 200 m × 200 m square region, and the effective destruction radius of the predator robot is 10 m. The perceptive radius is r_s = 40 m, the conventional communication radius is $r_c = r_l$ = 80 m, the strong communication radius is $r_c = r_h$ = 60 m, and the EMI radius of the interfering robot is r_o = 40 m. Moreover, we set the predator robot speed to 10 m/s, the prey robot speed as 12 m/s, and the interfering robot speed as 8 m/s. The simulation platform is a desktop computer equipped with a i7 CPU and a NVIDIA Geforce RTX 2080Ti GPU. We adopt Tensorflow 1.18, Open Gym 0.15 and Python 3.5 for experimental simulation verification.

6.1 Multi-robot in Multi-domain Scenario

In this subsection, we describe the simulation of the scenario described in Sect. 2, and determine the optimal collaboration strategy model between predator robots under confrontation with opponents. For validating our conclusion, we compared the win rates (Fig. 7) and reward results (Fig. 8) obtained by various DRL algorithms applied for this scenario.

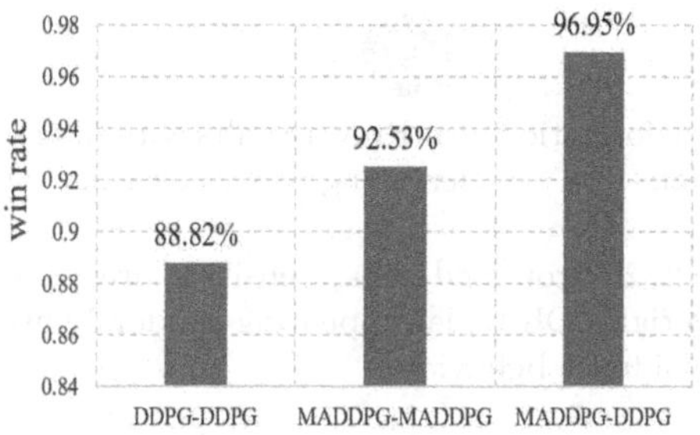

Fig. 7. Win rates for robots trained using various algorithms.

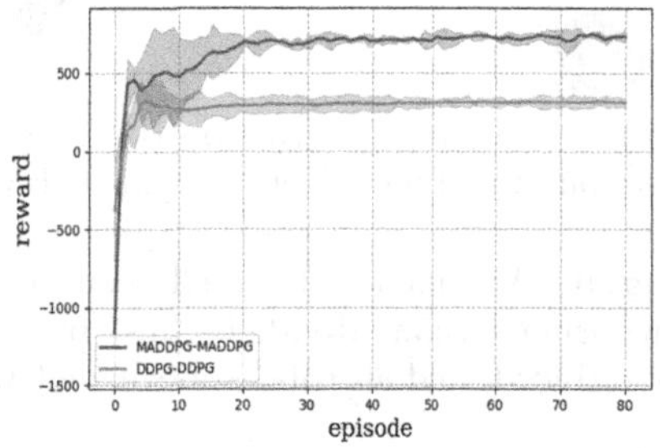

Fig. 8. Reward scores obtained through various DRL algorithms in scenario 4.

We define the win rate as that in 30,000 rounds of confrontation scenarios in Gym simulation environment. For the predator robots winning, they could capture the prey robot which has not reached the protected area.

After 80,000 iterations of training, the left and middle results shown in Fig. 7 indicate that both DDPG and MADDPG algorithms could train the predators to obtain an effective collaborative strategy. The MADDPG model provided a higher win rate than DDPG. Moreover, as shown in Fig. 8, MADDPG trained predator robot acquires more reward in a multi-domain environment.

In Fig. 7, the rightmost results are obtained by a cross-comparison experiment. Here, we trained the network model by both DDPG and MADDPG algorithms: the predator robot imported the model trained by MADDPG, and the opponent robot imported the model trained by DDPG. Compare with the middle result in Fig. 7, for the same DDPG algorithm trains the prey robot, the predator robot trained by MADDPG showed better encircling policy for the intrusion policy of the prey trained by DDPG.

In the simulation results presented in Fig. 9, the red balls represent predator robots, and their size indicates the effective attack range. The green balls represent the opponent robot. The smaller one is prey robot, whereas the larger one is interfering robot. Interfering robot size represents the signal jamming scope. The predator robot can cooperate in the effective range of communication. At the same time, the prey robots wait for the opportunity to assault the protected area under the cover of the interfering robots. All robots can identify and avoid obstacles well. Therefore, it can be proved that MADDPG can solve the networked multi-robot in competitive scenario under interference.

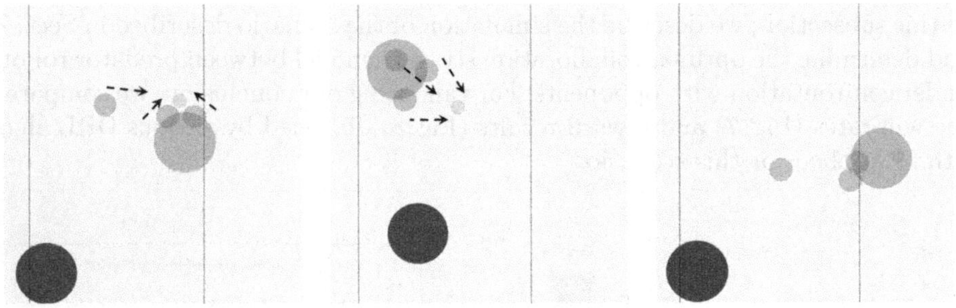

(a) Three predator robots surrounding prey robot. (b) Pursuit as a formation under EMI signal. (c) Prey assaults with interferencing robot covering.

Fig. 9. We simulated scaled scenario of Sect. 2. Protected area, sensitive area and intruding area are divide by lines from left to right. Obstacles appearing randomly are introduced, and all robots movement with avoidance behavior.

6.2 SCL Optimization

The goal of SCL is to improve the learning performance of an agent in the target task. The evaluation indicators of the learning performance can be measured with three aspects [25] that are learning speed improvement, improvement of jumpstart and asymptotic performance. Since the simulation involves the confrontation scenario between the two sides, we should not only compare the above three indicators but also compare the win rate under different scenarios, so as to prove the stability of scenario transfer. In addition, we prove the effectiveness of the proposed method in terms of the computation time.

Based on the above 5 indicators, we design two different scenario transfer cases to compare the impact of different scenario transfer methods on the transfer results. According to the sequence of substep training and the size of replay buffer, symmetric SCL (S-SCL) and asymmetric SCL (A-SCL) are proposed to verify the impact of those indicators on the transfer effect. Meanwhile, from the simulations on the scenarios described in Subsect. 5.4, we can observe the effect of different agent numbers on the transfer effect.

Symmetric SCL. In this subsection, the curriculum is designed as scenarios {1, 2, 3, 4} according to the agents constraints in the scenarios. In each scenario task, the size of training step is 20,000, and the size of replay buffer is 1000.

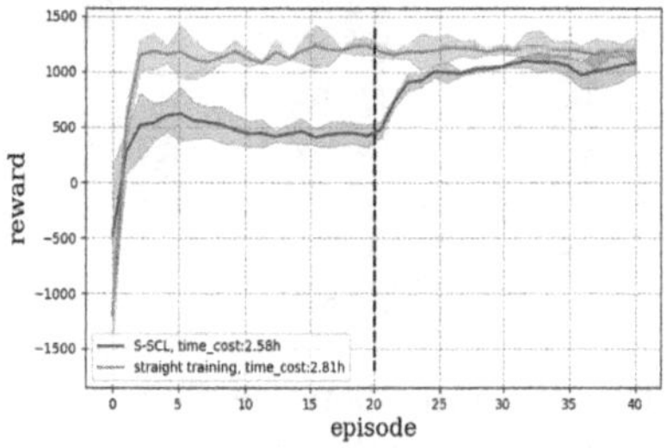

Fig. 10. Straight training and S-SCL in scenario 2.

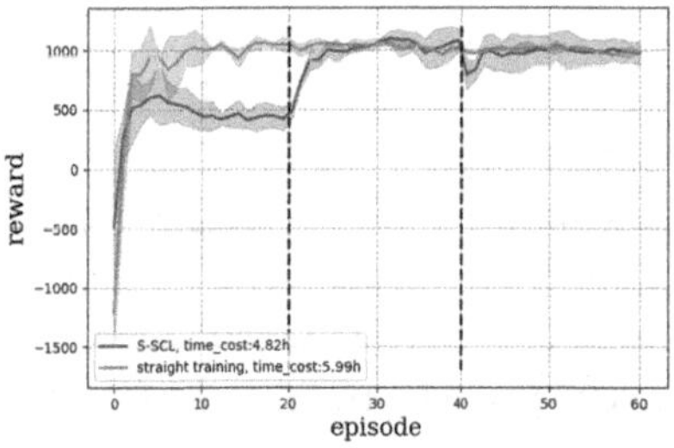

Fig. 11. Straight training and S-SCL in scenario 3.

First, the task curriculum includes scenario {1,2}. Figure 10 shows the comparison of two stages S-SCL and straight training reward in scenario 2. From the above-mentioned evaluation indicators, jumpstart and computation time of S-SCL are superior to those of straight training. Moreover, learning speed improvement and asymptotic improvement of S-SCL are worse than those of straight training. Figure 13 shows the win rate comparison of the above two methods. We can see that straight training is more effective. The reason is that the scene is relatively simple and the reward functions are relatively sparse. Therefore, the training step size of straight training is larger for scenario 2, and then more effective policy can be explored to obtain higher reward scores than pursuit policy trained in S-SCL. Thus, the method of straight training has more advantages in the reward and win rate indicators.

Next, we will verify the effect between two methods in scenario 3 as shown in Fig. 11. The jumpstart improvement and computation time indicates of S-SCL still have advantage than those of straight training. The learning speed of S-SCL is worse than that of straight training, while the asymptotic improvement is basically the same. The poor effect of learning speed improvement is due to the fact that the reward about interfering robots is not considered in the starting stage. The total number of agents is different, so the total score is not comparable. According to Fig. 11 and Fig. 13, the training time is reduced by 19.5% and the win rate is increased from 92.5% to 95%.

Figure 12 describes the advantage of S-SCL in scenario 4. The computation time of S-SCL is reduced by 13%, and the win rate is increased from 85.69% to 94.4% in Fig. 13. Therefore, the S-SCL method is superior to straight training. For the corresponding stages of scenario 3 and scenario 4, since the constraints of these two parts are negative reward feedback, there is a reward level decline for the curve in Figs. 11 and 12.

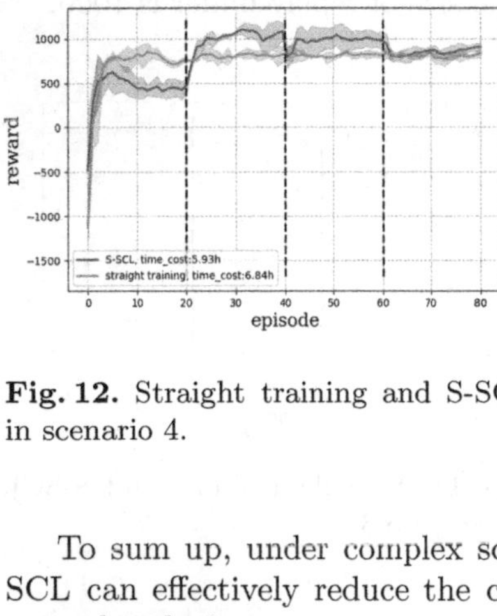

Fig. 12. Straight training and S-SCL in scenario 4.

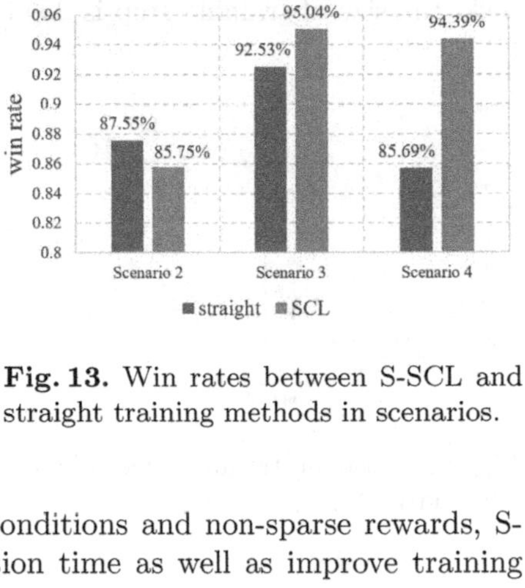

Fig. 13. Win rates between S-SCL and straight training methods in scenarios.

To sum up, under complex scenario conditions and non-sparse rewards, S-SCL can effectively reduce the computation time as well as improve training reward and win rates.

Asymmetric SCL. We observe that the results in the previous model can be used as the beneficial initial conditions for the next scenario. In this part, we explore the effect of different curriculum order on the experimental results. We designed the task curriculum as scenario 1, scenario IRAS (Interfering Robot Added into Scenario 1), scenario 3, and scenario 4. Based on the condition of global perception and communication in Subsect. 5.1, the IRAS introduces interfering robots, and each robot has a global observation perspective. Then, the constraint of the electromagnetic domain is introduced in scenario 3.

In scenario 1, the training step size is 20,000 and the replay buffer is 500, and the objective is to better train the collaborative pursuing ability of predator robots. In scenario IRAS, the replay buffer is 700, and the objective is to further train the confrontation ability of both sides in the electromagnetic field under the collaborative pursuing behavior. In Fig. 14, we do not consider the learning speed improvement due to the inconsistent number of agents. From the aspects of jumpstart improvement and asymptotic improvement, we observe A-SCL to show a great improvement in both initial and final rewards, with a slight advantage in the computation time. The left-side results in Fig. 17 indicates that A-SCL increases the win rate of straight training method from 93% to 97%.

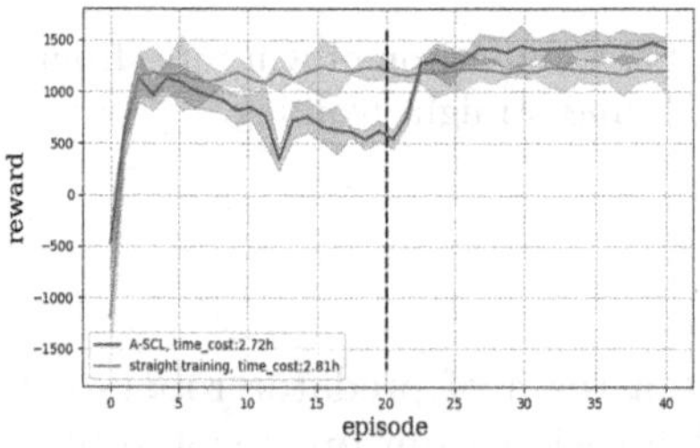

Fig. 14. Straight training and A-SCL in scenario 2.

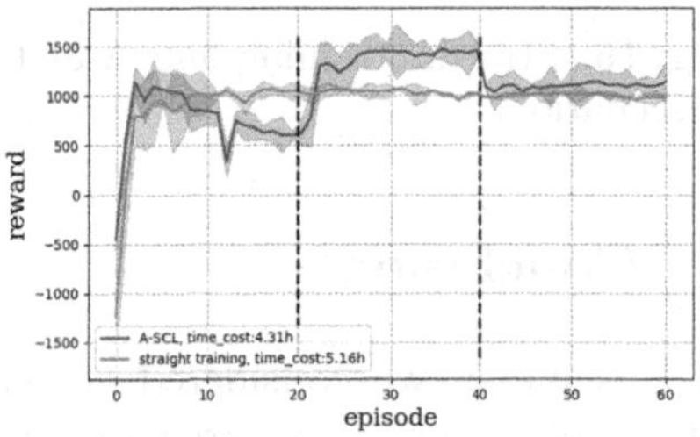

Fig. 15. Straight training and A-SCL in scenario 3.

Based on simulations in the previous part, the size of the replay buffer in scenario 3 is 900. Its purpose is to learn the strategy of maintaining communication cooperation in the situation of signal jamming. From the reward comparison of the three-stages A-SCL and straight training shown in Fig. 15, we observe that reward of A-SCL is slightly higher than those of straight training, and the computation time is reduced by 16.4%. Figure 17 indicates that A-SCL increases the win rate of straight training from 90% to 91%.

The size of the replay buffer in scenario 4 is 1100, and it aims to learn the obstacle avoidance function in the geographic domain in the electromagnetic domain condition. Figure 16 shows that the reward of A-SCL is higher than that of straight training, and the computation time is reduced by 11.4%. Figure 17 indicates that A-SCL increases the win rate of straight training from 87% to 97%.

In conclusion, we can design the size of the replay buffer manually, which will affect the weights of the trained agent models. The trained models are used as the initial condition for the subsequent scenario. Therefore, A-SCL has greater effect on the computation time, jumpstart improvement, and asymptotic improvement.

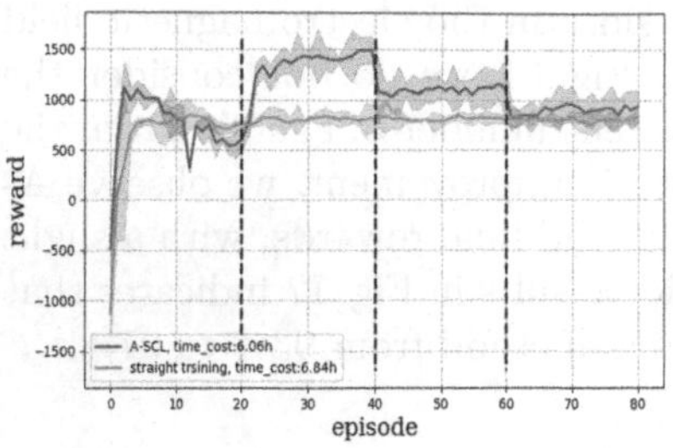

Fig. 16. Straight training and A-SCL in scenario 4.

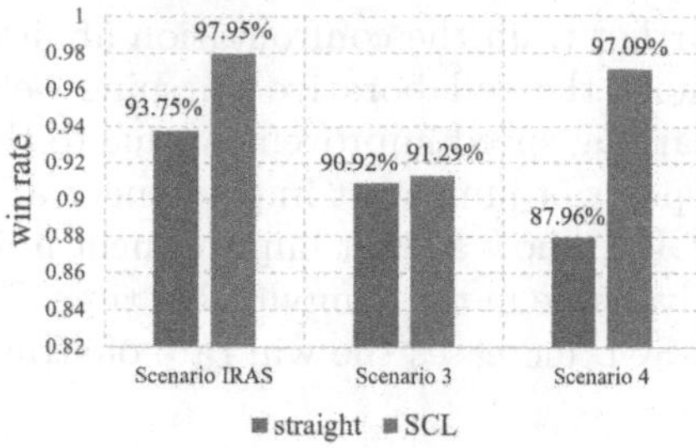

Fig. 17. Comparison of win rates from A-SCL and straight training.

7 Conclusion

In this paper, we considered a scenario where a team of predator robots collaboratively survey an area for prevention of invasion from opponent robots. To maximize the odds in a "cooperative–competitive" scenario, we adopted the cumulative reward as the performance metric. We modified the DRL algorithms for the scenarios for obtaining the optimal model of collaborative surround strategy under the constraint of maintaining communication quality and maximum sensor scope. We tackled the complex scenarios to further improve the SCL training method. Finally, the simulation results showed the effectiveness of the solution model in complex multi-domain problems and that the SCL method could improve efficiently and reduce the training time by 13%.

References

1. Bowling, M., Veloso, M.: Multiagent learning using a variable learning rate. Artif. Intell. **136**(2), 215–250 (2002)
2. Wu, Y., Ren, X., Zhou, H., Wang, Y., Yi, X.: A survey on multi-robot coordination in electromagnetic adversarial environment: challenges and techniques. IEEE Access **8**, 53484–53497 (2020)
3. Usunier, N., Synnaeve, G., Lin, Z., Chintala, S.: Episodic exploration for deep deterministic policies: an application to StarCraft micromanagement tasks. arXiv preprint arXiv:1609.02993 (2016)
4. Buşoniu, L., Babuška, R., De Schutter, B.: Multi-agent reinforcement learning: an overview. In: Srinivasan, D., Jain, L.C. (eds.) Innovations in Multi-Agent Systems and Applications - 1. Studies in Computational Intelligence, vol. 310. Springer, Heidelberg (2010). https://doi.org/10.1007/978-3-642-14435-6_7
5. Li, Y.: Deep reinforcement learning: an overview. arXiv preprint arXiv:1701.07274 (2017)
6. Hernandez-Leal, P., Kartal, B., Taylor, M.E.: Is multiagent deep reinforcement learning the answer or the question? A brief survey. arXiv preprint arXiv:1810.05587 (2018)
7. Lillicrap, T.P., et al.: Continuous control with deep reinforcement learning. arXiv preprint arXiv:1509.02971 (2015)

8. Lowe, R., Wu, Y., Tamar, A., Harb, J., Abbeel, O.P., Mordatch, I.: Multi-agent actor-critic for mixed cooperative-competitive environments. In: Advances in Neural Information Processing Systems, pp. 6379–6390 (2017)
9. Foerster, J.N., Farquhar, G., Afouras, T., Nardelli, N., Whiteson, S.: Counterfactual multi-agent policy gradients. In: Proceedings of AAAI Conference on Artificial Intelligence (2018)
10. Papoudakis, G., Christianos, F., Rahman, A., Albrecht, S.V.: Dealing with non-stationarity in multi-agent deep reinforcement learning. arXiv preprint arXiv:1906.04737 (2019)
11. Machado, M.C., Bellemare, M.G., Talvitie, E., Veness, J., Hausknecht, M., Bowling, M.: Revisiting the arcade learning environment: evaluation protocols and open problems for general agents. J. Artif. Intell. Res. **61**, 523–562 (2018)
12. Samvelyan, M., et al.: The StarCraft multi-agent challenge. In: Proceedings of International Conference on Autonomous Agents and Multi-Agent Systems, pp. 2186–2188 (2019)
13. Bard, N., et al.: The Hanabi challenge: a new frontier for AI research. Artif. Intell. **280**, 103216 (2020)
14. Littman, M.L.: Markov games as a framework for multi-agent reinforcement learning. In: Machine Learning Proceedings 1994, pp. 157–163. Elsevier (1994)
15. Wu, Y., Zhang, B., Yi, X., Tang, Y.: Communication-motion planning for wireless relay-assisted multi-robot system. IEEE Wirel. Commun. Lett. **5**(6), 568–571 (2016)
16. Wu, Y., Zhang, B., Yang, S., Yi, X., Yang, X.: Energy-efficient joint communication-motion planning for relay-assisted wireless robot surveillance. In: Proceedings of IEEE Conference on Computer Communications, pp. 1–9. IEEE (2017)
17. Kshemkalyani, A., Ali, F.: Fast graph exploration by a mobile robot. In: Proceedings of International Conference on Artificial Intelligence and Knowledge Engineering (2018)
18. Mordatch, I., Abbeel, P.: Emergence of grounded compositional language in multi-agent populations. In: Proceedings of AAAI Conference on Artificial Intelligence (2018)
19. Narvekar, S., Sinapov, J., Leonetti, M., Stone, P.: Source task creation for curriculum learning. In: Proceedings of International Conference on Autonomous Agents & Multiagent Systems, pp. 566–574 (2016)
20. Bengio, Y., Louradour, J., Collobert, R., Weston, J.: Curriculum learning. In: Proceedings of Annual International Conference on Machine Learning, pp. 41–48. ACM (2009)
21. Andreas, J., Klein, D., Levine, S.: Modular multitask reinforcement learning with policy sketches. In: Proceedings of International Conference on Machine Learning, pp. 166–175. JMLR. org (2017)
22. Wu, Y., Tian, Y.: Training agent for first-person shooter game with actor-critic curriculum learning. In: Proceedings of International Conference on Learning Representations (2016)
23. Wang, W., et al.: From few to more: large-scale dynamic multiagent curriculum learning. arXiv preprint arXiv:1909.02790 (2019)
24. Madden, M.G., Howley, T.: Transfer of experience between reinforcement learning environments with progressive difficulty. Artif. Intell. Rev. **21**(3–4), 375–398 (2004)
25. Lazaric, A.: Transfer in reinforcement learning: a framework and a survey. In: Wiering, M., van Otterlo, M. (eds.) Reinforcement Learning. Adaptation, Learning, and Optimization, vol. 12. Springer, Heidelberg (2012). https://doi.org/10.1007/978-3-642-27645-3_5

Adaptive Online Estimation of Thrashing-Avoiding Memory Reservations for Long-Lived Containers

Jiayun Lin[1], Fang Liu[1](✉), Zhenhua Cai[1], Zhijie Huang[1], Weijun Li[2], and Nong Xiao[1]

[1] School of Data and Computer Science, Sun Yat-Sen University, Guangzhou, Guangdong 510275, China
{linjy57,caizhh8}@mail2.sysu.edu.cn, {liufang25, xiaon6}@mail.sysu.edu.cn, jayzy.huang@gmail.com

[2] Shenzhen Dapu Microelectronic Co., Ltd., Shenzhen, China
liweijun@dputech.com

Abstract. Data-intensive computing systems in cloud datacenters create long-lived containers and allocate memory resource for them to execute long-running applications. It is a challenge to exactly estimate how much memory should be reserved for containers to enable smooth application execution and high resource utilization as well. Current state-of-the-art work has two limitations. First, prediction accuracy is restricted by the monotonicity of the iterative search. Second, application performance fluctuates due to the termination conditions. In this paper, we propose two improved strategies based on MEER, called MEER+ and Deep-MEER, which are designed to assist in memory allocation upon resource manager like YARN. MEER+ has one more step of approximation than MEER, to make the iterative search bi-directional and better approach the optimal value. Based on reinforcement learning and rich data, Deep-MEER achieves thrashing-avoiding estimation without involving termination conditions. Based on the different input requirements and advantages, a scheme to adaptively adopt MEER+ and Deep-MEER in cluster life cycle is proposed. We have evaluated MEER+ and Deep-MEER. Our experimental results show that MEER+ and Deep-MEER yield up to 88% and 20% higher accuracy. Moreover, Deep-MEER guarantees stable performance for applications during recurring executions.

Keywords: Memory reservation estimation · Intelligent cluster scheduling · Cloud datacenter · Reinforcement learning · Long-live container

1 Introduction

With the recent technical breakthroughs in big data, a dramatically increasing number of heterogeneous in-memory computation workloads, including machine learning [13–15], streaming processing [16–18], interactive query [19–21], and graph computation [22–24], are being deployed on shared cluster in data centers. These kinds of workloads

H. Gao et al. (Eds.): CollaborateCom 2020, LNICST 349, pp. 620–639, 2021.
https://doi.org/10.1007/978-3-030-67537-0_37

dealing with huge amounts of data are called long running applications (LRAs) [11], which have become the dominant workloads.

In today's data centers, clusters rely on resource managers, such as YARN [5], Mesos [6], Omega [7], Borg [8] and Kubernetes [9], to allocate resources for applications. These managers schedule computing resources like CPU and memory by packaging them as "containers". Different from traditional short-lived containers built to deal with batch jobs, containers for LRAs stay alive until an application completes, so it is named "long-lived containers".

Since the computing resources are limited and valuable, a long-existing concern that still remains in the industry is how to quantify user's resource demand when building a long-lived container. Take memory for instance, if too much memory is allocated to an application, the application utilizes only part of the allocated memory and the rest is wasted. On the contrary, if too little memory is reserved, the performance of the application may not be guaranteed and more seriously, the application may crash due to lack of memory. The optimal memory reservation estimation issue becomes more challenging based on this context.

1.1 Performance Inflection Point and Motivation

From previous works, we learn the concept of memory reservation elasticity [10] that slight reduction of memory supply under over-provision has no serious impact on performance. This property benefits from memory manager's two protection mechanisms. When the memory is insufficient, garbage collection will be triggered to free memory, or the program will be spilt to the disk. Under this situation, the slight performance penalty caused by insufficient memory supply is acceptable.

To understand memory elasticity, we submit the same application under different memory reservation and record the execution time. We use four different workloads to prove that memory reservation elasticity is a universal phenomenon. The result is shown as Fig. 1. It can be clearly observed that the curve has a long tail. Under over-reservation, moderate reduction on memory reservation has almost no effect on performance. But when memory reservation drops to a certain point, even slight reduction of memory leads to a sharp performance degradation. We call this point "performance inflection point", which is circled in red. Let's describe performance inflection point in detail using formula. We sort reserved memory sizes from large to small, denote them by $M_0, M_1, \ldots, M_n$ and the corresponding execution time by $T_0, T_1, \ldots, T_n$, and then compute the average of the leading execution time

$$T_{avg}^{i} = \frac{1}{i} \sum\nolimits_{n=0}^{i-1} T_n \tag{1}$$

for each i, i = 1, . . . , n. There exists a set of index I that for each i ∈ I, T_i satisfies the following formula:

$$\frac{T_i - T_{avg}^{i}}{T_{avg}^{i}} > 0.2. \tag{2}$$

The minimum value in the set minus one is the subscript corresponding to the performance inflection point, which is denoted by M in this paper. We have tried a lot of

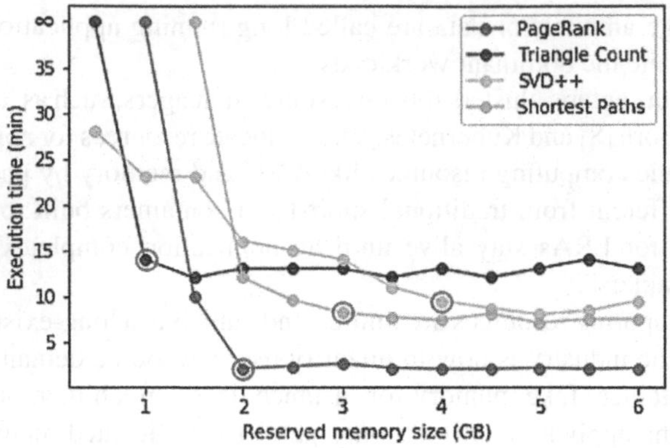

Fig. 1. Application execution time under various memory reservation.

minima. When it is set as 0.2, the calculated inflection points are most in line with people's intuitive feelings. Once we find the performance inflection point, we can find the best balance between memory utilization and application performance and this inflection point can be considered as optimal memory reservation.

A leading technique to estimate optimal memory reservation is MEER [10]. MEER uses histogram analysis to learn the rule of memory occupation and estimates an optimal memory reservation through iterative search. Although this work shows good results on estimation accuracy and saving memory resource, there is still room for improvement:

1. **The iterative search of MEER is monotonous.** MEER has the characteristic of self-decay, i.e., the estimated result decreases progressively and irrevocably before termination, which limits the opportunity for correction. Accordingly, MEER+ turns iterative search into bi-directional by adding a step of approximation. It achieves excellent accuracy in case that there is no enough historical data for detailed analysis and forecasting.
2. **Application performance is thrashing.** Both MEER and MEER+ have to experience inefficient execution required by the termination conditions, which may not be accepted by users. Deep-MEER exploits the potential of historical data generated in application execution with the help of reinforcement learning. Leveraging this unsupervised learning technique for data analysis, Deep-MEER wriggles out of termination conditions.

In the process of overcoming these limitations, we discovered an adaptive scheme to adopt MEER+ and Deep-MEER flexibly in different cluster life time to take full advantage of their own strengths: applying MEER+ to seek out optimal memory reservation quickly, then using thrashing-avoiding Deep-MEER for sustainable development of clusters as soon as reinforcement learning-based model is well trained with abundant historical data.

1.2 Main Contributions

The main contributions of this paper are as follows:

- To achieve higher estimation accuracy, we propose MEER+, which refines step size of iterative search to rectify the estimation result. MEER+ has an additional process of bidirectional finer search, where estimated reservation is tuned up by a more precise step size to approach the optimal reservation.
- To protect applications from performance thrashing, we propose Deep-MEER, which fully utilizes historical data by applying reinforcement learning-based estimation algorithm. Deep-MEER is not subject to termination conditions, so it avoids sever performance loss caused by the termination conditions.
- We show how to adopt MEER+ and Deep-MEER adaptively in cluster life cycle. Specifically, MEER+ makes a preliminary estimation when the cluster lacks data for reference in its infancy, while Deep-MEER enables stable application performance when the cluster has enough historical data in its maturity.
- We evaluate MEER+ and Deep-MEER on a cluster. Compared with MEER, MEER+ and Deep-MEER increase estimation accuracy by 88% and 20% respectively, and outperforms in resource utilization by 2×. Deep-MEER also has good generalization ability and incurs a smoother performance fluctuation curve.

The rest of this paper is organized as follows. Section 2 discusses the related work. Section 3 introduces accurate MEER+ and Sect. 4 describes thrashing-avoiding Deep-MEER, respectively. Further on, we present an adaptive strategy in Sect. 5. Section 6 gives the experimental results and the final section concludes this paper.

2 Related Work

Resource demand and system configuration forecasting has always been a research hotspot of resource management in datacenters. One of the most appealing research directions and challenges is that how resource elasticity can be leveraged to achieve trade-off between resource and efficiency. [25] first put forward the concept of memory elasticity in task level and proposes YARN-ME that leverages memory elasticity to reduce memory, which in turn reduces waiting time and makes task execution faster. Elasecutor [7] is present to be an elastic scheduler that dynamically allocates and explicitly sizes resources to executors. Then [10, 29] explore the same property in application level, which aims at finding out inflection point of application performance to make optimal memory reservation. More recently, Pufferfish [30] is a new manager to realize memory elasticity via lightweight virtualization.

People used to adopt default parameters or adjust parameters according to empiricism, which is rigid, manual and usually not the best. What algorithm has the best forecasting ability and what factor is the most influential is a problem worthy of discussion. MEER [10] and Prometheus [29] apply histogram analysis algorithm. Selecta [31] uses latent factor collaborative filtering to determine near-optimal configurations for cloud compute and storage resources. CherryPick [27] leverages Bayesian Optimization to build performance models for various applications. Elasecutor [32] relies on support vector regression (SVR) to train the model that predicts demand time series. When configuring large-scale systems with complex parameters, it is not wise to rely just on empiricism. With the rise of machine learning, scholars begin to consider how this adaptive and automated method can assist in decision making in the field of system parameter tuning [1, 2, 12]. Radical basis function neural network (RFBNN) and Swarm Intelligence Based Prediction Approach (SIBPA) [35] are developed to address the various complex factors in resource allocation. In [34], novel workload latent features are proposed and computed by applying unsupervised learning on the access logs. Deep dueling [28] is an advanced deep learning architecture that attains the optimal average reward much faster than Q-learning and is more applicable to resource slicing problem. Resource demand and system configuration forecasting algorithms integrated with machine-learning technique may be a major research status in the future.

There are various types of data and workloads [13–24] in data centers. In face of such diversity, how to optimize resource allocation adaptively for specific scenarios becomes an important concern. BestConfig [4] achieves automate configuration tuning under given application workloads, which shows improvement of performance on a diversity of workloads. Elastisizer [26] addresses cluster sizing problem automatically for different data-processing jobs. This issue remains to be further explored.

3 Estimation Under Insufficient Historical Data

In this section, we will first describe the overall architecture of MEER, point out the monotonicity of MEER, and then describe our design of MEER+ for achieving better estimation precision with one more step of approximation.

3.1 MEER Overview

MEER is a system that assists schedulers to efficiently estimate optimal memory reservations for diverse in-memory computing workloads. As seen in Fig. 2, MEER includes two stages:

Stage 1. Pilot Run (Blue Line). When the application is submitted for the first time, it runs under over-reservation M_0. This stage generates the original memory footprints for the following stage, which are recorded by the history server and the metrics system. Then MEER figures out the expectation of memory usage M_1 from memory footprints by using histogram analysis model and informs resource manager of the result.

Stage 2. Iterative Search (Red Line). When the application is submitted for the n th time, resource manager adopts the last estimated result M_{n-1} as memory reservation of this execution. Similarly, MEER records memory footprints and computes the expectation of memory usage M_n. Next, it evaluates the performance for judgement of termination. If the performance meets one of the termination conditions, iterative search is terminated and M_{n-2} is the ultimate estimated optimal reservation, which is the minimum estimated result that ensures application performance. Otherwise, MEER informs resource manager of the computed expectation M_n for next submission. In general, there are three termination conditions: 1) execution time is too long, 2) garbage collection is

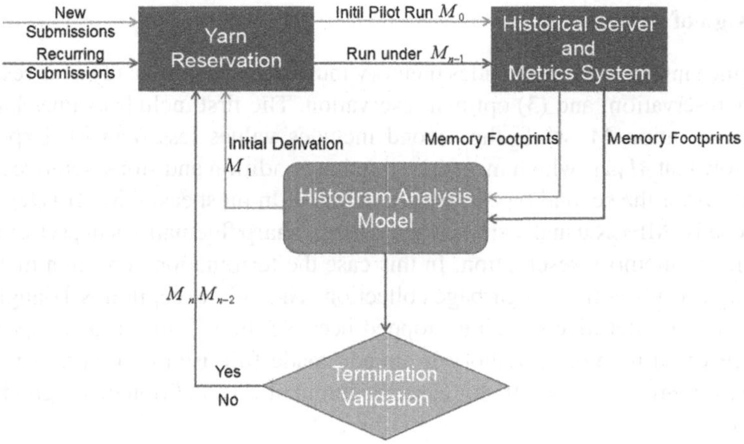

Fig. 2. Architecture and workflow of MEER. (Color figure online)

too time-consuming, and 3) memory utilization is satisfying. Except the third condition, for an iterative search to terminate, application must go through a time-consuming and inefficient run.

Limitation of Monotonicity. It is not hard to find that there is a certain gap between the optimal memory reservation estimated by MEER and the ground truth. We carefully observe the change of reserved memory size in the operation of MEER and find that in the whole process, the memory only decreases but not increases as shown in Fig. 3. Once the estimation result exceeds the inflection point, it stops immediately. This characteristic of monotonicity limits its opportunity of further approximation.

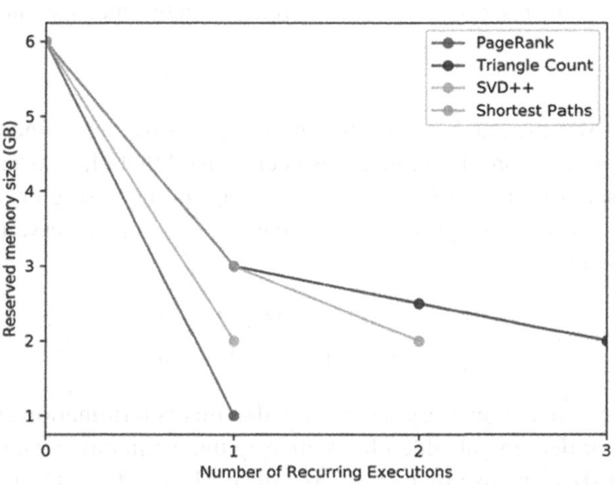

Fig. 3. Monotonic variations of memory reservation during recurring executions.

3.2 Design of MEER+

Performance inflection point divides memory into three categories: (1) over-reservation, (2) under-reservation, and (3) optimal reservation. The first includes values larger than optimal reservation M while the second includes values less than M. Experimental results show that M_{n-1}, which meets termination condition and stops iterative search, is either the first or the second type as shown in Fig. 4. In most cases, M_{n-1} is less than M. This is because MEER usually stops when it meets sharp fluctuations in performance out of insufficient memory reservation. In this case the termination condition met is time-consuming execution time or garbage collection. The only exception is Triangle Count. In this workload, iterative search is stopped because the memory usage expectation is close to reserved memory. Memory is already made full use of so there is no need to continue the search anymore. In this case, the termination condition met is good memory utilization.

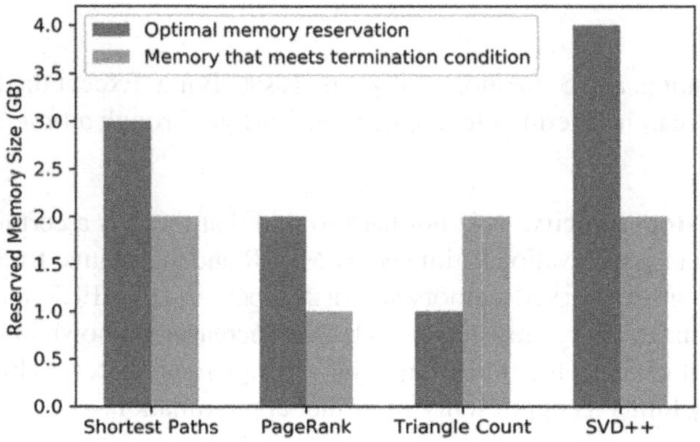

Fig. 4. Optimal memory reservation vs. Memory reservation that meets termination condition in MEER

MEER+ narrows the gap between M_n and M by destroying its monotonicity. It has an additional stage of approximation. It has been proved that M_{n-1} is either less than M or larger. Correspondingly, MEER+ reduces the gap by increasing or decreasing M_{n-1} within appropriate range. In approximation stage, the optimal reservation estimated by MEER+ is defined as:

$$\begin{cases} M_n = M_{n-1} + M_f, M_t < M \\ M_n = M_{n-1} - M_f, M_t > M \end{cases} \quad (3)$$

where M_t is the estimated memory reservation that meets termination condition and M_f is the increment or decrement added for rectifying the estimation result. Professionally, we call M_f step size. In case that $M_t < M$, since $M_t < M < M_{t-1}$, M_n is bound to nearer to M as long as:

$$M_f < M_{t-1} - M_t \quad (4)$$

And in case that $M_t > M$, M_n is certainly nearer to M as long as:

$$M_f < M_t - M \tag{5}$$

In a word, the estimation precision must be improved as long as the step size is appropriately set.

The workflow of MEER+ is shown in Fig. 5. There are three stages in total.

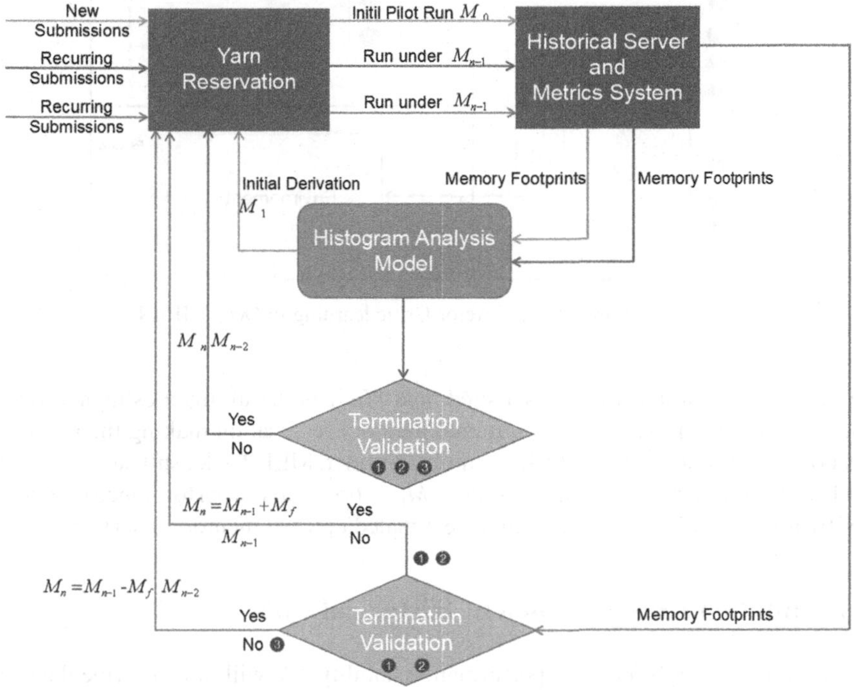

Fig. 5. Architecture and workflow of MEER+ (Color figure online)

Stage 1. Pilot Run (blue line). The same as Stage 1 of MEER.
Stage 2. Iterative Search (red line). The same as Stage 2 of MEER.
Stage 3. Approximation (green line). There is two branches according to termination conditions met in Stage 2. If condition 1 or 2 is met, Stage 3 is terminated if neither

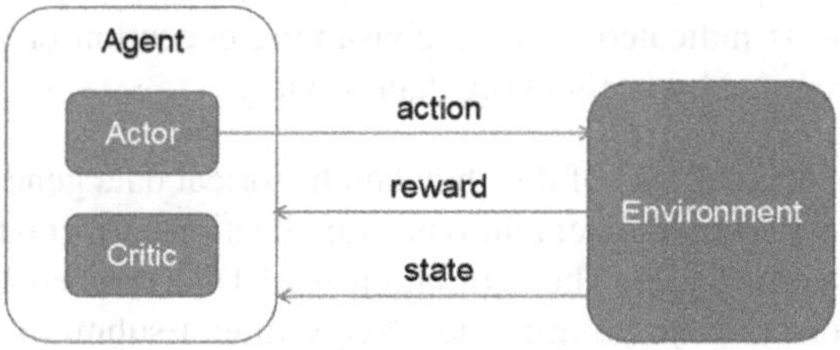

Fig. 6. Framework of actor-critic learning.

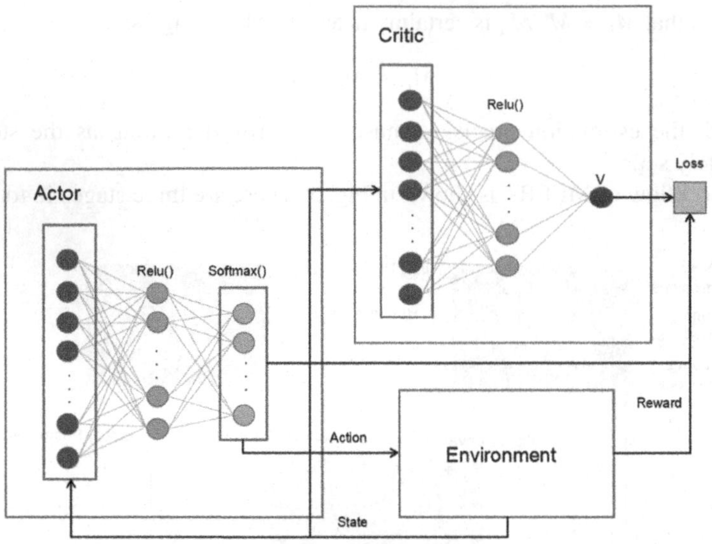

Fig. 7. Architecture of Actor-Critic learning in Deep-MEER.

termination condition 1 nor 2 is satisfied, and M_{n-1} is the ultimate estimated optimal memory reservation, which is the minimal memory reservation making the application runs normally again. Conversely, if condition 3 is met, MEER+ doesn't stop until condition 1 or 2 is met. Under this circumstance, M_{n-1} first leads to performance degradation so MEER+ selects M_{n-2} as the ultimate estimated optimal memory reservation.

4 Estimation Under Sufficient Historical Data

In this section, we will focus on performance stability. We will first describe the overall architecture of Deep-MEER, and then the design of reinforcement learning model and parameters for achieving thrashing-avoiding optimal memory reservation estimation with historical data.

4.1 Design of Deep-MEER

Limitation of Performance Trashing. When using MEER+ to look for optimal memory reservation, the application always suffers from a sharp performance degradation because termination is indicated by time-consuming execution or garbage collection. This may not be accepted by users in big data service.

Deep-MEER makes full use of the abundant historical data generated in MEER+ to eliminate the effect of termination conditions on application performance. Figure 8 shows the workflow of Deep-MEER. The estimation model is replaced by a reinforcement learning-based actor-critic learning model. Every time a submission comes, resource manager allocates memory resource with the model's guidance. Memory footprints generated are then used to further train the model. As we can see, the estimation is never limited by termination conditions. It is able to stop before the estimation result exceeds the performance inflection point so that users will never experience bad application performance.

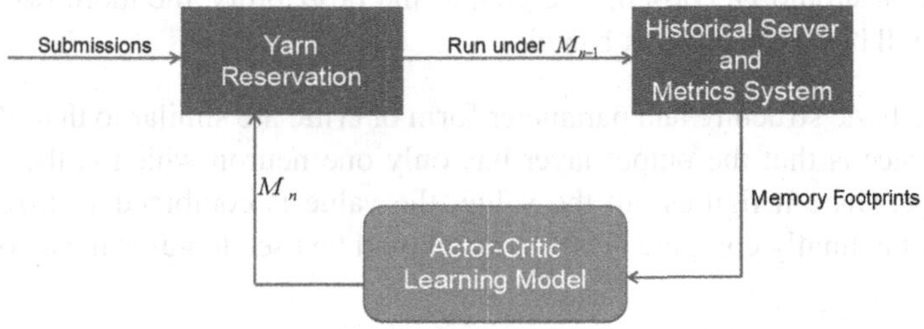

Fig. 8. Architecture and workflow of Deep-MEER.

4.2 Model Structure

Deep-MEER applies reinforcement learning by regarding the cluster as environment and seeing the memory estimator as agent. As shown in Fig. 6, the interaction between agent and environment is realized by the mutual transfer of three parameters: state, reward and action. Agent makes decisions under the guidance of a policy, i.e., provides action for the environment. Environment performs the action and responds to agent by returning reward and state. Then the agent learns from the return values, improves the policy to maximize the accumulated rewards. In actor-critic learning, agent is further separated into two units:

Actor. The actor acts as policy. It chooses an action based on probability and modifies the probability of each action according to the reward from environment and the judgement provided by critic. It implements a stochastic policy that maps the system state to the corresponding action.

Critic. The critic acts as value function that evaluates the policy. It passes judgement on the chosen action about how much it benefits the environment and provides useful reinforcement feedback for adjusting the policy.

More narrowly, actor and critic are both implemented by neural network. The whole structure is shown in Fig. 7.

Actor. Actor consists of three layers: input layer, hidden layer and output layer. The inputs of input layer are states coming from environment. Each circle represents a neuron and each line between two neurons represents a weight. Let ω denote the set of such parameters of weight. The value of each neuron is the weighted sum of the neurons in the previous layer except that the value of input neurons is provided by the environment. The mapping of neurons between each two layers is linear, which limits the expression ability of the model. Therefore, we introduce nonlinear factors by activation function to make the neural network approach any nonlinear function at well. The activation function of hidden layer is Relu(). The neurons in output layer correspond to the action that how much memory reservation is set for this application. They finally go through a Softmax() function, with the help of which the outputs are transformed to values between 0 and 1 and the sum of them is equal to 1. In other words, each output stands for the probability that the action should be chosen. The greater the probability, the more likely it is that the action will bring the greatest benefit.

Critic. The basic structure and parameter form of critic are similar to that of actor. The only difference is that the output layer has only one neuron which is the score given by the critic. Once it figures out the value, the value is combined with reward from environment to finally compute the loss, which will be used to guide the actor and critic to update.

Environment. Environment is the cluster to run the application. On the one hand, it runs the submitted application under memory reservation determined by the actor. On the other hand, it returns the state changed after executing the action as well as the reward that measures how much it benefits form the action.

Algorithm 1 Actor-Critic Model-based Estimation Algorithm

1: Initial state S_0
2: Operate the critic unit to compute $V(S_0)$
3: **for** $i = 1, 2, ..., N$ **do**
4:     Operate the actor unit to compute probability P_{i-1} of each action based on S_{i-1} and determine A_{i-1} with the highest probability
5:     Execute A_{i-1} to obtain S_i and R_i
6:     Operate the critic unit to compute $V(S_i)$
7:     Compute TD error δ_{i-1}
8:     Compute the loss $Loss(\delta_{i-1})$
9:     Update parameters ω of actor and critic guided by the loss
10: **end for**

Algorithm 1 is the pseudo code of the actor-critic model-based estimation algorithm. S_t, A_t and R_t refer to state, action and reward at time t respectively. MEER+ starts with running application under over-reservation. Accordingly, when initialing S_0, Deep-MEER reserves excessive memory. TD-error in the 7th line is commonly used to adjust the policy, which is defined as:

$$\delta_t = R_{t+1} + \gamma V(S_{t+1}) - V(S_t). \quad (6)$$

Loss function in the 8th line can be selected freely according to individual needs. Updating parameters in the 9th line refers to that following the chain rule, neural network operates back propagation that calculates the derivative of composite functions, propagates the gradient of the output unit back to the input unit, and adjusts the learnable parameters of the network according to the calculated gradient.

4.3 Model Parameters

We need to define the interaction information parameters to make the model run.

State. Table 1 shows states that are provided by the environment.

Table 1. Environment state parameters.

States	Meaning
Δt	Increased execution time compared to initial run
E	Expectation of memory footprints
$max_1 \sim max_n$	The n most frequent value of memory footprints in histogram analysis
$p_1 \sim p_n$	Corresponding frequency of $max_1 \sim max_n$

Reward. Reward is used to determine how good it is to execute a given action under the given state. The target of our model is to save memory and remain good performance at the same time. Therefore, the more memory is saved and the less time it takes to complete the application, we are closer to our expectations. Here we define reward at time *t* as a function:

$$R_t = \frac{M_0 - M_t}{T_t - T_0 + 1}, \tag{7}$$

where M_t and T_t refer to reserved memory size and execution time at time t respectively. We add one to the denominator to prevent the molecule from being divided by zero.

Action. In terms of action, each neuron of actor's output layer corresponds to a specific memory setting. For example, if neuron N_1 gets the highest probabilities, the action setting reserved memory as 0.5 GB will be chosen, and if neuron N_2 gets the highest probabilities, the action setting reserved memory as 1 GB will be chosen. And so on, for each of the neurons.

5 Adaptive Scheme

Since MEER+ and Deep-MEER has different superiority and condition for execution, they are suitable for different cluster stages and complementary to each other. We will describe how to choose strategy adaptively during cluster life cycle in this section (Fig. 9).

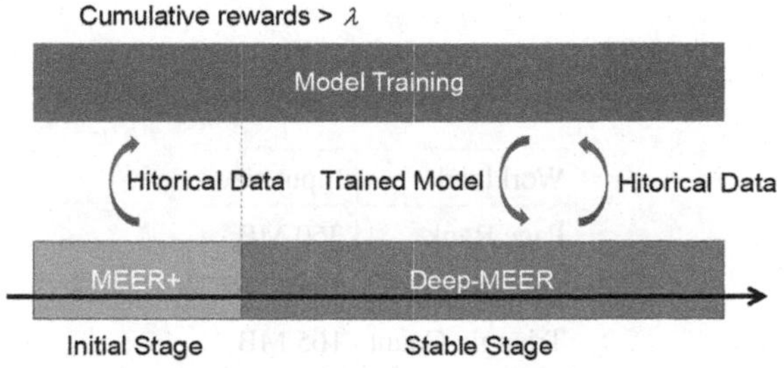

Fig. 9. Transition of strategy used during cluster life cycle.

Corresponding to estimation scheme, a cluster is divided into two stages:

Initial Stage. A cluster begins with initial stage. There is no prior experience when the cluster is just starting to work so MEER+ is adopted, sacrificing application performance several times to conduct the estimation. In the meantime, model training begins. A large amount of historical application execution data is produced in this stage, which is source material for model training.

Stable Stage. Stable Stage is the time when applications always perform well. The goal of reinforcement learning model is to maximize cumulative rewards. Once the cumulative rewards reach a threshold λ, we turn to adopting Deep-MEER and the cluster transforms into a stable stage. Threshold λ changes with application but it is easy to identify because it is the value cumulative rewards converging to. Of course it can be set a little smaller for fear of over fitting. Data from this stage can still be used to further train the model.

6 Evaluation Results

In order to evaluate our proposed schemes, we measure the performance of MEER+ and Deep-MEER from four aspects. (1) Estimation accuracy. Both schemes show nearer results to performance inflection point. (2) Generalization ability. The well-trained actor-critic model performs well even on workloads outside the training dataset. (3) Memory utilization. Idle memory is reduced by both of the proposed schemes to improve resource efficiency. (4) Application performance. Applications never suffer from performance plummeting under the estimation of Deep-MEER. In our experiments, we compare MEER+ and Deep-MEER with the state-of-the-art strategy MEER.

6.1 Experimental Setup

Through experiment in Sect. 1, we have proved the existence of performance inflection point in even 1-server-cluster, so we continue to conduct all the tests on this 1-server-cluster. The virtual machine is equipped with 1-core CPU, 12 GB memory and 200 GB virtual hard disks. We use Ubuntu 18.04 with Hadoop Yarn 2.7.3 and Spark 2.4.4 deployed. The only host acts as management node, HDFS data node and computing node simultaneously.

Table 2. Input size configuration of each workload.

Workload	Input size
Page Rank	360 MB
Shortest Paths	1900 MB
Triangle Count	165 MB
SVD++	170 MB

We run four representative workloads coming from SparkBench, Spark's benchmark performance test project for evaluation. In order to simulate the characteristics of application execution in industry, we configure the input size of each workload as shown in Table 2.

Both histogram analysis model and actor-critic learning model are implemented with Python. Execution time of the applications is obtained from history server of Spark and memory footprints are recorded in the metrics system. In the implementation of MEER+, we set $M_f = 0.5$, which ensures that formula (4) or (5) is met, but is not too conservative to limit the search speed. While computing loss value of Deep-MEER, we define the loss function as:

$$\text{Loss} = \frac{1}{n}\sum\nolimits_{n=1}^{N}(L + L_S), \tag{8}$$

where N is the number of state transitions. Both MEER and MEER+ are able to complete the estimation within five loops of iterations on all workloads. Accordingly, we set N as five.L and L_S are two typical loss functions. Specifically, L is defined as:

$$\text{L} = -\text{logP} \times \delta_t. \tag{9}$$

And L_S is defined as:

$$L_S = \begin{cases} 0.5x^2, |\delta_t| < 1 \\ |\delta_t| - 0.5, \textit{otherwise} \end{cases}. \tag{10}$$

To demonstrate the generalization ability of Deep-MEER, we run Page Rank and Triangle Count in initial stage to train the model and use the model to estimate on Shortest Paths and SVD++ in stable stage. What's more, we use MEER+ to estimate on Page Rank and Triangle Count independently for comparison.

For outstanding the advantages of our work, we take MEER as a contrast, showing and analyzing the experimental results from four aspects: estimation accuracy, generalization ability, memory utilization and application performance. Let's make agreement on some evaluation metrics before the analysis: (1) To describe estimation accuracy, we take ratio of differences between estimation result and the ground truth $\frac{M_n - M}{M}$ as a metric, which will be called relative error later for simplicity. Smaller error indicates higher accuracy. The generalization ability of model will be reflected by the relative error of the model's estimation results on the workloads outside the training dataset. (2) To describe memory utilization, we take the ratio of memory usage and estimated memory reservation $\frac{memory\_footprint}{M}$ as a metric. The higher the ratio, the higher the memory utilization. (3) To describe application performance, we use application execution time T as a metric, and there is no doubt that shorter execution time means better application performance.

6.2 Estimation Accuracy

An important goal of MEER+ and Deep-MEER is to find exactly the performance inflection point. Figure 10 shows the relative errors of final result of each strategy on the four typical workloads. Our results confirm that MEER+ has the best estimation ability, showing the best accuracy in three-quarters of the tested workloads. Although it doesn't perform well in workload Shortest Path, the difference between it and others is next to nothing with only 0.17. This situation is reasonable if the step size in Stage 3 is not accurate enough. Ordering reserved memory from largest to smallest, the theoretical inflection point of performance is the point where performance fluctuates greatly, not necessarily the first point that satisfies the termination condition or not. However, Stage 3 of MEER+ takes it as a signal to determine the ultimate estimated optimal reservation. This is the reason why if the step size is set too large or too small, it may cause some slight deviation from the actual value. As for Deep-MEER, it outstands in Shortest Path and SVD++. What's more, it performs far better than MEER in Page Rank. Despite the result of Deep-MEER is not satisfying in Triangle Count, the error is on a par with that of MEER's worst result in Page Rank. Since reinforcement learning learns memory occupation rule conforming to most workloads rather than a specific workload, errors are inevitable.

To have an overall evaluation of each scheme's estimative power, we compute average relative errors of all workloads for each strategy and show them in Fig. 11. It is observed

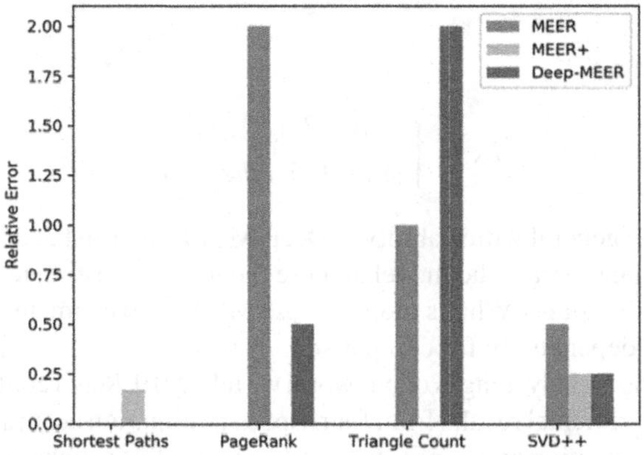

Fig. 10. Relative error of MEER, MEER+ and Deep-MEER on four benchmark workloads.

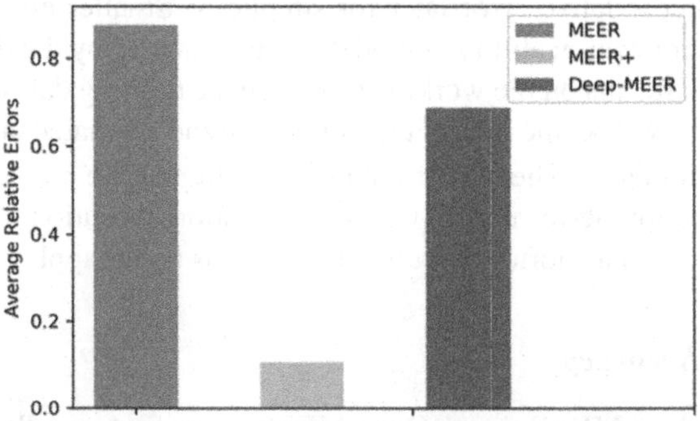

Fig. 11. Average relative error of MEER, MEER+ and Deep-MEER on four benchmark workloads.

that MEER+ shows the best preciseness with average relative error of 0.105. Deep-MEER comes after with average relative error of 0.688. They both perform better than MEER whose average relative error is 0.875. We can draw a conclusion that MEER+ and Deep-MEER reduce the relative error by 88% and 20% respectively.

6.3 Generalization Ability

The experimental results in Fig. 10 also indicates that reinforcement learning model based Deep-MEER has the advantages of excellent generalization ability. When adopting MEER or MEER+, the workload's estimated result is unique to itself, being not able to be applied to others. This is because the memory reservation is estimated only based on characteristics of memory usage of the current workload without any knowledge about that of other workloads. Whenever a new application is submitted, a new round of estimation will be activated. However, Deep-MEER learns the general rule of memory usage in multiple workloads, so the estimation model is applicable for any workloads. In our experiment, the actor-critic model is trained with data from workload Page Rank and Triangle Count but when we use the trained model to estimate on Shortest Paths and SVD++, the estimation result is still good with little relative error of 0 and 0.25. It is inspiring to know that once the model is trained, it is applicable for various workloads, which means for a cluster that runs millions of applications, the cost of training model can be ignored. Data centers may just need to pre-train a model using test data in trial operations of a small number of workloads, then the model can be applied directly in formal operations. Real-time data from new applications that use Deep-MEER to determine reserved memory size can also be utilized to train and perfect the model, making the model more inclusive.

6.4 Memory Utilization

The purpose of reducing memory reservation as much as possible is to improve memory utilization and save resource without secure impact on application performance. We calculate the memory utilization every second in application execution. Taking Page Rank as an example, we draw a diagram of the memory utilization changing with execution time. As shown in Fig. 12, the memory utilization of MEER+ and Deep-MEER is always greater than MEER, and their average utilization is about 21%, twice that of MEER. The peak memory utilization of MEER+ and Deep-MEER is 47% and 49% respectively, while the peak of MEER is only 23%. This is as expected, since MEER stops iterative search as soon as it meets the termination condition, which is rough and conservative, leading that the estimation result is often larger than the performance inflection point. However, the actual execution of the application does not need so much memory resource, so the memory utilization is not high. MEER+ effectively reduces the redundant memory reservation in approximation stage, and Deep-MEER improves the resource efficiency by learning the rule of memory occupation to make a closer estimation to the optimal value.

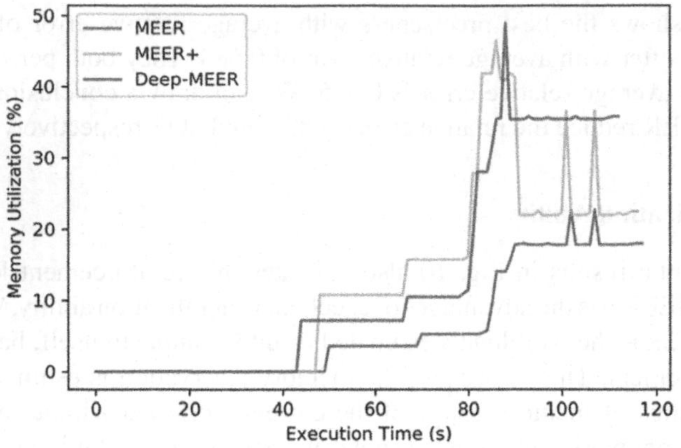

Fig. 12. Variants of memory utilization during the execution of Page Rank.

6.5 Application Performance

Thrashing-avoiding is another strength of Deep-MEER to ensure the stability of application performance, which means it is more user-friendly and more suitable for enterprises. We take an overall view of applications' performance in each recurring submission to illustrate how Deep-MEER protects applications from radical performance change.

Figure 13 is execution time in recurring executions of Page Rank and Shortest Paths. Both MEER and MEER+ experience a cliff of performance degradation in the process of estimation. This is caused by termination conditions, to meet which the application may go through a low performance run. MEER+ may experience more times of low performance for the stage of approximation may approach the optimal under inadequate reserved memory. In terms of Deep-MEER, its model always gets low rewards while

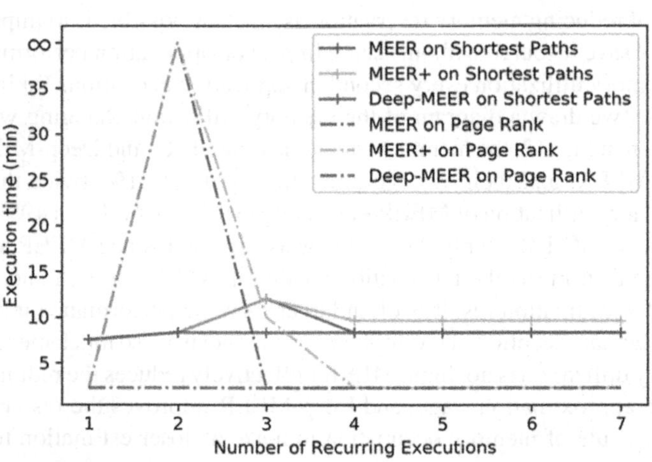

Fig. 13. Variations of execution time during recurring executions using MEER, MEER+ and Deep-MEER on benchmark workloads (Dotted lines refer to Shortest Paths and solid lines refer to Page Rank).

setting reserved memory size less than the performance inflection point in the process of training. Drawing a lesson from history, it mostly chooses memory settings that larger than inflection point to gain high rewards. This can explain why using Deep-MEER, workloads can always maintain satisfying performance.

7 Conclusion

In this paper, we present MEER+ and Deep-MEER, which are derived from MEER. MEER+ and Deep-MEER are designed to assist resource manager in estimating optimal memory reservation for long-lived containers accurately and effectively in data centers. By adding a process of approximation into MEER, MEER+ reverses research direction to rectify the estimated result. By combining fashionable reinforcement learning technique to explore the pattern of memory reservation, Deep-MEER exploits the potential of abundant historical data and prevents applications from being affected by termination conditions. The two schemes can be adopted adaptively in cluster life cycle to achieve trade-off between execution time and resource utilization. Both MEER+ and Deep-MEER have shown promising results in experiments.

Acknowledgments. This work is supported by The National Key Research and Development Program of China (2019YFB1804502), Key-Area Research and Development Program of Guangdong Province (Grant No.2019B010107001), and National Natural Science Foundation of China (Grant No.61832020, 61702569).

References

1. Zhang, W., Wang, L., Cheng, Y.: Performance optimization of Lustre file system based on reinforcement learning. J. Comput. Res. Dev. **56**(7), 1578–1586 (2019)
2. Zhao, T., Dong, S., March, V., et al.: Predicting the parallel file system performance via machine learning. J. Comput. Res. Dev. **48**(7), 1202–1215 (2011)
3. Boutin, E., Ekanayake, J., Lin, W., et al.: Apollo: scalable and coordinated scheduling for cloud-scale computing. In: 11th Symposium on Operating Systems Design and Implementation (OSDI), Broomfield, CO, pp. 285–300 (2014)
4. Zhu, Y., Liu, J., Guo, M., et al.: Bestconfig: tapping the performance potential of systems via automatic configuration tuning. In: 2017 Symposium on Cloud Computing (SoCC), Santa Clara, California, pp. 338–350 (2017)
5. Vavilapalli, V.K., Murthy, A.C., Dougla, C., et al.: Apache Hadoop YARN: yet another resource negotiator. In: 4th Annual Symposium on Cloud Computing (SoCC), Santa Clara, California, no. 5, pp. 1–16 (2013)
6. Hindman, B., Konwinski, A., Zaharia, M., et al.: Mesos: a platform for fine-grained resource sharing in the data center. In: 8th conference on Networked Systems Design and Implementation (NSDI), Boston, MA, pp. 295–308 (2011)
7. Schwarzkopf, M., Konwinski, A., Abd-El-Malek, M., et al.: Omega: flexible, scalable schedulers for large compute clusters. In: 8th ACM European Conference on Computer Systems (EuroSys), Prague, Czech Republic, pp. 351–364 (2013)
8. Verma, A., Pedrosa, L., Korupolu, M., et al.: Large-scale cluster management at Google with Borg. In: 10th European Conference on Computer Systems (EuroSys), Bordeaux, France, no. 18, pp. 1–17 (2015)

9. Burns, B., Grant, B., Oppenheimer, D., et al.: Borg, Omega, and Kubernetes. In: Communications of the ACM, New York, USA, vol. 59, no. 5, pp. 50–57 (2016)
10. Xu, G., Xu, C.: MEER: online estimation of optimal memory reservations for long lived containers in in-memory cluster computing. In: 39th IEEE International Conference on Distributed Computing Systems (ICDCS), Dallas, TX, USA, pp. 23–34 (2019)
11. Garefalakis, P., Karanasos, K., Pietzuch, P., et al.: Medea: scheduling of long running applications in shared production clusters. In: 13th EuroSys Conference, Porto, Portugal, no. 4, pp. 1–13 (2018)
12. Chen, H., Jiang, G., Zhang, H., et al.: Boosting the performance of computing systems through adaptive configuration tuning. In: 2009 ACM Symposium on Applied Computing (SAC), Honolulu, Hawaii, pp. 1045–1049 (2009)
13. Abadi, M., Barham, P., Chen, J., et al.: Tensorflow: a system for large-scale machine learning. In: 12th conference on Operating Systems Design and Implementation (OSDI), Savannah, GA, USA, pp. 265–283 (2016)
14. Meng, X., Bradley, J., Yavuz, B., et al.: Mllib: Machine learning in apache spark. J. Mach. Learn. Res. **17**(1), 1235–1241 (2016)
15. Zaharia, M., Chowdhury, M., Das, T., et al.: Resilient distributed datasets: a faulttolerant abstraction for in-memory cluster computing. In: 9th Conference on Networked Systems Design and Implementation (NSDI), San Jose, CA, p. 2 (2012)
16. Apache flink. http://flink.apache.org. Accessed 30 Mar 2020
17. Toshniwal, A., Taneja, S., Shukla, A., et al.: Storm@twitter. In: 2014 ACM SIGMOD International Conference on Management of Data, Snowbird, Utah, USA, pp. 147–156 (2014)
18. Zaharia, M., Das, T., Li, H., et al.: Discretized streams: fault-tolerant streaming computation at scale. In: 24th ACM Symposium on Operating Systems Principles (SOSP), Farminton, Pennsylvania, pp. 423–438 (2013)
19. Armbrust, M., Xin, R.S., Lian, C., et al.: Spark SQL: relational data processing in spark. In: 2015 ACM SIGMOD International Conference on Management of Data, Melbourne, Victoria, Australia, pp. 1383–1394 (2015)
20. Kornacker, M., Behm, A., Bittorf, V., et al.: Impala: a modern, open-source SQL engine for hadoop. In: 7th Biennial Conference on Innovative Data Systems Research (CIDR) (2015)
21. Saha, B., Shah, H., Seth, S., et al.: Apache Tez: a unifying framework for modeling and building data processing applications. In: 2015 ACM SIGMOD International Conference on Management of Data, Melbourne, Victoria, Australia, pp. 1357–1369 (2015)
22. Gonzalez, J.E., Xin, R.S., Dave, A., et al.: Graphx: graph processing in a distributed dataflow framework. In: 11th Conference on Operating Systems Design and Implementation (OSDI), Broomfield, CO, pp. 599–613 (2014)
23. Low, Y., Bickson, D., Gonzalez, J., et al.: Distributed graphlab: a framework for machine learning and data mining in the cloud. VLDB Endow. **5**(8), 716–727 (2012)
24. Malewicz, G., Austern, M.H., Bik, A.J.C., et al.: Pregel: a system for large-scale graph processing. In: 2010 ACM SIGMOD International Conference on Management of Data, Indianapolis, Indiana, USA, pp. 135–146 (2010)
25. Iorgulescu, C., Dinu, F., Raza, A., et al.: Don't cry over spilled records: memory elasticity of data-parallel applications and its application to cluster scheduling. In: Annual Technical Conference (ATC), pp. 97–109 (2017)
26. Herodotou, H., Dong, F., Babu, S.: No one (cluster) size fits all: automatic cluster sizing for data-intensive analytics. In: 2nd ACM Symposium on Cloud Computing (SoCC), Cascais, Portugal, no. 18, pp. 1–14 (2011)
27. Alipourfard, O., Liu, H.H., Chen, J., et al.: Cherrypick: adaptively unearthing the best cloud configurations for big data analytics. In: 14th Symposium on Networked Systems Design and Implementation (NSDI), Boston, MA, pp. 469–482 (2017)

28. Huynh, N.V., Nguyen, D.N., Dutkiewicz, E.: Optimal and fast real-time resource slicing with deep dueling neural networks. IEEE J. Sel. Areas Commun. **37**(6), 1455–1470 (2019)
29. Xu, G., Xu, C.: Prometheus: online estimation of optimal memory demands for workers in in-memory distributed computation. In: The ACM Symposium on Cloud Computing (SoCC), Santa Clara, California, p. 655 (2017)
30. Chen, W., Pi, A., Wang, S., et al.: Pufferfish: container-driven elastic memory management for data-intensive applications. In: the ACM Symposium on Cloud Computing (SoCC), Santa Cruz, CA, USA, pp. 259–271 (2019)
31. Klimovic, A., Litz, H., Kozyrakis, C.: Selecta: heterogeneous cloud storage configuration for data analytics. In: 2018 USENIX Annual Technical Conference (ATC), Boston, MA, USA, pp. 759–773 (2018)
32. Liu, L., Xu, H.: Elasecutor: elastic executor scheduling in data analytics systems. In: The ACM Symposium on Cloud Computing (SoCC), Carlsbad, CA, USA, pp. 107–120 (2018)
33. Peng, G., Wang, H., Dong, J., et al.: Knowledge-based resource allocation for collaborative simulation development in a multi-tenant cloud computing environment. IEEE Trans. Serv. Comput. (TSC) **11**(2), 306–317 (2018)
34. Erradi, A., Iqbal, W., Mahmood, A., et al.: Web application resource requirements estimation based on the workload latent features. IEEE Trans. Serv. Comput. (TSC), 1(2019)
35. Kholidy, H.A.: An intelligent swarm based prediction approach for predicting clou computing user resource needs. Comput. Commun. (CC) **151**, 133–144 (2020)

A Novel Probabilistic-Performance-Aware Approach to Multi-workflow Scheduling in the Edge Computing Environment

Yuyin Ma[1], Ruilong Yang[1](✉), Yiqiao Peng[2], Mei Long[4], Xiaoning Sun[1], Wanbo Zheng[3], Xiaobo Li[5], and Yong Ma[6]

[1] School of Computers, Chongqing University, Chongqing, China
yangrl@cqu.edu.cn
[2] Bashu Secondary School, Chongqing, China
[3] Kunming University of Science and Technology, Kunming, China
[4] ZBJ NETWORK Co. Ltd., Chongqing, China
[5] Chongqing Animal Husbandry Techniques Extension Center, Chongqing, China
[6] School of Computer and Information Engineering, Jiangxi Normal University Nanchang, Nanchang, China

Abstract. Edge computing is a decentralized computing infrastructure in which data, calculation, storage and applications are located somewhere between the data source and the computing facilities. While the edge servers enjoy the close proximity to the end-users to provide services at reduced latency and lower energy costs, we use from limitations in computational and radio resources, which calls for smart, quality-of-service (QoS) guaranteed and efficient task scheduling methods and strategies. For addressing the edge-environment-oriented multi-workflow scheduling problem, in this paper, we propose a probabilistic-QoS-aware approach to multi-workflow scheduling over edge servers with time-varying QoS. Our proposed method leveraged a probability-mass function-based QoS aggregation model and a discrete firefly algorithm for generating the multi-workflow scheduling plans. In order to prove the effectiveness of our proposed method, we conducted an experimental case study based on varying types of workflows and a real-world dataset for edge server positions. It can be seen that our method clearly outperforms its competitors in terms of completion time, cost, and deadline validation rate.

Keywords: Edge computing · Workflow scheduling · Probabilistic model · Quality-of-service (QoS)

This work is supported in part by the Graduate Scientific Research and Innovation Foundation of Chongqing, China (Grant No. CYS20066 and CYB20062), and the Fundamental Research Funds for the Central Universities (China) under Project 2019CDXYJSJ0022.

H. Gao et al. (Eds.): CollaborateCom 2020, LNICST 349, pp. 640–655, 2021.
https://doi.org/10.1007/978-3-030-67537-0_38

1 Introduction

The edge computing paradigm is evolving towards a highly efficient computing infrastructure. Different from traditional solutions, it is featured by a ubiquitous, heterogeneous collection of elastic computational entities in terms of, e.g., edge servers. It provisions the platforms for building complex and on-demand applications, in terms of, e.g., workflows with reduced prices compared to traditional parallel computing techniques, e.g., grids. Consequently, great growth in in the number of active research work regarding performance-aware scheduling methods for edge-environment-based workflows can be seen recently scheduling multi-task-based processes [7, 11, 18] over the edge platform refers to assigning workflow tasks into appropriate edge nodes or servers for execution. A workflow can usually be described as a Directed-Acyclic-Graph (DAG) with multiple tasks that meets the constraint of execution orders.

It is widely believed that to assign tasks within multi-workflows to distributed platforms is an NP-hard problem. It is thus impractical to yield optimal scheduling solutions by using traversal-based strategies. Recently, as novel bio-inspired and genetic algorithms are becoming increasingly versatile and powerful, a great deal of research efforts are paid to applying them in dealing with edge-environment-oriented workflow scheduling problem. However, for simplicity, most existing contributions in this direction consider that edge servers are with static and invariable performance. However, edge and cloud servers in real-world can show unstable and time-varying performance. For example, Schad *et al.* [12] obversed that Amazon EC2 cloud services are subject to performance variations of 24%, 20% and 19% for CPU performance, I/O performance and network performance, respectively. Jakson *et al.* [4] showed that the difference between the maximum and minimum runtime of servers is 7,900 s, or approximately 42% of the mean runtime within EC2.

As can be seen from the above analysis, existing heuristic and bio-inspired algorithms with static and time-invariant performance models can be ineffective in dealing with real-world edge-environment-oriented workflow scheduling requirements, where performance of edge servers and platform-level infrastructures themselves are with highly unstable and time-varying performance. To overcome this limitation, in this work, we propose a probabilistic-performance-aware approach to edge-environment-oriented multi-workflow scheduling. Instead of considering single-point and static performance, our proposed method captures the dynamics of performance of edge servers by leveraging the probability mass functions (PMF) of historical performance data and utilizes a firefly algorithm for optimizing the workflow scheduling plans via maximizing the probability that the process completion duration and cost meets the deadline constraint.

To validate our proposed method, we perform extensive simulative studies based on various widely-used scientific workflow templates and a position dataset for urban edge servers. Simulative results show that our proposed method beats its peers in terms of multiple performance metrics.

2 Related Work

The main objective of workflow scheduling algorithms is to identify the best resources in the edge environment for the applications (tasks) of workflows. To achieve this, the pertinent objectives for satisfying users performance constraints include reducing execution time and total execution cost. Recently, considerable research works were carried out in this direction.

For instance, Zhang *et al.* [18] developed a Two-stage Cost Optimization algorithm to schedule workflows on edge clouds. The algorithm first leveraged a BF algorithm for obtaining the initial scheduling strategy and then further optimized the scheduling plans by the first stage. Their algorithm aims to minimize the system cost while meeting the delay requirements of workflows. Kim *et al.* [7] studied the trade-off between execution cost and workflow delays in the mobile computing system and proposed a intelligent-control-based algorithm for achieving near-optimal trade-offs. The trace-driven simulation showed that the algorithm can achieve 71% saving of execution cost and 82% gain of as opposed to its peers. Pandey *et al.* [11] proposed a particle swarm optimization (PSO) algorithm for load-balancing of cloud servers, while minimizing the execution cost, i.e., communication cost plus cloud resource cost of workflows.

kaur *et al.* [6] leveraged a multi-objective bio-inspired procedure (MOBFOA) by augmenting the traditional BFOA with Pareto-optimal fronts. Their method deals with the reduction of ow-time, completion duration, and operational expenditure. Zhang *et al.* [17] considered a multi-objective genetic optimization (BOGA) and optimized both electricity consumption and DAG reliability. Casas *et al.* [1] considered an augmented GA with the Efficient Tune-In (GA-ETI) mechanism for the optimization of turnaround time. Verma *et al.* [13] employed a non-dominated-sorting-based Hybrid PSO approach and aimed at minimizing both turnaround time and expenditure. Zhou *et al.* [19] introduced a fuzzy dominance sort based heterogeneous completion time minimization approach for the optimization of both cost and turnaround time of DAG executed on IaaS clouds.

3 System Model and Problem Formulation

3.1 The System Model

As shown in Fig. 1, an edge computing environment can be seen as a collection of multiple edge servers usually deployed near base stations. By this way, users are allowed to offload compute-intensive and latency-sensitive applications, e.g., Augmented Reality (AR), Virtual Reality (VR), Artificial Intelligence (AI), to edge servers. With in an edge computing environment, there are m users, denoted by $U = \{u_1, u_2, \ldots, u_m\}$, and n edge servers stations, denoted by $ES = \{e_1, e_2, \ldots, e_n\}$. Each user has an application, in terms of a batch of tasks organized by a workflow, to be executed, and users mobile device is allowed to offload tasks to nearby edge servers.

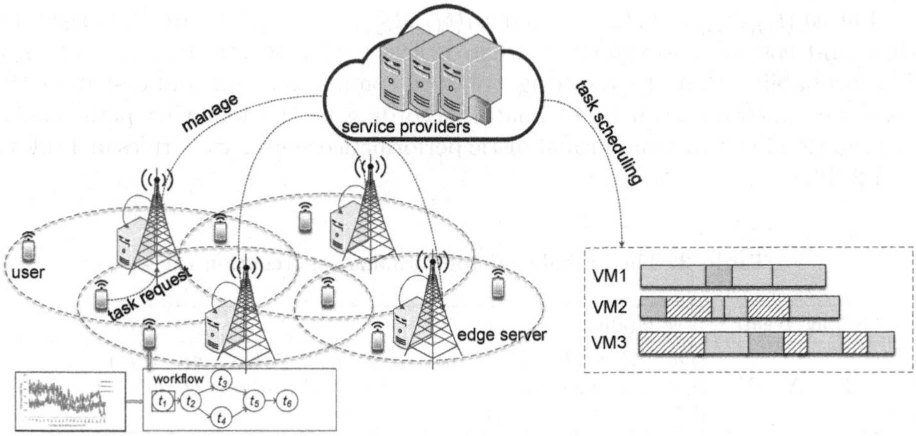

Fig. 1. The system architecture

3.2 The Probabilistic Performance Model

In this work, instead of considering static and time-invariant performance of edge servers, we consider time-varying performance of them. To be specific, we consider that the historical execution time of a certain workflow task upon an edge server, i.e., a discrete random variable X with *Dom(X)*, can be described by an empirical probabilistic distribution.

Consequently, the cumulative distribution function (CDF) of execution time and cost can be calculated as follows:

$$P(X \leq c) = P(X \leq floor(c)) + f_X(ceil(c)) \times \frac{c - floor(c)}{ceil(c) - floor(c)} \quad (1)$$

where c is deadline value, $Min(Dom(X)) < c \leq Max(Dom(X))$, $ceil(c) = Min\{c \mid c \in Dom(X) \text{ and } c \geq x\}$ and $floor(c) = Max\{c \mid c \in Dom(X) \text{ and } c < x\}$.

Table 1. The performance aggregation function of different workflow structural patterns

Workflow Patterns	Response Time	Cost
Sequence	$\sum_{i=1}^{n} RT(t_i)$	$\sum_{i=1}^{n} C(t_i)$
Parallel	$max_{1 \leq i \leq n} RT(t_i)$	$\sum_{i=1}^{n} C(t_i)$
Loop	$k \times RT(t_i)$	$k \times C(t_i)$

Let $w_i(t_{1j_1}, t_{2j_2}, \ldots, t_{nj_n})^{et}$ and $w_i(t_{1j_1}, t_{2j_2}, \ldots, t_{nj_n})^{cost}$ are the execution time and cost of a workflow w_i when it select edge servers of $(j_1, j_2, \ldots, j_n)$. The probability that the resulting workflow completion time and cost meet the deadline constraint, can be estimated according to the following performance aggregation functions and probabilistic performance aggregation rules in Tables 1 and 2 [3].

Table 2. The probabilistic performance aggregation rules

QoS aggregation	Probability
$Z = X + Y$	$Dom(Z) = \{z_1, z_2, \ldots, z_k\}, Max(m,n) \le k \le mn^1$, Each $z_i, 1 \le i \le k$, is the sum of some $x \in Dom(X)$ *and* $y \in Dom(Y), f_z(z_i) = \sum_{x+y=z_i} f_X(x) f_Y(y)$
$Z = X \cdot Y$	$Dom(Z) = \{z_1, z_2, \ldots, z_k\}, Max(m,n) \le k \le mn^1$, Each $z_i, 1 \le i \le k$, is the product of some $x \in Dom(X)$ *and* $y \in Dom(Y), f_z(z_i) = \sum_{x \cdot y = z_i} f_X(x) f_Y(y)$
$Z = MAX(X,Y)$	$Dom(Z) = Dom(X) \cup Dom(Y)$. $f_Z(z) = f_X(z) \cdot \sum_{y<z, y \in Dom(Y)} f_Y(y)$ if $z \in Dom(X)$ *and* $z \notin Dom(Y)$; $f_Z(z) = f_Y(z) \cdot \sum_{x<z, x \in Dom(X)} f_X(x)$ if $z \in Dom(Y)$ *and* $z \notin Dom(X)$; $f_Z(z) = f_X(z) \cdot \sum_{y \le z, y \in Dom(Y)} f_Y(y) + f_Y(z) \cdot \sum_{x<z, x \in Dom(X)} f_X(x)$ if $z \in Dom(X)$ *and* $z \in Dom(Y)$

[1] $m = |Dom(X)|, n = |Dom(Y)|$.

3.3 Problem Description

Based on the above analysis, the problem of probabilistic-performance-aware multi-workflow scheduling can be described as follows: given multiple workflows $w_{i_{1 \le i \le m}}$, we are interested to identify an edge server assignment plan $(t_{1j_1}, t_{2j_2}, \ldots, t_{nj_n})$ of w_i, with the highest probability that the workflow completion time and cost meet the deadline constraint.

$$max \;\; f = \prod P(w_i(t_{1j_1}, \ldots, t_{nj_n}))^{et_i} \le C_i^{et_i} \times \prod P(w_i(t_{1j_1}, \ldots, t_{nj_n}))^{cost_i} \le C_i^{cost_i} \quad (2)$$

s.t.

$$d_{ij} \le cov_j, i \in \{1, \ldots, m\} \; and \; j \in \{1, \ldots, n\} \quad (3)$$

$$x_{ij} \le 1, \; x_{ij} = \begin{cases} 1, & if \; e_j \; is \; selected \; for \; task \; t_i \\ 0, & otherwise \end{cases} \quad (4)$$

where $C_i^{et_i} (C_i^{cost_i})$ is the deadline constraint for completion time and cost of each workflow, d_{ij} is the distance between e_j and w_i, and cov_j is the coverage area of the j_{th} the server.

4 Firefly Algorithm

The firefly algorithm (FA) [15,16] is a meta-heuristic searching technology proposed, which simulates the luminous characteristics and attraction behavior of the fireflies. In this algorithm, fireflies are considered the sample points in the problem domain, and each firefly moves towards a brighter firefly that ultimately finds the optimal location. The individual renewal equation in FA consists of two parts: (1) the full attraction model; and (2) a randomly disturbed searching step size. The full attraction model requires every firefly to learn from all superior individuals.

In this work, we leverage the discrete derivative of FA, i.e., DFA, for solving the probabilistic-performance-aware workflow scheduling problem. The details are described as follows.

4.1 Encoding

In DFA, a schedule is an individual, described as a vector of integer values. The length of a vector is the same as the number of tasks in a workflow. The i_{th} element of the individual indicates to which server the i_{th} task of the workflow is scheduled to execute. Figure 2 gives a sample of an individual coding and a given workflow deployment, assuming that the workflow consists of eight tasks and is within the coverage range of by e_2, e_3, and e_4. In this schedule, t_1, t_4, t_6 are scheduled to be executed on e_3, t_2, t_3, t_7 are scheduled on e_2, and t_5, t_8 are scheduled on e_4, respectively.

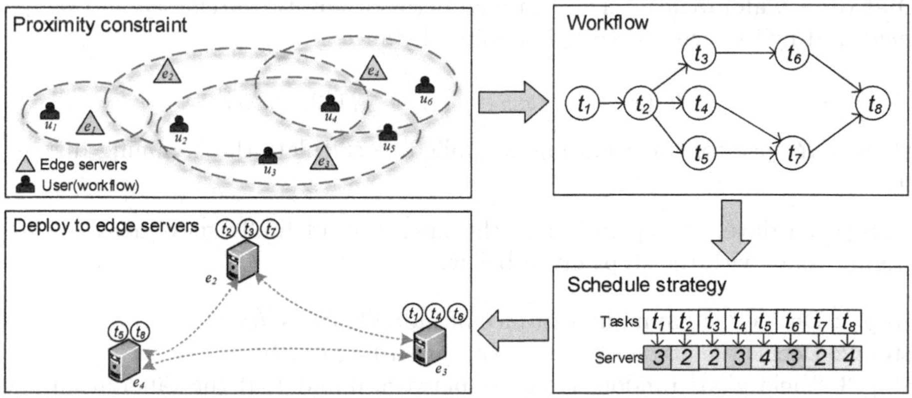

Fig. 2. An example of encoding

4.2 Population Initialization and Firefly Evaluation

The initialization of population is affected by the coverage range constraints due to the fact that, users can only offload tasks to reachable, in terms of the coverage range edge server.

As described in Sect. 3, our goal is to find the highest probability that the deadline constraint is met. The constraint is interpreted as a penalty function, where the highest probability is integrated into (5) as an evaluation of the fitness value of an individual. In case that the user is not reachable to servers, the fitness value is decided by the number of tasks that violate the deadline constraint.

$$F = \frac{1}{m} \times \begin{cases} \sum_{i=1}^{m} f_i, & if\ d_{ij} \leq cov_j \\ \sum_{i=1}^{m} f_i \times (\sqrt[n]{f_i})^{num}, & otherwise \end{cases} \tag{5}$$

where n denotes the total number of tasks in w_i, and $num \in \{1, 2, \ldots, n\}$, is the number of tasks that are unreachable by edge servers.

4.3 Individuals Update

The stipulation of update is that the darker firefly moves towards the brighter one. The population is updated iteratively and the scheduling strategy keeps being optimized. An example showing the update process is given in Table 3. The movement process is as follows.

Distance Calculation. The distance between any two fireflies p and p_{best} is measured by its corresponding hamming distance.

$$dis(x_i, x_j) = x_{id} \oplus x_{jd},\ d \in \{1, 2, \ldots, n\} \tag{6}$$

where OR operator for exclusion is applied to calculate the hamming distance [9].

β-step Update. β-step indicates the operation of firefly movement towards brighter ones with the steps given below.

Step 1: Decide the hamming distance of individuals as dis_1;
Step 2: Decide the attractiveness, $\beta(r)$, according to (7);
Step 3: Yield $|dis_1|$ random numbers between 0 and 1. If the random number is smaller than $\beta(r)$, the element of the p is substituted by the corresponding element of the brightest firefly;
Step 4: move towards the brightest one.

$$\beta(r) = \frac{\beta_0}{(1 + \gamma \times dis^2)} \tag{7}$$

α-step Update. The α-step update must be after the β-step update, according to (8), which is a process of random disturbance to avoid the solution space falling into the local optimization. Algorithm 1 presents all the operations of the discrete firefly algorithm.

$$x_i = x_i + \alpha(rand_{int}) \tag{8}$$

Table 3. Solution updation

Updation	Edge server assignment
Current firefly p	$\{2, 3, 2, 2, 3, 3, 4, 2\}$
Best firefly p_{best}	$\{3, 2, 2, 3, 4, 3, 2, 2\}$
Distance dis_1	5
Attractiveness $\beta(r)$	0.24
$rand(\ )between(0,1)$	{[0.13], 0.29, [0.03], [0.11], 0.67}
firefly p after β-step	$\{3, 3, 2, 3, 4, 3, 4, 2\}$
firefly p after α-step	{3, 3, 2, [4], [3], 3, 4, 2}

Table 4. Resource configurations and the price-per-minute of edge servers

Edge server types	Vcpu	Memory	Unit-price/minute
tp1	1 core	1g	0.0558 cents
tp2	1 core	2g	0.1262 cents
tp3	2 core	4g	0.1675 cents

5 Performance Evaluation

To evaluate the effectiveness of our proposed method, we conduct simulative experiments based on multiple workflow templates [5], namely, ***CyberShake***, ***Inspiral***, and ***Sipht***, as shown in Fig. 3.

We consider that all edge servers are with 3 different types of resource configurations and charging plans, i.e., ***tp1***, ***tp2***, and ***tp3***, as shown in Table 4. We tested the completion time of tasks on three types of edge servers at different periods, i.e., (a), (b) and (c), as shown in Fig. 4. As can be seen, the period of 4(a) shows the weakest performance fluctuations while 4(c) shows the greatest. The positions for edge servers and users are based on the dataset given in [2,8,14] and illustrated in Fig. 5.

Algorithm 1: Firefly Algorithm

Input: Algorithm related parameters: α, β_0, γ, $iter_{max}$
Output: Global QoS probability F_{best}, and schedule strategy S
Initial population of firelies: x_i, $i \in \{1,2,.,n\}$;
while t < $iter_{max}$ **do**
  for i=1:n **do**
    for j=1:i **do**
      if $F(x_i) > F(x_j)$ **then**
        move firefly j towards firefly i;
      else
        move firefly i towards firefly j;
      end
      Update the attractiveness of all fireflies;
      Evaluate new solution and update $F(x_i)$;
    end
    Rank the fireflies and find the current Global QoS probability F_{best}, and schedule strategy S;
  end
end
return F_{best}, S;

For comparison, we consider *pure* FA, GA, Greedy, and Random as baseline algorithms:

- ***pure* FA** [10]: the method is a heuristic algorithm, it is used to solve the multi-workflow scheduling problem with proximity constraint. It is noted that the QoS value in the method is constant. We use the mean QoS value as the static value of the firefly algorithm.
- **GA** [20]: the method is a heuristic search algorithm for workflow scheduling in cloud.
- **Greedy:** the method schedules unassigned tasks to the available lowest cost edge servers.
- **Random:** the method randomly selects a edge server for unassigned task.

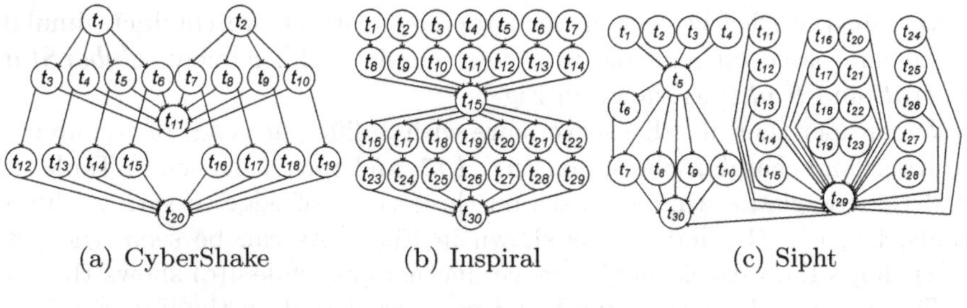

(a) CyberShake (b) Inspiral (c) Sipht

Fig. 3. Workflow process models

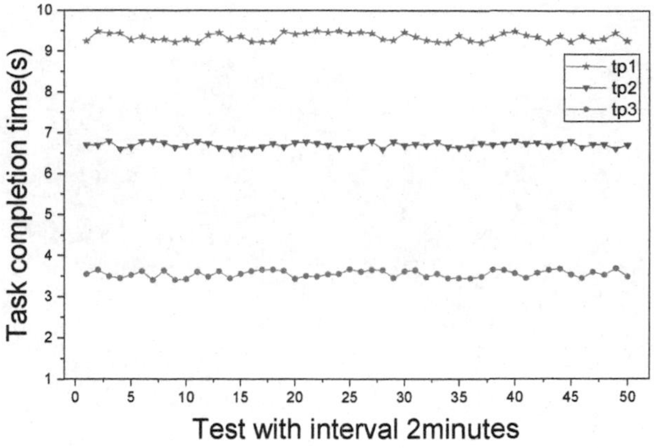

(a) Stable

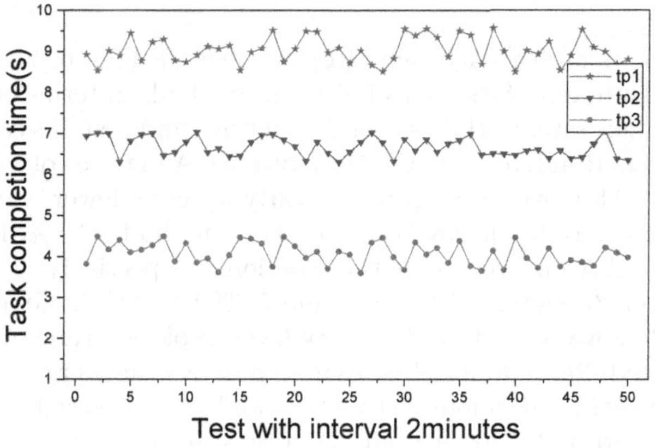

(b) Moderate

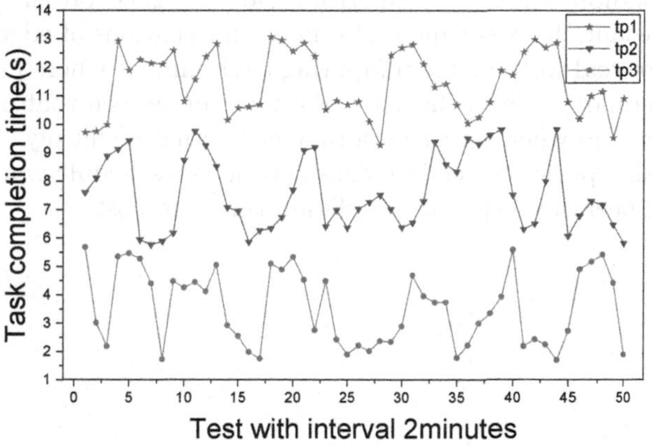

(c) Fluctuant

Fig. 4. The completion time of workflow tasks at different types of edge servers in different periods

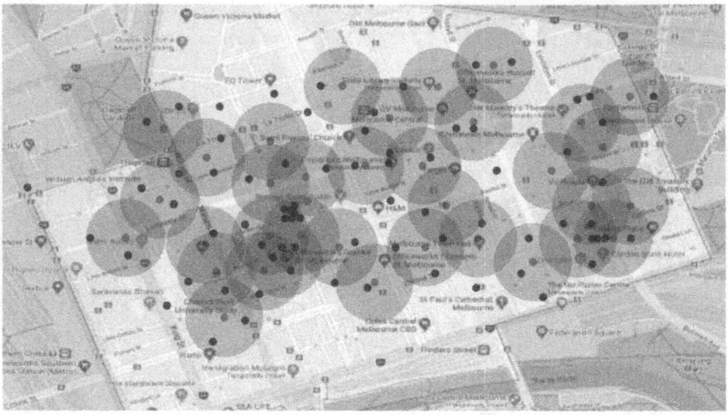

Fig. 5. Edge servers and users deployment

At the periods of 4(a), 4(b) and 4(c), we show in Figs. 6, 7 and 8 the comparison of scheduling performance of different methods in terms of deadline constraint satisfaction rate, workflow completion time, and cost. As can be observed, our method beats Random, *pure* FA, Greedy and GA in terms of average deadline validation rate. Moreover, our method clearly achieves lower workflow completion time and cost. To be specific, the cost of our method is 1.7%, 1.4%, and 3.4% lower than *pure* FA on average at three periods, respectively; 5.7%, 5.4%, and 8.4% lower than Random; 5.3%, 6.1%, and 5.6% lower than Greedy; and 2.2%, 1.4%, and 3.5% lower than GA. The workflow completion time of our method is 4.5%, 45.8%, and 49.7% lower than that of *pure* FA on average; 12.8%, 26.4%, and 62.3% lower than Random; 4.4%, 23%, and 34.7% lower than Greedy; and 14.5%, 51.4%, and 37.1% lower than GA, respectively.

The advantage of our proposed method is achieved because of the fact that traditional workflow scheduling algorithms are designed on the cloud without taking into account the real-time performance fluctuations of edge servers. However, this is not real in the edge computing environment where each server fluctuates in performance. From the above data, it can be seen that our method has a greater advantage when server performance fluctuates sharply. This is because our method takes performance into consideration to schedule workflow makes it superior to its peers in terms of completion time and cost.

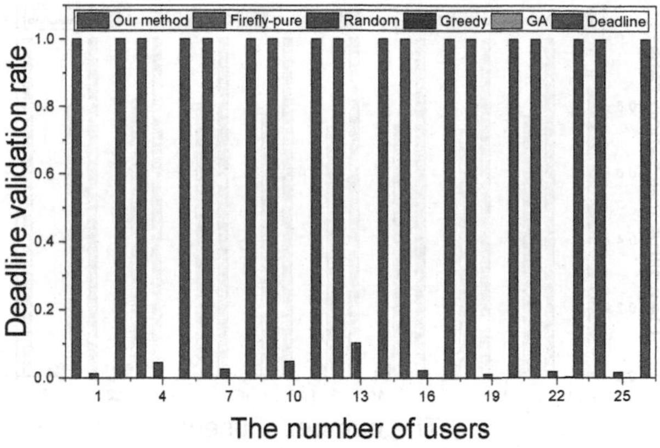

(a) deadline validation rate

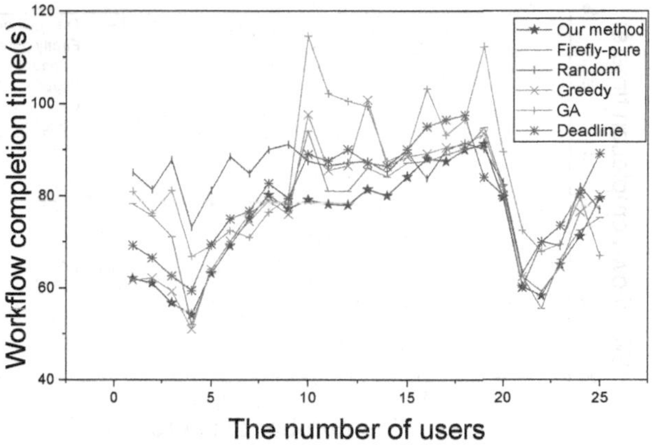

(b) completion time(s)

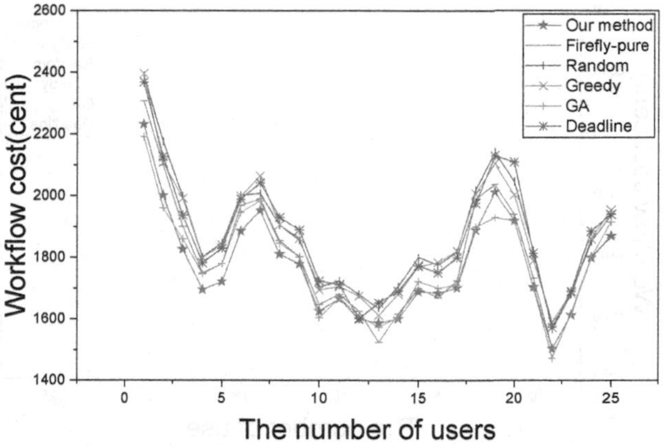

(c) cost(cent)

Fig. 6. The comparison of different methods at the period of 4(a) as the input performance data.

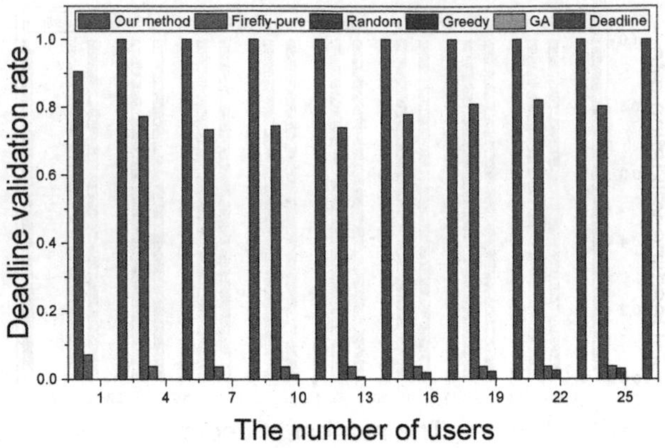

(a) deadline validation rate

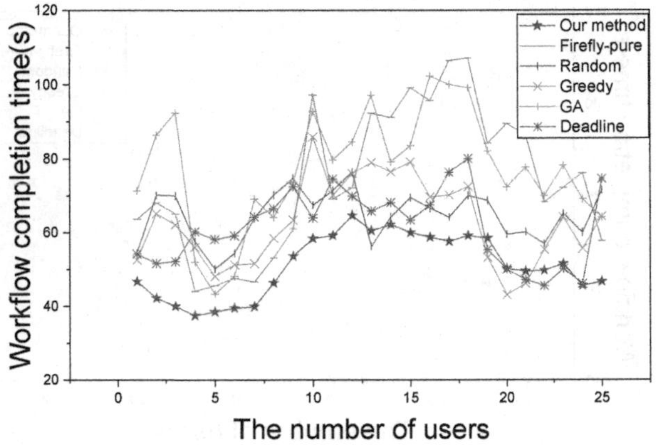

(b) completion time(s)

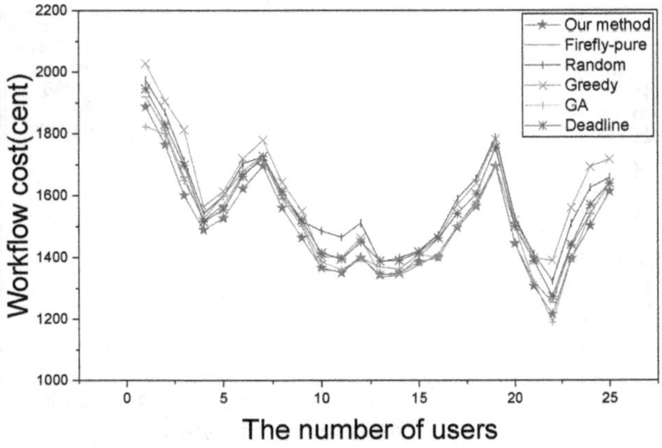

(c) cost(cent)

Fig. 7. The comparison of different methods at the period of 4(b) as the input performance data.

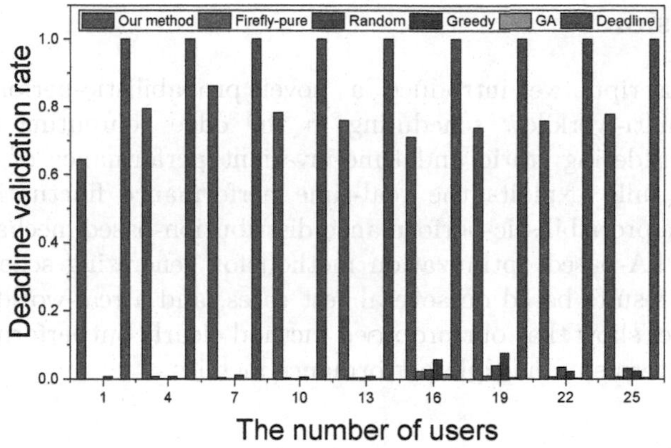

(a) deadline validation rate

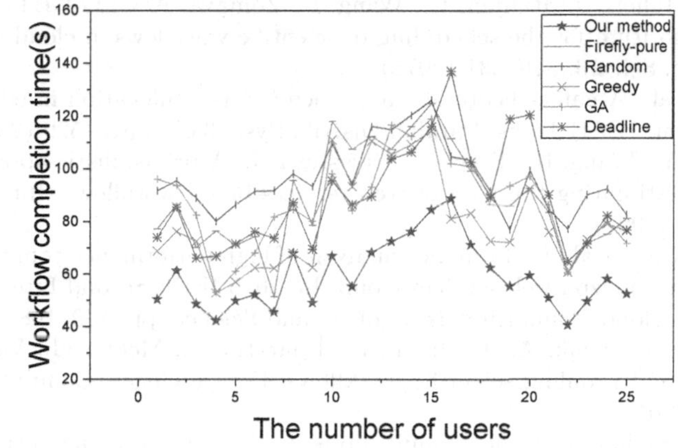

(b) completion time(s)

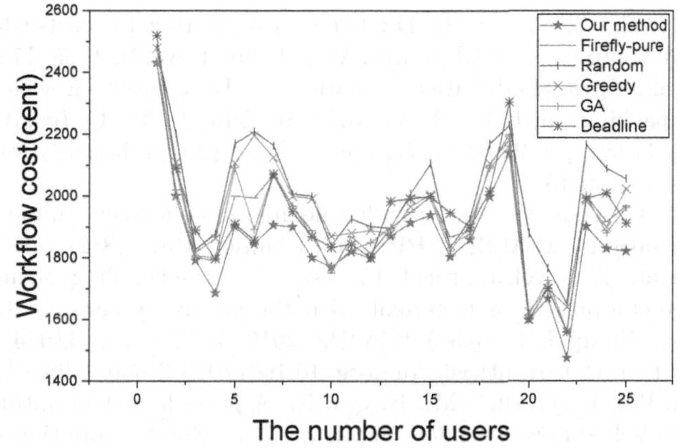

(c) cost(cent)

Fig. 8. The comparison of different methods at the period of 4(c) as the input performance data.

6 Conclusion

In this manuscript, we introduce a novel probabilistic-performance-aware method to multi-workflow scheduling in the edge computing environment. Instead of considering static and time-invariant performance of edge servers, our approach fully exploits the real-time performance fluctuations of them and leverages a probabilistic-performance-distribution-based mechanism in feeding a discrete-FA-based optimization method for generating scheduling plans. Experimental results based on several test cases, and a real-world edge-server-location dataset show that our proposed method clearly outperforms traditional approaches in terms of multiple performance metrics.

References

1. Casas, I., Taheri, J., Ranjan, R., Wang, L., Zomaya, A.Y.: GA-ETI: an enhanced genetic algorithm for the scheduling of scientific workflows in cloud environments. J. Comput. Sci. **26**, 318–331 (2018)
2. He, Q., et al.: A game-theoretical approach for user allocation in edge computing environment. IEEE Trans. Parallel Distrib. Syst. **31**(3), 515–529 (2020)
3. Hwang, S.Y., Wang, H., Tang, J., Srivastava, J.: A probabilistic approach to modeling and estimating the QoS of web-services-based workflows. Inf. Sci. **177**(23), 5484–5503 (2007)
4. Jackson, K.R., et al.: Performance analysis of high performance computing applications on the amazon web services cloud. In: 2010 IEEE Second International Conference on Cloud Computing Technology and Science, pp. 159–168. IEEE (2010)
5. Juve, G., Chervenak, A., Deelman, E., Bharathi, S., Mehta, G., Vahi, K.: Characterizing and profiling scientific workflows. Future Gener. Comput. Syst. **29**(3), 682–692 (2013)
6. Kaur, M., Kadam, S.: A novel multi-objective bacteria foraging optimization algorithm (MOBFOA) for multi-objective scheduling. Appl. Soft Comput. **66**, 183–195 (2018)
7. Kim, Y., Kwak, J., Chong, S.: Dual-side optimization for cost-delay tradeoff in mobile edge computing. IEEE Trans. Veh. Technol. **67**(2), 1765–1781 (2017)
8. Lai, P., et al.: Optimal edge user allocation in edge computing with variable sized vector bin packing. In: Pahl, C., Vukovic, M., Yin, J., Yu, Q. (eds.) ICSOC 2018. LNCS, vol. 11236, pp. 230–245. Springer, Cham (2018). https://doi.org/10.1007/978-3-030-03596-9_15
9. Lunardi, W.T., Voos, H.: An extended flexible job shop scheduling problem with parallel operations. ACM SIGAPP Appl. Comput. Rev. **18**(2), 46–56 (2018)
10. Ma, Y., et al.: A Novel approach to cost-efficient scheduling of multi-workflows in the edge computing environment with the proximity constraint. In: Wen, S., Zomaya, A., Yang, L.T. (eds.) ICA3PP 2019. LNCS, vol. 11944, pp. 655–668. Springer, Cham (2020). https://doi.org/10.1007/978-3-030-38991-8_43
11. Pandey, S., Wu, L., Guru, S.M., Buyya, R.: A particle swarm optimization-based heuristic for scheduling workflow applications in cloud computing environments. In: 2010 24th IEEE International Conference on Advanced Information Networking and Applications, pp. 400–407. IEEE (2010)
12. Schad, J., Dittrich, J., Quianeruiz, J.: Runtime measurements in the cloud: observing, analyzing, and reducing variance. Proc. VLDB Endow. **3**(1), 460–471 (2010)

13. Verma, A., Kaushal, S.: A hybrid multi-objective particle swarm optimization for scientific workflow scheduling. Parallel Comput. **62**, 1–19 (2017)
14. Xia, X., Chen, F., He, Q., Grundy, J.C., Abdelrazek, M., Jin, H.: Cost-effective app data distribution in edge computing. IEEE Trans. Parallel Distrib. Syst. **32**(1), 31–44 (2021)
15. Yang, X.-S.: Firefly algorithms for multimodal optimization. In: Watanabe, O., Zeugmann, T. (eds.) SAGA 2009. LNCS, vol. 5792, pp. 169–178. Springer, Heidelberg (2009). https://doi.org/10.1007/978-3-642-04944-6_14
16. Yang, X.S.: Firefly algorithm, stochastic test functions and design optimisation. arXiv preprint arXiv:1003.1409 (2010)
17. Zhang, L., Li, K., Li, C., Li, K.: Bi-objective workflow scheduling of the energy consumption and reliability in heterogeneous computing systems. Inf. Sci. **379**, 241–256 (2017)
18. Zhang, Y., Chen, X., Chen, Y., Li, Z., Huang, J.: Cost efficient scheduling for delay-sensitive tasks in edge computing system. In: 2018 IEEE International Conference on Services Computing (SCC), pp. 73–80. IEEE (2018)
19. Zhou, X., Zhang, G., Sun, J., Zhou, J., Wei, T., Hu, S.: Minimizing cost and makespan for workflow scheduling in cloud using fuzzy dominance sort based HEFT. Future Gener. Comput. Syst. **93**, 278–289 (2019)
20. Zhu, Z., Zhang, G., Li, M., Liu, X.: Evolutionary multi-objective workflow scheduling in cloud. IEEE Trans. Parallel Distrib. Syst. **27**(5), 1344–1357 (2016)

RCFC: A Region-Based POI Recommendation Model with Collaborative Filtering and User Context

Jun Zeng(✉), Haoran Tang, and Xin He

School of Big Data and Software Engineering, Chongqing University, Chongqing, China
{zengjun,tanghaoran,hexin}@cqu.edu.cn

Abstract. In the past few years, mobile application has been innovated by leaps and bounds, which leads the prevalence of location-based social networks (LBSNs). Point of interest (POI) recommendation aims to recommend satisfactory locations to users in mobile environment and plays an important role in LBSNs. However, there are still two challenges to be solved. One is the data sparseness caused by users who just visit a few POIs. The other is that it's hard to make reasonable explanation of recommendation from the perspective of real world. Hence, firstly we propose a region-based collaborative filtering to alleviate the data sparseness by clustering locations into regions. Secondly, we model the impact of two kinds of user contexts like geographical distance and POI category to make POI recommendation more reasonable. Finally, we present a joint model called RCFC which combines the two parts mentioned above. Results of experiments on two real-world datasets demonstrate the model we propose outperforms the popular recommendation algorithms and is more in line with the situation in real world.

Keywords: Recommendation system · Point of interest · Region · Context · Collaborative filtering

1 Introduction

Recent years have witnessed the rapid prevalence of location based social networks (LBSNs) such as Facebook, Twitter and Foursquare, which facilitate people's outdoor activities by recommending nearby POIs in real time [1]. LBSNs encourage users to share their experiences and locations by check-ins [20]. These check-ins embed abundant hints of users' preferences on locations. Implicit user preferences can provide better recommendation diversity [2]. Moreover, utilizing these check-ins helps users to explore new locations and bring more benefits to the third-parties like advertisers [12]. Therefore, it's still worth studying POI recommendation system in terms of its value [11].

Collaborative filtering (CF) is extensively investigated in recommendation system and widely used in industry [3]. It's based on a simple intuition that if users have similar rating records in the past, they are likely to rate new items similarly in the future [4]. It takes the full advantages of the similarity or correlations among user behaviors [5].

H. Gao et al. (Eds.): CollaborateCom 2020, LNICST 349, pp. 656–670, 2021.
https://doi.org/10.1007/978-3-030-67537-0_39

In most real-world applications, there are a huge number of users and locations. However, most users only visit a few of these locations, which causes thorny data sparseness [18]. Facing the data sparseness, CF can't calculate the similarity between users appropriately. Even though matrix factorization is an effective way to solve data sparseness [6], its mathematical nature makes explain recommendation from real-world perspective become much difficult. The key of reasonable explanation is to make full use of user context, as it greatly influences the user's decision on visiting a location [7]. Geographical distance is extremely important in user context. Users are more willing to visit locations closer to their current location [8]. For instance, if a user wants to watch a new movie that just comes out, he may tend to choose a nearby cinema, instead of going to one 20 kilometers away from home. In addition to the geographic distance, user daily activities usually present category-level transition patterns [9]. For example, user who loves stage plays may often go to locations classified as theater.

In this paper, inspired by CF and user context, we propose a novel region-based CF with the impact of context, which simulates the situation of users visiting POIs from real-world perspective. Our region-based CF is different from grid-based model which divide map into rectangular grids. We focus on the geographical relevance that can be regarded as irregular regions. For simplicity, we also call POI location.

The main contributions of this paper can be summarized as follows.

- Firstly, in order to alleviate the negative impact caused by data sparseness of CF, we cluster locations into many regions and figure out the preferences of users on regions instead of just locations. Then, we incorporate region factor into the CF.
- Secondly, since the full use of user context can explain recommendation well, we construct a probability model of geographic distance in a region and capture category level transition pattern between locations by a new category pairing method.
- Thirdly, we integrate our region-based CF with the impact of context by combining them in a linear way and propose our model called RCFC.
- Experimental results on two real-world datasets collected from Foursquare demonstrate our proposed model RCFC is superior to other popular recommendation models.

The rest of the paper is organized as follows. Related works are reviewed in Sect. 2. Section 3 describes in detail the design of RCFC model. Experiments are presented in Sect. 4 to analyze RCFC model and demonstrate its effectiveness compared to other recommendation algorithms. Finally, the Sect. 5 gives conclusion of this paper.

2 Related Work

POI Recommendation aims to obtain users' personal preferences according to their historical records such as check-ins, so as to recommend satisfactory locations to users in the future for saving time in making choices. To the best of our knowledge, there are many studies that extend CF or adopt user context for the sake of improving POI recommendation.

CF that is widely-used in industry is a classic popular algorithm in recommendation system. Liu [8] defines three types of friends in LBSNs and develop a two-step framework to leverage information brought by friends, which improves the quality of CF. Furthermore, Yang [10] proposes a general principled semi-supervised learning framework to alleviate data sparseness in CF via regularizing user preference and smoothing among neighboring users and POIs. The methods mentioned above focus on positive preferences of users. Tran [6] constructs a joint model which combines user embedding, user positive preference embedding and user negative preference embedding to complete recommendation from different perspectives. Therefore, it can be seen that the CF still plays a significant role today. However, the data sparseness is a great challenge for CF. Matrix factorization is an effective method. Xue [5] presents a deep matrix factorization structure to calculate the similarity between users and recommendations by learning a common low dimensional space for them.

User context always indicates the preferences of users explicitly and it can constrain predictions correctly no matter in recommendation system or other fields. Ye [13] argues that influence of geographical distance among two POIs is able to decide users' check-in behaviors and model it by power law distribution. Many geo-based POI recommendation models refer to his work so far. In contrast, Liu [14] considers the geographical influence from the location perspective rather than preferences of users and then models geographical neighborhoods of locations. Liu [15] proposes an adversarial learning model based on geographic information to dig deeper geographic representations. While most existing works discover the spatial, temporal and social patterns of user behavior, the information of context itself is often ignored [16]. To a large extent, location's category that represents its attribute can be regarded as expression of context itself. He [9] uses a third-rank tensor to predict the next preferred category that user may visit and then fuse distance to filter out POI candidates. These studies on user context are consistent with real world and give us great inspiration to our work.

In our previous work, we propose a deep recommendation framework by combining restricted Boltzmann machine with non-negative matrix factorization [18]. Furthermore, we construct a recurrent neural network based on self-attention to predict the next locations of users [20]. We also try some explorations by modeling geographical and social influence [19]. In this paper, different from our previous work, we focus on the region factor that could improve CF in POI recommendation. Moreover, we still consider the user context such as geographical distance and location category.

3 Region-Based POI Recommendation Model with Collaborative Filtering and User Context

3.1 Framework of RCFC

POI recommendation aims to recommend satisfactory locations to users by mining users' historical records of visiting locations. Now we introduce our recommendation model RCFC on the whole. RCFC is a joint model that combines region-based CF and the impact of user context. The whole framework of RCFC is shown in Fig. 1. Firstly, we cluster locations into different regions for capturing users' preference on regions and then incorporate that into CF. The goal of first step is to alleviate the negative impact caused by data sparseness. Secondly, after obtaining regions, we search the activity areas for users in order to find limited candidate locations and next use a non-linear model to calculate the probability of geographic distance. Thirdly, we capture the category-level transition pattern between locations by a novel category pairing method. The goal of second and third steps is to make recommendation more reasonable by adopting user context. Finally, we are able to recommend suitable locations to users by combination of all the steps mention above. The details of each step will be presented in following subsections.

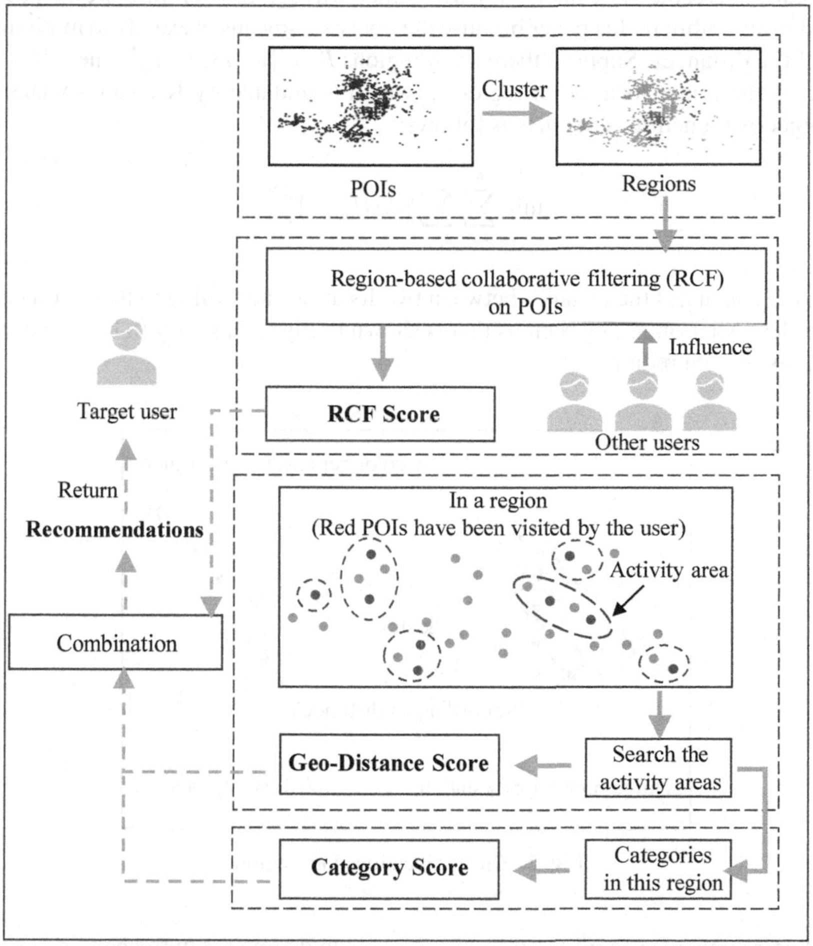

Fig. 1. The framework of RCFC

3.2 Region-Based Collaborative Filtering

For widely distributed locations, clustering them into regions is benefit for alleviating the data sparseness since there is a huge number of locations. We have a set U of users $\{u_1, u_2, \ldots u_m\}$, a set L of locations $\{L_1, L_2, \ldots L_n\}$,and a set C, of categories $\{c_1, c_2, \ldots c_p\}$. Each location is denoted by a triple $l = (lon, lat, c)$, which includes the longitude and latitude of the location and the category it belongs to. Before clustering locations, we use $\boldsymbol{u^L} = (u^{l_1}, u^{l_2}, \ldots, u^{l_n})$ to denote the corresponding vector of user u, where u^{l_j} represents the number of times user u has visited the location l_j.

It's intuitive and realistic to cluster locations in terms of distance. The more adjacent two locations are, the more likely they will gather to a region. Therefore, we adopt classic K-means algorithm in this paper because the goal of K-means is exactly to minimize the sum of the distances. Suppose there are q regions $R = \{r_1, r_2, \ldots r_q\}$ where $|R| \ll |L|$ and μ_i is the center location of region r_i, which is updated by K-means dynamically. The objective function is defined as follows:

$$\min \sum_{i=1}^{q} \sum_{l_j \in r_i} \left\| dis(l_j, \mu_i) \right\|^2 \tag{1}$$

where *dis* calculates the distance between two locations according to their latitudes and longitudes. After obtaining regions that is shown in Fig. 2, it's easy to know the overall check-ins of each region.

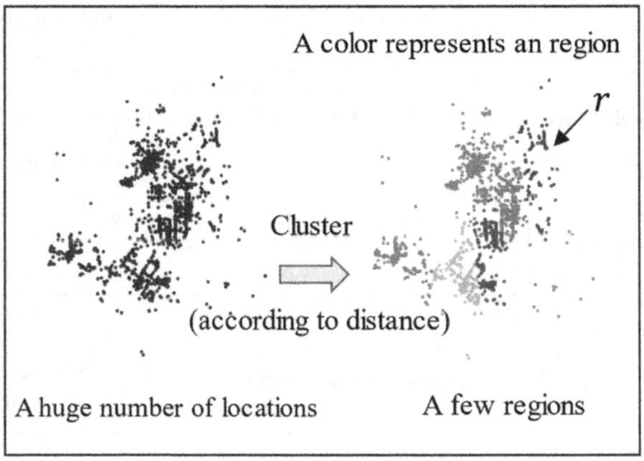

Fig. 2. Cluster locations into regions

Based on region, we use $\boldsymbol{u^R} = (u^{r_1}, u^{r_2}, \ldots, u^{r_q})$ to denote the corresponding vector of user u, where u^{r_j} represents the number of times user u has visited the region r_j. In order to compare with the overall regional visits, we have adopted a normalization on $\boldsymbol{u^R}$ in advance. The region similarity between a user and the whole is shown as follows:

$$sim(u, R) = \frac{\sum_{z=1}^{q} u^{r_z} \times cnt(r_z)}{\sqrt{\sum_{z=1}^{q} (u^{r_z})^2} \times \sqrt{\sum_{z=1}^{q} cnt^2(r_z)}} \tag{2}$$

where $cnt(r_z)$ denotes the ratio of the total number of check-ins in the region r to the maximum of all regions. $sim(u, R)$ indicates whether the user will follow others while choosing a region. If $sim(u, R)$ is large, it means the user is more willing to visit the regions that other users tend to visit. Otherwise, the user may have his own unique taste on visiting regions that is different from others. Then, the score $u^{r_{z'}}$ of user visiting region r_z is defined as follows, which considers both the public and the user himself:

$$u^{r_{z'}} = sim(u, R) \times cnt(r_z) + (1 - sim(u, R)) \times u^{r_z} \tag{3}$$

Our goal is to enhance CF by clustering all locations into regions. The original CF is shown as follows:

$$cf(u_i, l_j) = \frac{\sum_{z=1}^{SU(u_i)} sim(u_i, u_z) \times u_z^{l_j}}{\sum_{z=1}^{SU(u_i)} sim(u_i, u_z)} \tag{4}$$

$$Score_{u_i, l_j}^{RCF} = u_i^{r_{z'}} + \frac{cf(u_i, l_j)}{\max_{l_z \in L}(cf(u_i, l_j))} \tag{5}$$

where r_z is corresponding region of l_j. Unlike geographical and category contexts, region is a further strategy for CF so we don't adopt weighted calculation for them. Note that the user vectors in (4) and (5) are based on location and region respectively. r_z is corresponding region of l_j. Next, we will explore the impact of user context.

3.3 Geographical Distance Context

The major user context for POI recommendation is geographical distance. Users tend to visit locations that are closer to their current location. For the sake of modeling geographical distance, inspired by [13], we adopt a power law distribution that is shown as follows:

$$p(l_i|l_j) = a \times (dis(l_i, l_j))^b \tag{6}$$

where $p(l_i|l_j)$ denotes the probability of visiting location l_i while the current location is l_j. a and b are the parameters of the power law distribution.

For obtaining suitable a and$b$, firstly we convert it to a linear model by using logarithmic representation, which is defined as follows:

$$\log p(l_i|l_j) = \log a + blogdis(l_i|l_j) \tag{7}$$

$$y(x, \mathbf{w}) = a' + bx \tag{8}$$

where $y(x, \mathbf{w})$ denotes $\log p(l_i|l_j)$ and $\mathbf{w}$ is the parameter set. a' is equal to $\log a$. x denotes the pair of two adjacent locations l_i and l_j.

We adopt the least squares method in terms of its simplicity and the loss function that needs to be minimized is defined as follows:

$$\min_w \frac{1}{2} \sum_{x \in D} (y(x, \mathbf{w}) - t(x))^2 + \frac{\lambda}{2} \|\mathbf{w}\|^2 \tag{9}$$

where D is our real-world dataset and $t(x)$ is the logarithmic value of the true distance probability derived from D. Moreover, the last term is a regularization that is controlled by its weight λ.

Generally, a user has visited a few locations in a region but they denote the user's activity areas that are shown in Fig. 3. Hence, in order to filter out candidate locations, we set 0 to those locations that is out of user's activity areas. We only calculate geographical score for locations belong to user's activity area.

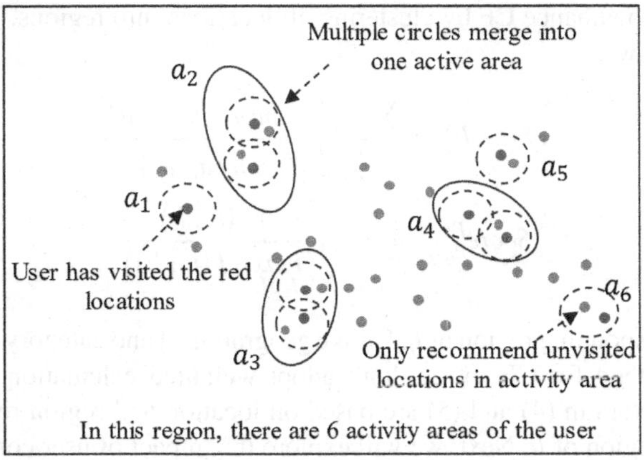

Fig. 3. Activity areas

In a region r, centered by location the user has visited, the circle with a radius of 0.5 kilometers composes an activity area. If two circles overlap, then merge them into one activity area. For a user, we have a set A_u of his activity areas $\{a_1, a_2, \cdots, a_z\}$, where a_0 represents the locations that the user has visited in this activity area. The geographical probability of a candidate location l_j is calculated as follows:

$$p(l_j|a_o) = \prod\nolimits_{l_i \in a_o} p(l_i, l_j) \tag{10}$$

where a_o is the corresponding activity area of l_j and l_i is the location user has visited in a_o. Next, the geographical score of unvisited l_j is defined as follows:

$$S^{Geo}_{u_i,l_j} = \frac{count(a_o)}{\max\limits_{a_x \in A_{u_i}} count(a_x)} \times \frac{p(l_j|a_o)}{\max\limits_{l_x \in a_o} p(l_x|a_o)} \tag{11}$$

where $count(a_o)$ is the function that counts the total number of check-ins of the user in a_o. Even in the same region, user has different check-in behaviors in different activity areas. In order to make all candidate locations comparable, we decide to normalize the geographical scores of them according to activity areas they belong to and also take the weight of each activity area into account.

3.4 Category Context

Categories usually have semantic information, reflecting functions of locations. In this paper, we present a novel pair-wise method to model the impact of category. If a user has visited both category c_i and c_j, then we can say there is a transition relationship between c_i and c_j. This kind of transfer has symmetry. Let T denote the category transition matrix, where its size is $p \times p$. T_{ij} represents value of the transition between c_i and c_j, which means the number of users who has visited both the two categories.

We use C_{u_i,r_z} to denote the set of categories that user u_i has visited in region r. For the category c_x of a candidate location l_j, its category score is defined as follows:

$$S_{u_i,l_j}^{Cate} = \begin{cases} \frac{count(c_x)}{\sum_{c_v \in C_{u_i,r}} count(c_v)}, c_x \in C_{u_i,r} \\ f(c_x), c_x \notin C_{u_i,r} \end{cases} \tag{12}$$

If user has visited the category before, then we calculate its probability according to user's check-ins in this region. Otherwise, we adopt function f to get its category score, which is shown as follows:

$$f(c_x) = \frac{\frac{1}{|C_{u_i,r}|} \sum_{c_v \in C_{u_i,r}} T_{c_v,c_x}}{\sum_{c_s \in C_{u_i,r}} f(c_s)} \tag{13}$$

Note that in both (12) and (13), c_v denotes the category user has visited and c_s denotes the unvisited category. After obtaining geographical distance score and category score, the final context score of location l_j is defined as follows:

$$Score_{u_i,l_j}^{Context} = \beta \times S_{u_i,l_j}^{Geo} + (1-\beta) \times \frac{S_{u_i,l_j}^{Cate}}{\max_{l_z \in r} S_{u_i,l_z}^{Cate}} \tag{14}$$

where l_z and l_j belong to the same region r and $\beta \in [0, 1]$. We have adopted normalization when calculating geographical score, so now we only have to normalize category score.

3.5 Joint Recommendation

We have introduced the region-based CF and obtain $Score_{u_i,l_j}^{RCF}$. RCF alleviates the data sparseness by clustering locations into regions and incorporating them into CF. Then, we combine the geographical and category context to get $Score_{u_i,l_j}^{Context}$, which models the impact of two major user context for explaining the recommendation more reasonable. Now we propose our final model called RCFC. It contains the $Score_{u_i,l_j}^{RCF}$ and $Score_{u_i,l_j}^{Context}$. RCFC is defined as follows:

$$Score_{u_i,l_j}^{RCFC} = \alpha \times Score_{u_i,l_j}^{RCF} + (1-\alpha) \times Score_{u_i,l_j}^{Context} \tag{15}$$

where $\alpha \in [0, 1]$. Finally, we sort the scores of all candidate locations and recommend top-K locations to users.

4 Experiments

In this section, we choose the optimal parameters of our proposed model and evaluate that with other popular POI recommendation models on two real-world datasets.

4.1 Datasets

We employ two real-world datasets that are collected from two cities on Foursquare, one is Los Angeles, the other is London. There are 48460 check-ins made with Los Angeles. They are made by 4746 users on 7135 POIs and the average check-ins of each user is 10. As for London dataset, there are 43912 check-ins. They are produced by 3470 users on 7941 POIs and the average check-ins of per user is 12. Both two datasets are much sparse. We randomly select 70% of the locations of each user as training data and the remaining 30% as test data. In addition, for the effectiveness of our experiments, the users who have visited less than 5 POIs and the POIs that have been visited by less than 5 users are removed.

4.2 Evaluation Metrics

We adopt precision, recall and F1-score to evaluate the performance. F1-score combines precision with recall and we use that to find the best values of parameters of our model. Precision and recall are used in comparing with other algorithms. All evaluation metrics are defined as follows:

$$Precision@K = \frac{1}{|U|}\sum_{u \in U}\frac{|rec^u \cap test^u|}{|rec^u|} \tag{16}$$

$$Recall@K = \frac{1}{|U|}\sum_{u \in U}\frac{|rec^u \cap test^u|}{|test^u|} \tag{17}$$

$$F1 - score@K = 2\frac{Precision@K \times Recall@K}{Precision@K + Recall@K} \tag{18}$$

where F1-score is based on the overall precision and recall. K is the number of recommended POIs, which is always set to 10, 15 and 20. rec^u is the recommendation list for user u and $test^u$ is the test data of user u.

4.3 Study on Parameters

There are three parameters of our model. Q determines how to cluster locations into regions. α weights the region-based CF and the impact of context and β weights the geographical distance and category. We do experiments on different parameter combinations.

The results of α are shown in Fig. 4. For the LA dataset, the optimal $\alpha = 0.7$ while $\alpha = 0.6$ is most suitable for the LON dataset. From the figure, it's observed that all curves first go up and then go down in both datasets. This indicates that with the increase of α, the performance of our model can be improved by region-based CF but then excessively ignoring the user context will cause the performance to decrease. $\alpha = 0.0$ means our model depends only on the impact of user context while $\alpha = 1.0$ means there is only region-based CF. However, the latter is better than the former on both datasets, which proves the effectiveness of clustering locations into regions for POI recommendation.

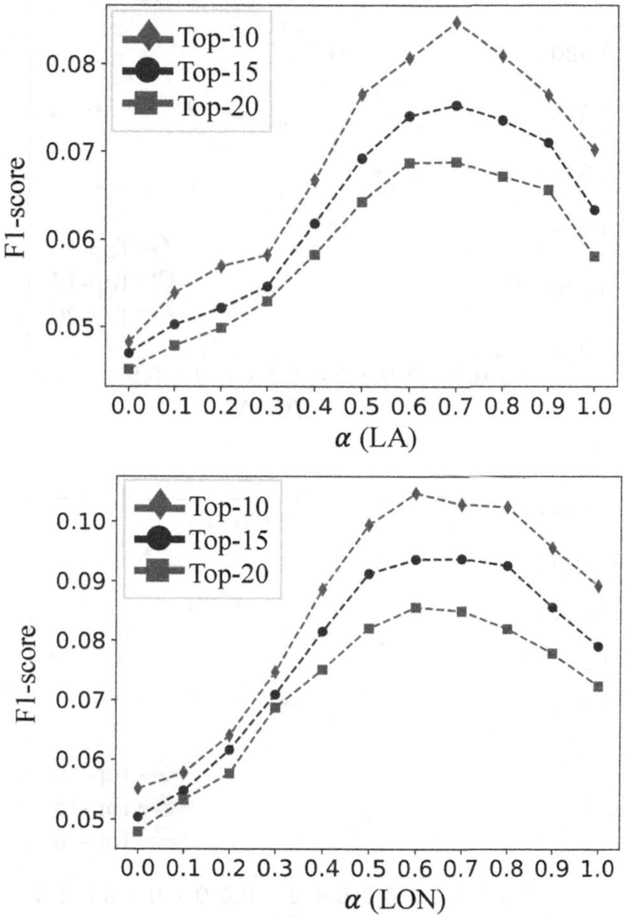

Fig. 4. F1-score under different α

The results of β are shown in Fig. 5. For the LA dataset, the optimal $\beta = 0.9$ while $\beta = 0.5$ is most suitable for the LON dataset. On one hand, the curves of LA dataset have clear upward trends before $\beta = 0.5$ but then become flat gradually, which reveals that geographical distance has a limited improvement. On other hand, while Top-K = 10 and15, the curves of LON dataset go down significantly after reaching their peaks, which shows that too much consideration of geographical distance will bring negative effect. However, for both datasets, region-based CF with only geographical context where $\beta = 1.0$ outperforms that with only category context where $\beta = 0.0$. This fully demonstrates that geographical distance is the most important context for POI recommendation and it's worth studying. Moreover, even though category context dose not play the equal impact as geographical distance, it can still improve the performance of POI recommendation based on geographical distance.

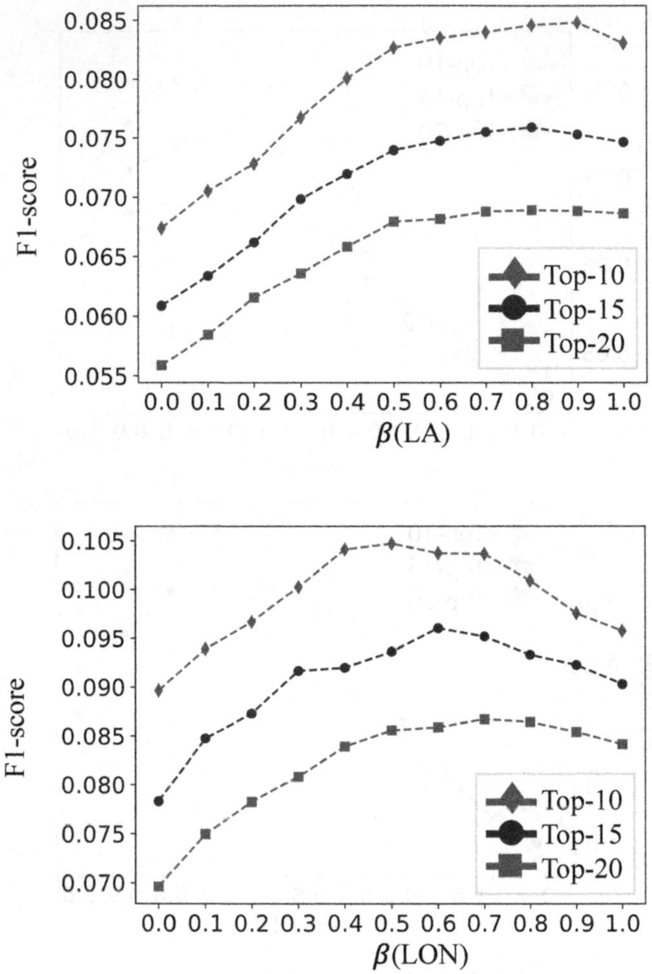

Fig. 5. F1-score under different β

As shown in Fig. 6, for the LA dataset, the optimal $Q = 11$ while $Q = 16$ is most suitable for the LON dataset. There are significant differences between two datasets. A huge decline happens for LA dataset when$Q = 2$, which indicates that few regions may result in insufficient clusters especially at the beginning. With the increase of Q, the advantage of region comes into play gradually. In contrast, for LON dataset, $Q = 2$ doesn't destroy the performance. That may be due to characteristics of the dataset because each city has different distributions of locations. Overall, there are many obvious fluctuations on both datasets. One possible reason is that, for each Q, the K-means method clusters locations into regions dynamically. In general, a larger region number can improve our model even though fluctuations exist. Meanwhile, a larger region number help us understand the regional behaviours of users.

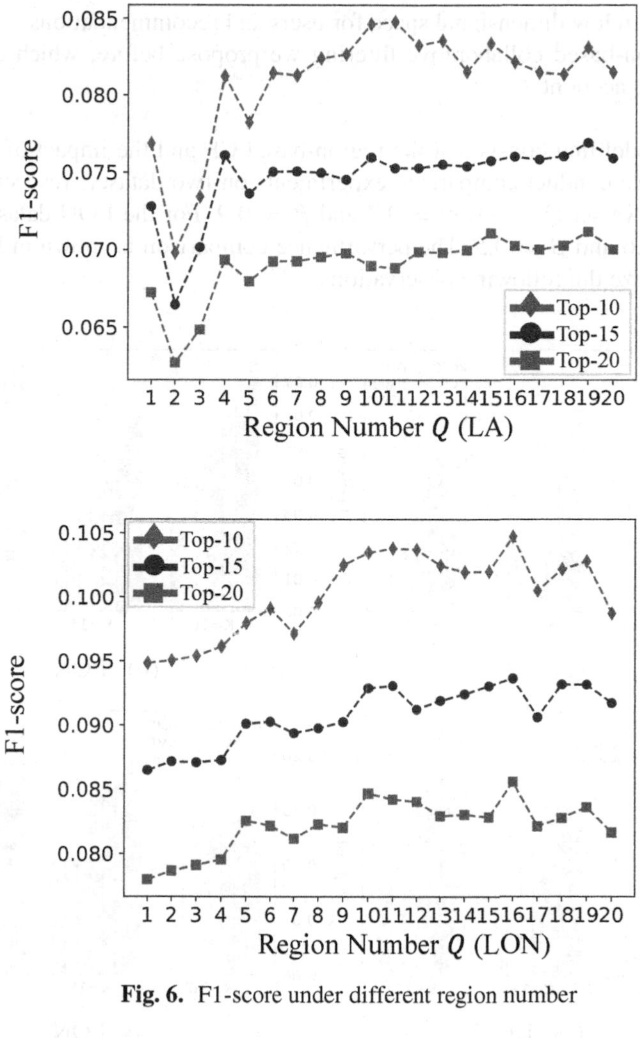

Fig. 6. F1-score under different region number

4.4 Performance Comparison

To comprehensively demonstrate the effectiveness of our model, we compare them with following popular models:

POP: A basic model that recommends popular POIs.
CF: User-based collaborative filtering, which is widely-used in industry.

NMF: A classic non-negative matrix factorization. It aims to fill in unknown items in user-POI matrix.
BPR [17]: Bayesian personalized ranking via optimizing the ordering relationship of users and POIs.
DMF [5]: A novel deep matrix factorization for recommendation system, which aims to learn a common low dimensional space for users and recommendations.
RCF: The region-based collaborative filtering we propose before, which only takes region factor into account

Our final model that consists of the region-based CF and the impact of context is called **RCFC**. We conduct comparison experiments on two datasets respectively. For the LA dataset, we set $Q = 11$, $\alpha = 0.7$ and $\beta = 0.9$. For the LON dataset, we set $Q = 16$, $\alpha = 0.6$ and $\beta = 0.5$. The performance comparison is shown in Fig. 7 and then we summarize the following observations.

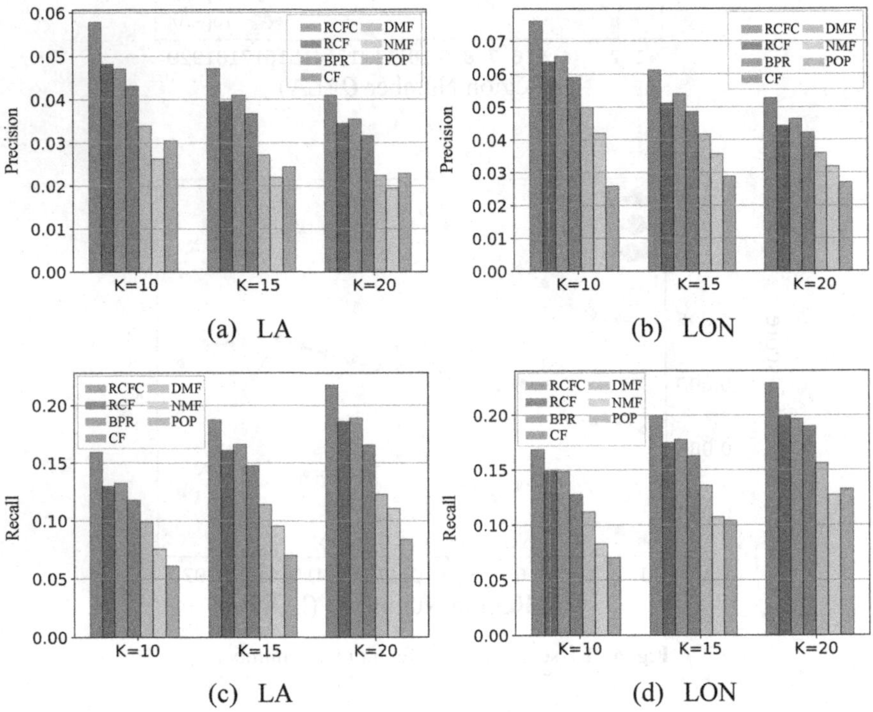

Fig. 7. Performance comparison in terms of precision and recall

Firstly, from a whole perspective, RCFC is obviously superior to other recommendation models in terms of both precision and recall. This result is due to clustering locations into regions and modeling the impact of context. RCF doesn't outperform BPR, which indicates BPR still has an excellent performance at present. BPR that aims to match the users and POIs can mine the correlations between them effectively. Although DMF is a state-of-art deep model for recommendation system, it's not as good as CF, only better than NMF and POP. One possible explanation can be that DMF depends on the structure of neural network to a great extent and it's also very sensitive to different datasets. Besides, DMF does not consider any unique characteristics of POI recommendation, such as geographical distance.

Secondly, the classical CF model performs well on both datasets, which fully proves why it can become the widely-used recommendation model in industry. This is why we are interested in it and proposed our own model. Our RCFC achieves excellent improvement by incorporating region and context into it. RCF only considers the region factor and ignores the impact of user context. Hence, it doesn't outperform RCFC. In terms of precision, NMF outperforms POP on LON dataset while POP is better than NMF on LA dataset. However, the recall of NMF basically is superior to POP on both datasets. In short, it's obvious that RCFC outperforms the popular recommendation models since it is based on region and user context. Meanwhile, RCFC can make reasonable explanation for recommendation according to region and context.

5 Conclusion

In this paper, we put forward two challenges in POI recommendation. One is the data sparseness caused by users who just visit a few locations and the other is that it's hard to make reasonable explanation for recommendation from the perspective of real world. Therefore, for the sake of alleviating data sparseness, we firstly present a region-based CF by clustering locations into different regions. Secondly, we explore the impact of two user context like geographical distance and location category to enhance the explanation of recommendation. Finally, we propose our joint model RCFC that combines region-based CF and user context. We construct experiments on two real-world datasets collected from two cities on Foursquare and the results demonstrate RCFC is superior to some popular POI recommendation models.

For future work, on one hand, we will compare different clustering methods to make the regions more accurate. On the other hand, we will take more contexts into account since there are still many useful contexts such as time and weather. We will continue our work of the model RCFC.

Acknowledgements. This research is Sponsored by Natural Science Foundation of Chongqing, China (No.cstc2020jcyj-msxmX0900) and the Fundamental Research Funds for the Central Universities (Project No.2020CDJ-LHZZ-040).

References

1. Li, H., Ge, Y., Lian, D., Liu, H.: Learning user's intrinsic and extrinsic interests for point-of-interest recommendation: a unified approach. In: The 26th International Joint Conference on Artificial Intelligence, IJCAI 2017, pp. 2117--2123 (2017)
2. Fletcher, K.K.: Regularizing matrix factorization with implicit user preference embeddings for web API recommendation. In: IEEE SCC, pp. 1-8 (2019)

3. Xue, H., Dai, X., Zhang, J., Huang, S., Chen, J.: Deep matrix factorization models for recommender systems. In: The 26th International Joint Conference on Artificial Intelligence, IJCAI 2017, pp. 3203--3209 (2017)
4. Sarwar, B.M., Karypis, G., Konstan, J.A., Riedl, J.: Item-based collaborative filtering recommendation algorithms. In: ACM WWW, pp. 285–295 (2001)
5. Hu, S., Tu, Z., Wang, Z., Xu, X.: A POI-sensitive knowledge graph based service recommendation method. In: IEEE SCC, pp. 197–201 (2019)
6. Tran, T., Lee, K., Liao, Y., Lee, D.: Regularizing matrix factorization with user and item embeddings for recommendation. In: ACM CIKM, pp. 687–696 (2018)
7. Liu, Q., Wu, S., Wang, L., Tan, T.: Predicting the next location: a recurrent model with spatial and temporal contexts. In: The 30th AAAI Conference on Artificial Intelligence, AAAI 2016 (2016)
8. Li, H., Ge, Y., Hong, R., Zhu, H.: Point-of-interest recommendations: learning potential check-ins from friends. In: ACM KDD, pp. 975–984 (2016)
9. He, J., Li, X., Liao, L.: Category-aware next point-of-interest recommendation via listwise bayesian personalized ranking. In: The 26th International Joint Conference on Artificial Intelligence, IJCAI 2017, pp. 1837–1843 (2017)
10. Yang, C., Bai, L., Zhang, C., Yuan, Q., Han, J.: Bridging collaborative filtering and semi-supervised learning: a neural approach for POI recommendation. In: ACM KDD, pp. 1245–1254 (2017)
11. He, J., Li, X., Liao, L., Song, D., Cheung, W.K.: Inferring a personalized next point-of-interest recommendation model with latent behavior patterns. In: The 30th AAAI Conference on Artificial Intelligence, AAAI 2016, pp. 137–143 (2016)
12. Cheng, C., Yang, H., King, I., Lyu, M.R.: Fused matrix factorization with geographical and social influence in location-based social networks. In: The 26th AAAI Conference on Artificial Intelligence, AAAI 2012, pp. 17–23 (2012)
13. Ye, M., Yin, P., Lee, W., Lee, D.L.: Exploiting geographical influence for collaborative point-of-interest recommendation. In: ACM SIGIR, pp. 325–334 (2011)
14. Liu, Y., Wei, W., Sun, A., Miao, C.: Exploiting geographical neighborhood characteristics for location recommendation. In: ACM CIKM, pp. 739–748 (2014)
15. Liu, W., Wang, Z., Yao, B., Yin, J.: Geo-ALM: POI recommendation by fusing geographical information and adversarial learning mechanism. In: The 28th International Joint Conference on Artificial Intelligence, IJCAI 2019, pp. 1807–1813 (2019)
16. Gao, H., Tang, J., Hu, X., Liu, H.: Content-aware point of interest recommendation on location-based social networks. In: The 29th AAAI Conference on Artificial Intelligence, AAAI 2015, pp. 1721–1727 (2015)
17. Rendle, S., Freudenthaler, C., Gantner, Z., Schmidtthieme, L.: BPR: Bayesian personalized ranking from implicit feedback. In: The 25th Conference on Uncertainty in Artificial Intelligence, UAI 2009, pp. 452–461 (2009)
18. Zeng, J., Tang, H., Li, Y., He, X.: A deep learning model based on sparse matrix for point-of-interest recommendation. In: The 31st International Conference on Software Engineering & Knowledge Engineering, SEKE 2019, pp. 379–492 (2019)
19. Zeng, J., Li, F., He, X., Wen, J.: Fused collaborative filtering with user preference, geographical and social influence for point of interest recommendation. Int. J. Web Serv. Res. (IJWSR) 16(4), 40–52 (2019)
20. Zeng, J., He, X., Tang, H., Wen, J.: A next location predicting approach based on a recurrent neural network and self-attention. In: The 15th International Conference on Collaborative Computing: Networking, Applications and Worksharing, CollaborateCom 2019, pp. 309–322 (2019)

Delay Constraint Energy Efficient Cooperative Offloading in MEC for IoT

Haifeng Sun[1(✉)], Jun Wang[2], Haixia Peng[3], Lili Song[1], and Mingwei Qin[4]

[1] School of Computer Science and Technology, Southwest University of Science and Technology, Mianyang 621010, China
dr_hfsun@163.com

[2] Guangdong Key Laboratory of Intelligent Information Processing, Shenzhen University, Shenzhen 518061, China

[3] Department of Electrical and Computer Engineering, University of Waterloo, Waterloo N2L 3G1, Canada

[4] School of Information Engineering, Southwest University of Science and Technology, Mianyang 621010, China

Abstract. Mobile edge computing (MEC) is a promising approach to execute delay-sensitive and computation-intensive applications in the resource-limited IoT mobile devices (IMDs) by offloading computing tasks to MEC servers. In this paper, we propose a neighbor-aided cooperative offloading scheme with delay constraint to improve the energy efficiency in MEC-based Internet of Things (IoT) networks. The network consists of a target IMD with some IMD neighbors, and an access point integrated with an MEC server. The latency-constrained tasks in the IMD can be partially offloaded to and executed by the selected neighboring IMD or the MEC server. Different from other works, our proposed offloading scheme selects the most energy-efficient one from all the neighboring IMDs as the offloading helper. Specifically, we formulate an optimization problem to minimize the total energy consumption while satisfying the computation delay constraint of each task, and obtain the most energy-efficient neighbor with the optimized division of tasks by solving the formulated problem. Moreover, we design an easy neighbor selection scheme with lower time complexity by the weighted value of the transmission rate for each neighbor. Numerical results show that the proposed scheme outperforms benchmark schemes significantly in terms of energy consumption and the supported maximum task length.

Keywords: Cooperative offloading · Delay constraint · Energy efficiency · Mobile edge computing · Neighbor selection

This work was supported in part by Doctoral Scientific Research Foundation of SWUST(16zx7106), Applied Basic Research Programs of Science & Technology Committe Foundation of Sichuan Province (2019YJ0309), Foundation of Sichuan Educational Committee (18ZB0611), and National Key Research & Development Project (2016YFF0104003).

H. Gao et al. (Eds.): CollaborateCom 2020, LNICST 349, pp. 671–685, 2021.
https://doi.org/10.1007/978-3-030-67537-0_40

1 Introduction

Internet of Things (IoT) applications have been growing explosively, which are usually delay-sensitive and computation-intensive, that require high processing capacities within sustainable time constraints [14]. Although IoT mobile devices (IMDs) have been becoming more and more powerful in the central processing unit (CPU), which may still not capable of handling computation tasks within limited time constraints. On the other hand, computation-intensive applications require more energy consumption, which shortens the lifetime of MIDs obviously, and high energy consumption in IMDs poses a significant obstacle for user experience. The traditional solution is to introduce cloud computing by offloading computation tasks to the cloud platform [1], but which imposes tremendous traffic load and induces high latency since data are sent to the cloud platform from plenty of IMDs located in different zones.

Mobile edge computing (MEC), a key technology for the 5th generation (5G) mobile networks, can relieve the shortcoming of centralized cloud computing by pushing the computation capabilities to MEC servers located at network edges, more closer to IMDs, that can provide sufficient computation capacities for IMDs, as well as reduce the traffic burden in core networks [15]. Researchers from academia to industry have been widely promoting MEC technologies to save energy consumption while reduce execution delay of applications in IMDs. Typically, an IMD can offload its computation tasks to an access point (AP) integrated with an MEC server providing rich computation resources.

On the other hand, task offloading is not always beneficial in saving energy of IMDs according to some research results [7]. If the IMD is far away from the AP with worse communication links, some tasks even consume more energy if they were offloaded to the MEC server than processed locally due to the communication overhead in energy. But some works show that the relay node situated between the IMD and the AP can save the communication energy in total [3,8,10].

For a common scenario of an IMD with some neighboring IMD nodes, selecting cooperative nodes for offloading from the neighbors with minimized energy consumption is of great realistic meaning. In some cases, tasks can be partitioned into many different segments for execution, thus many neighbor nodes can be selected as offloading helpers. But in other cases, tasks are indivisible or can only be partitioned into two segments for offloading, then only one neighbor node can be selected as the offloading helper.

In this paper, we consider a scheme of selecting the most energy-efficient cooperative neighbor node as the offloading helper with the execution delay constraint. Helped by the selected neighbor node, the minimized energy consumption is realized among all the neighbor nodes of the IMD. To achieve the goal, we first build a system model to introduce the neighbor-aided cooperative task offloading processes, then formulate the energy consumption and the delay constraint in each process at local, at the cooperative node and at the MEC server, respectively. Afterwards, we derive the optimization problem to minimize the total energy consumption and prove it is convex, so it can be effectively tackled by conventional methods like toolbox CVX [4]. At last, we perform the numerical results to demonstrate the efficiency of the proposed scheme. In particular, the main contributions of this paper can be summarized as follows.

1) We consider a scenario of a group of randomly located IMDs, and an AP integrated with an MEC server providing rich computation resources. In which an IMD can select one neighbor as the offloading helper due to the very limited number of partitioned segments of its applications.
2) In order to prolong the lifespan of the IMDs, a cooperative scheme of selecting the most energy-efficient neighbor as the offloading helper with execution delay constraint is investigated. The system model and the corresponding optimization problem are presented accordingly.
3) By solving the problem using CVX toolbox, the neighbor with the minimized energy consumption is confirmed and the segment length for task offloading in each process is decided. Extensive numerical experiments validate the advantage of our proposed scheme in neighbor selection and segmentation of tasks for energy saving.

The rest of this paper is organized as follows. In Sect. 2, we review the related work. We specify the system model and problem formulation in Sect. 3. Section 4 describes the proposed energy-efficient problem and then we get the optimal solution. Performance evaluations are illustrated in Sect. 5, and Sect. 6 concludes this paper.

2 Related Work

Saving energy with low execution latency to satisfy the user quality of experience is of great value in IMDs. Executing programs slowly in IMDs can save energy. If the clock speed of the CPU is reduced by half, the execution time doubles, but only one quarter of the energy is consumed [7]. However, it hardly satisfies the delay constraint by many delay-sensitive applications. Sending computation to another machine is not a new idea. For example, cloud computing can save energy for mobile users through computation offloading [13]. But, in the IOT circumstance with tremendous IMDs, the remote centralized cloud server will induce bandwidth congestion and thus longer delay. The idea of MEC moves cloud servers to the network edge, which is smart and effective in dealing with the shortcoming of centralized cloud server schemes, and has been attracting tremendous attention of researchers from academia to industry.

On the other hand, according to various cost/benefit studies, only if the energy consumption for transmitting and receiving data is less than the execution cost in the IMD system locally. Since the communication distance, the wireless channel states, the length of task input-bits, the time constraint and so forth will affect the offloading cost/benefit jointly, energy efficient offloading to optimize the resource allocation is one of the key problems in MEC.

Some research works reveal that, cooperative offloading can save the energy consumption aided by neighbor nodes than offloading the partial task directly

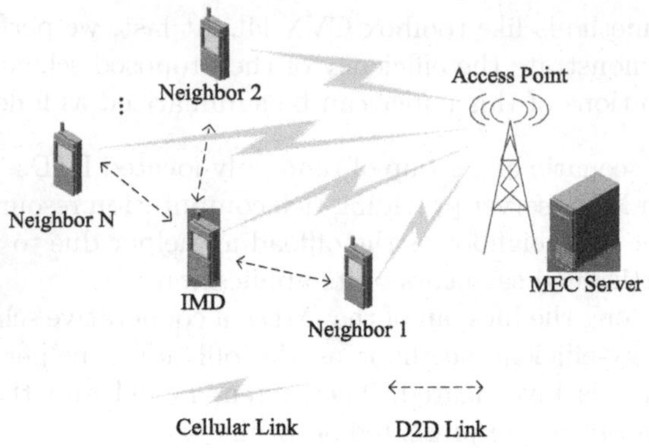

Fig. 1. System model.

to the MEC server especially in bad communication conditions. The authors in [10] study an energy-efficient computation offloading scheme with specialized cooperative nodes, in which N IMDs, M cooperative edge nodes and one cloud server are considered. Tasks in IMDs can be offloaded to one of the edge nodes or forwarded to the cloud server directly decided by the formulated energy optimization problem. Cao *et al.* [3] introduce another scenario of one IMD and an AP with a neighboring node located between them as the helper to assist the computation task offloading with energy efficiency, in which time is divided into four slots for cooperation over the whole block of task computation process. Different from these works, in this paper, we consider a scenario of an IMD surrounded by many neighboring IMD nodes, and the best energy efficient one will be selected as the cooperative offloading helper due to the tasks can only be partitioned into limited segments in the applications.

3 System Model

In this section, we introduce a scenario of a target IMD surrounded by many neighboring IMD nodes in the MEC-based IoT network, then formulate the energy consumption and delay constraint in local processing, neighbor node processing and MEC processing, respectively.

As shown in Fig. 1, we consider a network consisting of an AP integrated with an MEC server providing rich computation resources, one IMD and a set of K neighboring IMDs of it as cooperative candidates for helping offload computation tasks to the MEC server. Since some computation tasks are indivisible or can only be partitioned into a very limited number of segments for execution, we consider the case that only one neighbor node is selected as the offloading helper, achieving the minimized energy consumption with delay constraint. Each segment of the task can be processed locally, offloaded to one of its neighbor nodes through D2D links or to the MEC server directly through cellular links for execution. Further more, segments offloaded to the neighbor node can continually be offloaded to the MEC server through cellular links. After the task is performed at the neighbor or the MEC server, the result will be sent back to the IMD.

In the system, the D2D links and the cellular links are considered to be deployed over different frequencies, so the D2D links and the cellular links do not interfere with each other [6,12]. Same with many other research works, we suppose the D2D links and the cellular links are working with pro-defined fix bandwidth [9]. We also suppose channels from IMDs to the AP follow the quasi-static block fading. That is, the channel state remains unchanged during the offloading period of one computational segment [5]. In addition, costs of delay and energy consumption for result downloading are not taken into consideration since the result has much less length [11].

We suppose a set of neighbor nodes of the IMD in the network are denoted by $\mathbb{K} = \{1, 2, \ldots, K\}$, and the IMD has the computation tasks of $L > 0$ input-bits with the delay constraint $T \geq 0$.

Consider the case of partial offloading, a task of L input-bits of the IMD can be partitioned into four different segments. Let $l_u \geq 0$, $l_u^s \geq 0$, $l_k \geq 0$ and $l_k^s \geq 0$ denote the number of segment input-bits for local computing in the IMD, offloading to the MEC server from the IMD, offloading to a neighbor node k, $k \in \mathbb{K}$ of the IMD, and offloading to the MEC server from k, respectively. Then we have

$$L = l_u + l_u^s + l_k + l_k^s. \tag{1}$$

Note that the segment of l_k bits executed on the node k, and the other segment of l_k^s bits offloaded from k to the MEC server are both offloaded from the IMD to its neighbor node k. Then, the total offloaded bits from the IMD to its neighbor node k is $l_k + l_k^s$.

The whole processes of task computation and offloading in the system can be summarized into three stages. The first stage is local processing, the second stage is neighbor node processing, and third stage is MEC server processing. In the following section, we will formulate the energy consumption and delay constraint for each stage in detail.

3.1 Local Processing

In the stage of local processing, it includes computation for the partial task of l_u bits at the local IMD, the offloaded partial task of $l_k + l_k^s$ bits to its neighbor node k, and the offloaded partial task of l_u^s bits to the MEC server from the IMD.

Local Computing. Let c_u denote the number of CPU cycles for computing each task input-bit, f_u denote the maximum computation capability (in CPU cycles/s) at the IMD, and f_u' denote the computation capability on demand, respectively. Then the computation latency T_u^C in the IMD is

$$T_u^C = \frac{c_u l_u}{f_u'}. \tag{2}$$

Since the execution of computation tasks is constrained within the delay constraint T, each part of the execution tasks is requested to be finished within T. As a result, the time for executing the computation task with l_u input-bits need to satisfy the requirement $T_u^C \leq T$. Substituting (2) into the requirement and we get $\frac{c_u l_u}{f_u'} \leq T$. Considering $f_u' \leq f_u$, then $\frac{c_u l_u}{f_u} \leq \frac{c_u l_u}{f_u'}$, and thus we get the delay constraint as

$$\frac{c_u l_u}{f_u} \leq T. \tag{3}$$

Let γ_u denote the effective capacitance coefficient of CPU at the IMD [2]. Then the energy consumption E_u^C for local computing at the IMD is [11,16]

$$E_u^C = \gamma_u c_u {f_u'}^2 l_u. \tag{4}$$

From (2) and (4) we get, the longer computation latency, the less energy consumption for local computing. In order to get the minimized energy consumption for the local execution of l_u bits, we set T_u^C with the delay constraint T, that is $T = T_u^C = \frac{c_u l_u}{f_u'}$, and we get $f_u' = \frac{c_u l_u}{T}$, then (4) can be denoted by

$$E_u^C = \frac{\gamma_u {c_u}^3 {l_u}^3}{T^2}. \tag{5}$$

Computation Offloading to Neighbor Node. Before a neighbor node k computing the offloaded task bits, the IMD will firstly offload $l_k + l_k^s$ task bits to the node k with transmit power $P_{u,k} \geq 0$ through the D2D link. Let $h_{u,k} \geq 0$ denote the channel power gain from the IMD to k, and B_1 the channel bandwidth. Accordingly, the achievable data rate (in bits/s) for task offloading from the IMD to its neighbor node k is

$$r(P_{u,k}) = B_1 \log_2 (1 + \frac{P_{u,k} h_{u,k}}{\sigma_k^2}), \tag{6}$$

where σ_k^2 denotes the noise power at k.

Then the offloading delay $T_{u,k}^O$ and the energy consumption $E_{u,k}^O$ from the IMD to its neighbor node k are given respectively by

$$T_{u,k}^O = \frac{l_k + l_k^s}{r(P_{u,k})}, \tag{7}$$

$$E_{u,k}^O = \frac{P_{u,k}(l_k + l_k^s)}{r(P_{u,k})}. \tag{8}$$

Computation Offloading to MEC. Suppose the partial task offloaded from the IMD to the MEC server through the cellular link with transmit power $P_{u,s} \geq 0$, and $h_{u,s} \geq 0$ denote the channel power gain from the IMD to the MEC server, then the achievable data rate (in bits/s) for task offloading from the IMD to the MEC server is given by

$$r(P_{u,s}) = B_2 \log_2 (1 + \frac{P_{u,s}h_{u,s}}{\sigma_s^2}), \tag{9}$$

where σ_s^2 denotes the noise power at the MEC server and B_2 denotes the cellular channel bandwidth, respectively.

Accordingly, The offloading delay and the energy consumption from the IMD to the MEC server are thus

$$T_{u,s}^O = \frac{l_u^s}{r(P_{u,s})}, \tag{10}$$

$$E_{u,s}^O = \frac{P_{u,s}l_u^s}{r(P_{u,s})}, \tag{11}$$

3.2 Neighbor Node Processing

The stage of neighbor node processing includes computation for the partial task of l_k bits in the neighbor node k and offloading l_k^s bits to the MEC server from k.

Computing at Neighbor Node. Let c_k denote the number of CPU cycles for computing each task input-bit, γ_k denote the CPU effective capacitance coefficient, f_k denote the maximum computation capability (in CPU cycles/s), and f_k' denote the computation capability on demand at the neighbor node k, respectively. Thus, the computing time T_k^C at the neighbor node k is expressed as

$$T_k^C = \frac{c_k l_k}{f_k'}. \tag{12}$$

The delay constraint on the neighbor node includes the transmit delay and the computing delay. Then, we have

$$T_{u,k}^O + T_k^C \leq T. \tag{13}$$

Substituting (7) and (12) into (13), and because $f_k' \leq f_k$, similarly as in (3), we get the delay constraint as

$$\frac{l_k + l_k^s}{r(P_{u,k})} + \frac{c_k l_k}{f_k} \leq T. \tag{14}$$

Similarly as in (5), the energy consumption E_k^C for local computing at the neighbor node k is

$$E_k^C = \frac{\gamma_k {c_k}^3 {l_k}^3}{(T - T_{u,k}^O)^2}. \tag{15}$$

Substituting (7) into (15) we get

$$E_k^C = \frac{r(P_{u,k})^2 \gamma_k {c_k}^3 {l_k}^3}{(Tr(P_{u,k}) - (l_k + l_k^s))^2} \tag{16}$$

Computation Offloading. Suppose the partial task is offloaded from a neighbor node k through the cellular link with transmit power $P_{k,s} \geq 0$, and $h_{k,s} \geq 0$ denote the channel power gain from the node k to the MEC server. The achievable data rate (in bits/s) for task offloading from k to the MEC server is thus

$$r(P_{k,s}) = B_2 \log_2 (1 + \frac{P_{k,s} h_{k,s}}{\sigma_s^2}), \tag{17}$$

Accordingly, the offloading delay and the energy consumption for task offloading from k to the MEC server can be separately expressed as

$$T_{k,s}^O = \frac{l_k^s}{r(P_{k,s})}, \tag{18}$$

$$E_{k,s}^O = \frac{P_{k,s} l_k^s}{r(P_{k,s})}. \tag{19}$$

3.3 MEC Server Processing

Since the purpose of proposed scheme is to minimize the energy consumption in the MEC-based IoT network and the MEC server has sufficient computation resources in general, the energy consumption of the computation for the offloaded segments at the MEC server is not take into consideration. Therefore, the stage of MEC server processing only includes the offloaded task computing.

We also suppose the partial tasks offloaded to the MEC server from the IMD and its neighbor node do not need to wait for execution in queue. Let c_s denote the number of CPU cycles for computing each task input-bit, γ_s denote the CPU effective capacitance coefficient, f_s denote the maximum computation capability (in CPU cycles/s) at the MEC server, and f_s' denote the computation capability on demand, respectively. Then, we have the time that the MEC server computes the partial tasks offloaded from the IMD and k as

$$T_{u,s}^C = \frac{c_s l_u^s}{f_s'}, \tag{20}$$

$$T_{k,s}^C = \frac{c_s l_k^s}{f_s'}. \tag{21}$$

Similarly as in (3) and (13), the delay constraint at the MEC server includes the offloading delay and the computing delay for tasks offloaded from the IMD and from the neighbor node k, respectively. Then, we have

$$T_{u,s}^O + T_{u,s}^C \leq T. \tag{22}$$

$$T_{u,k}^O + T_{k,s}^O + T_{k,s}^C \leq T. \tag{23}$$

Substituting (10) and (20) into (22), and substituting (7), (18) and (21) into (23), respectively. Thus we get

$$\frac{l_u^s}{r(P_{u,s})} + \frac{c_s l_u^s}{f_s} \leq T, \tag{24}$$

$$\frac{l_k + l_k^s}{r(P_{u,k})} + \frac{l_k^s}{r(P_{k,s})} + \frac{c_s l_k^s}{f_s} \leq T. \tag{25}$$

4 Problem Formulation and Proposed Offloading Approach

In this section, we formulate an energy-efficient problem with delay constraint for the neighbor-aided cooperative MEC-based IoT network, then we get the optimal solution for the problem, and propose an offloading scheme in selecting the cooperative neighbor with minimized energy consumption. We also present a simple weighted value cooperative neighbor selection scheme to reduce the complexity as a comparison.

4.1 Problem Formulation and Optimal Solution

As the MEC server has reliable power supply, we focus on minimize the total energy consumption caused by offloading and computation at IMDs subjected to the task's delay constraint T. The formulation also optimizes the task partition.

We denote the total energy consumption cooperated by a neighbor node k as E_k, then

$$E_k = E_u^C + E_k^C + E_{u,k}^O + E_{u,s}^O + E_{k,s}^O. \tag{26}$$

Substituting (5),(8),(16),(11) and (19) into (26), we get

$$\begin{aligned} E_k =& \frac{\gamma_u {c_u}^3 {l_u}^3}{T^2} + \frac{r(P_{u,k})^2 \gamma_k {c_k}^3 {l_k}^3}{(Tr(P_{u,k}) - (l_k + l_k^s))^2} \\ &+ \frac{P_{u,k}(l_k + l_k^s)}{r(P_{u,k})} + \frac{P_{u,s} l_u^s}{r(P_{u,s})} + \frac{P_{k,s} l_k^s}{r(P_{k,s})}. \end{aligned} \tag{27}$$

We design a variable of the IMD's task partition vector $\boldsymbol{l} \triangleq [l_u, l_u^s, l_k, l_k^s]$. The delay constraint energy minimization problem is then formulated as

$$(\text{P1}): \quad \min_{\boldsymbol{l}} E_k \tag{28a}$$

$$\text{s.t.} \quad T \geq 0, \tag{28b}$$

$$l_u \geq 0, l_u^s \geq 0, l_k \geq 0, l_k^s \geq 0, \tag{28c}$$

$$(1), (3), (14), (24) \text{ and } (25),$$

where (28c) is the constant of delay constraint required by tasks at the IMD.

Since $\frac{{l_k}^3}{(l_k+l_k^s))^2}$ is convex with $l_k \geq 0$ and $l_k + l_k^s > 0$, then the term $\frac{r(P_{u,k})^2 \gamma_k {c_k}^3 {l_k}^3}{(Tr(P_{u,k})-(l_k+l_k^s))^2}$ in the objective function is jointly convex with respect to $l_k \geq 0$ and $\frac{l_k+l_k^s}{r(P_{u,k})} < T$. So we get (P1) is convex. Thus (P1) can be optimally solved by the using convex optimization toolbox CVX [4].

By solving problem (P1), we get the minimized energy consumption of each neighbor node for the neighbor-aided cooperative offloading progress, and then we can get one neighbor node with the minimized energy consumption among them. But as we know, the computation complexity is high to solve the convex optimization problem. We hereby design a weighted value cooperative neighbor selection scheme.

4.2 Weighted Value Cooperative Neighbor Selection Scheme

For neighbor-aided cooperative offloading, the energy consumption for computation in each neighbor differs little since the computing power of IMDs are similar for many cases, while the main difference for energy consumption is in the process of offloading, so we can only consider the energy consumption for the offloading process. From (8) and (19) we conclude that the offloading energy consumption is proportional to the transmit power and the length of offloading bits, while inversely proportional to the data rate. Suppose the IMDs have identical transmit power, then we can define the weighted value cooperative neighbor selection scheme as

$$k^* = \arg\min_{k \in \mathbb{K}} (\alpha r(P_{u,k}) + (1-\alpha) r(P_{k,s})), 0 \leq \alpha \leq 1, \tag{29}$$

where α is the coefficient to evaluate the importance between the communication rates from the IMD to its neighbor node k, and from the neighbor node k to the MEC server. The time complexity of the scheme is $O(n)$.

Apparently, the vector $\boldsymbol{l}$ and energy consumption E^* of the neighbor node k^* confirmed by (29) can also be got by solving problem (P1).

4.3 Maximum Task Length

The maximum task length discloses how many bits of a task are supported within the given delay constraint T in the MEC-based IoT network. The maximum length of the task input-bits for the neighbor node k of the IMD will be derived when the task is executed at local, at the neighbor node and at the MEC server, simultaneously. Because in this case, the three nodes can fully take advantage of their available communication and computation resources. We then formulate the problem as

$$\begin{aligned} (\text{P2}): \quad & \max_{l} \quad l_u + l_u^s + l_k + l_k^s && \text{(30a)} \\ \text{s.t.} \quad & (3), (14), (24), (25), (28\text{b}) \text{ and } (28\text{c}) \end{aligned}$$

Note problem (P2) is a linear program and can thus be efficiently solved for every neighbor node k via standard concave optimization techniques, and hereby we can get a neighbor with the maximum task length. The maximum length of the task input-bits for the neighbor k^* confirmed by (29) can also be got.

5 Numerical Results

In this section, we provide the numerical results to evaluate the performance of the proposed two neighbor-aided cooperative offloading schemes compared with two benchmark schemes. The proposed offloading schemes include

1) Offloading with the best neighbor: The cooperative neighbor with the minimized energy consumption and the maximum task length in the system among all the neighbor nodes of the IMD are selected by solving problem (P1) and problem (P2) for each node, separately.
2) Offloading with the neighbor of weighted value: The cooperative neighbor k^* is selected by solving (29), and the minimized energy consumption as well as the maximum task length in the system are got by solving problem (P1) and problem (P2) by setting $k = k^*$, separately.

The benchmark schemes are

1) Local computing: The IMD executes whole computation tasks locally by itself. We can get the maximum length of task input-bits by solving $\frac{Tf_u}{c_u}$ from (3), and get the energy consumption $E_u^C = \gamma_u c_u f_u^2 L$ by Eq. (5).
2) Offloading without neighbor: The IMD offloads computation tasks to the MEC server without corporation of a neighbor. Like the scheme of offloading with the best neighbor, this scheme corresponds to solving problem (P1) and problem (P2) by setting $l_k + l_k^s = 0$ to get the minimized energy consumption and the maximum task length, separately.

In the simulation set-up, we consider the neighbors of the IMD are randomly located within the area of a rectangular coordinate system in order to get the influence of the different locations of the neighbor nodes. Let (c_k^x, c_k^y) denote the location coordinates in X-axis and Y-axis of a node k in the rectangular coordinate system. Then we set $0 \le c_k^x \le 500m$ and $0 \le c_k^y \le 200m$. Therefore, the whole rectangular area is evenly divided into a grid network of 26∗11 with the IMD located at $(100, 0)$ and the AP at $(400, 0)$, separately. At last, we randomly

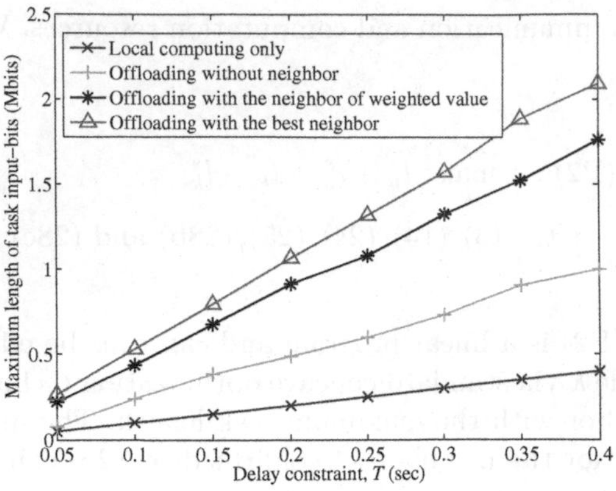

Fig. 2. Average maximum length of task input-bits versus the delay constraint T.

set 20 neighbor nodes located at different intersections in the whole rectangular areas.

Let d denote the distance from the transmitter to the receiver, then the path-loss between any two nodes is $\beta_0(d/d_0)^{-\xi}$, where $\beta_0 = -60$ db corresponds to the path-loss at the reference distance of $d_0 = 10$ m, and $\xi = 3$ is the path-loss exponent. Furthermore, we set $B_1 = B_2 = 1$ MHz, $\sigma_k^2 = \sigma_s^2 = -70$ dBm, $c_u = c_k = c_s = 10^3$ cycles/bit, $P_{u,k} = P_{u,s} = P_{k,s} = 40$ dBm [3], $\gamma_u = \gamma_k = 10^{-26}$, $f_u = f_k = 1$ GHz and $f_s = 5$ GHz.

Figure (2) shows the average maximum length of task input-bits (Mbits) versus the delay constraint T. It is observed that the maximum length of task input-bits increases as the delay constraint becomes longer. In detail, the scheme of offloading with the best neighbor supports the maximum length of task input-bits compared with other schemes. The scheme of the neighbor selected by the weighted value is better than the other two benchmark schemes of local computing and offloading without neighbor aiding. The reason is the adopted offloading strategy can assist the IMD by executing more extra parts of the tasks within the delay constraint. Compared the scheme of local computing only with the scheme of offloading without neighbor, the latter one can execute more data within the delay constraint. Tasks can not be finished within the delay constraint when the length of task input-bits exceeds the supported maximum length, which means that the proposed scheme of offloading with the best neighbor supports the longer maximum length as well as the bigger computation capacities.

Figure (3) shows the average minimum energy consumption versus the delay constraint T with $L = 0.02$ Mbits. The results demonstrate that the minimum energy consumption decreases with the prolonged delay constraint. The proposed scheme of offloading with the best neighbor consumes the least energy for all cases, while the scheme of local computing only consumes the most energy especially when the delay constraint is smaller. The differences between schemes become smaller as the delay constraint becomes longer, and eventually almost

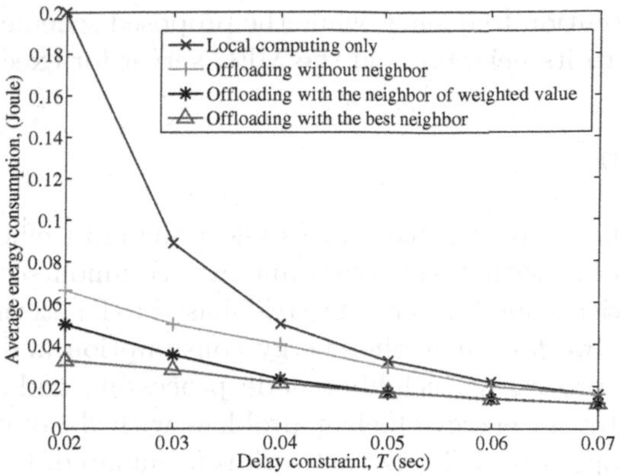

Fig. 3. Average minimum energy consumption versus delay constraint T.

keep in with the same scale, due to more tasks will be executed locally for the slack long delay constraint.

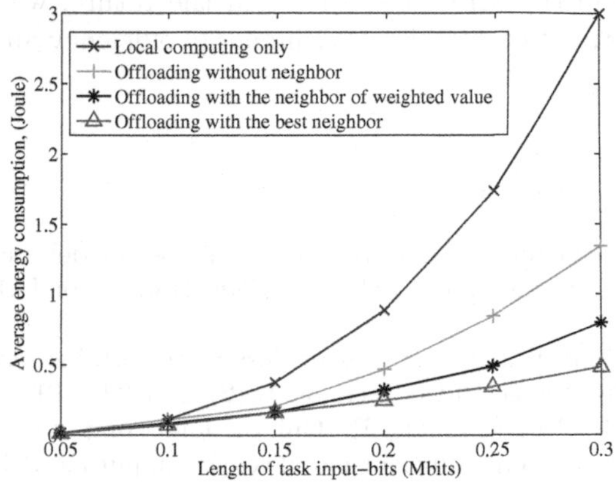

Fig. 4. Minimum energy consumption versus the length of task input-bits (Mbits).

Figure (4) shows the minimum energy consumption versus the length of task input-bits (Mbits) with the delay constraint $T = 0.3s$. The simulation results demonstrate that the minimum energy consumption grows exponentially with the prolonged length of task input-bits, and the proposed scheme of offloading with the best neighbor out performs the others at all cases with lower energy consumption. When the task length is small, the scheme of local computing achieves the similar performance compared with others due to the task can be finished with lower CPU execution frequency within the delay constraint. But when the task length increases, local computing will cost more energy for the increased CPU execution frequency, while the proposed scheme can offload some parts of the tasks to its neighbor and the MEC server for execution.

6 Conclusion

In this paper, we have investigated how to select the most efficient neighbor for cooperative offloading with delay constraint in a common scenario of a target IMD neighbored with some IMDs in the MEC-based IoT network. After building the system model, we formulate the energy consumption and delay constraint problems in local processing, neighbor node processing and MEC processing, respectively. Hereby, an energy-efficient problem with delay constraint for the neighbor-aided cooperative IoT edge network is formulated. By solving the optimization problem, the neighbor of the target IMD with the least energy consumption is selected. We also present a simple weighted value cooperative neighbor selection scheme to reduce the complexity as a comparison. Moreover, in order to get the supported maximum task length in the MEC-based IoT network, we formulate the task length optimization problem. Numerical results indicate that the proposed scheme outperforms the benchmark schemes significantly in terms of the supported maximum task length and the energy efficiency, which is feasible and effective in the MEC-based IoT. In the future, we will research the scheme of selecting a set of neighbors with energy efficiency and execution delay constraint.

References

1. Barrameda, J., Samaan, N.: A novel statistical cost model and an algorithm for efficient application offloading to clouds. IEEE Trans. Cloud Comput. **6**(3), 598–611 (2018)
2. Burd, T.D., Brodersen, R.W.: Processor design for portable systems. J. VLSI Sig. Process. Syst. Sig. Image Video Technol. **13**(2), 203–221 (1996)
3. Cao, X., Wang, F., Xu, J., Zhang, R., Cui, S.: Joint computation and communication cooperation for energy-efficient mobile edge computing. IEEE Internet Things J. **6**(3), 4188–4200 (2019)
4. Grant, M., Boyd, S.: CVX: Matlab software for disciplined convex programming, version 2.1 (2014). http://cvxr.com/cvx
5. Guo, S.T., Liu, J.D., Yang, Y.Y., Xiao, B., Li, Z.T.: Energy-efficient dynamic computation offloading and cooperative task scheduling in mobile cloud computing. IEEE Trans. Mob. Comput. **18**(2), 319–333 (2019)
6. Hu, G., Jia, Y., Chen, Z.: Multi-user computation offloading with D2D for mobile edge computing. In: 2018 IEEE Global Communications Conference (GLOBECOM), pp. 1–6 (2018)
7. Kumar, K., Lu, Y.H.: Cloud computing for mobile users: can offloading computation save energy? Computer **4**, 51–56 (2010)
8. Ning, Z., Dong, P., Kong, X., Xia, F.: A cooperative partial computation offloading scheme for mobile edge computing enabled internet of things. IEEE Internet Things J. **6**(3), 4804–4814 (2019)

9. Peng, H., Ye, Q., Shen, X.: Spectrum management for multi-access edge computing in autonomous vehicular networks. IEEE Trans. Intell. Transp. Syst. 1–12 (2019)
10. Vu, T.T., Huynh, N.V., Hoang, D.T., Nguyen, D.N., Dutkiewicz, E.: Offloading energy efficiency with delay constraint for cooperative mobile edge computing networks. In: IEEE Global Communications Conference. IEEE, New York (2018)
11. Wang, J., Feng, D., Zhang, S., Tang, J., Quek, T.Q.S.: Computation offloading for mobile edge computing enabled vehicular networks. IEEE Access **7**, 62624–62632 (2019)
12. Wu, Y., Chen, J.C., Qian, L.P., Huang, J.W., Shen, X.: Energy-aware cooperative traffic offloading via device-to-device cooperations: an analytical approach. IEEE Trans. Mob. Comput. **16**(1), 97–114 (2017)
13. Yang, K., Ou, S., Chen, H.H.: On effective offloading services for resource-constrained mobile devices running heavier mobile Internet applications. IEEE Commun. Mag. **46**(1), 56–63 (2008)
14. Zhang, N., et al.: Physical layer authentication for internet of things via WFRFT-based Gaussian tag embedding. IEEE Internet Things J. https://doi.org/10.1109/JIOT20203001597
15. Zhang, N., Wu, R., Yuan, S., Yuan, C., Chen, D.: RAV: relay aided vectorized secure transmission in physical layer security for internet of things under active attacks. IEEE Internet Things J. **6**(5), 8496–8506 (2019)
16. Zhang, W., Wen, Y., Guan, K., Kilper, D., Luo, H., Wu, D.O.: Energy-optimal mobile cloud computing under stochastic wireless channel. IEEE Trans. Wireless Commun. **12**(9), 4569–4581 (2013)

Author Index

Printed in the United States
By Bookmasters